Power System Protection

IEEE Press
445 Hoes Lane
Piscataway, NJ 08854

Power System Protection

Fundamentals and Applications

John Ciufo
Ciufo and Cooperberg Consulting Inc. (CCCI)
Mississauga
Greater Toronto Area
Ontario
Canada

Aaron Cooperberg
Ciufo and Cooperberg Consulting Inc. (CCCI)
Mississauga
Greater Toronto Area
Ontario
Canada

IEEE Press Series on Power and Energy Systems
Ganesh Kumar Venayagamoorthy, Series Editor

IEEE PRESS
WILEY

Published by John Wiley & Sons, Inc., Hoboken, New Jersey.
Published simultaneously in Canada.

Library of Congress Cataloging-in-Publication Data:

Names: Ciufo, John, 1953- author. | Cooperberg, Aaron, 1952- author.
Title: Power system protection : fundamentals and applications / John Ciufo, Ciufo and Cooperberg Consulting (CCCI), Aaron Cooperberg, Ciufo and Cooperberg Consulting (CCCI).
Description: First edition. | Hoboken, New Jersey : John Wiley & Sons, Inc., [2022] | Series: IEEE press series on power and energy systems | Includes bibliographical references and index.
Identifiers: LCCN 2021048461 (print) | LCCN 2021048462 (ebook) | ISBN 9781119847366 (cloth) | ISBN 9781119847373 (adobe pdf) | ISBN 9781119847380 (epub)
Subjects: LCSH: Electric power systems–Protection.
Classification: LCC TK1010 .C58 2022 (print) | LCC TK1010 (ebook) | DDC 621.31–dc23/eng/20211122
LC record available at https://lccn.loc.gov/2021048461
LC ebook record available at https://lccn.loc.gov/2021048462

Cover Design: Wiley
Cover Image: © Patrick Jennings/Shutterstock

Set in 9.5/12.5pt STIXTwoText by Straive, Chennai, India

10 9 8 7 6 5 4 3 2 1

Contents

About the Authors

John Ciufo, P. Eng.

Is a professional Electrical Engineer with over forty years of electric utility experience with a focus on protection and control (P&C) engineering. He has worked for Hydro One Inc., formerly Ontario Hydro from 1976 to 2010. Over the years, he has held many different positions in the P&C power system discipline. His experience includes engineering, power system reliability compliance, smart grid, asset management, development of strategies, policies, developing functional and design standards, cybersecurity, and digital stations. He completed his career with Hydro One as a Senior Manager – Protection & Control Strategies and Standards in Asset Management. In 2011, he became a principal owner of Ciufo & Cooperberg Consulting Inc. a consulting company that specializes in power system protection.

John has an extensive background in protection and control systems in the electric industry. He is a registered Professional Engineer in the Province of Ontario and was a member of the Northeast Power Coordinating Council (NPCC), Task Force on System Protection for 10 years, and a Vice-Chair from January 2008 to January 2010. Was a member of the North American Electric Regulatory Corporation (NERC), System Protection and Control Subcommittee since May 2004, and was the Chair between the periods June 2008 to September 2010. He was on several NERC drafting teams developing protection and control-related standards and has co-authored several industry papers.

During his career at Ontario Hydro/Hydro One, he was instrumental in developing protection engineering application standards; protection philosophies and processes; transitioned the company from electromechanical designs to microprocessor-based designs that resulted in significant cost savings; developed protection and control health indices and asset management processes; developed reliability compliance programs and processes; developed the company's cybersecurity compliance program, introduced and piloted digital station designs, transitioned the company from microwave to Digital Synchronous Optical Network (SONET) teleprotection systems, established, developed, and conducted internal company protection training, among other achievements.

John continues to be active in the industry and provides engineering services to many North American electric utilities. He is a Technical Advisor for the Centre for Energy Advancement Technological Innovations (CEATI) for the Protection and Control Group.

Aaron Cooperberg, Licensed Protection Engineer

Is professionally licensed to practice Power System Protection in the Province of Ontario. Upon graduation having specialized in Power Systems he began his career with Ontario Hydro in1977. He was assigned there to the protection engineering group responsible for the specification and design of protection systems. He spent 21 years working alongside Ontario Hydro's top experts in Power

System protection engineering carrying out many protection system designs during the period of rapid expansion of Ontario's 500 kV transmission system and the construction of large multi-unit Nuclear, Thermal and Hydraulic generation sites.

His last 13 years with Ontario Hydro and then Hydro One was focused on the Asset Management of the province's Protection Systems. Here he developed industry-recognized Asset Management design and business case strategies for the proactive replacement of end-of-life protection systems. He has also led the team responsible for developing the technical requirements to connect multi-tapped generation to Hydro One's Transmission and Distribution systems. He completed his career with Hydro One as a Senior Manager – Protection & Control Planning. In 2011, he became a principal owner of Ciufo & Cooperberg Consulting Inc. a consulting company that specializes in power system protection.

Aaron possesses an in-depth knowledge of protection systems for transmission as well as generation having designed protection systems for nuclear and hydroelectric generation as well as transmission for the former Ontario Hydro. He provided expert testimony to the U.S–Canada Committee on the August 14th, 2003, blackout. He was on the NERC drafting team developing Protection and Control Standard PRC-001 for Protection Coordination. He was a speaker at conferences on the topics of protection systems as well as Asset Management and has presented at CEATI and Electric Power Research Institute (EPRI). Aaron continues to be active in the industry providing engineering services to many North American electric utilities.

Preface

This book contains the accumulated experiences and practices used by the authors who have each, practiced protection engineering for over forty years.

Protection engineering is a specialty within the study of power system engineering. It is generally, not taught in engineering programs except for some specialized post-graduate programs. Considering that every power system big, and small, regardless of voltage level, requires the application of protection systems; the authors felt that there was a gap in the industry, and there is a need for more protection system information and guidance for new protection practitioners.

Protection is a highly complex discipline requiring several years of specialized engineering development following graduation. Utilities typically resort to the recruitment of graduate power system engineers into the field of protection engineering. Historically, these recruits would gain the necessary experience and training while working along with seasoned engineers over many years as they gain confidence. This mentoring approach is becoming more difficult to implement. Specifically, this mentoring approach relies on several years of overlap which is becoming more difficult to attain as many experienced staff with lifelong knowledge have, or are retiring, leaving fewer, and fewer experienced mentors.

New protection practitioners to this field require resources, and the means to gain the necessary know-how. It is for this reason that we felt compelled to write this book, to provide new protection practitioners with a book they can relate with for Power System Protection Fundamentals and Applications. It is the intent of the authors, that this book facilitates knowledge transfer via the use of a structured set of fundamental protection principles, explanatory illustrations, and applications of these principles.

The authors appreciate the challenges for new protection practitioners. It is a complex field requiring knowledge of electrical engineering, power systems, power equipment, protection engineering, telecommunications, power system analysis, control, and more recently, computer programming, and networks as the industry transforms into a digital world. Protection practitioners are tasked with designing, maintaining, operating, compliance, managing, and diagnosing protection system applications. As such, they are accountable to make these systems work and function per design; they represent the process metaphorically, where "the rubber meets the road."

This book is written with the approach that in this new and dynamic digital transformation, the understanding of the underlying protection principles is key to the successful development of a protection practitioner. Fundamentally, protection practitioners are held accountable to design, operate, maintain, and implement workable solutions to support the reliable operation of the power system. It is for this purpose; we wrote this book to be a balance between theory and practical applications for the intent of being relatable.

Acknowledgements

John Ciufo

An undertaking of this nature requires a passion for the practice of protection engineering. It also requires dedicating personal time to its development, and as such, I would like to thank my family for their cooperation and understanding.

This book was made possible with the encouragement and support of my dear wife Maria. I would like to also thank my children, Vanessa Lynn and Mark Joseph, for their inspiration during the writing of this book.

Additionally, I would like to thank my colleague and friend, Aaron Cooperberg, for his shared interest and dedication to this subject and for co-authoring this book.

Aaron Cooperberg

During my career at Ontario Hydro/Hydro One, I was always passionate about sharing my engineering knowledge with co-workers and particularly with junior protection engineering staff. This passion for sharing knowledge has lead me to co-author this book.

I would like to acknowledge the encouragement, support, and patience of my wife Rina without whom this book would not be possible.

Additionally, I would like to thank my colleague and friend, John Ciufo for whom I have the utmost respect. John's commitment and determination were instrumental while co-authoring this book.

John and Aaron

We would like to express our sincere thanks to Ontario Hydro/Hydro One for providing the opportunities to learn and practice power system protection and control engineering. This has allowed us to contribute to Hydro One's success and advancements to the Ontario power grid and ultimately for the betterment of the people of Ontario.

We would like to convey our sincerest gratitude to Murched Ajami, Ian Bradley, Mark Ciufo, and Miroslav Kostic for reviewing our manuscript, providing their direction, and their continued support.

1

What Is Power System Protection, Why Is It Required and Some Basics?

1.1 What Is Power System Protection?

Our modern human civilization is dependent on the electric power system to enable all of its critical functions: food, health, sanitation, security, commerce, and progress. The electric power system is dependent on protections. By electric power system, we are referring to power generation and a network of wires that connect generation to the load locations where it is utilized to power the functions above. Protections consist of an assembly of electric components, and consequently, are better referred to as protection systems. Protection systems continuously monitor the equipment that the power system itself is comprised of for abnormal operating conditions. Protections are automatic systems that once an abnormal condition is detected, quickly as possible isolates the abnormal condition by the tripping of circuit breakers or the operation of fuses.

Power system protection systems are referred to as secondary equipment, as the primary equipment is transformers, lines, buses, generators, capacitors, breakers, disconnectors, etc. Primary equipment is directly involved with electric energy supply and delivery. Protection systems are designed and installed to oversee and "protect" primary equipment and the integrity of the power system.

In essence, power system protections "protect" power system primary equipment and, thereby, maintain system integrity and safety.

Protection systems are to a power system as a panel circuit breaker/fuse is to a household electrical circuit panel.

In addition to protecting power system primary equipment, power systems also employ remedial action schemes (RASs), previously known as special protection systems (SPSs), to protect the integrity of the power system. RAS/SPSs can monitor frequency, voltage, and operating contingencies that require immediate system correct actions, among others.

Power system protections are classified as "mission-critical" assets, as failure to operate or, if they do not operate as intended, have grave consequences to the continued operation of the power system.

A protection system itself is comprised of Individual devices, sub-systems, and numerous pieces of equipment as follows:

- Protection relays that monitor the power system for abnormal conditions.
- Communication systems that are used as part of the overall protection system functionality.
- Voltage and current sensing equipment that steps down high-power system values to much lower values capable of being input into the protection relays.

Power System Protection: Fundamentals and Applications, First Edition. John Ciufo and Aaron Cooperberg.

- Direct current (DC) auxiliary supply including batteries and their chargers used to power protection relays, auxiliary devices, communication systems and trip circuit breakers.
- Control circuitry working with protections to trip circuit breakers or other interrupting devices such as circuit switchers.

Most reliability organizations that oversee the adequacy of protections include the above-listed components as part of an overall protection system. Batteries are not included just the battery circuits. Also, circuit breakers are not included just the breaker trip coils are. However, batteries and breakers are key components of protection systems but fall under the jurisdiction of station engineering. The consequence of such definitions only impacts compliance and organizational accountability.

A typical protection system consisting of these components is illustrated in Figure 1.1 showing that a protection system consists of many components, or sub-systems: CTs, PTs, protective relays, auxiliary relays, control wiring, equipment mounting panels, DC power supplies, telecommunications, and breaker trip coils. A protection system, in the general case, is not just one device, or subsystem, it consists of several sub-systems, each containing several devices that represent the whole. To function correctly, each of the components or sub-systems must themselves operate correctly … it is a serial operation. Each of these sub-systems and their functions will be discussed in more detail in Chapter 2, Section 2.1

It is not possible to design an electric power system that is immune to equipment failures and abnormal operating conditions. Therefore, all power systems must deploy highly reliable protection systems that can quickly detect abnormal conditions and take appropriate actions to mitigate abnormalities.

In the normal state of a power system, there is a balance of electric energy sufficient to meet the needs of the connected load, in real-time, and the power system operating quantities such as voltages, currents, and frequency, are all within the design ratings of the primary equipment.

Abnormal conditions result when system faults occur that cause these operating quantities to deviate beyond equipment ratings. Protection systems are designed to monitor power system quantiles for such abnormalities and operate to isolate these fault events that cause abnormal quantities. One prominent operating quantity that is drastically impacted by such events is current. System faults also referred to as disturbances, can cause normal load current to increase from several hundred amps to 70,000 A which can cause major damage if not cleared in fractions of a second. Currents of such high quantities can cause thermal damage, mechanical damage, forces are so high that metal bus bars can bend, equipment failures, fires, safety issues, and a collapse of the power system if not cleared within the short-time ratings of primary equipment.

Some examples of system events that cause abnormal conditions are as follows: lightning strikes (Figure 1.2), wind, ice storms, animal contact, equipment failures (Figure 1.3), car accidents knocking down electrical poles/equipment, etc., that cause short circuits or broken connections. Such events are also referred in the industry, as faults. Faults and their types, causes, and how to calculate fault values will be further discussed in Chapter 6.

1.2 Why Is Power System Protections Required?

Power systems are designed, planned, and constructed to limit failure modes and equipment damage and thereby enhance overall system reliability.

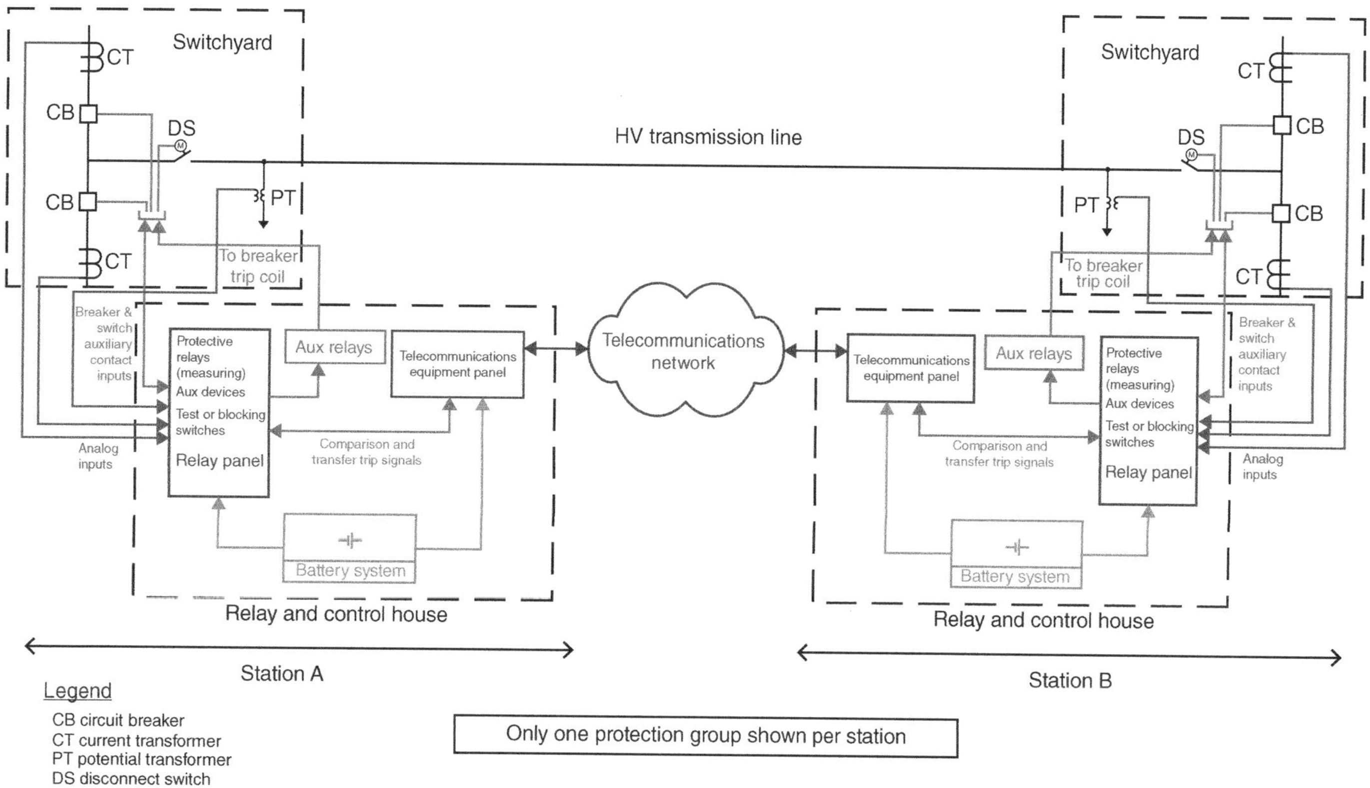

Figure 1.1 Illustration of a protection system for a transmission line [1].

Figure 1.2 Environmental risk lightning strike – Dallas Tx. Source: NOAA Photo Library / Flickr / CC BY 2.0.

Figure 1.3 Failed transformer on fire – Thessaloniki Greece. Source: Konstantinos Stampoulis / Firefighters.gr / Wikimedia Commons / CC BY-SA 3.0 GR.

The power system is designed to balance performance and minimize the cost of energy delivery. The planning, design, and implementation of a power system is a balance of initial capital costs and ongoing maintenance costs with the potential cost impact of power system equipment failure.

Power systems are exposed and subjected to environmental elements such as rain, snow, ice, lightning (Figure 1.2), storms, and other such environmental risks. These risks cause, power system's primary equipment components to make unwanted contact with other components, referred to as faults which result in fault currents in the order of 10–100 times normal load currents. Transmission lines have the highest risk of environmental elements due to their increased natural exposure to the environment.

It should be noted that protection systems **cannot** prevent faults or equipment failures. They detect an abnormality by monitoring quantities such as increased currents, depressed voltages resulting from failures. A limited number of protection devices can respond to failures without directly monitoring electrical quantities example are gas, temperature, and light-sensing devices.

1.2.1 Minimize Primary Equipment Damage

Power system equipment is designed and constructed to limit failure modes. However, power system's primary equipment can and does fail for the following reasons:

(1) Soon after installation, due to either a design or manufacturing flaw.
(2) Equipment failures due to prolonged operation beyond the equipment's rated design parameters.
(3) Equipment failures due to adverse environmental conditions such as salt pollution, animal contact, or high wind and lightning strikes during storms.
(4) Equipment operated beyond their normal expected life span.

Primary equipment is designed to withstand a certain level of fault exposure. Protection systems operating within that exposure period will minimize damage to the equipment and possibly prevent catastrophic failure thereby, also decreasing equipment outage time. It should be noted the cost and installation of some primary equipment such as generators and large power transformers are in the order of US$10's of millions of dollars each. More significant than the cost is the time needed to manufacture as it can take up to one to two years as this type of equipment is only made by custom order.

1.2.2 Provide Continuity of Service by Minimizing Outage Time and Service

1. Power systems are classified as critical infrastructure due to the 24/7 dependence on electric power for modern-day life. Automatic protection systems are designed to detect faulted equipment and remove the minimal primary equipment required to remove the abnormal condition and maintain a continued supply of electric power to as many customers as possible.
2. Protection systems are automatically designed to restore power system elements to service following the abnormal condition being removed. The automatic restoration of equipment back into service generally takes no more than one second. Fast fault clearing will promote power quality and prevent equipment with ride-through capability from powering down. However, the resultant momentary loss of supply voltage still affects many loads without such capability.

1.2.3 Promote Safety

Working with electricity can in general be dangerous. Engineers, electricians, and other workers especially those working around high voltage equipment such as in switchyards or near metal-clad switchgear equipment are at added risk should the protection equipment not perform correctly.

While the electrical shock is indeed a hazard, there are also more immediate hazards due to explosions because of the rapid expansion of very hot air resulting in exploding equipment with resultant flying shrapnel akin to an exploding hand grenade. Therefore, protection systems are fundamental to the monitoring, detection of abnormalities, and quick removal of such abnormalities to promote safety.

1.2.4 Maintaining Power System Integrity

Protection systems mitigate damage and maintain system integrity. An integrated power system can only tolerate fault conditions for a very short time. Faults must be removed very quickly by isolating the faulted power system equipment. This is usually done by tripping circuit breakers or by the opening of other dedicated isolating devices such as circuit switchers.

Typical clearing times for high voltage system faults are in the order of less than 200 ms. When faults are not isolated and eliminated quickly enough, power system instability usually results, affecting the integrity of the entire interconnected power system.

Fault clearing times are always specific to each power system with such factors as: voltage level, generation inertia, interconnectivity, and transmission topology are all involved and usually require sophisticated power system studies to determine.

1.3 Some Basic Protection System Terms and Information

1.3.1 Relay

In the most simplistic functional terms, it is an electrically operated switch. Relays monitor electrical quantities in a low voltage circuit environment to operate output contacts that are used to control/energize independent circuits, normally operating at much higher voltages.

A relay is defined as an electric device that operates to close contacts or by some other means to complete a circuit following specific predetermined input conditions being met.

A relay may consist of a single unit or multiple units all performing to achieve the same desired results. Most of the time the inputs are electrical but could also be actuated by heat, mechanical vibration, gas accumulation, etc.

For the official definition of a relay refer to the Institute of Electrical and Electronic Engineers (IEEE) Standard C37.90-2005 [2].

1.3.2 Protective Relays

In the electrical power industry, protective relays monitor power system quantities such as current, voltage, impedance, and frequency. They do so, to quickly identify a fault/abnormal condition, and isolate it by triggering a circuit breaker to open. They are intended to provide the "last line" of defense for the power system. The selection and applications of protective relays shall achieve reliability, selectivity, speed, and must coordinate/be selective with other protective relays.

Protective relays have been developed over the last 125 years and have gone through significant advancements. Initially, they were electromechanical devices and have transitioned from there to solid-state devices, to modern-day multifunctional microprocessor devices offering advanced functions and capabilities.

Protection systems and their components are utilized in all parts of electric power systems for the detection of abnormal conditions.

1.3.3 Protective Relaying

The electrical engineering science and discipline of protection system design and operations are generally referred to as Protective Relaying. This term has been used, due to the dominant use of protective devices called relays, in protection systems. The protective relay provides the measuring and intelligence of the protection system as shown in Figure 1.1 above. The various protective relay types and operations are described in Chapter 4.

1.3.4 Protection Engineering

Power system protection engineering is a specialty within the discipline of electrical engineering. Except for some post-graduate study courses, protection engineering is normally learned on the job and takes several years of application to become a protection practitioner.

Protective relays and auxiliaries and the systems architecture have evolved dramatically in comparison to other power system technologies. With the migration of protective relays from electromechanical to digital devices, modern-day protection system design and maintenance require knowledge of telecommunications, computers, and computer networking in addition to electrical engineering, power systems, power system equipment, power system operations, protective relay operations, and monitoring. It's for this reason that this specialized branch of power system engineering has gone from what was once viewed before the 1990s as being somewhat limited to being dynamic and career rewarding.

1.3.5 Protection System Objectives

1. Clear the fault fast to minimize damage, maintain the stability of the generators, and minimize shock to customer loads.
2. Clear selectively by only tripping those breakers closest to the fault. It is relatively easy to clear a fault fast; the challenge is to do it selectively.
3. Keep it as simple as possible without unwarranted complications. The cost of installing protection systems is relatively small compared to the capital investment in other system equipment. However, regardless of initial costs, ongoing operating and maintenance costs can be significant. Also, simple and well-understood protection schemes are less prone to misoperations, are more reliable and easier to maintain.
4. To allow the free flow of normal and emergency power without imposing load limitations.
5. Be effective under all credible operating conditions.

1.3.6 Protection System Characteristics

Protective relays in particular and protection systems, in general, must possess specific attributes to be effective. The required characteristics for the protective relay to perform its function properly are defined in terms of selectivity, sensitivity, and speed leading to dependability, security (see Section 1.3.7), and overall reliability as shown in Figure 1.4.

1.3.6.1 Selectivity

A key objective in protection design is that, for a given abnormal condition, the least number of protection devices operate to trip circuit breakers, so that the impact on the system and customers is minimized.

To achieve this, protective relays must be designed to differentiate between conditions for which they must operate and conditions for which they must not operate. This is essential to achieve the above objective and also to coordinate with other protection systems. This characteristic of protection design is referred to as selectivity where protection systems coordinate their operation and tripping of circuit breakers with respect to each other.

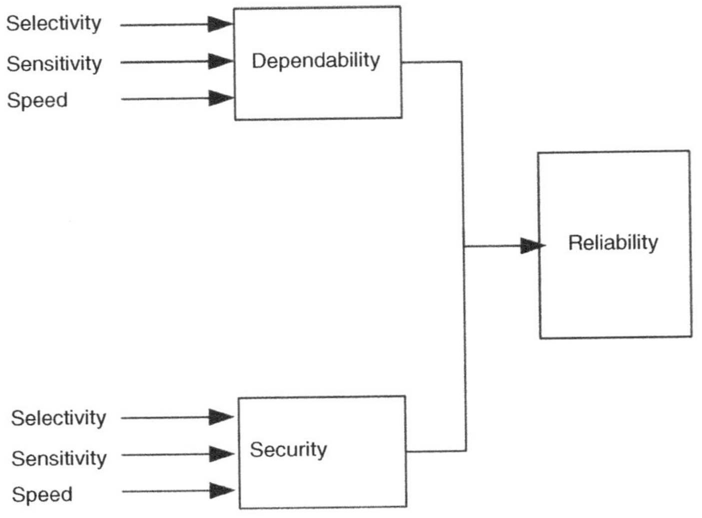

Figure 1.4 Protection system characteristics leading to overall reliability.

Different types of relays have different operating characteristics; therefore, selectivity must be established over the full range of fault and operating conditions. Discrimination methods, such as measured quantities, time, combinations of magnitude and time, distance, are used to provide selectivity.

1.3.6.2 Sensitivity

Sensitivity is a relay's ability to operate under the required minimum operating condition. In normal day-to-day operations of a power system, generation, lines, and other system elements are switched in and out to support maintenance and other operating requirements. The relay must be able to operate when power equipment is out of service for maintenance or otherwise leading to minimum operating conditions.

1.3.6.3 Speed

This refers to the time it takes for the protection system which includes the relay and associated auxiliary devices to operate and effect a change of output state upon the detection threshold conditions having been met.

Power system planners specify the required clearing times for abnormal conditions. These times are among the fundamental requirements used by protection practitioners to determine and design appropriate protection systems.

Protection clearing time, which depends on the speed of operation of the relay, among others, is important in clearing faulted or damaged power system equipment. Faulted or damaged equipment leads to dangerously high short circuits. Speed of protections impacts on the damage caused by short-circuit currents and maintaining power system stability.

Protection systems are required to isolate faulty element(s) within the power system parameters assigned to it. Overall system stability can be affected by faulted power system equipment that fails to be isolated within predetermined power system stability limits. The power system stability limits are dictated in part by voltage level and proximity to sources of generation.

System Planners conduct system studies taking into consideration operating contingencies to determine the maximum short-circuit clearing times necessary to ensure the power system remains stable.

System Planners also conduct system studies to ensure the adequacy of equipment short circuit withstands values such as circuit breakers that are rated for maximum short-circuit interrupting capacities.

It should be noted, that by reducing the margin of intentional time delay for protections that are intended to coordinate with respect to each other, loss of selective tripping of breakers could happen in certain circumstances.

There are relays in certain situations where intended built-in time delay is necessary to ensure selective tripping of breakers.

1.3.7 Protection System Reliability

Reliability refers to the ability of the protective system to operate correctly at all times. Overall protection reliability is defined by how dependable and how secure it is under all possible operating scenarios.

Dependability is a measure of the protective system to trip when required to do so.
Security is a measure of the protection system not to trip when not required to do so.

1.3.7.1 Dependability

Protection system dependability is achieved by ensuring that the protective relays and their application within the overall protection system operate when required.

Consider the following simple power system with a focus on protecting the transmission line, Line1. An abnormal event occurs on Line1 as shown below in Figure 1.5. When the protection systems operate correctly as designed; Line1, with the abnormal condition, will be isolated automatically by protections R1 at Station A, and R3 at Station B, thereby; tripping circuit breakers CB1 and CB3, respectively, at each line terminal station. When this happens, these protections are seen to have operated correctly as intended. In other words, the protections that were relied upon by design to isolate Line1 did so correctly and dependably.

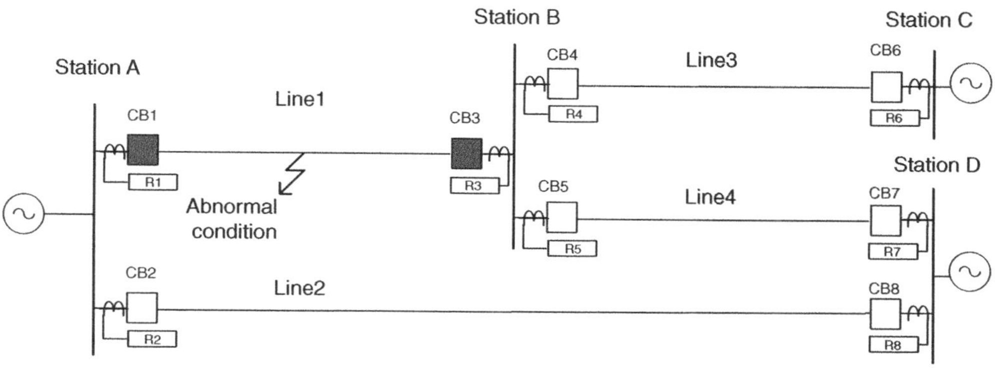

Figure 1.5 Example of a dependable protection.

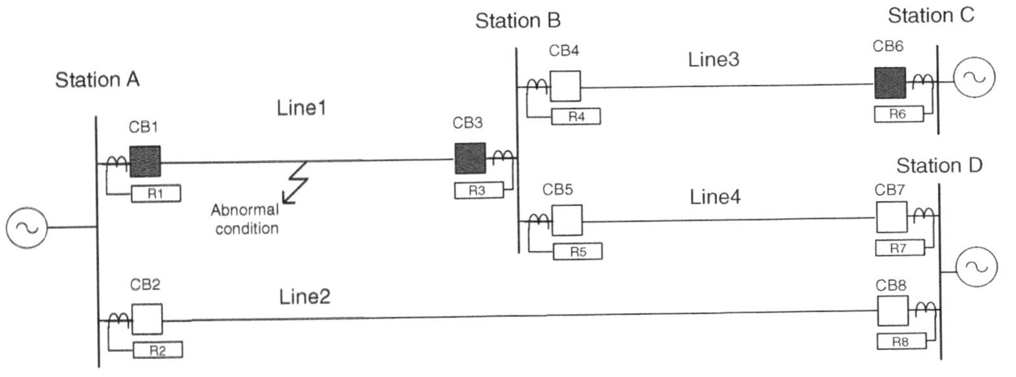

Figure 1.6 Example of a protection that is not secure.

1.3.7.2 Security

Protection system security is achieved by ensuring that the protective relays and their application within the overall protection system do not operate when not required. When they do operate correctly, the minimum numbers of circuit breakers are tripped to isolate the power system equipment with the abnormal condition.

It is vitally important that for power system integrity, that the minimum number of breakers only be tripped as necessary via the protections. Otherwise, more power system equipment is removed from service than is necessary. When this happens, it negatively affects load flows leading to higher operating costs. It may also lead to local blackouts and certainly makes it more difficult for line maintenance crews to determine the nature of and the location of the abnormal condition thereby, prolonging power system equipment to be out of service longer than necessary.

Consider the same example, an abnormal event occurs on Line1 as shown above in Figure 1.6. As in the previous example, protections, R1 and R3 will trip circuit breakers CB1 and CB3 at each of the line terminals. However, what if protection R6 at Station C also operates simultaneously with protection R3 at Station B to trip its CB6?

When this happens, protection R6 is seen to have operated incorrectly and not as intended. In other words, the protections that were relied upon by design to isolate only the abnormal condition on Line1 were not the only protections to operate. Protection R6 at Station C should only operate for an abnormal condition on Line3 and operated incorrectly for an abnormal condition on Line 1. Protection R6 has shown itself to not be secure protection as it should not have operated. When protection operates in this manner, not according to its intended design for an abnormal condition on another line, it does so incorrectly, and not securely.

1.3.8 Protection System Backup

It is of the utmost importance to ensure protection systems are dependable. It is impossible to guarantee that any system is 100% dependable as components fail from time to time in any system. Therefore, to ensure that circuit breakers are tripped there needs to be another independent protection that functions as its backup. There are two fundamentally different approaches to provide backup. Since these two approaches are so fundamentally different from each other with huge differences in installed costs, it can be said that each approach can be regarded as being a philosophy of backup. One philosophy is known as a remote backup while the other is known as a local backup.

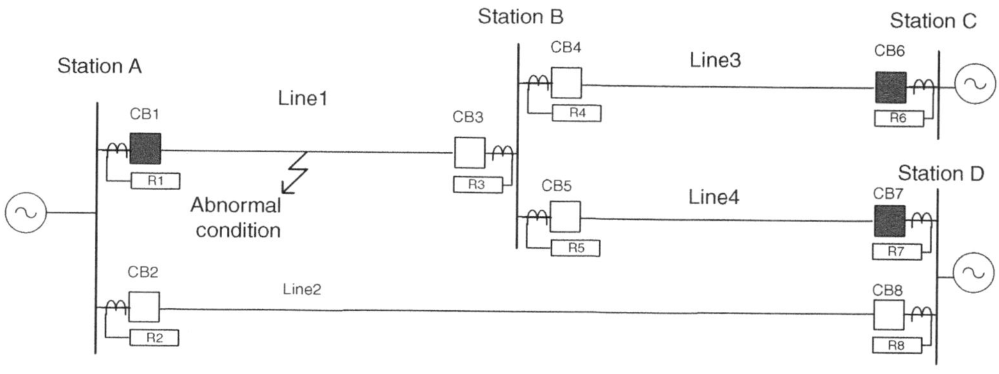

Figure 1.7 Remote backup.

1.3.8.1 Remote Backup

Consider the following simple power system where an abnormal condition occurs on Line1 as shown in Figure 1.7. Protection system R1 at Station A operates correctly to trip circuit breaker CB1. However, the protection system R3 at Station B does not operate at all leaving circuit breaker CB3 not tripped and unable to isolate the abnormal condition. Protection systems R6 and R7 at Station C and Station D are designed with backup systems to correctly identify an abnormal condition on Line1. This is done by intentionally time delaying the operation of protection systems R6 and R7 to wait for protection system R3 at Station B to operate first to trip circuit breaker CB3. When this does not happen after a predetermined set time, remote protection systems R6 and R7 operate to trip circuit breakers CB6 and CB7 at Station C, thus isolating the abnormal condition on Line1.

1.3.8.2 Local Backup

Consider the following simple power system where an abnormal condition occurs on Line1 as shown in Figure 1.8. Protection system R1 at Station A operates correctly to trip circuit breaker CB1. However, the protection system R3 at Station B does not operate at all leaving circuit breaker CB3 not tripped and unable to isolate the abnormal condition.

However, instead of relying on protections R6 and R7 at Station C to backup failed protection R3 at Station B two independent protections are applied at Station B. These two protections are designated R3A and R3B. These two independent protection systems when they operate similarly

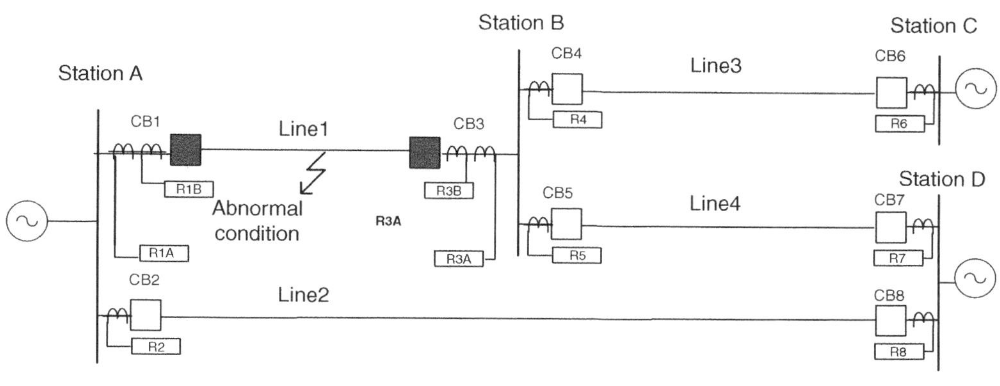

Figure 1.8 Local backup.

with respect to each other are known as being redundant protection systems. Operating similarly means if one is instantaneous without intentional time delay, they both are instantaneous. This means that for a common abnormal condition being detected they both operate simultaneously.

1.3.9 Protection System Redundancy

When applying local backup, overall protection system reliability is achieved by designing protection systems with adequate redundancy of equipment and functional adaptability to mitigate the risk of single component failure. Redundancy ensures adequate protection even when an unexpected failure of a protection system occurs. It is common practice to apply two functionally similar protection systems to achieve dependability. To apply three similar and independent protections would increase overall dependability but with an increased exposure to operation thereby reducing security. Overall, redundant protection systems are seen to strike the correct balance between reliable protections on the one hand and cost on the other.

Another advantage of redundancy is to allow temporarily taking one protection system out of service, let's say the A system to maintain and check it, and to rely on the other B protection system, all this being done while allowing the power system element to be protected and to remain in service. This is allowed since each protection system operates without intentional time delay and is done when system operating conditions allow for it when incremental weather permits, as faults primarily occur when there is lightning or high wind.

Some design features to provide redundancy include the following:

- Two, or more, independent protection systems
- Independent protective relay current sources in the form of separate current transformers
- Independent protective relay potential sources in the form of separate potential devices or independent secondary windings on a single device. The second method is considered acceptable when the digital relay itself recognizes there is a loss of potential supplied to it and adaptively applies other protections not requiring potential to operate.
- Duplicate power supplies from station batteries or redundant battery systems.
- Physical separation of protection and control equipment in substation control houses to minimize the chance of common mechanical damage affecting all the protection functions

References

1 Protection System Maintenance, Technical Reference, NERC, Sept. 13, 2007 (Originally drawn by J. Ciufo).

2 IEEE Standard C37.90 2005, Standard for Relay Systems Associated with Electric Power Apparatus.

2

Basic Power System Protection Components

2.1 General Description

Power system protections are an assembly of electric components, collected together as functional subsystems, with a simple goal to "protect" power system equipment and thereby, the integrity of the power system.

The application of power system protection is a specialized discipline that utilizes many components all interconnected to achieve the common goals of detecting and isolating abnormal system conditions. A well-designed and applied protection system should be as simple as possible while achieving its performance requirements. Those performance requirements very often dictate which components or sub-systems are utilized. Regardless of how it is achieved, there are basic components and sub-systems native to all power system protections.

The illustration in Figure 2.1 depicts the various components and sub-systems for two dual redundant transmission line protections at one terminal of a typical transmission line protection where communication systems are used.

Computerized protective relays and components offer the integration of some sub-systems functions such as trip auxiliary relays, telecom interfaces, and scheme logic. For the protection system as a whole to function correctly, each of the components or sub-systems must themselves operate correctly.

2.2 Power System Protection Components

2.2.1 Instrument Transformers

Protection systems continuously monitor the power system for faults and abnormal conditions. Primarily, they do so by measuring, at the protective relay location, the voltage, current, and possibly other electrical quantities. Transmission systems operate at voltages that range from 115,000 to 750,000 V, with fault currents that can reach up to 70,000 A or higher.

Protection system components are classified as secondary type devices that are designed and built to operate at voltages of 120/69 V and 5 A (in North America) or 1 A (Europe) nominal.

Therefore, instrument transformers are used as interfaces between the high-power transmission network and the lower power protection system, serving to protect personnel and protection devices from high voltage and currents. Furthermore, this allows for the use of the same relays independent of the voltage level. They also permit the use of reasonable insulation levels and current carrying capacity in relays and meters at a common base.

Power System Protection: Fundamentals and Applications, First Edition. John Ciufo and Aaron Cooperberg.

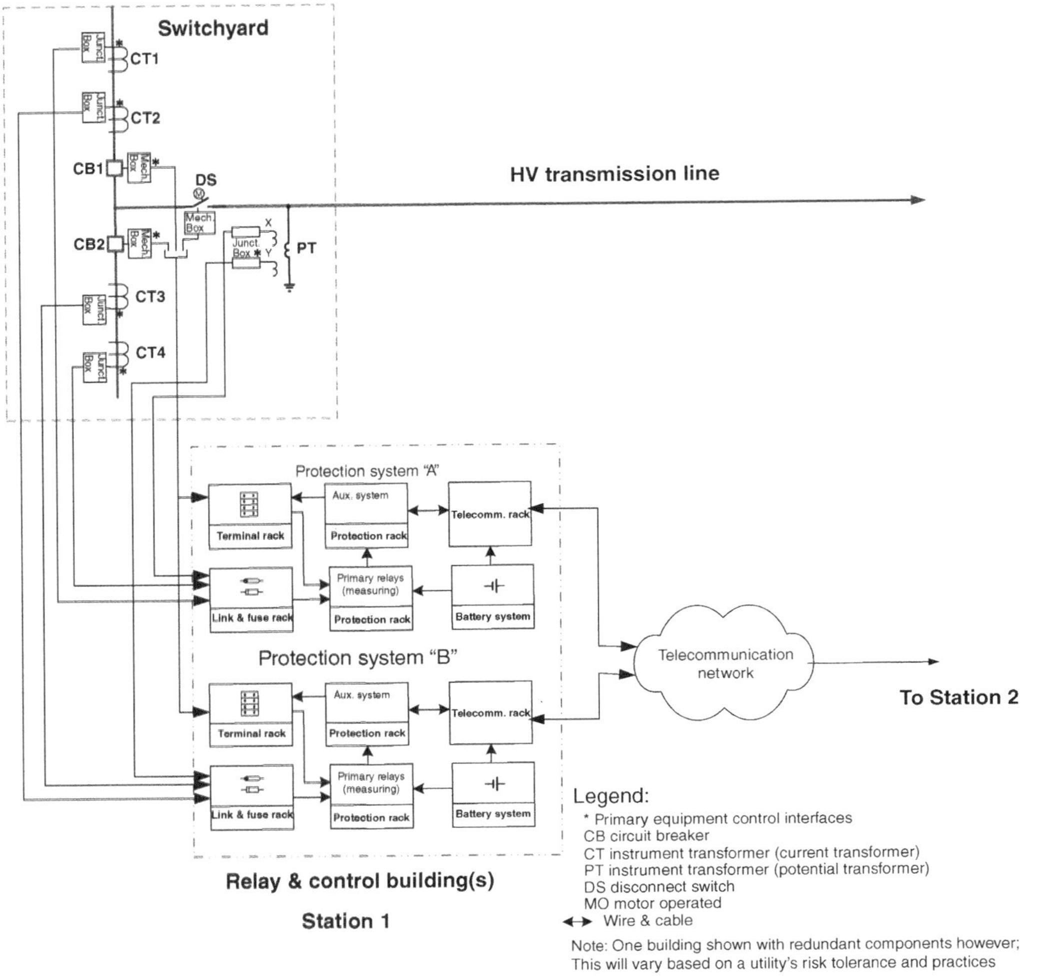

Figure 2.1 Components and sub-systems for a redundant transmission line protection at Station 1.

Voltages and currents are measured by relays only after they are reduced to levels that can be suitably used for both cost and safety considerations. Power system voltages and currents are stepped down using instrument transformers that are specifically designed and constructed for this application.

Current transformers (CTs) are built to withstand nominal load current continuously and high primary fault currents that in modern systems can exceed 70 kA temporarily. Many CTs are rated to withstand 100 A secondary winding current for three seconds. A CT with a current step-down ratio of 800:1, for example, will produce a secondary current of 87.5 A with a primary fault current of 70 kA.

Potential transformers (PTs) are built to withstand nominal power system voltages continuously and over-voltages temporarily.

Standard secondary voltages are typically 120 V phase-to-phase or 69 V phase-to-neutral. Standard secondary currents are 5 A in North America and 1 A in Europe while measuring power system full-load current. CTs are usually manufactured with several ratios taps to allow for the matching of full-load currents, short circuit values, relay setting sensitivities, and CT saturation levels.

Instrument transformers are required to provide, continuously, protective relays with power system current and voltage information. It is via these power system quantities that protection systems continually monitor the state of the power system. Instrument transformers, also commonly known as CTs, PTs, and CVTs (capacitor voltage transformers), reduce normal, power system higher operating voltages (500/230/115/44/27.6 kV) and high load currents (thousands of amps) to 120 V/69 V and 5 A or 1 A, respectively.

The performance of current measuring instrument transformers is different for protections compared to those used for metering. Instrument transformers supplying metering are intended to accurately measure load currents while those used for protections are intended to accurately measure fault currents. For this reason, the design and building of each type are fundamentally different from each other. Voltage measuring instrument transformers are usually also dedicated to either metering or protection applications. Metering applications in this context mean revenue metering. However, telemetry used in power system operator control usually obtains measured secondary quantities either directly from the same instrument transformers as those used by the protections or via the protective relays themselves when they are digital devices. This is discussed in more detail in Section 3.1.

2.2.2 Protective Relays

Protective relays also referred to as primary relays, represent the intelligence of the protection system. Modern protective relays are referred to as intelligent electronic devices (IEDs), and they integrate legacy sensing, scheme logic functions, and more recently, telecom interfaces, among others.

The power system instrument transformer's secondary low-level voltages and currents are inputs to the measuring functions of the protective relays – via the AC control cables. These monitored quantities or some combination thereof are compared against thresholds (termed "protection settings") that are pre-determined and programmed/set into protective relays. If the comparison indicates that a threshold is exceeded, it will trigger the decision block, where it will consider other possible factors before initiating a trip decision (breaker open, etc.). If a trip decision is warranted, then the protective relay will assert a trip output contact(s). The protective relay output contacts are wired into the control circuits for the associated protection zone's isolation breakers, and or disconnect switches where applicable, via their Mechanical (Mech.) Boxes, through a set of auxiliary relays, termination devices, and isolation devices. The result is that the closure of the protective relay's contact causes the isolation device(s) to open as shown in Figure 2.2 trip output block.

2.2.3 Auxiliary Logic

Electromechanical relay-based protection systems always required a lot of panel space due to their fundamental property of essentially being one relay per function. Up until the age of digital relays, all logic had to be implemented via discrete auxiliary relays and timers. Much design effort that took place from approximately 1928 till 1995 revolved around the application of auxiliary relay logic using discrete auxiliary relays and timers.

Following the advent of digital relays, the function of auxiliary relays and associated timers began to be replaced by the inherent property of digital relays to develop and implement logic all internal

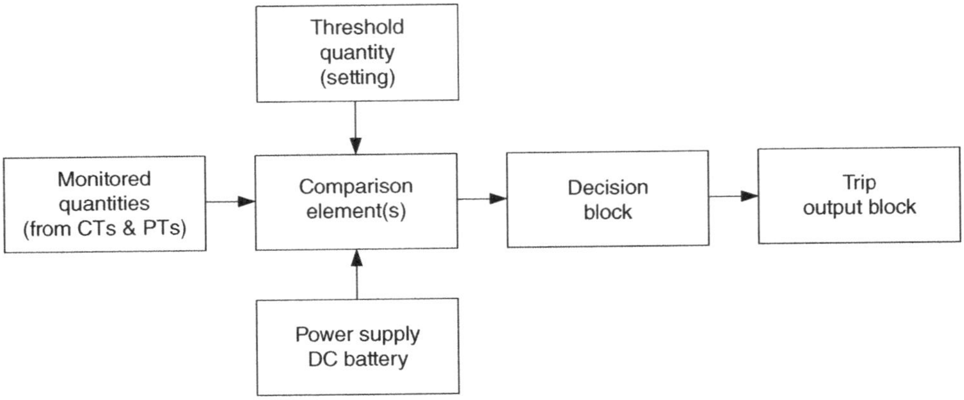

Figure 2.2 Simplified block diagram of protective relay.

to the digital relay itself. As the development of digital relays matured, the internal logic became more sophisticated and replaced most of all the auxiliary functions that were the domain of discrete auxiliary relays and timers.

2.2.3.1 Auxiliary Relays

Auxiliary relays, powered by DC, provide high-speed tripping, heavy-duty breaking, targeting, and latching functions. They offer numerous contact combinations per relay that may be used for the creation of relay logic. Electromechanical measuring relays such as overcurrent and distance relays would be of limited use without external auxiliary relay functions. Within this context, timers are included as being part of the auxiliary relay system.

Discrete auxiliary relays and timers were directly mounted on panels, either flush or surface mounted, up until about 1970. In the following decades, auxiliary relays began to be mounted in relay cases that were themselves either flush or surface mounted refer to Figure 2.6 and Figure 2.7 below.

Auxiliary relay cases were typically directly rack mounted on 19-in. racks. As a result, individual relays and other modular plug-in components such as target relays were combined to provide all auxiliary protection functions.

2.2.3.2 Application of Auxiliary Relays

Measuring relays in most cases have light-duty-rated tripping contacts. However, the making capacity of these tripping contacts is sufficient to trip a breaker. It's the breaking capacity that they cannot handle. Breakers universally are manufactured with a heavy-duty pallet switch wired in series with the trip coil. When the relay's light-duty contact trips the breaker, it draws current below the maximum making capacity of this initiating contact. The measuring relay trip contact will stay closed as long as the fault exists, (until the breaker is tripped and interrupts the fault current); typically, less than 100 ms for systems operating > 200 kV. Since the breaker pallet switch follows the breaker position, it will not open until the breaker is opened. Following fault clearance, the initiating relay contact will open. However, the breaker tripping current is already zero as the pallet switch has opened.

As protection systems became more sophisticated. Protections in addition to just tripping the isolating breakers, also initiated breaker failure, automatic reclosing, and the communication keying of remote trip, transfer trip, permission, etc. Therefore, tripping auxiliary relay schemes were developed.

The measuring relay initiating contact picks up a set of auxiliary relays in parallel, all connected to and energized by what is known as a tripping bus. The light-duty measuring relay contact did not have the breaking capacity to break the auxiliary relay load on the tripping bus. This was especially true for larger auxiliary relays whose coils were mainly inductive. The function that allowed this to function adopted by some utilities is the automatic seal-in circuit as shown in Figure 2.3.

Figure 2.4 illustrates the tripping of a breaker directly from a light-duty relay. All breakers come wired from the factory with a breaker "A" pallet in series with the trip coil.

Under normal circumstances, the measuring relay trip contact stays closed until the fault is cleared and will open last. The breaker "A" pallet will be the trip coil current breaking contact for which it is suitably rated. In summary, the breaker "A" pallet contact opens first before the measuring relay contact and interrupts the trip coil current; not subjecting the measuring relay contact to the trip coil current.

Refer to Figure 2.5 illustrating a circuit breaker trip control module – Two auxiliary relays are energized in parallel from protection trip contacts. A high-speed auxiliary relay is picked up to trip the breaker as quickly as possible, in 2.5 ms. A slow-speed auxiliary relay picks up in 50 ms, rated to

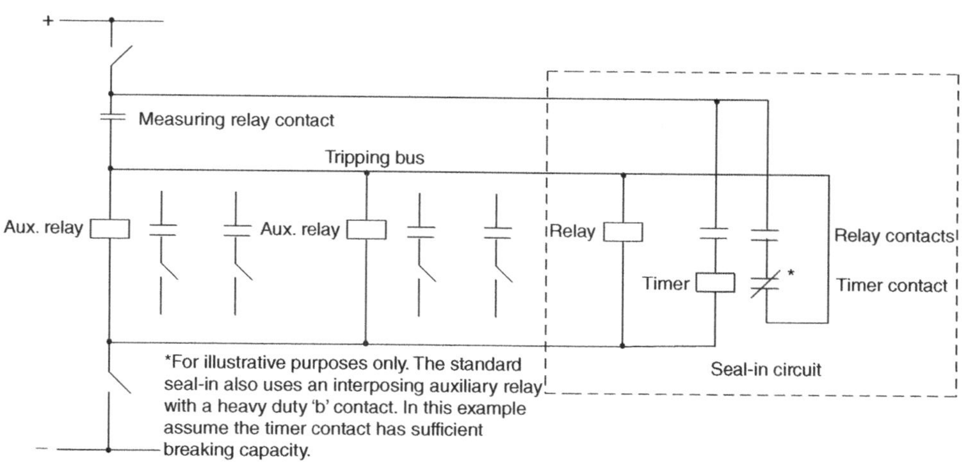

Figure 2.3 A conceptual illustration of a seal-in circuit.

Figure 2.4 Simple breaker trip circuit from a protection relay.

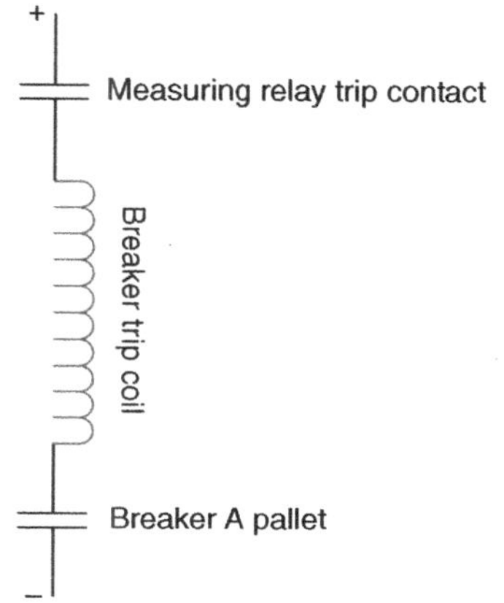

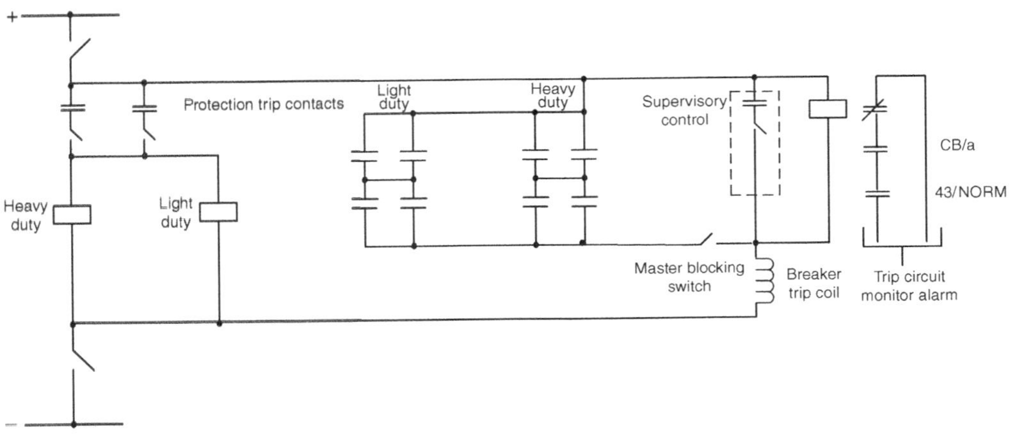

Figure 2.5 Conceptual illustration of a circuit breaker trip control module.

open the trip coil current. When the protection trip contacts open the light-duty contact relay will immediately drop out. The heavy-duty relay will drop out after another 50 ms. The four contacts in an "H" pattern for each are designed for maximum make (contact closing)/break (contact opening) current capacity.

A master blocking switch can be opened to block all protection trips. A supervisory control open breaker contact is in parallel with the master blocking switch from the protections.

The breaker trip coil is normally monitored. One such scheme, others are available, monitors the trip coil for continuity works by using a solid-state auxiliary relay with a high impedance that essentially operates on voltage only. It stays continuously picked up via receiving negative polarity (back negative) through the breaker trip coil itself. Its form "B" contact will close should the breaker trip coil circuit be open due to the breaker trip coil being open-circuited. The trip circuit monitor alarm is supervised by the breaker "A" pallet and the breaker test normal switch. The "A" pallet will disable the alarm whenever the breaker is open.

With the advent of digital relays, trip coil monitoring has been developed using an optically isolated digital input to the digital relay that tripped the breaker. More recent applications use features built into the trip-rated solid-state relay (SSR) outputs. The logic that uses the breaker pallet switch is implemented via a simple truth table that comes standard with those relays to further simplify wiring, testing, and maintenance.

2.2.4 Panels and Racks

Protection systems are assembled from many different types of components such as protective relays, auxiliary devices, wires/cables, terminations, and isolation devices. These components are all interconnected to form a protection system. The general industry-based philosophy of power system protection application is to divide the power system into protective zones which can be isolated electrically by disconnecting a minimum number of system elements. This logically divides the power system into the following power system-based equipment protective zones: lines, transformers, busses, capacitors, feeders, breakers, etc. Therefore, protection systems are developed, assembled, and organized into power system-based protection systems such as a line protection system and transformer protection systems. These systems are predominately

Figure 2.6 A set of legacy protection panels on several racks for a line protection system.

Figure 2.7 Mounting of legacy auxiliary logic.

assembled onto 19/24-in.-wide, by 84-in.-high, metal panels or racks. Components can be surface, or flush mounted onto these panels and racks; all of the current designs are rack mount. A single protection system can consist of a single rack or several racks (Figures 2.6 and 2.7).

Legacy auxiliary devices such as relays and timers are usually mounted in auxiliary cases that in turn are mounted on panels or racks. Older style electromechanical protective relays are also mounted in cases specifically made for them that then mount on panels or racks.

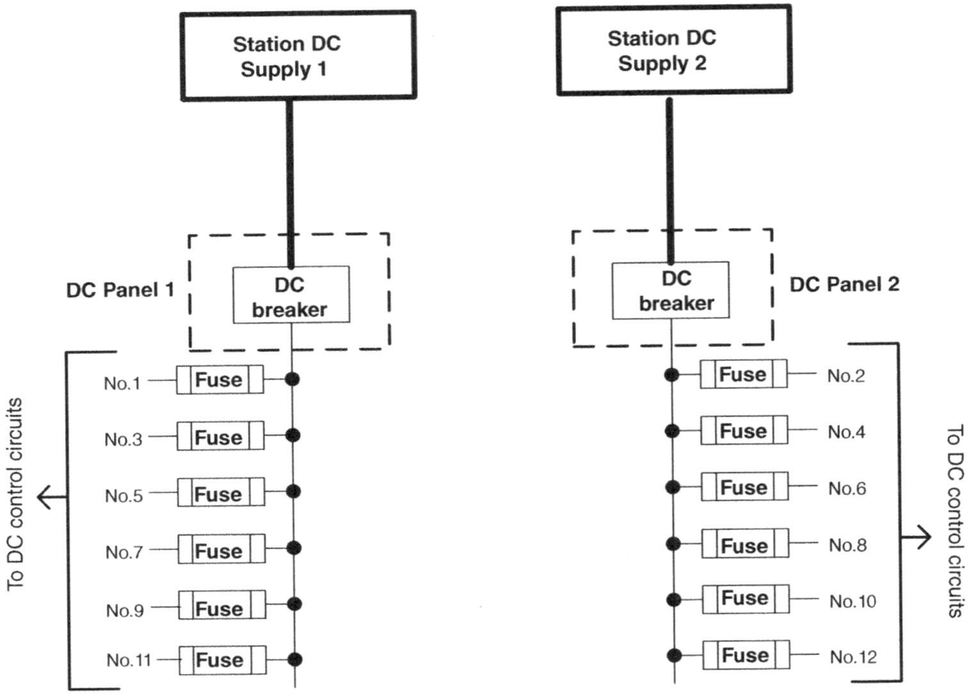

Figure 2.8 A basic battery system (not all components shown).

2.2.5 Battery Systems Used for Protections

Most protection components require a power source to function. Due to the mission-critical nature of these systems, DC power is used, as the protection systems must still function if there is a total loss of AC power to the station. Standalone battery systems are employed to power protection system components and the connected circuitry of isolation devices (breakers). Transmission stations typically use AC Station Service transformers connected to the power system to supply station AC requirements including the battery chargers.

A typical battery system consists of a battery (125, or 250 V DC) which in itself consists of hundreds of individual battery cells all interconnected to achieve the appropriate voltage and power requirements; a battery charger; DC distribution panels, similar to a residential electrical distribution panel; wiring, fuses, fuse monitoring devices, battery ground fault detectors, and a transfer scheme (Figure 2.8).

2.2.6 Telecommunications

Most protective relays operate based on intelligence gathered from measurements taken at the relay location. However, for some schemes, this is not sufficient information to maintain selectivity. This is the case for most high voltage transmission lines. It is common practice to use at least two zones of protection to ensure full line protection coverage. The first zone is set short of the remote end of the line and is designed to trip instantaneously. The second zone is set beyond the remote end and is time delayed before tripping, to provide selectivity, thereby, removing the minimum number of system elements from service. Most power systems cannot tolerate time-delayed tripping for line

Figure 2.9 Digital telecommunication equipment used for line protections.

faults. High voltage lines carry large power transfers; delayed tripping would cause the system to become unstable, and would result in a power system collapse.

To overcome delayed tripping and ensure power system integrity, more complex transmission protection schemes are used. They all rely on high-speed communications links between the line end terminals. These systems are normally referred to as pilot relaying systems, that is, these systems utilize communications paths to send signals from the relaying systems at one end on the line to that at the other end, thereby, resulting in instantaneous tripping for all line faults.

Telecommunications are critical in linking protection systems between stations, for the transfer tripping of remote power circuit breakers to isolate faulted elements such as transmission lines, and for the transfer of sophisticated data between systems necessary for secure operation (Figure 2.9).

2.3 Physical Implementation

2.3.1 Relay and Control Building

Unlike power system primary equipment designed for outdoor operation, protection and control systems are designed and manufactured to be housed in buildings. These buildings must be environmentally controlled for such elements as temperature, dust, and electromagnetic interference. Operation of these systems outside their normal intended operation environments will lead to misoperations. All of the above-mentioned sub-systems, except for instrument transformers, Mech. boxes, circuit breakers including the associated control circuits and trip coils and switchyard cables, are housed in Protection and Control Buildings or rooms.

In many locations, Protection and Control Buildings are air-conditioned and heated. Digital protection and control equipment are usually rated for a maximum ambient temperature of 55 °C.

2.3.2 Location of Instrument Transformers

Instrument transformers can either be installed in an outdoor station switchyard or indoor metal-clad compartment cells. The CTs are mostly an integral part of the high voltage transformer or breaker bushings when in outdoor switchyards and independent when located in an indoor metal-clad. Potential devices, whether in switchyards or indoor metal-clad, are almost always "free-standing" dedicated equipment. The instrument transformer outputs are connected via suitable cables/wiring first into a dedicated Junction Box adjacent to the instrument transformers in the switchyard. More cables are connected from the Junction Box into a relay and control room building, also referred to as a relay or P&C building (or room). Here, the cables are terminated onto a terminal link and fuse racks within the P&C building from which more cables and wires are distributed to the ac inputs of protective relays mounted on relay panels. Modern indoor metal-clad are built to arc-proof standards that allow the protection equipment to be installed directly into arc-proof instrument compartments that are integral to the metal-clad cells themselves. For metal-clad switchgear applications, the CTs and PTs are connected to the protective relays via intervening terminal blocks within the metal-clad compartment cells.

2.3.3 Terminations

Terminations include such items as terminal blocks, crimp lugs, and CT links and fuse holders. A typical protection system is mounted on either a single or on multiple panels. Each one of these panels has its own set of terminal rails holding typically anywhere from 50 to several hundred terminal blocks. The various protection panels interface with each other utilizing these terminal blocks.

Protection relays have their own terminations, which are typically blocks, with screw-in terminations found on the back of each relay. Wires are connected to these blocks via crimps at the end of each wire that is screwed in.

Current links are sometimes used to terminate AC cables coming from CTs typically located in the switchyard to facilitate isolation of the protections from CTs which must be shorted while being disconnected from the primary protection relay. An alternative to current links adopted by many utilities is to terminate CTs directly onto dedicated switches designed for this purpose. These dedicated switches work by shorting the incoming CT secondary leads using automatic spring action shorting devices whereas current links require the manual placing of shorting bars onto the incoming CT links. CTs as opposed to voltage transformers cannot be operated with primary current flowing and the secondary windings being open-circuited. To do so is to create extremely high and dangerous voltages across the CT secondary leads.

The AC cables coming from potential transformers PTs whether they are located in an outdoor switchyard in metal-clad cells are typically terminated onto fuse holders. Some means of determining whether the potential is lost due to open fuses are used. In legacy systems, simple pilot lights are used for this purpose. Modern digital relays come with a native loss of potential type logic used for this same purpose.

2.3.4 Protection Isolation Devices

DC blocking switches are used for the isolation of all types of DC inputs and outputs that interface with the protection system. For example, all outputs that cause tripping of circuit breakers are wired through blocking switches connected in series with the tripping path.

Blocking switches provide two important maintenance functions. By allowing outputs to be blocked, testing the protection system's correct response to a simulated fault condition can be performed without actually tripping the protection zone. The second one is to provide a metallic point for applying DC voltage to various inputs to artificially simulate inputs to the relay representing the change of state. Examples are the position of breakers disconnects either open or closed, telecommunication signals for tripping or blocking as well as signals from other primary relays such as breaker failure relays vital to the logic used by the protection system in its operation. The reason is that digital relays almost always receive an external change of logic state by applying DC voltage to what is known as digital inputs.

As modern systems which use digital signals over fiber optic cable are deployed, the mechanisms for assured blocking and testing of input signals become more sophisticated and complex.

2.3.5 Wiring and Cable (Control Wiring)

There are two broad categories of cables used, AC wiring and DC wiring:

AC wiring includes cables from current and voltage measuring instrument transformers CTs and PTs that are located in outdoor switchyards or indoor metal-clad. Individual AC cables typically include six conductors for three-phase systems to allow for Delta Connections. Where it is a certainty that only Wye or Wye-Ground connections will be made four conductors may be used.

DC wiring includes all wiring associated with the interfacing of protections with other protections or systems such as telecommunications as well as the final tripping of circuit breakers. Included within this category is the auxiliary supply of DC voltage from station batteries used by the protection system for all of its various functions. Individual cables usually have many conductors with numbers such as 12, 24, and 48 conductor color-coded wire pairs being common in a single-bundled cable (Figure 2.10).

2.4 Power System Isolation Devices and Control Interfaces

2.4.1 Isolation Devices

Protective relays are the "brains" of any protection system that must work with isolation devices such as circuit breakers or sometimes circuit switchers. They also automatically isolate equipment via motorized disconnectors. These devices are rated for and capable of providing the necessary isolation of power system equipment. An isolation device such as a circuit breaker is required to safely open and isolate the faulted power system equipment, as these types of isolators are high-energy devices specifically designed and manufactured for this purpose.

Protection systems are connected to breaker trip or open coils within the circuit breaker. Once a circuit breaker trip coil is energized, it initiates the tripping/opening of the breaker isolator to open up the electric circuit. Modern breakers are typically designed to open, once a trip control is initiated to it, within 2–3 cycles for 500/230 kV type SF6/air type breakers, respectively.

Figure 2.10 An illustration of relays, panels, terminations, wiring, and isolation switches.

2.4.2 Control Interfaces

All power system primary equipment such as circuit breakers, disconnect switches, and load switchers, have complex interfaces used to connect and implement primary equipment control logic and schemes. They interact with the protection and remote supervisory control systems and they provide local equipment operation. Primary equipment is installed within the switchyard and they interface via their mechanical wiring box, normally referred to as the Mech. Box. The DC-powered interface circuitry is contained within the Mech. Boxes. Control cables are wired from the Mech. Boxes in cable pans to Protection and Control Buildings, where they are terminated onto terminal racks for further distribution.

2.5 Redundancy Arrangements

2.5.1 Instrument Transformers

To ensure reliability, it is common in the Grid Codes of most jurisdictions to require at least two separate Instrument Transformer (IT) sources for protection systems.

Figure 2.1 shows a common arrangement that addresses the requirement for current measurement redundancy and independence requirements. CTs are required to provide separate secondary

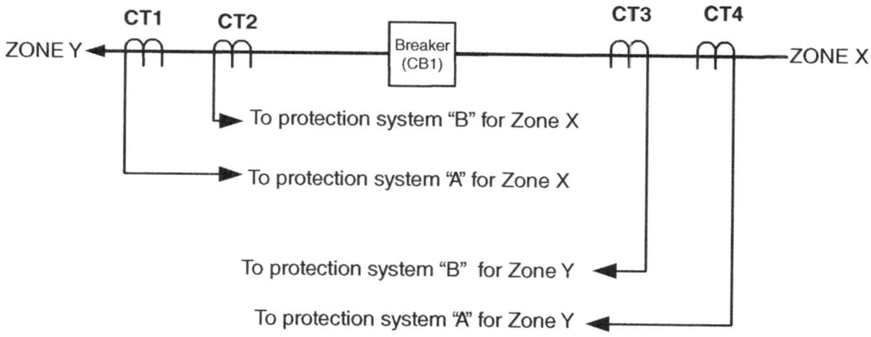

Figure 2.11 Example of redundant CTs.

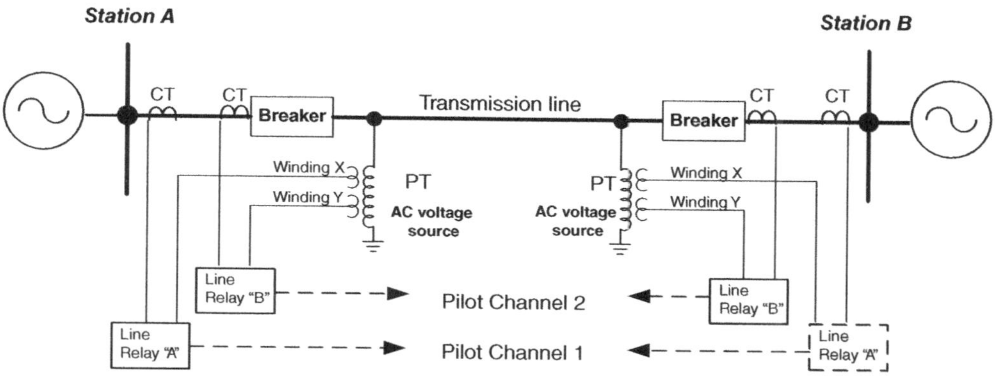

Figure 2.12 AC voltage inputs.

AC current sources for each redundant protection system. This is required so that a shorted, open, or otherwise failed CT circuit will not result in the failure of both redundant protection systems. Figure 2.11 shows the use of four CTs from a breaker with bushing CTs to separate the current measurement for the two protection systems for zones A and B. Protection zoning will be discussed later in the text in Chapter 5.

For potential measurement redundancy, it is typical to require at least two separate secondary windings supplying voltages for Protection Systems. This is required so that a shorted, open, or otherwise failed voltage circuit will not result in the failure of both redundant protection systems.

Figure 2.12 shows a potential device with two independent secondary voltage windings. The two secondary voltage sources are utilized independently by the two protective relay systems. Both Protection Systems in Figure 2.12 requires voltage measurements to perform their protective functions and must have separate secondary sources as illustrated.

It should be noted that fully redundant potential sources whether PT or CVT would require additional space and would incur higher costs; therefore, it is acknowledged that the single primary winding, representing a single common mode failure. This practice is normally acceptable in most jurisdictions.

To ensure that protections are dependable, typical protection design practice is to utilize two independent protective relays for each of the transmission primary equipment being protected. These relays may be located at the same terminal to provide what is known as redundant A and B protections or may be located at two different terminals in what is known as local primary and

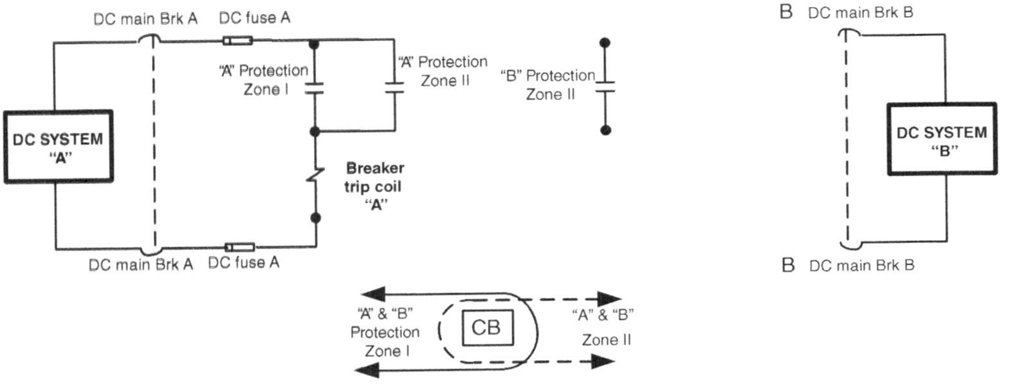

Figure 2.13 Breaker trip circuits illustration.

with remote backup protection. The specific scheme used whether redundant or local with timed backup is dependent on meeting the power system stability limits in terms of maximum tolerable fault clearing times for that power system. Redundancy is more expensive to implement.

2.5.2 Dual Breaker Trip Coils

Most transmission circuit breakers are purchased with two independent trip coils to meet reliability standard requirements. The protection systems and each trip coil must be operated from independent DC control circuits to avoid a common mode failure disabling the protection completely (Figure 2.13).

The breaker trip coil initiates the action that operates the breaker to clear a fault. Therefore, its failure to operate will cause delayed clearing times and possible loss of coordination and selectivity. Redundant protection system output trips are issued to independent trip coils, for breakers that are fitted with dual trip coils.

3

AC Signal Sources

3.1 Introduction

Ideally, protection relays would measure voltages and currents directly on the power system. In this manner, no errors in magnitude or shift in phase relationship would exist. This of course is simply not possible. There is no choice but to resort to the use of instrument transformers that reduce power system voltages and currents to levels that can be used safely and effectively by protection devices. Unfortunately, the laws of physics apply and errors are introduced. Nevertheless, instrument transformers, properly selected and configured into a protection system are capable of reproducing a replica of system electrical quantities within acceptable limits of error.

Various standards organizations have created criteria or accuracy classifications that manufacturers must abide by. A higher accuracy class comes at a higher cost. Consequently, there is always a cost–benefit analysis to be done when specifying instrument transformers. The protection system designer will specify the accuracy class that achieves the right balance between the quality and cost for the given application.

3.2 Current Transformers

Current transformers (CTs) are used for two main applications:

1. to supply measurements for operating telemetry and revenue metering, and
2. to supply measurements to power system protective relays.

These two applications impose different requirements. Metering demands high-accuracy transformation under normal steady-state conditions, and the performance during fault conditions is unimportant. Protective relaying requires reasonably accurate information up to the highest fault current levels possible, and this accuracy should be maintained whatever the shape of the primary current waveform. The focus in this book is on protection system applications.

The primary winding of a high current rated CT usually consists of a single turn, obtained by running the power system's primary conductor through the CT core. The normal rating of CT secondaries is standardized at 5 A in North America, whereas 1 A is the standard in Europe and other parts of the world. A CT is also called a series transformer because the primary, usually one turn, is connected in series with the line or equipment.

Power System Protection: Fundamentals and Applications, First Edition. John Ciufo and Aaron Cooperberg.

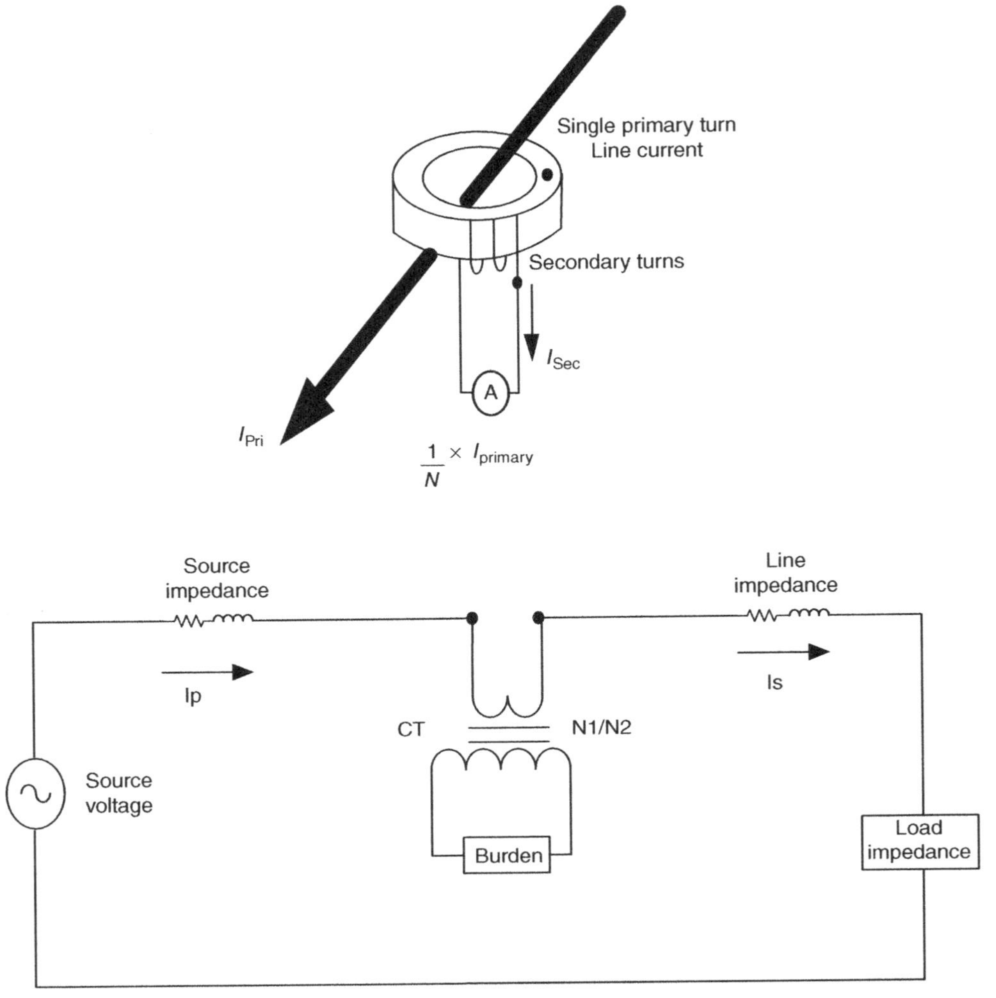

Figure 3.1 Typical series connected CTs.

Referring to Figure 3.1, the primary windings of CTs are connected in series with the power circuit and in series with the system impedance which governs the current through the primary windings. The secondary windings are connected in series with current elements of protective relays.

A CT is fundamentally similar to voltage transformers (VTs). The main parts of a CT are its primary and secondary windings. These windings are linked by an electromagnetic flux contained within a magnetic core made with silicon steel of high magnetic permeability. The primary in most designs does not have an actual winding at all but a single current-carrying conductor. Rather than a time-varying voltage applied across a primary winding inducing an electromagnetic flux in the core, the alternating through current in the primary single conductor creates a time-varying magnetomotive force and magnetic flux density corresponding to the material permeability. Figure 3.2 below depicts a simplified equivalent circuit of a CT.

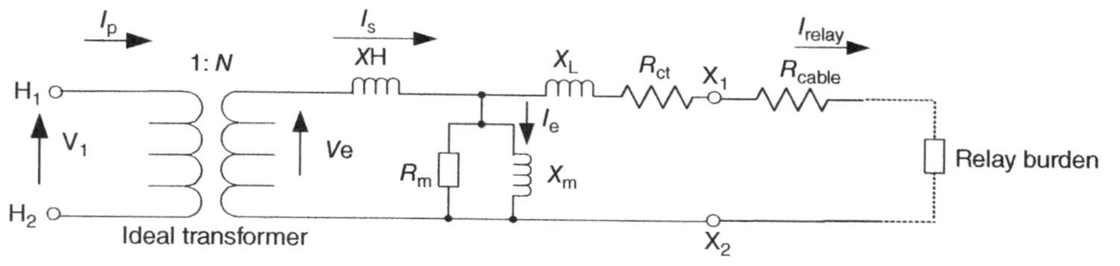

X_m is the magnetizing branch reactance and Rm represents the I^2R heating loss

X_H and X_L are the primary and secondary leakage reactance

R_{ct} is the CT internal resistance as specified by the CT manufacturer

R_{cable} is the resistance of the cable from the CT typically in a switchyard to the relay location usually in a control house

Figure 3.2 CT equivalent circuit.

As noted, CTs are connected in series as depicted in Figure 3.1. This is as opposed to VTs that are connected in parallel. A CT is wired in series with the source such that the CT primary current is the phase current of the equipment that is being measured. Since the purpose of a CT is to measure current, the impedance of the CT is much less than the impedance of the circuit being measured (ideally zero). The CT secondary current is not determined by the impedance of the load connected across the secondary, as is the case of a VT, but is entirely dependent upon the amount of current flowing through the primary, which is in a series circuit. Therefore, CTs can be viewed as a constant current source; CT secondary current will flow as long as the measured equipment phase current is flowing, and the CT does not significantly saturate (see Section 3.2.8).

As shown in Figures 3.1 and 3.2, the primary current passes through the center of the core creating a magnetomotive force that induces a magnetic flux in the core. This in turn induces a voltage across the secondary winding leads at the terminal block (`X1`, `X2`). This voltage will rise in magnitude until a secondary current begins to flow into whatever device is connecting to the secondary leads which can produce the countering magnetomotive force from the primary. The current transformation is inversely proportional to the number of winding turns.

Most high current rated CTs that are used for protection systems are classified as low leakage reactance type L or C. The secondary windings, for these types, are uniformly distributed around the core so that any magnetic flux produced by the current links all turns of all windings thereby making the leakage reactance negligible. On multi-ratio types of CTs, each of the taps is also uniformly distributed. Therefore, the CT equivalent circuit can be further simplified by removing the X_H and X_L components as shown in Figure 3.3.

$$V_e = (\mathrm{N2}/\mathrm{N1})/\mathrm{V1}; I_S = (\mathrm{N1}/\mathrm{N2})/I_p; I_{relay} = I_S - I_e; \mathrm{N1} = 1 \text{ for most except for auxiliary CTs.}$$

If there was no magnetizing branch, all the secondary current of the CT would flow into the burden. However, the non-linear magnetizing branch draws a current I_e to magnetize the core. It is the value of this current that causes the difference between the actual output of the CT, I_S, and the current into the relay, I_{relay}, that produces the error seen as an apparent ratio error as discussed further in Section 3.2.6.

The CT accuracy is proportional to the magnetizing current. As the CT operating point moves toward the saturation area of the excitation characteristic, the error of the CT will increase. It should be understood that the high exciting current of the larger core at low load is not considered significant for protective relaying.

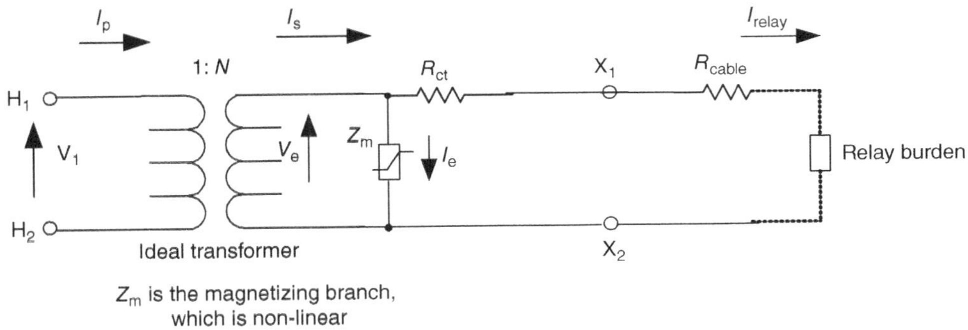

Figure 3.3 Simplified CT equivalent circuit.

The flux density level, with which CTs operate, differs widely, from load currents to fault currents. Fault currents are many orders of magnitude larger than the load current. CTs, as a current source, are expected to operate up to the system fault currents within the linear range of the magnetization curve. To facilitate this performance in the entire range of its operation without significant errors, the core of the CT is designed to operate at low flux density at rated current.

Devices supplied by CTs are known as burdens. The term burden is used for CTs instead of the more familiar term load used for devices supplied by VTs. This highlights the fundamental cause and effect difference between the two types of transformers. The voltage across the secondary of an energized VT is the causative factor and will drive current through the load connected to it according to the load impedance. As load impedances tend to be in a range from large to infinite, the current effect is relatively small and the secondary voltage can remain fixed according to its voltage transformation ratio and transformer regulation characteristic. There is no intrinsic tendency to push current into a connected device.

However, in an energized CT, the current is the causative factor and will drive a voltage across the admittance connected to it. As admittances tend to range from small to zero, the voltage effect for a given current can be extremely large. It is due to this electrical inverse relationship that devices connected to VTs are known as loads (impedances) and devices supplied by CTs are known as burdens (admittances). A large load is a low impedance, and a large burden is a low admittance.

The magnetic and other forces of high fault current can destroy CTs. CTs should be capable of withstanding currents of up to 20 times their normal continuous secondary current rating which is experienced typically under fault conditions for a short duration. This duration of time must be considerably longer than it takes to operate relays and trip breakers. The CT manufacturer will specify this maximum time.

A CT is a constant current source; when a CT is open-circuited, all the current flowing from the primary winding must be magnetizing current. Since the magnetic circuit is designed for low magnetizing current when the transformer is loaded, this large increase in magnetizing current will build up an enormous flux in the magnetic circuit and cause the transformer to act as a step-up transformer, inducing a very high voltage across the terminals of the secondary. It will continue to induce a voltage across the secondary leads until extremely high spikes of voltage, whose peak can reach several thousand volts, could occur. This can cause an electrical arc across the secondary leads much like an arc welder.

It is extremely important to strictly adhere to the concept that a CT must never be left open-circuited when primary current is flowing through it. A VT of course is just the opposite, where the secondary must never be shorted when voltage is applied to the primary winding unless done under very controlled conditions usually to test the transformer to calculate the percent impedance.

CTs are designed and built to withstand secondary fault currents up to 100 A for the short time required to clear a fault including the extreme conditions when first-line breakers fail and backup systems need to operate. However, not only the CTs are at risk from high sustained fault currents but so are the relays. Protection relays are also designed to withstand 100 A for a short period of time before they are either seriously damaged or even catch fire.

3.2.1 Current Transformer Secondary Burden

There are two types of CT connections widely used in protections applied to three-phase systems. The first is the wye connection whereby the CTs are connected in wye and so are the relays. Refer to CT connections in Tables 3.1 and 3.2 showing such a connection. In this example should the three-phase fault current be 100 A the relays will also be exposed to 100 A until they operate to trip the breakers. The second is the delta connection also shown in Tables 3.1 and 3.2 whereby the current into the relay is $\sqrt{3}$ times the individual phase current through the CT primary. As an example, when subtracting I_A vectorially from I_B, the resultant $I_A - I_B$ is $\sqrt{3}$ larger than I_A. The secondary burden of a three-phase connected CT for various phase connections and different types of faults can be calculated as shown in Table 3.2.

Digital relays are always supplied by wye-connected CTs. Where delta-connected CTs are required, for example when zero-sequence current needs to be blocked from entering the relays, it can be done internally in the digital relay algorithm. The only way to do this with electromechanical relays is by actually connecting the CTs in delta, (see chapters 7, 8).

There are significant advantages to supplying relays from wye-connected CTs with the star point being made in the P&C Control House where the relays are located. The first and most obvious reason is that only a four-conductor CT cable is required from the CT location to the P&C Control House. CTs connected in a delta at the P&C Control House, where the relays are located, require a six-conductor CT cable. It is more expensive to purchase and install six-conductor over four-conductor cable. A second and less obvious reason is that digital relays also record the oscillography of the fault currents for analysis. The digital relays record the phase currents before the wye connection is made. Where delta connections are required for reasons described later in the text, the digital relays arithmetically compute the phase currents into delta thereby preserving the individual phase currents for oscillography.

Typically, utilities bring all the CT leads into the P&C Control House where they are connected either in wye or in a delta at that location depending on the application. For digital relay applications, the CT leads are always connected in wye.

Definitions:

R_b is the total relay burden seen by the CT
R_s is the CT internal secondary winding resistance
R_{cable} is the resistance of the cables connecting the CT to the relay
R_{relay} is the resistance of the relay

Since protections must cater for all possible types of faults, when calculating the total burden, the single phase to ground fault is used.

$$\text{Relay burden } R_b = R_s + 2R_{cable} + R_{relay} + RG_{relay}$$

Table 3.1 Summary of CT secondary burden by CT connections.

CT connection	3-PH fault	PH-PH fault
Wye Connected at P&C Control House	$R_b = R_s + R_{cable} + R_{relay}$ [a)]	$R_b = R_s + R_{cable} + R_{relay}$
Delta Connected at P&C Control House	$R_b = R_s + 3R_{cable} + 3R_{relay}$	$R_b = R_s + 3R_{cable} + 3R_{relay}$

CT connection	Phase – ground fault
Wye Connected at P&C Control House	$R_b = R_s + 2R_{cable} + R_{relay} + RG_{relay}$
Delta Connected at P&C Control House	$R_b = R_s + 2R_{cable} + 2R_{relay}$

a) This equation assumes four-wire cable is brought into the P&C Control House. Where six-wire cable is brought into the P&C Control House, then this equation becomes $R_b = R_s + 2\,R_{cable} + R_{relay}$

RG_{relay} is mainly applicable to electromechanical applications where there is an independent ground relay exclusive of the phase relays. For digital relay applications, where multiple elements exist in one device, only R_{relay} is applicable.

3.2.2 Current Transformer Types

There are various types of CTs manufactured to match specific applications. Some applications are high to medium voltage located in switchyards or in metal-clad switchgear, while others are lower voltage usually located in on or near relay panels.

Most CTs are a variation of a "donut"-shaped iron core with windings wrapping around the core whose leads are taken to a common set of terminal blocks. The iron core itself can be made of degrees of quality in terms of magnetic permeability. As a general rule, the better the magnetic characteristic of the iron core, the higher its accuracy rating and the higher the cost. The current passing through the center of the iron core is known as the primary, and the winding wrapped around the iron core is the secondary.

3.2.2.1 Bar-Type CT

A bar-type CT has an insulated straight conducting bar that passes through the iron core and is fixed in the manufacturing of the device where external primary connections are made. A secondary winding wraps around the iron core. This bar then becomes the primary conductor of the CT. This type of CT is usually used for outdoor applications at distribution voltage levels typically at 13.8 or 27.6 kV with suitable basic impulse level (BIL) for outdoor use (Figure 3.4).

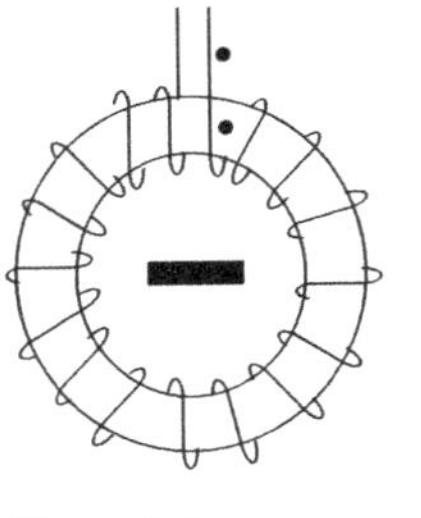

Figure 3.4 Bar-type current transformer.

The bar-type CT is mainly used when ratio matching is required for electromechanical relays. These types are being phased out as digital relays no longer require external ratio matching as ratio matching is achieved in the relay itself via a programmable setting.

Table 3.2 CT secondary burden by CT connections.

Connected CTs	Burden	Fault type	Connection drawing
Wye @ P&C Control House	A, B, C CTs each have a burden R_b Total burden per CT $R_b = R_s + \text{Rcable} + R_{relay}$	3PH	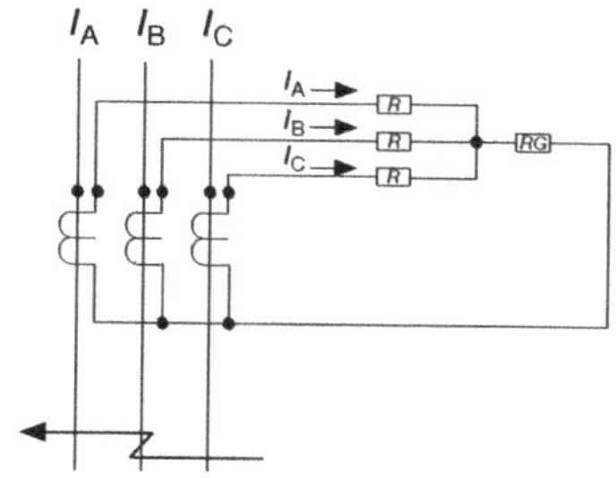 Three-phase
Wye @ P&C Control House	A, B CTs each have a burden R_b. Total burden per CT $R_b = R_s + R_{cable} + R_{relay}$	PH-PH	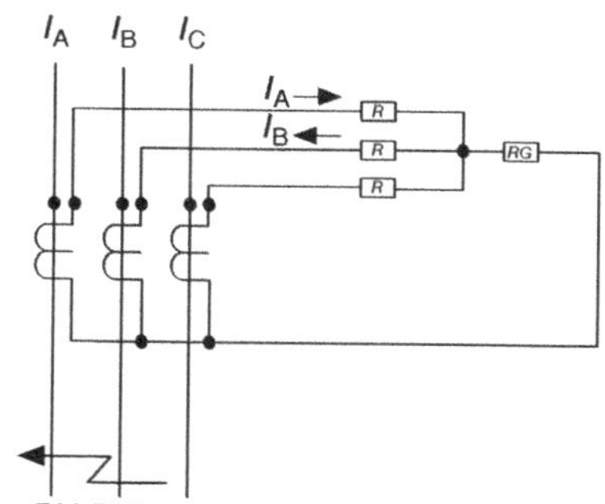 PH-PH
Delta @ P&C Control House	A, B, C CTs each have a burden R_b. Total relay burden $R_b = R_s + 3R_{cable} + 3R_{relay}$	3PH	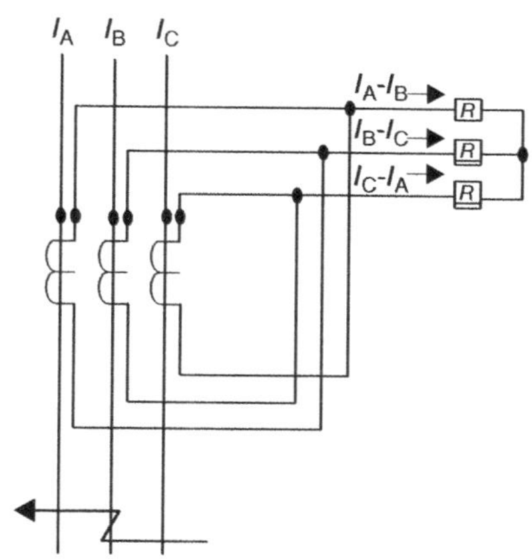 Three-phase

(Continued)

Table 3.2 (Continued)

<table>
<tr><th>Connected CTs</th><th>Burden</th><th>Fault type</th><th>Connection drawing</th></tr>
<tr><td>Delta @ P&C Control House</td><td>A, B CTs each have a burden R_b.
Total relay burden
$R_b = R_s + 3R_{cable} + 3R_{relay}$</td><td>PH-PH</td><td>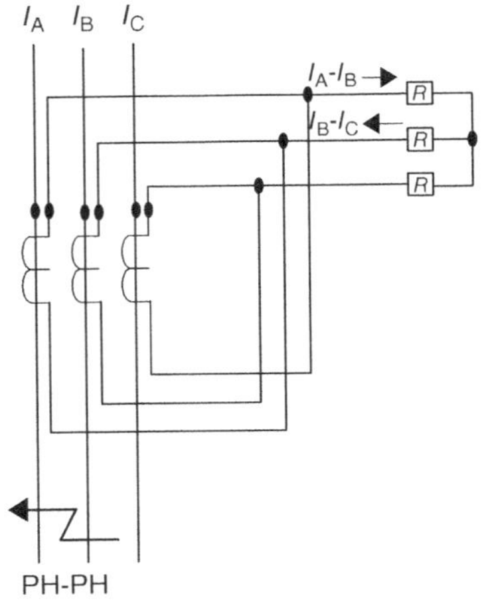
</td></tr>
<tr><td>Wye @ P&C Control House</td><td>A CT has a burden R_b
Total relay burden
$R_b = R_s + 2R_{cable} + R_{relay} + RG_{relay}$</td><td>PH-G</td><td>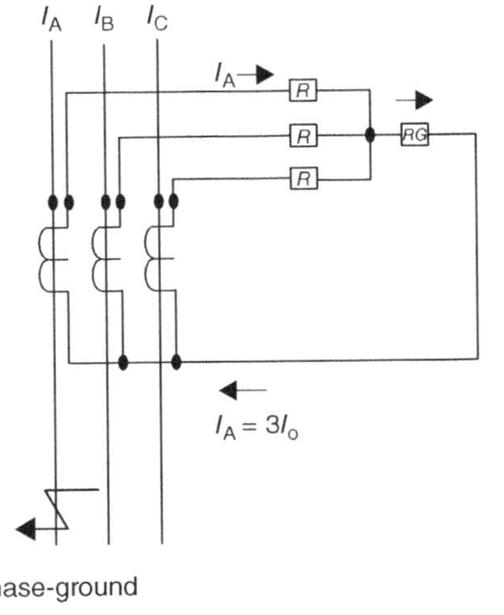
</td></tr>
<tr><td>Delta @ P&C Control House</td><td>A CT has a burden R_b
Total relay burden
$R_b = R_s + 2R_{cable} + 2R_{relay}$</td><td>PH-G</td><td>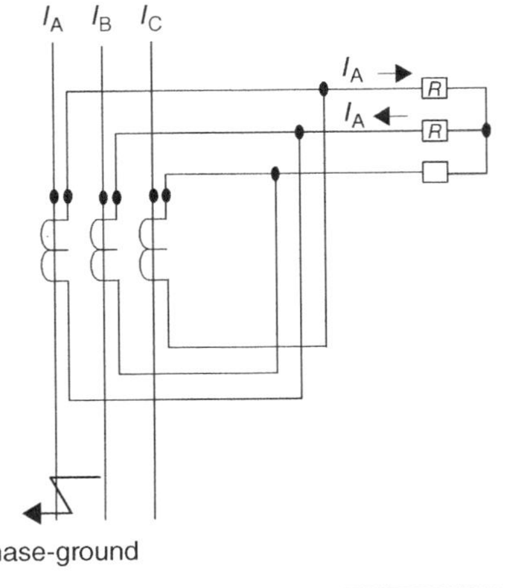
</td></tr>
</table>

3.2.2.2 Bushing-Type CT

A bushing-type CT has an iron core with nothing in its center. A secondary winding or typically windings wrap around the iron core. Any primary conductor passing through the center of the core is not integral to the CT itself and is independent of it. No actual primary winding connections are being made. This type of CT is installed in the oil-insulated bushings of transformers, breakers, and generators (Figure 3.5).

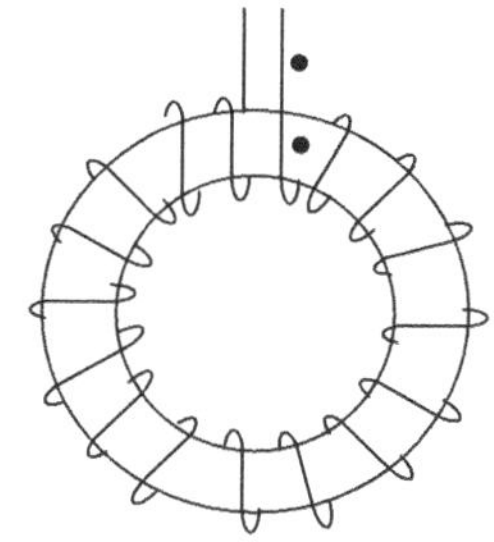

Figure 3.5 Bushing-type current transformer.

3.2.2.3 Window-Type CT

A window-type CT has an iron core with nothing in its center. A secondary winding or typically windings wrap around the iron core. Any primary conductor passing through the center of the core is not integral to the CT itself and is independent of it. These types of CTs are usually free-standing and have more than one conductor passing through the iron core.

A typical application for this type of CT is what is known as split-phase protection of hydraulic generators where the stator winding for each phase is split into two parts with each part passing through a window-type CT in opposite directions. The magnetic fluxes induced by each primary conductor are neutralized by the other with the net effect that only a residual flux is induced should one of the primary conductors carry more in-phase current than the other (Figure 3.6).

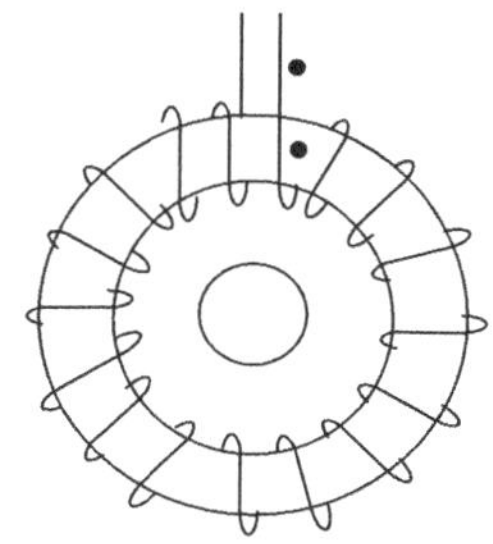

Figure 3.6 Window-type current transformer.

3.2.2.4 Wound-Type CT

A wound-type CT has a primary winding of usually multiple turns wrapping around an iron core. The secondary winding is wrapped around the same iron core. Both the primary and secondary windings are insulated from each other. The entire device is manufactured as one integral structure.

Figure 3.7 Wound-type CT.

This type of CT is used where current transformation ratios need to be matched or where galvanic isolation is required between protection devices or groups are required. This type of CT is therefore referred to as an auxiliary unit (Figure 3.7).

3.2.3 Current Transformer Polarity

As AC current reverses direction every half cycle, when current is the only electrical quantity supplied to a protection relay, CT polarity would have no significance. However, in many protection relays, the directional flow of current is compared to either another current or a fixed voltage. Also, CTs are regularly interconnected to interact with each other where the instantaneous direction of current flow determines whether a relay operates or not.

Correct relay operation, therefore, depends on the CT polarity being respected.

CT polarity is defined as instantaneous current into a designated spot on the primary with a corresponding instantaneous current out of a designated spot on the secondary, refer to Figure 3.8 below.

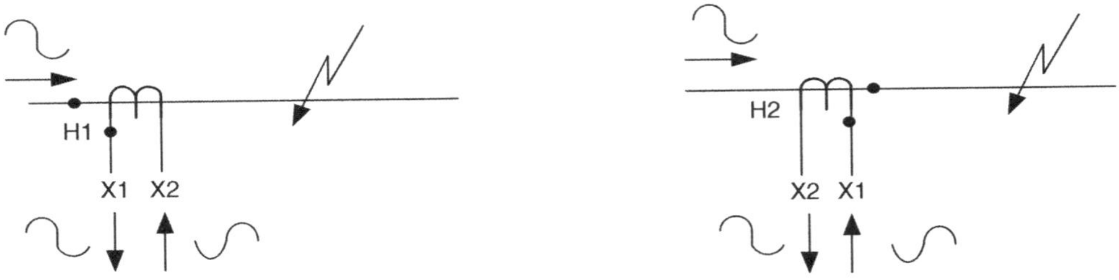

Figure 3.8 The polarity of current transformers.

The CT primary and secondary terminals are identified either by painted polarity marks or by the symbols H1 and H2 for the primary terminals and X1 and X2 for the secondary terminals.

When instantaneous primary current enters the CT input defined as H1, instantaneous secondary current leaves the output defined as X1. When instantaneous primary current leaves the CT input defined as H2, instantaneous secondary current enters the output defined as X2. In simple terms, "into the spot out of the spot," into the non-spot out of the non-spot, "out of the spot into the spot," and "out of the non-spot into the non-spot" as shown in Figure 3.8.

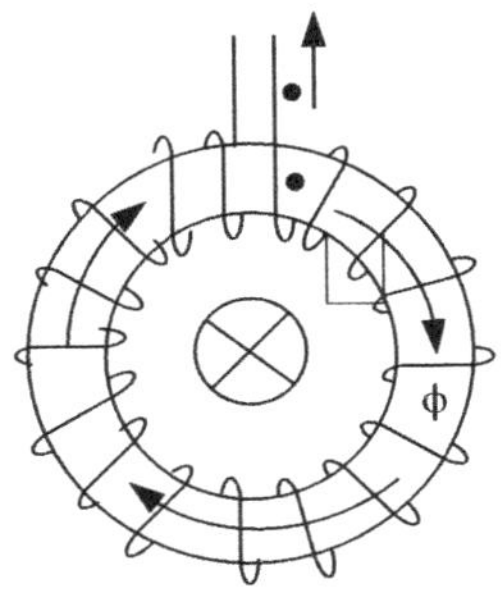

Figure 3.9 Bushing CT showing primary current entering the spot.

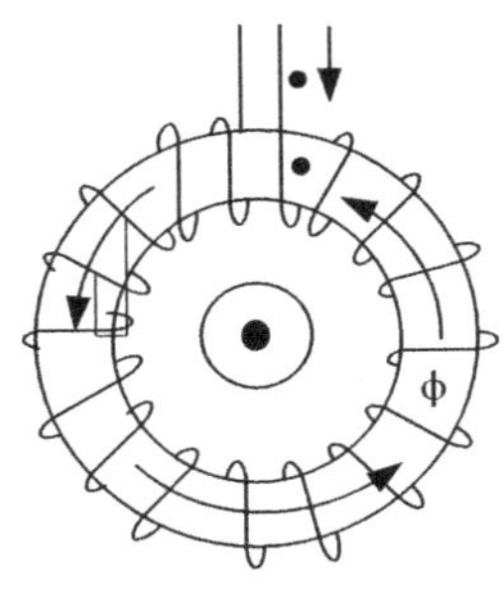

Figure 3.10 Bushing CT showing primary current leaving the spot.

The universally accepted symbols for currents entering and leaving are adopted for these illustrations. In simple terms, symbols represent what an arrow would look like to an observer if an arrow would enter or leave the CT. The head of the arrow is just a sharp point symbolized by a single dot on the drawing. The tail of the arrow is made of two feathers at right angles to each other symbolized by a cross. To visualize current entering a CT, a single dot is used just like the tip of an arrow entering it. Similarly, to visualize the current leaving a CT a cross is used just like the feathers of an arrow observed when it is leaving the CT.

Refer to Figures 3.9 and 3.10 showing primary currents first entering then leaving the spot side of a bushing CT. The magnetic flux changes instantaneous direction depending on the instantaneous direction of the primary current. In the first case, instantaneous current will leave the secondary spot; in the latter, it will enter the secondary spot.

The secondary leads of two CTs may be connected in series such that the secondary current will circulate as shown in Figure 3.11. This principle is the basis for differential type protection as shown in Figure 3.12 for illustrative purposes as this topic will be covered fully in a later chapter under differential protection.

3.2.4 Current Transformer Ratios

CTs used for protection usually have multiple windings to provide various transformation ratios. Some transmission companies specify CTs with multiple ratios that are provided by separately wound windings that are fully distributed about the core, while other transmission companies do not due to the expense. For CTs with fully distributed windings for each of the specified sets of rated ratios, the CT is constructed with a single primary conductor that passes through the center of the transformer core upon which all secondary windings are uniformly and symmetrically

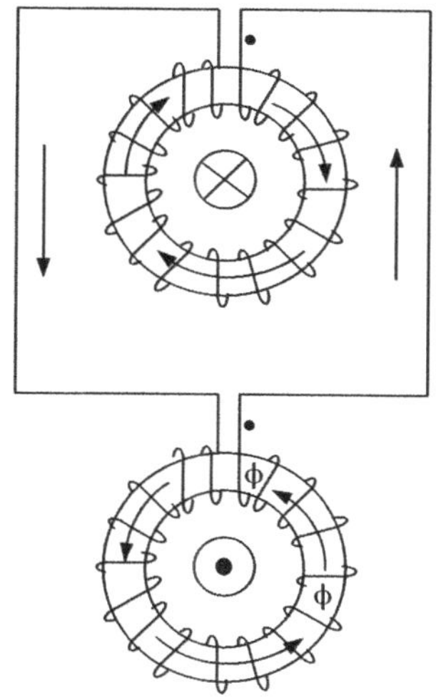

Figure 3.11 Two CTs with the secondary windings connected in series.

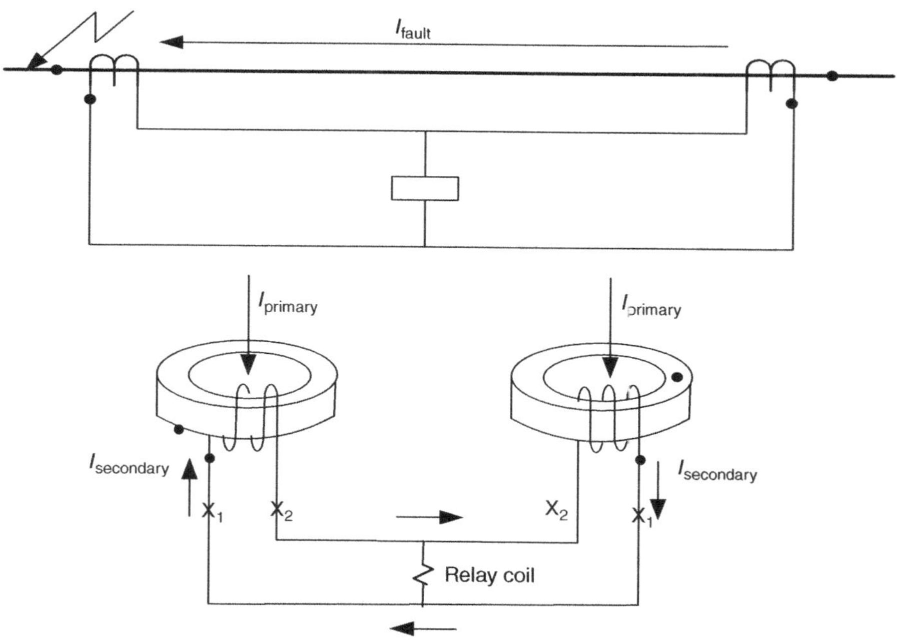

Figure 3.12 Bushing CTs connected differentially with an external fault.

wrapped. Essentially zero leakage flux exists. All flux which links the primary conductor also links the secondary winding.

The equivalent circuit for a CT with uniform geometry is shown in Figure 3.13. X_H and X_L are negligible because leakage flux is negligible. For CTs that do not have fully distributed windings for each ratio, a single fully distributed winding is wound around the iron core and this winding is tapped according to the turn ratio being used. These multi-ratios tapped CTs have leakage flux as not all the magnetizing of the iron core is being used to transform primary to secondary.

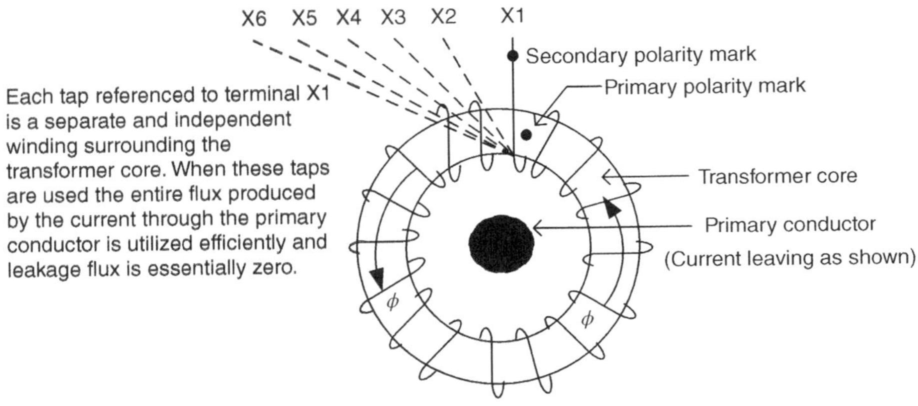

Figure 3.13 Construction of multi-ratio CTs.

Table 3.3 Fully distributed CT winding and tapped ratios.

Current ratio	Turn ratio	Secondary taps
400-5	80:1	X3-X4
600-5	120:1	X2-X3
800-5	160:1	X5-X6
1000-5	**200:1**	**X1-X2**
1200-5	240:1	X4-X6
1400-5	280:1	X2-X5
1600-5	**320:1**	**X1-X3**
2000-5	**400:1**	**X1-X4**
2200-5	440:1	X2-X6
2400-5	**480:1**	**X1-X5**
3200-5	**640:1**	**X1-X6**

CT manufacturers apply ratio correction factors in the design and construction of CTs that are premised on minimum leakage flux being present. This is only applicable when the CTs have completely distributed secondary windings for which secondary leakage reactance is so small that it may be assumed to be zero.

In this respect, even though the total secondary winding is completely distributed, tapped portions of this winding may not be distributed as depicted in Table 3.3 as non-bolded ratios along with Figure 3.14. The resulting secondary leakage reactance, which may be significant, could introduce significant errors if an undistributed tapped portion is used.

The secondary winding consists of sections comprising turns that are symmetrically distributed around the core to permit each tap section to be uniformly distributed around the core circumference. A polarity mark is assigned to terminal `X1` indicating the polarity for the marked primary. If this is a breaker bushing CT, the orientation is at the end of the breaker terminal opposite to the breaker contacts. A spot painted black on the core indicates the physical orientation of the core with

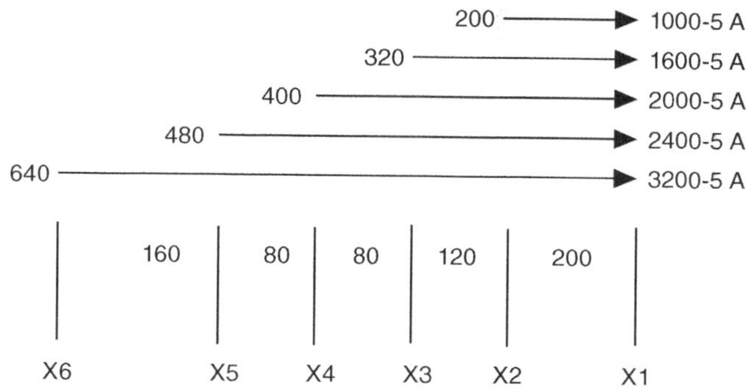

Figure 3.14 Typical multi-ratio bushing CT.

respect to the secondary designated spot. When intermediate taps are used, the tap numerically nearest to X1 has the same polarity as X1.

The ratios in Table 3.3 of the fully distributed windings are all referenced to terminal X1 and are bolded in the table. The actual ratios by rating for this CT which are typical of many transmission companies are **3200/2400/2000/1600/1000-5 A** as listed in Table 3.3. Each one of these ratios represents a separate and independently wound winding with the number of turns as shown in the table in the turn ratio column. Other possible ratios may be obtained by using taps other than those referenced to X1. These ratios are obtained by using partial windings that are not fully distributed on the core. The ratios are accurate but leakage flux is introduced that can be shown as X_H and X_L being other than equal to zero in Figure 3.15.

Referring to Figure 3.13, when the X2 – X5 intermediate taps are chosen, the overall number of winding turns on the core is $120 + 80 + 80$ or 280 turns to a single primary. The overall ratio is then 280×5 or 1400-5 A. However, not all the magnetic flux produced by the current through the single primary is used efficiently in producing current in the secondary windings as trapped leakage flux is introduced. This concept is best described as shown in Figure 3.13.

CT error results from the excitation current shunting through the magnetizing branch. The secondary current into the relay is no longer a direct function of the turn ratio. Refer to Figures 3.13 and 3.15 illustrating where the secondary current is shunted through reactance X_M and resistance R_M. These two impedances make up the magnetizing branch. Current I_e through X_M is used in the production of the magnetizing flux, and current through R_M represents heat losses in the magnetic core such as eddy current losses.

Any impedance connected to a CT is referred to as a burden. The burden includes the impedance of the cables from CTs located in the switchyard, where the breakers and transformers are located,

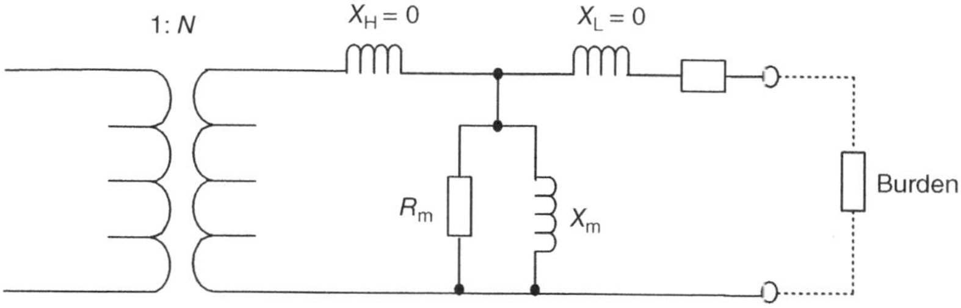

Figure 3.15 Electrical equivalent model of a CT with fully distributed windings.

to the relay building where the protection relays are located. It also includes the impedance of the relays that in the case of electromechanical relays can be significant. The resistance of the leads and the relay are summed to determine the overall burden on the CTs. The typical lead burden is in the range of 1–2 Ω.

Relay burden for electromechanical relays is specified in VA as the resistance varies with relay current pickup. For example, a relay with a specified VA burden of 1 VA will represent a burden of 1 Ω when set for a 1 A pickup and will represent a burden of 4 Ω for a 0.5 A pickup.

Digital relays represent a constant burden of typically 0.02 Ω or approximately 2% of that for most electromechanical devices. Also, since electromechanical devices have a single protection function, a few of them are usually connected in series supplied from the same CT's significantly adding to the overall CT burden. Digital relays have many protection functions in one device that do not add to the overall burden. The burden of the CT leads would be the same unless the digital relay is mounted in the breaker or transformer mechanism box in the switchyard which is possible.

3.2.5 Auxiliary Transformers

In many situations involving electromechanical relays, CT ratios (CTRs) must be matched to either other CTs or to available taps on relays. In either case, auxiliary CTs of the wound type are used to convert one ratio to another. For example, to match 1200-5 A CTs with 2400-5 A in a differential scheme would require a 10-5 A ratio matching auxiliary CTs.

Refer to Figure 3.16 showing where the conventional connection of 10-5 A CTs could result in a high reflected relay burden depending on the application. Also, shown is an example where a conventional 5-5 A auxiliary CT can be used to achieve the same 2:1 ratio without changing the overall burden. This is an effective method to deal with reflected burdens. High burdens can affect the overall transient response of CTs and their accuracy depending on the application.

Digital relays can be programmed to accept mismatched relay ratios and have drastically reduced the need for auxiliary CTs. Digital relays represent very low burdens compared to electromechanical in the order of 0.02 Ω which is 1/50 of that of electromechanical. Nevertheless, the CT leads still represent a significant burden that could be in the order of 1 Ω or higher.

For a 10-5 A ratio matching transformer, the 5 A winding will have twice as many turns as the 10 A winding

When the 10-5 A auxiliary CT is used for increasing the overall CTR of 1200-5 A CT's to 2400-5 A overall, the 5 A winding will be connected to the relay and the 10 A winding to the 1200-5 A source CTs. Then, the reflected relay impedance from the 5 A side to the 10 A side appears as $0.5^2 = 0.25$ times the actual relay impedance.

When the 10-5 A auxiliary CT is used to decrease the overall CTR of 1200-5 A CTs to 600-5 A overall, the 10 A winding will be connected to the relay and the 5 A winding to the 1200-5 A source CTs. Then, the reflected relay impedance from the 10 A side to the 5 A side appears as $2^2 = 4$ times the actual relay impedance.

3.2.6 Current Transformer Accuracy Classifications

CT accuracy classifications are according to IEEE Standard C57.13-216 [1].

It is recognized that where the leakage flux (refer to Figure 3.13) is significant, there is no simple method of assigning a legitimate accuracy to these CTs. Their performance can only be determined by testing them. CTs with significant leakage flux within the transformer core is designated as class *T* or class *H* before 1968 where *H* stood for high secondary impedance.

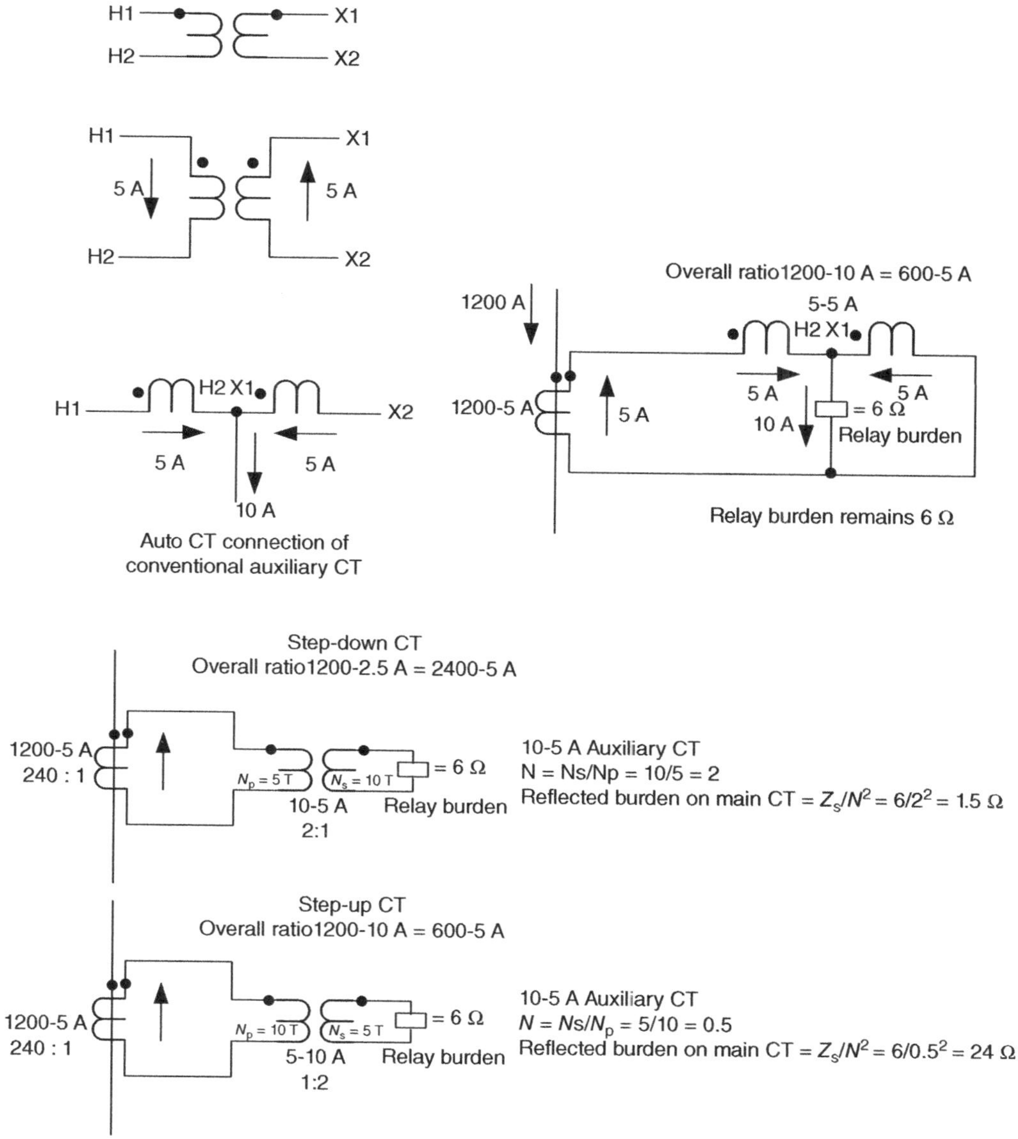

Figure 3.16 Auxiliary current transformers and reflected burdens.

Other CTs are designed to minimize the leakage flux in the core such a bushing-type CTs with fully distributed symmetrical windings surrounding the core. For these CTs, the leakage reactance can be paralleled with the magnetizing impedance, and with this, their performance can be predicted. These CTs are designated as class C or class *L* before 1968 where *L* stood for low secondary impedance [2].

The *C* designation is followed by a number indicating the secondary terminal voltage which the CT can deliver to a standard burden at 20 times rated secondary current without exceeding 10% ratio correction.

For relaying CTs, the ratings are by voltage class, like C100, C200, C400, and C800, corresponding to standard burdens of 1.0, 2.0, 4.0, and 8.0 Ω, respectively. Therefore, with the 800 V rating and its 8 Ω standard burden: 8 Ω × 5 A × 20 = 800 V. For simplicity, relay burdens are universally treated as purely resistive even though it is not. It is industry standard therefore for relay burdens and cable lead resistances to be added together without regard to power factor.

If the secondary voltage is less than the capability of the CT, then I_e will be small and have a minimum effect on I_S. If the secondary voltage exceeds the CT capability, i.e. will be significant, and the secondary current will be less than expected, and the CT performance impaired. Waveform distortion is extremely high when CTs operate over the knee of the saturation curve.

The excitation characteristic of a CT is expressed as a curve similar to that shown in Figure 3.17 and is a plot of secondary exciting voltage vs. secondary exciting current plotted on a log–log graph with square decades.

The knee point voltage shown in Figure 3.17 is defined as the sinusoidal voltage of rated frequency applied to the secondary terminals of the transformer, all other windings being open-circuited, which, when increased by 10% causes the exciting current to increase by 50%. For IEEE *C* classified CTs, the knee point voltage is the point on the excitation curve where the tangent is at 45° to the abscissa [1]. The excitation curve is plotted on log–log paper with square decades. This definition is for non-gapped CTs. When the CT has a gapped core, discussed later, the knee point voltage is the point where the tangent to the curve makes an angle of 30° with the abscissa.

The International Electrotechnical Commission (IEC) differently defines the knee point as the intersection of straight lines extending from the non-saturated and saturated portions of the exciting curve. The IEC [3] knee point is at a higher voltage than defined by IEEE [1].

As shown in Figures 3.17 and 3.18, a C800 classified CT will be capable of producing a secondary voltage of up to 800 V without saturating when subjected to symmetrical ac excitation. Past the knee point where the CT is fully saturating with the curve flattening out, where no amount of increased excitation current results in a higher secondary voltage the amount of excitation current required to produce the required magnetic flux in the iron core is up to 10 A. At this secondary voltage, the CT

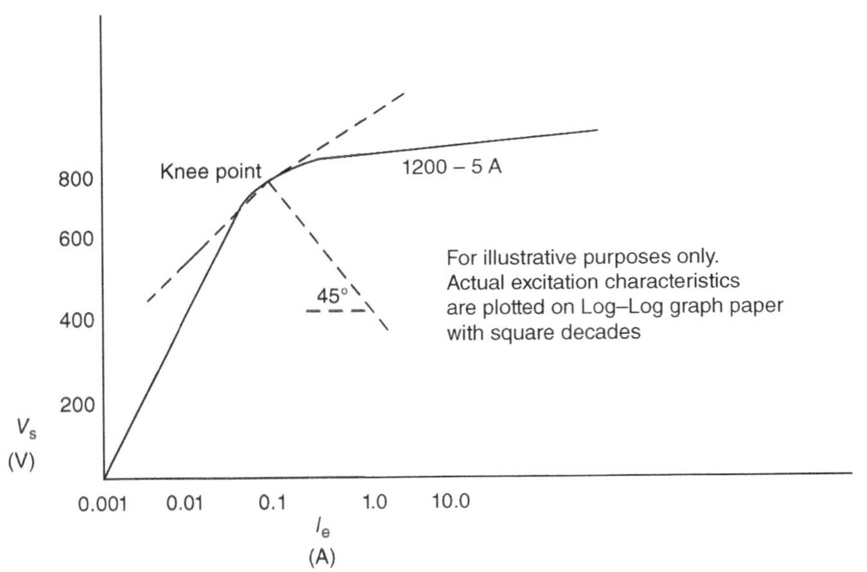

Figure 3.17 Typical excitation characteristic for a CT-rated C800.

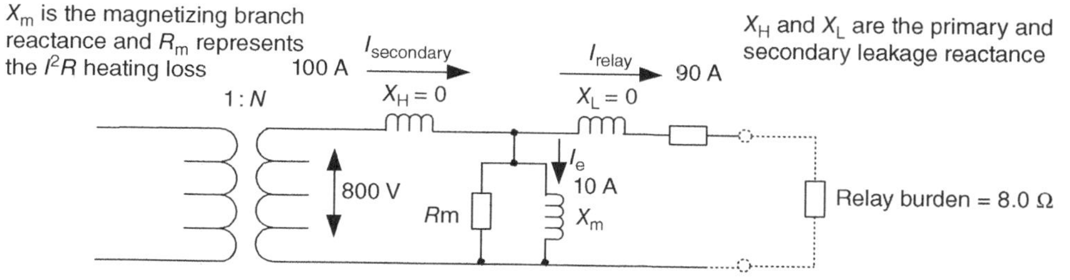

Figure 3.18 Electrical equivalent model of a C800 class CT.

is rated to develop 20 × 5 A or 100 A secondary current with no more than 10 A or 10% of 100 A as shunted by the excitation branch. It is for that reason that a C800 classified CT will exhibit an overall ratio error of 10%. Beyond this value of secondary voltage, the CT continues to provide secondary current but requires increasing amounts of excitation current thereby significantly adding to the overall error. The error remains below 10% up to 10 A excitation current where the CT core is fully saturated as shown in Figure 3.17. This industry standard is an accepted simplification that does not account for the excitation current being in quadrature with the through secondary current to the mainly resistive burden.

With the advent of lower burden digital relays, many transmission utilities specify C800 accuracy class instead of 2.5L800 where the rated error was kept to only 2.5%. In general, digital relays represent lower burdens and therefore require lower secondary voltages to produce the same amount of secondary current into the burden. The higher accuracy class before 1985 was necessary for electromechanical relays as they represented a significant burden. For illustrative purposes, the typical burden of an electromechanical overcurrent or distance relay is approximately 1 Ω. Since electromechanical relays provide a single function, it was not unusual to have three or four relays in the current loop each representing approximately 1 Ω burden for a total burden of approximately 4 Ω. Add to this a cable lead burden of 1 Ω, and the overall burden is 5 Ω. At a fault current of 100 A secondary, the secondary voltage required to be produced by the CT is 500 V. At $V_S = 500$ V, the excitation current starts becoming significant enough to affect CT accuracy.

The better quality core iron for a 2.5L800 rating guarantees that even with the much higher burden and resultant V_S, the overall accuracy remains similar to a C800 CT when using a digital relay whose burden remains constant at 0.02 Ω. The difference in construction between the two accuracy classes is in the physical size of the magnetizing core. The C800 CT is physically smaller and less expensive to produce.

3.2.7 General Characteristics of CTs

Common notation for CTs as indicated on their rating plates is shown below:

2400/1600/400-5 A 4000/3200/1600/1200-5 A

The hyphen separates the primary and secondary windings, and the slashes indicate available taps on the secondary. Typical convention is to use a hyphen the actual rating referred to 5 A secondary current and a colon when referred to a ratio to 1. For example, a 2400-5 A rating has a ratio of 480:1. The secondary rating of 5 A is standard in North America although there are some exceptions such as CTs used in the protection of autotransformer tertiary windings.

Electromechanical relays used in North America beginning in the 1920s were built to withstand a continuous current of 5 A representing typical load current in the CT primary. It was felt by the major manufacturers at the time that overcurrent relays, in particular, could more dependably operate at higher secondary currents. However, European manufacturers felt otherwise and decided to standardize on 1 A secondary current representing full load for their relays. CTs were designed with 1 A secondary ratings to match. Whether 5 A or 1 A rating they are known as CT nominal current I_N. Many transmission utilities specify each available tap on the secondary to be brought out to terminal blocks to be a fully distributed winding surrounding the core of each CT to keep the leakage flux to a minimum.

Some transmission utilities also adopt the use of air-gapped core vs. closed core CTs in protective relaying applications at 230 kV and above to keep remanence to a minimum. Remanence is the magnetic flux that remains in a magnetic circuit after the removal of an applied magnetomotive force (current).

Under no circumstance, should a CTR be chosen for a relay, be relied upon to operate, that results in higher than 100 A ($20 \times I_N$) secondary for a maximum symmetrical system fault.

For example, for a maximum fault (phase or ground) of 48 kA, the CTR should be no lower than 2400-5 A (480:1) so that a 48 kA primary fault current results in 100 A secondary current into the relay. For a 63 kA maximum fault current, the CTR should be 3200-5 A (640:1) to give a secondary current just under a 100 A. For other relays that are not relied upon to operate, they may accept considerably more current for a short time according to their manufacturer specifications. The distinction as to whether a given relay is relied upon to operate will be covered in Chapter 5. The general rule is that the relay that is relied upon to operate should not be made to accept more than 100 A symmetrical under worse case fault conditions.

3.2.8 Response of CTs Under Transient Power System Conditions

It must be considered when applying protections that the asymmetrical nature of power system fault current causes CTs to saturate. As saturation develops, the CT secondary current will start showing evidence of distortion and will no longer be sinusoidal and symmetrical. The performance of protection devices, depending on the type of protection and equipment type, will experience and be affected by distorted current waveforms. The greater the effect of system asymmetry on a given CT based on the CTs own characteristics, the faster a given CT goes into saturation. The saturation and the severity of that saturation are directly related to the time it begins to saturate.

3.2.8.1 The Effect of CT Saturation on Protections

Electromechanical relays when used in differential protection applications tend to have natural filtering that makes them less likely to operate incorrectly. Digital relays in identical differential applications require sophisticated and complex algorithms to match the natural security of electromechanical. These algorithms depend on CT minimum times to saturate within specific time parameters to ensure protections remain secure. Typically, these algorithms require at least a half good cycle of CT secondary current to make the correct decision to operate or not. Also, severely saturating CT secondary currents will slow down the operation of digital relays compared to that of electromechanical. This is most evident in the application of impedance relays that protect transmission lines.

3.2.8.2 Causes of CT Saturation

The main cause of CT saturation is the exponentially decaying DC component of the primary fault current. All fault currents in three-phase power systems consist of two components, AC and DC. The AC components vary sinusoidally with time, also referred to as symmetrical current, refer to

Chapter 6 for a more detailed discussion on faults. The DC component is non-periodic and decays exponentially with a time constant of L/R that is proportional to X/R. The DC component makes the symmetrical current asymmetrical. The DC current decays over approximately 6–12 cycles. The magnitude of the DC component and its duration is significantly dependent on the L/R ratio of the source impedance feeding the fault.

A higher L/R ratio leads to a more severe more slowly decaying DC component. Interconnected transmission system source impedances tend to represent lower L/R ratios compared to those at or close to generating stations where the value of inductance is highest due to the proximity of the generators that are almost exclusively inductive.

Distortion of the secondary current begins whenever conditions are such that the core flux density enters the region of saturation. The factors influencing the core flux density are the physical parameters of the CT, the magnitude, duration, and waveform of the primary currents and the nature of the secondary burden. Saturation of the core can be produced by excessive symmetrical fault currents and by lower magnitude asymmetrical (offset) fault currents. By far, DC offset is the most significant factor affecting the degree of CT saturation.

When a fully offset current is impressed on the primary of a CT, the DC offset will, in general, cause a rise in flux in the core several times greater than that required to transform the 60 Hz component of the current.

The variation in the transient component of flux is a function of both the primary system DC time constant and the time constant of the CT secondary circuit. The CT secondary circuit time constant is a function of CT secondary resistance (internal resistance and lead resistance), burden impedance, and the CT magnetizing impedance. The total flux required to reproduce the offset current is considerably greater than that required to simply reproduce symmetrical fault current.

A CT will reproduce the offset fault current completely provided that the induction does not reach its saturation flux density level. If the induction does greatly exceed the saturation flux density, the CT will produce a distorted secondary current similar to that shown in Figures 3.19 and 3.20 using the IEEE CT saturation calculator [4]. In this example, the CT secondary current is initially

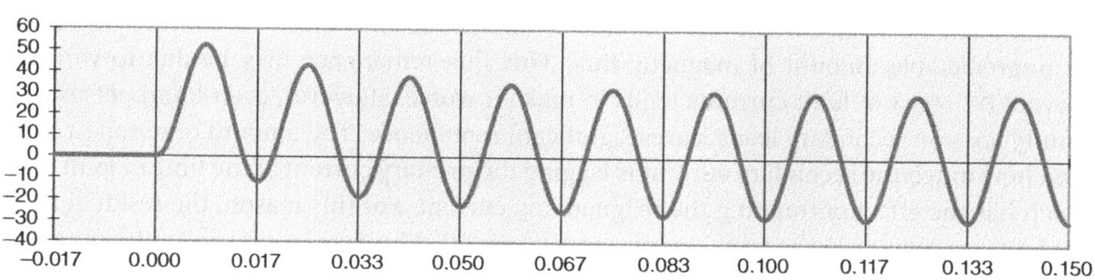

Figure 3.19 Low primary fault current.

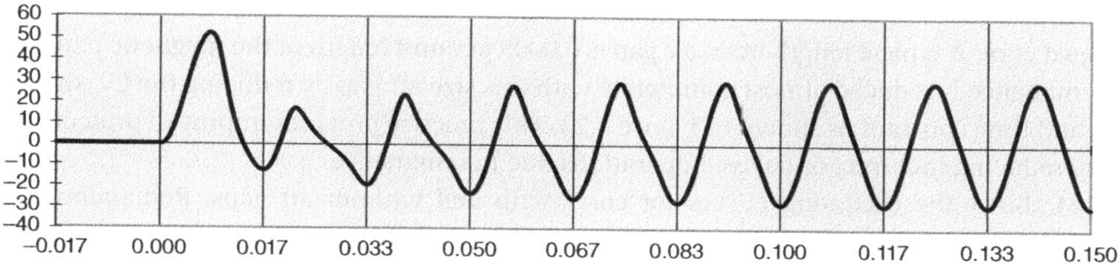

Figure 3.20 Secondary CT fault current during moderate saturation.

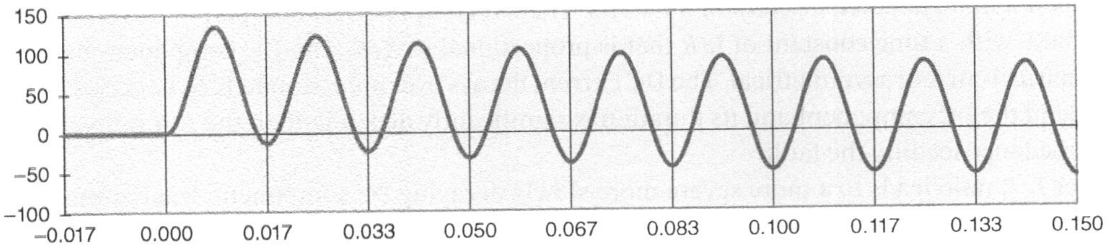

Figure 3.21 High primary fault current.

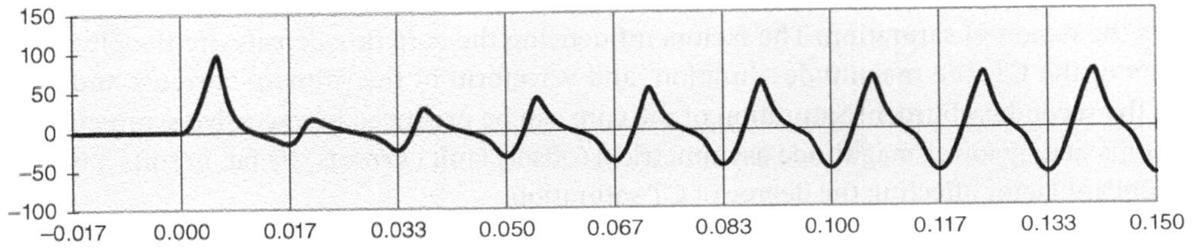

Figure 3.22 Secondary CT fault current during severe saturation.

fully offset due to the DC component of the primary fault current. The secondary current does not severely distort in the first cycle and only begins to do so starting in the second cycle. This is referred to as a secondary current with one good cycle.

In many cases, the CTs will be capable of accurately reproducing offset currents for several good cycles before starting to saturate. However, under certain conditions, severe saturation cannot be avoided and CTs will produce distorted secondary currents even during the first half cycle as shown in Figures 3.21 and 3.22 [4] where the secondary is a true replica of the primary current for only one-quarter cycle.

3.2.8.3 Flux Remanence in the CT Core

Another significant cause of CT saturation is the ferromagnetic character of a CT core that may retain an unpredictable amount of magnetic flux. This flux remanence may be due to various causes. Severe DC offset of fault currents tends to make it worse. However, even improper use of DC continuity tests on secondary leads leaves significant remanence. The amount of remnant flux depends on how much the secondary current is lagging the primary current at the time of fault isolation which has the effect of trapping the magnetizing current. For this reason, the residual flux will either improve or worsen the transient response of the CT. Flux remanence will either oppose or aid the buildup of core flux caused by the DC offset fault current.

3.2.8.4 Use of Air Gaps to Reduce Remanence

The time to saturation that is shortened by remanence in the core may be remedied with the use of an air-gaped core. A typical length of an air gap is 0.0005 per unit length of the magnetic path in the core. Remanence is reduced almost completely with this size air gap by reducing the CT shunt impedance and time constant as shown in Figure 3.23. This practice provides improved protection reliability by reducing incorrect or delayed operations due to remanence.

Figure 3.24 shows the excitation curves for cores with and without air gaps. Remanence is reduced to 6% or less for smaller gaps and almost completely for larger gaps. However, the unique characteristic of slowly decaying air gap CTs can at times be the cause of protection misoperations. For example, the flux may not have decayed to the remanence level fast enough so

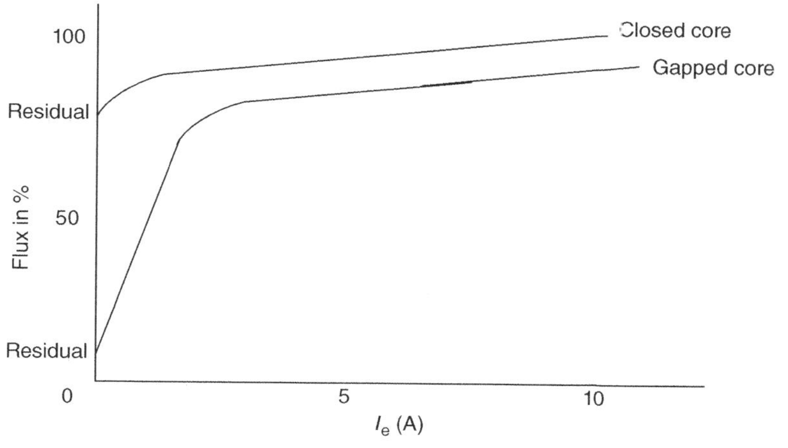

Figure 3.23 Comparison of residual flux with a closed core and gapped core.

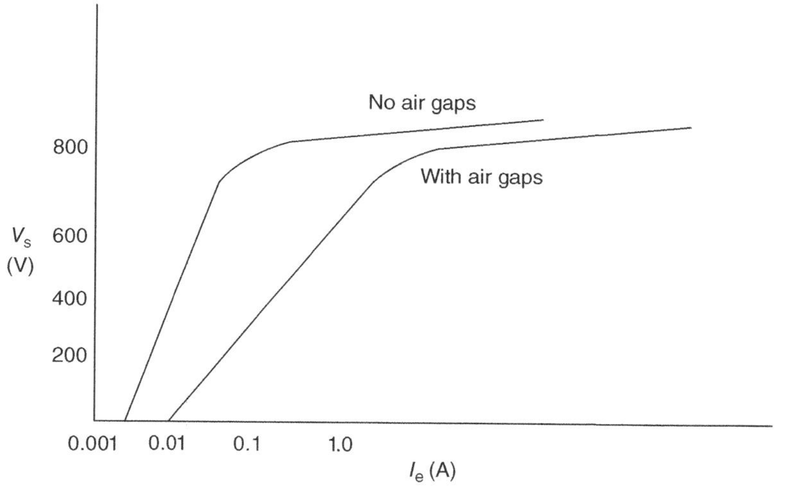

Figure 3.24 CT excitation characteristic with and without air gaps.

that secondary current continues to flow into a relay for a short time following breaker operation. This phenomenon should be taken into account when applying settings to relays with assumed resetting times skewed by this characteristic.

3.2.8.5 Methods to Ensure Correct CT Performance

In most cases, while applying protection systems, the first practical approach of ensuring appropriate CT performance is to calculate how the CT behaves at steady-state conditions. If possible, the CT should be sized to have a knee point voltage well above that required for a maximum calculated fault current and CT secondary burden. The calculated steady-state voltage for the given application should preferably be no more than 50% of the standard classification rating of the CT. The 50% safety margin in most cases will provide a suitable margin taking into account the DC component of the waveform as well as possible CT remanence.

3.2.8.5.1 Steady-State Performance

Steady-state conditions mean when the DC offsetting current in the power system has had an opportunity to decay sufficiently to allow the CT secondary current to be transformed is no longer distorted or asymmetrical. In most cases using this method, it is simple to ensure that the steady-state error of the CT will not be exceeded. It does so by having the CT operate within its defined accuracy class:

(a) The maximum symmetrical fault through the CT shall not exceed 20 times the CT-rated current at the connected tap. For a CT with 5 A secondary rating, $I_S < 100$ A.
(b) The voltage across the CT burden does not exceed the accuracy class voltage of the CT at the selected tap. For a C800 CT: $V_S < 800$ V.

Therefore, by following the above practice, the CT steady-state error will be within its rating, for example, a C800 CT steady-state error will be less than 10%.

Good utility practice based on experience is that selecting a CT with steady-state secondary terminal voltage rating at least greater than twice the secondary voltage that is developed across the CT terminal for the maximum symmetrical fault current the CT should perform adequately under dynamic conditions:

$$V_S \geq 2 * I_{max} * Z_b$$

I_{max}: Maximum symmetrical fault current (secondary)
Z_b: Secondary burden (cable and protection devices)

For a closed core CT with a 10% accuracy class and 5 A secondary as shown in Figure 3.17, the knee point voltage is at approximately 50% of the excitation voltage at 10 A excitation current (10% error). So, if the secondary voltage at maximum fault is less than half of the CT secondary terminal voltage rating, the operating point of the CT would be at or below the knee point voltage for the worst-case fault conditions.

The above practice does not consider the internal CT resistance, which can become a factor when it is relatively large compared to the secondary burden Z_b. So, if the CT excitation characteristics and the CT resistance are available, a better and more realistic criterion for CT secondary voltage rating would be: $V_S \geq 2 * I_{max} * (Z_b + R_{ct})$, where R_{ct} is the CT internal resistance usually found on the CT excitation curve sheet provided by the manufacturer of the CT.

3.2.8.5.2 Transient Performance

In some situations, it is not possible to rely only on steady-state calculations to ensure correct CT response in a given application. There are a several reasons for this. First and foremost, it may not be possible to obtain a safety margin of 50%. Another reason is the introduction of higher than previous fault currents with the expansion of the network or with a new generation being built nearby. A new generation station introduces another problem as the L/R ratio that determines the severity of DC offset increases.

As previously described, the severity of CT saturation does not occur immediately, and depending on various factors, a CT will replicate the primary without distortion for a short time before entering saturation. In almost all cases, the CT will come out of saturation after a finite number of cycles provided that the CT does not exhibit ac saturation as the DC component decays. The time to saturation, i.e. the time preceding the onset of saturation during which the CT output is a faithful replica of the primary current, is most critical for the reliable operation of digital relays. The faster the time to saturate for a given CT application is also a reliable indicator of the severity of waveform distortion. In other words, the shorter the time that the secondary current is no longer a replica of the primary current the greater the amount of secondary current distortion directly impacting correct relay performance. When doubt exists whether the steady-state CT performance calculation is an adequate indicator, detailed time to saturate calculations are necessary.

3.2.8.5.2.1 Time to Saturate The time to saturate of a CT is determined by the following parameters:

1. Fault current magnitude.
2. Degree of fault current offset.
3. The time constant of the DC component of fault current.
4. Remanence flux in the core.
5. Secondary circuit resistance including the effect of secondary winding resistance.
6. Secondary excitation impedance of the CT (magnetic permeability of the core).
7. CT turns ratio.
8. Is the secondary winding fully distributed on the core or does leakage flux exist.

When fault levels are very high and/or the CT secondary burdens are large, it may not be possible to size the CT so that it does not saturate during such faults.

Depending on the CT saturation factor and the primary system time constant, it takes some time for the CT to go into saturation.

In critical applications where CT steady-state performance calculations are insufficient to ensure confidence in the overall performance of the protections, complex calculations may be necessary to determine the time to saturate. In this situation, it is recommended that the following excellent publications be referred to determine whether a CT in the given application is suitable from a dynamic performance perspective.

IEEE Power Engineering Society, "Transient response of current transformers," IEEE Special Publication, IEEE Publication #76 CH1130-4-PWR, Jan. 1976 [5].

IEEE Guide for the Application of Current Transformers Used for Protective Relaying Purposes IEEE Std C37.110™-2007 [4].

CT saturation calculator (PSRC) Prepared for IEEE Std C37.110™-2007. The CT saturation calculator provides a simple indication of whether a CT will saturate in a specific application and also provides an indication of actual waveshape of the secondary current [6].

3.2.9 General Requirements for CT Sizing

An important consideration for CTs is the selection of a suitable ratio for the provision of appropriate current signals to the relay. The ratio must not be too high, so that the secondary current at minimum fault levels is adequate to be above minimum setting thresholds (typically 0.25 A or 0.5 A secondary). On the other hand, the ratio must not be too low to avoid subjecting the relay to excessive continuous current under normal conditions or excessive secondary currents exceeding the short-time rating. Also, it is best not to exceed halfway up the linear range of the CT on high fault levels (i.e. exceeding halfway to the knee point voltage, leading to saturation of the current signal).

3.2.9.1 Maximum Expected Load Current

The first step in selecting an appropriate CTR is to determine the maximum expected load current. For some lines, future line loadings can be predicted fairly accurately, e.g. lines supplying radial loads. For other lines, particularly those located in the network where there are many number parallel paths, some of which can be external to the utility, it is often not possible to predict future line loadings, except for a relatively short period. In these cases, the maximum loading could be based on extreme conditions. For example, a maximum load angle of 40° across the line and a difference of 15% between line end voltages could be used as a reference condition.

For shorter lines, this condition could produce excessive load current levels that exceed the current-carrying capacity of the line and/or station equipment; in some cases, it is the equipment rating that defines the maximum continuous loading. A simple means of finding the maximum load current of a given line is determined by the type of conductor used. This information is typically available either from planners or design groups.

In determining the CTR, the approach that may be used is to define the derived maximum load level as the 5 A level on the secondary of the CT. This leads to choosing the primary tap that is the nearest, but not less, to the maximum load current.

3.2.9.2 Maximum Symmetrical Fault Current

A fault study can be conducted to determine the maximum symmetrical fault current and the related X/R ratio of the primary system for the specific application of the CT under consideration. The X/R ratio may be needed should time to saturate need to be determined using previous references.

The secondary current for the maximum primary symmetrical fault shall not exceed 20 times the CT current rating, at the selected tap. Choose the closest higher ratio. This will ensure that the CT error will remain within its accuracy class (10% for C), and the mechanical forces on the CT windings due to high fault currents can be tolerated by the CT physical construction. Typically, 20 × 5 A or 100 A for a 5 A rated CT and a few short seconds allowing for protection operation and circuit breaker tripping.

3.2.9.3 Maximum CT Burden

Since protections must cater to all possible types of faults, calculate the total burden for a single phase to ground fault (refer to Section 3.2.1).

$$\text{Relay burden } R_b = R_s + 2R_{cable} + R_{relay} + RG_{relay}$$ [1]

Definitions:

R_b is the total relay burden seen by the CT
R_s is the CT internal secondary winding resistance
R_{cable} is the resistance of the cables connecting the CT to the relay
R_{relay} is the resistance of the relay

3.2.9.4 Calculate the Steady-state CT Secondary Voltage (V_S)

For a given symmetrical line to ground fault current and overall CT, burden calculate the CT secondary voltage V_S. Then, compare V_S to the CT saturation characteristic evaluating whether V_S is lower or higher than the CT saturation voltage.

If V_S is lower than halfway to the knee point, this specific CT application should be satisfactory. CT saturation should not affect the performance of the protection system. Should V_S be higher than halfway to the knee point, if using electromechanical relays, digital relays with considerably lower burden could be used to replace the electromechanically relays. Should that not be possible or digital relays are already being used refer to Section 3.2.8.5.2.1.

1 RGrelay is only applicable when calculating overall electromechanical relay burdens.

3.2.9.5 CT Application Example

This example is intended to illustrate using electromechanical relays to determine the CT performance of an existing installation of a transmission line protection.

Given typical values for this type of application (Figure 3.25):

R_L lead resistance – CTs located in the switchyard at the breakers to the relays in the control house 50 ft of 12 AWG wire @ 1.98 Ω/1000 ft. R_L burden is 1.98/1000 × 50 = 0.09 Ω, one direction, 0.18 Ω loop resistance.

CT Ratio (CTR) = 1200/5 = 240:1

CT secondary winding resistance $R_s = 0.51\ \Omega$

Auxiliary CT secondary winding resistance R_s aux = 0.25 Ω

L-G fault: $I_S = 19{,}000\ \text{A}/240 = 79.2\ \text{A}$, three-phase fault

$I_S = 18{,}000\ \text{A}/240 = 75\ \text{A}$

21 Distance relay burden: 1.9 Ω

51*N* Time overcurrent relay burden: 3.5 Ω

51 Phase time overcurrent relay burden: 0.7 Ω

Calculate V_S for a 3-Phase fault:

The three 5-5 A auxiliary CTs are connected Y-D giving an overall step-up ratio of 1: $\sqrt{3}$

$N = N_s/N_p = (\sqrt{3})^2 = 3$

The reflected burden on the main CT, therefore, $= 0.7\ \Omega \times (\sqrt{3})^2 = 2.1\ \Omega$

$V_S = \text{3-Phase fault current} \times (R_{ct} + R_{ct}\ \text{aux} + R_L + R_{21} + R_{51\ \text{reflected}})$

$V_S = 75\ \text{A} \times (0.51 + 0.25 + 0.09 + 1.9 + 2.1)$

$V_S = 364\ \text{V}$

Calculate V_S for a L-G fault:

Since the auxiliary CT are connected in Y-D, the zero-sequence current will circulate in the delta connection and will not enter the 51 relays. Therefore, they represent no burden to the main CT.

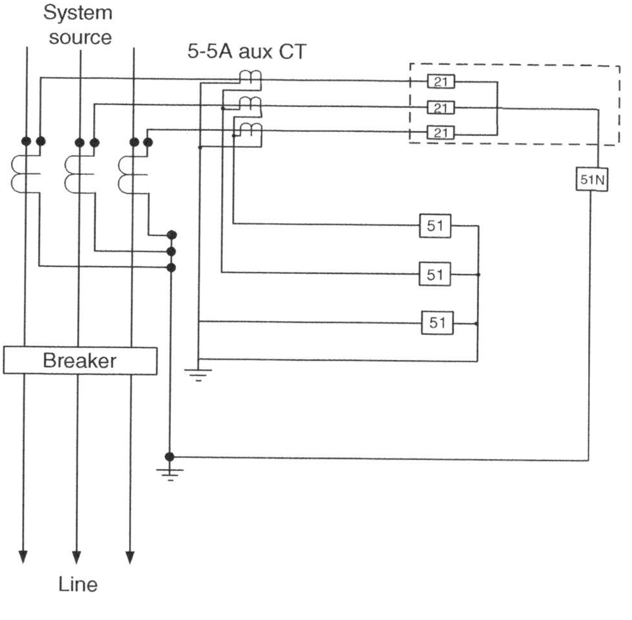

Figure 3.25 CT and relay connections – electromechanical.

However, their internal winding resistance must be multiplied by 3 as the zero-sequence current circulates in all three aux. CT.

$$V_S = \text{L-G fault current} \times (R_{ct} + 3 \times R_{ct}\ \text{aux} + 2R_L + R_{21} + R_{51N})$$

$$V_S = 79.2\ \text{A} \times (0.51 + 0.75 + 0.18 + 1.9 + 3.5)\ \Omega$$

$$V_S = 542\ \text{V}$$

Referring to Figure 3.26, the calculated value for V_S being 542 V is more than halfway up the knee point. This voltage should be no greater than half the knee point being 400 V.
Solution: Replace the electrometrical relays with a single digital relay (Figure 3.27).

R_L lead resistance – CTs located in the switchyard at the breakers to the relays in the control house 50 ft of 12 AWG wire @ 1.98 Ω/1000 ft. R_L burden is 1.98/1000 × 50 = 0.09 Ω, one direction, 0.18 Ω loop resistance.
CT Ratio (CTR) = 1200/5 = 240:1
CT secondary winding resistance R_s = 0.51 Ω
L-G Fault: I_S = 19,000 A/240 = 79.2 A
21 Digital distance relay burden: 0.02 Ω (51*N* ground element is internal to the relay).

$$V_S = \text{L-G fault current} \times (R_{ct} + R_L + R_{21})$$

$$V_S = 79.2\ \text{A} \times (0.51 + 0.18 + 0.02\ \Omega)$$

$$V_S = 56.2\ \text{V}$$

The replacement of the electromechanical relays with digital has drastically increased the CT performance for this application.

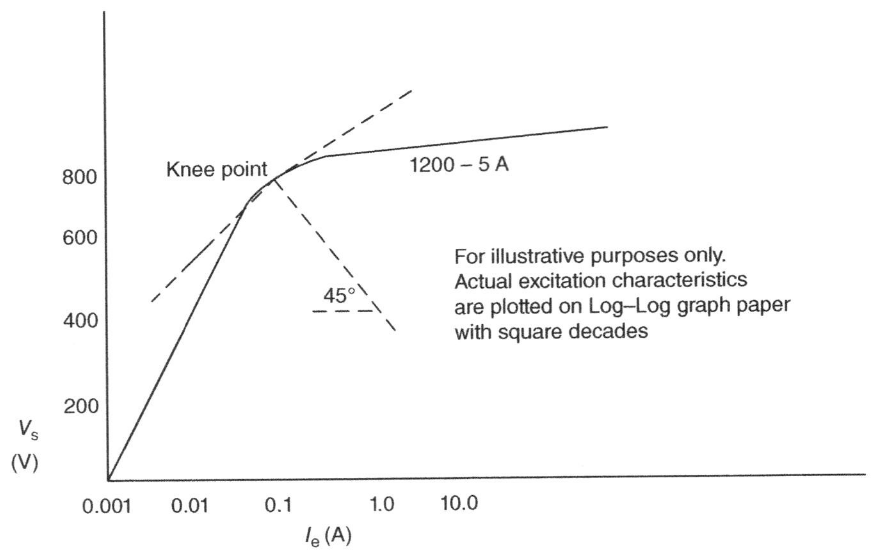

Figure 3.26 Typical CT excitation characteristic curve rated C800.

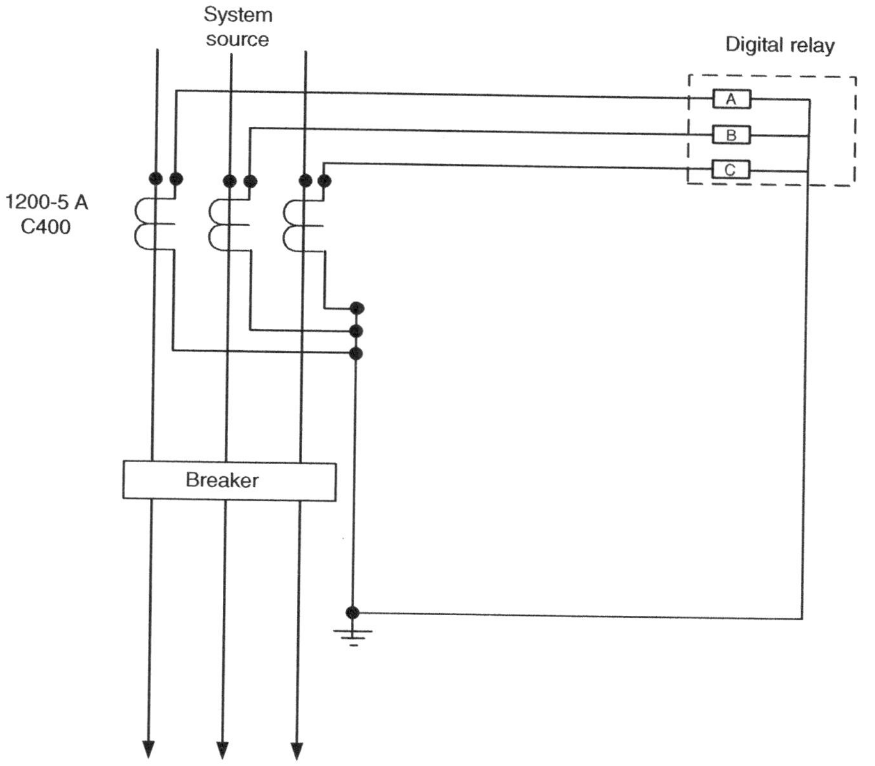

Figure 3.27 CT and relay connections – digital relay.

3.3 Voltage Sources

Voltage sources are used to provide input signals to voltage-dependent relays such as distance and voltage relays and other devices such as synchronizing equipment. Distance relays are used for line protections, and voltage relays are used for several applications including automatic breaker reclosing.

Typically, VTs are used as voltage sources for most applications. There are two types of VTs: electromagnetic type and capacitor voltage divider type – coupling capacitor voltage transformers (CCVTs or CVTs).

Voltage sources are shunt-connected onto the power circuit to measure voltage. Their secondaries are shunt-connected with voltage coils or elements of relays or instruments. They are normally designed in North America to have secondary windings rated at 115 V L-L (66.4 V L-N) or 120 V L-L (69.3 V L-N), regardless of their primary voltage rating.

Electromagnetic VTs are similar to small power transformers but differ only in details of design that control ratio accuracy over the specified range of output. Capacitor voltage transformers (CVTs) are very popular on high-voltage power systems due to their lower cost compared with the cost of VTs at higher voltages.

Most utilities do not have performance issues with VTs and CVTs compared with CTs. Most applications have limited burdens/loads connected to their secondary that are well within their typical 200 VA accuracy burden limit.

CVTs and electromagnetic VTs are designed to give an accurate reproduction of the 60 Hz primary system voltage to secondary voltage. However, CVTs may have a poor transient response characteristic during faults. CVTs are designed with energy charge and storage elements such as capacitors and inductors, and there is a time lag for changing the energy and or charge states appropriate to the magnitude of the primary fault voltages. The consequence of the time lag is that distance measuring relays applied for line protection may suffer reach errors in calculating the distance to the fault. This will be discussed in more detail in the chapter devoted to line protection.

3.3.1 Magnetic Voltage Transformers

Magnetic VTs are used to provide input signals for relays almost exclusively where the primary voltage is 115 kV or less.

3.3.1.1 Magnetic Voltage Transformers Equivalent Circuit

Figure 3.28 shows a magnetic (inductive) VT circuit representation. However, in general, it is also applicable to capacitor VTs as far as accuracy and measuring errors.

3.3.1.2 Protection of VTs

VTs can be protected with fuses on the primary side for sub-transmission voltages. Fuses are not normally available that have sufficient interrupting ratings for use at higher voltages. It is common not to fuse the primary. For high-voltage applications, the primaries of the VTs are generally protected by the transmission line relays if the VTs are connected on the line or by the bus relays if VTs are connected to the bus.

VTs should always be protected by secondary fuses or by miniature circuit breakers located as near the transformer as possible. The secondary fuses protect the secondary control cables and connected burden against a short circuit. However, the primary fuses, if installed or miniature circuit breakers, will not clear the secondary short circuit because of low primary current for secondary circuits.

Each of the secondary windings is normally separately fused at the VT/CVT mechanism box and each relay location. For electromechanical relays, it was common practice to use three pilot lights at each relay panel as a visual indication of fuse health. For digital relays, utilities no longer use pilot lights to indicate loss of potential as digital relays have a native loss of potential logic that can be used to alarm a blown fuse condition.

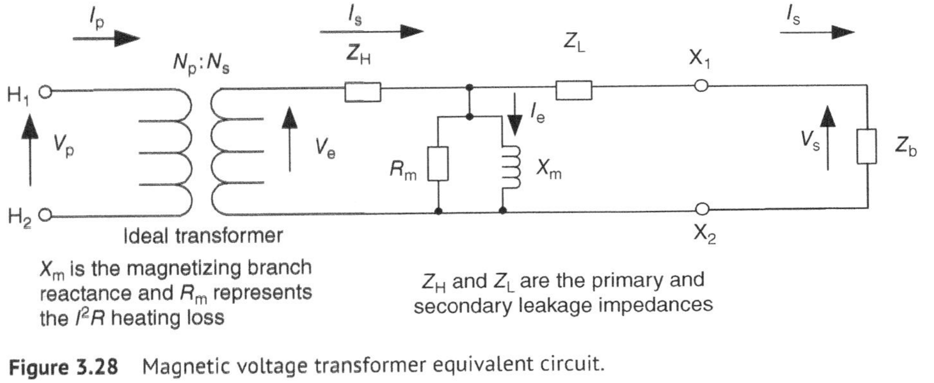

Figure 3.28 Magnetic voltage transformer equivalent circuit.

One of the problems encountered by some utilizes while changing from electromechanical to digital is that older fuses seemed to open-circuit for no apparent reason when used for digital relay replacements. Without the use of pilot lights and with extremely low burdens, very little current flowed normally through the older fuses. This after time tends to damage the fuses probably due to moisture in the air.

3.3.2 Capacitive Voltage Transformer (CVT)

CVTs are used to provide voltage input to relays almost exclusively where the primary voltage is 230 kV and above.

The physical size of a VT for higher voltages is proportional to the rated voltage. Therefore, the cost is much higher for such applications, and at higher voltages, the CVT is often more economical. The CVT is a capacitor voltage divider, and although the construction of a VT and CVT is different, from a protection point of view, they are functionally the same.

A CVT is a device that utilizes a capacitive voltage divider to reduce the high primary voltage to a medium voltage (e.g. 15 kV), which in turn is reduced to the common secondary voltages of 66.4 V or 69.3 V by magnetic VTs as shown in Figure 3.29. In addition, the CVT consists of auxiliary components in the form of a compensating reactor and a ferroresonance suppression circuit. The accuracy of the phase angle of the output voltage is adjusted by using a tuning reactor which is adjusted to match the capacitive reactance of the capacitive voltage divider. Protective gaps are provided to prevent over-voltage, and adjustment is made in the transformer ratio to provide a precise overall voltage ratio.

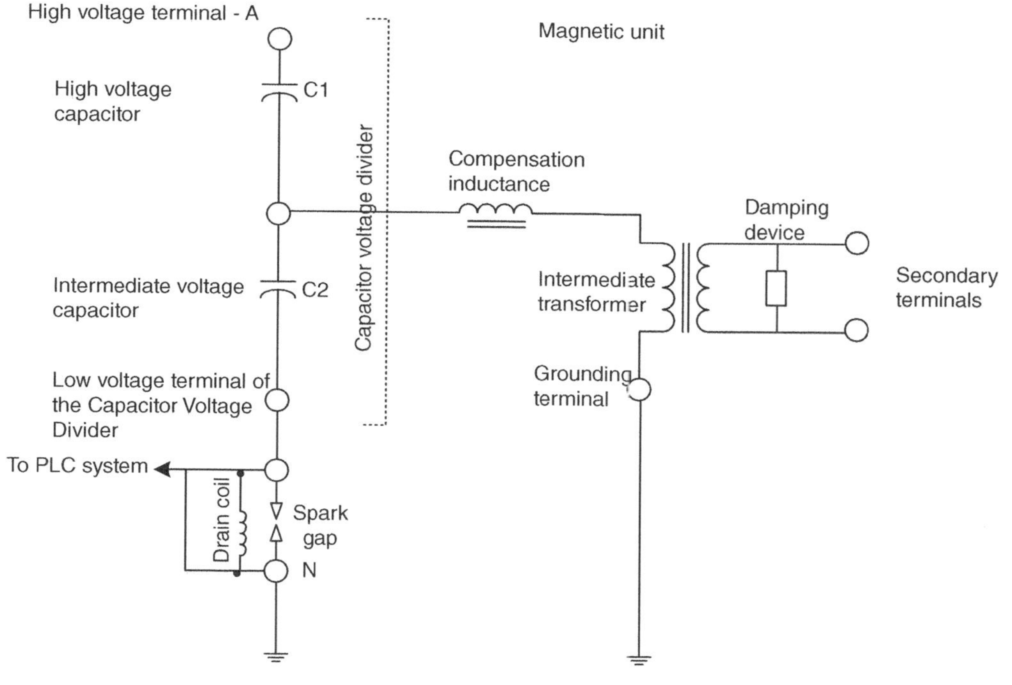

Figure 3.29 Typical CVT diagram with a carrier interface.

3.3.3 Bushing Potential Devices

A simple capacitive voltage divider can also be created with a bushing of a breaker or a transformer, utilizing the capacitance effect of the bushing and tapping off on a point within the insulation. While the transient characteristics of this potential device are somewhat different from the CVT above, the concerns are the same concerning the performance of distance relay elements.

The most common application of this type of potential device is the bushing potential device (BPD) that can be specified for most power transformers at substations. These BPDs are used for phase and ground line backup protections and automatic reclosing of transformer LV breakers on voltage presence.

BPDs have shown a tendency to produce significant amounts of third harmonics. Electromechanical and solid-state voltage relays that operate on unfiltered RMS voltage are vulnerable to mal-operation. Third harmonics in three-phase power systems are all in phase, when added directly can be of sufficient magnitude to operate sensitive voltage sensing protection typically used to detect ground faults by measuring zero-sequence voltage. A possible solution is to replace those relays with digital that possess arithmetic filtering techniques that filter out all but 60 Hz.

References

1 IEEE C57.13-2016, Standard Requirements for Instrument Transformers.

2 C. Russell Mason, *The Art and Science of Protective Relaying*. Schenectady, New York: John Wiley & Sons, Inc. 1956, General Electric Company.

3 IEC Standard 61869-1 Instrument Transformers General Requirements (IEC International Electrotechnical Commission).

4 IEEE Standard C37.110, 2007, Guide for the Application of Current Transformers Used for Protective Relaying Purposes.

5 IEEE Publication #76 CH1130-4-PWR, Jan. 1976, Transient Response of Current Transformers.

6 CT Saturation Calculator (PSRC) Prepared for IEEE Std C37.110™-2007. The CT saturation calculator provides a simple indication of whether a CT will saturate in a specific application and also provides an indication of actual waveshape of the secondary current.

4

Basic Types of Protection Relays and Their Operation

4.1 General

Protective relays are the building blocks used to develop protection systems. In this chapter, the term relay refers to those relays known as protective relays. These are the relays that receive power system electrical quantities, typically voltages and/or current signals, and use them to compare those calculated values against set points known generically as relay settings. When a set point is exceeded, an output contact in the relay closes. The closed contact in turn initiates what is known as tripping commands.

Examples of tripping commands include tripping a circuit breaker or breakers, initiating the breaker failure protection of those breakers, initiating automatic breaker reclosing, initiating remote trip, or keying transfer trip (see Chapters 10 and 14). From the 1920s till the late 1960s, the only type of protective relays available was electromechanical. In the late 1960s, solid-state protective relays were introduced and were adopted by many manufacturers as an alternative to the labor-intensive electromechanical relays. However, solid-state design while less expensive to manufacture, was still limited in function.

In the mid-1990s, a new type of protective relay was introduced by all the major manufacturers including some start-up companies. These relays were built around central processor units (CPUs), that by that time became powerful enough to provide multi-functionality not found in solid-state designs.

Whereas solid-state designs compared voltages across replica impedances that were purely analog electrical quantities, digital relays convert the analog signals received from current transformers (CTs) and potential transformers (PTs) into digital equivalent values in front-end digital signal processors (DSPs), then apply that digital information to mathematical algorithms to operate the relay.

Digital relays held an enormous advantage over any of their predecessors with the new ability to add multi-functionality to the device. For example, electromechanical and solid-state relays could only accept and operate according to one group of settings while digital relays typically were capable of accepting multiple setting groups and to adaptively apply any one of them automatically or by operator control according to changing power system topologies and operating conditions.

Also, many advantageous functions were newly possible such as oscillography, storage of history of events, and of course relay logic that was until then implemented external to the protective relay using auxiliary relays and timers. All these new functions dramatically reduced the overall installed cost of installation and made the replacement of end-of-life electromechanical-based systems economically feasible.

Power System Protection: Fundamentals and Applications, First Edition. John Ciufo and Aaron Cooperberg.

The algorithms applied to digital relay firmware by the various manufacturers always had to mimic the fundamental governing equations that described electromechanical relay response. However, as will be seen, whereas electromechanical devices had by their very nature inherent filtering, digital relays were susceptible to the transient responses of CTs and PTs. Sophisticated algorithms were developed to overcome these inherent limitations.

Nevertheless, the underlying digital relay responses always mimicked the electromechanical devices they were replacing.

> *It is for this reason that the best way to understand how digital relays respond and operate is to review the governing equations that electromechanical relays are built upon.*

4.2 Fundamental Principles and Characteristics

All protective relays, whether electromechanical, solid-state, or digital, are built to respond in a predetermined way upon the receipt of specific electrical quantities. These relays are user-settable within a given range to suit given power system protection criteria.

The fundamental principles of operation are similar for all relay types. An appreciation for the methods used in the development of electromechanical relays is invaluable in understanding how to apply all types of modern digital relays as well.

The algorithms used in digital relays mimic the principles and characteristics of electromechanical relays.

An understanding of the methods used to match the characteristics of electromechanical relays to those of the power system provides a solid understanding of all types of relay applications. For these reasons, this chapter focuses primarily on the operating principles of electromechanical relays.

4.2.1 Non-directional Induction Disk Overcurrent

This type of relay was one of the first to be developed some 100 years ago. An induction disc relay is nothing more than a very simple split-phase induction motor with contacts being closed when the disc rotates in one direction, and opens contacts when rotating in the opposite direction due to the spring action applied to it.

4.2.2 Induction Principle of Operation

A torque-producing force is developed in a disc formed of non-magnetic current-conducting material, usually aluminum but could also be made of copper. The torque is produced by the interaction of electromagnetic fluxes with eddy currents that are induced in the disc by these fluxes [1].

Figure 4.1 illustrates how a torque-producing force may be developed in a section of the disc when two adjacent fluxes pass through the disc at right angles. Flux ϕ_1 and ϕ_2 are shown when both are increasing in magnitude. Each flux induces a voltage around itself in the disc. The induced voltages cause currents to flow in the disc.

In Figure 4.1, the current produced by flux ϕ_1 is represented as $i\phi_1$ and the current produced by flux ϕ_2 is represented as $i\phi_2$. These currents and fluxes act on each other to produce rotational forces that act on the disc as follows:

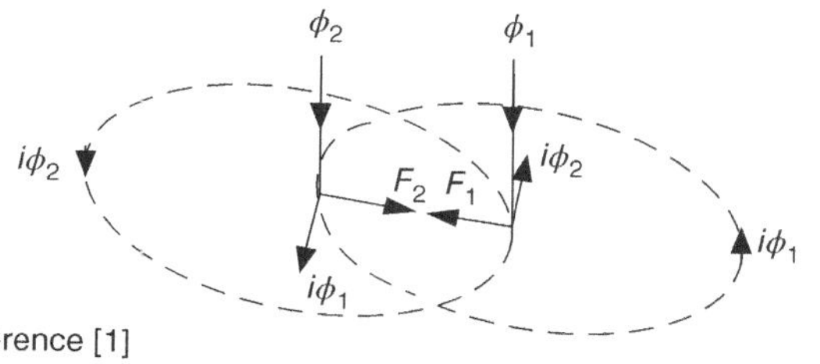

Figure 4.1 Torque production in an induction disc relay. Source: Based on Russell Mason [1].

As the fluxes are alternating and out of phase with respect to each other by phase angle θ, they can be represented as the following:

$$\phi_1 = \phi_1 \ \sin \omega t$$
$$\phi_2 = \phi_2 \ \sin (\omega t + \theta)$$

As the eddy currents produced as a result of these fluxes are also alternating, but out of phase with the fluxes by 90° they can be represented as the following:

$$i\phi_1 = k\,\phi_1 \ \cos \omega t$$
$$i\phi_2 = k\,\phi_2 \ \cos (\omega t + \theta)$$

The two forces F_1 and F_2 in Figure 4.1 are represented as per reference [1]:

$$F_2 = k\phi_2\, i\phi_1$$
$$F_1 = k\phi_1\, i\phi_2$$
$$F = F_2 - F_1 = k(\phi_2\, i\phi_1 - \phi_1\, i\phi_2)$$

Substituting these values gives the following:

$$F = k\{\phi_1\, \phi_2[\sin (\omega t + \theta) \ \cos \omega t - \sin \ \omega t \ \cos (\omega t + \theta)]$$

This can be reduced to $F = (F_2 - F_1)$ being expressed as $F = K\,\phi_1\,\phi_2 \sin(\theta)$
Therefore, the rotational net force or torque applied to the disc can be expressed as:

$$T \propto \phi_1\phi_2 \ \sin(\theta) \ [1] \tag{4.1}$$

When ϕ_1 and ϕ_2 are in phase, no amount of flux magnitude produces any amount of force on the disc. However, when ϕ_1 and ϕ_2 are out of phase, force is produced on the disc. When this happens, the speed at which the disc rotates is a function of the magnitude of the two fluxes ϕ_1 and ϕ_2. Also, the direction of the force and hence the direction of motion of the disc depend on which flux would be leading the other.

4.3 Overcurrent

4.3.1 Induction Disc Time-Overcurrent

A shaded-pole structure overcurrent relay only needs to accept a single current to operate. Normally, the magnetic flux produced by this single current input produces a single magnetic flux in phase with the current. In that case, no torque would be applied to the disc according to Eq. (4.1).

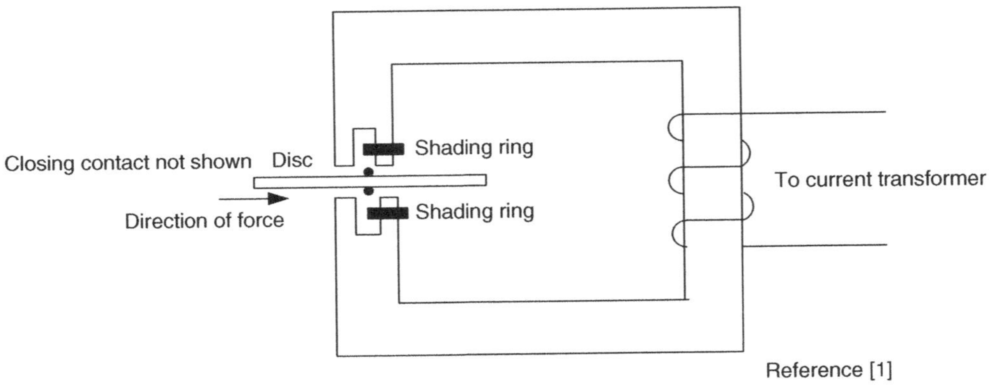

Figure 4.2 Shaded-pole structure. Source: Based on Russell Mason [1].

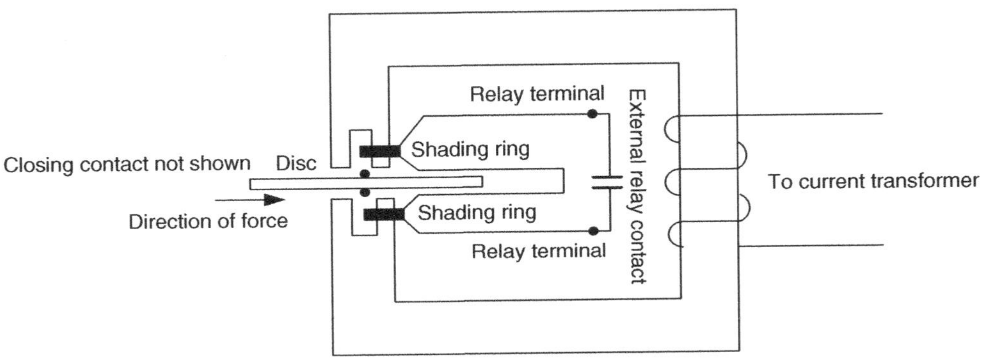

Figure 4.3 Torque control shaded-pole structure. Source: Based on Russell Mason [1].

The method universally adopted to cause two out-of-phase magnetic fluxes to be produced is by placing what is known as a shading ring on one half of each pole as shown in Figure 4.2. The magnetic flux in the air gap where the disc resides is split into two 90° out-of-phase components by the size and design of the shading ring generally of copper that encircles a portion of the pole.

Contacts are attached to the shaft of the disc as to close contacts in one direction only as one of the 90° out-of-phase fluxes always leads the other. The name used to describe this type of structure is aptly called a "shaded-pole structure."

It should be noted that just adding a shading ring around one-half pole will produce two magnetic fluxes out of phase but not necessarily out of phase by the required 90°. The method that produces the requisite 90° phase shift is to view the magnetic flux in the core consisting of two components. Then, apply the superposition principle in the linear magnetic circuit, assuming that the core is not saturated.

The shading ring can be opened with two leads taken to terminal blocks such that torque is only applied when the two leads are shorted as shown in Figure 4.3. This method allows for external torque control of the relay disc. An example where it would be useful to external torque control an overcurrent relay is by a distance relay or directional relay. The purpose of torque control will be covered in Chapter 15.

4.3.2 Inverse Time-Overcurrent Relay

An inverse time-overcurrent relay adopts the shaded-pole structure where maximum torque is always produced as θ always equals 90° by design. The torque is directly proportional to the product of the magnitude of the two fluxes. The magnitude of these fluxes is directly proportional to that of the current. Fast relay operation is produced at high current and slow relay operation at low current as shown in Figure 4.4.

4.3.2.1 Inverse Time-Overcurrent Characteristics

A feature in the design and manufacture of inverse time-overcurrent relays is the ability to alter the inverse time-characteristic using a permanent magnet that encompasses the disc as can be seen in Figure 4.5.

Specific inverse time-overcurrent characteristics can be created by the geometry of the permanent magnet. The industry has assigned categories to a family of characteristics as shown in Figure 4.6 where the relay basic pickup starts just above infinity.

Figure 4.4 Inverse time-overcurrent characteristic.

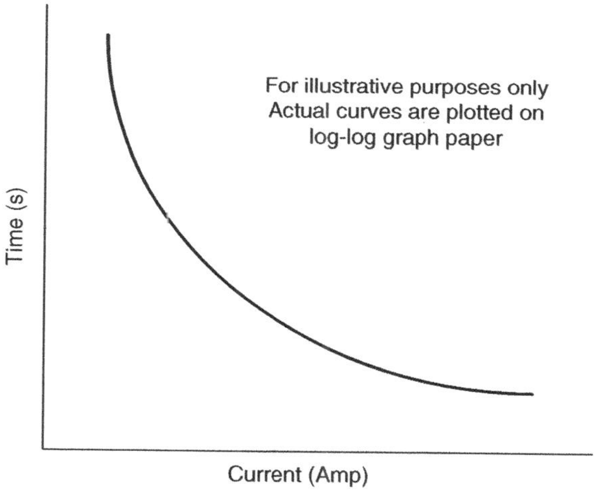

Figure 4.5 Typical inverse time-overcurrent relay of the shaded-pole structure. Source: Courtesy of GE.

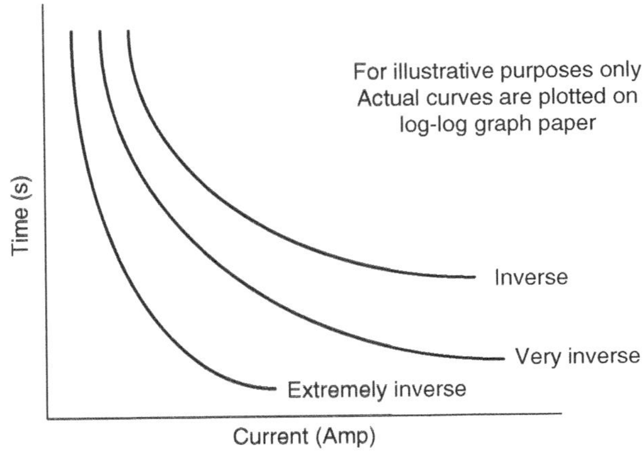

Figure 4.6 Various inverse current-time characteristic types.

There are defined functions that describe the various inverse time-overcurrent curves. One would have to refer to the relay manufacturer for the specific set of curves applicable.

Equations for IEC Operate Curves as illustrated in Figure 4.6 are:

Curve C1 Inverse

$$t = \mathrm{TD}\,[0.14/(M^{0.02} - 1)]$$

where:

t = operating time in seconds
TD = time dial setting
M = multiples of pickup

For a basic pickup of 2A and time dial (TD) of 0.5, the operate time at 20A (10 × pickup) is as follows (Figure 4.7):

$$t = 0.5\,[0.14/(10^{0.02} - 1)] = 1.5 \text{ seconds}$$

Each set of curves also has a corresponding function that defines the time to reset the disc to its normal position after it has closed contacts. This happens when the actuating current has disappeared as would happen when the breaker trips to eliminate the fault current. The time delay to reset is significant as will be described in a future chapter.

$t_r = (-\mathrm{TD} \times 1.08)/(\mathrm{M}^2 - 1)$ where t_r is the time delay to reset

Curve C2 Very Inverse

$$t = \mathrm{TD}\,[13.5/(\mathrm{M} - 1)]$$

Curve C3 Extremely Inverse

$$t = \mathrm{TD}\,[80.0/(M^2 - 1)]$$

Curve C4 Long Time

$$t = \mathrm{TD}\,[120/(M - 1)]$$

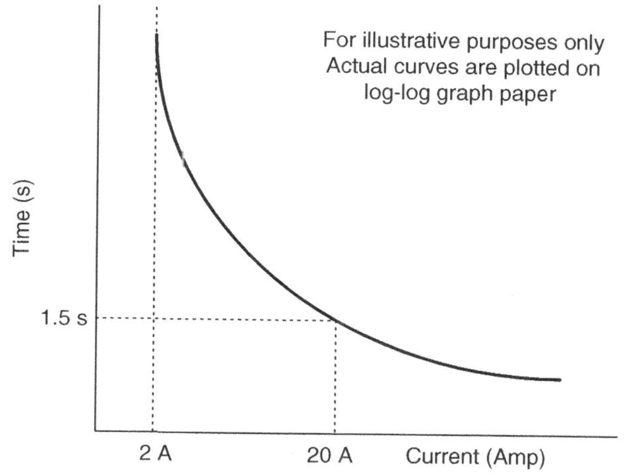

Figure 4.7 Operating time at 10x pickup in the example.

Equations for Typical US Curves as illustrated in Figure 4.6 are as follows:

Curve U1 Moderately Inverse

$$t = \text{TD}\ [0.0226 + 0.0104/(M^{0.02} - 1)]$$

Curve U2 Inverse

$$t = \text{TD}\ [0.180 + 5.95/(M^2 - 1)]$$

Curve U3 Very Inverse

$$t = \text{TD}\ [0.0963 + 3.88/(M^2 - 1)]$$

Curve U4 Extremely Inverse

$$t = \text{TD}\ [0.0352 + 5.67/(\text{M}^2 - 1)]$$

The purpose for the wide variation in time-current characteristics is to allow coordination with other downstream (further away from the source) devices whether fuses or relays. For example, fuse time-current characteristics are typically Extremely Inverse. Also, relays must coordinate with equipment damage curves which are typically Very Inverse. In general, time-current characteristics provide for faster operating times as shown in Figure 4.7. The greater the magnitude of fault current the faster it operates reducing the risk to equipment damage. This is compared to definite time delayed overcurrent operation where the time delay is independent of fault current level.

4.3.2.2 Basic Pickup Current Setting

The definition of the term basic pickup current is defined as the amount of actuating current into the inverse time-overcurrent relay where that current and the curve meet at infinity as shown in Figure 4.8a. At this current, the disc would just begin to physically vibrate and start to rotate. The multiples of pickup refer to the actual fault current measured by the relay divided by its basic pickup current. This number is the variable M used in the previous equations to calculate the relay operational time for any given setting.

The basic pickup current is adjusted by using taps on the winding to the CT as shown in Figure 4.8b. The basic current pickup taps can be seen in the picture in Figure 4.5 directly on the top of the relay. The taps are chosen by a simple screw-in lug for a given pickup value as shown in Figure 4.9.

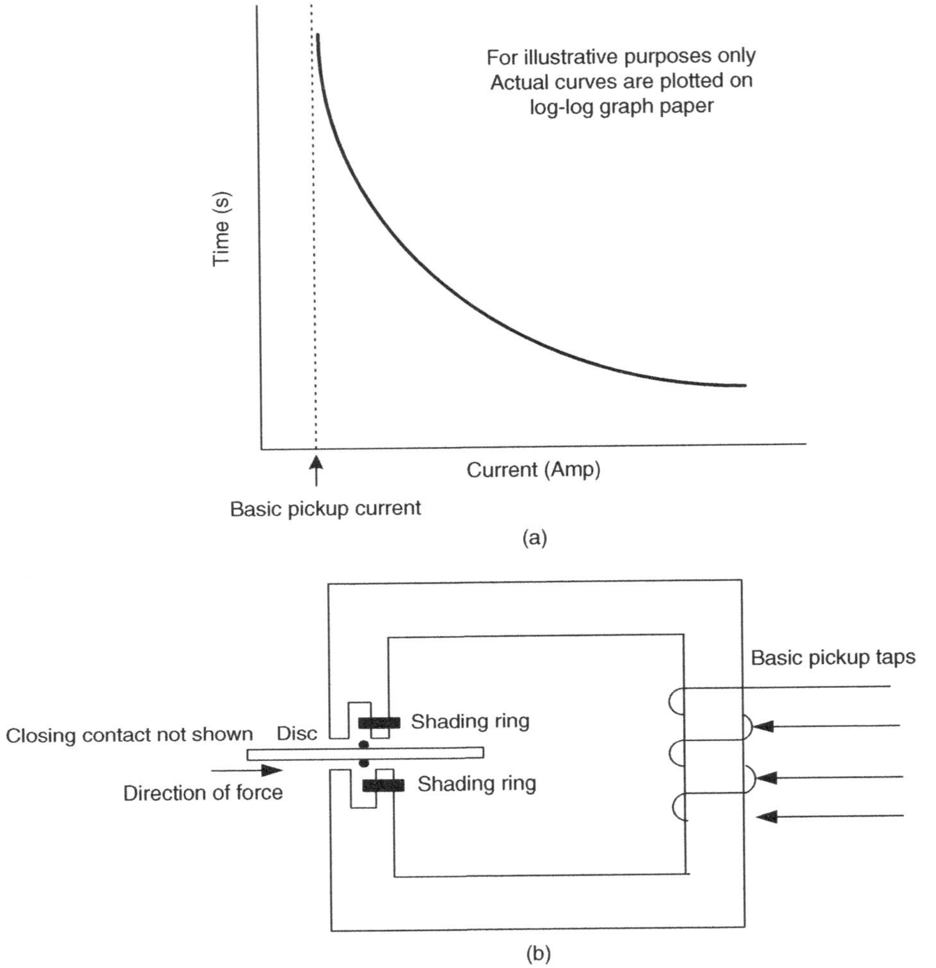

Figure 4.8 (a) Basic pickup current. (b) Basic pickup current taps. Source: (b) Based on Russell Mason [1].

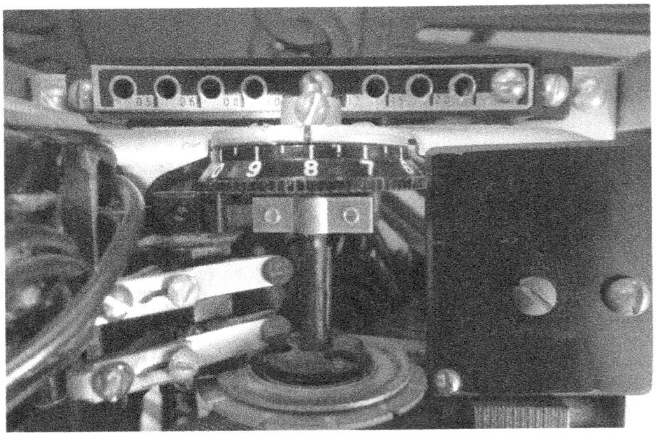

Figure 4.9 Close-up showing pickup taps and time dial. Source: Courtesy of GE.

4.3.2.3 Time Dial Adjustment

The TD is an adjustment that alters the distance the disc needs to rotate before closing contacts. The physical adjustment that alters this distance is known as the TD. The TD can be seen in the picture in Figure 4.9 sitting on top of the shaft. As the TD is moved to a higher number, the de-energized resting position of the disc is moved to a new resting position further from the contact closing position. The disc has more distance to travel and will take more time before closing contacts for the same multiple of basic pickup current.

4.3.2.4 Setting Adjustments for an Inverse Time-Overcurrent Relay

A given inverse time-overcurrent characteristic may be moved around the time-current plane vertically by increasing or decreasing the TD and horizontally by increasing or decreasing the current pickup as shown in Figure 4.10.

4.3.3 Time Coordination with Overcurrent Relays

It is relatively simple to apply overcurrent protection to a radial line. There is usually no reason to intentionally time delay the overcurrent protection. In this case, even a basic instantaneous overcurrent relay would suffice as shown in Figure 4.11.

The need for protection coordination becomes apparent when one considers two lines in a network as shown in Figure 4.12. When more than one relay is designed to sense a fault condition,

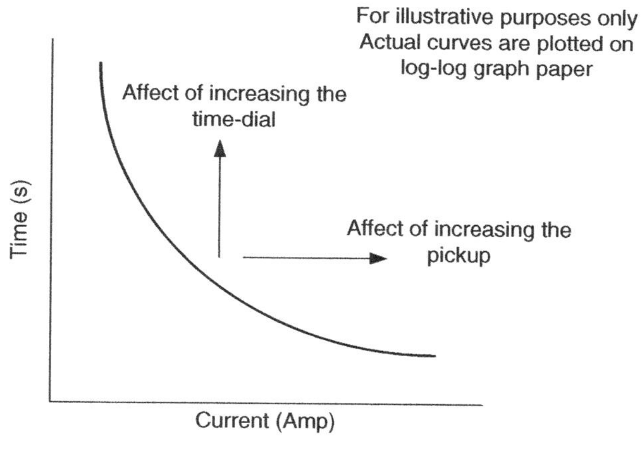

Figure 4.10 Method of setting inverse time-current characteristics.

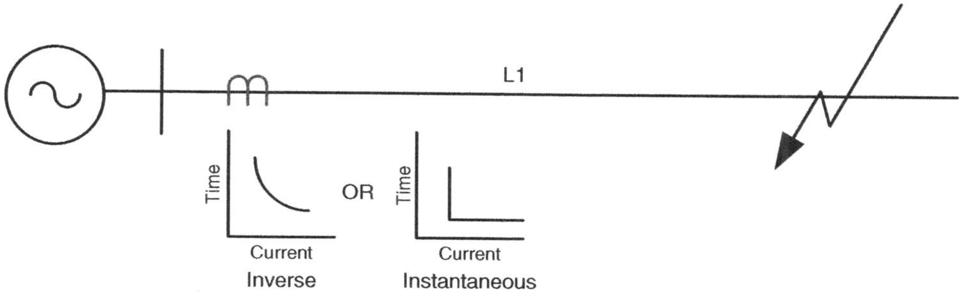

Figure 4.11 Overcurrent protection of a simple radial line.

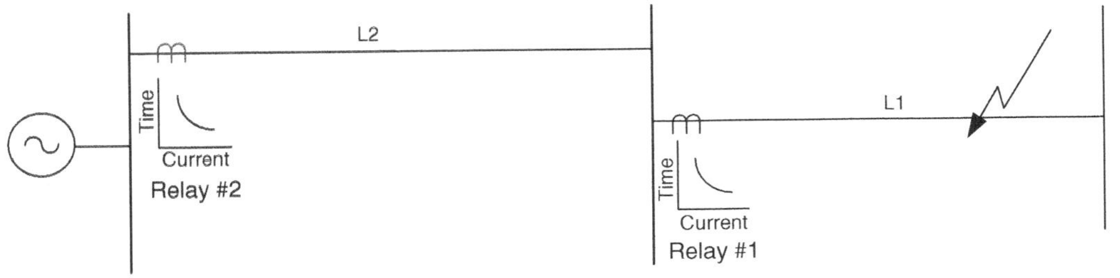

Figure 4.12 Example of the need for relay coordination.

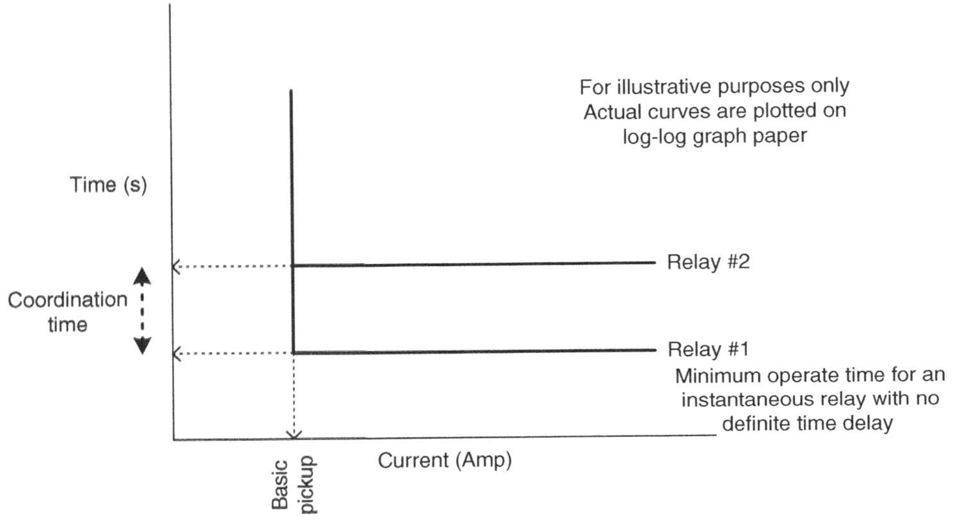

Figure 4.13 Step-timed coordination with definite time delays.

but trip different breakers one of the methods used to achieve selective tripping is through time coordination. Referring to Figure 4.12, there may be two reasons why Relay #2 might sense a fault on line L1. The most obvious reason is that Relay #2 backs up Relay #1 by design and therefore is set to cover both lines. Another less obvious reason is that Relay #2 is mainly set to only sense faults on line L2 and not on line L1. In either case, by assigning an intentional time delay to Relay #2, it is possible to achieve coordination between the two relays. This is shown conceptually for two lines for simple illustration. In practice, there are usually several lines in parallel with all the lines shown that need to coordinate with respect to each other.

Coordination in this example may be achieved by making Relay #1 instantaneous and Relay #2 time delayed by enough time to ensure that the breaker isolating line L1 has opened to eliminate the fault and allow Relay #2 to reset itself. This method of coordination is known as step-timed as shown in Figure 4.13.

In this example, Relay #1 and Relay #2 are set for the same pickup. However, Relay #2 is timed to operate later than Relay #1. These two relays coordinate with each other.

Timed overcurrent coordination may also be achieved by time coordinating inverse time-overcurrent relays as shown in Figure 4.14. In this example, both Relay #1 and Relay #2 may be set for the same pickup current yet the TD of each can be set differently to achieve the minimum

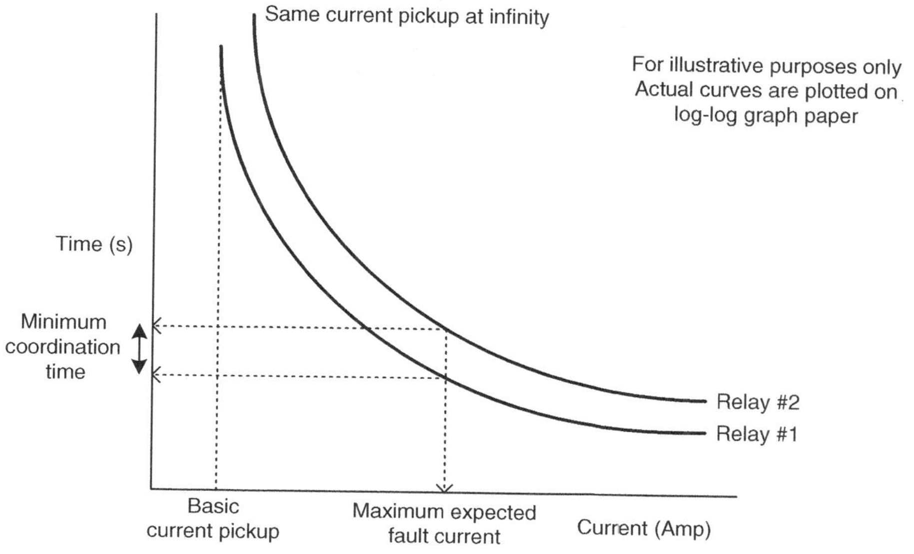

Figure 4.14 Time coordination with inverse time-overcurrent.

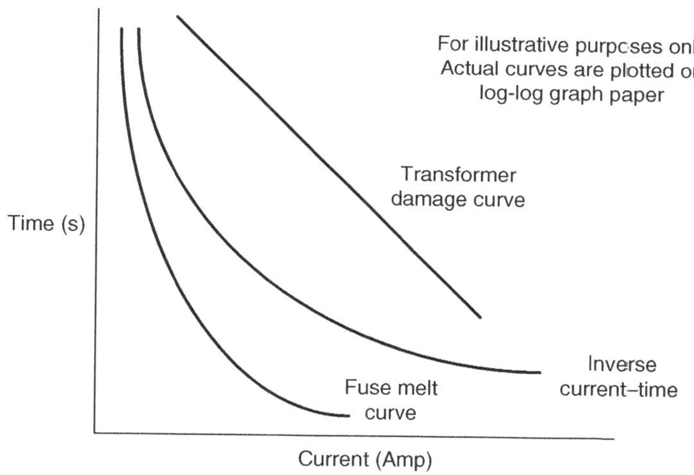

Figure 4.15 Transformer damage curve, fuse melt curve, and inverse time-overcurrent relay characteristics.

coordination margin at the maximum expected fault current. Typically, relay pickup is defined in multiples of basic pickup current for that specific relay. In Figure 4.14, the relay is set to pickup at the maximum expected fault current which is at specific multiples of basic pickup current.

Inverse time-overcurrent relays have an advantage over definite time instantaneous overcurrent relays. They can coordinate with fuses and power equipment damage curves such as transformers whose characteristics are also inverse time-current as shown in Figures 4.15 and 4.16 further discussed in Chapter 15.

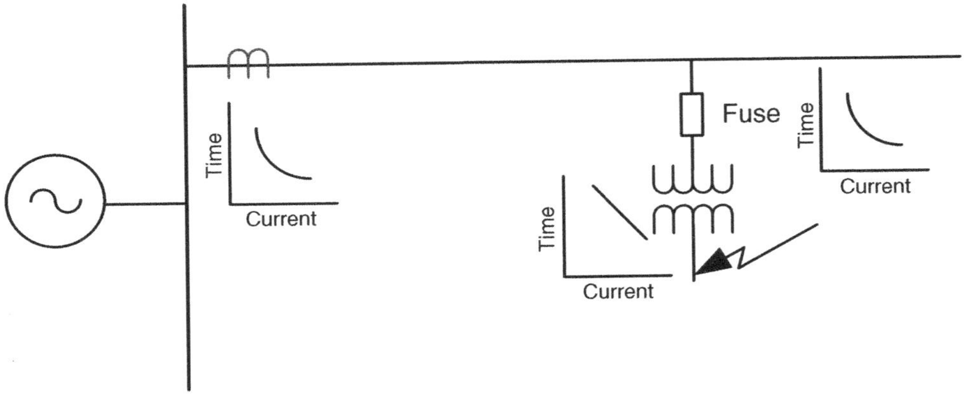

Figure 4.16 Single line diagram for the coordination example in Figure 4.15.

4.3.3.1 Coordination Time for Electromechanical Relays

The minimum coordination time for electromechanical inverse time-overcurrent relays is 400 ms calculated as follows:

- Relay #1 and Relay #2 both see the fault on line L1 as shown in Figure 4.12.
- Relay #1 closes contacts and trips the line L1 breakers to isolate the fault.
- Relay #2 induction disc is a mechanical device with disc over-travel.
- Breaker time 8 cycles or 135 ms.
- Disc rotational over-travel is 4 cycles or 65 ms.
- The minimum coordinating time is therefore 200 ms.
- Add an equal amount of time as for a margin applied coordination time for a total of 400 ms.

4.3.3.2 Coordination Time for Digital Relays

The minimum coordination time for digital inverse time-overcurrent relays is 300 ms calculated as follows:

- Relay #1 and Relay #2 both see the fault on line L1 as shown in Figure 4.12.
- Relay #1 closes contacts and trips the line L1 breakers to isolate the fault.
- Relay #2 being a digital device can be set for instantaneous reset.
- Breaker time 8 cycles or 135 ms.
- Time to dropout output relay approximately 5 ms.
- The minimum coordinating time is therefore 140 ms.
- Add an equal amount of time as margin and the coordination time can be as short as 280 ms. The minimum coordination time is typically 300 ms for digital relays at many utilities.

4.3.4 Directional Overcurrent Relays

Directional overcurrent relays provide a dual function. Firstly, they respond to a given sensed current above a predetermined set threshold. Secondly, they only respond to that current set threshold being exceeded for a fault in one direction and not in the other for a particular direction from the relay location.

Simple induction disc overcurrent relays that receive an actuating signal from a single current source are by definition non-directional. There is no way of defining the current direction in an AC system without a polarizing quantity used as a reference since the current is changing direction every half cycle at system frequency.

4.3.4.1 Method of Directioning

The method used to direction overcurrent relays is to introduce a polarizing quantity such as voltage or an unidirectional current to compare the actuating current with. In the example of Figure 4.17, voltage is used as the polarizing quantity. The voltage measured by the relay will not change direction with fault location. The current measured by the relay will change direction with fault location by 180°. In a power system, the angle between the voltage and current would be less than 90° for faults in the forward direction but greater than 90° for faults in the reverse direction. For simplicity, refer to Figure 4.18 where the system current and polarizing voltages are assumed to be in phase.

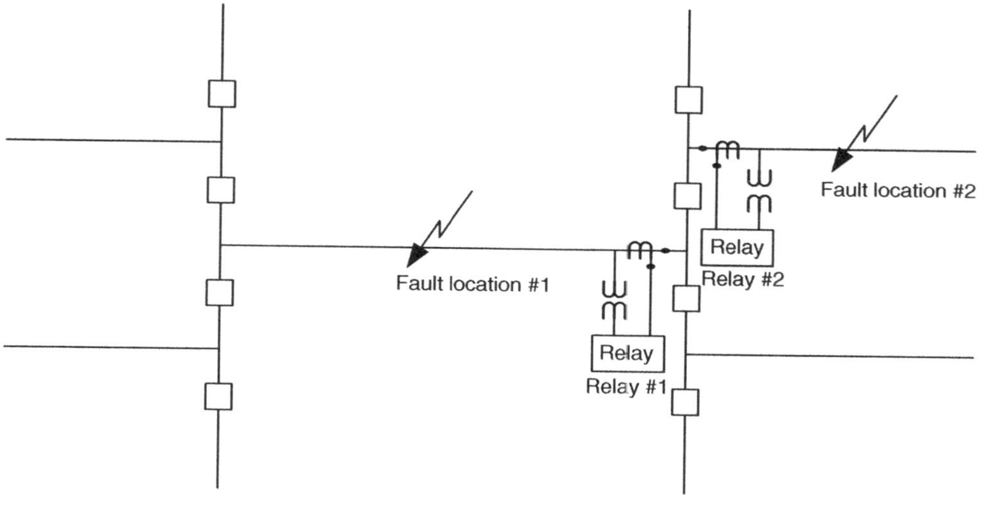

Figure 4.17 Need for directioning.

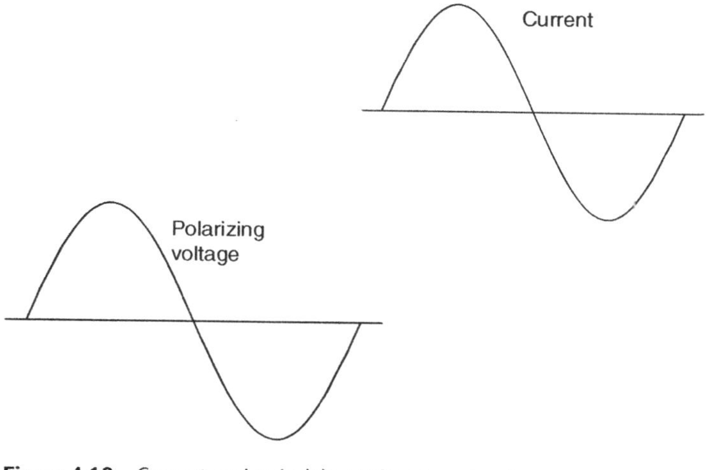

Figure 4.18 Current and polarizing voltage are in phase.

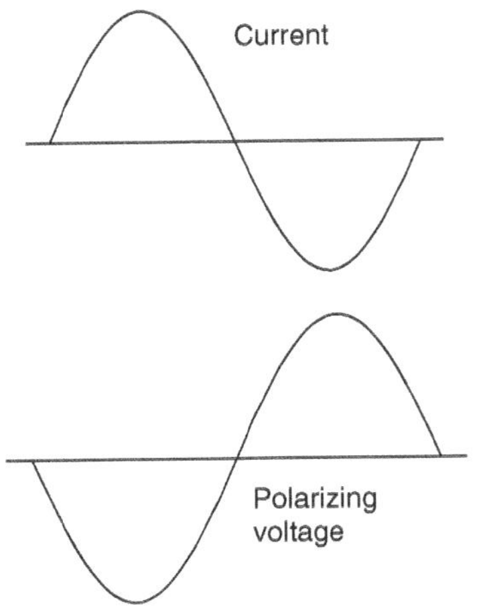

Figure 4.19 Current and polarizing voltage are out of phase.

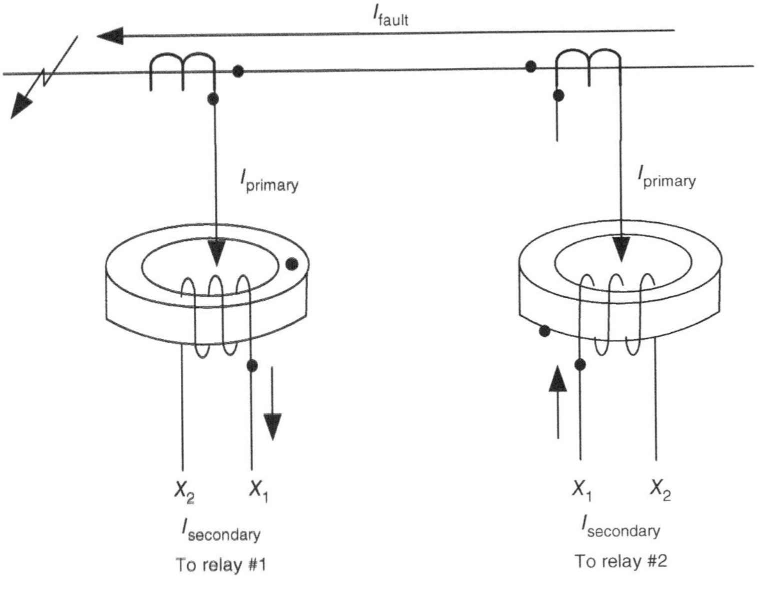

Figure 4.20 Current transformer polarity and connections.

The relay will operate for faults in the forward direction when the voltage and current are in phase and will restrain when they are not as shown in Figure 4.19.

Refer to Figure 4.20 which illustrates the concept covered in Chapter 3 on CT polarity. These are the CTs shown earlier in Figure 4.17. For a fault in fault location #1, the instantaneous current flows into relay #1 and out of relay #2. With the common polarizing voltage, relay#1 will operate and relay #2 will restrain.

4.3.4.2 The Watt-Hour Structure

The watt-hour meter structure in Figure 4.21 receives its name from the fact that the same construction is used for watt-hour meters. It contains two separate coils on two different magnetic circuits, each of which produces one of two necessary fluxes for driving the rotor which is also a disc.

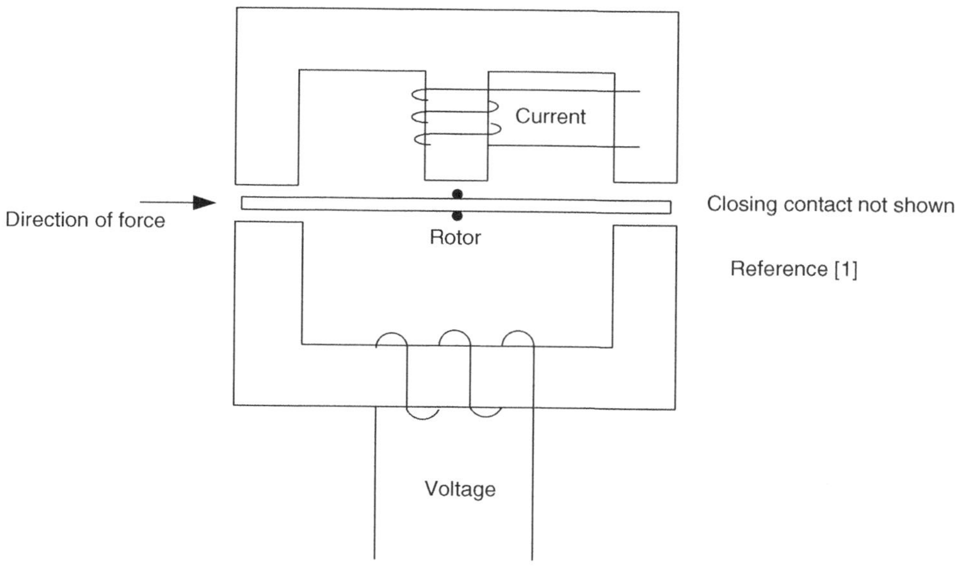

Figure 4.21 Watt-hour meter structure. Source: Based on Russell Mason [1].

A watt-hour-meter structured induction disc relay receives one actuating quantity from a current transformer source and the other actuating quantity from a voltage transformer source.

The torque developed by the induction disc is proportional to the product of the two fluxes which cross it at a right angle and the sine of the angular displacement of those fluxes or $T = K\,\phi_1\phi_2 \sin(\theta)$ (Eq. (4.1)). One flux is a function of the relay current and the other is a function of the current through the voltage coil produced by the voltage to the relay. Thus, maximum torque is developed when the two fluxes are displaced by 90° similar to an induction disc relay. Note, that the flux is always in phase with the current through the coil not the voltage across the coil.

4.3.4.3 The Induction Cup Structure

The induction cup resembles an induction motor, except that the rotor iron is stationary, while the rotor conductor portion is free to rotate. The cup structure employs a hollow cylindrical rotor. The induction cup, being smaller and lighter, is a more efficient torque producer than either the shaded-pole or the watt-hour-meter structure and is used in high-speed relays (Figure 4.22).

4.3.4.4 Relay Phase Relationship of Voltage and Current in a Directional Relay

As shown in Figures 4.23 and 4.24, the phase relationship between the voltage and current associated with the voltage coil is a function of the overall impedance of the voltage coil itself which may be altered by inserting a combination of resistance and capacitance in series with the voltage coil. By inserting a combination of resistance and capacitance in series with the voltage coil, the angle between the applied voltage and voltage coil current $I_{\text{voltage coil}}$ may be changed to almost any value either lagging or leading V without changing the magnitude of $I_{\text{voltage coil}}$.

Figure 4.24 illustrates the principle of operation. In this phasor diagram, the reference point is when the polarizing voltage and the current through the relay are in phase at unity power factor. Given the usual impedance characteristic of the voltage coil, the current lags behind the applied voltage by angle θ. Flux ϕ_2 is in phase with this current. Flux ϕ_1 is in phase with the current through the current coil. Maximum torque is applied to the disc when ϕ_1 and ϕ_2 are out of phase by 90°.

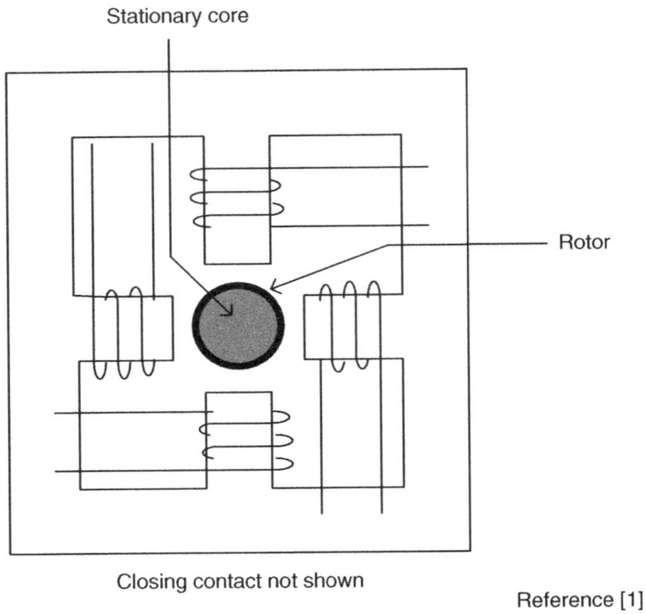

Figure 4.22 Induction cup structure. Source: Based on Russell Mason [1].

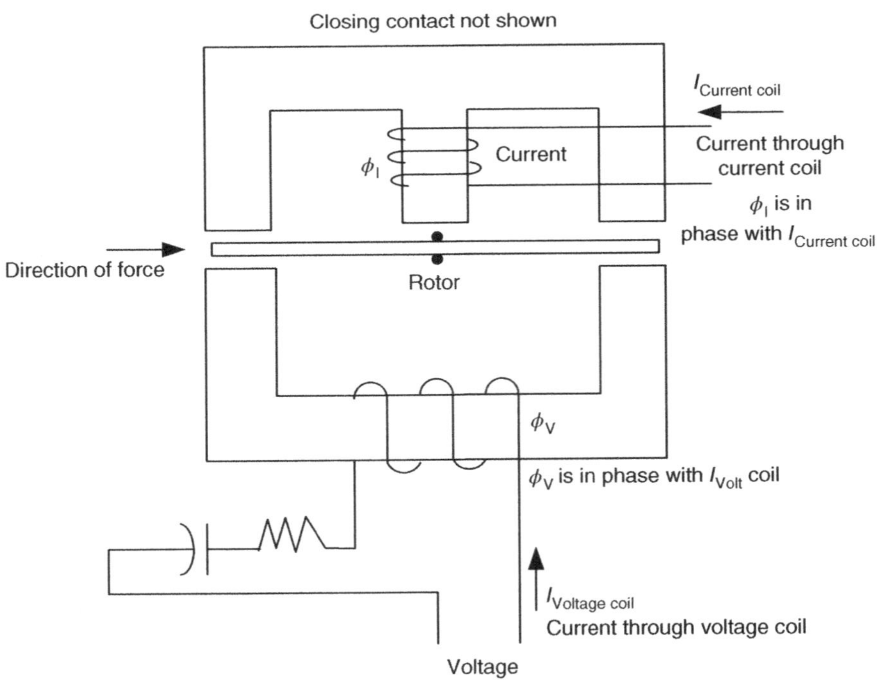

Figure 4.23 Watt-hour meter structure as a directional overcurrent relay.

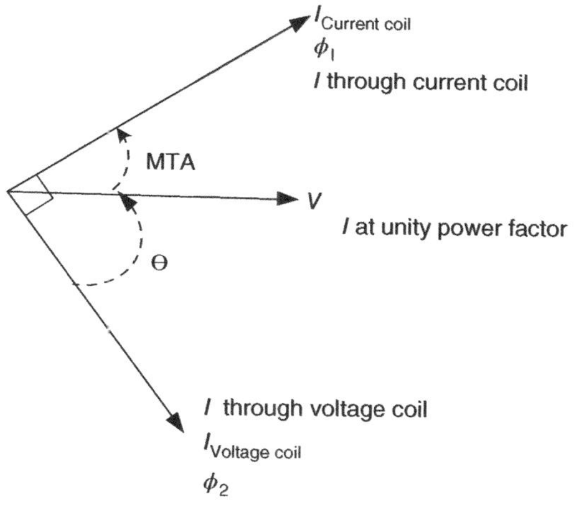

Figure 4.24 Phasor diagram showing the relationship of voltage and current for maximum torque in the example above.

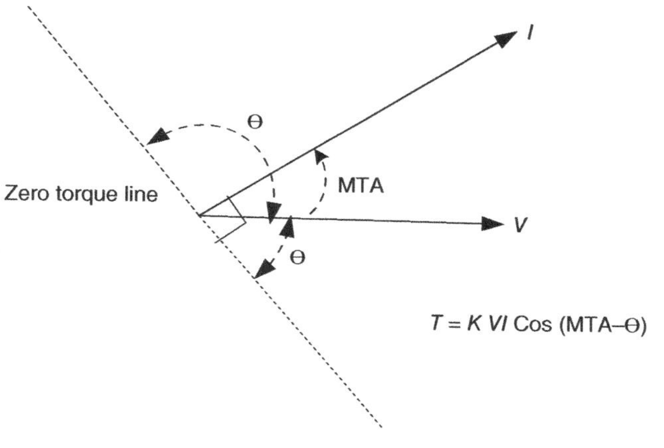

Figure 4.25 Generalized phasor relationship of voltage and current.

The maximum torque angle (MTA) is a characteristic value built into the relay by the manufacturer such that maximum torque is applied to the disc when the $\text{MTA} + \theta = 90°$.

Figure 4.25 shows a generalized phasor diagram for maximum torque in a current–voltage induction disc directional relay.

The torque equation representing this relay is $T = K\,VI\cos(\text{MTA} - \theta)$ where the MTA is the maximum torque angle and θ is the angle described below. Torque is the rotational force that causes the disc to rotate. When the disc rotates in one direction, it causes contact closure and when rotated in the other direction does not result in contact closure. In an electromechanical relay, the rotational force is a function of the cosine of the relay MTA, and θ, the angle the current deviates from that of unity power factor, with the polarizing voltage as reference. When current either lags the polarizing voltage or leads it as shown in Figure 4.25 for the two extreme values of θ such that no torque is developed defines the Zero Torque Line. When the current reverses itself with respect to the polarizing voltage beyond these values, the disc begins to rotate in the opposite direction. In a digital relay, the algorithm governing directioning would adopt this same equation. However, since the concept of actual torque is not applicable, contact closure is simply determined by the phase relationships of polarizing voltage and current. For digital relays, the term MTA is therefore replaced with the term relay characteristic angle or RCA.

Directional relays are used to recognize the distinction between current being supplied from one direction or the other in a power system. A directional relay recognizes differences in phase angle between two sensed quantities. This recognition is reflected in contact action and is limited to differences in phase angle exceeding 90° from the phase angle at which maximum torque is developed.

In general, for directional relays, the quantity that produces one of the fluxes is called the polarizing quantity. It is the reference against which the phase angle of the other quantity is compared. Consequently, the phase angle of the polarizing quantity must remain more or less fixed when the other quantity goes through wide changes in phase angle. The polarizing quantity is usually the voltage to the relay and is the reference. Its angle and magnitude are assumed to be constant. Sometimes in the case of current into the grounded neutral of autotransformers, this current is used for polarizing instead of voltage, as the direction of this current is most of the time constant regardless of fault location. It should be noted, that in some rare cases this current could reverse depending on the system impedances and should be studied.

4.3.4.5 Typical Application of Directional Phase Overcurrent Relays

Refer to Figure 4.26 showing a phasor diagram for a single-phase directional relay with an MTA of 30° lead.

Of the various connections for single-phase directional relays possible in a three-phase system is the 90° or quadrature connection, whereby the current coil is connected in one phase and the voltage is taken from a pair of other phases as depicted in Figure 4.27.

Refer to Figure 4.28 showing the phasor diagram for three single-phase directional overcurrent relays, each with an MTA of 30° lead, where quadrature connection is used for the voltage input to each relay.

The purpose of quadrature connection of voltage input to a relay is to guarantee a healthy polarizing voltage for any combination of phase–phase faults. This would not apply to three-phase faults where the polarizing quantity may be too low to operate the relay. For three-phase faults farther away from the relay location, the voltage drop to the fault is usually sufficient to polarize the relay. For three-phase faults close to the relay location, the voltage drop to the fault could be too low and not be sufficient to polarize the relay. In this case, a high set instantaneous overcurrent relay/element can be used to supplement the directional overcurrent relay. Only a single-phase

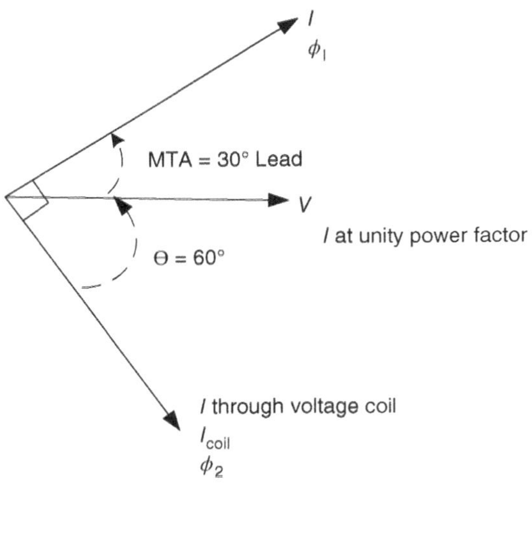

Figure 4.26 Phasor diagram – single phase directional relay with MTA 30° lead.

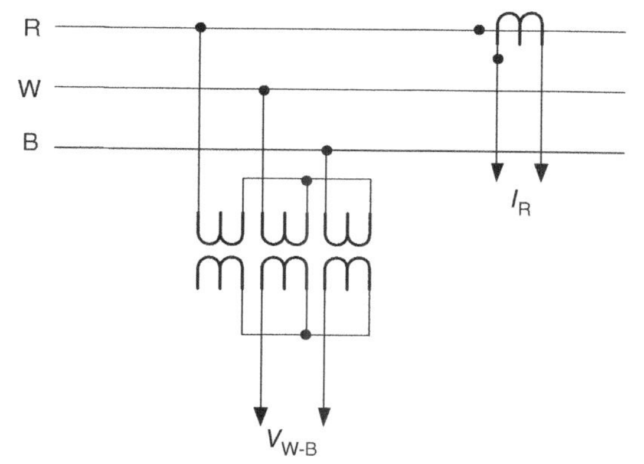

Figure 4.27 Quadrature connection of a single red phase directional relay.

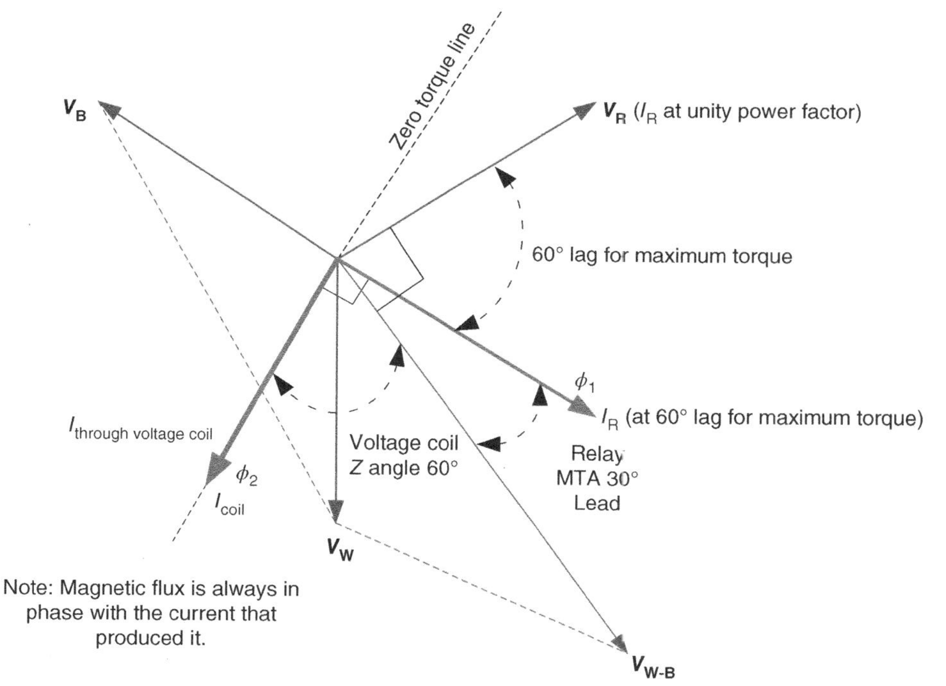

Figure 4.28 Phasor diagram for three single-phase directional overcurrent relays connected in quadrature.

non-directional overcurrent relay is required to protect for a three-phase fault. A high pickup current setting assures that it does not operate for external faults.

When three single-phase relays each with an MTA of 30° lead are connected with polarizing voltage not taken from the same phase as the phase current, but the quadrature voltage the overall MTA of the three relays becomes 60° lag. That means that the three relays will provide coverage for phase–phase and three-phase faults so that maximum torque is developed when the protected line impedance angle is 60° as can be seen in Figure 4.28. Most 115 kV lines have a characteristic impedance angle of 60°. The same can be done for 230 kV lines where the characteristic impedance is approximately 75°–85°.

4.3.4.6 Typical Application of Ground Directional Overcurrent Relays

Directional overcurrent relays designed for the detection of phase faults cannot be relied upon to detect ground faults. Phase overcurrent relays must be given a high basic pickup current setting to ensure that they don't operate for a three-phase load. Ground relays are generally given a low basic pickup current setting to make them as sensitive as possible. Faults involving ground very often include high fault resistance which limits the fault current thereby requiring that the relays be made sensitive as possible. Ground overcurrent relays measure ground or neutral current directly. They are optimized to operate for low ground-fault current and not operate for a balanced three-phase load.

Ground directional overcurrent relays are typically dual-polarized with both voltage and current polarization. Voltage polarizing never changes polarization for all fault locations regardless of system conditions. However, the magnitude of voltage required to operate the relay depends on the location of a ground fault. On occasion, it may be too low to provide sufficient actuating quantity to operate an electromechanical relay. Therefore, as a general rule, most utilities use current polarizing also where possible to supplement voltage polarization. Current polarizing can on occasion give a false indication of fault location, such as autotransformer neutral ground current reversal. Therefore, it is prudent where possible to provide dual polarization.

Refer to Figure 4.29 showing the potential transformer connection used to obtain zero-sequence voltage for voltage polarization (see Chapter 8). Current polarization can be obtained from a CT in the neutral to the ground connection of an autotransformer as shown in Figure 4.30 or a CT in the closed delta tertiary of an autotransformer as shown in Figure 4.31.

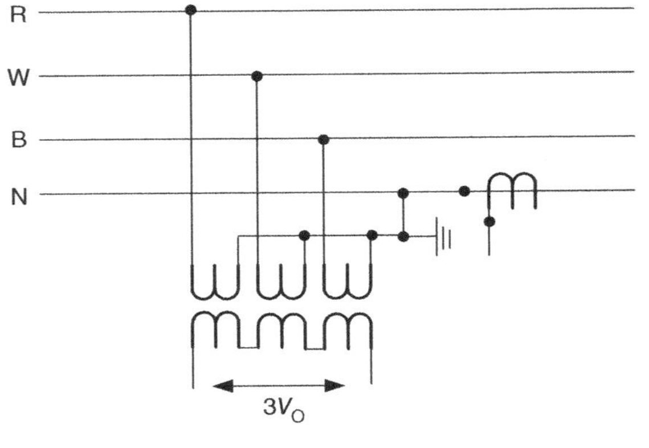

Figure 4.29 Zero-sequence voltage ($3V_0$) derived for polarization of ground directional overcurrent relays.

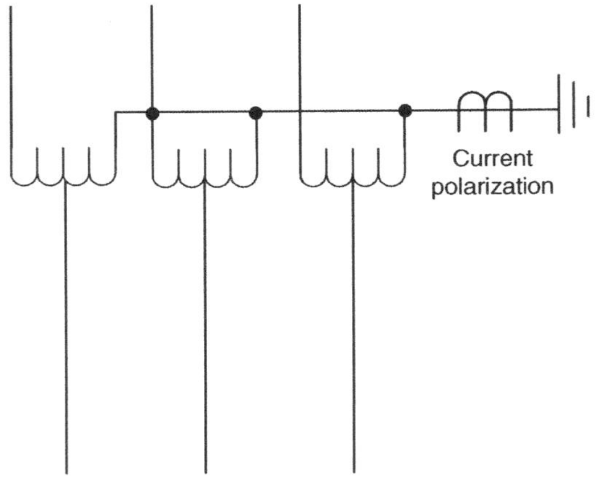

Figure 4.30 Zero-sequence current derived for polarization of ground directional overcurrent relays from auto neutral-gnd.

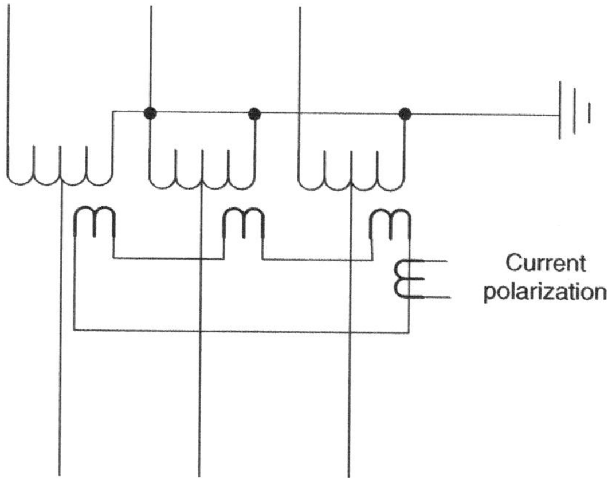

Figure 4.31 Zero-sequence voltage derived for polarization of ground directional overcurrent relays from delta tertiary.

Digital directional ground overcurrent elements apply an entirely different principle using negative sequence polarizing quantities instead. All a digital relay requires is phase currents and phase-neutral voltages to direction correctly.

4.4 Differential

4.4.1 General

The most efficient technique that provides high-speed discrimination between in-zone and out-of-zone faults is differential protection (see Chapter 7 for the definition of zones). The electrical quantities entering and leaving the protected zone are compared via current transformers. When added, if the net value between all sensed circuits is sufficiently close to zero, no fault is assumed to exist. However, if the net value when added exceeds the set pickup value, an internal fault exists with the difference current being able to operate differentially connected relays.

Almost any type of relay can be made to operate as a differential relay. It is not so much the relay construction as the way the relay is connected in a CT circuit that makes it possible to operate in a differential protection mode.

This chapter covers the basics of differential protection as applied to buses and transformers. This topic will be covered extensively in Chapter 8 for transformers and in Chapter 9 for buses.

4.4.2 Differential Principle Used in Bus Protection

Ideally, the most straightforward application of differential protection is for the protection of a simple stretch of the bus bar. In practice, the biggest problem is the performance of CTs for of out of zone faults. In this case, multiple infeeds can saturate one CT to the exclusion of others as will be seen later.

4.4.2.1 Fundamental Principle of Operation

CTs are located at all incoming and outgoing nodes of the protected bus. The CT secondary windings are then interconnected, and the coil of an overcurrent relay is connected across the CT secondary circuit. The CT polarities are critical to the correct operation of this type of connection. In this application, the current relay could be an instantaneous or induction disc type relay. The effective CT ratios at either end of the differential connection must be the same for a simple two node bus protection as illustrated in Figure 4.32. Without the CT ratios being the same, there is a

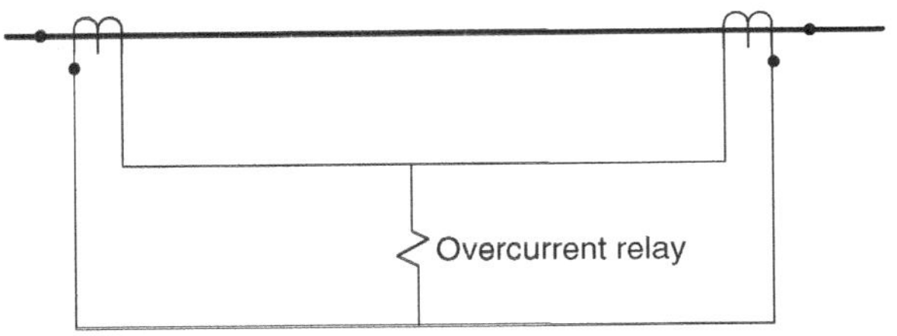

Figure 4.32 Simple differential connection using an overcurrent relay.

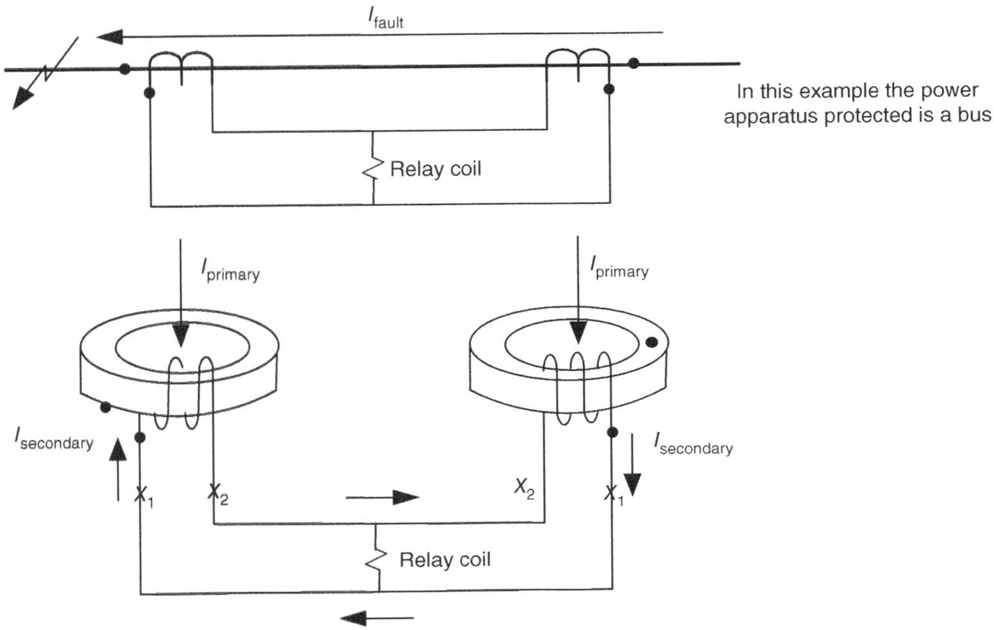

Figure 4.33 Simple differential connection method of operation – external fault.

mismatch between secondary currents that would flow into the relay at the relay node. In an ideal situation, the secondary current representing the fault current would all circulate around the CT connection or all flow into the relay coil at the relay node.

For an external fault, the instantaneous primary current entering the CT causes a secondary excitation voltage resulting in instantaneous secondary current to leave X_1. At the same instant, the primary current is leaving the other CT causing a secondary excitation voltage resulting in the secondary current to enter X_1. The secondary currents simply circulate around the CT connection with no current entering the relay coil (Figure 4.33).

For an internal fault, the opposite is true. The instantaneous primary current entering the CT causes a secondary excitation voltage resulting in instantaneous secondary current to leave X_1. At the same instant, the primary current is also entering the other CT causing a secondary excitation voltage resulting in the secondary current leaving X_1. The secondary currents meet at the relay node and both flow through the relay coil.

An internal fault located anywhere between the two CTs causes current flows to the fault from both sides. The sum of the CT secondary currents will flow through the overcurrent relay (Figure 4.34).

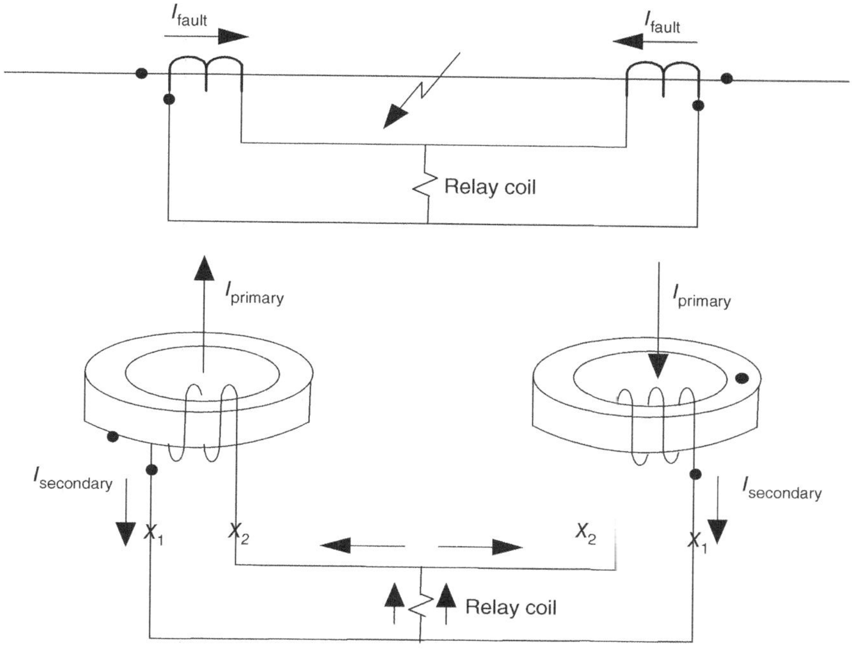

Figure 4.34 Simple differential connection method of operation – internal fault

It is not necessary that fault current flow from both sides to result in secondary current flow through the relay. A flow on one side only, or even some current flowing out of one side while a larger current enters the other side, will cause a differential current.

The differential current will be equal to the vector difference between the currents entering and leaving the protected zone. Should the differential current exceed the relay pickup value, the relay will operate.

It is a simple step to extend the principle to a system element such as a bus having several connections as shown in Figure 4.35. It is merely necessary that all CTs have the same ratio and that they be connected so that the relay receives no current when the total current leaving the protected element is equal in magnitude and phase to the total current entering the element. In this case, the secondary exciting voltages of all the CTs are such that all the currents flow around the differential circuit and no current flows through the relay.

4.4.2.2 Security for Out-of-Zone Faults

The performance of CTs is a significant factor for bus differential protection. As shown in Figure 4.36, infeeds to a high magnitude bus fault can be through multiple sources. The one CT closest to the fault needs to transform the entire fault current while the others only need to transform a portion of it. The result of seeing the entire infeed is to possibly cause severe waveform distortion in that one CT due to severe saturation (Chapter 3). This issue is common for high voltage buses at terminal switching stations as well as low voltage buses at substations. There are two types of bus differential protection using electromechanical relays. One is known as low impedance bus differential, and the other is high impedance bus differential protection. Digital relay is a third type using a different method of ensuring security for out-of-zone bus faults.

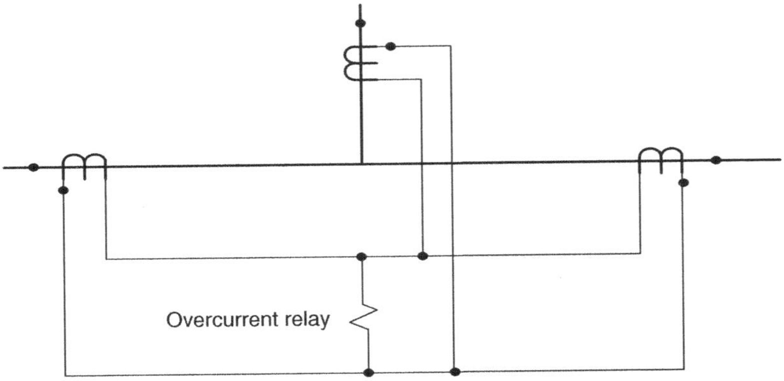

Figure 4.35 Differential bus protection.

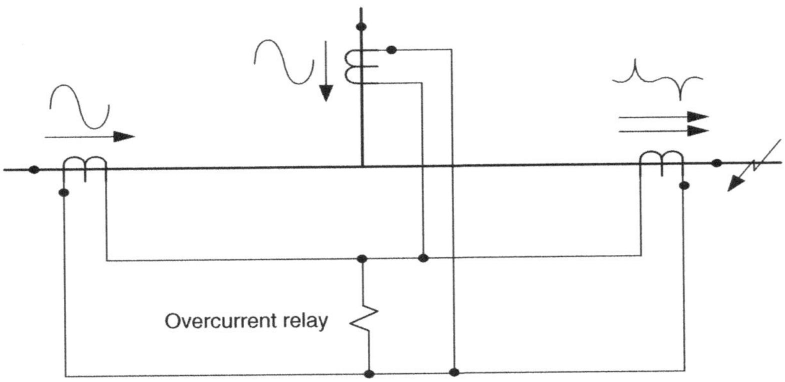

Figure 4.36 Waveform distortion for out-of-zone faults.

4.4.2.3 Low Impedance Differential Protection

All differential protections that use straight overcurrent relays with low burden fall into the category of low impedance differential. Induction disc inverse time-overcurrent relays used in bus differential protection were found to be highly stable for applications up to 115 kV system typical short circuit levels.

The induction disc inverse time-overcurrent relay characteristic known as "short time," (Figure 4.37) even when set sensitively, was found to provide superior performance by remaining stable even for severe out-of-zone faults.

Some utilities that used inverse electromechanical relays for bus protection tried experimenting with digital relays using the same inverse characteristic. It was surmised that the reason electromechanical relays were secure for external faults was a function of the inverse–time relationship. However, when digitally duplicated inverse–time characteristics were applied the differential protection tended to operate for severe close-in external faults.

The reason electromechanical relays were so stable was due to their fundamental operating property. Torque is applied to the disc when two sinusoidal fluxes out of phase by 90° are directed downward into the disc. When CTs severely saturate, the secondary current loses its sinusoidal character. The additional harmonic components in distorted waveforms tend to interfere with torque production from power frequency components in a way that tends to retard disc rotation. With severely distorted waveforms, the two fluxes simply cannot react with one other in time. Without

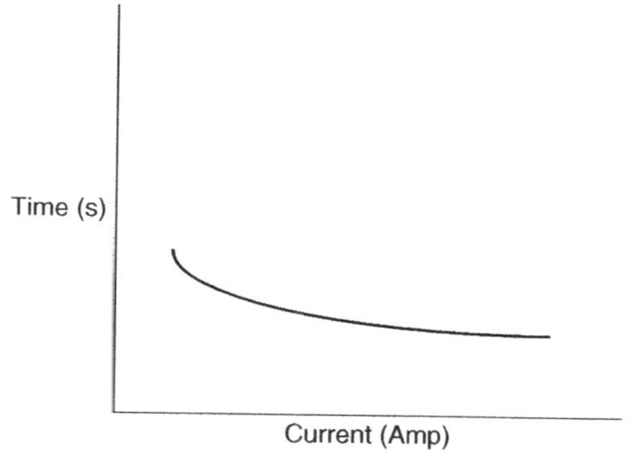

Figure 4.37 Short-time inverse characteristic.

the ability to produce torque, the relay remains stable. The simple digital overcurrent relays convert the current via an analog to digital conversion and apply the digital representation of the current to the short-time inverse equation. Whereas the electromechanical relay provides natural filtering for severe waveform distortion and restrains, the digital relay does not, therefore it trips. More sophisticated digital techniques are employed for bus differential relays covered later in Chapter 9 on bus protection.

4.4.2.4 High Impedance Differential Protection

Whereas low impedance differential protection is effective at lower system voltages, it becomes unstable for applications at higher voltages such as 230 kV and above. System short circuit levels at this voltage are considerably higher. For this reason, low impedance differential schemes tend to misoperate for out of bus zone faults. A different type of differential protection known as high impedance differential protection is effective and stable at voltages of 230 kV and above.

Low impedance differential protection works on the principle that the relay burden in parallel with all the CTs is of low magnitude with operational stability calculated based on current. In high impedance differential protection schemes, the overall relay burden in parallel with the CTs is of high magnitude with operational stability calculated based on voltage. Refer to Figure 4.38 showing a typical high impedance differential connection.

A maximum three-phase external fault is considered along with the assumption that the CT closest to the fault needs to transform the entire infeed and goes completely into saturation. A fully saturated CT looks electrically to the other CTs as a dead short. The maximum voltage across the relay node is the sum of the secondary currents from all CTs (excluding the saturated CT) times the resistance of the CT leads and the internal resistance of the shorted CT. A stabilizing resistor is added such that the current seen by the relay resulting from the stabilizing voltage will be just below its pickup setting. This protection is effective, as the worst-case fully saturated CT is assumed in calculating the value of the stabilizing voltage. The value for the stabilizing resistor is chosen to ensure the relay does not operate for that value. The other option is to use a voltage protective relay with high impedance.

4.4.3 Differential Principle Used in Transformer Protection

The performance of CTs under fault conditions is not a significant factor for transformer differential protection. Fault currents are restricted by the transformer impedance. Also, there is no real possibility for multiple infeeds saturating a single CT. Nevertheless, there are other issues unique to transformers that are dealt with using the percent differential relay.

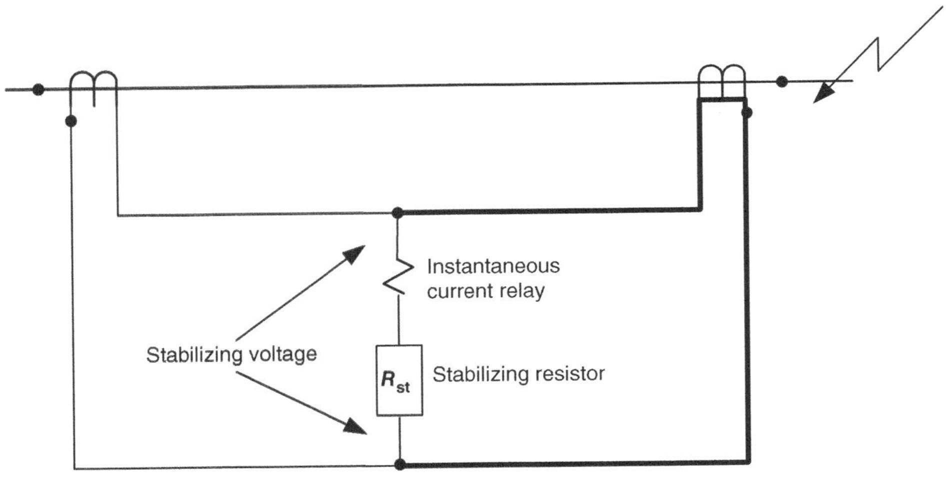

Figure 4.38 High impedance differential protection.

4.4.3.1 Percent Differential Relay

The basic differential principle can still be applied where a power transformer is involved. The ratios and connections of the CTs on opposite sides of the power transformer must be such as to compensate for the magnitude and phase angle shift between the power transformer currents on either side. Chapter 8 on transformers will show this can either be done externally for electromechanical relays with CT ratios and connections or internally for digital relays via matrix manipulation – internally mathematically compensated algorithms.

The most extensively used form of differential relay is the percentage-differential type. This is essentially the same as the overcurrent type but the differential current required to operate this relay is a variable quantity, owing to the effect of a restraining coil.

The ratio of the differential operating current to the restraining current is a fixed percentage, which explains the name of this relay. As will be covered in Chapter 8 on transformers, there are three main factors contributing to a certain amount of out-of-zone fault current also "spilling" into the operating coil. These factors are CT ratio mismatch, CT error, and the effect of tap changing. Since the percent differential relay has a rising pickup characteristic as the magnitude of through current increases, the relay is restrained against operating improperly while still being sensitive for low magnitude faults.

Percent differential relays are usually instantaneous or high speed. Time delay is not required for selectivity because these relays are virtually immune to the effects of transients when applied properly (Figure 4.39).

Figure 4.40 represents the basic construction and operation of a typical electromechanical percent differential relay. The relay depicted in this figure has three coils wound around a magnetically permeable iron bar. The CT secondary leads are connected to restraint windings. The two restraint windings are then connected in parallel with each other and then in series with an operating coil. The other side of the operating coil is connected in parallel with the two non-spot sides of the CTs.

The operating coil has N number of turns while each restraint winding has half the number of turns or $N/2$ turns. The operating winding is wound in the opposite direction to the two restraint windings. The induced magnetic flux ϕ_{operate} produced by the operating coil is in the opposite direction to the magnetic flux ϕ_{restrain} produced by the two restraint windings. A net magnetic flux in the operate direction will pull the plunger-type solenoid relay contact to the closed or typical trip direction.

The primary current is shown in the direction of supplying load where the CT secondary currents circulate around the two CTs. Any difference in value between them will flow through the operating coil.

The electrical circuit representation of the percent differential relay in Figure 4.40 is shown in Figure 4.41.

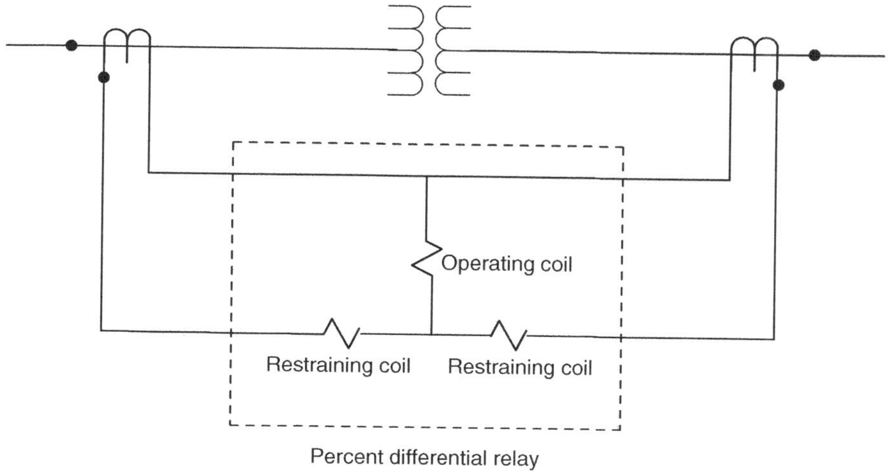

Figure 4.39 Percent differential relay.

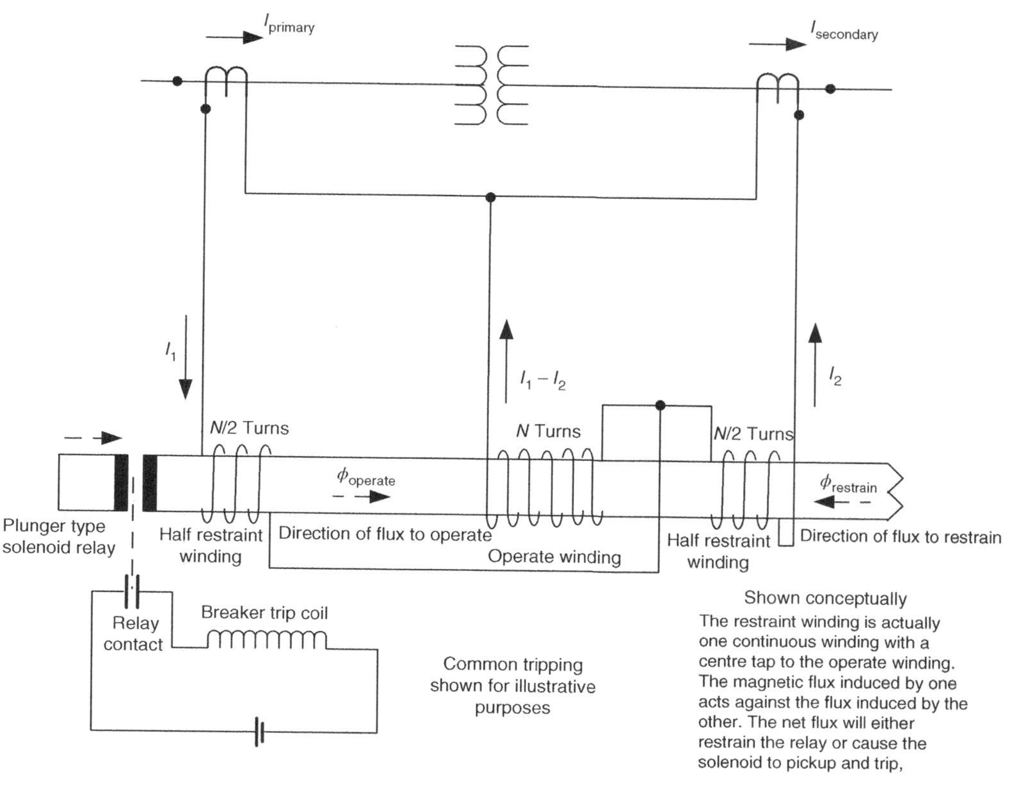

Figure 4.40 Electromechanical percent differential relay basic construction and operation.

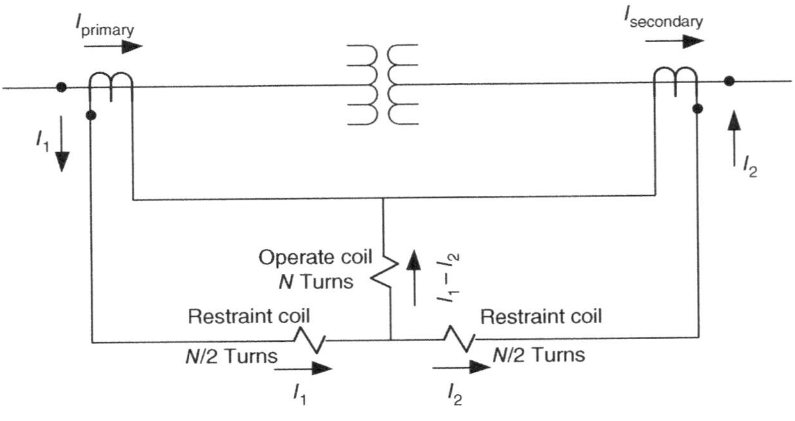

Figure 4.41 Electrical circuit representation of the percent differential relay.

Referring to Figure 4.42, the differential current in the operating coil is equal to $I_1 - I_2$, and the equivalent current in the restraining coil is equal to $(I_1 + I_2)/2$ since the operating coil is connected to the midpoint of the two half restraining coils. If N is the number of turns on the operating restraining coil, the total ampere-turns are $I_1N/2 + I_2N/2$ which is the same as if $(I_1 + I_2)/2$ were to flow through the entire coil.

The percent slope shown in Figure 4.42 is derived in terms of competing for magnetic fluxes via the tendency to operate in Ampere-Turns and the tendency to restrain in Ampere-Turns. The percent slope is the tendency to Operate in Ampere-Turns divided by the tendency to Restraint in Ampere-Turns.

Example 4.1 The percent slope shown in Figure 4.42 is derived in terms of competing for magnetic fluxes via the tendency to operate in Ampere-Turns and the tendency to restrain in Ampere-Turns. The percent slope is the tendency to Operate in Ampere-Turns divided by the

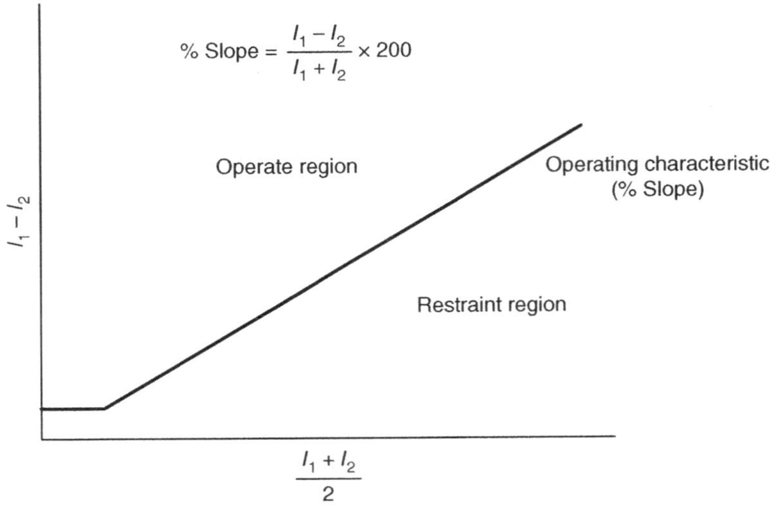

Figure 4.42 Operating characteristic for the percent differential relay (typical of electromechanical relays).

tendency to Restraint in Ampere-Turns.

$$\text{Operate in Ampere-Turns} = (I_1 - I_2)\,N_o$$

$$N = \text{number of turns in the operate winding}$$

$$\text{Restraint in Ampere-Turns} = \frac{I_1 N_r}{2} + \frac{I_2 N_r}{2}$$

$$N_r = \text{number of turns in the restraining windings}$$

$$\text{Slope in Ampere-Turns} = \frac{(I_1 - I_2)\,N_o}{\frac{I_1 N_r}{2} + \frac{I_2 N_r}{2}}$$

$$= \frac{(I_1 - I_2)N_o}{\frac{(I_1 + I_2)N_r}{2}} \text{ when } N_o = N_r$$

$$\%\text{Slope} = \frac{(I_1 - I_2)200}{(I_1 + I_2)} = 45\%$$

The degree of percent slope is a function of the relative values of N_o for the operate coil and N_r the two restraint coils. When N_o and N_r are equal, the percent slope is fixed at 45%. When the number of turns N_o making up, the operated coil is higher than the number of turns N_r making up the two restraint windings the percent slope will be lower than 45%. When the opposite is true and there are more relative turns in the restraint windings, the percent slope will be higher than 45%. The percent slope is therefore adjustable by means of adding inter-turn taps to the operate and restraint windings.

The operating characteristic of a typical digital relay is shown in Figure 4.43. As will be covered in Chapter 8 on transformers, the percent slope derived for electromechanical relays is equally applicable to digital relays. Digital relay manufacturers mimic the known behavior of electromechanical relays for simplicity of use.

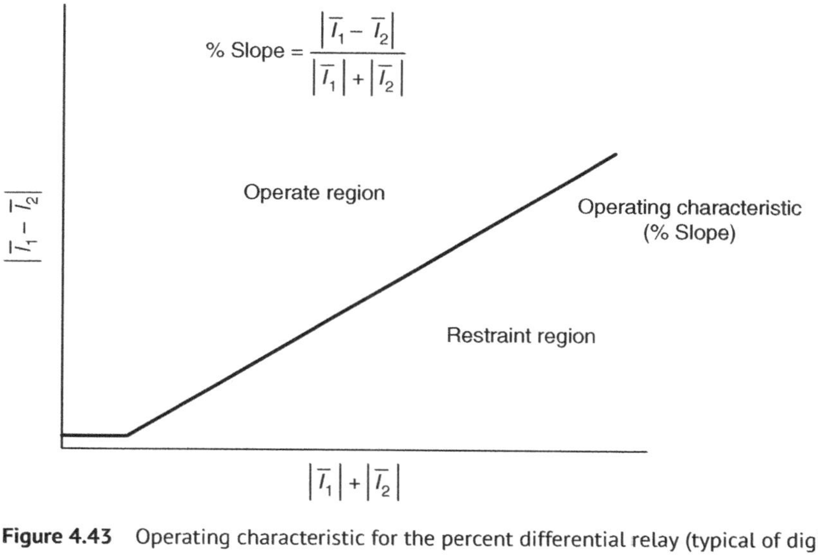

Figure 4.43 Operating characteristic for the percent differential relay (typical of digital relays).

4.5 Distance

4.5.1 General

Overcurrent relays respond to current and, if directional, receive polarization via an additional sensed quantity either voltage or current. Distance relays are fundamentally different as they balance voltages and currents, the ratio of which corresponds to an impedance. Impedance is an electrical measure of distance along a transmission line that usually has a linear relationship with line length. This explains the generic name distance applied to this group of relays. The applications of distance relays are covered in Chapter 14.

4.5.2 Need for Distance Protection

The source impedance (Z_s) is the Thevenin equivalent impedance behind the bus where the protection relays are located. In an ideal network where the source impedance is constant, overcurrent relays could be used whenever there is a need to underreach the remote end of a line. Line impedance (Z_L) is the impedance of the line from where the relay measures voltage to the location of the fault. The fault current is simply $I_F = V/Z_L$ with Z_s not affecting the set reach of the relay, refer to Figure 4.44. In reality, however, Z_s is not constant and changes with the number of lines in service as well as the amount of upstream generation in service at any given time that make up the Thevenin equivalent network impedance, refer to Figure 4.45. This matters as the relay setting need constant updating as the source impedance changes with time.

Distance relays measure the impedance between the relay location and the fault and if this impedance falls within the relay operating characteristic, which it is set to reach, it will operate to trip. The impedance to the fault is the ratio between the voltage (IZ_L) and the current (I) supplied to the relay. In other words, for a fault along the transmission line, the voltage at the relaying location is the IZ_L voltage drop of the line from the relaying location to the point of fault. Therefore, the voltage to the current ratio for this fault will be $V/I = Z_L$. For a fault inside the line section, $V/I < Z_L$ and conversely, outside the line section $V/I > Z_L$. Therefore, a protective relay designed to operate if it measures a ratio $< Z_L$ and restrains for ratios $> Z_L$, provides a very predictable form of protection usually independent of any source impedance consideration.

4.5.3 Impedance Relay Principle of Operation

An impedance relay is a device that responds to any component of impedance, resistance, reactance, or usually a combination of both. In an impedance relay, the torque produced by a current

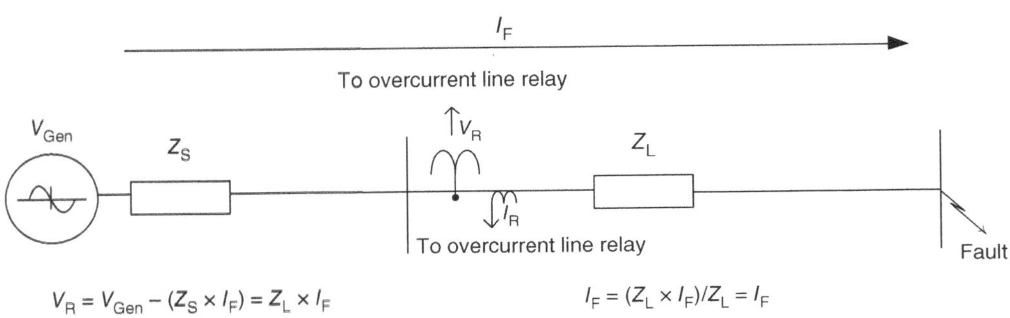

Figure 4.44 Fault detection using a simple overcurrent relay.

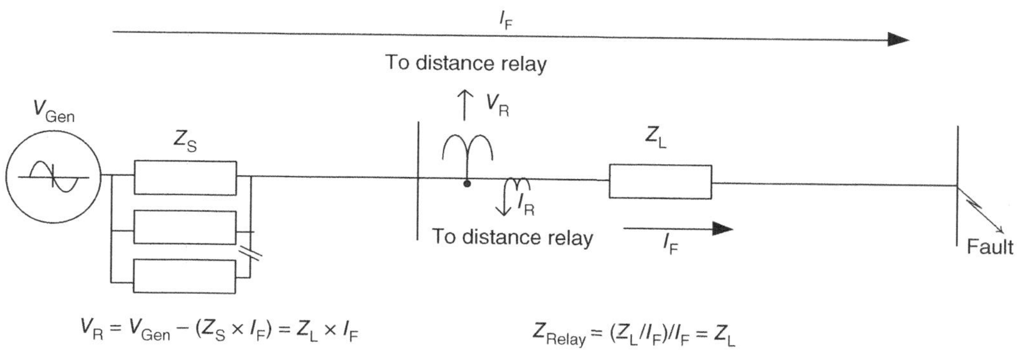

Figure 4.45 Fault detection using an impedance relay.

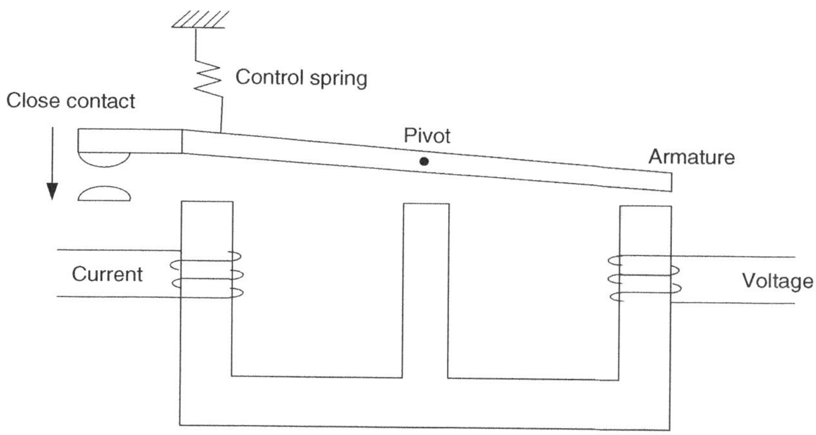

Figure 4.46 Simple impedance relay basic construction and operation. Source: Based on Russell Mason [1].

element is balanced against the torque of a voltage element. The current element produces positive torque whereas the voltage element produces negative torque. An impedance relay is therefore a voltage-restrained overcurrent relay (Figure 4.46).

The equation which describes relay torque is as follows:

$$T = K_1 I^2 - K_2 V^2 - K_3$$

where I and V are RMS magnitudes of the current and voltage, respectively, and K_3 represents the control spring.

At the balance point, when the relay is on the verge of operating, the net torque is zero:

$$K_2 V^2 = K_1 I^2 - K_3$$

Dividing by $K_2 I^2$:

$$\frac{V^2}{I^2} = \frac{K_1}{K_2} - \frac{K_3}{K_2 I^2}$$

$$\frac{V}{I} = Z = \sqrt{\frac{K_1}{K_2} - \frac{K_3}{K_2 I^2}}$$

Neglecting the effect of the control spring whose effect is significant only at low currents:

$$Z = \sqrt{\frac{K_1}{K_2}} = \text{constant}$$

In other words, an impedance relay is on the verge of operating at a given value of the ratio of V to I, which may be expressed as impedance [1].

Refer to Figure 4.47 showing that the relay will pick up for any combination of V and I represented by a point above the characteristic in the positive torque region. In other words, the relay will pick up for any value of Z less than the constant value represented by the operating characteristic. By adjustment via taps on the current and voltage coils, the slope of the operating characteristic can be changed so that the relay will respond to all values of impedance less than any desired upper limit.

A much more useful way of showing the operating characteristic of distance relays is using an impedance diagram or R-X diagram as shown in Figure 4.48.

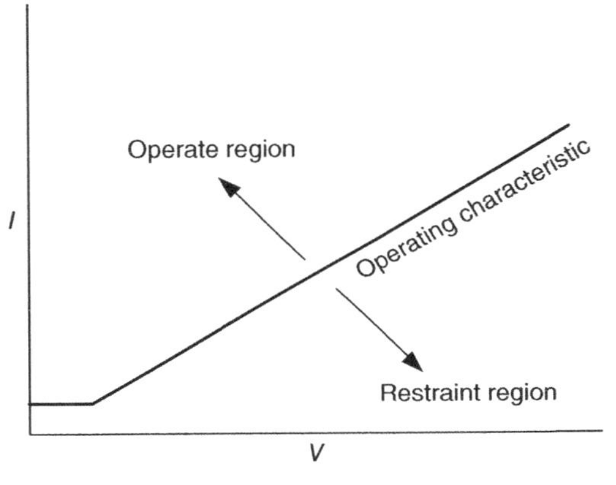

Figure 4.47 Operating characteristic of an impedance relay.

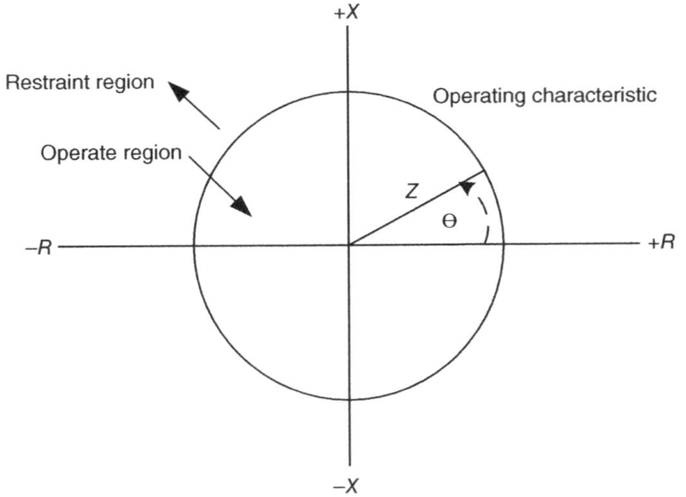

Figure 4.48 Operating characteristic of an impedance relay on an *R*-*X* diagram.

The numerical value of the ratio of V to I is shown as the length of a radius phasor, such as Z, and the phase angle θ between V and I determine the position of the phasor.

Since the operation of the impedance relay is independent of the phase angle between V and I, the operating characteristic is a circle with its center at the origin. Any value of Z less than the radius of the circle will result in the production of positive torque and any value of Z greater than this radius will result in negative torque, regardless of the phase angle between V and I.

An impedance relay is non-directional. It operates for a specific measurement of impedance thus ensuring that a predetermined amount of line is protected and not beyond. For applications, when directioning is required, it can be achieved via directional current supervision.

When impedance relays were originally invented around 1928, the phase relays that are used to detect phase-to-phase and three-phase faults, measured phase-to-phase voltage and single-phase current. For example, a red phase relay would by convention measure $V_R - V_W$ phase-to-phase voltage and I_R phase current. As shown in Figure 4.49, this type of relay connection leaves the relay with two different reaches depending on whether the phase fault is phase–phase or three-phase. A relay calibrated to be set for a certain line impedance reach for three-phase faults would underreach by a factor of 0.866 for a phase–phase fault. For a relay calibrated to be set for a certain line impedance reach for a phase–phase fault would overreach by a factor of 1.15 for a three-phase fault.

Some manufacturers chose to set calibration according to phase–phase faults while others preferred to calibrate it for three-phase faults. Both approaches are acceptable provided that the

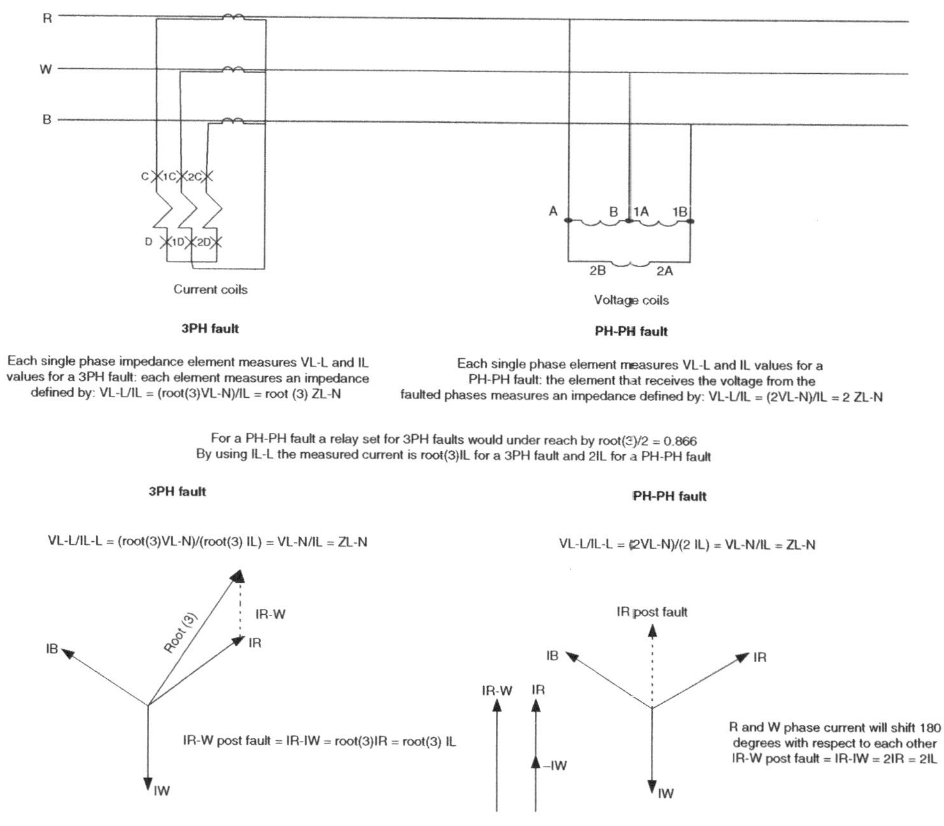

Figure 4.49 Two fault detecting techniques used for impedance relays.

protection practitioner recognizes what type of fault the relay was calibrated for. As will be covered in Chapter 14 for line protection, a zone 1 distance relay is set to cover 80% of the line and the zone 2 relay is set at 125% of the line. Therefore, there is a material difference as to whether the relay was a zone 1 underreaching relay or a zone 2 overreaching relay. If it was a zone 1 underreaching relay, the protection practitioner would not mind that the relay is underreaching by a factor of 0.866 for some other types of faults. Conversely, if it was a zone 2 overreaching relay, the protection practitioner would not mind that the relay over reach by a factor of 1.15 for some other types of faults.

To measure the same distance or impedance ratio for all faults involving more than one phase, it was recognized by the industry around 1950, that a distance relay supplied with the voltage between the two faulted phases and the vectorial difference of their currents could achieve this. For example, for an R-W phase fault, the relay measures their currents, in this case, the relay measures $V_{R\text{-}W}/(I_R - I_W) = Z_1$, the impedance of the line between the relay and the point of fault. For a three-phase fault, the relay measures current as $\sqrt{3}\,I_L$ and measures voltage as $\sqrt{3}\,V_{L\text{-}N}$, the $\sqrt{3}$ multipliers cancel. For a phase–phase fault, the relay measures current as $2\,I_L$ and measures voltage as $2\,V_{L\text{-}N}$, the 2 multipliers cancel. In this manner, the relay will have the same response to a phase-to-phase and three-phase fault as shown in Figure 4.49.

4.5.4 MHO Relay Principle of Operation

The admittance or mho relay is a form of impedance relay that is inherently directional. The mho relay has a voltage-restraining element that opposes a directional overcurrent relay using voltage as a polarizing quantity. In other words, the device is a voltage-restrained directional relay. There is nothing magical about the term admittance or mho for this relay type. It is essentially an impedance relay whose characteristic is offset in the forward direction such that the characteristic passes through the origin. The term mho was given to this type of relay only as a means of separating its unique operating characteristics from that of the simple impedance relay. Mho or admittance has little further meaning.

The electromechanical mho relays used by most utilities were of the induction cup construction. It resembles an induction motor, except that the rotor iron is stationary, only the rotor conductor portion being free to rotate. The cup structure employs a hollow cylindrical rotor. The induction disc is a very efficient torque producer that operates at high speed (Figure 4.50).

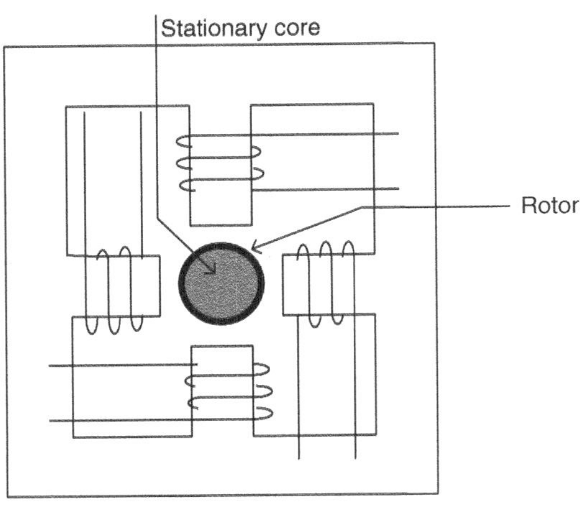

Figure 4.50 Induction cup structure. Source: Based on Russell Mason [1].

The equation which describes the relay torque as a voltage-restrained directional relay is as follows: $T = K_1\ VI \cos(\text{MTA} - \theta) - K_2V^2 - K_3$

At the balance point, the net torque is zero and hence:

$$K_2V^2 = K_1VI \cos(\text{MTA} - \theta) - K_3$$

Dividing both sides by K_2VI:

$$\frac{V}{I} = Z = \frac{K_1}{K_2}\cos(\text{MTA} - \Theta) - \frac{K_3}{K_2VI}$$

Neglecting the control spring effect:

$$Z = \frac{K_1}{K_2}\cos(\text{MTA} - \Theta)$$

This equation which describes the mho relay is like that of a directional relay but with no voltage term [1]. The relay has but one circular characteristic such that the diameter of the circle is shown when $\theta = \text{MTA}$. Every other cord on the circle is defined similarly (Figure 4.51).

4.5.4.1 Offset-MHO Relay Principle of Operation

The offset-mho relay is similar to the mho characteristic except that the operating characteristic is shifted such that it no longer passes through the origin. The shift is accomplished by using what is known as a current bias. This consists of introducing a voltage into the voltage-restraining circuit (Figures 4.52 and 4.53).

4.5.4.2 Reactance – Type Distance Relay

The reactance distance relay is an overcurrent element with a directional element that either opposes or aids in relay operation. In other words, it is an overcurrent relay with directional relay restraint. This is contrasted with a mho relay which is a directional overcurrent relay with voltage restraint.

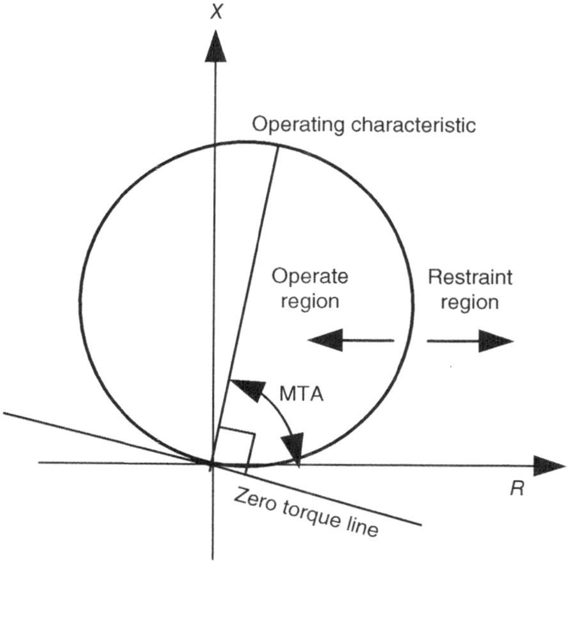

Figure 4.51 Operating characteristic of a Mho relay on an *R-X* diagram.

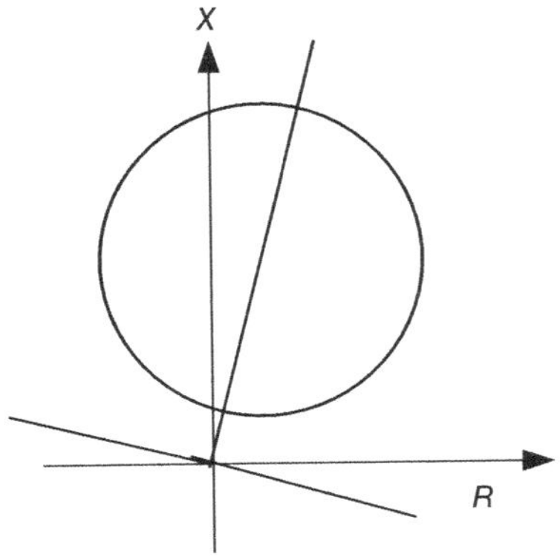

Figure 4.52 Forward offset MHO characteristic.

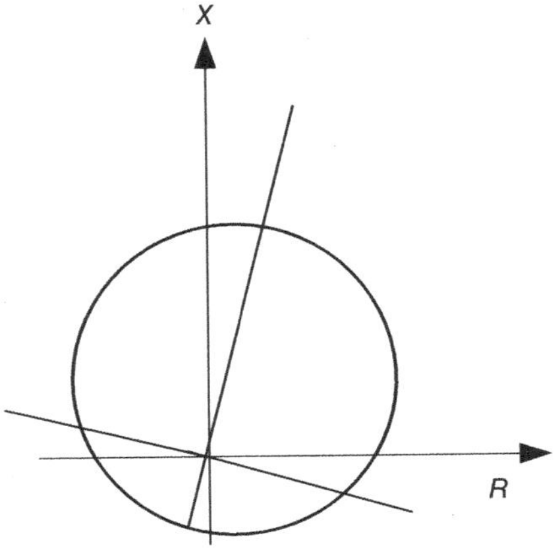

Figure 4.53 Reverse offset MHO characteristic.

The equation which describes the relay torque is as follows:

$$T = K_1 I^2 - K_2 VI \sin \theta - K_3$$

At the balance point, the net torque is zero and hence:

$$K_1 I^2 = K_2 VI \sin \theta - K_3$$

Dividing both sides by I^2:

$$K_1 = K_2 \frac{V}{I} \sin \Theta - \frac{K_3}{I^2}$$

Neglecting the control spring affect:

$$\frac{V}{I} \sin \Theta = Z \sin \Theta = X = \frac{K_1}{K_2}$$

$$X = \text{constant}$$

This means the relay has an operating characteristic such that all the impedance values that lie on this characteristic have a constant X component [1]. This characteristic is shown in Figure 4.54 which is a straight line on the R-X plane. The significance of this characteristic is that the resistance component of the measured impedance does not affect the relay operation.

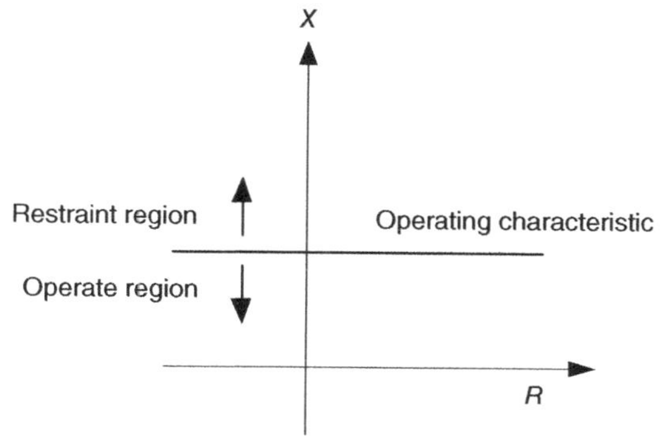

Figure 4.54 Operating characteristic of a reactance relay.

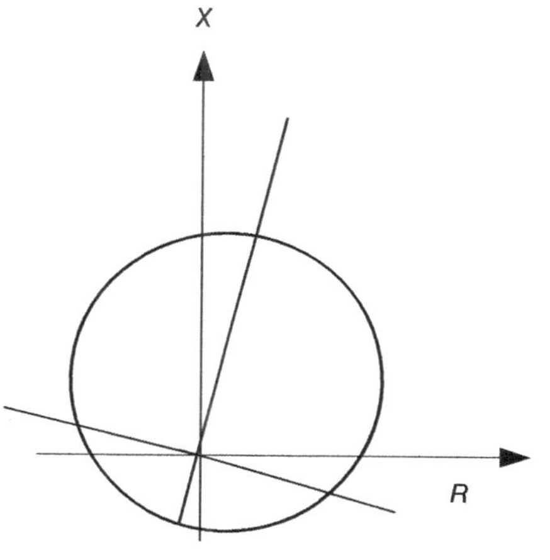

Figure 4.55 Offset Mho relay.

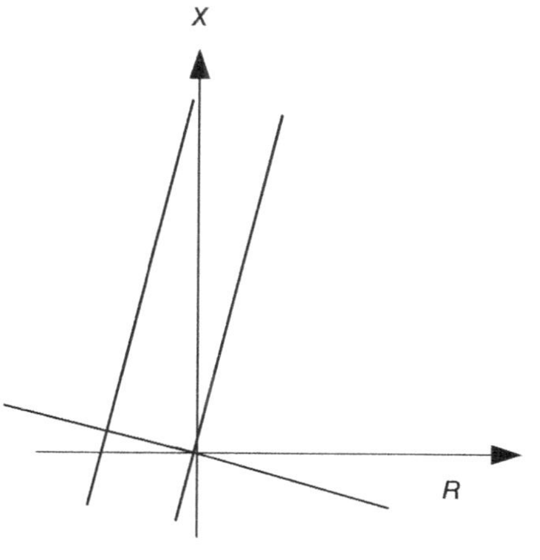

Figure 4.56 Blinder relay.

4.5.4.3 Blinder-Type Distance Relay

Blinder-type distance relays are used for two purposes. The predominant use for blinders is to block an impedance type relay from operating under load. The other use is for generator out-of-step protection. A blinder may be viewed as essentially an offset mho characteristic with an extremely large reach (something akin to the earth looking flat to an observer at its surface). Refer to Figure 4.55

showing an offset mho characteristic with typical reach. Refer to Figure 4.56 showing the same relay but with a much larger reach, such that, the characteristic looks like a straight line.

Reference

1 C. Russell Mason, *The Art and Science of Protective Relaying*. Schenectady, New York: John Wiley & Sons, Inc. 1956, General Electric Company.

5

Protection Information Representation, Nomenclature, and Jargon

5.1 General

The culmination of the work of a protection practitioner for a given protection system design is represented by an industry-standard set of documents. These documents are mainly in the form of a set of drawings; a set of protective relay settings and logic; and setting calculation documentation, which includes supporting fault calculation results.

For a given simple protection system, there could be 20 protection drawings, a relay setting and logic document that could result in 20–50 pages, and a setting calculation document that can be in the order of 10–40 pages.

Protection design, construction, and maintenance heavily rely on the use of several different types of drawings to depict and communicate the specifics of a protection system. An up-to-date set of documentation must be retained at the local site, and a master copy is retained, managed, and backed up, as these documents represent critical company records.

This chapter focuses on protection system information that is developed and provides and communicates protection system specifics. Engineering drawings and the various types of engineering drawings developed by protection practitioners to represent their design, support operations, and maintenance are described.

It should be noted that the terms drawings and diagrams are used interchangeably within the text. There is an insignificant difference between them except that drawings tend to be more detailed.

5.2 Protection Drawing Types

Protection systems are represented in various drawings. The purpose of a drawing is to transfer engineering information or requirements to one or several users so that productive work can be performed. This work may consist of such actions as estimating, planning, manufacturing, construction, commissioning, maintenance, and troubleshooting.

Interpretation of drawings requires knowledge of circuit diagram graphics, numeric symbols, and basic circuit practices.

The protection discipline relies on the various drawings to safely complete its work, and therefore, accuracy and maintenance of the protection drawings are critical. There are several types of drawings depicting electrical circuits. The most basic representation is the single-line diagram (SLD) which also shows in conceptual form, the alternating current (AC) connections of the protections.

The comprehensive AC connection for protection is shown in the three-phase diagram otherwise, known as the AC elementary wiring diagram (EWD). The comprehensive direct current

Power System Protection: Fundamentals and Applications, First Edition. John Ciufo and Aaron Cooperberg.

(DC) connection for protection is shown in the DC, EWD, or sometimes also referred to as, a DC control drawing. It should be noted that sometimes these types of drawings are combined into a single EWD.

The detailed instructions describing the specific connections between relays, terminal blocks, current links, fuses, current transformers (CTs), voltage transformers (VTs), etc., are all shown in connection wiring diagrams (CWDs). These are the drawings used by panel electricians to construct relay panels and interconnections between panels and apparatus in the switchyard.

The physical layout of protection equipment on racks and panels is shown on electrical arrangement (EA) drawings.

These protection drawings have a hierarchy in terms of logical information flow and project development:

1. The SLD is typically the first drawing that is developed. It depicts the largest scope, includes the power system equipment, a high-level representation of the breaker isolation devices, protection relays and their associated protection zones, and the specific CTs and PTs that are used for the protection relays.
   This drawing provides an overview of the power equipment and its associated protection systems. It has a broad width, with the least amount of detail. The information from this drawing is used to develop the protection system engineering details, described in the AC and DC EWDs/control drawings.
2. The AC and DC EWDs/control drawings are subsequently developed. These are developed for each protection system used and for each of their primary and backup systems. The EWDs are detailed and depict all AC and DC engineering information.
3. For microprocessor-based relays, logic diagrams are created that provide information about the specifics of the protection system, its operation, inputs, and outputs, etc. It is implemented via algorithms in the firmware of the digital protection relay. In legacy installations, logic is developed via the use of discrete auxiliary relays, and timers that are hard-wired.
4. The wiring diagrams and the electrical arrangement drawings are developed last. These drawings are very detailed and represent information to permit the physical arrangement of equipment, the installation and routing of cables, and the termination of wires onto correct terminals.

5.2.1 Single-line Diagram

The SLD depicts the electrical presentation and the flow of power for the system under consideration. It includes all the interconnections and basic components, such as the power system equipment (lines, buses, transformer, etc.), and the protection system components such as the circuit breakers, CTs, PTs, protection relays, and their associated protection zones. Figure 5.1 shows an example SLD of a load-substation and a partial representation of the actual station.

The purpose of the SLD is to supply, in concise form, significant information about the power system equipment and its associated protection systems. The SLD, also referred to as the Simplified One Line diagram, provides information about current and potential transformers which connect relays to the system or which are installed for metering.

Since a balanced three-phase system is always solved as a single-phase circuit composed of one of the three lines and a neutral return, it is convenient to show no more than one phase and the neutral return when drawing a diagram of the circuit.

Often the diagram is further simplified by omitting the completed circuit through the neutral and by indicating the component parts by standard symbols. Such a simplified diagram of an electric

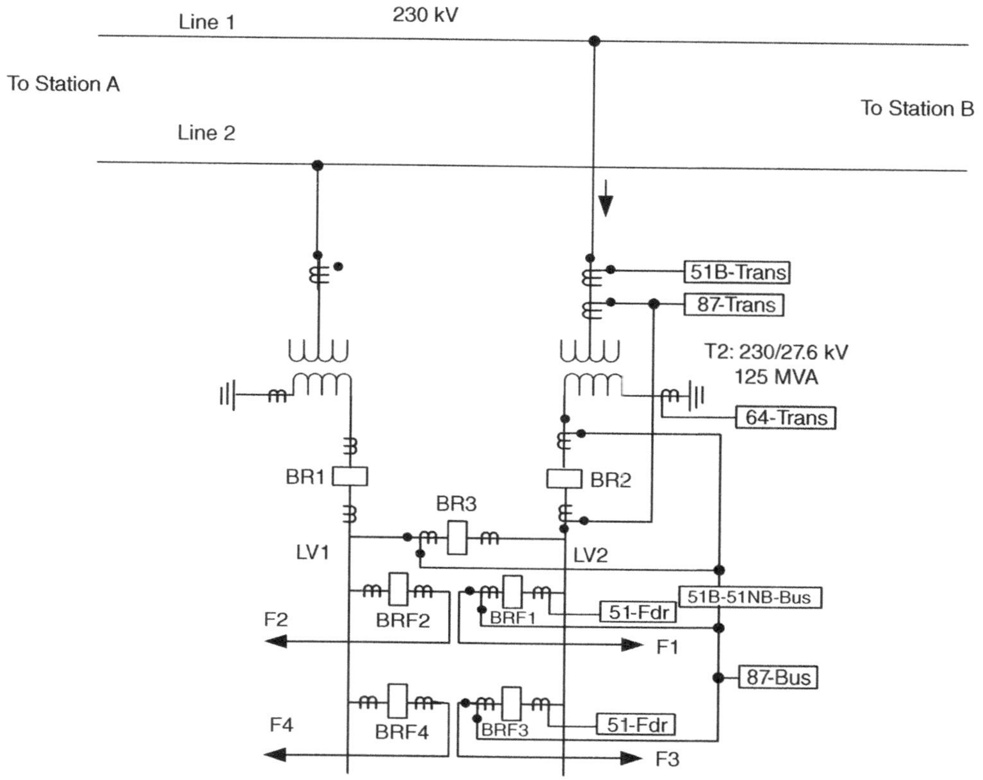

Note: For Simplicity, Only Partial Protections Shown (Right-Side)

(a)

(b)

Figure 5.1 (a) An example of a single-line diagram. (b) An example actual station (partial) for the single-line diagram.

system is called a SLD. It indicates by a single-line and standard symbols and nomenclature the network and associated apparatus including the protections.

The industry uses a set of apparatus symbols used in the SLD that is useful for an understanding of the protections. Specifically, different symbols are used for the types of transformer connections.

The symbol for a protective relay is either a rectangle or a circle with a number inside. The number refers to a protective function designated by an industry-standard (IEEE C37.2) [1] and addressed later in Section 5.3.1.

5.2.2 Three-Phase Diagram (AC EWD)

The three-phase diagram shows all AC connections between power system elements as well as between CTs, VT's, and relays. A protection practitioner who is proficient in interpreting the SLD can usually correctly envision the three-phase connections but not always. When in doubt, the AC, EWD should be consulted. It is not unusual to find some information that is not represented on the SLD as it is sometimes quite difficult to correctly transcribe some information on the AC, EWD to a SLD.

5.2.2.1 Transcription of Information from AC EWD to Single-Line Diagrams

Refer to Figures 5.2–5.4 showing typical three-phase connections including protections and the single-phase equivalent representations.

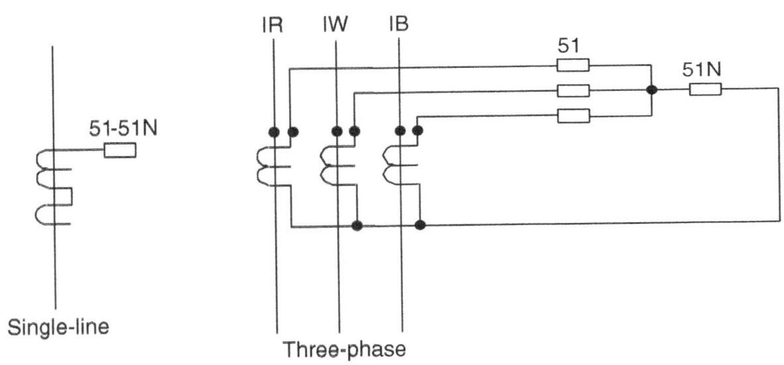

Figure 5.2 Three single-phase and ground overcurrent relays and wye CTs.

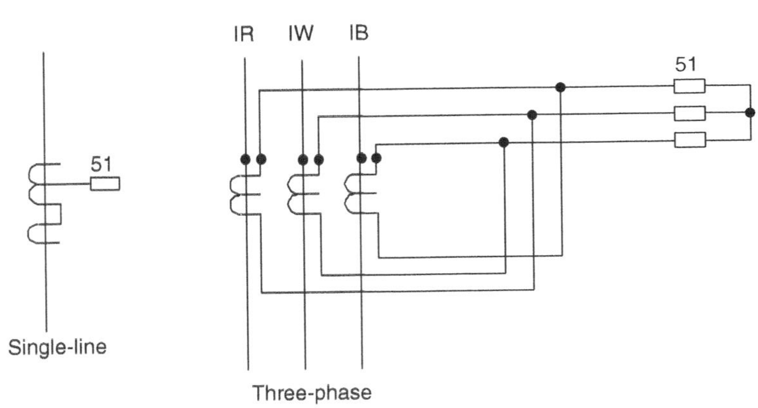

Figure 5.3 Three single-phase overcurrent relays and delta CTs.

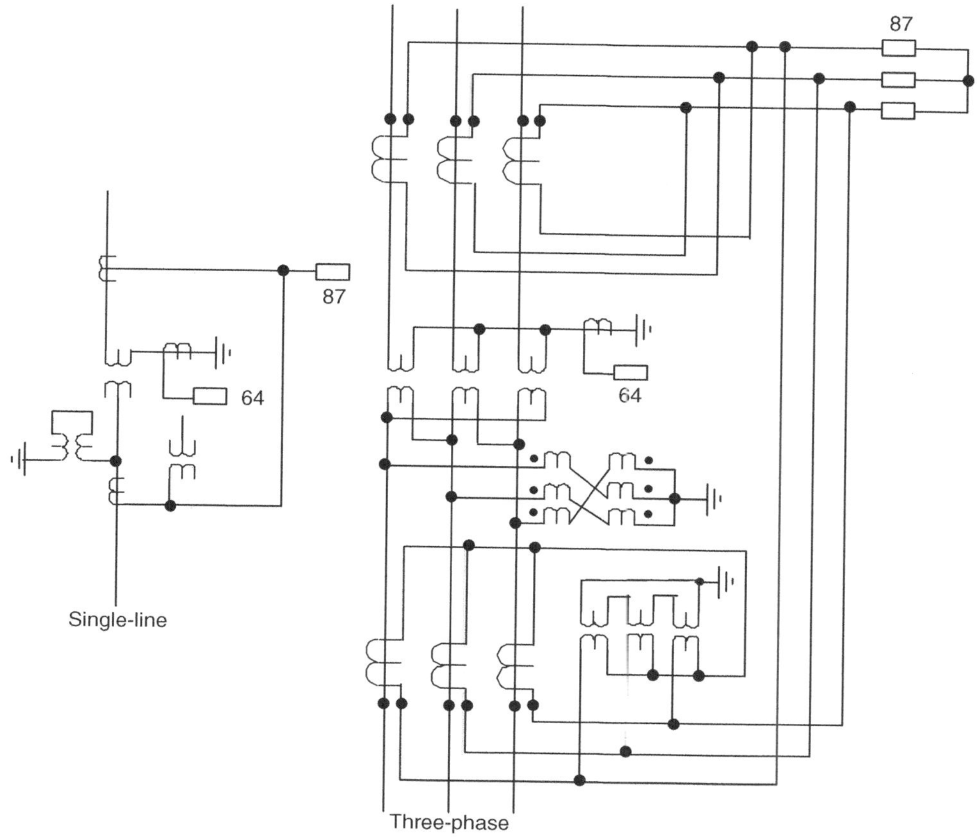

Figure 5.4 An example transformer connection and protection.

5.2.3 DC Elementary Wiring Diagrams – Also Known as DC Control Diagrams

All protection systems rely on DC supplies for many purposes. DC: powers digital relays, is used for auxiliary relay schemes, to power system breakers – via the breaker tripping coils for both protection and control functions – and other isolation type power equipment. DC is also used to automatically reclose breakers from protections, and to close breakers by system operators. All monitoring and recording systems are powered by DC.

The source of DC is large banks of station batteries. The typical DC supply voltage is, 125 or 250 V. The station batteries are continuously kept charged via charging systems. The reason that AC supply is not used for any of these protection and control systems is dependability. Protection and Control systems are classified as mission-critical assets and are also required to function when the power system AC goes down.

DC connections of relays and associated auxiliary relay systems are represented as DC EWD, or as DC Control Drawings. Refer to Figure 5.5 showing an example DC, EWD type drawing.

DC elementary drawings are very useful when trying to determine the operation of certain circuitry. DC elementary drawings, also sometimes referred to as schematics, provide a detailed description of circuit operation, but not physical layout information.

In this example, the circuits represented in this drawing are all powered from the station DC battery, via a branch circuit feed from the battery DC distribution panel, not shown.

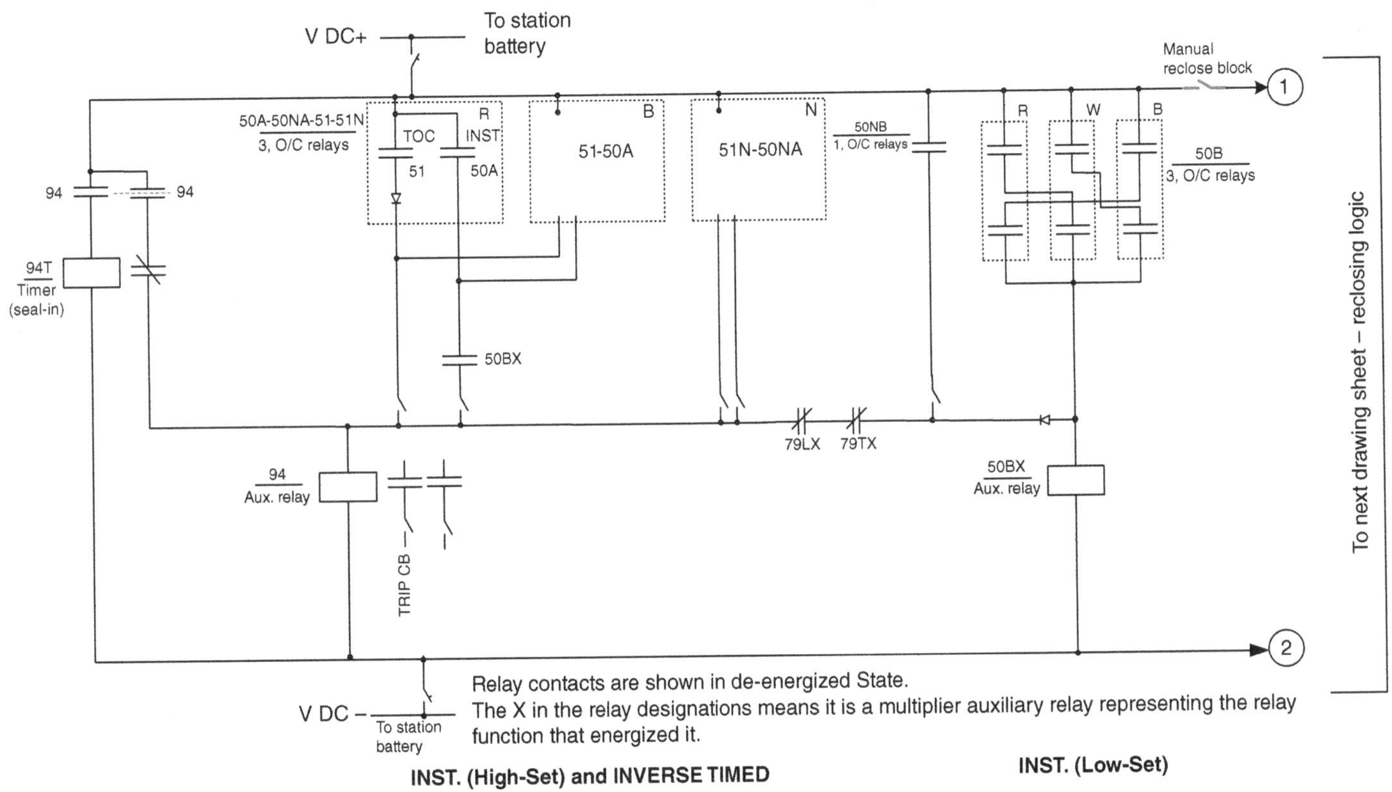

Figure 5.5 Simple example DC, EWD, or DC control drawing.

The overcurrent measuring relays' output contacts are connected in parallel. Once anyone of the overcurrent contacts closes, it will cause the energization of auxiliary DC relays 94 and 50BX. One of the 94 auxiliary relay contacts is wired/connected to the associated isolation breaker's trip coil, to open the breaker.

The overcurrent relays are AC connected as per Figure 5.2 to CTs. Once the measured relay's current exceeds the overcurrent threshold setting, it will pick up its output contact, depicted in this DC elementary, to initiate the tripping sequence, amongst other logic functions.

5.2.4 Electrical Arrangement (EA)

The electrical arrangement diagram shows the physical location of the protection devices such as relays, terminal blocks, block switches, etc. on a free-standing structure. The structure is also referred to as a rack. It is typically, steel construction and 19 or 24 in. wide and 80 in. in height.

A protection rack is a free-standing steel rectangular-shaped frame that typically measures 19 in. wide that is bolted to the floor, refer to Figure 5.7 below. Protection relays and other devices are either mounted directly between the left and right rails of the rack, or on metal panels that are mounted to between the rack rails. The rack, therefore, consists of several vertical 19-in. panels with panels of varying heights.

The electrical arrangement drawing shows on which rack and panel the protection devices are located. The protection devices have associated sets of terminal blocks which are also mounted on the rack. Typically, all rack/panel cables/wiring, are first connected to the rack-mounted terminal blocks. Separate wiring is used from the terminal blocks to the back of the protection devices to complete the circuits.

Each protection system has its own electrical arrangement drawings. Protection racks are typically mounted in rows, and back-to-back, within a dedicated protection and control building. The control building also has an electrical arrangement drawing showing the location of the protection system racks within the control building.

These drawings are also used to construct the panels, so they are often drawn to scale.

Figure 5.6 shows an example of protection and tele-protection electrical arrangement diagrams for mounting equipment on racks.

5.2.5 Connection Wiring Diagrams (CWD)

CWDs indicate where each particular wire of a protection system is terminated. The CWD shows the terminal blocks and components with the interconnecting wiring locations. Typically,

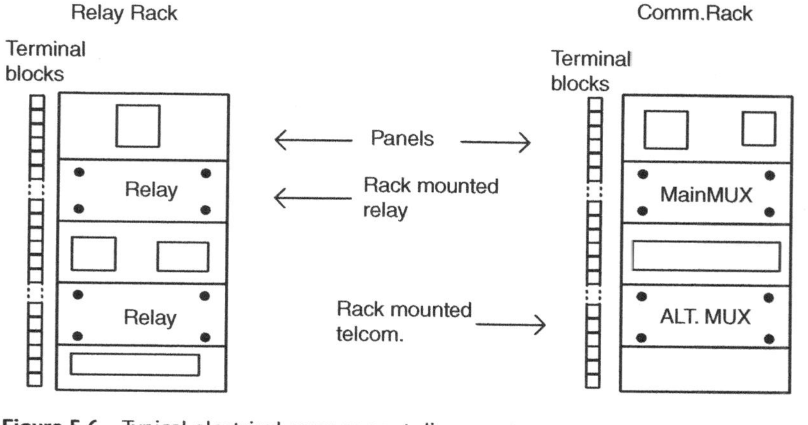

Figure 5.6 Typical electrical arrangement diagrams.

the actual wires and cables are not shown since this would result in a mass of lines across the drawing. The CWD also includes outlines of the racks, panels, and terminal blocks, depicting the appearance of the equipment as closely as possible. This is challenging due to a large number of cable and wire connections.

The industry may use slightly, different approaches to simplify and provide the necessary required wiring information. Some generator plants use wiring tables. CWDs are used mostly for construction, commissioning, and testing.

There can be hundreds of thousands of meters of cables, wires, terminal blocks (connection points for wires), refer to Figure 5.8c, racks, panels, and protection system relays and devices, installed at a station. Many cables run from power equipment mechanical/control boxes located in the switchyard such as breakers, disconnects, transformers, CTs, and PTs. These cables are installed in protected cable trays, in the switchyard, and terminate into one or more protection and control buildings.

Each cable is uniquely labeled and contains many pairs of wires. Each wire and terminal block is labeled, and wire routing is defined by the termination labels. The wires in each cable are terminated typically, onto building terminal racks. From there, wires are run to individual protection racks and panels, where they terminate onto the rack's terminal blocks. From there, they are wired to individual protection relays and other devices; refer to Figure 5.8c.

Figure 5.7 shows a digital relay on the left that interfaces with telecommunication equipment shown on the right. The rear of the digital relay is shown in Figure 5.8a.

Figure 5.7 Digital relay interfacing with telecommunication equipment.

The CWD will indicate the terminal designation on the digital relay connecting to a terminal block on the relay rack that in turn is connecting to a cable running between racks to a designated terminal block on the telecommunication rack and then from this terminal block to a terminal connection on the rear of the telecommunication equipment.

Figure 5.8b shows the inside of an arc-proof metal-clad instrument compartment where the protection relay is mounted on the swing door connecting to terminal blocks mounted inside. From there, cables leave this compartment connecting to CTs, PTs, breaker trip coil, disconnect and breaker pallet switches, etc.

5.2.6 Protection Logic Diagrams

Protection systems are actioned once one or more of their measuring relays, or actuating devices, operate an output, refer to Figure 5.5. This starts a series of actions that ultimately can cause a trip signal to be sent to the protection's isolation devices/breakers, and in some cases, also the automatic restoration process of reclosing.

After initiation and before the tripping of the breaker(s), the protection system ensures operational dependability and security by incorporating several conditions and assurances before sending a trip signal. The actions that occur between initiation and the final decision to proceed with a trip decision are referred to as the protection logic. Depending on the application, it can be simple, or very complicated with hundreds of conditions that need to be satisfied.

Protection logic for traditional electromechanical protection systems was developed by using many DC auxiliary relays and timers, and it was shown on protection DC, EWDs as depicted in the simple example case of Figure 5.5. The protection practitioner is required to interpret the DC circuits and understand the intended protection logic.

(a)

Figure 5.8 (a) Rear connections of the digital relay. (b) Example relay connections on metal-clad compartment door. (c) Example set of terminal blocks. (d) Example connection wiring diagram CWD.

(b)

(c)

Figure 5.8 (*Continued*)

With modern microprocessor/digital relays, protection system logic and controls, have transitioned from hard-wired using discrete DC devices to using software logic. This has transformed the information contained on traditional DC, EWD.

Referring to Figure 5.9, it should be noted that the drawing's main function is to provide information on the relay model used, and its associated inputs and outputs. The details regarding what logic is being used and how, are contained within the relay, and is no longer readily available from the DC, EWD.

Digital relays are computers with dedicated inputs and outputs, and with embedded software. They offer a library of protection functions and logic capabilities, all user-configured, via a relay software configuration tool provided by the relay manufacturer. These devices, upon being received

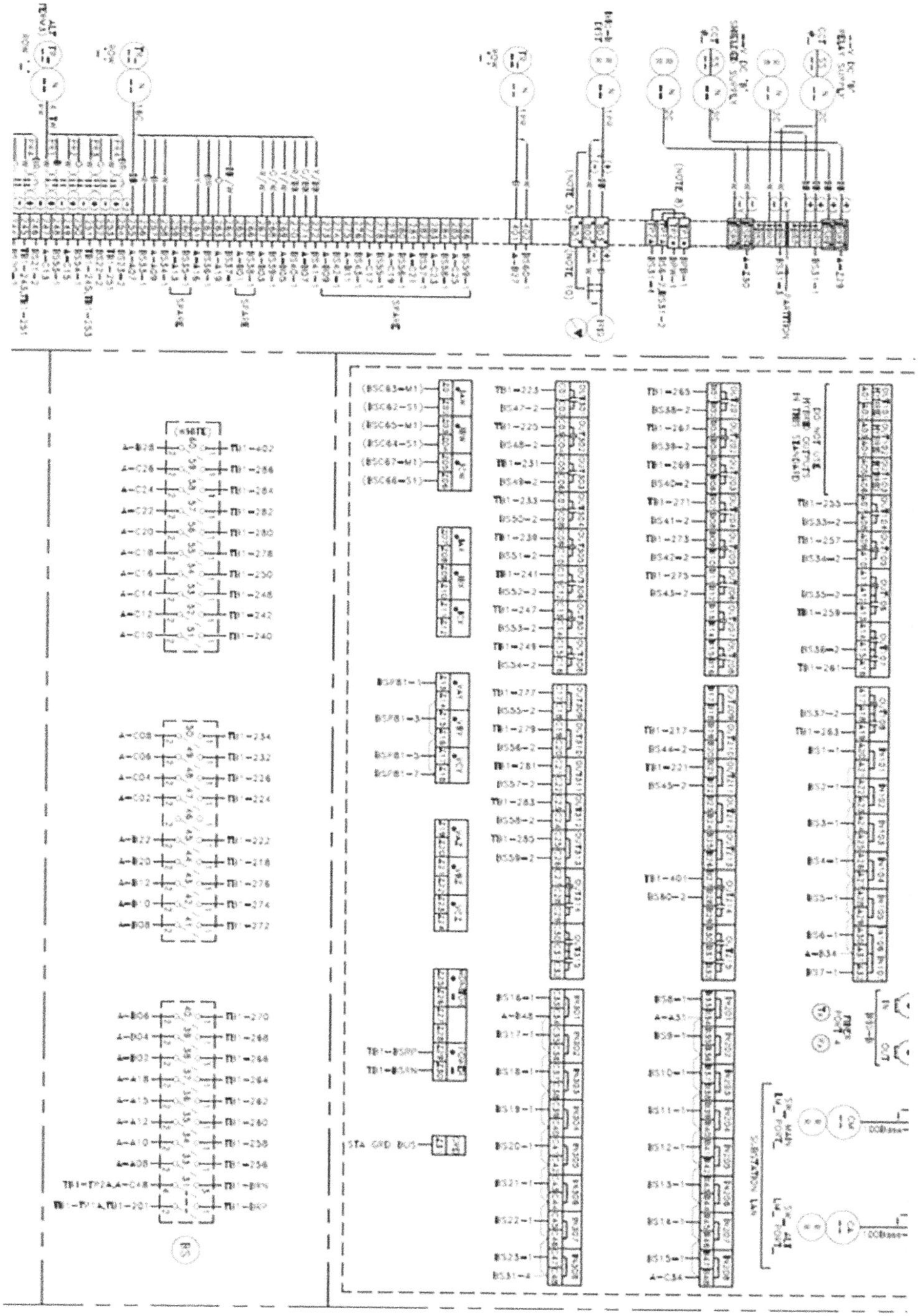

(d)

Figure 5.8 (*Continued*)

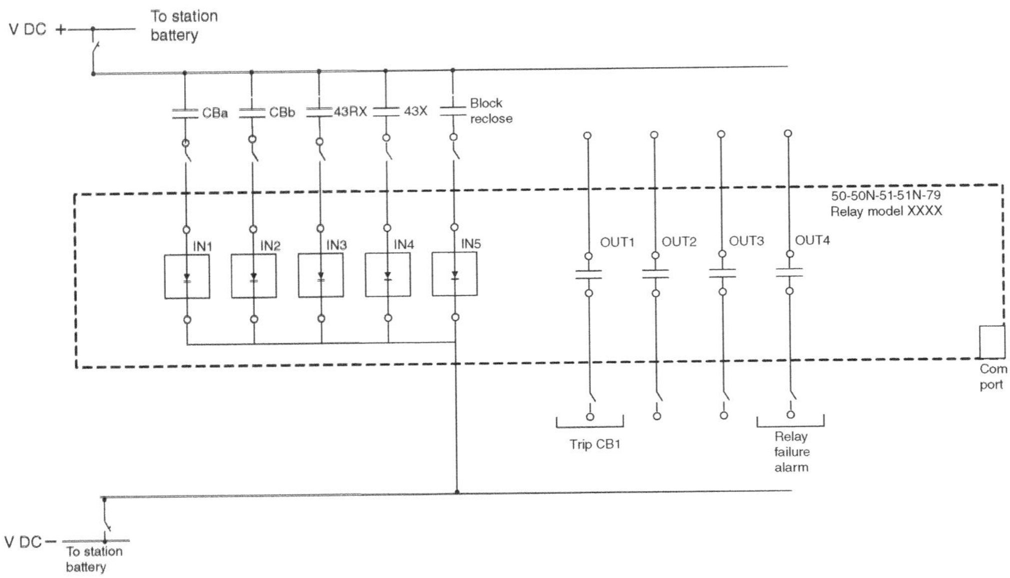

Figure 5.9 An example DC, EWD for a digital relay (conceptual).

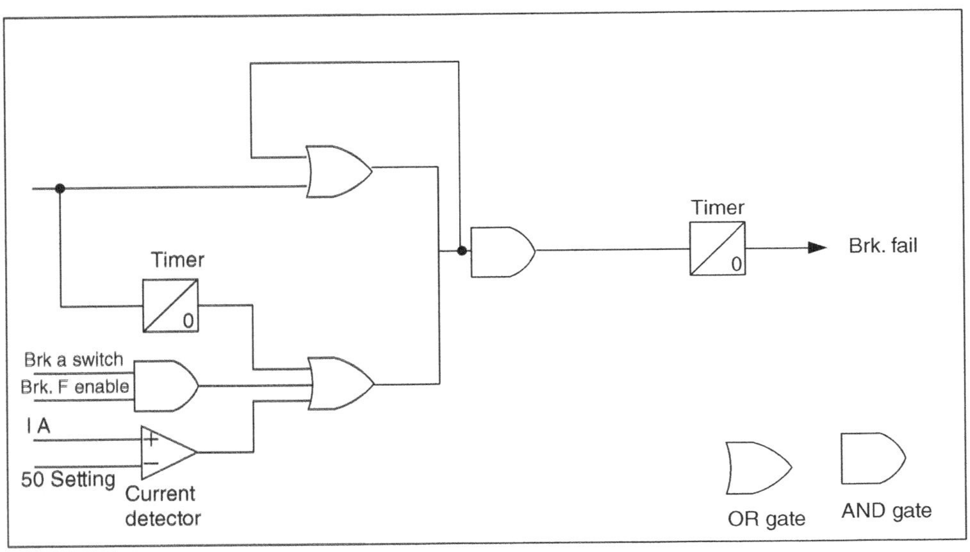

Figure 5.10 An example of a simple protection logic drawing.

from the manufacturer, must be customized by configuring, programming and setting them, specific for their intended application.

As a consequence of transitioning from electromechanical to digital protection systems, there was an industry need to develop protection logic diagrams. In most cases, the relay manufacturers provide a mechanism to easily derive and capture the relay logic information once configured.

Protection logic drawings allow protection practitioners to readily review and analyze protection system logic, without the need to access the relay, and the configuration tool. This type of drawing, refer to Figure 5.10, completes the series of required drawings records for the protection system application.

5.2.7 Protection Settings Records and Support Information

Upon a protection system's equipment and models having been selected, protection functions determined, and the logic and family of associated diagrams are complete; the protection practitioner determines the appropriate relay settings.

Two broad setting tasks are required, the measuring relay protection functions threshold settings (pickups), and the relay configuration settings. Both of these relay settings information processes should be documented, and retained for the life of the protection system asset.

5.2.8 Protection Relay Threshold/Pickup Settings

A protection system can invoke the use of several protection functions such as overcurrent elements, distance elements, voltage elements, timers, etc. For each one of these elements, the protection practitioner is required to calculate and determine the appropriate threshold/pickup settings. To do so, system fault studies need to be conducted that includes the protected power system equipment in the study area. A series of fault calculations, including different fault types, at several locations, and under normal and minimum conditions need to be conducted. Faults calculations are discussed in Chapter 6.

With available fault information, the protection practitioner determines the appropriate settings for each of the protection elements deployed. This information and process must be documented and retained for record purposes. An example of a simple manual setting process is depicted in Chapter 15, Section 15.8.

To aid protection practitioners, some utilities develop and use setting tools such as customized spreadsheets per protection function type, to facilitate, and document this derivation process. This results in a uniform and consistent set of records.

5.2.9 Relay Configuration and Settings File

Most modern protection systems use microprocessor/digital-based protection relays. These types of protection relays, although they have advanced the industry, have also increased the complexity of protection relay configuration and settings. They offer a library of protection functions, configuration, and logic capabilities, resulting in approximately 50 pages of settings. Moreover, these relays can be configured with several setting groups, each group, representing a different configuration in logic, settings, and relay behavior. This can result in multiplying the hundreds of settings by five. The net result is that a relay setting file can result in hundred pages of information.

Digital relay manufacturers, provide clients with its relay configuration interface software tool, for their suite of relays. This configuration software is installed on the protection practitioner's computer. The software permits users to configure, create logic, set all relay settings, amongst other tasks, "off-line" (not connected to the relay), and to save all of this information which is unique for each relay, as a computer setting file.

The protection practitioner, via the configuration tool, selects the protection functions that are required to achieve the protection schemes from the functions that are available within the relay. They develop logic settings and settings to configure the relay to act and perform as the intended design. The threshold settings are also entered derived from the calculation process described above, for each used protection function.

Once completed, the software setting file is typically peer-reviewed, as part of a quality check. Relay settings are critical, and due to the volume of information, they can be prone to input error. The reviewed setting file once finalized and approved, should contain the names of the protection practitioner, and the reviewer, to be provided to the field staff, where it is download into the relay, and then tested and commissioned into service. The relay setting file should be retained, and managed, as an official record, and can be used to reconfigure the relay if ever required to do so.

Since there are many hundreds of possible configurations and settings that are possible with a digital relay, many utilities develop and standardize a relay standards template. The relay setting template contains a standard set of configured settings requiring the protection practitioner to enter application specific and threshold settings, which represents a subset of data, thereby; reducing errors.

It should be noted that legacy electromechanical protection relays, also have relay configuration and threshold settings. They are typically less than a dozen or so of settings, per relay, and can be described on 1–2 sheets of paper, that were referred to as relay setting sheets.

Regardless of relay technology type, the protection system fault study results, threshold calculations, and the specific relay configuration and setting should be retained, and managed, as an official record.

5.3 Nomenclature and Device Numbers

The functions and standard device function numbers of various relays and equipment are identified by the ANSI/IEEE Standard C37.2-2008 [1]. Such function numbers are used on various protection information systems such as drawings, settings, etc. as shown in Figure 5.11.

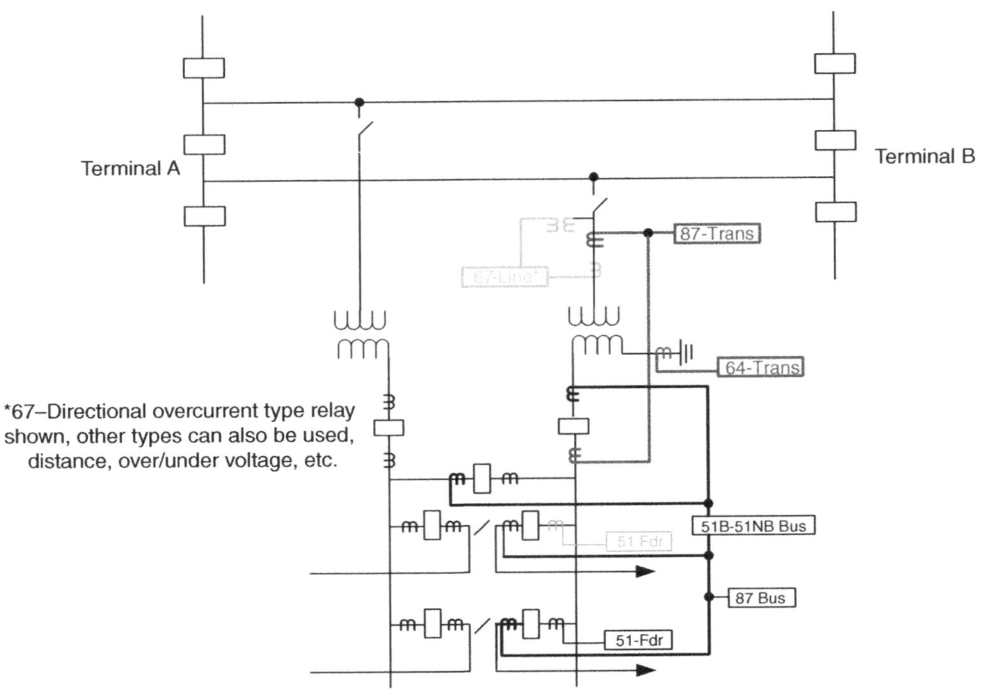

Figure 5.11 Single-line depicting protection function numbers.

5.3.1 Commonly Used Protection Device Numbers

The following is a list of commonly used protection device numbers. For a complete definition, refer to ANSI/IEEE Standard C37.2-2008 [1].

21	Distance relays
25	Synchronizing or Synchro check device
27	Undervoltage relays
32	Directional power relays
37	Undercurrent relays
40	Field Relays
41	Field circuit breakers
43	Manual transfer devices
46	Reverse-phase relays
50	Instantaneous overcurrent relays
51	AC time overcurrent relay
52	AC circuit breaker
59	Overvoltage relays.
62	Breaker Failure Time-Delay
63	Pressure switches
64	Ground detecting relays
67	AC directional overcurrent relays
68	Blocking relays
74	Alarm relays
79	AC reclosing relays
81	Frequency relays
86	Locking-out relays
87	Differential protective relays
89	Line Switch
94	Tripping or Auxiliary relays
50N	Instantaneous ground overcurrent relay
51N	AC time ground overcurrent relays
87N	Differential ground protective relays

5.3.2 Prefix and Suffix Meaning

Suffix Letters are used with device function numbers for various purposes. To prevent possible conflict, each suffix letter should have only one meaning for individual equipment. The letters should generally form part of the device function designation. The two suffixes typically used are **G** for Ground and **N** for Neutral.

Suffix **N** is generally used in preference to **G** for devices connected in the secondary neutral of CTs. Suffix **G** is used by most utilities in the secondary of a CT whose primary winding is located in the neutral of a machine or power transformer.

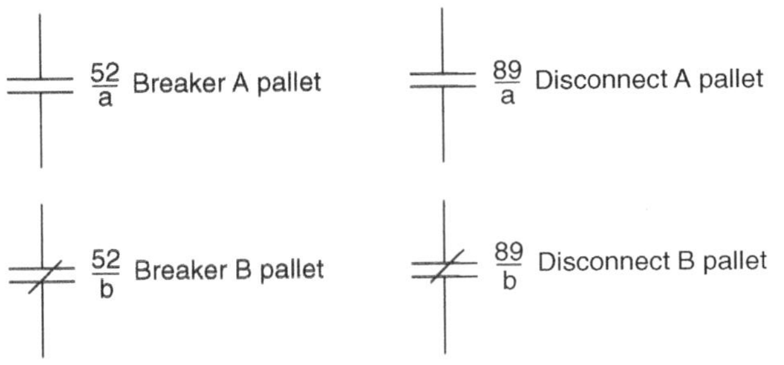

Figure 5.12 Breaker and disconnect pallet switch representation.

The suffix letters which denote parts of the main device and those which cannot or need not form part of the device function designation are generally written directly below the device function number on drawings. It is common practice to represent breaker and disconnect pallet switches in this manner as shown in Figure 5.12.

Breaker "A" and "B" pallet switches and disconnect "A" and "B" pallet switches are represented on drawings as shown in Figure 5.9. Pallet switches follow the position of the device itself. A-breaker "A" pallet contact will be open when the breaker is open and closed when the breaker is closed. A-breaker "B" pallet contact will be closed when the breaker is open and open when the breaker is closed. The designation X suffix is used when a pallet multiplier is being used. The same holds for disconnect switches. That is, the pallet switch energizes an auxiliary latching relay, whose contacts are used to represent the pallet switch; it is used to "multiply" the pallet switch function for applications where more breaker pallets are required than available.

The simple designation "a" or "b" is used in most cases where the part of the travel where the contacts change position is of no significance in the protection, control or operating scheme. However, where there is a need to monitor the contacts to change position at any particular point in the travel of the main device such as, for the so-called "fast breaker pallet" used in breaker failure for the 62A path, the designation "aa" is usually used instead. Most breakers provide access to such intermediate pallet position indications.

The designations for auxiliary relay contacts are as follows:

a. Contact that is open when the main device is in the standard reference position, which means "de-energized" or with the actuating quantity removed, commonly referred to as the normal position and that closes when the device assumes the opposite position.
b. Contact that is closed when the main device is in the standard reference position commonly referred to as the normal position and that opens when the device assumes the opposite position.

On electrical diagrams, the "b" contacts of all devices, including those of relays and those with suffix letters should be shown as normally closed contacts and all such "a" contact should be shown as normally open contacts.

For those devices that have no de-energized position the preferred method of representing these contacts is an "a" switch. Each contact should be identified on the elementary diagram as to when it closes.

Any devices such as latching relays typically used for contact multiplication, that have no de-energized position should have their contacts shown in the position most suitable for the ready

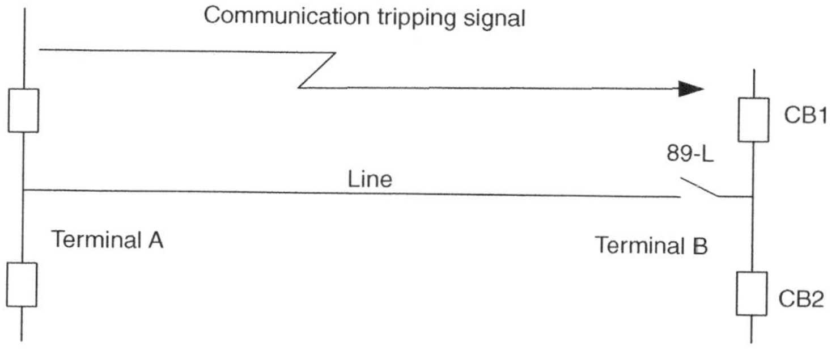

Figure 5.13 Illustration of pallet switches used as interlocks.

understanding of the operation of the device in the equipment and sufficient description should be used as necessary on the EWD to indicate the contact operation.

5.3.3 Interlocks

The purpose of using pallet switch interlocks in protection schemes is to allow protection or tele-protection-related operations only when certain conditions are met. The most notable condition is to only trip breakers that need to be tripped to isolate a fault. There are also other condition types requiring interlocks such as to block tripping out-of-service power system equipment during maintenance. Another condition type requiring interlocking is to block automatic breaker reclosing until motor-operated isolating switches are fully open thereby isolating the faulted equipment first. These and others all require interlocks. Refer to Figure 5.13 where a simple communication signal is generated at Terminal A and transmitted to Terminal B to trip breakers CB1 and CB2.

In this situation, breakers CB1 and CB2 should only be tripped when the line is in-service at Terminal B as when the 89-L line isolating disconnector switch is closed. Therefore, in logic an 89-L pallet contact would be used to allow breaker tripping. When using electromechanically operated auxiliary logic, an 89-L "A" pallet would be placed in series with the communication receive tripping signal so that breakers CB1 and CB2 could only be tripped provided 89-L is closed. When a digital relay or digital logic controller is used instead of electromechanical, it makes little difference whether a disconnect "A" pallet or a "B" pallet is used as an interlock. The logical state of any input into a digital device can be converted to either "A" or "B" normally open or normally closed position using simple logic inverse commands. This is convenient when there are a limited number of spare "A" pallet contacts to use even when multiplied.

Refer to Figure 5.14 showing a digital relay or logic controller with three contacts connecting to what is commonly known as digital inputs. Contact 89a is the line "A" pallet, 89b is the line "B" pallet and TTR is the transfer trip receive communication contact. When a transfer trip signal is received from station A, the TTR contact closes. However, CB1 and CB2 are only allowed to trip provided the line disconnector "A" pallet is closed indicating the line is in-service at station B. As previously mentioned, a line disconnector "B" pallet can be used instead provided it's catered for by logically inverting the input within the relay or logic controller.

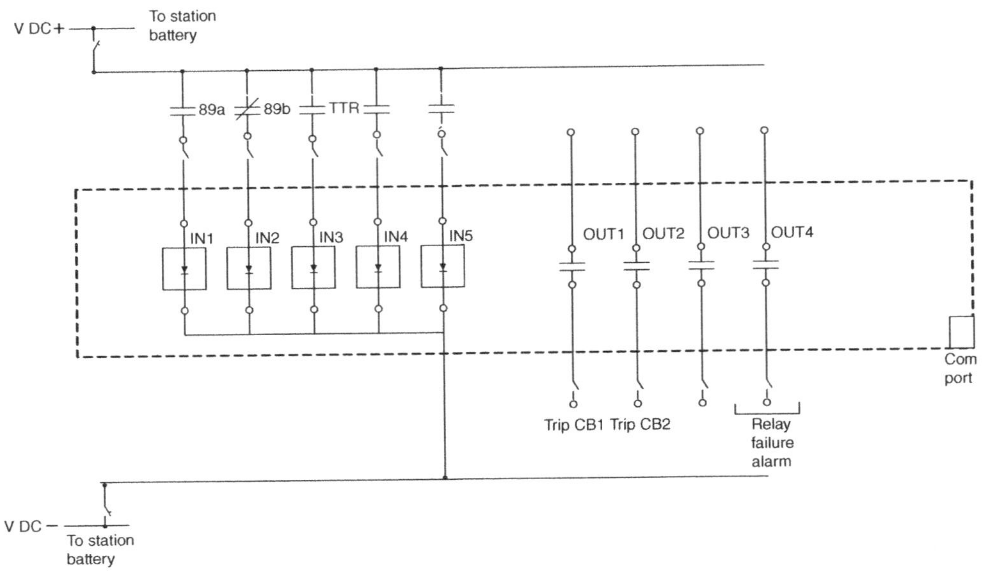

Figure 5.14 Illustration of pallet switch inputs to a relay or logic controller.

5.4 Classification of Relays

Relays are classified in broad categories by methods of operation and by response characteristics.

5.4.1 Methods of Operation

Electromechanical relay operates electromechanically.

Gas Pressure relay responds to gas pressure.

Latching relay changes state and latches usually mechanically and then stays that way until changed again.

Percentage differential relay operates due to a difference in two currents with a linear increase in relay operation with a magnitude of through current.

Solid-state relay uses solid-state components to compare measured values to set replica values.

Digital relay uses digital means to compare algorithms and preset values to measured values.

5.4.2 Response Characteristics

Blinder relay measures impedance and is generally used to block another impedance relay from tripping.

Definite time relay closes contacts after a definite time delay.

Differential relay measures the value of electrical quantities and closes contacts when the difference between them exceeds a specified amount.

Directional overcurrent relay measures current and closes contacts when the current exceeds a predetermined value and is in a predetermined direction with respect to a polarizing reference current or voltage.

Directional power relay measures current and voltage and closes contacts when the two are in a predetermined phase relationship above a set threshold and is in one direction.

Directional relay operates when the voltage and current are in a phase relationship sufficient to cause relay operation for current in one direction.

Distance relay operates when measured impedance including magnitude and angle is above a set threshold.

Frequency relay operates when the measured system frequency is either above or below a set threshold.

Ground relay is designed to measure either zero-sequence voltage or zero-sequence current.

Impedance relay operates when measured impedance in magnitude only is above a set threshold.

Inverse time relay operates when measured voltage or current is exceeded and operates with increasing speed according to a set function as the magnitude of either the voltage or current rises.

Lens relay is similar to a mho relay except that the characteristic is lens-shaped instead of circular to reduce the likelihood of operation under load.

Mho relay is an impedance relay that can be offset such that the center of the characteristic does not encompass the origin on a *R-X* plane. The angle of the circular characteristic's diameter is adjustable with the angle known as the maximum torque angle (MTA) or relay characteristic angle (RCA).

Negative phase sequence relay operates when either negative sequence voltage or negative sequence current exceeds preset values which occur during unbalanced conditions.

Neutral relay is designed to measure either zero-sequence voltage or zero-sequence current when either by direct measurement or by digital calculation with the resultant being above a preset value.

Open phase relay responds to an open phase or phases.

Overcurrent relay measures current and closes contacts when the current exceeds a predetermined value due to a fault.

Overload relay measures current and closes contacts when the current exceeds a predetermined value due to an overload.

Overvoltage relay measures voltage and closes contacts when the voltage exceeds a predetermined value.

Phase comparison relay monitors the currents entering and leaving a given protection zone such as a line and operates when the phase relationship between those currents exceeds a predetermined value.

Rate of change relay measures typically frequency but can be other electrical quantities and operates when the rate of change accelerates beyond a predetermined set value.

Reactance relay is an impedance relay with an impedance characteristic running in parallel with the resistive axis and operates purely to a change of measured reactance regardless of impedance angle.

Residual relay operates when either zero-sequence voltage or zero-sequence current are vector subtracted from each other either by direct measurement or by digital calculation with the resultant being above a preset value.

Time overcurrent relay operates when measured current is exceeded and operates with increasing speed according to a set function as the magnitude of current rises.

Undercurrent relay operates with a drop in measured current below a predetermined threshold.

Undervoltage relay operates with a drop in measured voltage below a predetermined threshold.

5.5 Protection Jargon

In addition to classifying relays according to their response characteristics, protection jargon or a special vocabulary had to be created to allow protection practitioners to communicate effectively and to describe fully the detailed operation and performance of either a relay or a protection.

5.5.1 Relay Operation and Performance

Assertion refers to digital relay protection elements or inputs that change state from logical state 0 to logical state 1. These are synonymous with the electromechanical relay term pickup used for protection elements.

Block refers to preventing a relay from operating for a given situation.

Block short is to implement blocking via a blocking signal or by setting an impedance relay to see only so far into impedance and no more. An example of this is an impedance relay covering a line but blocking short of a tapped or terminating transformer.

Contact opening time the time taken for relay output contacts to open following relay de-energization.

Dropout contact opening following relay de-energization.

Dropout time the time taken for a relay output contact to change state such as open when de-energized.

Instantaneous no intentional time delay has been applied to either a relay or a protection system.

Memory action when voltage is stored for a short time for a relay to use when voltage is required to operate correctly and the voltage is no longer available.

MTA is the maximum torque angle of an electromechanical relay. This is achieved when the electrical angle between two electrical quantities such as current and voltage produces maximum torque on a disc or induction cup.

Offset a Mho characteristic usually passing through the origin that is offset either in the forward or reverse direction.

Operating characteristic any relay characteristic plotted on a graph.

Operating time, the time taken for a measuring relay to close contacts following an internal decision to trip.

Overtravel refers to an induction disc type relay whose disc continues to turn despite measured current being removed.

Pickup refers to a relay that closes its output contacts following a measured electrical quantity exceeding a predetermined value.

Polarization refers to when a measured electrical quantity either voltage or current that does not change direction regardless of fault location is used as a relay directioning reference.

Poly-phase refers to a relay where more than one phase is involved.

RCA is the relay characteristic angle of a digital relay This is the electrical angle between two electrical quantities producing directioning in the intended direction. The RCA is similar to MTA in an equivalent electromechanical relay.

Reach the location based on impedance measurement whether on a line or for a transformer, etc., where a given protection responds to a fault and beyond that point it does not.

Reset refers to when relay output contacts return to their normal state either normally open or normally closed following relay de-energization.

Reset time the time taken for relay output contacts to change state following the removal of initiating electrical quantities.

Setting instructions used to calibrate relays to respond in a predetermined manner. These instructions typically would include electrical quantities as well as relay internal logic.

Torque control refers to the ability to prevent an induction disc from turning altogether. This is typically done in an electromechanical relay with a contact from another relay opening up the shading ring on one of the relay poles. In a digital relay it is done via the setting software. It is usually used to maintain coordination following breaker automatic reclosing downstream.

Transient response refers to how a given protection-related device responds to an abrupt change of input state. Devices where transient response typically can be an issue are relays, CTs, and capacitor voltage transformers (CVTs).

Voltage restraint is typically used to prevent overcurrent relays from operating to differentiate between load and fault current.

5.5.2 Protection System Operation and Performance

Backup protection, protection that backs up a primary protection.

Breaker failure protection, protection dedicated to a breaker should the breaker fail to operate as intended.

Dependability an aspect of protection reliability associated with protection operation when required to do so.

Directional comparison protection any protection scheme that depends on the direction of current to a fault.

Distance protection refers to impedance relays used to cover lines whose impedance is usually in a linear relationship with length of line.

False tripping refers to when protection trips a breaker unnecessarily.

Ground protection that operates on zero-sequence quantities either zero-sequence voltage or zero-sequence current.

Guard signal a continuous signal that is received under normal conditions indicating that telecommunications are functioning correctly.

Incorrect relay operation when a relay operates not as designed.

Line terminal refers to where a line terminates onto a bus.

Load shedding a power system protection that removes degrees of load from service due to stressed power systems as measured by either under voltage or under frequency to maintain power system stability.

Local backup protections are duplicated locally to provide backup as opposed to backup systems that are located remotely.

Loss of excitation when a machine loses its field.

Open phase protection that responds to an open phase or phases by various means of detection.

Out of step when a generator loses synchronism with other system generators and begins to slip poles.

Overreach a distance protection that operates for a fault beyond its intended reach according to its setting.

Overreaching protection either an overcurrent or impedance-based protection that is meant to reach into another zone according to its setting.

Permissive a protection scheme where permission to trip usually received via telecommunication from another location is required.

Pilot protection any protection that uses telecommunications as part of the protection scheme.

Primary protection that is intended by design to operate first before any backup protection.

Reliability protection that is dependable and secure.

Remote backup protections and related systems that are located remotely provide backup as opposed to protections duplicated locally.

Security that aspect of reliability associated with protection operation when not required to do so.

Zone of protection apparatus that is intended to be protected by a given set of protections usually bounded by breakers.

Reference

1 ANSI/IEEE Standard C37.2-2008, Standard for Electrical Power System Device Function Numbers, Acronyms, and Contact Designations.

6

Per-Unit System and Fault Calculations

6.1 General

A protection practitioner, to set protective relay pickup thresholds and to design and implement protection designs, must be able to develop and analyze power system electrical circuits. They do so mainly to calculate faults, also known as short circuit quantiles that are in turn, used to derive protective relay settings and designs.

Fundamental to this process is the development of power system electrical equipment circuit representations, or models, for such equipment as generators, lines, transformers, etc. The electrical equipment representations are interconnected to form electric circuits that mimic the actual connectivity of the power system equipment being studied and protected.

The practice is to take the power system network being studied and represent it by a one-line diagram. This diagram is then converted into an impedance diagram listing the equivalent circuit of each component. An interconnected power system uses several different voltages to transfer power via the use of transformers. The electric circuit solution analysis is more onerous due to transformers operating at different voltages. The Per-Unit Theorem/method was introduced many years ago with the main focus to provide a less onerous method, among other reasons.

Fault calculations pertain to the determination of current and voltages with their relative phase angles at various points in the power system network; following the occurrence of an abnormal condition such as a fault or open circuit. These calculations are useful in determining the power system equipment ratings and for protection systems. The focus for protection practitioners is on the latter – to determine appropriate protective relay settings and behavior.

This chapter will expand on the per-unit concept and will focus on the application of fault studies to determine adequate relay settings. The use of symmetrical component theory will be used for fault calculations, and a summary of this method will be provided.

The method of symmetrical component theory was presented by Charles L. Fortescue in 1918 [1]. There are many references available in the industry that provides detailed information on symmetrical components and their applications. It is the intention of the authors of this book to only summarize the symmetrical components method and to provide examples of its use. It is recommended that readers reference other, more detailed resources, on this subject, and there are many available. The inclusion of more detailed discussions would be too voluminous to include in this type of book. One such reference is J. Lewis Blackburn, "Symmetrical Components for Power System Engineering," 1993 by Marcel Dekker Inc. [2]. Another such reference is Paul M. Anderson, "Analysis of Faulted Power Systems" 1995 by Wiley-IEEE Press [3].

Power System Protection: Fundamentals and Applications, First Edition. John Ciufo and Aaron Cooperberg.

6.2 Per-Unit

It is an electric power industry standard practice to use the Per-Unit method to represent power system information. As such, a protection practitioner must become familiar with the use of per-unit.

Per-Unit (PU)

Electrical quantities, whether voltage, current, kVA, or impedance, may all be expressed as a unit of a base value provided that the base value itself is derived from two of the other electrical quantities assigned specifically as base values. In simple terms, a per-unit value of any electrical quantity is the ratio of a given quantity to its base value. The base value itself is derived from two other selected base or reference quantities.

In the per-unit system, various physical quantities such as kVA, voltage, current, and impedance are expressed as a decimal fraction of base quantities. In the per-unit system, different voltage levels disappear and a power network reduces to a system of simple impedances expressed as per-unit.

Electric power equipment manufacture's nameplate and test data are provided to utilities in per unit. Therefore, the protection practitioner needs to understand it, be able to use the per-unit (PU) system and able to covert from PU to natural units (e.g. volts, Ampere, Ohm) and vice versa, as required. It should be noted that some manufacturers may also use percent instead of PU; however, PU is nothing but percent divided by 100.

There are many benefits in using the PU system which can be found in many power system engineering textbooks. Some of the more salient benefits are as follows:

(1) Simplifies calculations especially when one or more transformers are included in the models/calculations. The PU impedance of a transformer is the same when referred to either side of the transformer … no need for the "Turns" (a^2) factor and is the same regardless of winding connections. This results in the ability to use math operations without the need to cater for multiple transformer ratios and windings.
(2) The PU impedance of a three-phase transformer is the same regardless of its winding configurations, i.e.: wye–delta, wye–wye, and delta–delta.
(3) The PU impedance of equipment lies in a narrow range, while the actual ohmic values vary widely; this is useful for estimating values when no data are available.

For ease of calculations, it is necessary to choose one voltage level and refer the impedances of all the components to this voltage referred to as the base voltage.

It is common practice to select 115/220/500 kV as the base voltages for the transmission system but others can be used. The system networks invariably operate at different voltage levels at various points due to the use of transformers.

Having established the voltage of one part of the system as the base voltage, the transformer turns ratio determines the base voltages for other parts.

Since the electrical quantities are interrelated, it would suffice to establish two base quantities. The other quantity normally chosen is the kVA/MVA apparent power.

The industry-accepted units are the kilovolt (kV) and megavolt-ampere (MVA) for the base quantities of voltage and power.

The base quantities of the other two quantities, current and impedance, are easily determined as they are bound by fundamental laws. The base quantities are scalar as far as the equations linking the base magnitudes are concerned.

6.2.1 Base Quantity Equations

Traditionally, the two electrical quantities chosen as base values, that the other quantities are referenced to, are kV and MVA.

6.2.1.1 Establish the Base Voltage (kV) and Power (MVA)

$$\text{Base voltage} = \text{kV}_\text{B} \ \left(\text{Phase-to-Phase Voltage in kV for a three-phase system}\right)$$

$$\begin{aligned}\text{BaseMVA} &= \text{MVA}_\text{B}\ (\text{Three} - \text{Phase Apparent Power})\\ &= \sqrt{3}\,\text{kV}_\text{B} \times I \ \text{ where } \ I = \text{Line Ampere}\end{aligned}$$

6.2.1.2 From Base Voltage and Power, Calculate the Base Current and Impedance

Since three-phase MVA $= \sqrt{3}\,\text{kV} \times I$ where kV is line-line voltage and I is line current

$$\text{Base current} = \frac{\text{Base MVA}}{\sqrt{3}\ \text{Base kV}}\ \text{A}$$

$$\text{Base impedance} = \frac{\text{Base kV}^2}{\text{Base MVA}}\ \Omega$$

It should be noted that the base impedance is in Ohm per phase.

Generators with different MVA ratings can be synchronized to the system at different points. The usual practice is, therefore, to select 100 MVA for transmission applications or 10 MVA for distribution as the common power bases for all the sub-systems, although other base values can be used.

With a voltage base chosen for one sub-system, the base voltage for the adjacent sub-systems connected through a transformer is determined by the transformer turns ratio. This means that base Ohm and base Ampere for the sub-systems will thus be correspondingly different.

6.2.2 Per-Unit General

A per-unit system is a method of expressing numbers for ease of comparison. By definition, per-unit value of any quantity is the ratio of its value to a chosen base value.

$$\text{Per-Unit value} = \frac{\text{Actual value}}{\text{Base value}} \tag{6.1}$$

The "Actual value" is a scalar or complex value of a quantity expressed in its proper units, namely, volts and amps. The "Base value" is the arbitrarily chosen or derived scalar value of the same quantity expressed in the same units.

It is simpler to conduct calculations using integer values of per unit. However, equipment data are normally provided as a percent on the equipment rating. Percent can be easily converted to PU by dividing by 100 or vice versa by multiplying by 100.

$$\text{Percent value} = 100 \times (\text{per-unit value})$$

$$\text{Per-unit voltage kV}_\text{PU} = \frac{\text{Actual kV}}{\text{Base kV}} = \frac{\text{kV}_\text{a}}{\text{kV}_\text{B}} \tag{6.2}$$

The per unit and percent are dimensionless ratios, which may be a scalar or complex number.

For example, the system nominal voltage of 230 kV is 1.045 PU or 104.5% of the chosen base value of 220 kV. If 100 MVA is chosen as the base, then 50,000 kVA is 0.5 PU or 50%. Similarly, 10 Ω compared to a base of 5 Ω is 2 PU or 200%.

Consider the case where the normally generated voltage of 13.8 kV is stepped up to 230 kV for transmission and then stepped down to 27.6 kV for sub-transmission. If for this system, the base voltages selected are 13.8, 230, and 27.6 kV. These are each 1 PU voltage in their respective parts of the power system. A measured value of 28.4 kV at the sub-transmission level is, therefore, equal to 1.029 PU (28.4/27.6).

The voltages above are the L–L or delta voltages; the related line-to-neutral voltages or star voltages are also 1 PU This means for the 230 kV system, 1 PU line-to-neutral voltage is $230/\sqrt{3} = 132.8$ kV.

The product of two PU quantities yields a value in PU However, the product of two quantities in percent must be divided by 100 to obtain the result in percent.

To summarize, for three-phase power systems, it is normal to choose three-phase MVA and line-to-line voltage for the base values.

6.2.3 Per-Unit Impedance

Having selected base volts (kV) and base power (kVA or MVA), then base impedance can be calculated:

$$\text{MVA} = \frac{V^2}{Z}$$

$$Z = \frac{V^2}{\text{MVA}}$$

$$Z = \frac{(V_{\text{base}})^2}{\text{MVA}_{\text{base}}}$$

For example, where base voltage = 220 kV and base MVA = 100 MVA

$$Z\,\text{base} = 220^2/100 = 484\,\Omega$$

It is desirable to have the impedance of a system component such as lines or transformers converted directly from Ohm to per-unit value without first determining the base Ohm given the base values of voltage and power for the system concerned.

$$\text{PU Ohm} = \frac{\text{Ohm}}{\text{Base Ohm}}$$

It should be noted that a calculated fault MVA can be directly converted to a system impedance as follows:

$$\text{PU system impedance} = \frac{\text{Base MVA}}{\text{Fault MVA}}$$

6.2.4 Conversion of PU Values To Different Bases

The sub-systems of a network are normally interconnected through transformers. The chosen MVA will be common to all the sub-systems but the base volts will be the rated voltage of each sub-system. The base Ohm and base Ampere will be therefore correspondingly be different.

As long as the ratio of the base voltages on the two sides is equal to the transformer L–L voltage ratio for 3φ systems (turns ratio for 1φ system), PU values of system components will be the same and the two interconnected systems can be treated as a single system with the appropriate PU impedance of transformer being shown. Only if the actual voltage and current values are needed, then the base values of the sub-systems are required.

Normally, the manufacturer specifies the PU impedance value based on the equipment ratings. These may be different from the standardized power system bases. For instance, the reactance of transformers, generators, and motors is based on their MVA and voltage ratings. But their reactance is required to be converted to the chosen system base values such as 100 MVA and 220 kV. The conversion is facilitated by expressing the same impedance in Ohm on two different sets of PU bases.

Let $Z_{PU\ old}$ be the PU impedance of Z_a Ohm on the old set of bases represented by the subscript 1.
Let $Z_{PU\ new}$ be the PU impedance of Z_a Ohm on the new set of bases represented by the subscript 2.

$$Z_{PU\ old} = \frac{Z_a}{Z_{B1}} = \frac{Z_a X\ MVA_{B1}}{(kV_{B1})^2} \quad (6.3)$$

$$Z_{PU\ new} = \frac{Z_a}{Z_{B2}} = \frac{Z_a X\ MVA_{B2}}{(kV_{B2})^2} \quad (6.4)$$

$$\frac{Z_{PU\ new}}{Z_{PU\ old}} = \frac{MVA_{B2}}{MVA_{B1}} \times \frac{(kV_{B1})^2}{(kV_{B2})^2} \quad (6.5)$$

Therefore, the general equation is

$$Z_{PU\ new} = Z_{PU\ old} \times \frac{MVA_{B2}}{MVA_{B1}} \times \frac{(kV_{B1})^2}{(kV_{B2})^2} \quad (6.6)$$

or

$$Z_{PU\ new} = Z_{PU\ old} \times \frac{MVA_{PU\ new2}}{MVA_{PU\ old}} \times \frac{(kV_{PU\ old})^2}{(kV_{PU\ new2})^2}$$

It should be particularly noted that the voltage-square factor of conversion is used only when the same voltage level is involved, but with slight differences for the voltage bases. It should never be used where the base voltages in the two windings are proportional to the transformer turns ratio, such as going across a step-up or step-down transformer.

The PU impedance of a transformer is the same on each side of the transformer.

Consider a transformer with a voltage ratio of kV_2/kV_1, as shown in Figure 6.1.

Let the actual impedance of the transformer viewed from the low voltage (LV) side be Z_{a1}.
Let the actual impedance of the transformer viewed from the high voltage (HV) side be Z_{a2}.

$$Z_{PU1} = \frac{Z_{a1}}{Z_{B1}} = Z_{a1} \times \frac{MVA}{(kV_1)^2}$$

$$Z_{PU2} = \frac{Z_{a2}}{Z_{B2}} = Z_{a2} \times \frac{MVA}{(kV_2)^2}$$

But $Z_{a2} = Z_{a1} \times \dfrac{(kV_2)^2}{(kV_1)^2}$

Therefore, $Z_{PU\ 2} = \dfrac{Z_{a1} \times MVA}{(kV_1)^2} = Z_{PU1}$

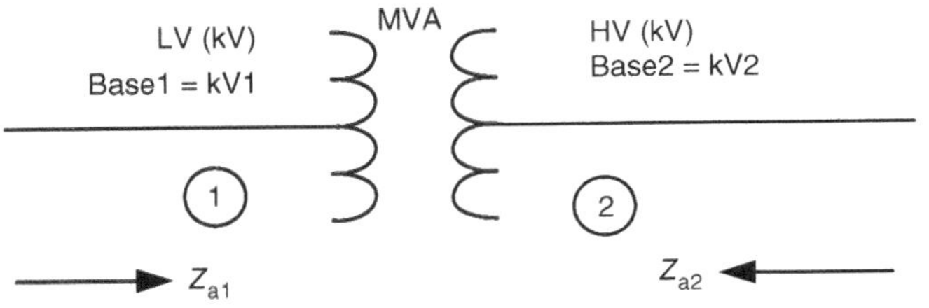

Figure 6.1 Two winding transformer.

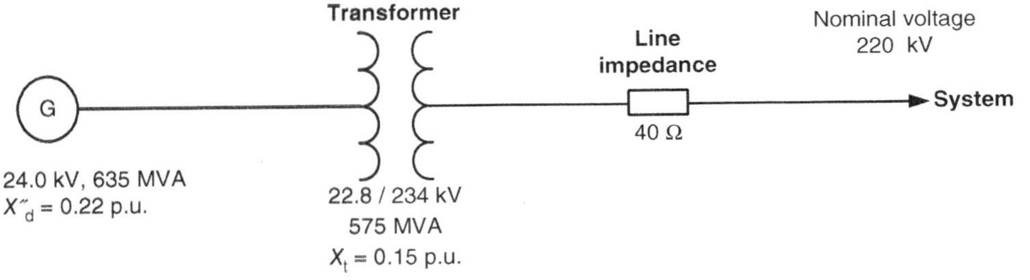

Figure 6.2 An example case.

The above conversion is illustrated by considering the following example, Figure 6.2
Let the system bases be 100 MVA and 220 kV.

$$\text{System base impedance} = \frac{\text{Base kV}^2}{\text{Base MVA}} = \frac{220\text{ kV}^2}{100\text{ MVA}} = 484\,\Omega$$

$$\text{Line impedance in PU} = 40/484 = 0.08265\text{ PU}$$

The given transformer reactance PU value needs to be converted to the new MVA base and the new voltage base of 220 kV from 234 kV.

$$\text{Transformer } X_T = 0.15 \times \frac{100}{575} \times \left(\frac{234}{220}\right)^2$$

$$\text{Transformer } X_T = 0.0296\text{ PU}$$

Based on the transformer turns ratio, the base voltage on the generator side of the transformer is $(220/234) \times 22.8 = 21.4\text{ kV}$

$$\text{Generator } X_d'' = 0.22 \times \left(\frac{24}{21.4}\right)^2 \times \left(\frac{100}{635}\right)$$

$$= 0.22 \times \frac{100}{635} \times \left(\frac{24}{21.4}\right)^2$$

$$\text{Generator } X_d'' = 0.0436\text{ PU}$$

$$\text{Total impedance in PU} = 0.08265 + 0.0296 + 0.0436 = 0.1558\text{ PU}$$

$$\text{The impedance as seen by the HV 220 kV system:} 0.1558 \times 484 = 75.4\,\Omega$$

The impedance diagram for the network with PU values is as follows (Figure 6.3).

For comparison, the following represents the same system with impedances provided in Ohm (Figure 6.4).

Figure 6.3 Impedance diagram in PU for Figure 6.2.

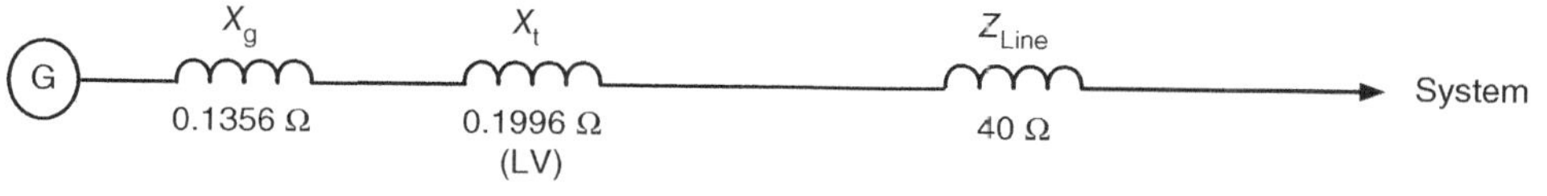

Figure 6.4 Impedance diagram in Ohm for Figure 6.2.

Similarly, the impedance as seen by the HV, 220 kV system, without using PU

$$(0.136 + 0.2) \times (234/22.8)^2 + 40 = 75.4\ \Omega$$

6.2.5 A Summary for Solving a Problem Using Per Unit

(1) Choose a convenient set of base quantities at one point in the network. Normally start at the system source.
(2) Select the Base MVA and Base kV. Typical transmission values are 100 MVA, 500/220/115 kV, respectively, and for distribution, 10 MVA, and 44/28/14 kV.
(3) Find the corresponding voltage base quantities for other parts of the network using the "turns" ratio to determine the set of base quantities applicable… retain the same MVA base throughout the network.
(4) Calculate the base impedance and base current once the Base MVA and Base kV have been selected as per Step 2.
(5) Clearly document your base quantities for which the PU quantities have been derived
(6) Covert equipment impedances from Ohm to per unit.
(7) Solve the problem in per-unit quantities.
(8) Convert back to actual quantities at the location where required, by multiplying per-unit quantities by the base quantities applicable at that network location (Table 6.1).

Table 6.1 Base quantities summary.

Base Quantities Summary Table		
Quantity	**Three-phase base**	**1-PH base**
Voltage	$V_{L\text{-}L}$	$V_{L\text{-}N} = V_{Base\ L\text{-}L/1.732}$
Current	I_{Line}	I_{Line}
Power	kVA total (3PH)	kVA per phase
Base current	Base kVA/[1.732 × Base Voltage (kV)]	Base kVA/Base Voltage (kV)
Base impedance Note: Is in Ohm per phase	Base voltage(kV)²/Base MVA or Base Voltage ($kV_{L\text{-}N}$) × 1000/Base current	(Base voltage(kV)²/ Base kVA) × 1000 or Base voltage/Base current

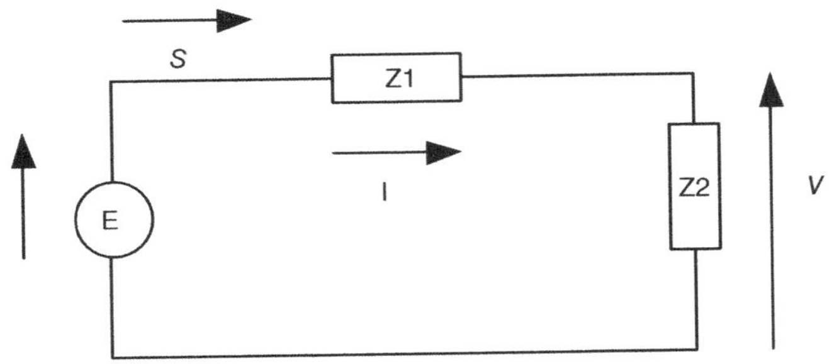

Figure 6.5 Simple circuit example 1.

6.2.6 Example

Figure 6.5 below shows a circuit which though more easily solved directly, will serve as a simple example of per-unit application:

Solve for I, S, and V

Nominal System Source: 10,000 kVA, 230 kV

$E = 268$ kV @ 0°; $Z1 =$ j800 Ω; $Z2 =$ 8000 + j800 Ω

Select the following base quantities: V base = 230 kV; S base = 10,000 kVA or 10 MVA.
Therefore:

$$I_{\text{base}} = 10{,}000\ \text{kVA}/230\ \text{kV} = 43.5\ \text{A}$$

$$Z_{\text{base}} = 230\ \text{kV}/43.5 = 5290\ \Omega \text{ or } 230^2\ \text{kV}/10\ \text{MVA} = 5290\ \Omega$$

Converting actual circuit quantities to per-unit quantities:

$$E = 268\ @\ 0°/230 = 1.165\ \text{PU}$$

$$Z_1 = \text{j}800/5290 = \text{j}\,0.151\ \text{PU}$$

$$Z_2 = 8000 + \text{j}800/5290 = 1.51 + \text{j}\,0.151$$

Solving for I:

$$I = E/(Z1 + Z2) = 1.165\ @\ 0°/1.51 + \text{j}\,0.302$$
$$= 1.165\ @\ 0°/1.54\ @\ 11.3°$$
$$= 0.757\ @\ -11.3°$$

$$S = E \times I^* = (1.165)\ \left(0.757\ @\ 11.3° = 0.883\ @\ 11.3° = 0.865 + \text{j}\,0.173\ \text{PU}\right)$$

Converting back to actual quantities:

$$I = I \times I_{\text{base}} = \left(0.757\ @\ -11.3°\right) \times 43.5 = 33\ @\ -11.3°$$

$$S = S \times S_{\text{base}} = \left(0.865 + \text{j}0.173\right) \times (10{,}000)$$

$$P = 8{,}650\ \text{kW};\ Q = 1{,}730\ \text{kVAR}$$

6.3 Fundamental Need for Fault Information

Fault calculation information is primarily needed for protection practitioners to understand, design, and set protective relays. Maximum and minimum values are required to ensure adequate protection performance under all expected operating system conditions.

Protective relays represent the intelligence of the protection system. The power system instrument transformers secondary low-level voltages and currents are inputs to the protective relays – via the AC control cables. These monitored quantities or some combination thereof are compared against thresholds – protection settings – that are computed by protection practitioners, and are set, into the protective relays.

If the relay comparison element computes that a threshold is exceeded, it will trigger the decision block shown in Figure 6.6, where it will consider other possible factors before initiating a trip decision (breaker open or closed, etc.). If a trip decision is warranted, then the protective relay will assert a trip output contact(s). The protective relay output contacts are wired as inputs to the associated protection elements' isolation breakers, via their Mech. Boxes (electrical interfaces), normally through a set of auxiliary relays, termination devices, and isolation devices. The result is that the protective relay closes a contact that has a net effect of causing the isolation device(s) to trip/open.

Fault calculations are used by protection practitioners to determine abnormal voltages, currents, and relative phase angles and to enable them to determine settings for protective relay threshold values that assure protection trip dependability and security.

Protection systems must operate under system normal conditions (all elements in service) and are also expected to operate under credible system contingencies such as some transmission lines and generators out of service for maintenance or others.

Protection practitioners need to calculate and set protective relays for maximum as well as credible minimum system elements in service, to assure power system reliability. Therefore, for any given fault location, two fault values are normally calculated for each fault type, i.e. three-phase and L-G. One under normal conditions, all power system elements in service; and a second, with the credible number of power system elements removed from service, to minimize the fault quantities at the relay location.

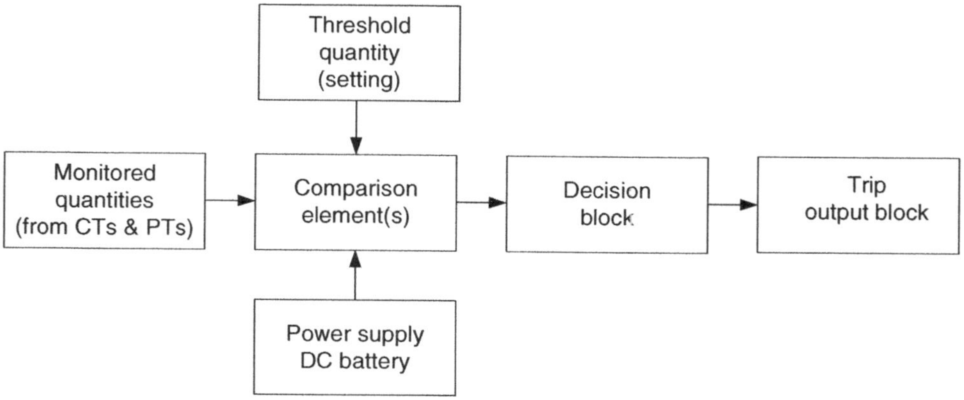

Figure 6.6 Simplified block diagram of a protective relay.

6.3.1 Types of Faults

Faults are abnormal conditions that can occur on power systems due to lightning strikes, equipment damage, wind and ice storms, animal contact, and general vehicle accidents, and the like.

When a fault occurs, it causes a short circuit of some sort depending on how many phases are involved and whether the ground is involved. Regardless, the normal load current increases to fault values which are 100+ times load and affect system voltages.

Recently, power system faults can reach levels of 63,000 A or more. Such magnitudes of fault current can cause severe damage and forces, sufficient to bend solid bus bars. These high currents caused by the fault must be cleared fast to mitigate damage to equipment, safety and to maintain the integrity of the system. Balanced and unbalanced faults occur on the power system, and protections must operate to minimize equipment damage, maintain stability, and safety.

In order of normal increasing severity, the various faults for which protections must operate are as follows:

1.	Single phase-to-ground fault (L-G)	Unbalanced type fault
2.	Phase-to-phase fault (PH-PH)	Unbalanced type fault
3.	Double phase-to-ground fault (PH-PH-G)	Unbalanced type fault
4.	Three-phase fault (three-phase)	Balanced type fault

The three-phase type of fault typically produces the highest fault level and the line–ground fault type which is typically lower is the most dominant type at 80%. Therefore, for protection setting applications, the three-phase and L-G faults are the two types that are typically conducted to derive protection relay settings. For that reason, these two fault types are covered in this chapter. Other sources may be referred to for a discussion and analysis of other fault types [4].

6.3.1.1 Balanced vs. Unbalanced

6.3.1.1.1 Balanced

In a balanced three-phase power system, each of the phases draws the same amount of current. The magnitudes of currents are the same; however, each of the phase currents is displaced with respect to each other, by 120 electrical degrees. Similarly, a balanced three-phase fault implies that fault current in each of the phases is equal in magnitude and displaced to each other by 120 electrical degrees.

In a balanced system, one need only to model and analyze one phase, normally the "A" phase, and from the phase A, currents and voltages one can derive the other B and C phase quantities as the counter-clockwise rotation ABC is universally standard, and each phase is displaced by 120° apart (Figure 6.7).

In terms of phasors, one can write the same balanced set as follows. Note that the phasors are RMS quantities:

$$V_a = V_m \angle \phi_m; V_m = \text{Vector Magnitude}, \angle \phi_m \text{ is the vector angle}$$

$$V_b = V_m \angle \phi - 120^\circ$$

$$V_c = V_m \angle \phi - 240^0 = V_m \angle \phi + 120^0$$

For a balanced system, one needs only to model and analyze one phase and to derive the other two-phase quantities based on the rotation and 120° phase displacement.

For balanced circuit analysis, normal circuit theorems apply – it is not necessary in this case to use symmetrical components.

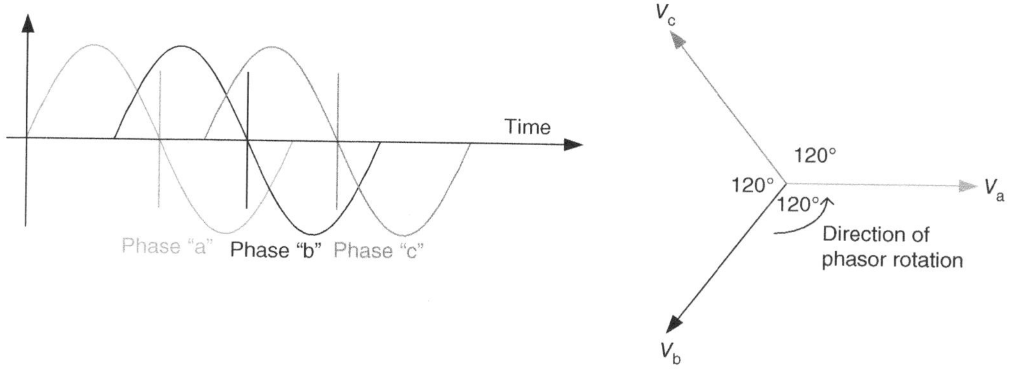

Figure 6.7 A balanced three-phase set.

6.3.1.1.2 *Unbalanced*

Unbalanced type faults such as the L-G fault are difficult to analyze using conventional circuit analysis.

Depicted in Figure 6.8 is an unbalanced set of currents, it should be noted that the magnitudes and phase angles for the three vectors I_A, I_B, and I_C are not the same. Therefore, one cannot model and analyze Phase A and derive Phase B and C quantities directly from Phase A as the vectors are not balanced as opposed to a balanced set.

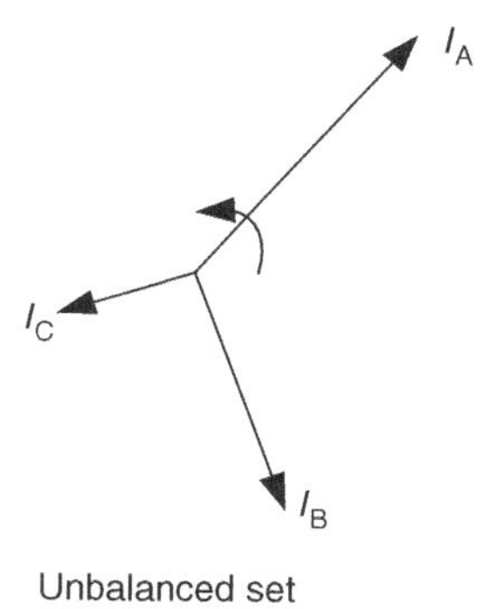

Figure 6.8 An unbalanced three-phase set.

The analysis of unbalanced faults is more challenging. However, fortunately for the industry at the time, Charles Fortesque in a 113-page paper [1] presented at the 1918 annual AIEE convention provided the industry with a theory and methodology that is still presently being applied today.

Unbalanced fault analysis requires the use of symmetrical components. The symmetrical component theory states, for any three-phase system, any unbalanced fault condition can be solved by superposition of three distinct sets of positive, negative, and zero-sequence sets, as shown below in Figure 6.9. The sets of symmetrical components (or subsets) of the original unbalanced set are assigned by naming convention, positive, negative and zero-sequence components. The first two component sets are balanced and symmetrical while the zero-sequence set are only symmetrical as shown in Figure 6.9.

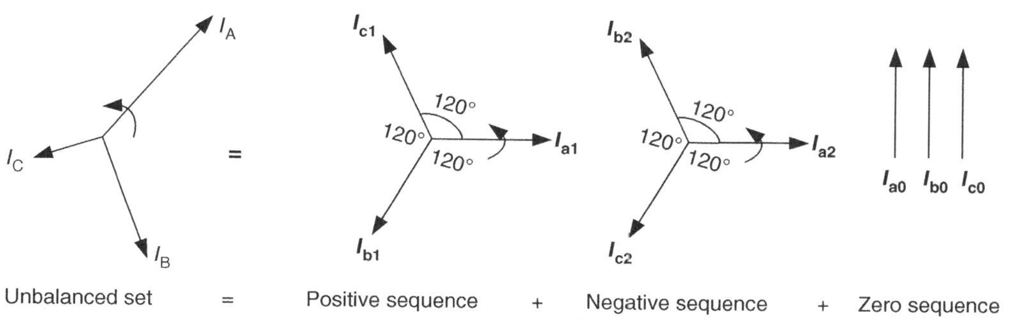

Figure 6.9 Unbalanced 3PH set represented by symmetrical components.

In summary, more details are discussed below; the unbalance set of the three phases can be resolved as follows:

$$I_A = I_{a1} + I_{a2} + I_{a0} = I_1 + I_2 + I_0$$
$$I_B = I_{b2} + I_{b2} + I_{b0} = a^2 I_1 + aI_2 + I_o$$
$$I_C = I_{c1} + I_{c2} + I_{c0} = aI_1 + a^2 I_2 + I_0$$

Positive	Negative	Zero
$I_{a1} = I_1$	$I_{a2} = I_2$	$I_{a0} = I_{b0} = I_{c0}$
$I_{b1} = a^2 I_{a1}$	$I_{b2} = aI_{a2}$	
$I_{c1} = aI_{a1}$	$I_{c2} = a^2 I_{a2}$	

where $a = 1\angle 120° = -0.5 + j0.866$; $a^2 = 1\angle 240° = -0.5 - j0.866$.

6.4 Symmetrical Components

The need for symmetrical component theory was derived from the need to provide a systematic way of calculating unbalanced faults or short circuits to properly understand, design, and implement protection systems. Unbalanced faults, such as phase-to-phase and line-to-ground, are complex, and symmetrical component theory provides a methodology to calculate such faults.

The theory is based on the ability to analyze any system fault through the application and superposition of three individual three-phase networks:

(a) the positive phase sequence network
(b) the negative phase sequence network
(c) the zero-phase sequence network

The advantage of this approach is that a three-phase network can be represented by a single-phase equivalent circuit; this simplifies the analysis.

Before the use of computers to conduct fault studies, ac network analyzers were extensively used. Network analyzers provided a direct and physical analog model of the system under study. The analyzer only represented one phase. Though it was theoretically possible to represent all three phases of the system on the analyzer to simulate an unbalanced fault, it would have been very cumbersome.

Simpler methods were devised to continue the use of single-phase analyzers. The one which gained extensive and universal application was the method of symmetrical components.

The power system normally operates in a balanced three-phase steady-state mode; however, there are under certain situations, faults, that can cause unbalanced operations.

Note that in the following sections, ABC and RWB (Red White Blue) phase designations are used interchangeably.

6.4.1 Theory of Symmetrical Components

The basic theory of symmetrical components is that any unbalanced three-phase system of current or voltage phasors can be resolved into three systems. Each with three phases but with different phase sequence. The first two sets (positive, negative) are balanced and symmetrical while the third set (zero) is only symmetrical.

Figure 6.10 Positive sequence phasors.

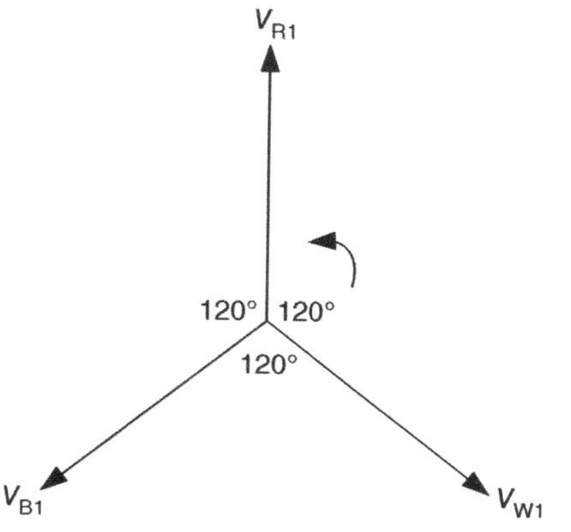

Limiting the discussion to three-phase systems only, three distinct sets of components are involved. The positive, negative, and zero-sequence components, are applicable to both currents and voltages. The direction of rotation for the sequence components is counter-clockwise, however rotates with a different phase sequence – defined by the order in which the phasors pass a given position.

6.4.1.1 Positive Sequence Phasors

It is a trio of phasors, balanced in that they are equal in magnitude and 120° apart. The phase sequence is RWB (assuming red, white, and blue labeled phases), the same as that of the original unbalanced phasors and that of the power system. It should be noted that the set refers to phase currents and line-to-neutral (or line-to-ground) voltages supplied by the source generators.

Figure 6.10 shows a set of positive sequence voltage phasors.

The subscript "1" is used to identify the positive sequence components. The "a" operator is a unit phasor with an angle of 120°, used extensively as a convenient shorthand. For example, a phasor can be rotated 120° in the counter-clockwise direction by multiplying it by an "a" operator.

$$a = \angle 120 = -0.5 + j0.866$$

$$a^2 = \angle 240 = -0.5 - j0.866$$

$$a^3 = \angle 360 = 1.0 + j0$$

Therefore, the previous positive sequence voltage set can be designated as follows:

$$V_{R1} = V_1$$

$$V_{W1} = a^2 V_1 = V_1 \angle 240$$

$$V_{B1} = a V_1 = V_1 \angle 120$$

6.4.1.2 Negative Sequence Phasors

It is also a balanced set of three phasors, all equal in magnitude and 120° apart but with the phase sequence reversed as shown in Figure 6.11. The phase sequence is opposite to that of the positive sequence and is RBW.

The subscript 2 is used to identify the negative sequence components. The negative sequence phasors can be designated as follows:

$$V_{R2} = V_2$$

$$V_{W2} = a V_2 = V_2 \angle 120$$

$$V_{B2} = a^2 V_2 = V_2 \angle 240$$

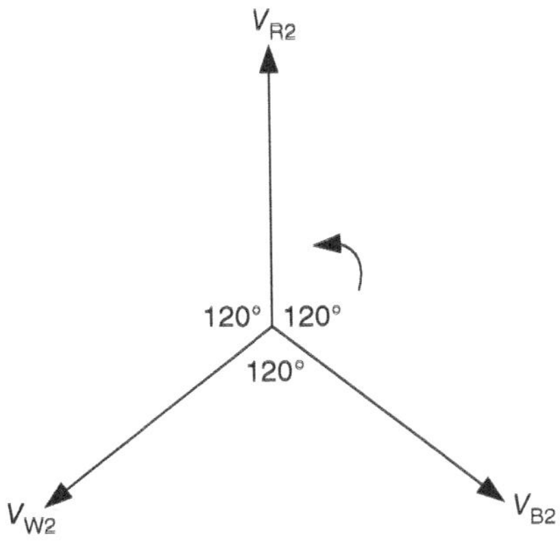

Figure 6.11 Negative sequence phasors.

6.4.1.3 Zero-Sequence Phasors

This set consists also of three phasors, equal in magnitude and always in phase. The set is denoted by the subscript 0 (Figure 6.12).

$$V_{R0} = V_{W0} = V_{B0} = V_0$$

All sequence components exists only as a set of three phasors, and a phasor within a sequence component set never exists alone. When one phasor is determined, the other phasors of the set are reference to it.

6.4.2 Phase Quantities In Terms of Sequence Components

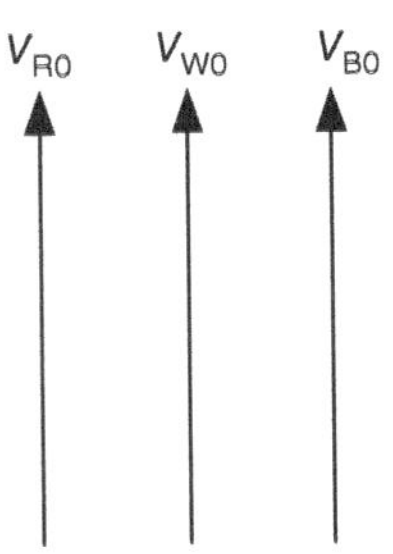

Figure 6.12 Zero-sequence phasors.

An unbalanced three-phase system can be resolved into three-component systems, one of positive, one of negative, and one of zero-phase sequence. It should be noted that the reference phase (usually the red phase or A phase) quantities of the three-component systems need not necessarily be equal in magnitude nor be in phase. This means V_{R1}, V_{R2}, and V_{R0} (and by extension, the two-component sets of the other two phases) are not necessary equal in magnitude or angle with respect to each other.

Knowing the sequence components, the original unbalanced system can be determined by the following equations:

$$I_R = I_1 + I_2 + I_0; \qquad V_R = V_1 + V_2 + V_0 \tag{6.7}$$

$$I_W = a^2 I_1 + aI_2 + I_0; \qquad V_W = a^2 V_1 + aV_2 + V_0 \tag{6.8}$$

$$I_B = aI_1 + a^2 I_2 + I_0; \qquad V_B = aV_1 + a^2 V_2 + V_0 \tag{6.9}$$

Based on these equations, it is shown that the sequence quantities, in terms of phase quantities of an unbalanced system, can be derived as follows [2–4]:

$$I_0 = \frac{1}{3}\left(I_R + I_W + I_B\right); \qquad V_0 = \frac{1}{3}\left(V_R + V_W + V_B\right) \tag{6.10}$$

$$I_1 = \frac{1}{3}\left(I_R + aI_W + a^2 I_B\right); \qquad V_1 = \frac{1}{3}\left(V_R + aV_W + a^2 V_B\right) \tag{6.11}$$

$$I_2 = \frac{1}{3}\left(I_R + a^2 I_W + aI_B\right); \qquad V_2 = \frac{1}{3}\left(V_R + a^2 V_W + aV_B\right)$$

6.5 Sequence Impedances of Power Apparatus

Power system equipment represents different values of impedance to positive sequence (PS), negative sequence (NS), and zero-sequence (ZS) voltages. These impedances are known as sequence impedances and vary according to the type, construction, and connections of the apparatus.

It is typical for non-rotating equipment that the PS and NS impedances are the same, for example, the leakage impedance of the transformers and the phase to the neutral impedance of transmission lines.

The transformers present different zero-sequence impedances depending upon the type of core, winding connections, and grounding details.

The PS and NS impedances of a transmission line is the straight line impedance from the source end to the terminating end. But, the zero-sequence impedance is loop impedance; line impedance plus the return path for the zero-sequence currents. The return path can be either through the ground or a parallel combination of ground and grounded sky-wire or cable sheath.

The following sections provide background information on the impedances of power apparatus.

6.5.1 Synchronous Machinery

The equipment covered are generators, synchronous motors and condensers, and to a lesser extent induction motors designated usually by using a circle symbol in sequence diagrams. The rotating magnetic fields in a generator are the main source of positive sequence voltages in a system network. The symmetrical fault currents supplied by this type of apparatus decrease exponentially with time from a relatively high initial value to a low steady-state value.

The DC offset of the initial fault current is generally not used for protection considerations but is taken into consideration to determine the rating of the interrupting device such as a circuit breaker. It is also considered when analyzing CT transient performance. The PS impedance can have three values during the transient period immediately following the occurrence of a fault. These are the direct-axis sub-transient reactance (X''_d), the direct-axis transient reactance (X'_d), and the direct-axis synchronous reactance (X_d).

The sub-transient reactance value is normally used for short circuit calculations as this yields the highest initial fault current which is important while considering high-speed relays and circuit breakers.

The contribution to fault currents by induction motors is most often ignored for power system protection purposes because of their very short time constants.

The values of generator reactance vary with the designs of the machines, and the specific values are provided by the manufacturer. Typically, X''_d in per unit of the machine rating is 0.07–0.3 PU with time constants (t_c) of 20–50 ms, X'_d is about 1.5–2.5 times X''_d with t_c. of 0.6–4 seconds, and X_d is about 4–15 times X''_d for sustained faults.

However, the applicability of using generator sub-transient reactance values in determining the source impedance has no practical value when determining fault currents for feeders at distribution voltage levels where slower-speed relays are used for protection. The slower speed ensures that they operate only after the generator reactance decays to that of transient/synchronous reactance. Furthermore, the system network typically represents a much greater component than the generator reactance in making up the overall source impedance thereby having little effect on the fault current.

Though the generators do not produce negative sequence voltages, the NS sequence current can flow through the windings when impressed with unbalance voltages as a result of an unbalanced fault or open phase. The MMF associated with these currents in the three phases will pass flux alternately between and under the poles at double frequency. For non-salient-pole machines, the negative sequence impedance is the same as X''_d. However, for salient-pole machines particularly without damper windings, the value is a bit higher than X''_d. The difference is ignored unless the fault under calculation is very near the machine terminals.

Generators and sources are represented by impedance in the negative sequence networks without a voltage source.

Zero-sequence currents in the stator do not cause a rotating flux in the rotor field, and the effective magnetic path is confined to the leakage paths in the stator, dependent upon the machine construction. The ZS reactance is generally low and variable. Since generators are generally not solidly grounded, the contribution of ZS currents from generators is minimal.

It should be noted that the resistance of the armature windings is so small that it is ignored. However, it is important in determining the dc time constant of an asymmetrical fault current.

A large power system consists of many generators, interconnecting transmission lines, transformers, and other equipment; when considering a fault in a part of such a system, an equivalent source with appropriate impedance values is normally used to represent the equivalent of all of the system up to the point of fault except for the immediate element supplying the fault current. This is generally the case when one is conducting fault calculations manually or "by hand." For such cases, it is usual to reduce a system network to two equivalent sources one at each end of an area to be studied with an equivalent transfer impedance.

Most utilities today use fault calculation software tools to model the utilities' power system and to conduct fault studies, and in some cases, to conduct protection relay coordination studies. Therefore, the need to simplify the power system to the local area of study by deriving system equivalents and transfer impedances is not needed. The whole power system is modeled and is used to derive the fault studies. The use of computerized software tools has greatly aided the efficiency of fault analysis and protection coordination. A system model including thousands of buses can calculate results in seconds, allowing one to analyze many more operating contingencies and fault types than ever before.

6.5.2 Transmission Lines

The PS and NS impedances are the same. The ZS impedance is affected by the coupling between the phase conductors, the overhead ground wire, if it exists, and the ground through which the ZS current returns to the source.

The ZS impedance is generally higher than the PS impedance, and the reactive component is usually 2.5–3 times greater. For very short lines it can as low as 2.

For cables, the ZS impedance is dependent upon the cable construction, the position and grounding of the cable sheaths, and the resistivity of the soil. Cable charging is as much as 10 times greater than overhead lines of the same length.

Some typical line positive sequence and zero-sequence impedances are shown in Table 6.2.

6.5.3 Transformers

Since transformers are symmetrically wound and stationary devices, the positive and negative sequence reactance are identical being the transformer leakage impedance between the respective windings.

Table 6.2 Typical line impedances.

Voltage (kV)	Typical positive sequence (PS) impedance (PU/km)	Typical zero-sequence impedance
115	0.003	Estimated at 2.5–3 × PS
230	0.001	Estimated at 2.5–3 × PS
500	0.00013	Estimated at 2.5–3 × PS

The ZS impedance varies considerably according to the construction of the transformer and the presence, or otherwise, of a delta winding. The zero-sequence impedance of a star winding will be very high if no delta winding is present. The actual value will depend on whether there is a low reluctance return path for the third-harmonic flux.

For three-limb designs without a delta, where the return flux path is capacitance to ground with air being the dielectric, the determining feature is usually the grounded tank and possibly the core support framework, where this flux creates a circulating current around the tank and/or core framework. The impedance of such winding arrangements is likely to be in the order of 75–200% of the positive-sequence impedance between primary and secondary windings.

For five-limb cores and three-phase banks made of three single-phase units, the zero-sequence impedance will be the magnetizing impedance for the core configuration. Should a delta winding exist, then the third-harmonic flux will create a circulating current around the delta, and the zero-sequence impedance is determined by the leakage field between the star and the delta windings. Again, the type of core will influence the magnitude of the impedance because of the effect it has on the leakage field between the windings.

Five-limb transformers have their zero-sequence impedances substantially equal to their positive-sequence impedance between the relative star and delta windings (Figure 6.13).

It should be noted that if the connections are such as to block zero-sequence currents like an ungrounded neutral of a star winding, the ZS impedance is infinite and a break will be shown in the zero-sequence network diagrams.

6.5.3.1 Equivalent Circuit: Positive and Negative Sequence Impedances

The transformer reactance is normally measured by the manufacturer and given on the nameplate. It is normally based on the self-cooled rating of the transformer bank. The magnetizing impedance is several orders of magnitude higher than the leakage reactance of about 6–15%. It is common practice to ignore, except for special situations, the magnetizing impedance, and the transformer is represented in the positive and negative sequence networks as a series impedance/reactance refer to Figure 6.13.

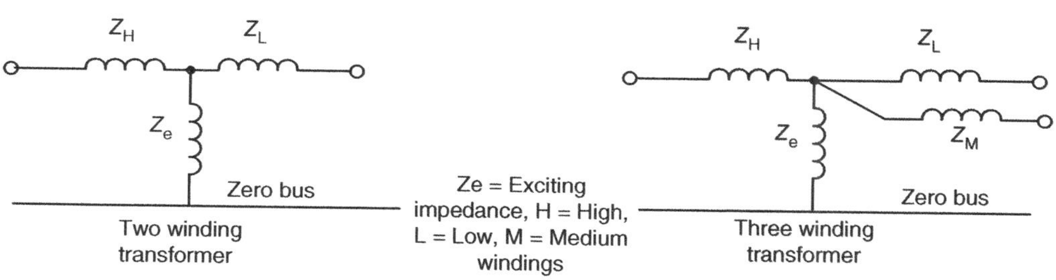

Figure 6.13 Transformer equivalent circuits.

It should be noted from Figure 6.13 that, Z_H, Z_L, and Z_M are components of leakage reactance represented as an equivalent Y rather than an equivalent delta. For example, consider the 3-Wdg. transformer for which the manufacturer provides the leakage impedances between the windings as Z_{HM}, Z_{HL}, and Z_{ML}, normally on different MVA ratings and at one of the winding voltages. They are converted to the equivalent Y values as follows:

$$Z_H = \frac{1}{2}(Z_{HM} + Z_{HL} - Z_{ML}) \qquad Z_M = \frac{1}{2}(Z_{HM} + Z_{ML} - Z_{HL})$$
$$Z_L = \frac{1}{2}(Z_{HL} + Z_{ML} - Z_{HM}) \tag{6.12}$$

The neutral point of the Y mathematical representation of impedances has no physical meaning. It facilitates the determination of currents and voltages at the transformer terminals. It is possible to get a negative value for one of the components, usually for an auto-bank, but this does not mean that it is capacitive reactance.

6.5.3.2 Zero-Sequence Circuit and Impedance

In the zero-sequence network, although the leakage impedance is identical to the positive sequence value (when zero-sequence path is available), the zero-sequence magnetizing impedance is dependent upon the transformer core construction and can be much lower than the above-mentioned value of 200%.

In three-phase banks constructed of single-phase transformers and in three-phase shell cored transformers, the zero-sequence magnetizing impedance is large and can be ignored as in the positive and negative sequence networks. In three-limb core-type transformers, however, the zero-sequence flux must be completed through the oil or tank. Owing to the high reluctance of the flux path, zero-sequence magnetizing impedance is of the order of only 100–400%. It is for this reason that a single-phase fault can have a higher current than three-phase. Nevertheless, since it represents a high impedance ground source, it has negligible effect and is therefore neglected in most fault studies.

Zero-sequence currents have a path to flow if the neutral of the Y-connected windings is grounded with the presence of a delta winding. A grounded zig-zag winding also provides a path for zero-sequence currents. When zero-sequence currents flow in any winding, another winding on the same core must provide a path for the flow of balancing currents to nullify the amp-turns. The reactance associated with the leakage flux between these two windings constitutes the zero-sequence impedance.

The mode of connection of the transformer leakage impedance to the external circuit is determined by taking account of each winding arrangement and its connection or otherwise to the ground. As a method to modeling transformer zero-sequence connections, the industry has developed an aid.

Referring to the two winding transformer models in Figure 6.14 below, imaginary links "a" and "b" are used to derive the connections. If zero-sequence currents can flow into and out of a winding, for example, a solidly grounded Y-winding, the winding terminal is connected to the external circuit, that is link "a" is closed. If zero-sequence currents can circulate in the winding without flowing in the external circuit, for example, a delta winding, the winding terminal is directly connected to the zero bus that is, link "b" is closed.

Refer to Figure 6.15 for a grounded Y-delta transformer, for which the zero-sequence equivalent circuit is shown without the magnetizing impedance.

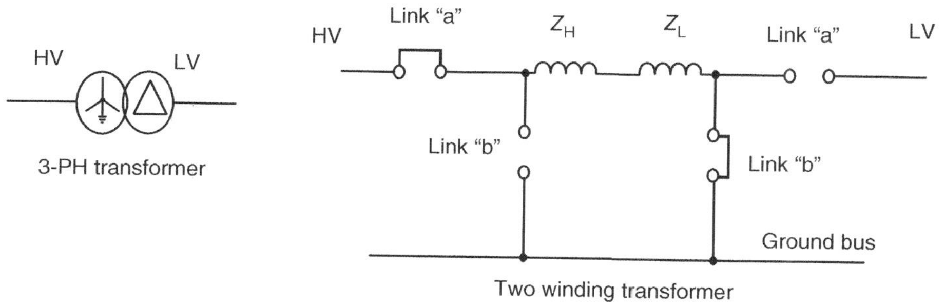

Figure 6.14 Two winding transformer zero-sequence equivalent circuit model.

Figure 6.15 Wye Grd–Delta winding transformer zero-sequence equivalent circuit.

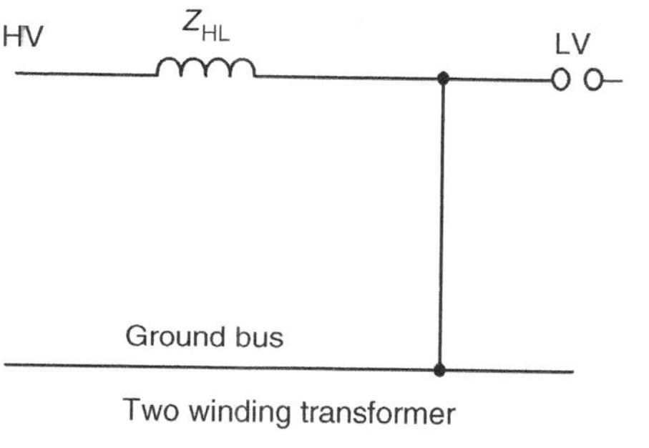

- The HV Link "a" is closed because the grounded neutral permits zero-sequence currents to flow up the neutral and is balanced by the circulating currents within the delta.
- The HV Link "b" is open because the HV winding is not a delta.
- The LV Link "b" is closed because the LV winding is delta connected.
- The LV Link "a" is open because zero-sequence currents cannot flow into a delta – no connection to ground.

From this circuit, it is seen that the zero-sequence impedance as viewed from the HV terminals is $Z_H + Z_L = Z_{HL}$. When viewed from the LV terminals, the impedance is infinite, and therefore, no current will be contributed from the LV side for an L-G fault (Figure 6.15).

The same method can be applied to three winding transformers, as illustrated by the following example shown in Figure 6.16.

6.5.3.3 Zero-sequence Impedance with Neutral Grounding Impedance

The zero-sequence impedance of a neutral grounding Z_n is $3\,Z_n$. The reason for this can be understood from Figure 6.17 below.

At the neutral point, the zero-sequence currents I_o in the three phases combine to give $3I_o$ in the neutral grounding impedance. The zero-sequence impedance voltage at the neutral point is given by:

$$V_o = 1/3\ (V_{an} + V_{bn} + V_{cn}) = V_n \text{ But } V_n = 3I_oZ_n$$

Therefore, $V_o = 3I_oZ_n$, and $Z_o = V_o/I_o = 3\,Z_n$

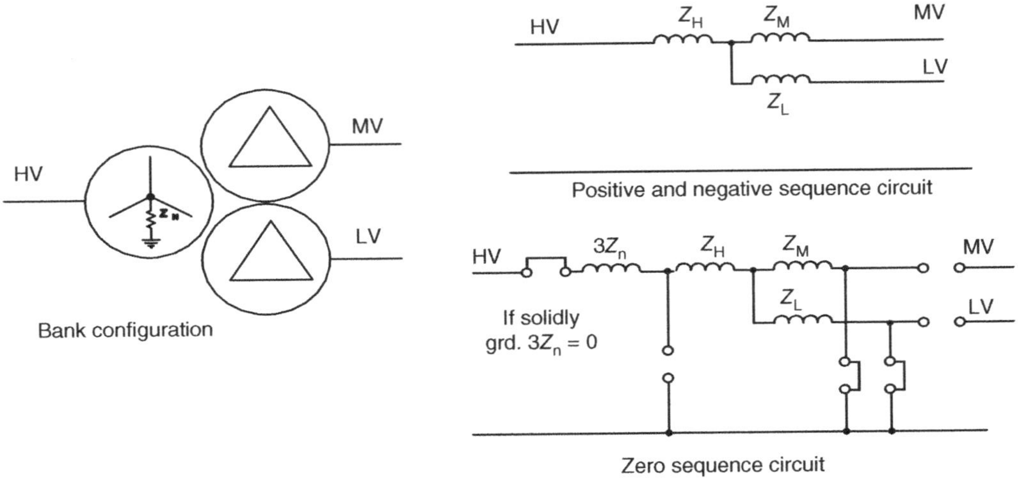

Figure 6.16 Three winding transformer zero-sequence equivalent circuit model.

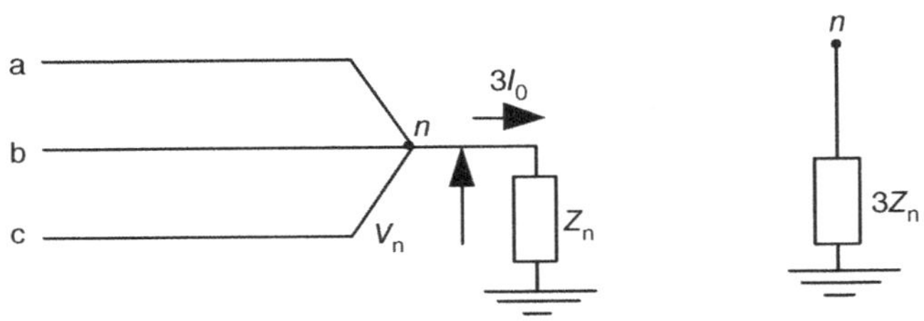

Figure 6.17 Neutral grounding impedance model.

As an example, refer to the above zero-sequence circuit for a Y grounded, through Z_n, delta–delta transformer.

6.5.3.4 Banks of Three, Single-Phase Transformers

When three individual 1-Ph transformers are connected to form a three-phase bank that is used in the power system, the individual nameplate PU or percent impedance will be the Z_T, the leakage impedance but on the three-phase MVA base and the system L-L kV.

For illustration, consider 3, 1-Ph. Transformers each nameplate rated as:

$$20\ \text{MVA}, 66.5\ \text{kV}: 13.8\ \text{kV}, X = 10\%$$

Converting into actual Ohm,

$$X_{TH}\ (\text{on the HV side}) = 0.1 \times Z_{base} = 0.1 \times \left(66.5^2/20\right) = 22.11\ \Omega$$

$$X_{TL}\ (\text{on the LV side}) = 0.1 \times Z_{base} = 0.1 \times \left(13.8^2/20\right) = 0.952\ \Omega$$

Let the transformers be connected in wye on the HV windings to a 115 kV system, while the LV windings are connected in delta to form a 13.8 kV system.

Then by definition, for three-phase transformer bank

$$X_{TH}\ (\text{on the HV side}) = 0.1 \times Z_{base} = 0.1 \times \left(115^2/60\right) = 22.11\ \Omega\ \text{Primary}$$

$$X_{TL}\ (\text{on the LV side}) = 0.1 \times Z_{base} = 0.1 \times \left(13.8^2/60\right) = 0.317\ \Omega$$

Table 6.3 Typical transformer impedances.

Transformer type	Typical PU impedance (ONAN)
Large load station transformers	0.2–0.25
Wind farm transformer	0.05 PU
Autos 230/115	9%@118 kV on its MVA rating
500/230	13%@240 kV on its MVA rating

Remembering that the reactance is phase-to-neutral Ohm, the individual reactance is identical on the primary side. But, on the LV side, the individual reactance of 0.952 Ω is across the delta (13.8 kV).

The equivalent wye impedance is given by:

$$(0.952 \times 0.952)/(3 \times 0.952) = 0.952/3 = 0.317\,\Omega \text{ same as derived for three-phase bank.}$$

Some typical transformer Impedances are provided in Table 6.3 above; however, actual values should be used.

6.5.3.5 Typically Used Transformers and Models

The modeling of transformers for representation in positive, negative, and zero-sequence circuits can be found in typical power system engineering books. Therefore, a complete list of such has been excluded. However, the following have been provided due to their predominant use in the industry (Figures 6.18–6.21).

6.5.3.5.1 Auto Transformers

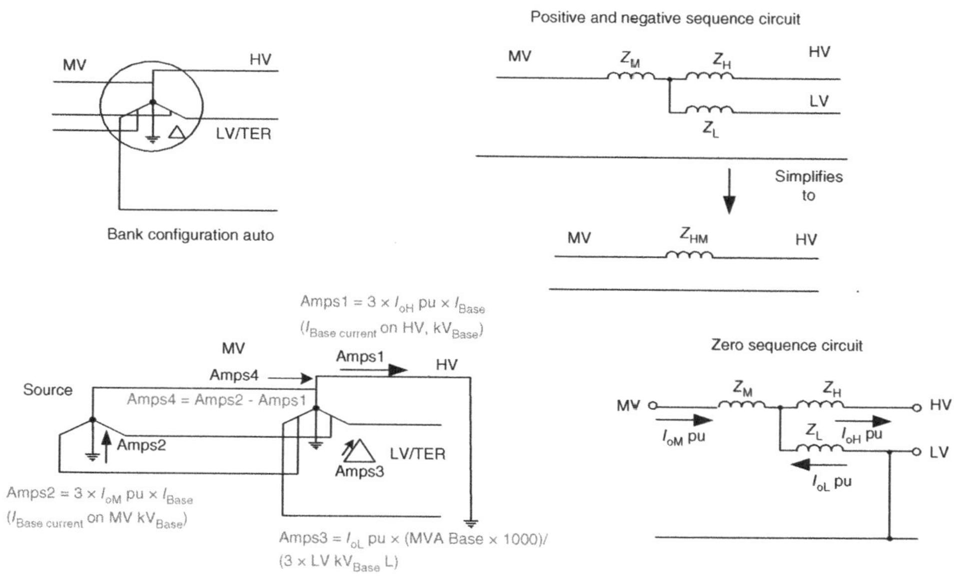

Figure 6.18 Auto transformer equivalent circuits.

6.5.3.5.2 Wye grd – Zig-Zag

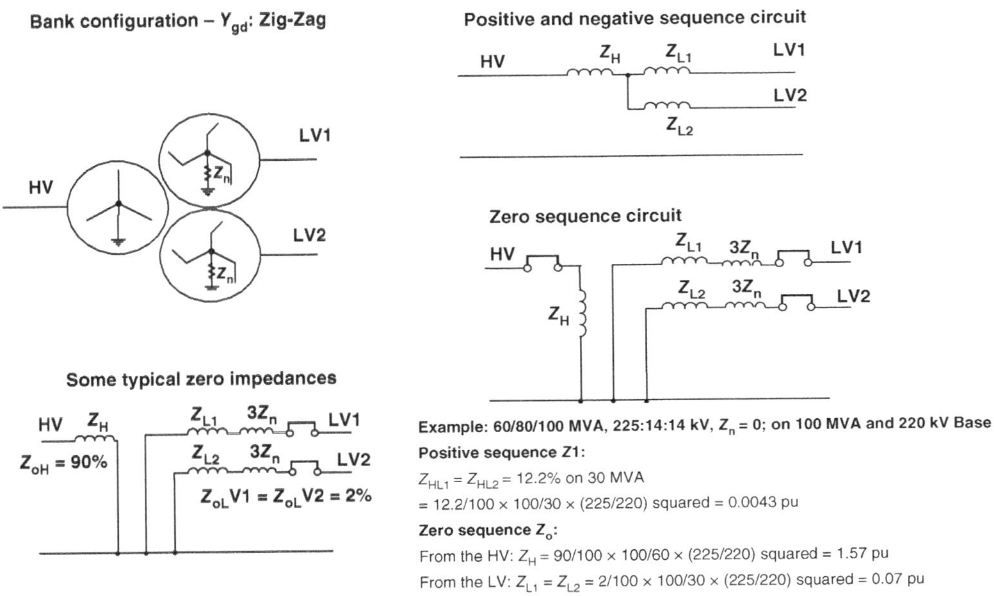

Figure 6.19 Wyegrd zig-zag transformer equivalent circuits.

6.5.3.5.3 Wye–Wye–Delta

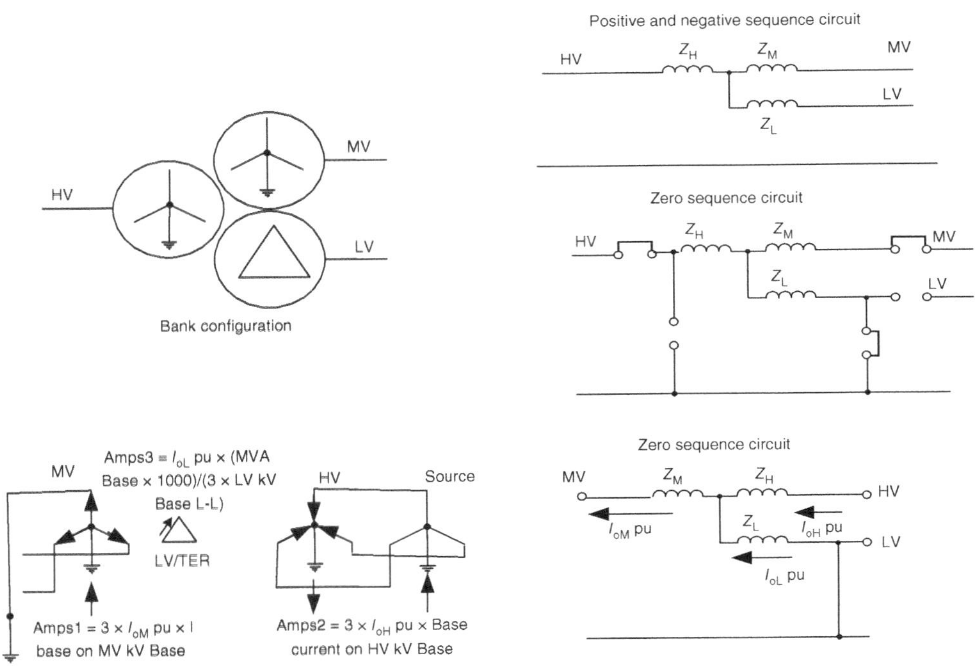

Figure 6.20 Wye–wye–eelta transformer equivalent circuits.

6.5.3.5.4 Grounding Transformer

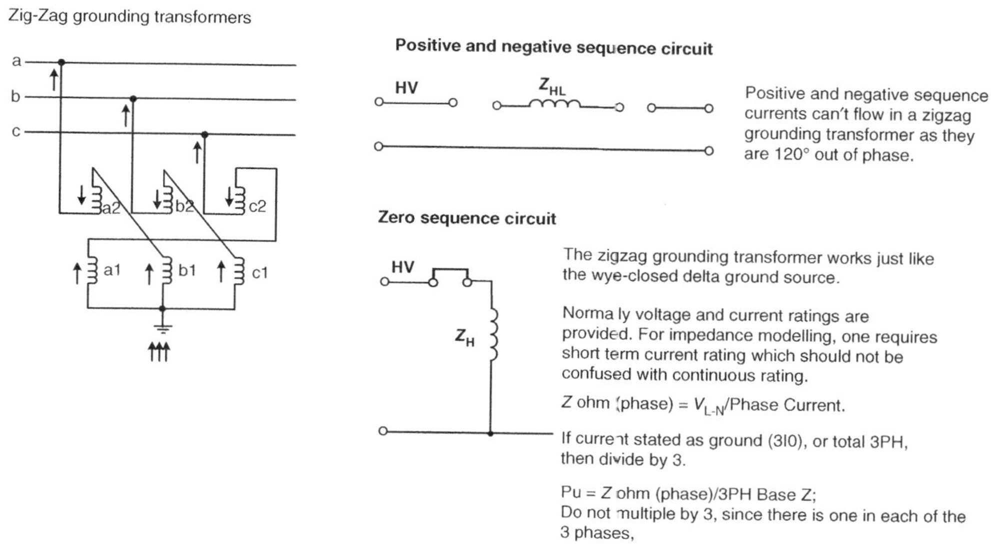

Figure 6.21 Grounding transformer equivalent circuits.

6.6 Balanced Fault Analysis

The severity of faults occurring on a power system network is usually measured in terms of short circuit level which is given by $3X$ (nominal system voltage in kV_{l-n}) × (maximum fault current in kA) MVA.

In order of increasing severity, the various faults are as follows:

1. Single phase-to-ground fault
2. Phase-to-phase fault
3. Double phase-to-ground fault
4. Three-phase fault

A three-phase fault is a balanced fault, while the others are unbalanced faults as they do not involve all the three phases. The single line-to-ground fault occurs most frequently, while the three-phase fault is the least frequent. However, the system should be designed based on the most severe conditions that it can be exposed to.

Protection practitioners are most concerned with three-phase faults, generally, the most severe, and the line-to-ground fault, being the most dominant fault type in the system; normally, 80% of all fault types are line-to-ground.

6.6.1 Balanced Fault Calculations

A balanced three-phase system can be represented by a single-phase equivalent circuit. It consists of one phase and the neutral return path which is not normally drawn in the circuit diagram. Practically, there may not be a neutral conductor since the sum of the three phases sum to zero at the neutral point.

With the total system represented by a one-line diagram and its component values expressed in per-unit values on a common base, one could solve for currents and voltages under a three-phase fault at any given location. This involves the reduction in the network to a single equivalent impedance to be placed in series with a voltage to calculate the current at the fault location. Working back with this fault, current through the network, currents, and voltages at other locations are determined. This retracing task is facilitated by the use of distribution factors.

6.6.2 Simplifying Assumptions

In the majority of fault studies, the following assumptions are made to simplify calculations for both balanced and unbalanced faults:

(a) Load currents are considered negligible compared to fault currents, and therefore, an unloaded system is considered except in case of an open phase fault condition.
(b) All generated voltages are assumed to be in phase and equal in magnitude to the nominal terminal voltage.
(c) Resistance in the network elements is neglected, for manual calculations.
(d) Transformer magnetizing currents and line charging currents are ignored.

6.7 Sequence Networks

The system is assumed balanced except at the point of a fault. This means the impedances of the network elements in all three phases including mutual impedances and generated voltages are equal, and the three sequence components are independent and do not react with each other.

The system can be therefore considered as being comprised of three networks, one for each sequence component. Currents and voltages of only one sequence will exist in each network. These are called sequence networks, and their interconnection represents the unbalance caused by faults.

The positive sequence network (PSN) includes the source voltages and the impedances of the network elements.

The negative sequence network (NSN) is a duplicate of the PSN but excludes voltages as no generator is operating with a reverse phase sequence. The other variant is that the NS reactance of the rotating machinery may be different from the PS reactance.

The zero-sequence network (ZSN) differs from the NSN due to the ground current contributions. The transformers present various values for their zero-sequence impedances depending upon the type and winding connections and the grounding arrangements or lack of grounding.

6.7.1 Sequence Network Interconnections

6.7.1.1 Principle of Interconnections

There are six quantities to be considered at the point of fault, V_R, V_W, V_B and I_R, I_W, I_B. From these, the relationship between the sequence components of the reference phase (V_1, V_2, V_0, I_1, I_2, I_0) can be determined if any one of the two following conditions is met: (i) if any three are known provided they are not all voltage or current and (ii) if any two are known and two others known to have a specific relationship.

The relationships are called circuit constraints. These constraints determine the type of interconnections among the sequence networks. It is the adopted practice to derive the relationships with the red phase (A Phase) as the reference phase and the faults are chosen to be balanced relative to the reference phase.

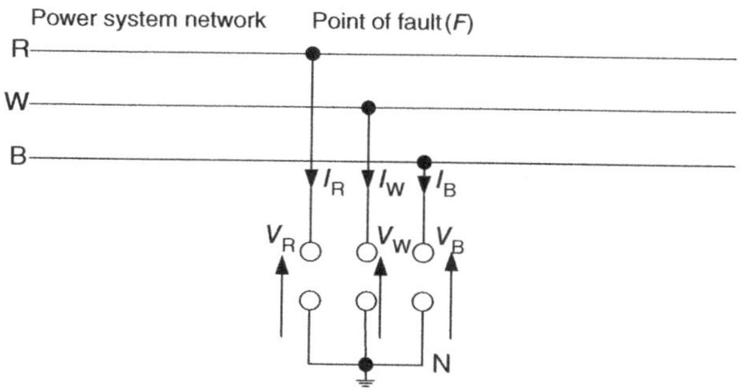

Figure 6.22 Network equivalent at the fault location.

6.7.1.1.1 *Representation Techniques*

Under three-phase fault analysis, the system in each of the networks behind the point of fault can be reduced to a single equivalent impedance representation. It is then convenient to show the sequence networks as blocks with fault terminals F and N for external connections.

To facilitate the expression of fault conditions, three imaginary leads of zero impedance are connected to the three phases at the point of fault. These leads are connected together or to ground to represent the various types of faults, and from the conditions set up by these connections, the relationships are derived. This is illustrated in Figure 6.22.

Consider a simple system as per Figure 6.23 with a fault at F, for which the positive, negative, and zero-sequence networks are shown.

The box representation of sequence networks is shown as follows (Figure 6.24).

6.7.1.1.2 *Three-Phase Fault*

Three-phase faults represent a balanced fault type. Therefore, a balanced three-phase network can be replaced by a single-phase equivalent circuit to determine the three-phase fault currents, and symmetrical components are not needed to solve three-phase faults.

However, to continue with the concept of using the box representation of sequence networks which are further used for other unbalanced fault types below, the three-phase sequence connections are depicted below for a three-phase fault (Figure 6.25).

The conditions at the point of fault: $V_R = V_w = V_B, I_R + I_W + I_B = 0\, I_{a0} = 0, V_{a1} = 0, V_{a2} = 0$

This indicates that the positive and negative sequence networks are to be shorted without interconnecting them. The zero-sequence circuit network is open as the ground is not involved. Also, there is no negative sequence voltage so that the equivalent circuit is only the positive sequence. The fault current I_F in all the three phases will be equal in magnitude to I_1 but displaced in angle by 120°.

6.7.1.1.3 *Single Phase-to-Ground Fault*

The conditions at the fault point: $V_r = 0, I_w = I_b = 0$

$$V_r = V_1 + V_2 + V_0 = 0 \tag{6.13}$$

$$I_0 = \frac{1}{3}\left(I_r + I_w + I_b\right) = \frac{1}{3}I_r \tag{6.14}$$

$$I_1 = \frac{1}{3}\left(I_r + aI_w + a^2 I_b\right) = \frac{1}{2}I_r \tag{6.15}$$

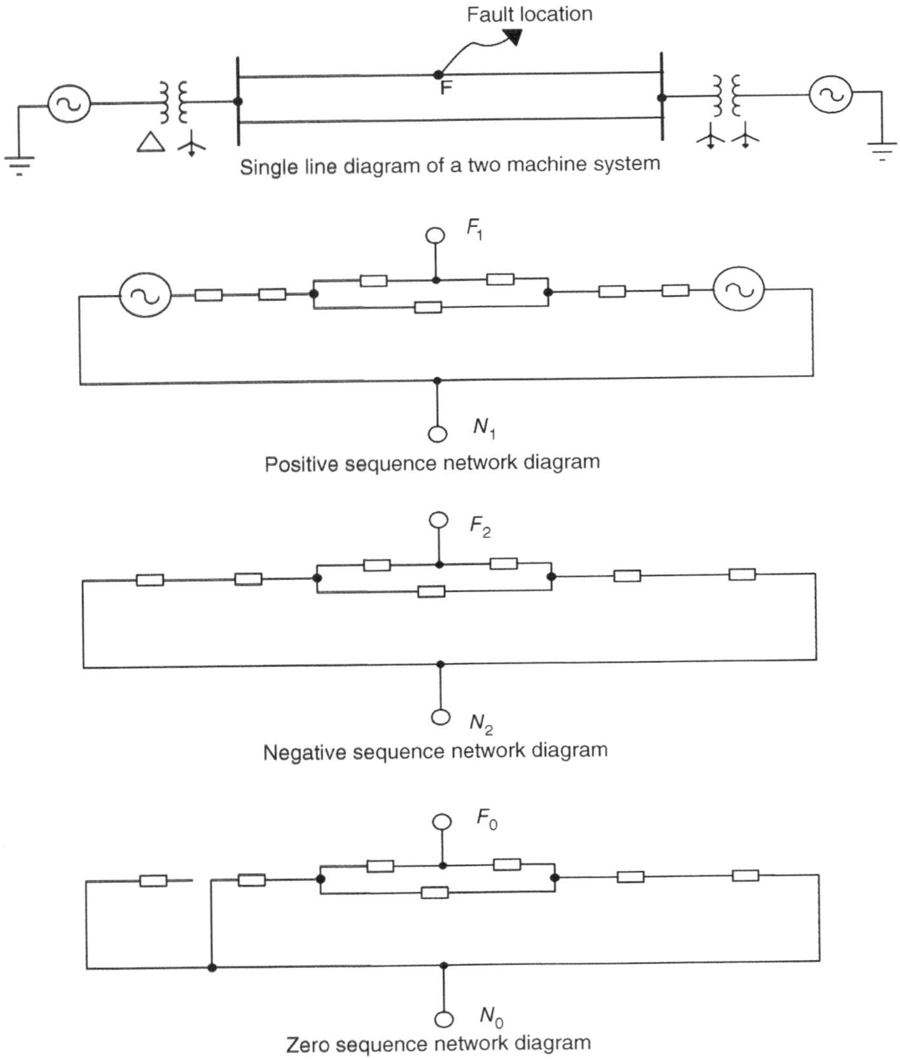

Figure 6.23 A simple system and its sequence network diagrams.

$$I_2 = \frac{1}{3}\left(I_r + a^2 I_w + a I_b\right) = \frac{1}{3} I_r \tag{6.16}$$

$$I_1 = I_2 = I_0 = 1/3\, I_r \tag{6.17}$$

Equations (6.13) and (6.17) are the circuit constraints that call for the sequence networks to be connected in series at the fault point as shown (Figures 6.26 and 6.27).

Now, consider a single phase-to-ground fault through fault impedance Z_F as shown below:

$$I_0 = \frac{1}{3}\left(I_r + I_w + I_b\right) = \frac{1}{3} I_r \tag{6.18}$$

The conditions at the fault point: $V_r = I_r\, Z_F\, I_w = I_b = 0$.

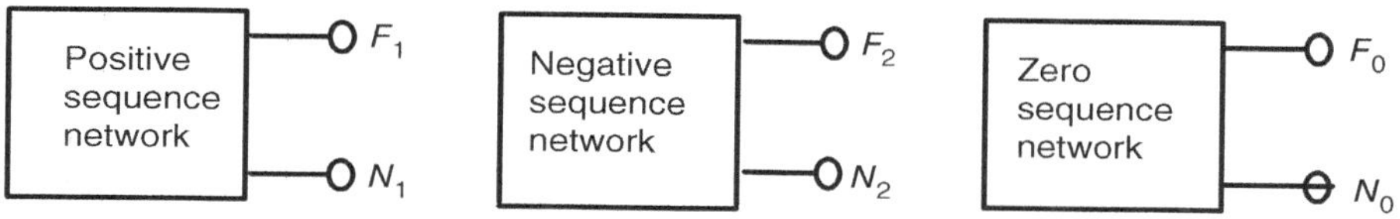

Figure 6.24 Sequence equivalent network blocks.

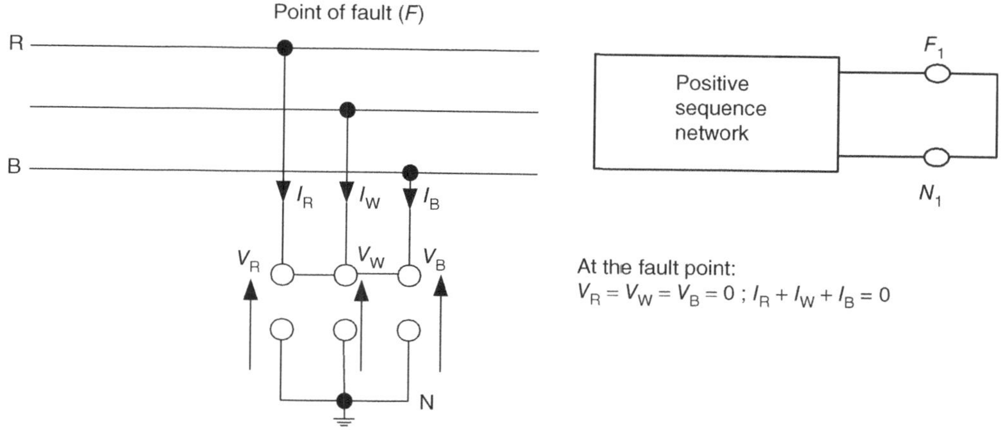

Figure 6.25 Three-phase sequence network connections.

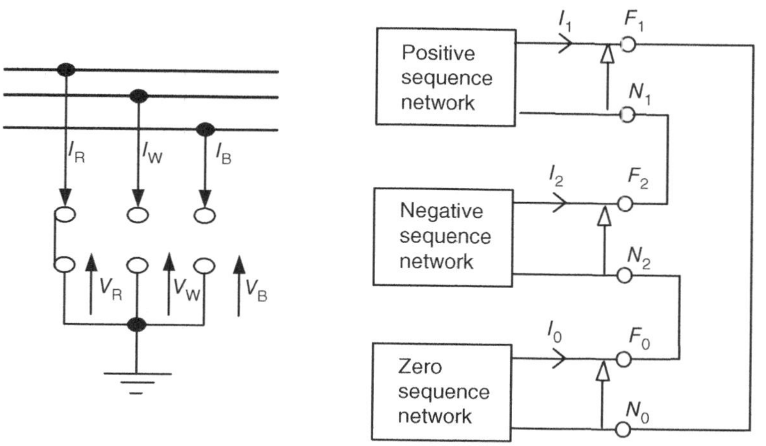

Figure 6.26 Line-ground sequence network connections.

Given that:

$$I_1 = \frac{1}{3}\left(I_r + aI_w + a^2 I_b\right) = \frac{1}{3}I_r \tag{6.19}$$

$$I_2 = \frac{1}{3}\left(I_r + a^2 I_w + aI_b\right) = \frac{1}{3}I_r \tag{6.20}$$

$$I_1 = I_2 = I_0 = 1/3\, I_r \tag{6.21}$$

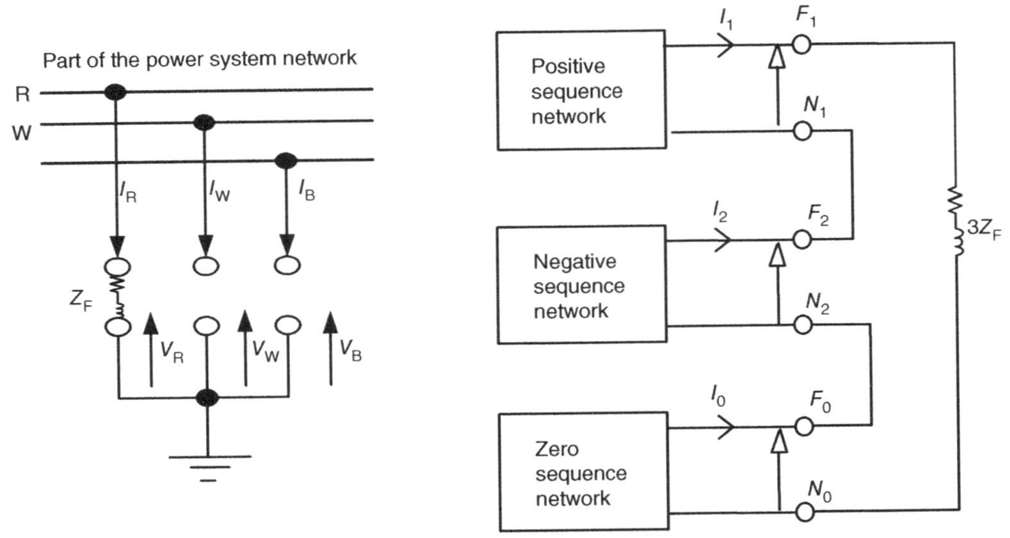

Figure 6.27 Line-ground fault through fault impedance, sequence network connections.

$$V_r = V_1 + V_2 + V_0 = I_r Z_F \text{ But } I_r = 3I_0$$

$$\text{Therefore, } V_1 + V_2 + V_0 = I_0 \left(3Z_F\right) \tag{6.22}$$

Equations (6.21) and (6.22) are the circuit constraints that call for the sequence networks to be connected in series at the fault point as shown above.

6.8 Summary of Unbalance Fault Calculations

A simple example will be used to describe the process.

Draw a One-Line Diagram or Single-Line (Figure 6.28):

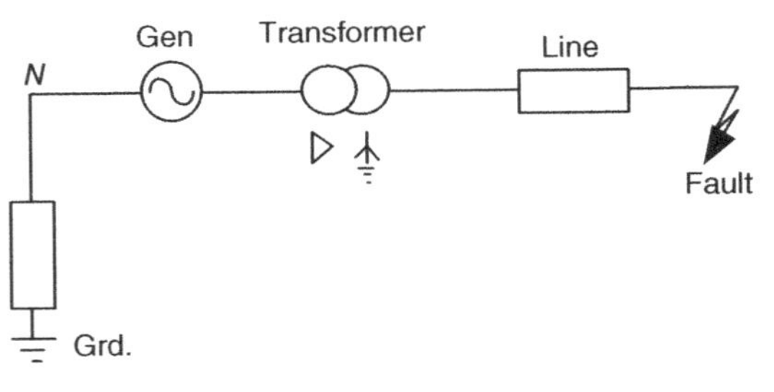

Figure 6.28 Example one-line diagram.

6.8.1 Positive Sequence Diagram

(1) Start with the neutral point, all generators and load neutrals are connected to N_1.
(2) Include all source voltage (phase-to-neutral voltages)
(3) Impedance network, positive sequence impedance per phase
(4) Diagram finishes at the fault point F_1 (Figure 6.29).

V_1 = Positive Sequence PH-N Voltage at Fault Point.
I_1 = Positive sequence Phase Current Flowing into F_1
$V_1 = E_1 - I_1\ (ZG_1 + ZT_1 + ZL_1)$

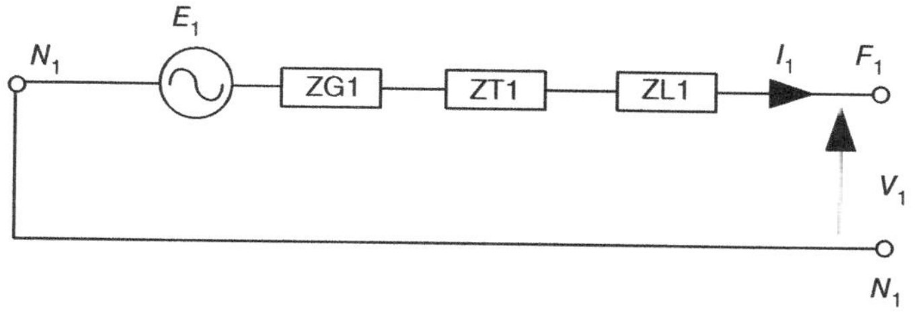

Figure 6.29 Positive sequence diagram.

6.8.2 Negative Sequence Diagram:

(1) Start with the neutral point, all generators and load neutrals are connected to N2.
(2) No voltage included, and no negative sequence voltage is generated.
(3) Impedance network, negative sequence impedance per phase
(4) Diagram finishes at the fault point F_2 (Figure 6.30).

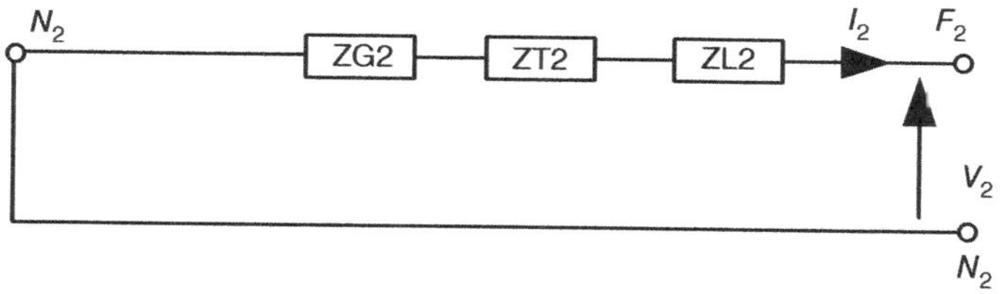

Figure 6.30 Negative sequence diagram.

V_2 = Negative Sequence PH-N Voltage at Fault Point.
I_2 = Negative Sequence Phase Current Flowing into F_2
$V_1 = -I_1\ (ZG_2 + ZT_2 + ZL_2)$

6.8.3 Zero-Sequence Diagram

For "In Phase" (zero-phase sequence) currents to flow in each phase of the system, there must be a fourth connection (this connection is typically the neutral or ground connection) (Figure 6.31).

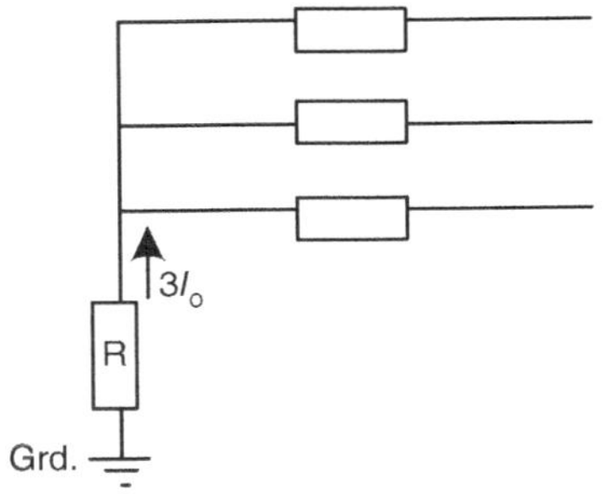

Special consideration required for transformers, e.g. delta–wye transformer
Thus, equivalent single-phase zero-sequence diagram (Figure 6.32):

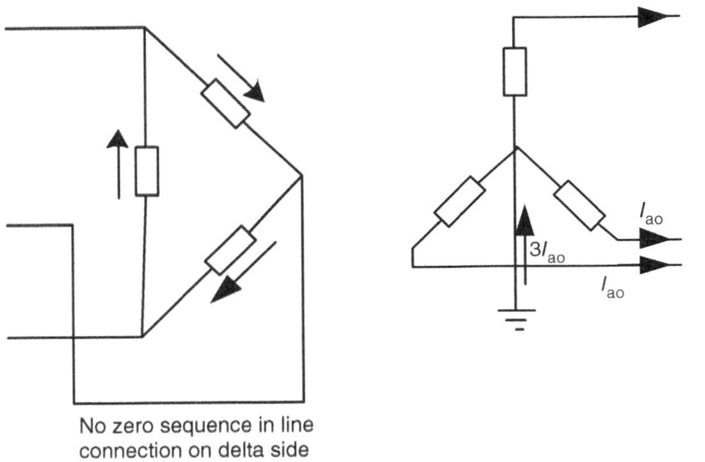

Figure 6.31 Delta–wye transformer, three-phase representation.

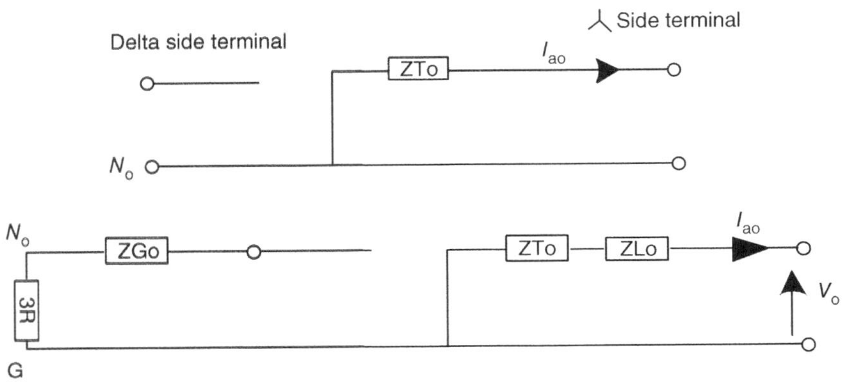

Figure 6.32 Zero-sequence diagram.

V_0 = zero-sequence Ph-G voltage at the fault point
I_0 = zero-sequence current flowing into F_0
$V_0 = -I_0\,(ZT_0 + ZL_0)$ (Figure 6.33).

6.8.4 Conduct the Fault Study

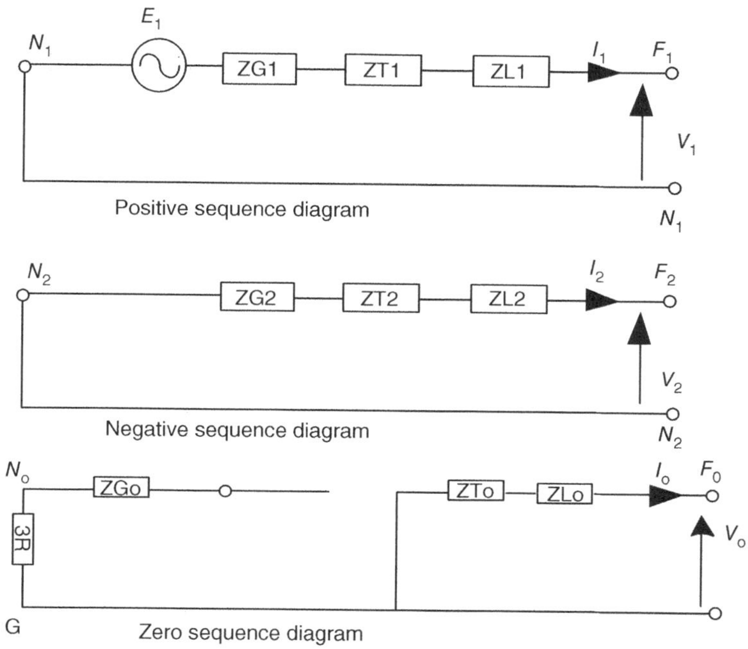

Figure 6.33 The sequence diagrams.

6.9 High-Level Summary of the Fault Calculation Process

The following is intended to illustrate the steps needed to conduct fault studies for the purpose of calculating relay thresholds. The following steps have been developed based on using a "hand" calculation method. Some of these steps may not be required if one is using software tools and an official system base case(model), where the system and model integrity has been deemed appropriate for such use.

6.9.1 Develop a Single-Line Diagram of the Studied Area

Develop/draw a single-line diagram of the area under study that includes the protected element(s). The number of elements included in any particular study will vary based on the topology of the system and the number of system equivalents needed to accurately represent the integrated system.

The single line should be labeled with the appropriate power system elements and depict the relay location(s).

6.9.2 Determine the Studied Power System Element Impedances

Collect and/or determine the power system element impedances – positive and zero-sequence, negative sequence is normally assumed to be equal to the positive sequence. Such data should be added to the single-line diagram or a separate impedance diagram.

Possible data sources: Utility equipment information database, equipment test sheets, manufacturer information, sometimes assumptions are based on similar or typical values.

Select and document on the single-line, the base system of voltage and the apparent power that will be used to determine the per-unit impedance of the power elements- normally 100 MVA and 118.05/230/500 kV are used for the transmission system. Convert all power system element impedances to the selected base values.

6.9.3 Develop Sequence Impedance Models

Develop and draw the three sequence models for the studied area – positive, negative, and zero-sequence networks. For each network, determine the appropriate system equipment model and sequence representation for it. Include for each sequence network, the appropriate sequence impedance for each of the respective sequence networks.

6.9.4 Determine the Fault Types and System Conditions

Using the above-developed single-line drawing, make multiple copies. Depict the fault type(s) and the system conditions for which the relay scheme electrical quantities need to be determined to assure trip dependability and security.

6.9.5 Conduct the Fault Studies and Determine Relay Quantities

Fault calculations are based on reducing the networks, whether for balanced or unbalanced faults, to a single impedance diagram, and then interconnecting them appropriately based on symmetrical component theory to calculate the various fault types.

Faults are normally calculated using one per-unit pre-fault voltage at 0° displacement from the horizontal (reference phase). Once the total fault current is determined, the current distribution is calculated to determine the electrical quantities where the relays are located.

Calculating fault studies for a small system where it can be reduced to a handful of buses by using system equivalents is very useful. However, for medium to large systems, it is very time-consuming and not practical to do so. Therefore, off-the-shelf computer applications that are optimized for fault calculation are generally used by larger utilities.

6.10 Useful Fault Calculation Formulas/Methods

6.10.1 Conversion from Short Circuit Values to System Impedances

Often, short circuit values are provided for three-phase and L-G faults at a bus. With that information it becomes possible to find system impedances at that given point to calculate faults for a system network further from this point.

Three-Phase Faults

$$I_{3\phi} = \frac{1000\ \text{MVA}_{\text{SC}}}{\sqrt{3}\ \text{kV}} \tag{6.23}$$

$$\text{MVA}_{\text{SC}} = \frac{\sqrt{3} I_{3\phi}\ \text{kV}}{1000} \tag{6.24}$$

$$Z_{\text{Ohms}} = \frac{V_{\text{L-N}}}{I_{3\phi}} = \frac{1000\ \text{kV}}{\sqrt{3}\, I_{3\phi}} = \frac{\text{kV}^2}{\text{MVA}_{\text{SC}}} \tag{6.25}$$

$$Z_{\text{PU}} = \frac{Z_{\text{Ohms}}}{Z_{\text{base}}} = \frac{\text{kV}^2 \text{MVA}_{\text{base}}}{\text{kV}^2 \text{MVA}_{\text{SC}}} = \frac{\text{MVA}_{\text{base}}}{\text{MVA}_{\text{SC}}} \tag{6.26}$$

The system positive sequence impedance $Z_1 = Z_2$ is thus given by:

$$\frac{\text{MVA}_{\text{base}}}{\text{MVA}_{\text{SC}}} \tag{6.27}$$

L-G Fault

$$\text{MVA}_{\phi\text{GSC}} = \frac{\sqrt{3} I_{\phi\text{G}} \text{kV}}{1000} \tag{6.28}$$

where $I_{\phi\text{G}}$ is the total L-G fault current and kV is the system L-L voltage in kilovolts.

$$I_{\phi\text{G}} = \frac{1000\ \text{MVA}_{\phi\text{GSC}}}{\sqrt{3}\ \text{kV}} \tag{6.29}$$

$$\text{But, } I_{\phi\text{G}} = I_1 + I_2 + I_0 = \frac{3V_{\text{L-N}}}{Z_{\text{G}}} \quad \text{where } Z_{\text{G}} = Z_1 + Z_2 + Z_0 \tag{6.30}$$

$$Z_{\text{G}} = \frac{3V_{\text{L-N}}}{I} = \frac{\sqrt{3}\text{kV}3V_{\text{L-N}}}{1000\ \text{MVA}_{\phi\text{GSC}}} = \frac{3\text{kV}^2}{\text{MVA}_{\phi\text{GSC}}}\Omega \tag{6.31}$$

The system zero-sequence impedance Z_0 is thus given by:

$$Z_0 = Z_{\text{G}} - Z_1 - Z_2 = Z_{\text{G}} - 2Z_2 \tag{6.32}$$

6.11 Fault Calculation Examples

6.11.1 Three-Phase Fault Example

Find the three-phase fault current in each section for the system depicted below in Figure 6.34.
Using the Steps as discussed in Section 6.9:

Step 1. Develop a single-line diagram of the studied area; it has already been developed and is depicted in Figure 6.34 below.

Step 2. Determine the studied power system element impedances

Covert all impedances to per unit, for simplicity, ignore all resistance values; select 50 MVA base and 132 kV base, resulting in the following:

Feeder:

$S_{\text{base}} = 50\ \text{MVA}$, $V_{\text{base}} = 27.5\ \text{kV}$, $Z_{\text{base}} = 27.6^2/50 = 15.24$

$I_{\text{Base}} = (50 \times 1000)/(1.732 \times 27.6) = 1046$

Feeder impedance $= 8\ \Omega/15.24 = \text{j}0.525$ PU

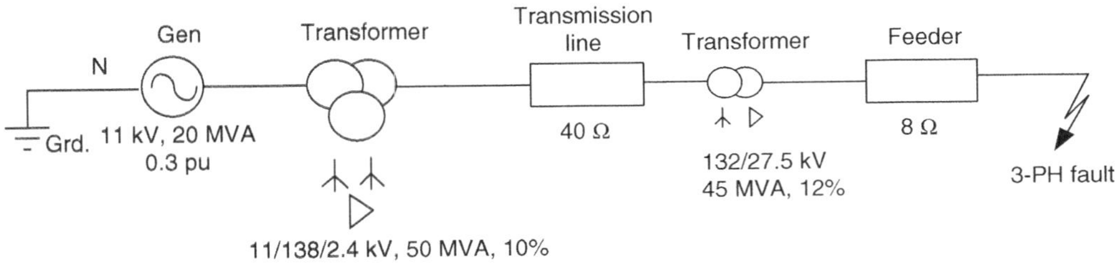

Figure 6.34 Single-line diagram of the studied area.

Feeder Transf.:

$S_{\text{base}} = 50$ MVA, $V_{\text{base}} = 132/27.5$ kV

Z_{base} LV $= 27.6^2 2/50 = 15.24$, Z_{base} HV $= 132^2/50 = 348.5$

I_{Base} LV $= (50 \times 1000)/(1.732 \times 27.6) = 1046$

I_{Base} HV $= (50 \times 1000)/(1.732 \times 132) = 218.7$

Feeder transformer impedance provided = 12% or 0.12 PU on rating

On 50 MVA base and since the voltage ratio is the same as the base voltage: $0.12 \times 50/45 = j0.133$ PU

Line:

$S_{\text{base}} = 50$ MVA, $V_{\text{base}} = 132$ kV, $Z_{\text{base}} = 1322/50 = 348.5$

$I_{\text{Base}} = (50 \times 1000)/(1.732 \times 132) = 218.7$

Line impedance $= 40\,\Omega/348.5 = j0.115$ PU

Gen Transf.:

1. $S_{\text{base}} = 50$ MVA, $V_{\text{base}} = 132$ kV
2. Z_{base} HV $= 1322/50 = 348.5$
3. I_{Base} HV $= (50 \times 1000)/(1.732 \times 132) = 218.7$
4. Source transformer impedance provided = 10% or 0.10 PU on rating
5. On 50 MVA base and on 132 kV: $0.10 \times 50/50 \times (138/132)2 = j0.109$ PU

   Gen:
6. $S_{\text{base}} = 50$ MVA, V_{base}, must maintain transformer ratio $= (11/138) \times 132 = 10.52$ kV
7. Z_{base} HV $= 10.52^2/50 = 2.21$
8. $I_{\text{Base}} = (50 \times 1000)/(1.732 \times 10.52) = 2744$
9. Gen impedance provided = 0.3 PU on rating; on 50 MVA and 10.52 kV $= 0.3 \times (50/20) \times (11/10.52)^2 = j0.82$ PU

Step 3. Develop sequence impedance models

Since the assignment is to calculate a three-phase fault, only the positive sequence network is required. A three-phase fault represents a balance type fault, and therefore, symmetrical components are not required.

Step 4. Determine the fault types and system conditions

For this example, a single three-phase fault will be conducted at the remote end of the feeder, under normal system conditions, that is, all power system elements in service.

Step 5. Conduct the fault studies and determine relay quantities

The system impedance network of Figure 6.35 is reduced to a single voltage source and impedance, Thevenin equivalent, as shown in Figure 6.36 below.

The total fault current equals $V_{\text{PU}}/Z_{\text{total PU}} = 1.0/j1.7 = -j0.59$ PU

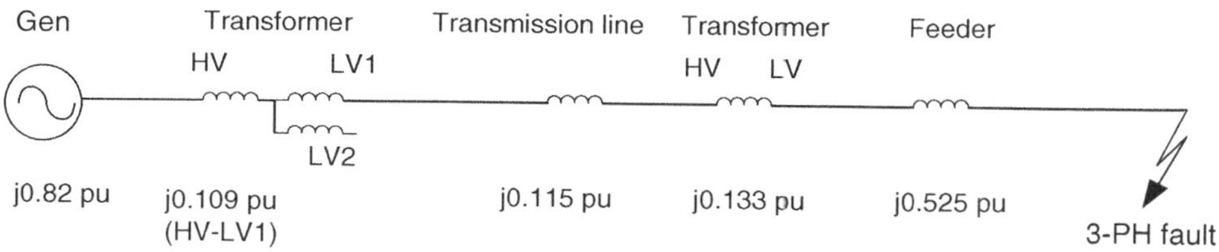

Figure 6.35 Impedance diagram of the studied area.

Figure 6.36 Simple three-phase balanced fault.

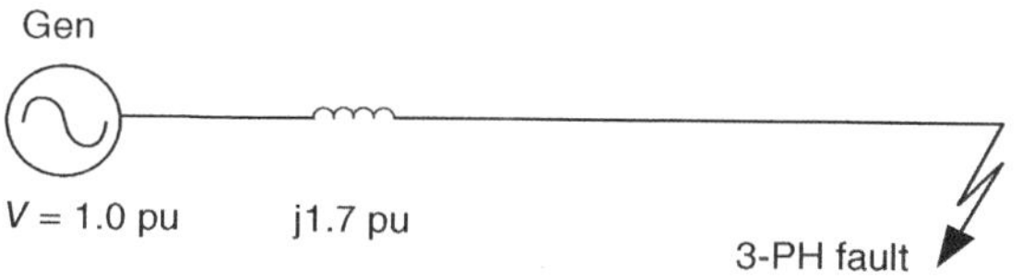

Fault current in amps: $(I_{PU} \times I_{Base})$
Feeder: $0.59 \times 1046 = 617$ A
Line: $0.59 \times 218.7 = 129$ A
Gen: $0.59 \times 2744 = 1619$ A
As a check: The turns ratio of generator transformer is $138 : 11 = 12.55 : 1$
Amp-turns balance needs $12.55 \times 129 = 1619$ A on the source side.

6.11.2 Line-to-Ground Fault Example

Given the following power system network: two, 230 kV terminals (A and B) with two transmission circuits as depicted below in Figure 6.37.

The task is to determine the impedance "seen" by a ground mho distance relay at Bus A, at Terminal A1, with a fault location at Bus B.

The fault calculations will be derived based on normal system conditions which translate to all power system elements remaining in service.

The impedance networks for this example network are as follows, with impedances presented in per unit on 100 MV and 230 kV base (Figure 6.38–6.41).

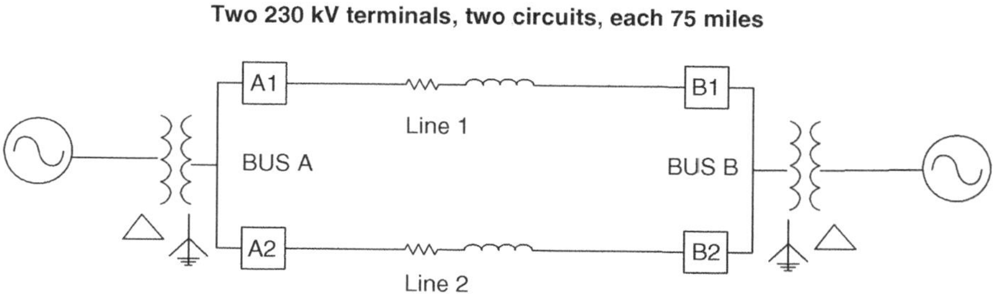

Figure 6.37 L-G fault example power system.

6.11.2.1 Positive Sequence

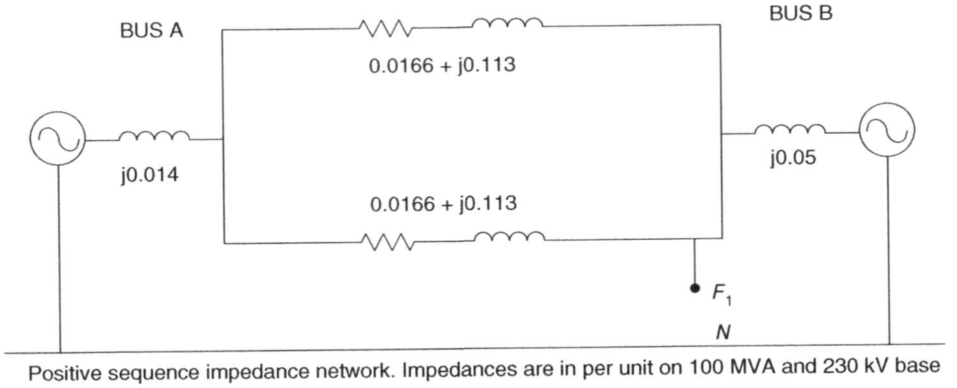

Figure 6.38 Positive sequence network for the L-G example.

6.11.2.2 Negative Sequence Network

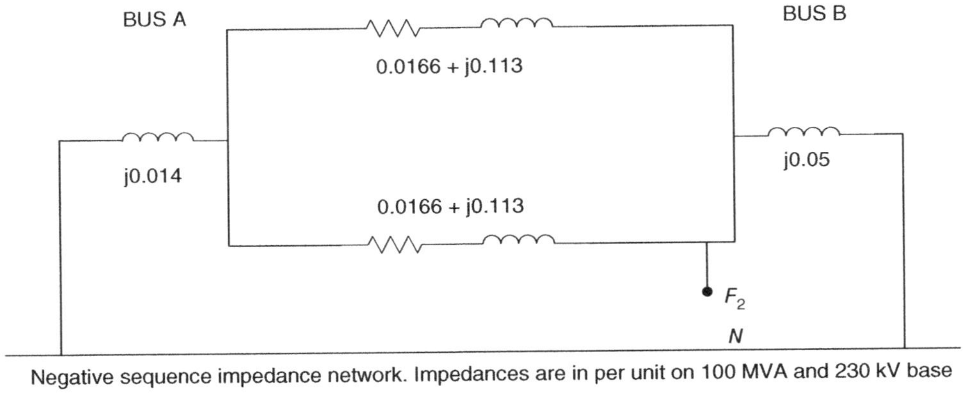

Figure 6.39 Negative sequence network for the L-G example.

6.11.2.3 Zero-Sequence Network

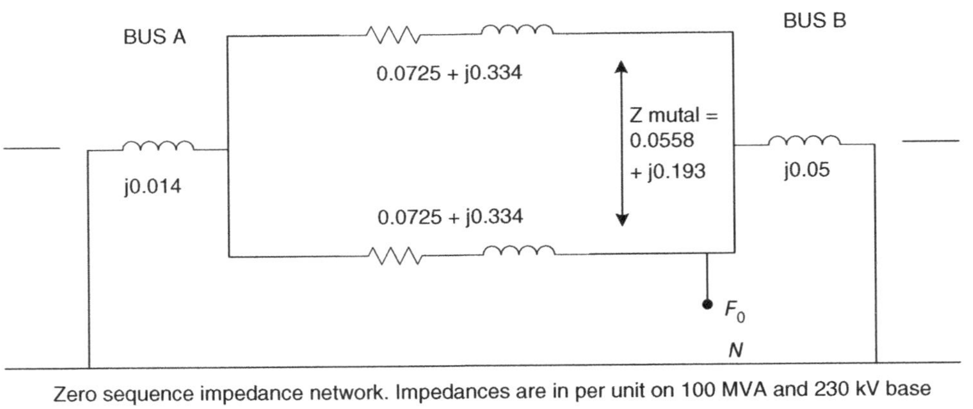

Figure 6.40 Zero-sequence network for the L-G example.

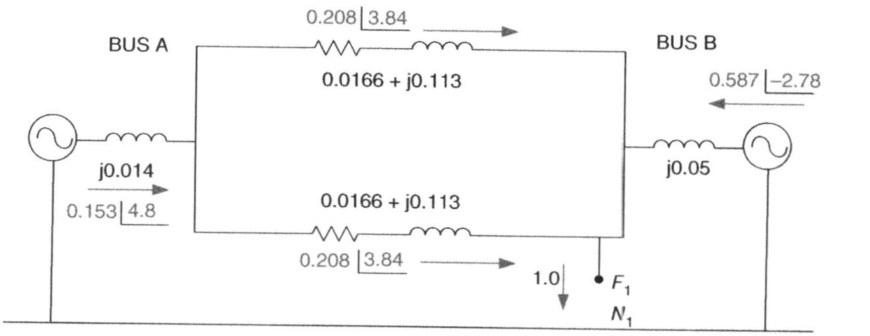

Positive sequence impedance network. Impedances are in per unit on 100 MVA and 230 kV base

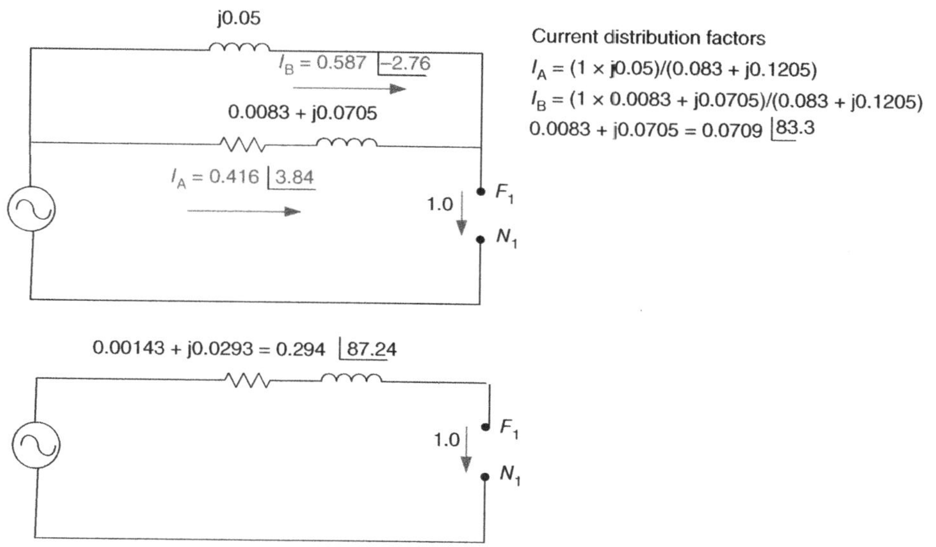

Figure 6.41 Reduction in the positive sequence network for the L-G example.

6.11.2.4 Reduction in the Positive Sequence Network

In strong interconnected systems the post fault source voltages remain at 1PU. Therefore it will be assumed that the two positive sequence voltages are 1.0 per unit so that the positive sequence network will reduce to 1.0 voltage in series with a single impedance.

The negative sequence distribution factors will be just the same as the positive sequence except there are no voltage sources. Also, the negative sequence current distribution factors will be the same as the positive sequence (Figure 6.42).

6.11.2.5 Reduction in the Zero-Sequence Network

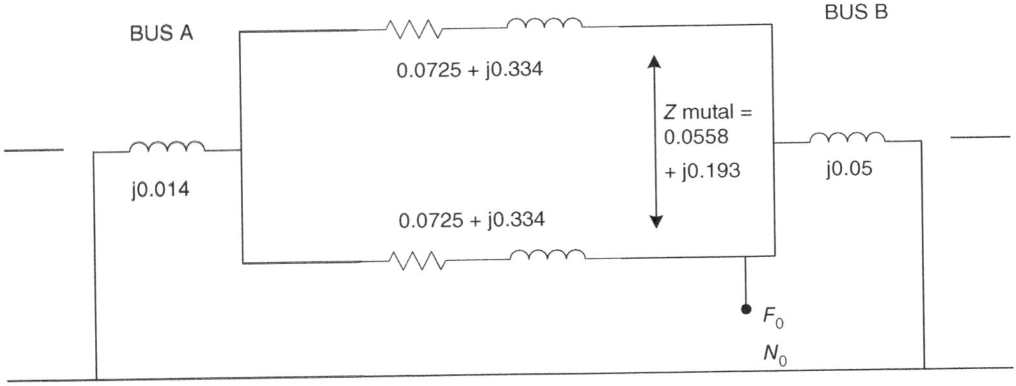

Zero sequence impedance network. Impedances are in per unit on 100 MVA and 230 kV base

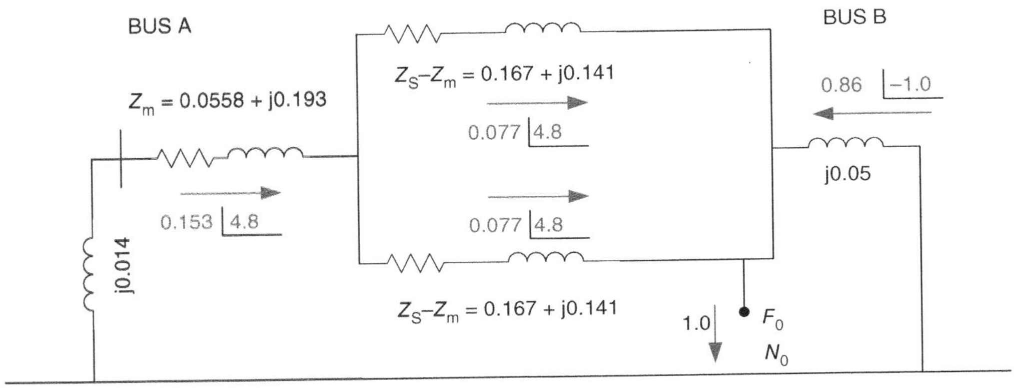

Zero sequence impedance network. Impedances are in per unit on 100 MVA and 230 kV base

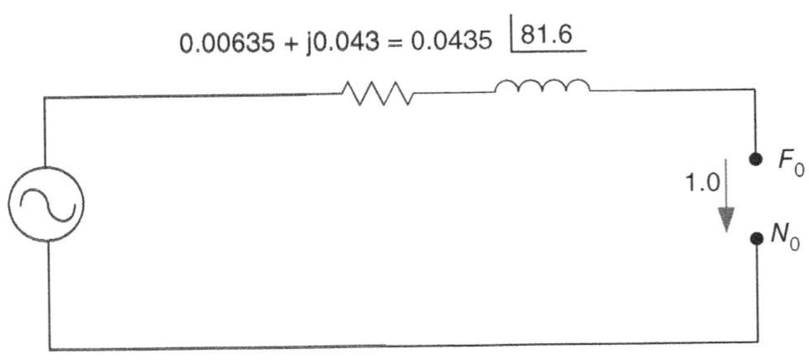

Figure 6.42 Reduction in the zero-sequence network for the L-G example.

6.11.2.6 Calculate the L-G Fault

As per Section 6.7.1.1.3, L-G -type faults require the series connection of the positive, negative, and zero-sequence networks, refer to Figure 6.43 as follows:

$I_1 = I_2 = I_0 = V/(Z_1 + Z_2 + Z_0) = 1.0/(0.00921 + j0.1016) = 9.8\ @ -84.8°$ PU

On 230 kV, $I_{base} = (100\ \text{MVA} \times 1000)/(1.732 \times 230\ \text{kV}) = 250$ A

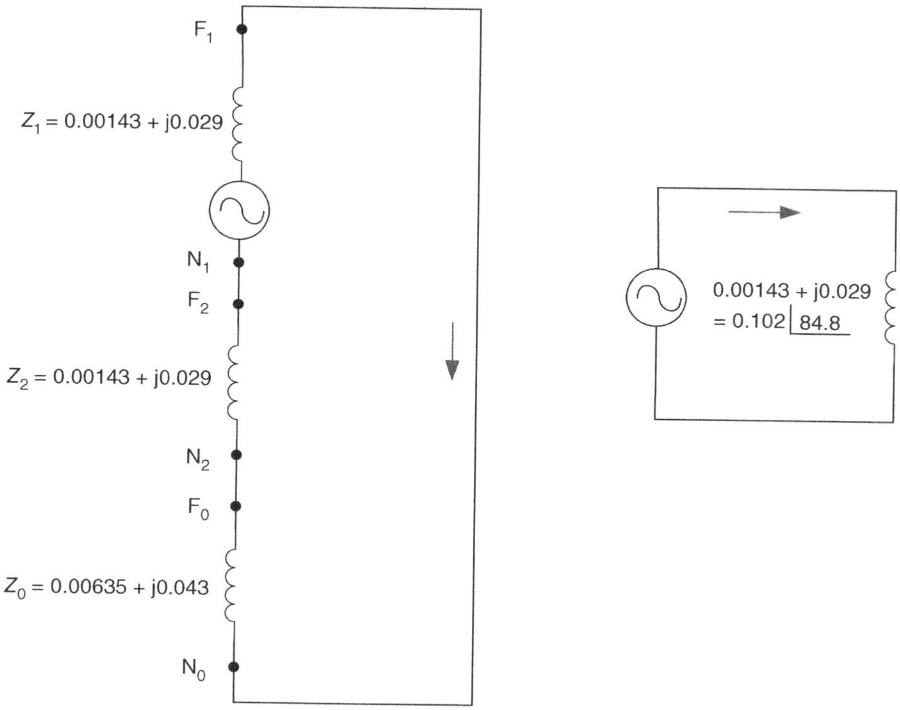

Figure 6.43 Connections of the sequence networks for the L-G example at Bus B.

Distribution factor = DF

Source A fault contribution

$I_A = \mathrm{DF}_1 \times I_1 + \mathrm{DF}_2 \times I_2 + \mathrm{DF}_0 \times I_0$

$I_A = 0.416@3.84 \times 9.8@{-84.8} + 0.416@3.84 \times 9.8@{-84.8} + 0.153@4.8 \times 9.8@{-84.8}$

$I_A = 0.985\ @4.0 \times 9.8\ @84.8 \times 250 = 2420@{-80.8}°$ amps

$I_B = a^2\, I_1 + a\, I_2 + I_0$

$I_B = [a^2\ \mathrm{DF}_1 + a\ \mathrm{DF}_2 + \mathrm{DF}_0] \times 9.8@{-84{,}8}°$

$I_B = [1@240 \times 0.416@3.94 + 1@120 \times 0.416@3.94 + 0.154@4.8] \times 9.8\ @\ {-84.8}$

$I_B = -0.262@4.0 \times 9.8@{-84.8} \times 250$

$I_B = -650@{-80.8}$ A

$I_C = aI_1 + a^2 I_2 + I_0$

$I_C = [a\mathrm{DF}_1 + a^2\ \mathrm{DF}_2 + \mathrm{DF}_0] \times 9.8@{-84{,}8}°$

$I_C = -0.262@4.0 \times 9.8@{-84.8} \times 250$

$I_C = -650\ @\ {-80.8}°$ A

The terminals A_1 and A_2 line flows will of course split evenly in this case as shown in Figure 6.44 below.From terminal B by similar reasoning:

$I_A = 2.03@{-2.5} \times 9.8@{-84.8} \times 250 = 4960@{-87.3}$ A

$I_B = 0.265@{-2.0} \times 9.8@{-84.8} \times 250 = 650@{-86.8}$ A

$I_C = 0.265@{-2.0} \times 9.8@{-84.8} \times 250 = 650@{-86.8}$ A

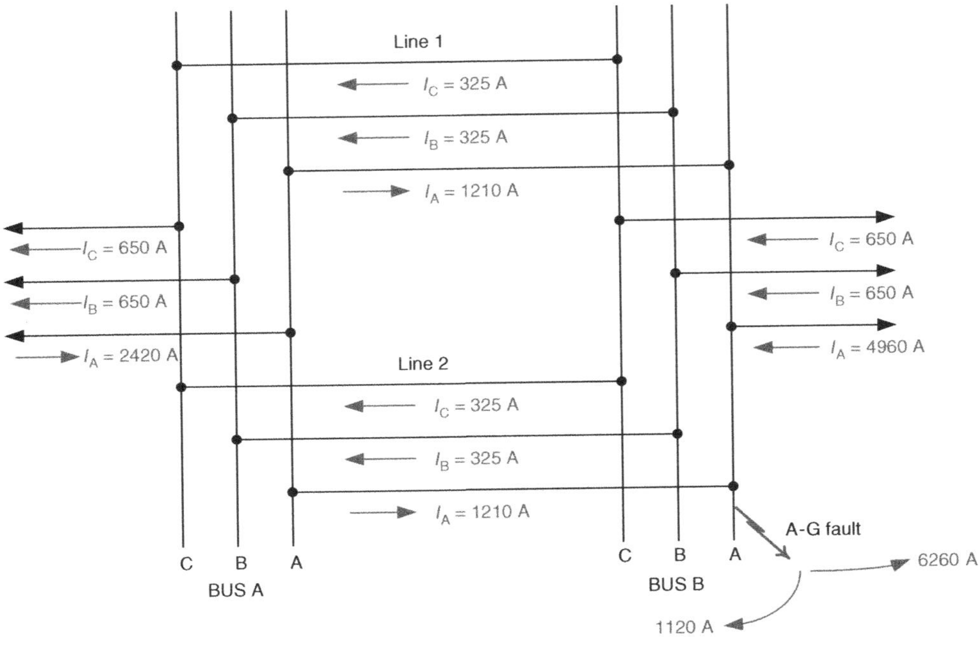

Figure 6.44 Bus A and B current distribution for an L-G at Bus B.

Voltages at Bus A

Ignoring the resistance

$V_{a1} = E_{a1} - I_1 Z_1$

$V_{a1} = 1.0 - (0.98@{-90} \times 0.415) \times 0.14@90$

$V_{a1} = 0.943 \text{ PU}$

$V_{a2} = -I_2 Z_2$

$V_{a2} = -0.057 \text{ PU}$

$V_{a0} = I_0 Z_0$

$V_{a0} = -(9.8@\text{-}90 \times 0.153) \times 0.014@90$

$V_{a0} = 0.021 \text{ PU}$

$V_A = V_{a1} + V_{a2} + V_{a0}$

$V_A = 0.943 - 0.057 - 0.021$

$V_A = 0.865 \text{ PU} = 199 \text{ kV}$

Impedance Seen by a Ground Distance Relay at Terminal A1.

Assuming the relay is feed by Y-connected PTs and line currents.

A relay that is supplied with line-to-neutral voltage and line current plus a proportion of the residual current will measure and can be set for the straight positive sequence impedance of the line. This concept is further expanded in Section 14.4.2.

Therefore, a ground distance relay measures the following, which permits the ground and phase distance elements to be set to cover the same percentage of the line:

$$
\begin{aligned}
Z_{\text{relay Grd.}} &= V_A / \left[I_A + I_{a0} \times \left(Z_0 - Z_1/Z_1\right)\right] \\
&= 0.865 \ @ \ 90 / \left[4.83 + 0.755 \times (0.334 - 0.113/0.113)\right] \\
&= 0.137 \ @ \ 90^\circ
\end{aligned}
$$

It should be noted that the positive sequence impedance of the line is 0.113, and the ground distance element measures an impedance of 0.137; this is due to the mutual impedance between the two parallel lines which causes an apparent effect.

References

1 Charles L. Fortescue Method of symmetrical co-ordinates applied to the solution of polyphase networks, AIEE Transactions, 37, part ll, 1027–1140, Presented at the AIEE in Atlantic City, N.J. 1918.

2 J. Lewis Blackburn, *Symmetrical Components for Power System Engineering*, New York: Marcel Dekker Inc. 1993

3 Paul M. Anderson, *Analysis of Faulted Power Systems*, Wiley-IEEE Press, 1995

4 William D. Stevenson, *Elements of Power System Analysis*, New York, San Francisco, Toronto, London: McGraw-Hill Book Company 1982

7

Protection Zones

7.1 Protection Zones General

The philosophy of protection design is to divide the power system into protective zones that should be protected adequately with the minimum amount of system disconnected when only the faulted zone trips. These protection zones should cover the entire power system completely, leaving no part unprotected. When a fault occurs, protection is required to select and trip only the circuit breakers needed for fault clearing, as in most applications, it is the breakers that define the zone.

7.2 Zones Defined

Power systems are designed with the understanding that protection systems will trip defined parts of the system for various fault locations. The defined parts of the system that are protected are what are termed protection zones. For example, a generator can be defined as being a single protection zone. Therefore, should a fault occur in the generator then the generator synchronizing breaker will be tripped and the generator itself will be shut down. If the generator and main output transformer are defined as a single protection zone, then for a fault in either element, the breakers on the high side of the main output transformer will be tripped and the generator will be shut down. From a protection perspective, it is not relevant how the system zones are defined as much as the need to ensure that the protections match those system zones. Properly designed protection systems achieve the selectivity that matches the zones conceived by the system planners.

A protection zone is a region of the power system which defines its protection jurisdiction. In the example of zones of protection in Figure 7.1, the generator, bus, and line protection zones each have their respective boundaries. It is essential that zones of protection overlap so that no portion of the power system ever remains unprotected. This is especially important for circuit breakers as breakers cannot identify a fault as opposed to fuses.

Protections are typically designed with two independent protection systems such that overall, the protections can be relied upon to isolate the faulted zone even when one of the protections fails to either detect the fault or attempt to trip the circuit breakers. For this purpose, one of two approaches may be adopted. In the first approach, the protected zone is covered at the same location with similar but independent protections that mirror each other. This is known as local backup also referred to as redundancy. In the second approach, there is primary protection intended to only cover the protected zone and backup protection intended to cover more than one zone. In many cases, the backup protection is located remotely from the local zone of protection and covers multiple protected zones, this is known as remote backup. For example, referring to Figure 7.1, the protection

Power System Protection: Fundamentals and Applications, First Edition. John Ciufo and Aaron Cooperberg.

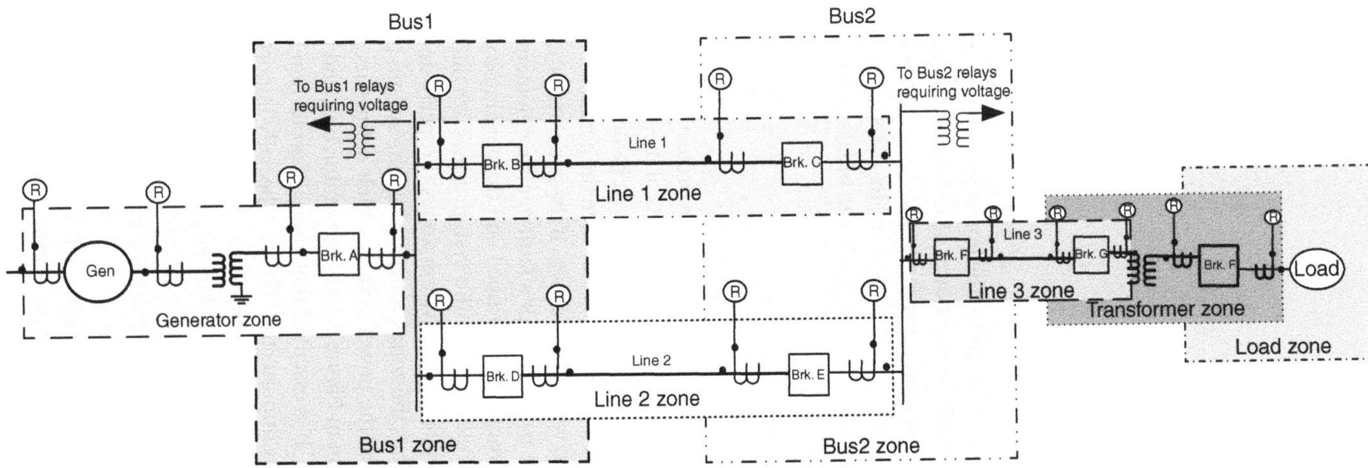

Figure 7.1 Example zones of protection.

system associated with Line 3 at Bus2, for Breaker F, is the local protection for Line 3; its remote backup protection(s) are the protection systems located at Bus1, for Breakers, B and D.

It is common practice, to meet reliability requirements, to use redundant protections to meet protection reliability performance for power system elements operating above 200 kV and with remote backup protections for 115 kV and lower system voltages.

The shaded zones indicated in Figure 7.1 are the primary protection zones. For a fault in the primary protection zone, all the breakers belonging to that zone would be tripped. Generally, the current transformers (CTs) providing current to the protection relays define the boundaries of the zones. Adjacent protection zones are made to overlap thereby providing complete coverage. A fault in the overlapping zones trips many more breakers as all the breakers associated with each zone are tripped. For this reason, overlapping protection zones should if possible be made smaller by adding more breakers. Where the remote breakers are included in the protection zone, they are tripped via communication.

The location of a current measuring relay defines the protection zone for directional overcurrent measuring relays. Voltage measuring relay location defines the protection zone for impedance measuring relays.

7.3 Zone Overlap Around Breakers

As circuit breakers define the boundaries of the protection zones, it is advantageous to place CTs on either side of each of the breakers and to overlap the protection zones around those breakers. For this reason, breakers almost always come equipped with CTs built into their bushings as shown in Figure 7.2.

It is useful from a protection perspective, to zone the protections around breakers while measuring current from CTs on those breakers. Other power system elements, such as generators and transformers, are also provided with bushing CTs for similar reasons (Figure 7.3).

For protections to cover all system elements to isolate them from the power system when they are faulted requires that protection zones overlap. Without zone overlap, what is known as blind spots are created for which no protection exists. Since breakers define the isolation points of the zone of protection, there is no better way to provide zone overlap then to overlap around the breakers that define the zone. Most utilities specify breakers and transformers with CTs located within their bushings. This application permits zoning the protections around the protected equipment with the

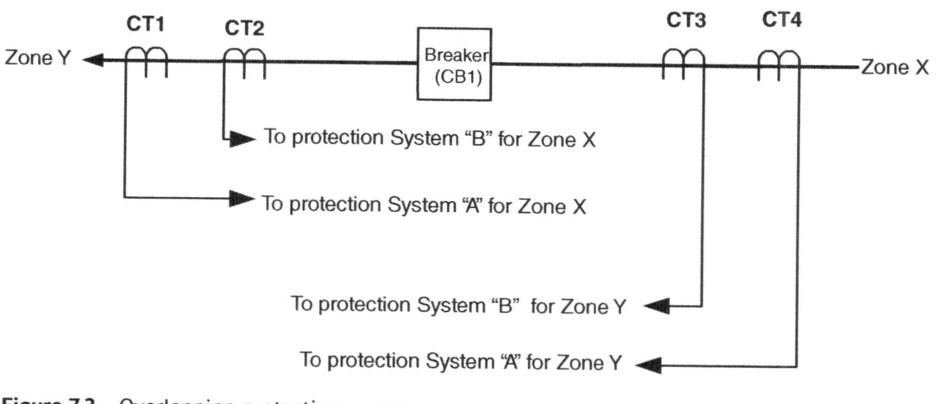

Figure 7.2 Overlapping protection zones.

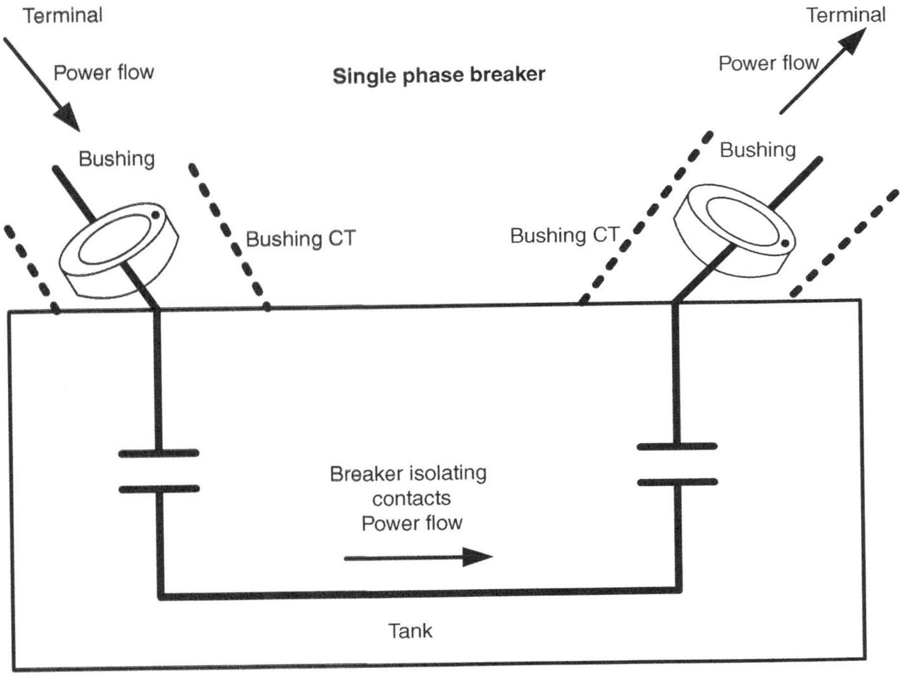

Figure 7.3 Circuit breaker with bushing CTs.

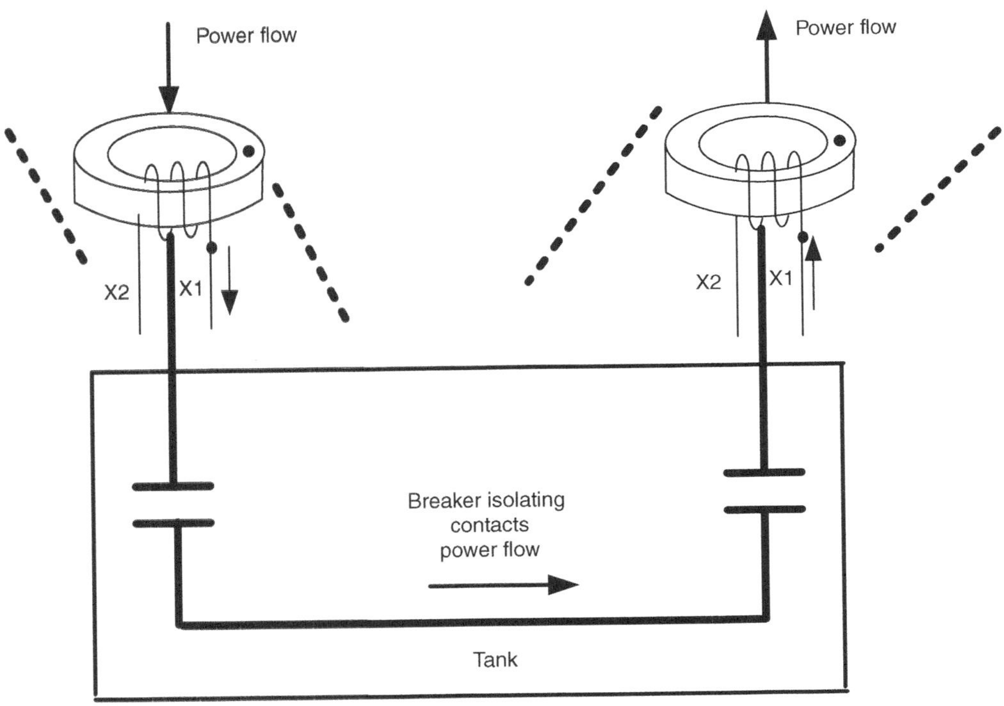

Figure 7.4 An illustration of breaker bushing CT polarity.

breaker or transformer bushing defining the protection zone. When zoned around breaker bushings, the breaker itself is protected with overlapping protection zones. A fault within the breaker itself produces the tripping of breakers in other zones.

When breakers define the zone of protection for protections that are intrinsically directional, such as directional overcurrent, differential or distance type protections, the spot polarity of the CTs is crucial to the correct operation of the protection. Figure 7.4 illustrates a typical breaker bushing CT polarity, where the spots are on the system side of the incoming and outgoing bushings. This is the industry standard followed by all breaker manufacturers to allow protections to be easily and consistently applied with regard to direction.

7.4 Protection Zoning at Stations

7.4.1 HV Switching Stations

Proper CT connections, taking into account polarities, are required to define the intended protection zone. Take for example Figure 7.5 showing zoning and overlapping zones at a typical switching station.

In this example also shown in Figure 7.6, the bus differential protection R1, having a protection designation number of 87, is zoned by the three diameter breakers and the one bus tie breaker. The CT polarities are such that all infeeds to an internal bus fault shown in Figure 7.6 will add vectorially at the differential node and flow into the relay, for the correct operation of the differential relay for internal bus zone faults.

Similarly, the infeeds and outfeeds to an external bus or line fault will add vectorially at the differential node such that the currents will flow around the CTs and none into the relay. Note that in Figure 7.7, the current outfeeds through the bus tie breaker are in the reverse direction from an internal bus fault in the previous example. This current leaving the bus zone prevents the differential relay from operating for an external fault.

The same principle can be applied to a line fault where the outfeed to a line fault is shown in Figure 7.8.

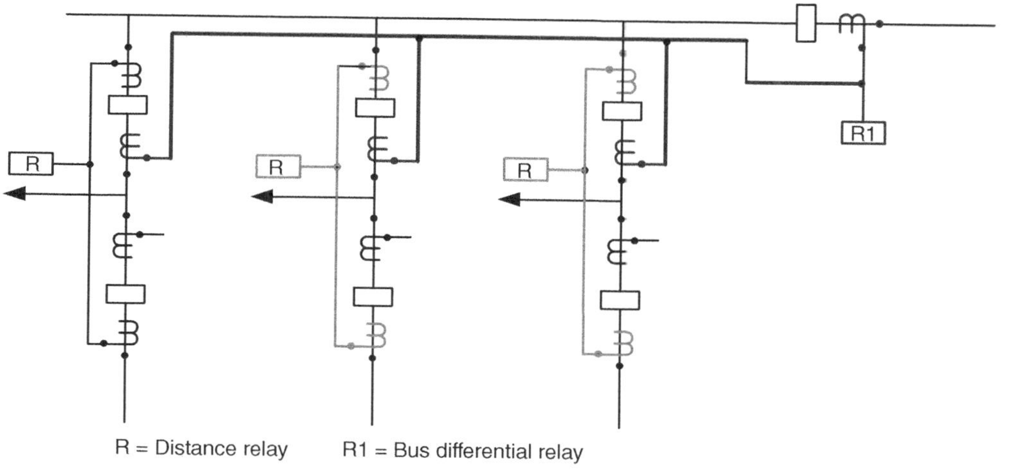

Figure 7.5 Zoning and overlapping zones at a typical HV switching station.

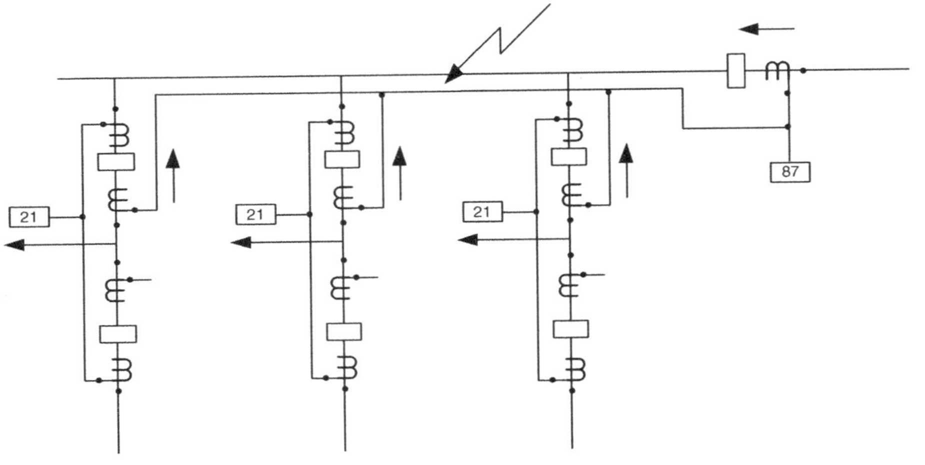

Figure 7.6 Example of current flows for an internal bus fault.

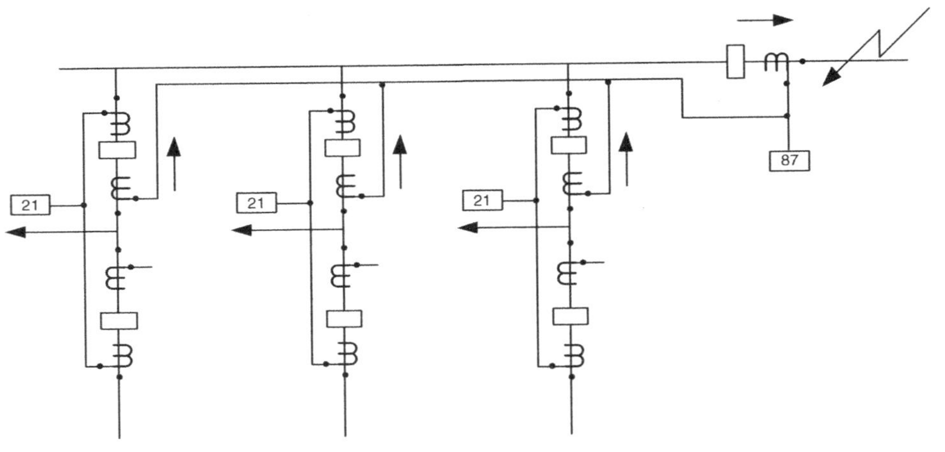

Figure 7.7 Example of current flows for an external bus fault.

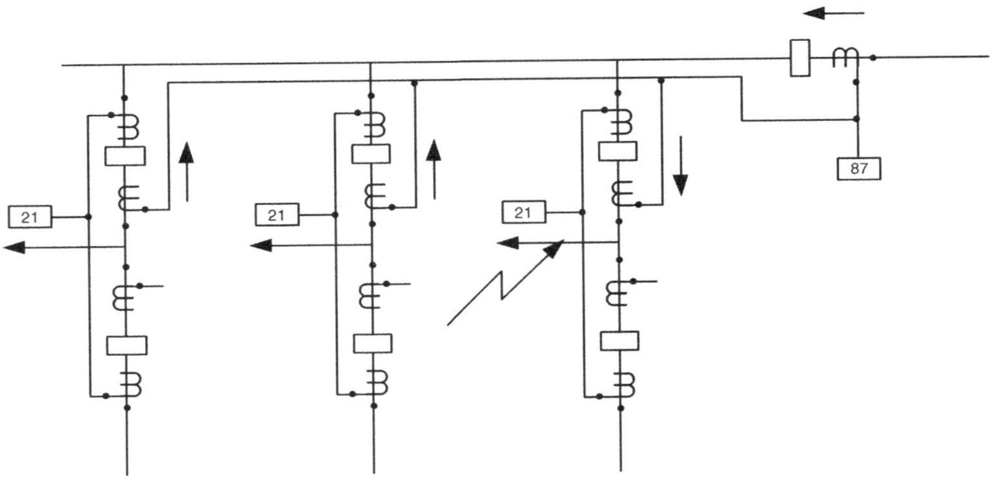

Figure 7.8 Example of current flows for an external line fault.

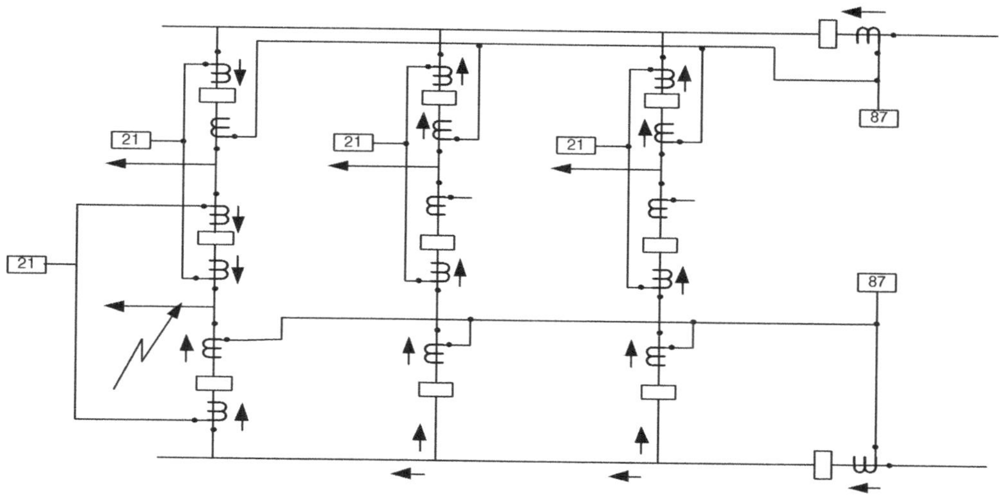

Figure 7.9 Example of current flows for an external line fault showing flows from all buses.

Figure 7.9 shows an external line fault at a switching station where a more complete picture of current flows is depicted. In this example except for the CTs supplying the faulted line protection, all other current flows are into and out of the various protection zones whether lines or the bus. Selective tripping of only the faulted zone is achieved in this manner at switching stations.

7.4.2 LV Distribution Stations

Figure 7.10 shows the various protection zones found at typical load substations. The protection zoning follows the CT connections as illustrated. In this illustration, there are three distinct protection zones: transformer, bus, and feeder.

7.4.2.1 The Transformer Zone

The transformer is a substation-protected zone. For a fault in the transformer, the transformer could be isolated by tripping the line breakers at the terminal station via communication and by tripping the transformer LV breaker. This is applicable where no dedicated transformer high voltage (HV) breaker or circuit switcher is located at the substation. The transformer motor-operated disconnect would then be opened upon completion of opening the remote station breakers. Thus, upon successful isolation of the transformer, the line can be restored to service via automatic reclosing. Similarly, without communication to the remote terminals, a dedicated transformer HV breaker or circuit switcher would be tripped to isolate the transformer from the HV line.

Refer to Figure 7.11 for the distribution of currents for a transformer fault at a substation where the bus tie breaker is operated closed. The transformer is protected using differential protection (87-Trans).

The infeeds to the faulted transformer shown in Figure 7.11 flow into the transformer differential protection via the CT on the HV transformer bushings and the CT on the transformer LV breaker as shown, such that the secondary currents combine to produce a spill current in the differential circuit to operate the relay.

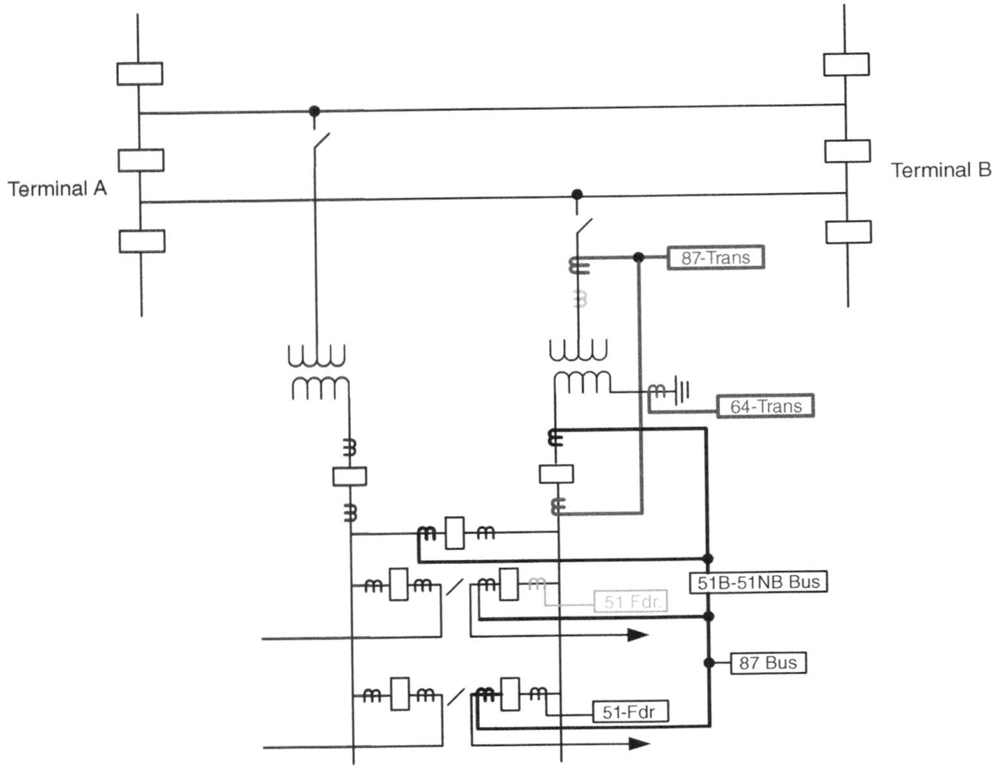

Figure 7.10 Load substation protection zones.

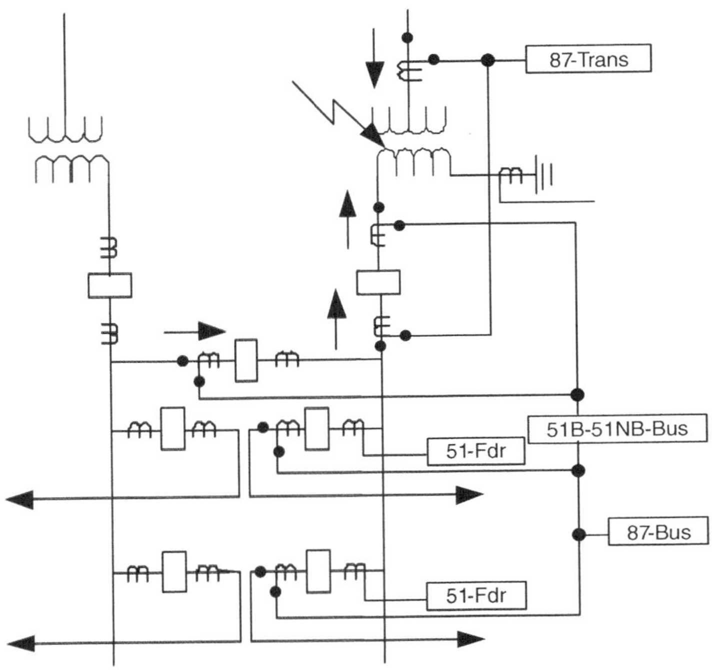

Figure 7.11 Current distributions for a transformer fault at a substation.

The infeeds to the faulted transformer shown in Figure 7.11 flow into the CT on the bus tie breaker but flow out of the CT on the transformer LV breaker such that the secondary currents circulate around the CTs. The transformer differential protection would operate for the transformer fault but not the bus differential protection or bus overcurrent backup protection.

7.4.2.2 The Bus Zone

The bus is another substation-protected zone. For a fault in the bus zone, the bus is isolated by tripping the transformer low voltage (LV) breaker, the bus tie breaker, and all the feeder breakers. Unless the feeders are non-radial or have distributed generation, they do not need to be tripped since they do not provide any fault infeed.

Refer to Figure 7.12 for the distribution of currents for a bus fault at a substation. The substation bus is protected by two protections, a differential (87-Bus) and an overcurrent backup (51B/51NB).

The infeeds to the faulted bus flow into the bus differential protection via the CT on the transformer LV breaker and the CT on the bus tie breaker. The bus overcurrent protection would also operate for a bus fault via the same infeeds as shown.

The infeed to the faulted bus flow into the CT on the transformer HV bushings and out of the CT on the transformer LV breaker such that the secondary currents circulate around the CTs. The bus differential protection would only operate for the bus fault but not the transformer differential protection.

An infeed to the bus fault from distributed generation may trip the feeder protection unless it has been directioned or the infeed is too low, below the pickup setting threshold, to operate the overcurrent relays.

7.4.2.3 The Distribution Feeder Zone

The distribution feeder is yet another substation-protected zone. For a fault on the feeder, the feeder breaker is tripped then automatically reclosed to allow for downstream reclosers or fuses to isolate the faulted portion of the feeder.

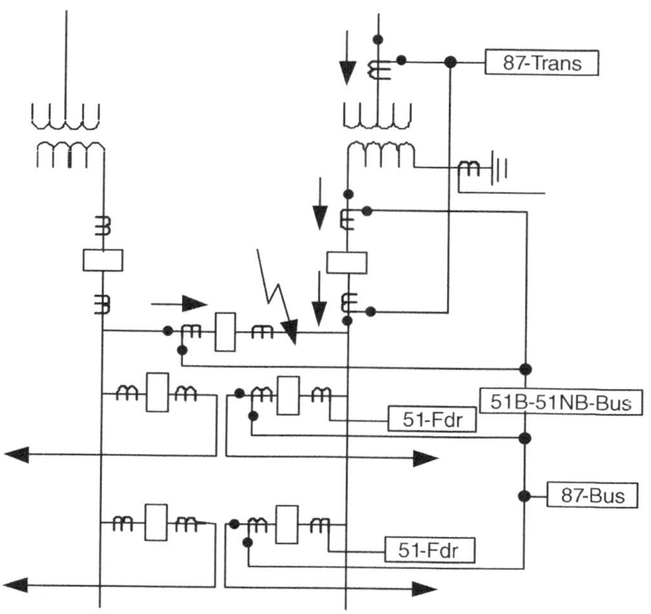

Figure 7.12 Current distributions for a bus fault at a substation.

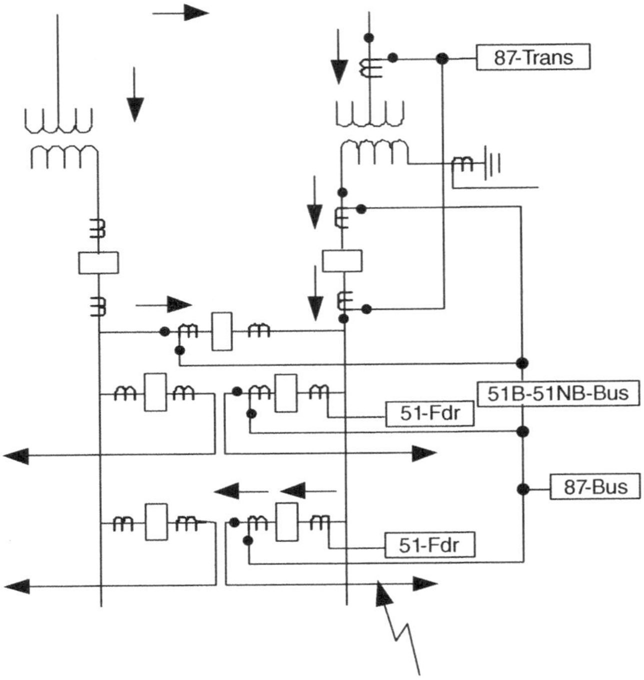

Figure 7.13 Current distributions for a feeder phase fault at a substation.

Refer to Figure 7.13 for the distribution of currents for a feeder fault at a substation. The infeeds to the faulted feeder flow into the feeder overcurrent protection via the CT on the feeder breaker as shown. The infeed to the faulted feeder flow into the CT on the transformer HV bushings and out of the CT on the transformer LV breaker such that the currents circulate around the CTs.

The infeed to the faulted feeder flow into the CT on the bus tie breaker and into the CT on the transformer LV breaker but flow out of the CT on the feeder breaker such that the secondary currents circulate around the CTs. Neither the transformer nor the bus differential protection operates but the bus overcurrent protection would see with the feeder fault and must coordinate with it as backup protection.

Refer to Figure 7.14 for the distribution of currents for a feeder ground fault outside the substation. The infeeds and outfeeds are similar to the previous example except that the transformer ground backup protection (64-Trans) and the bus ground backup (51NB-Bus) sees the fault. Transformer protection will be discussed in detail in Chapter 8, and the focus in this chapter is on protection zoning. For completeness, a series fuse is included in this example as fuses are normally used for distribution system sectionalizing of feeders. The portion of feeder beyond the fuse is in fact in the sectionalizing fuse zone. Figure 7.15 shows the time–current coordination between four distinct protection zones that also include the sectionalizing fuse. It should be noted each zone as it moves away from the location of fault covers a wider portion of equipment to be removed compared to the previous protection zone.

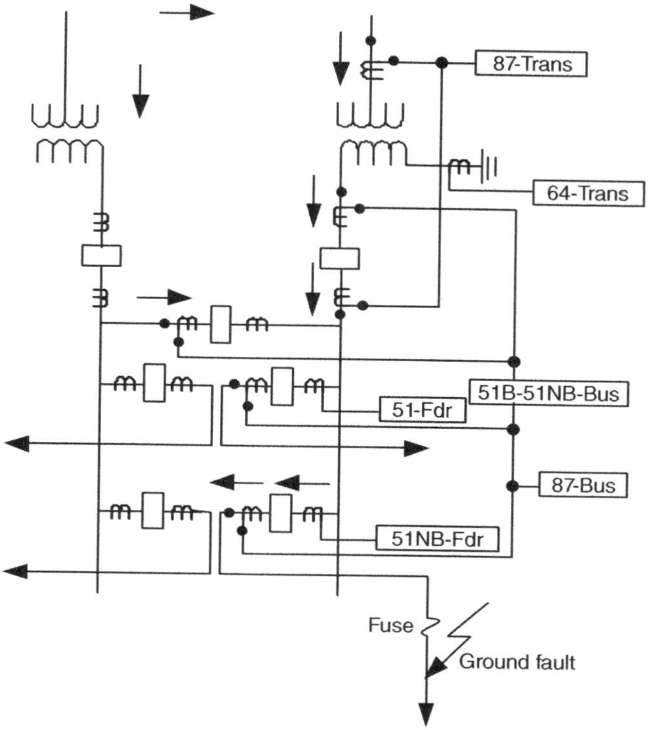

Figure 7.14 Current distributions at a substation for a feeder ground fault.

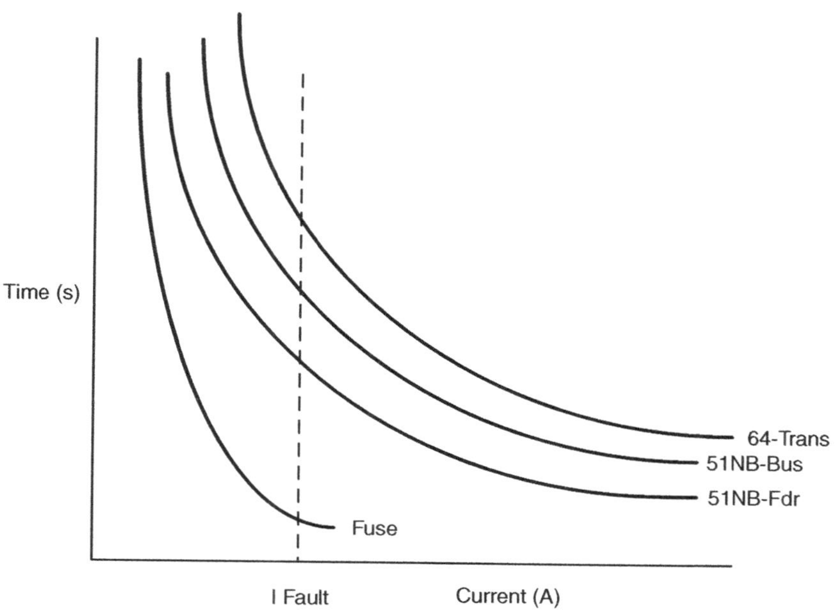

Figure 7.15 Substation protection zone coordination for a feeder ground fault.

7.5 Protection Zones in General

7.5.1 Lines

There are four distinct methods of achieving protection selectivity for lines. One method is to provide overlapping zones with timed coordination. A second method is to provide overlapping zones with communication and some form of sophisticated logic scheme. A third method is for the protection at each end of the line to cover most but not the entire line. The protections at each line end subsequently via communication transfer trips the remote line breakers. A fourth method is to provide some form of zone tight differential protection with communication such as pilot wire using telephone lines or line current differential using digital fiber.

HV lines traditionally have zones defined by line length, or distance from the relay location, using impedance relays. Either timed coordination in what is known as stepped time coordination or communication-based schemes may be used to achieve selectivity. The former is satisfactory where system fault clearing time is not a significant factor. Where system fault clearing times are critical, the latter tends to be used.

For lines at all voltage levels, zone tight line current differential protection can be used where suitable digital communication media is available. Packets of digital data representing the currents at the terminals are transmitted and received via either direct fiber or uploading onto a (synchronous optical network [SONET]) digital communication system whereby the data are compared with respect to each other at the line terminals. This works much like a simple differential protection scheme providing a zone tight type of protection.

LV lines that typically are radial supply tend to be protected with timed overcurrent relays where inverse characteristics covering the same portion of line are set to time coordinate. Where these lines are non-radial supply, pilot wire or line differential protection would be used.

7.5.2 Transformers

Power transformers are manufactured with bushing CTs on the high and LV sides as well as CTs connected in the star point neutral to ground connections. Autotransformers have in addition, CTs for tertiary windings that are sometimes brought out to bushings.

Figure 7.16 depicts a substation power transformer protection zone using the transformer HV bushing CTs and station LV winding breaker CTs.

Where there are dual secondary windings, the CTs on each of the two LV winding breakers may be connected in parallel in most circumstances as shown in Figure 7.17.

Autotransformers may have various types of zoning. In Figure 7.18, the autotransformer bushing CTs are used for the transformer differential protection and for the bus differential protection. There are two variations to this type of protection zoning. One is exactly as shown where there are no dedicated circuit breakers on the HV side. This is common to stations where transmission lines terminate directly onto autotransformers.

In other cases, where there are buses on both sides of the autotransformer and lines terminate onto breaker diameters, the zoning for both the HV and LV sides is as shown in Figure 7.19. Note that overlapping zones are achieved via overlapping CTs.

Yet, another variation of zoning used for autotransformers, regulators, and phase-shifting transformers is shown in Figure 7.20. The A and B groups, the two local protections, do not provide coverage for the same zone. In this method of zoning, the main protection covers the autotransformer, while there is an overall zone tight or unit protection covering the autotransformer and

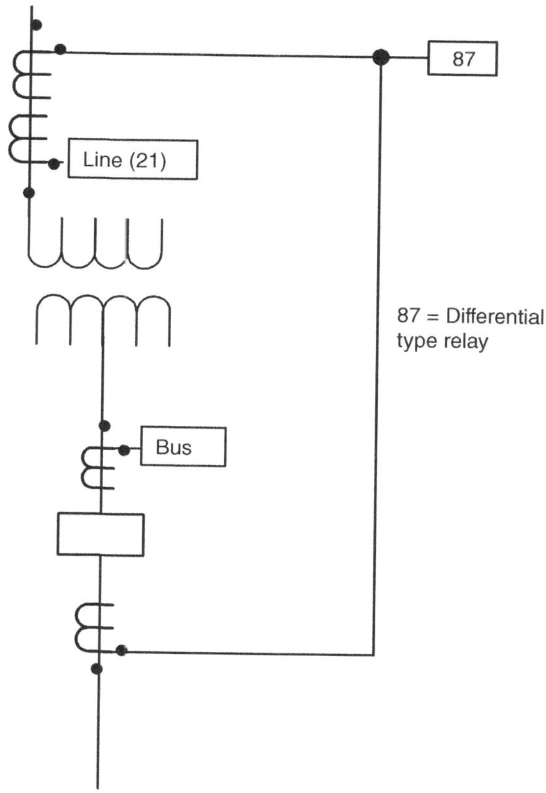

Figure 7.16 Zoning for a transformer with a single secondary winding.

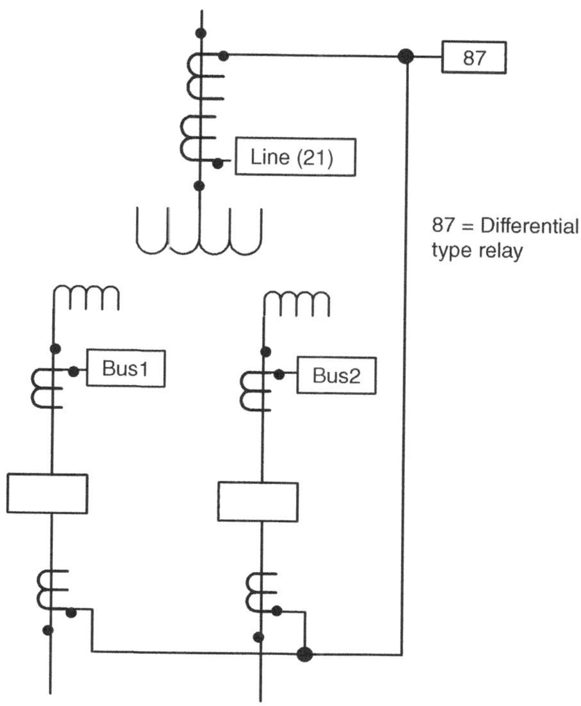

Figure 7.17 Zoning for a transformer with double secondary windings.

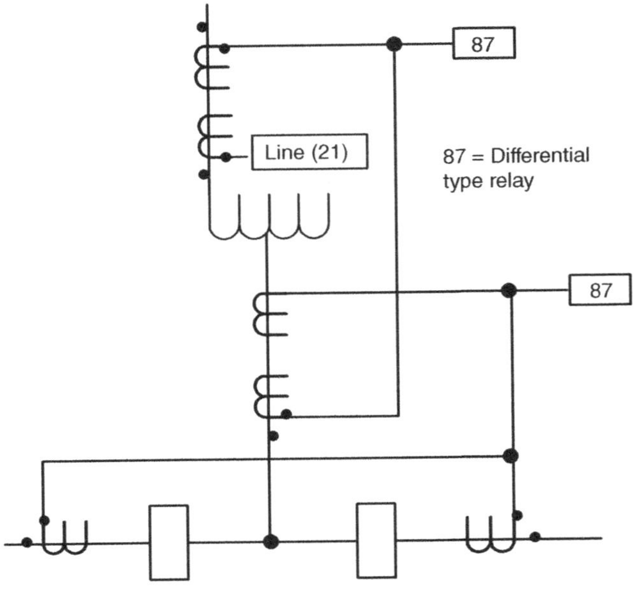

Figure 7.18 Zoning for lines terminating into an autotransformer.

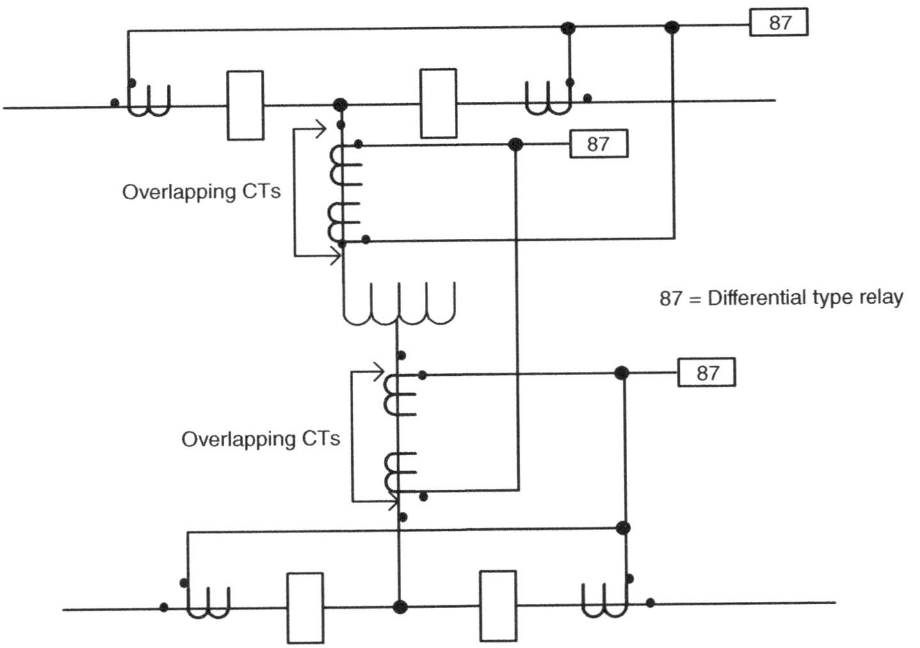

Figure 7.19 Zoning for autotransformers connected to buses.

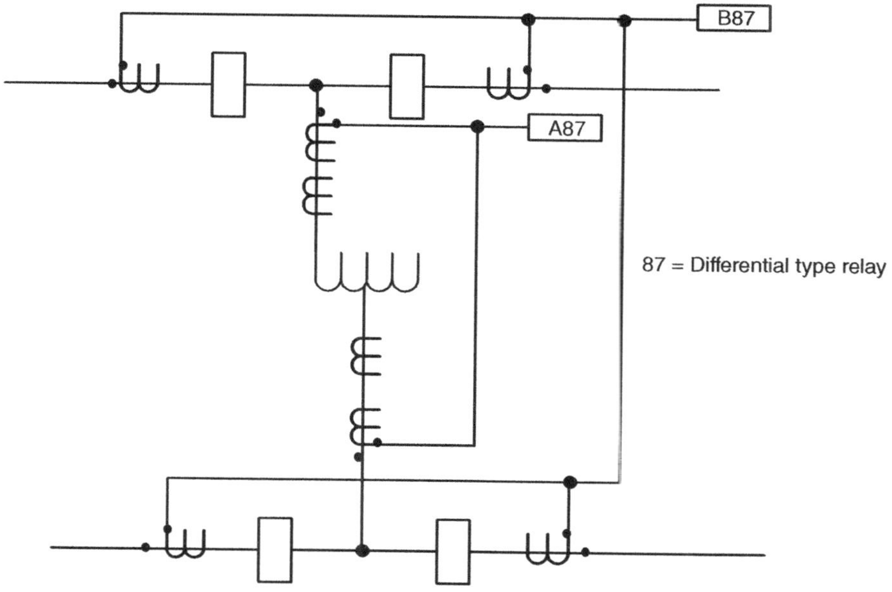

Figure 7.20 Zoning for autotransformer differential and overall bus differential.

both the HV and LV buses. Note that the A group bus protections that independently cover the HV and LV buses are not shown in this example. This zoning arrangement is driven by the availability of CTs.

7.5.3 Generators

The zoning for a generator and associated unit (step-up) transformer is shown in Figure 7.21. It is common practice to configure unit step-up transformers Delta on the LV side and solidly grounded Wye on the HV side as most transmission systems are grounded systems. There are many protections associated with the generator and the unit transformer which are a backup to the transmission system protections. Figure 7.21 shows the zoning for the generator and unit transformer where there are dedicated differential protections for each. In many applications, the dedicated differential protections are then duplicated.

It is also possible to apply dedicated differential protections as primary protections to the generator and unit transformer and overall backup differential protection encompassing both the generator and unit transformer as one overall generator – unit transformer set as backup protection as shown in Figure 7.22.

7.5.4 Protection Zones Overlapping Between Stations

There are systems where protection zones do not overlap via breakers and CTs but by impedance. Figure 7.23 shows such an example where a remote line distance protection located at Terminal B is set to see approximately half-way into the remotely located autotransformer impedance at Terminal A. The area of overlap between the line and autotransformer protection is bounded by the autotransformer bushings and approximately half the autotransformer impedance.

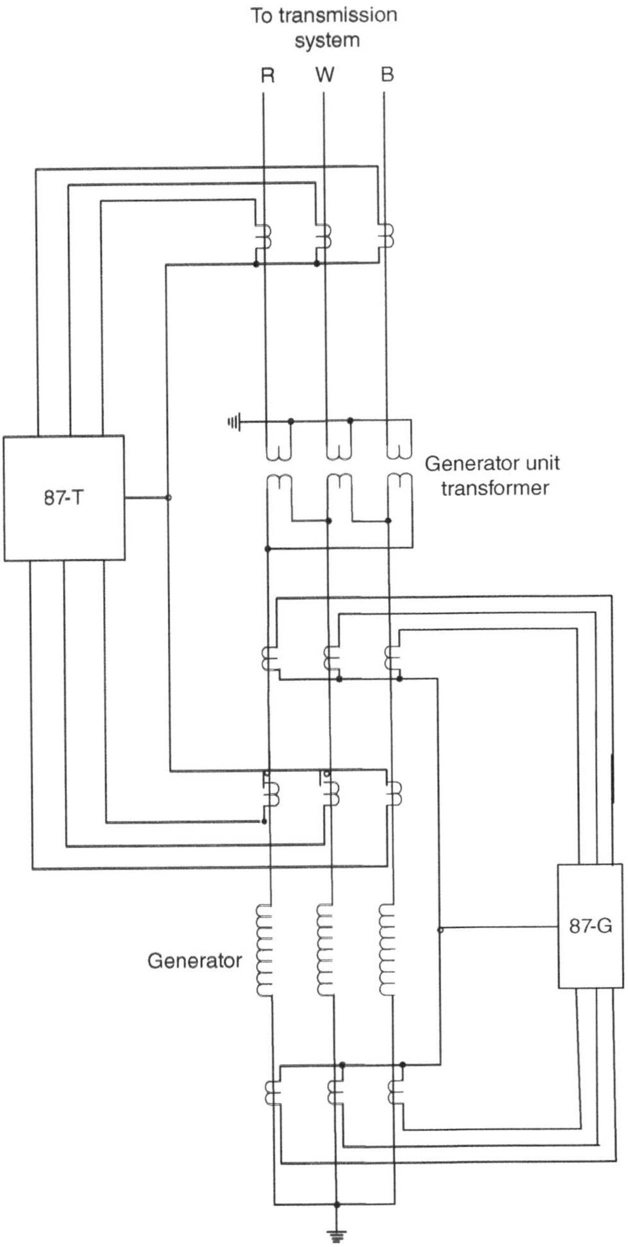

Figure 7.21 Generator and unit transformer dedicated protection zones three-phase connections shown.

Refer to Figure 7.24 below showing an example where the Zone 2 mho relays at the two switching stations see partially into the transformer impedance at a substation. The area of overlap between the two-line protections and the transformer differential protection is bounded by the transformer HV bushings and a portion of transformer impedance.

This protection zone overlap is of particular importance to some utilities. Substations tap directly to HV lines that are subjected to regional operating authority criteria. The criteria require that all

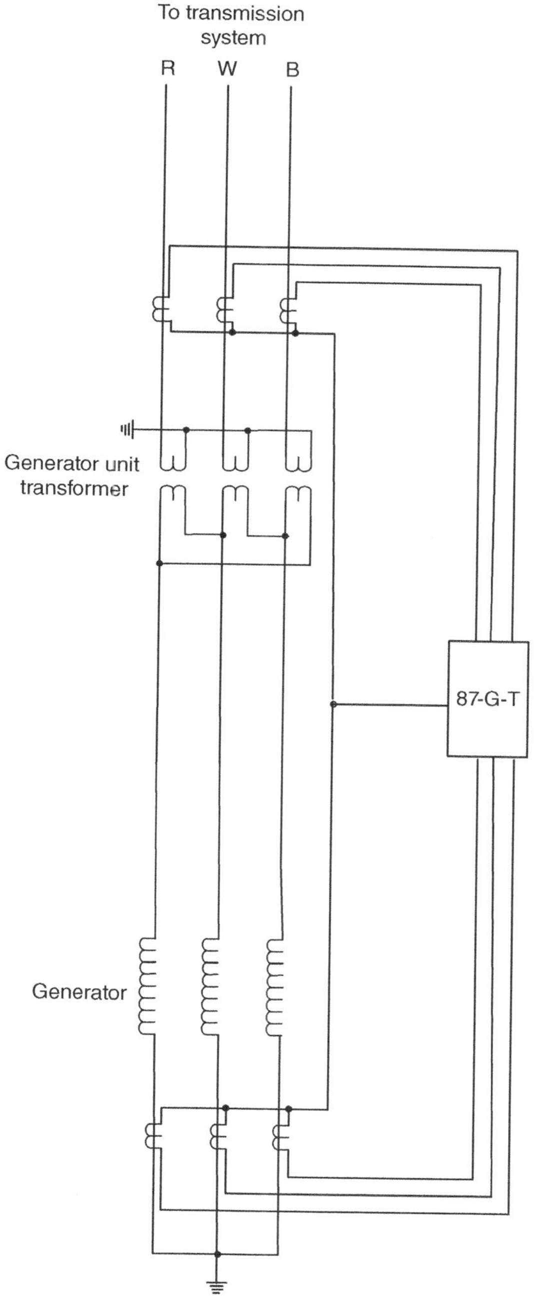

Figure 7.22 Generator and unit transformer overall protection zone three-phase connections shown.

bulk electricity system (BES) facilities have redundant protection systems. This means that BES facilities have two distinct protection and tele-protection systems; each fed from a separate set of batteries and configured to initiate circuit breaker tripping through separate (redundant) trip coils. None of this applies to substations even though these stations tap directly to BES facilities. The reason this is considered acceptable is due to the overlapping of line and substation protection

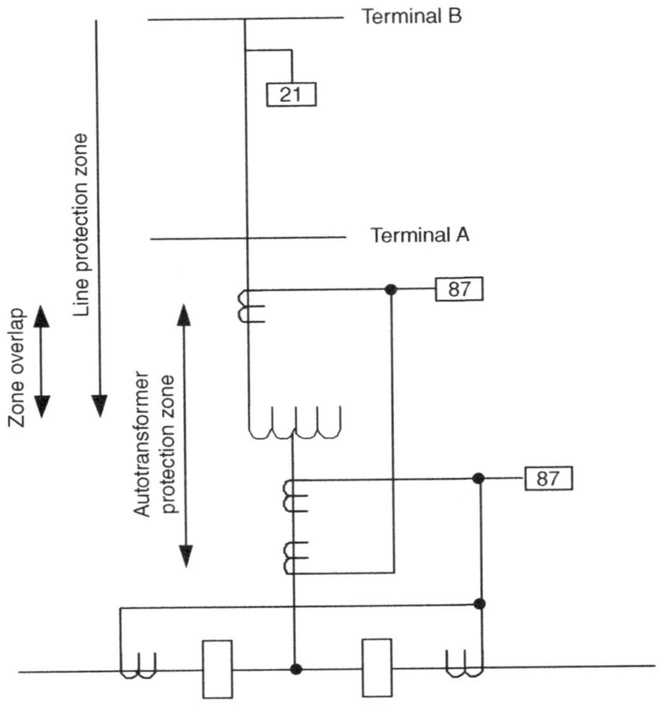

Figure 7.23 Example of overlapping line and autotransformer zones.

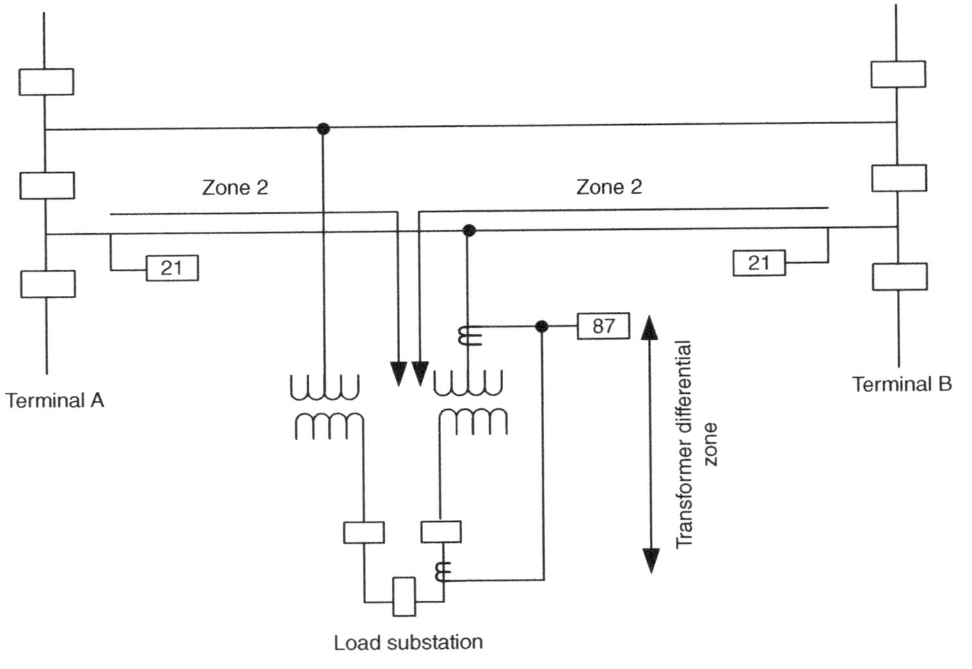

Figure 7.24 Example of overlapping line and substation transformer zones.

zones. The rationale is that any fault high enough in magnitude to be considered a BES event will be seen and cleared by at least one line protection at the switching stations where dual redundancy exists. If it were not for this, these substations would be mandated to have redundant transformer protections. Nevertheless, it is good utility practice to use redundant transformer protections for substation transformers for other reasons such as to facilitate maintenance of protections without removing the transformer from service.

7.6 Backup Protection

All protections need to be backed up to meet expected protection reliability performance. A single failure of protection equipment at the time of fault must always be catered for. The definition of protection equipment includes such items as DC circuits, protection relays, and breaker trip coils. This equipment can either be backed up remotely by the use of a protection zone extension or locally by redundant protection zones. In some cases, a combination of the two is used.

7.6.1 Remote Backup

Remote backup means that the protections at the remote station(s) are intentionally relied upon to adequately provide backup to a local station.

Refer to Figure 7.25 for a typical example of remote backup where station A backs up the protection equipment at station B. In this example, the Zone 2 distance protection tripping zone, as defined in Figure 7.25, at Terminal B, would normally cover all faults on the line up to Terminal C. However, since Terminal B protections are not redundant for example, with only a single battery, it must be assumed that any one of these could fail locally. In this case, the remote protection system at Terminal A labeled Zone 3, which is timed delayed to allow the Terminal B protection to trip first, will back up all protection equipment at Terminal B.

Figure 7.25 postulates that breaker CB1 has failed when the local distance protection (21) attempts to trip CB1. The remote timed Zone 3 protections (21) at Terminal A will see the fault at Terminal C and trips CB2 and CB3 to isolate the infeed from Terminal A. Remote backup

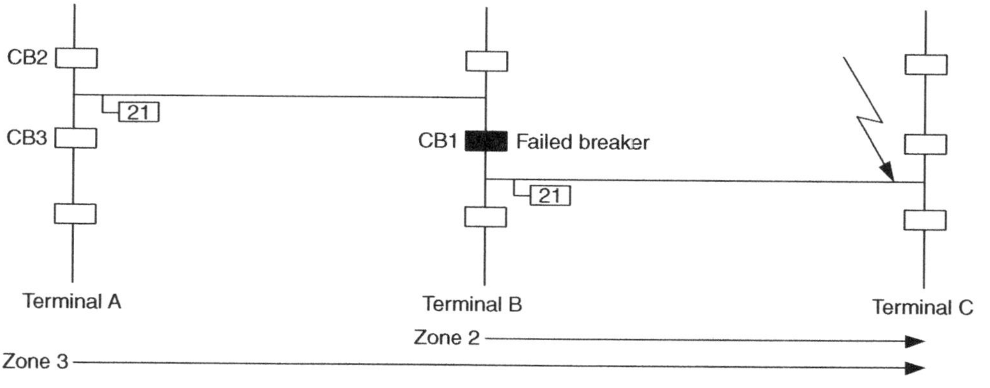

Figure 7.25 Example of remote backup for a failed breaker at Terminal B.

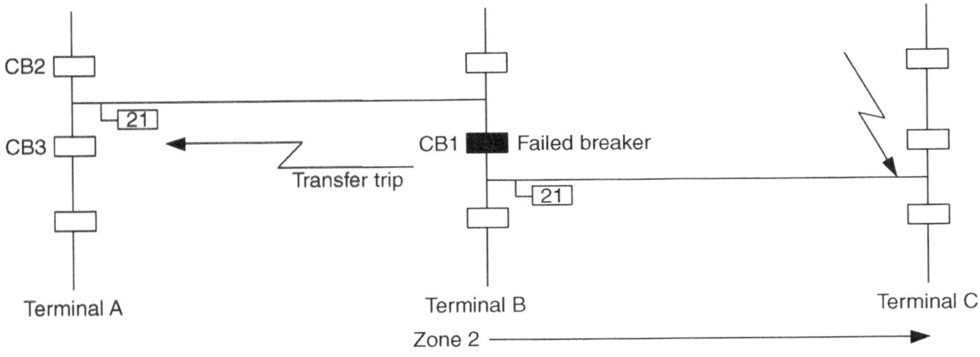

Figure 7.26 Example of local backup for a failed breaker at Terminal B.

protections have disadvantages such as longer clearing times, larger zones subject to tripping on load and swings, among others.

7.6.2 Local Backup

The protection philosophy that defines 230 and 500 kV line protections in general is local backup. This includes transfer tripping of remote breakers for a local breaker failure.

Reliability is achieved by using redundant protection systems with two groups of line protections, identical in function for each line. To achieve redundancy that matches the protections, geographically diverse main and alternate communication routes are used. Local breaker failure protection makes use of these communication routes for transfer trip of remote breakers to isolate a local breaker failure condition. The two groups of protections are self-contained and independent of each other. Each group is capable of providing complete high-speed protection for line faults.

The fundamental principle underpinning local backup is that only single contingency failure during a single event needs be catered for. For example, one may postulate that the A group protection could fail but never that the A group protection fail along with the failure of the B group for the same fault. Double contingency failure of protection equipment is therefore not expected or catered for at all. In general, utility practice is for power system elements operating above 200 kV, to use two local protection systems; the two groups of protections are designated the "A" group and the "B" group. Neither group is considered secondary to the other.

Figure 7.26 considers that breaker CB1 has failed when the local distance protection attempts to trip CB1 for the fault location shown. The dedicated breaker failure protection at Terminal B senses that CB1 has failed and transfer trips CB2 and CB3 to isolate the infeed from Terminal A.

7.7 CT Configuration and Protection Trip Zones

The secondaries of CTs in three-phase systems can be configured in either wye or a delta depending on the application. The decision to connect CTs in one configuration or the other determines whether the protection zone is "zone tight" to external ground faults or not. For digital relays, the CTs are always configured in wye in the analog circuit, and then, the relay's numerical processing algorithm is used to perform the translation to a delta configuration if needed.

The connection of CTs, whether connected in wye or delta, is used to define protection tripping zones. For example, relays having their CTs connected in delta can detect phase faults, but not ground faults.

7.7.1 CTs Connected in Wye

Refer to Figure 7.27 showing the method of connecting CTs in a three-phase system with the CTs configured in wye. In this example, the load is balanced or faults are assumed to be three-phase. The three-phase currents add vectorially from each other at the star point with the resultant equal to zero. No zero-sequence current, also referred to as residual current, flows.

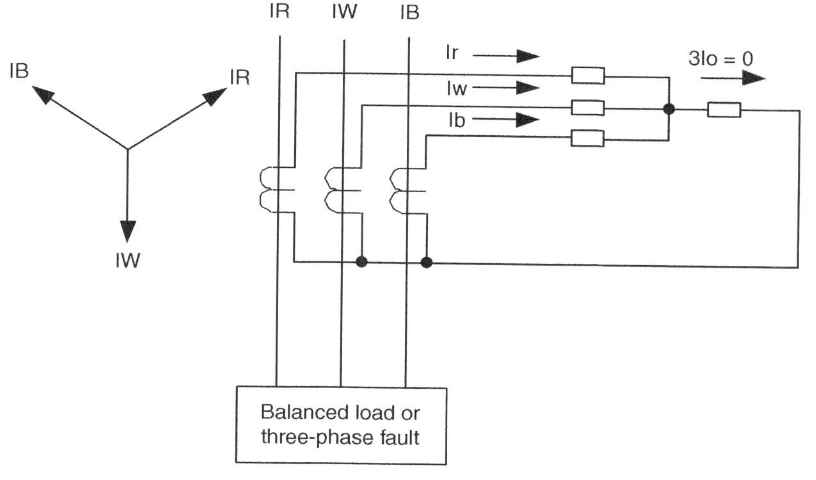

Figure 7.27 CTs configured in wye and a balanced load or three-phase fault.

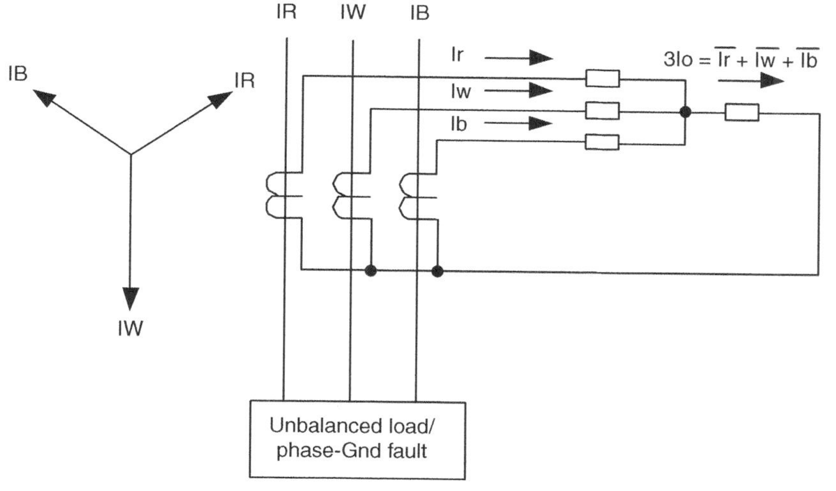

Figure 7.28 CTs configured in wye and unbalanced load or phase-ground fault.

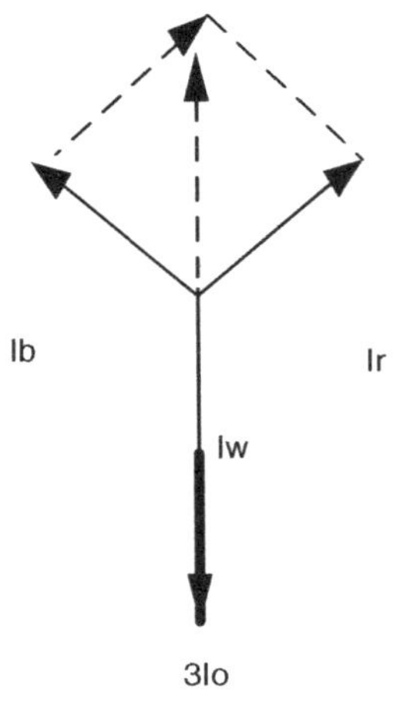

Figure 7.29 Derivation of residual current for a white phase-to-ground fault.

In this example, the load is unbalanced or faults are assumed to be phase-to-ground (Figure 7.28). The three-phase currents add vectorially from each other at the star point with the resultant not equal to zero. There is a net zero-sequence current – also referred to traditionally as residual current – flowing as shown in Figure 7.29.

7.7.2 CTs Connected in Delta

Refer to Figure 7.30 showing the method of connecting CTs in a three-phase system with the CT secondaries configured in delta. There is no path for zero-sequence current to flow into the relay, refer to Figure 7.31. This type of configuration is used to effectively create a zone tight differential protection for external line-to-ground faults. For example, autotransformer electromechanical differential protection must use delta-connected CTs; otherwise, it would operate for out-of-zone line-to-ground faults. The zero-sequence currents circulate around the delta and do not enter the relay. This topic will be discussed in detail in Chapter 8 on transformer protections.

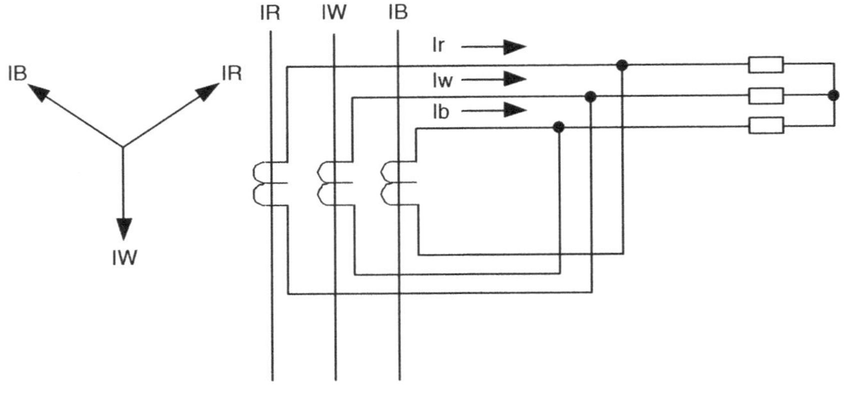

Figure 7.30 CTs configured in delta.

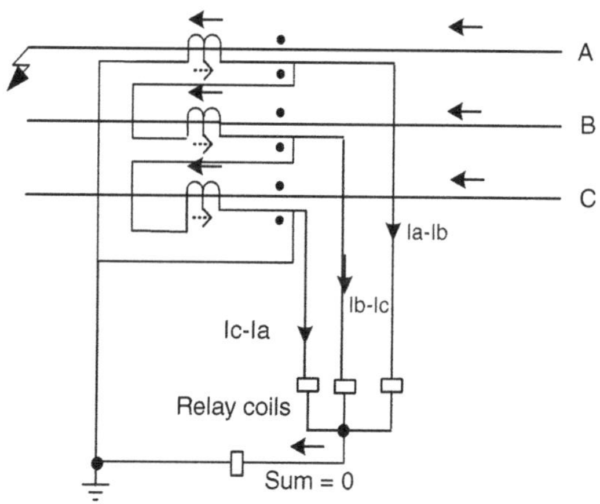

<u>Using symmetrical component method-refer to sub-module 1.10</u>

$Ia = Ia1 + Ia2 + Ia3$

$Ib = Ib1 + Ib2 + Ib3 = a^2Ia1 + a\,Ia2 + Ia0$

$Ic = Ic1 + Ic2 + Ic0 = a\,Ia1 + a^2Ia2 + Ia0$

$Ia + Ib + Ic = Ia0 + Ib0 + Ic0 = 3Ia0 = 3Ib0 = 3Ic0$

$a = 1@120$ degrees $= -0.5 + j0.866$

$a^2 = 1@240$ degrees $= -0.5 - j0.866$

$a^3 = 1@360$ degrees $= 1 + j0$

1,2,&0represents +,–& 0 sequence components respectively

a and a^2 are operators

$Ia\text{-}Ib = (Ia1\text{-}Ib1) + (Ia2\text{-}Ib2) = (1\text{-}a^2)Ia1 + (1\text{-}a)Ia2 = (1.5 + j0.866)Ia1 + (1.5 - j0.866)Ia2$

$Ib\text{-}Ic = (1\text{-}a^2)Ib1 + (1\text{-}a)Ib2 = a^2(1+a^2)Ia1 + a(1\text{-}a)Ia2 = (a^2\text{-}a)Ia1 + (a\text{-}a^2)Ia2 = -j1.732\,Ia1 + j0.1.732\,Ia2$

$Ic\text{-}Ia = (1\text{-}a^2)Ic1 + (1\text{-}a)Ic2 = a(1+a^2)Ia1 + a^2(1\text{-}a)Ia2 = (a\text{-}1)Ia1 + (a^2\text{-}1)Ia2 = (-1.5 + j0.866)Ia1 + (-1.5 - j0.866)Ia2$

It should be noted that zero sequence components are not present in the output circuits; they merely circulate in the delta connection.

For a phase-a-to-ground fault, and assuming $Ia2 = Ia1$, the output currents become:
$Ia\text{-}Ib = 3\,Ia1$; $Ib\text{-}Ic = 0$; $Ic\text{-}Ia = -3\,Ia1$, the sum = 0

Figure 7.31 Delta configured CTs and its response to an A-to-ground fault.

7.8 Where Protections Zones do not Overlap Around Breakers

7.8.1 Blind Spot Created by Non-Overlap of Protections Zones around Breakers

There are two types of 500 kV and above breakers typically used by utilities. There are the breakers that come with traditional bushings that house CTs; these are normally referred to as dead tank type; they are grounded requiring no insulated mounting posts. For these types of breakers, the CTs are mounted in the bushings on both sides of the breaker.

There are also live tank breakers that are mounted on insulated pedestals. For these types of breakers, the CTs are external to the breaker and are also mounted on insulated pedestals (also called free-standing CTs). However, since the cost of pedestal mounting is high, they are often mounted on only one side of the breaker. This means that the traditional method of overlapping protection zones around breakers is not achieved. This creates a blind spot such that faults even when detected by protections have no means of actually isolating the fault.

Refer to Figure 7.32 showing a fault in a blind spot on the bus between Breaker CB2 and the CT. This fault is seen by the Line 1 protection which then trips CB1 and CB2. However, the fault cannot be cleared by tripping CB2 as infeed to the fault continues to flow from bus B via breaker CB3 and by infeed from Line 2 remotely.

The solution is to rely on breaker failure or backup protection. To illustrate using the example given above, when breaker CB2 is initially tripped, breaker failure is also initiated. Since current continues to flow through the CTs to the fault, the breaker failure protection that was initiated will

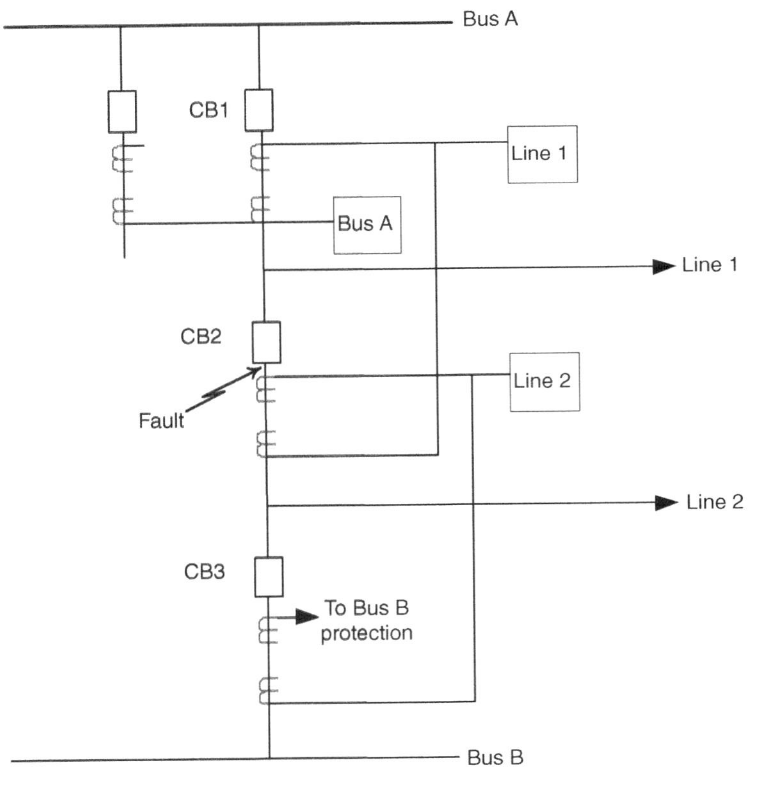

Figure 7.32 Blind spot on a 500 kV bus diameter.

operate. When breaker CB2 failure operates, it keys transfer trip to the remote end of Line 2 and also trips breaker CB3 locally.

Breaker failure protection is relied upon for the protection of this blind spot, and it has not been traditionally duplicated. Since most faults are line-to-ground, utility practice is to add frame leakage protection to sense ground faults in the blind spot area. Since its purpose is to back up the breaker failure protection, the frame leakage protection should have a battery supply separate from the main breaker failure protection.

Since the amount of exposed bus that constitutes the blind spot is not much larger than a meter long and since a phase fault in this area is extremely rare, industry practice is to not duplicate this protection.

7.9 Lines Terminating Directly on Buses at a HV Switching Station

Refer to Figure 7.33 showing a line terminating directly onto a bus at a switching station. Past utility protection practice was to include the bus within the line protection zone. Present utility protection practice is to keep the bus and line protection as independent protection zones. This could be achieved by creating separate protection zones with the creation of new diameters. Re-terminating the line onto a dedicated breaker diameter zone is very expensive. One much less expensive solution is to install free-standing CTs in the switchyard to service new bus protection and others to service the existing line protection as shown in Figure 7.34.

This is an example of two protections covering separate system elements that trip the same breakers. The two protections overlap at the free-standing CTs while sharing one common set of isolating points, provided by the respective circuit breakers.

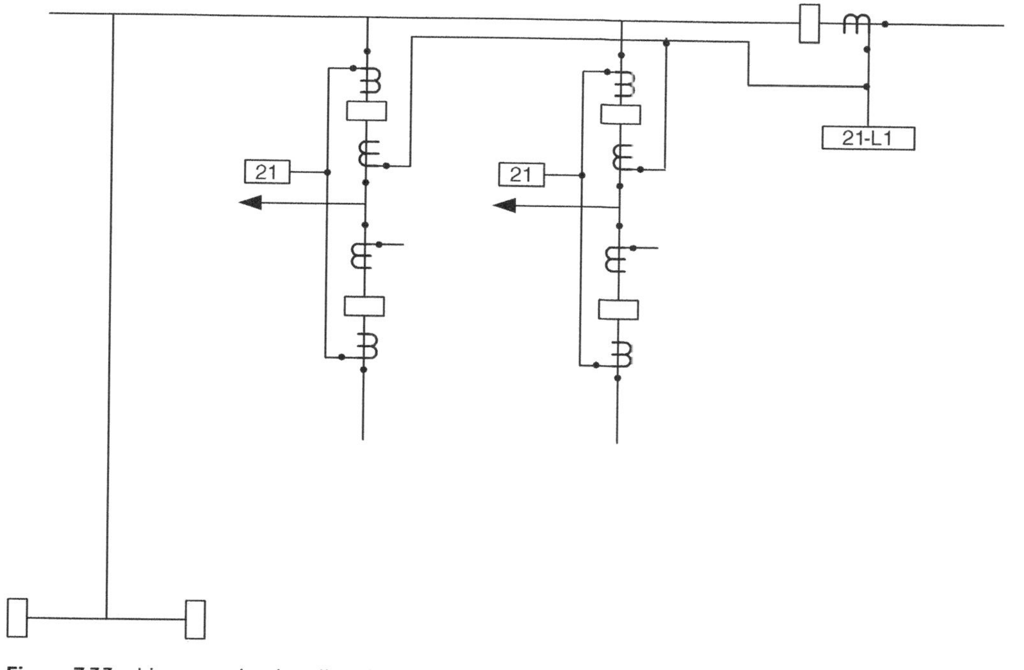

Figure 7.33 Lines terminating directly onto a bus at an HV switching station.

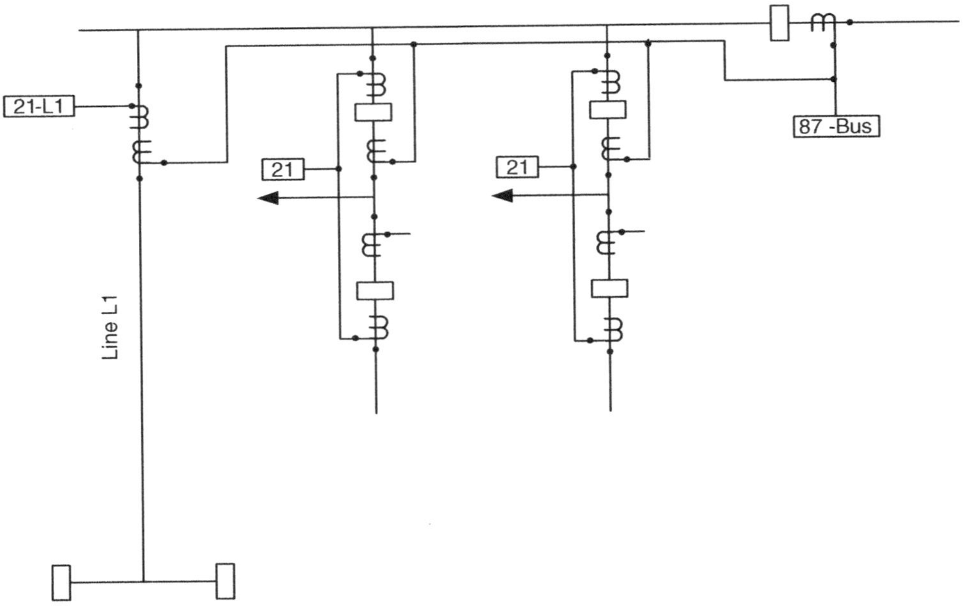

Figure 7.34 Lines terminating directly onto a Bus at an HV switching station.

There are two advantages to modifying the protection zones as in Figure 7.34. The first advantage is that bus protections that are optimized for bus faults are much more secure for out-of-zone bus faults. The second advantage is that dedicated bus and line protections facilitate the ability to determine fault location dependent on which protection operated.

Nevertheless, the line and bus constitute a single zone in terms of breaker tripping and zone isolation.

8

Transformer Protection

8.1 Introduction

There are four common abnormal conditions affecting transformers. They are short circuits, open circuits, overvoltages and overloads.

Specific overheating protection is not provided as there are controls to alarm to control pumps and banks of fans for cooling of the transformer without needing to trip the unit.

Overload protection is generally not provided. However, there are specific applications where it is important to automatically remove a transformer from service due to overload in which case it is provided (see Section 8.3.6.7).

Protection for transient overvoltage such as those caused by lightning strikes is provided mainly by horn gap protectors, lightning arresters, or transformer rod gaps.

The types of transformers covered in this chapter are power transformers, autotransformers, and grounding transformers (Figure 8.1).

8.2 General Principles

Several various types of protection are necessary to adequately protect a transformer bank against internal faults within the tank and within the transformer's overall protection zone. They include differential, overcurrent, and gas protection [1].

Whether to duplicate the differential protection usually depends on the MVA rating of the bank. The general utility adopted rule is that any transformer bank larger than 50 MVA requires "A" and "B" differential protection. For example, a 33 MVA transformer bank typically only requires single differential protection in the "A" group. The "B" group consists of phase and ground overcurrent backup. The criterion for double differential protection is to allow operators to maintain the transformer in service, while one protection group is being maintained.

In the case of a smaller transformer bank, the impedance of the bank limits the fault value sufficiently enough to allow timed fault clearance. For these banks, it is possible to conduct routine scheduled maintenance on the "A" group differential protection while still adequately protecting the in-service bank with the "B" group timed overcurrent protection. For larger banks, above 50 MVA, timed clearance as primary protection even during the short time for scheduled maintenance is not acceptable. For that reason, instantaneous "A" and "B" differential protections are required to allow the bank to stay in service during scheduled protection maintenance.

Power System Protection: Fundamentals and Applications, First Edition. John Ciufo and Aaron Cooperberg.

Figure 8.1 33 MVA 115–13.8 kV Power Transformer.

8.3 Differential Protection Power Transformers

In general, differential protection provides the best overall protection for both phase and ground faults, except where the current is limited by high impedance grounding. At substations that exclusively supply three-wire load, grounding reactors are used to limit the fault current to below differential protection detection. Restricted ground fault (RGF) protection which is a type of differential protection optimized for this purpose works very effectively independent of the main differential protection.

8.3.1 Factors Affecting Transformer Differential Protection

Four main factors influence the design of differential protection.

1. Phase shifting between the primary and secondary sides of the transformer due to configuration such as wye–delta or delta–wye.
2. The flow of zero-sequence current is determined by transformer configuration and whether an artificial ground source such as a grounding transformer is used and where it is electrically located.
3. Unbalances due to the following two reasons: Current transformers located on the primary and secondary of the transformer that comes close to but still do not exactly match the transformation ratio. Under-load tap changing and voltage taps on either the primary or secondary.
4. Inrush current upon placing a transformer on potential that appears momentarily as an internal transformer fault.

8.3.2 Phase Shifting from Primary to Secondary

Whenever a three-phase transformer is configured with different winding connections, a phase shifting of current and voltages from primary to secondary is created.

These are the main reasons that transformers are connected wye or delta:

1. Delta windings are desirable as they provide a path for third harmonics to flow for power quality.
2. Wye winding connections are usually used to provide a source of ground current either solidly grounded or through a reactor-connected star point to ground.

There are many types of phase shifting used by utilities for their transformers which take into account the transmission and distribution systems and the phase relationship between them. The predominant one is secondary lagging the primary by 30° as shown in Figure 8.2. As an example that is worth analyzing to show how many other types of phase shifting may be achieved is the secondary lagging the primary by 210° as shown in Figure 8.3.

The standard approach using non-digital differential relays is to correct for the 30° phase shift via the configuration of CTs. For example, phase shift correction is achieved by choosing CTs on the wye transformer side connected in delta and on the delta side in wye. Figure 8.4 shows a typical delta–wye 30° lag transformer with CTs connected in such a way as to correct the 30° phase shift. Note that the CT delta connection is identical to the transformer delta connection.

Figure 8.5 shows an example where 210° phase-shifting correction is achieved through the connection of CTs in delta and by rolling phases similar to that of the transformer itself.

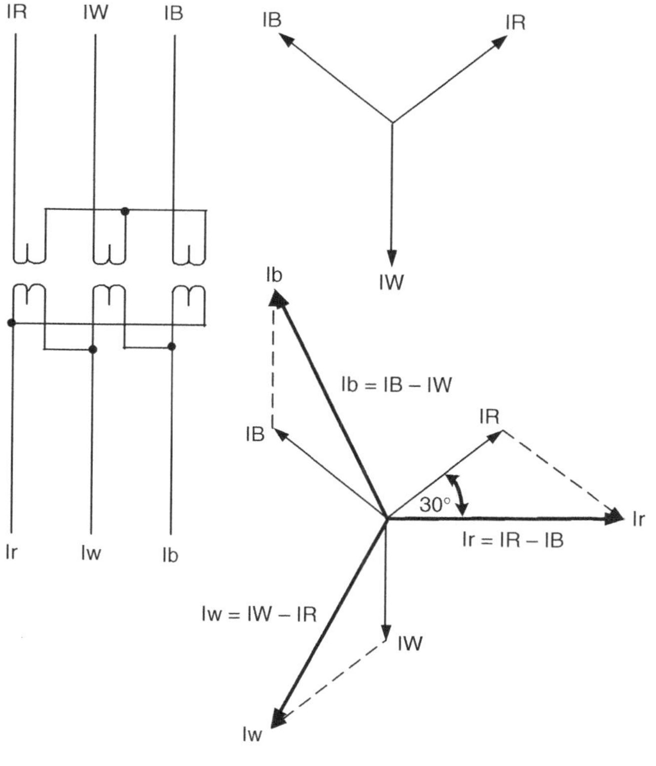

Figure 8.2 Example of a wye–delta configure transformer where the secondary lags the primary by 30°.

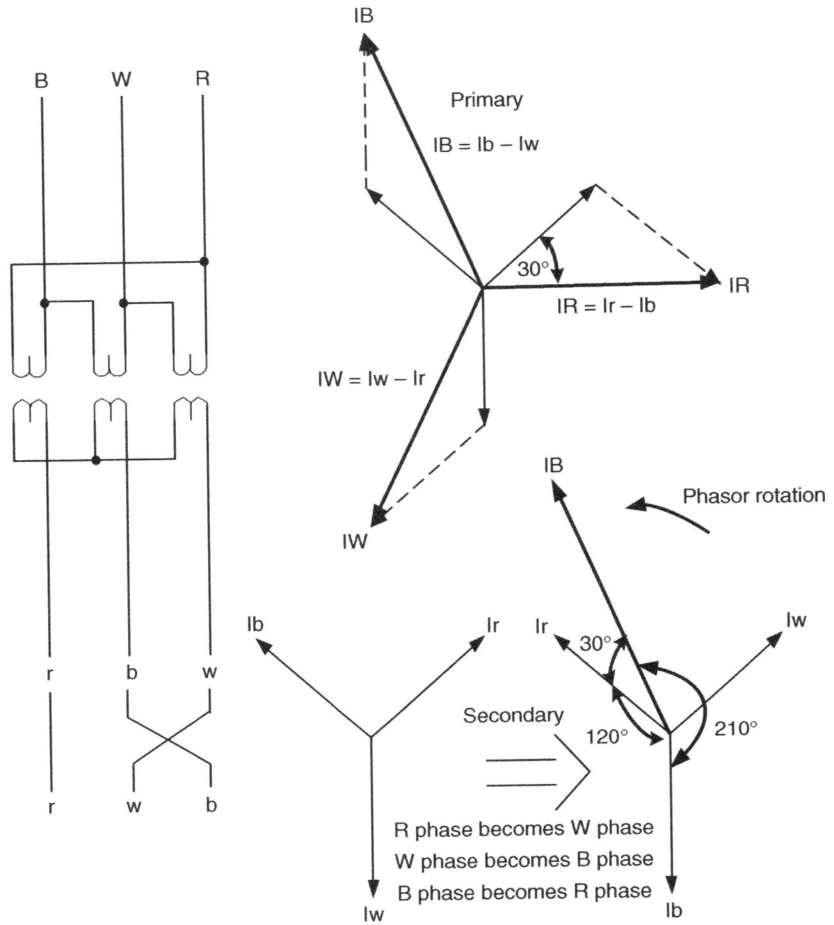

Figure 8.3 Example of a wye–delta configure transformer where the secondary lags the primary by 210°.

The application of modern digital transformer differential relays is significantly different than electromechanical or solid-state. The CTs on both sides of the transformer are connected in wye, and the required phase-shifting correction is implemented within the relay algorithm. For example, one of the achievable setup questions would be to choose a phase shift that fits the transformer such as 30° lag or 210°. There are two main advantages to connecting the CTs in wye instead of delta. Only a four-wire cable from the outdoor switchyard to the relay building is required instead of a six-wire, and a true measurement of each of the phase currents for fault recording is achieved (Figure 8.6).

8.3.3 The Flow of Zero-Sequence Current

8.3.3.1 Substation Grounding Requirements

The secondary rating of typical power transformers supplying sub-transmission distribution systems are 44, 27.6, 13.8 kV among other commonly used similar voltages.

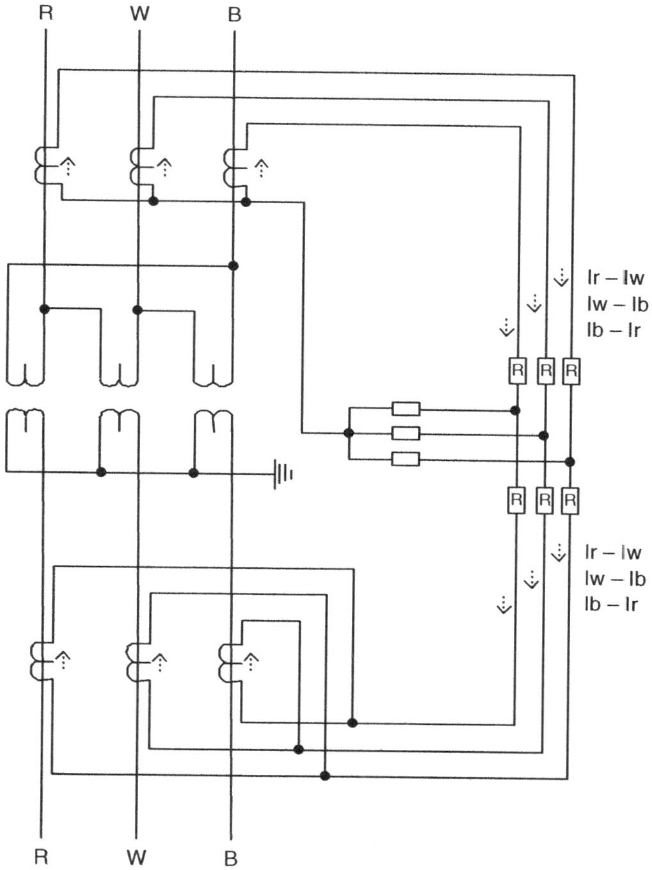

Figure 8.4 Typical 30° phase shift correction using CT connection.

A 44 kV distribution system is usually exclusively three-wire and is supplied from a wye–wye transformer. This transformer also has a delta tertiary to circulate third harmonic currents and eliminate third harmonic voltages. This arrangement permits grounding the sub-transmission neutral at the station. It is not, however, brought out to the overhead lines. The main purpose is to stabilize the neutral point and provide a source of ground current for relaying during line-to-ground faults. Only balanced three-phase loads are permitted on this type of distribution system.

A 27.6 kV distribution system is usually exclusively four-wire and is supplied from either a delta–wye transformer grounded through a low impedance grounding reactor, or from a wye-zigzag transformer grounded through a low impedance grounding reactor, or from a wye–delta transformer with an independent low impedance grounding transformer. In each case, the station neutral is brought out to the overhead lines and unbalanced loads, usually in the form of single-phase laterals, are permitted on this type of distribution system.

A 13.8 kV distribution system is usually either three-wire or four-wire and is supplied from either a delta–wye transformer grounded through a low or high impedance reactor, or from a wye-delta transformer with an independent low or high impedance grounding transformer. Where the station is designed to supply single-phase laterals, the station neutral is brought out; otherwise, there is no reason to.

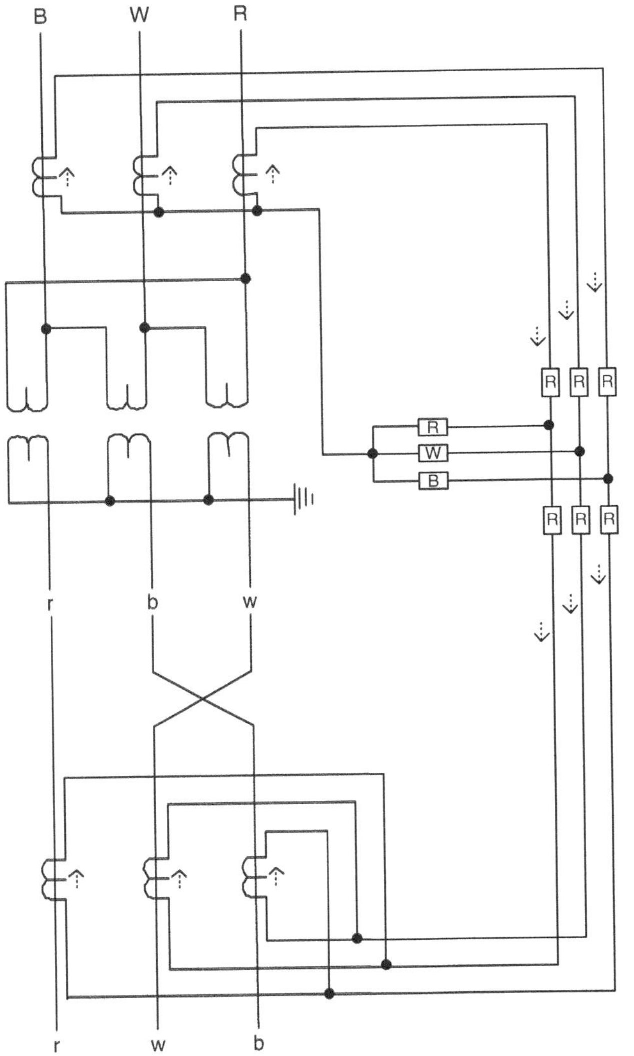

Figure 8.5 Typical 210° phase shift correction using CT connection.

8.3.3.2 Artificial Ground Sources

8.3.3.2.1 Wye–closed delta Transformer

A wye–closed delta configured transformer can make a suitable artificial ground source. In a wye–closed delta transformer with a grounded neutral, the closed delta offers a low impedance path for zero-sequence currents to flow. With a line-to-ground fault on the system, the zero-sequence currents flow into the ground at the point of fault and back through the neutral of the wye–delta transformer. In other words, it provides a low impedance path for zero-sequence currents to flow in the system. The wye–delta configuration can be either a three-phase transformer or three single-phase transformers either one of them configured as shown in Figure 8.7.

Figure 8.7 shows a ground fault on one of the phases. The total ground fault current is divided into three equal parts as it splits and flows in the system. These currents are all single-phase currents,

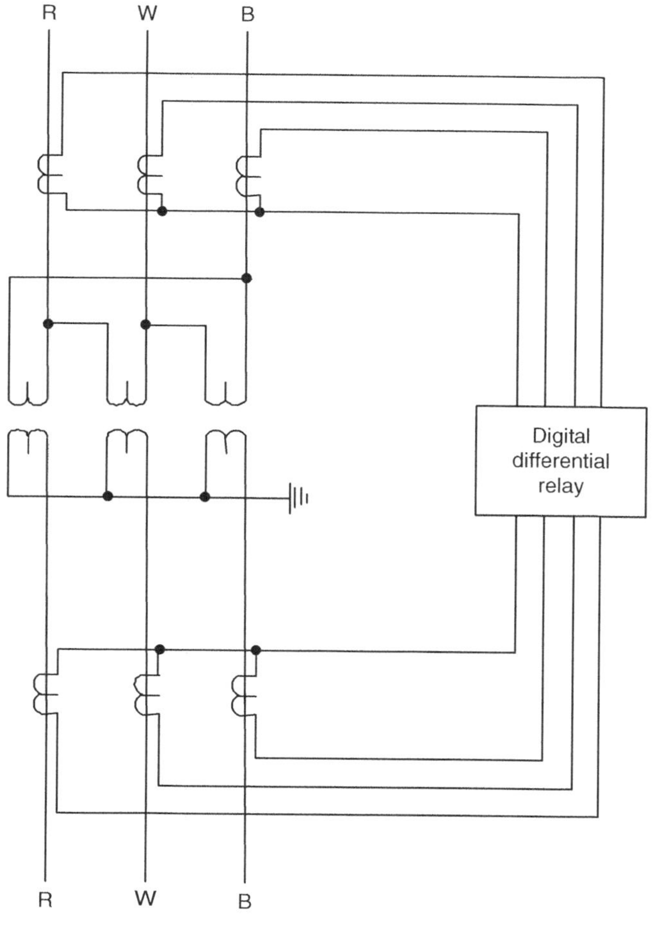

Figure 8.6 CT connections to a transformer differential relay.

and each arrow represents one-third of the total ground fault current. These are the zero-sequence components of the ground current as they are all equal in magnitude and phase relationship.

The closed delta permits essentially equal and opposite currents to flow in the two windings on each core section. This current only meets transformer leakage reactance rather than the high magnetizing reactance it would meet if the closed secondary delta connection were open. Each pair of windings on the same core section acts like a simple two-winding single-phase transformer.

What would happen if the three components of ground fault current flowing into the grounded Wye star-point attempted to divide into unequal parts? The secondary windings are connected in series forming a delta connection. Since they are connected in series, the currents flowing in each respective secondary winding are equal to each other by definition. Furthermore, the secondary winding on phase a in the delta is wound on the same iron core section as the primary wye a winding. Since the primary and secondary windings share the same iron core, these windings are affected by the same magnetic flux path. The windings of phase b and c are constructed the same way and respond similarly. In essence, the ampere turns on each core section are equal and opposite just like in a two-winding transformer under load. However, just like a standard two-winding transformer, the difference in ampere turns between the pair of windings on individual core section

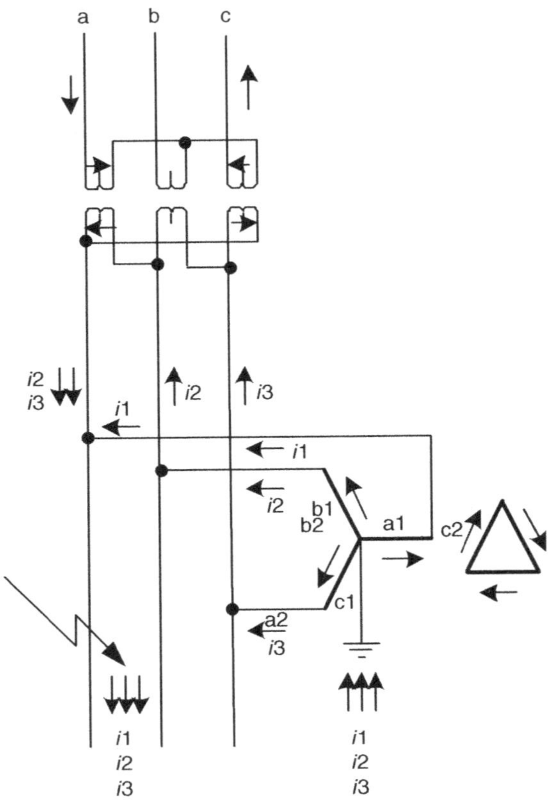

Figure 8.7 Wye–closed delta transformer used as a ground source.

is due to the ampere turns necessary to magnetize the core. The effect of the closed secondary delta with the currents in each of the secondary windings is to force the respective primary winding currents to all be the same magnitude. In other words, if the current in any wye primary phase winding tries to become smaller than that in the other two phases, there would be an additional magnetomotive force (MMF) available to produce additional magnetic flux and voltage in the deficient core section in the primary wye winding. This would be in a direction to tend to increase this current and therefore tend to restore normal current balance.

The fact that the series delta connection requires the secondary currents to be equal so too the currents in the wye primary connection have to be equal. This is similar to the primary and secondary ampere turns being equal on a regular two-winding transformer except for magnetizing current.

8.3.3.2.2 Zigzag Grounding Transformer

A more suitable method of providing an artificial ground source would be to install a zigzag grounding transformer. The previous method of using three single-phase transformers configured wye–closed delta works but is not as efficient under normal operating conditions.

Refer to Figure 8.8 showing a zigzag grounding transformer connection. The two windings marked a1, a2 and b1, b2 and c1, c2 are each wound on the same core section, respectively. The flow of ground fault currents is similar to that for a wye–closed delta grounding transformer. Since the three zero-sequence currents flowing into the star-point of the grounding transformer are all in phase, they divide equally.

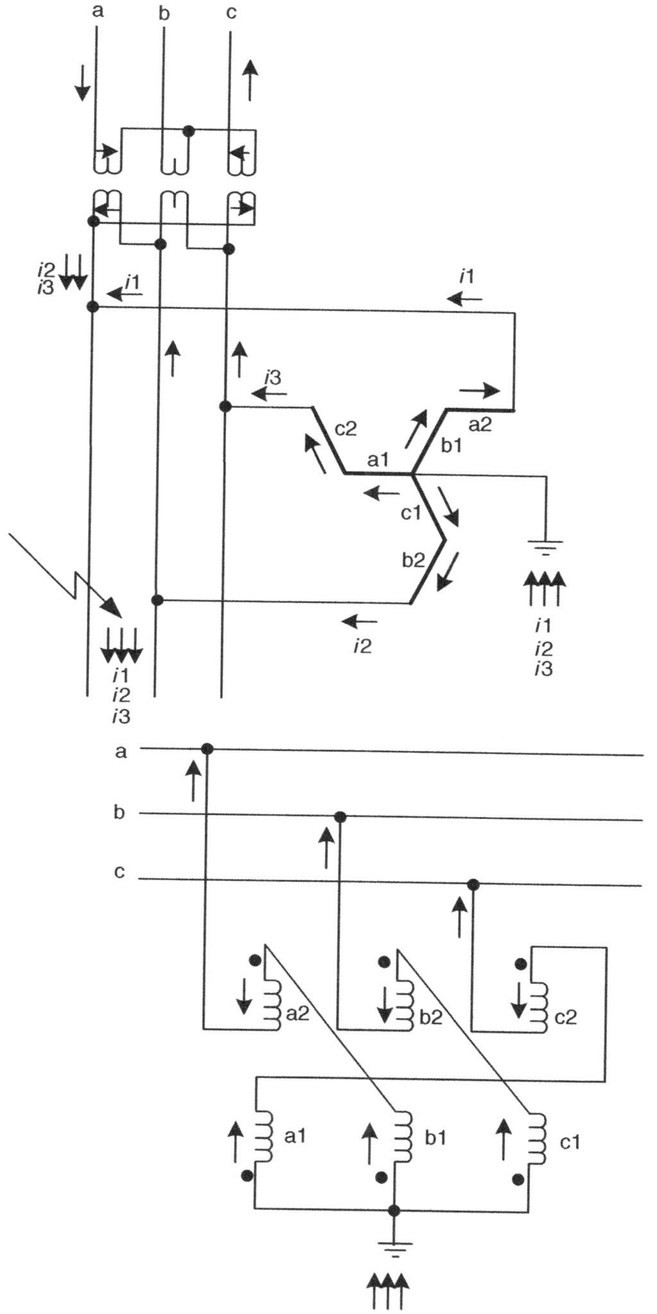

Figure 8.8 Zigzag grounding transformer as a ground source.

The zigzag grounding transformer works similarly to a wye–closed delta grounding source. The series connection between windings on adjacent core legs forces the currents to be equal. If any unbalance should appear, additional MMF will provide additional flux and therefore added voltage in the unbalanced core section in a direction that restores the current balance.

Positive and negative sequence currents cannot flow in a zigzag grounding transformer as they are 120° out of phase. Zigzag grounding transformers represent a very high impedance under normal balanced system conditions and a very low impedance for zero-sequence currents for unbalanced system conditions.

The zigzag configuration ensures that no magnetizing current will flow under normal system conditions. However, a wye–closed delta transformer will draw magnetizing current. It is for this reason that zigzag grounding transformers are the preferred choice when an external ground source, also referred to as an artificial ground, is required.

8.3.4 Flow of Zero-Sequence Currents in Differential Circuits

8.3.4.1 Delta–Wye Transformer

When using electromechanical relays, phase-shifting correction is achieved by connecting the CTs on the transformer delta side in wye and on the transformer wye side in delta. This connection ensures that the secondary currents will circulate between the CTs as required for any external fault or load and that the phase currents add vectorially to zero. For external phase faults on the wye side of the bank, the particular CT connection would not matter as long as phase-shifting correction is achieved.

For an external ground fault however, where zero-sequence current can flow in the wye winding, the delta CT connection circulates the zero-sequence components of the current inside the delta. This effectively prevents them from entering the external, to the delta, CT connections to the relay as shown in Figure 8.9. This is necessary as there are no zero-sequence components of current on the delta side of the transformer for a ground fault on the wye side. Therefore, there is no possibility for the zero-sequence currents to simply circulate between the two sets of CTs. Furthermore, if the CTs on the wye side were not delta connected, the zero-sequence components would flow in the operating coils and cause the relay to operate undesirably for external ground faults.

Figure 8.10 shows the same CT connections and the flow of the zero-sequence currents through the relay operating coils for an internal ground fault. In both examples, the zero-sequence current flows through the wye-connected load as shown. If not for the possibility of zero-sequence current flowing through the load thereby completing their path in theory they would not flow at all. In reality, some zero-sequence current would flow owing to the affects of capacitance to ground.

8.3.4.2 Wye–Delta Transformer

When using electromechanical relays, phase-shifting correction is achieved by connecting the CTs on the transformer wye side in delta and on the transformer delta side in wye. This connection ensures that the secondary currents would circulate between the CTs as required for any external fault or load and that the phase currents add vectorially to zero for any type of phase fault. For external phase faults on the delta side of the bank, the particular CT connection would not matter as long as phase-shifting correction is achieved.

When the transformer secondary is delta configured, most utilities typically introduce an artificial ground source such as a zigzag grounding transformer. Section 8.3.3.2.2 previously described the workings of this type of ground source. The effect of using an artificial ground on the delta secondary transformer side is to introduce a source of zero-sequence current for an external ground fault without any means of preventing it from entering the operating coils of the differential relay.

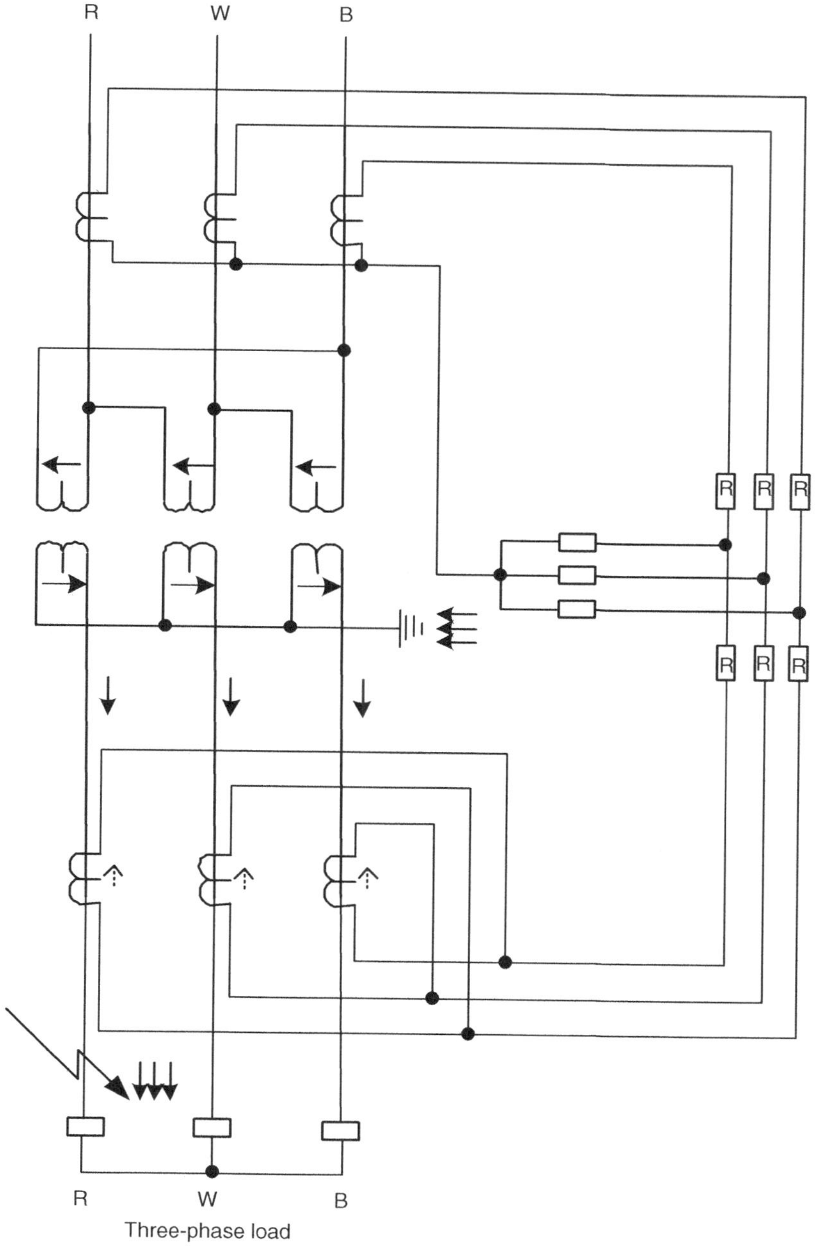

Figure 8.9 Zero-sequence current distribution for a delta–wye transformer and external ground fault.

Refer to Figure 8.11 where an external ground fault and consequent zero-sequence currents flowing are shown to operate the differential relay.

The universally accepted solution preventing relay operation in this given situation is to trap the zero-sequence currents thereby keeping them away from the relay operating coil while allowing them to circulate harmlessly for this situation. This is done by using a zero-sequence shunt where the zero-sequence components of current are kept out of the external secondary circuits

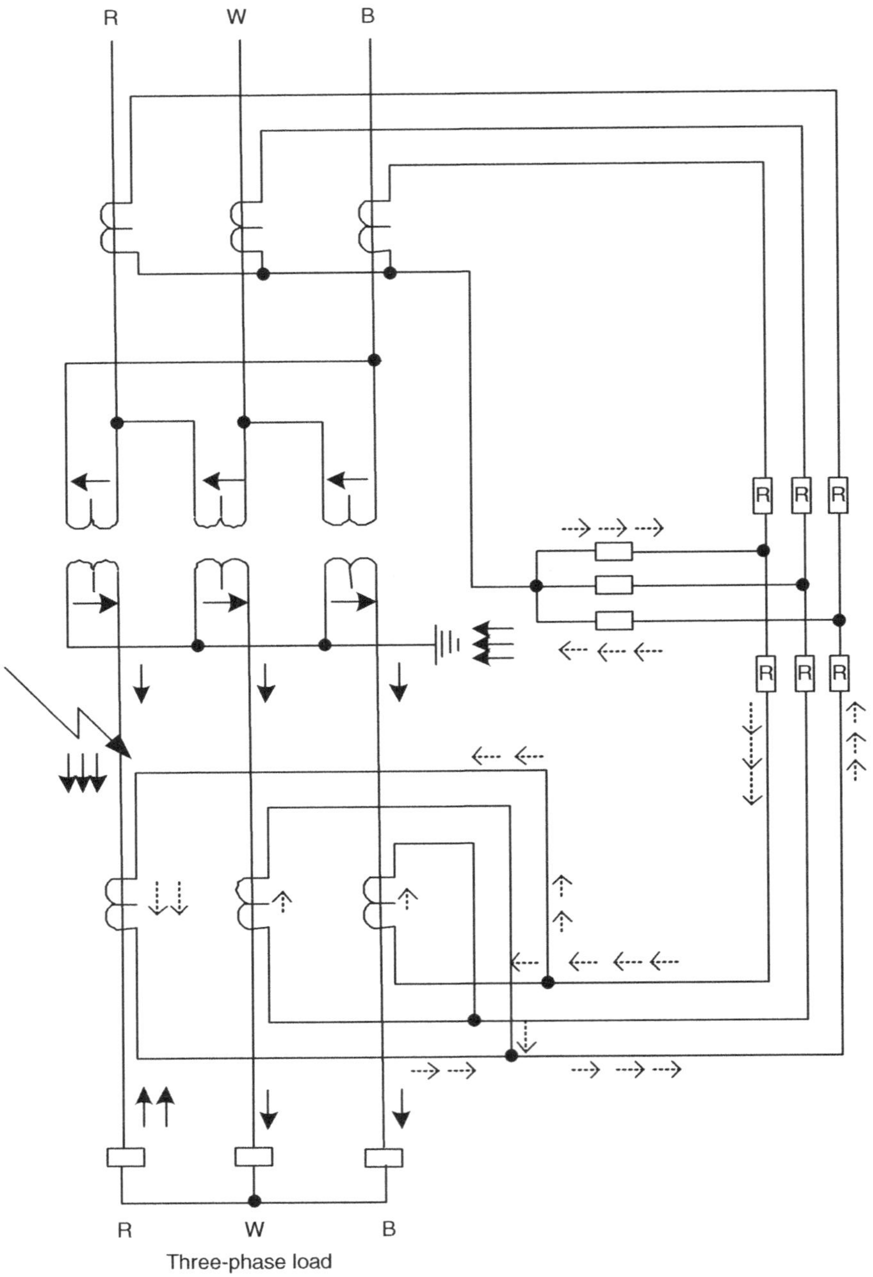

Figure 8.10 Zero-sequence current distribution for a delta–wye transformer and internal ground fault.

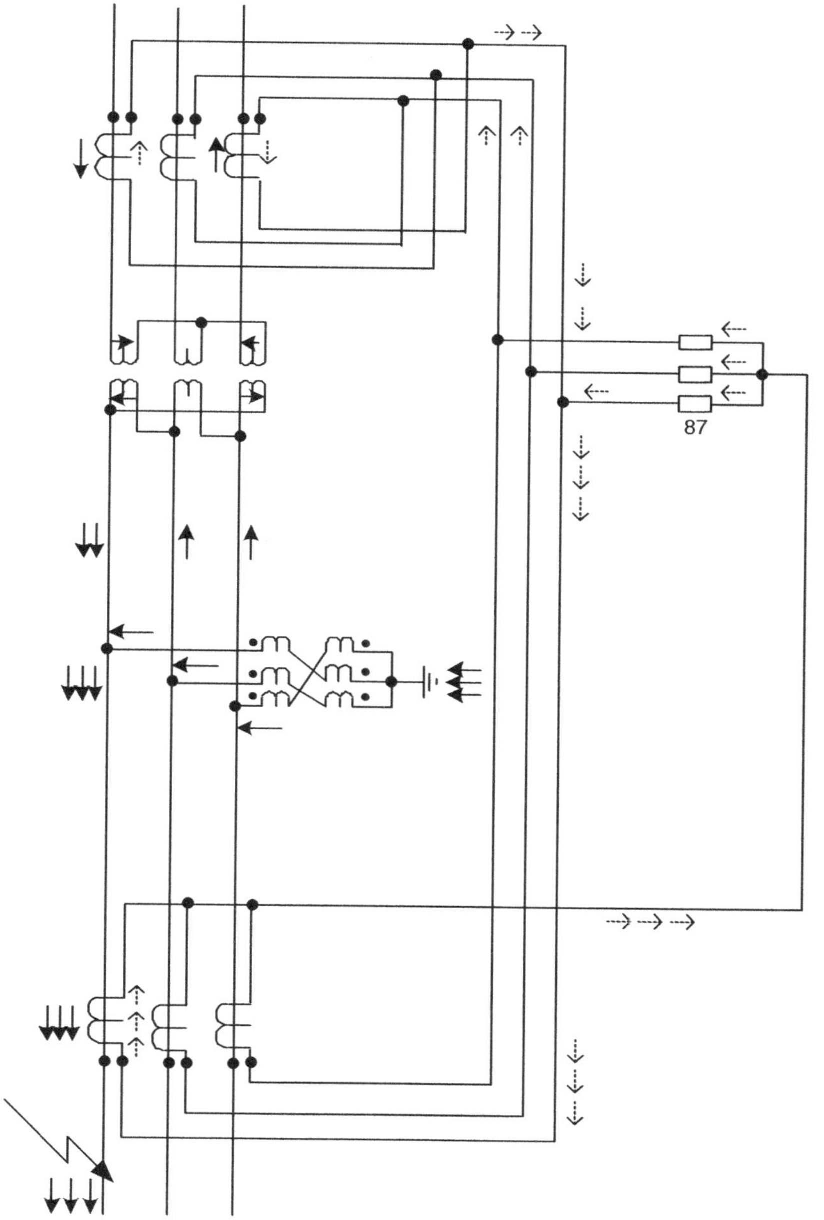

Figure 8.11 Zero-sequence current distribution for a wye–delta transformer with a grounding transformer and an external ground fault.

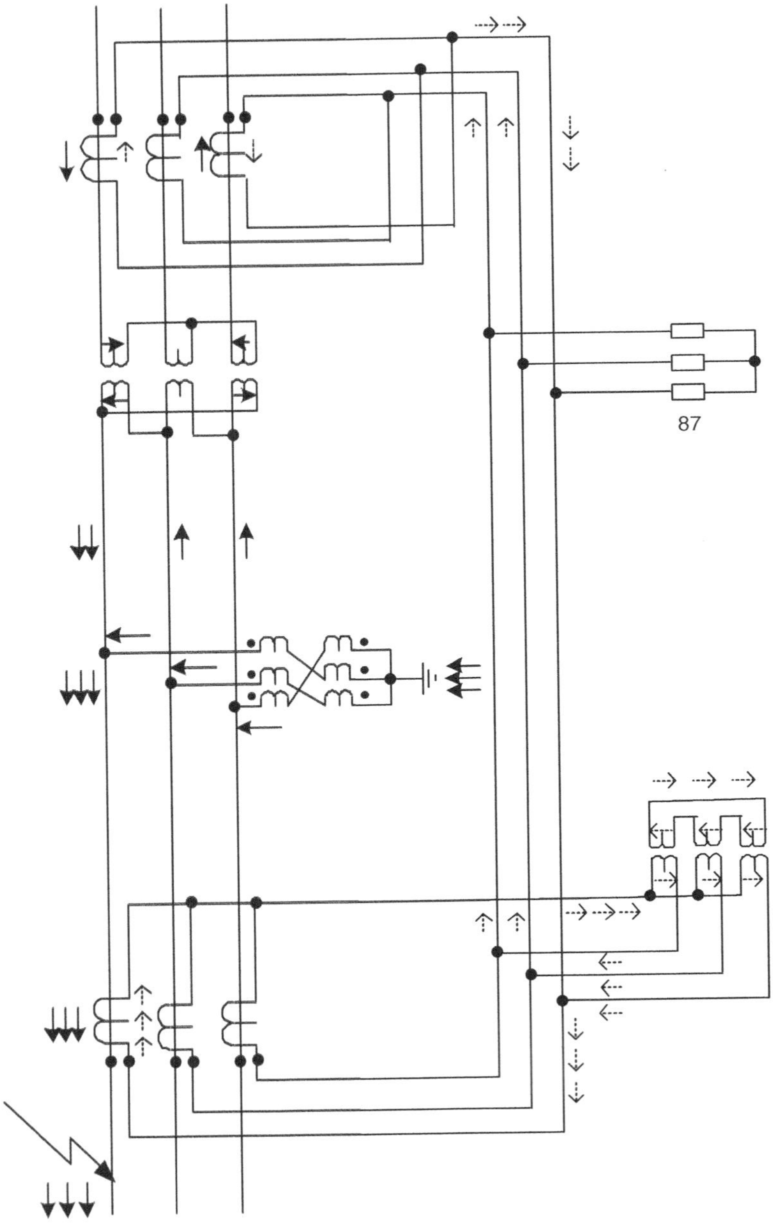

Figure 8.12 Zero-sequence current distribution for a wye–delta transformer with a grounding transformer and an external ground fault with a zero-sequence shunt.

of wye-connected CTs. Refer to Figure 8.12 for an external ground fault where a zero-sequence shunt is used when a grounding transformer is electrically located on the delta side of a wye–delta transformer within the transformer zone of protection. The shunt itself consists of three 5-5 A auxiliary CTs-connected wye–closed delta. Wherever a zero-sequence shunt is used, the neutral of the relay connection must not be connected to the neutral of the CTs; otherwise, it will not work as intended. When applied correctly, the transformer differential relay will not operate for external faults. Where modern digital relays are used, the zero-sequence shunting is done by eliminating the zero-sequence current matrixes from the calculation. The disadvantage of eliminating the zero-sequence currents from the matrix is that the protection becomes less sensitive for ground faults in the protected zone. Refer to Figure 8.13 for an internal ground fault for the identical application showing the transformer differential relay operating as intended.

Refer to Figure 8.14 showing the zero-sequence current distribution for an external fault where the grounding transformer is located outside the transformer differential protection zone. Also, refer to Figure 8.15 showing the zero-sequence current distribution for an internal fault where the grounding transformer is located outside the transformer differential zone. In both cases, the protections are dependable and secure for internal and external faults.

Where a zigzag grounding transformer is electrically located in the transformer differential zone, it is also covered for phase and ground faults by the same transformer differential protection. When located outside the transformer protection zone, an independent phase and ground protection are required. Figure 8.16 shows the phase protection of a zigzag grounding transformer. The CTs are connected in the delta so that zero-sequence currents circulate in the delta CT connection. This way the phase relays can be set very sensitively as they will not operate for ground faults and therefore can be set to pick up at a low value thereby increasing protection sensitivity.

Figure 8.17 shows the flow of positive, negative, and zero-sequence currents that normally flow in a transformer differential protection circuit. In this example, it is assumed that the load is unbalanced to show the flow of zero-sequence current as well.

Figure 8.18 shows an application used by some utilities where an overcurrent relay is placed in the neutral to the ground connection of the wye CTs and the differential relay.

Figure 8.19 shows the same ground overcurrent protection for an external fault where it does not operate.

While discussing grounding transformers as ground sources, it is important to keep in mind that for an overcurrent protection to operate correctly, the ground source must be electrically located upstream of the protection. Refer to Figure 8.20 where a zigzag grounding transformer is sourcing zero-sequence current to a ground fault. In this example, the ground source is electrically located between the 51N relays and the power transformer. The distribution of zero-sequence currents shows that the relay will operate correctly.

Refer to Figure 8.21 where a zigzag grounding transformer is sourcing zero-sequence current to a ground fault. In this example, the ground source is electrically located downstream of the 51N relays and the power transformer. The distribution of zero-sequence currents shows that the relay will not operate correctly. When using artificial ground sources, it is easy to make this mistake. The effect of this mistake is to leave power system equipment completely without protection and must be avoided at all costs.

8.3.5 Restricted Ground Fault Protection

Should the current for an internal line-to-ground fault be limited to a low value by high impedance grounding of the transformer neutral, it is possible that the differential relay may not receive

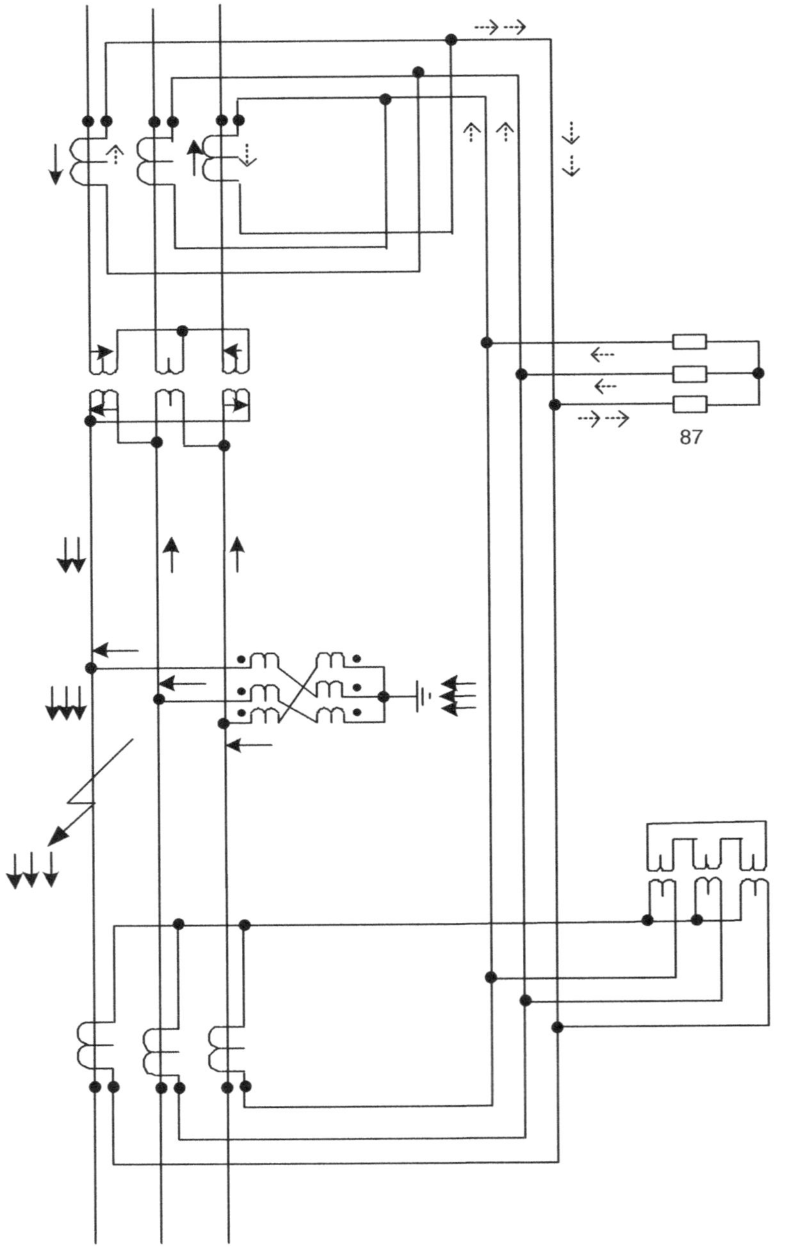

Figure 8.13 Zero-sequence current distribution for a wye–delta transformer with a grounding transformer and an internal ground fault with a zero-sequence shunt.

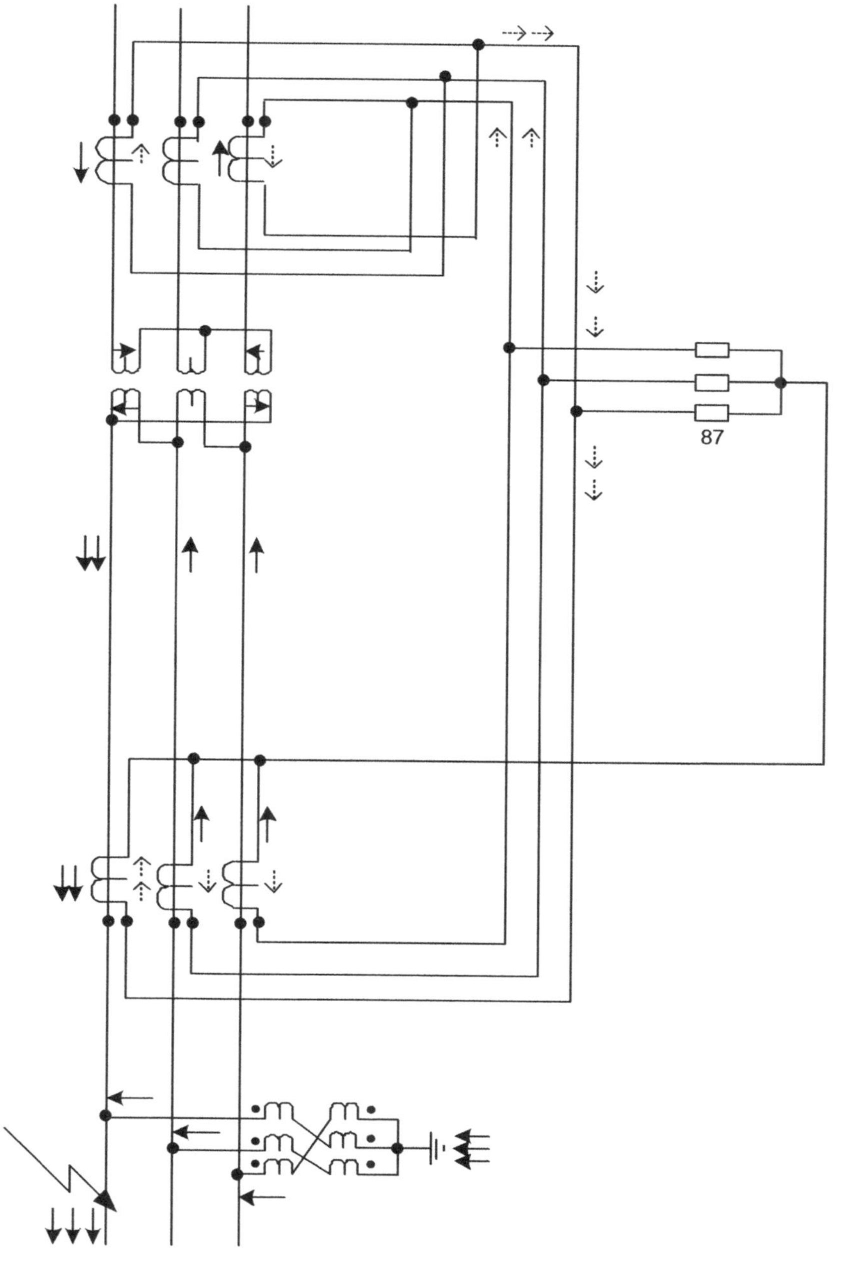

Figure 8.14 Zero-sequence current distribution for a wye–delta transformer with an out-of-zone grounding transformer and an external ground fault.

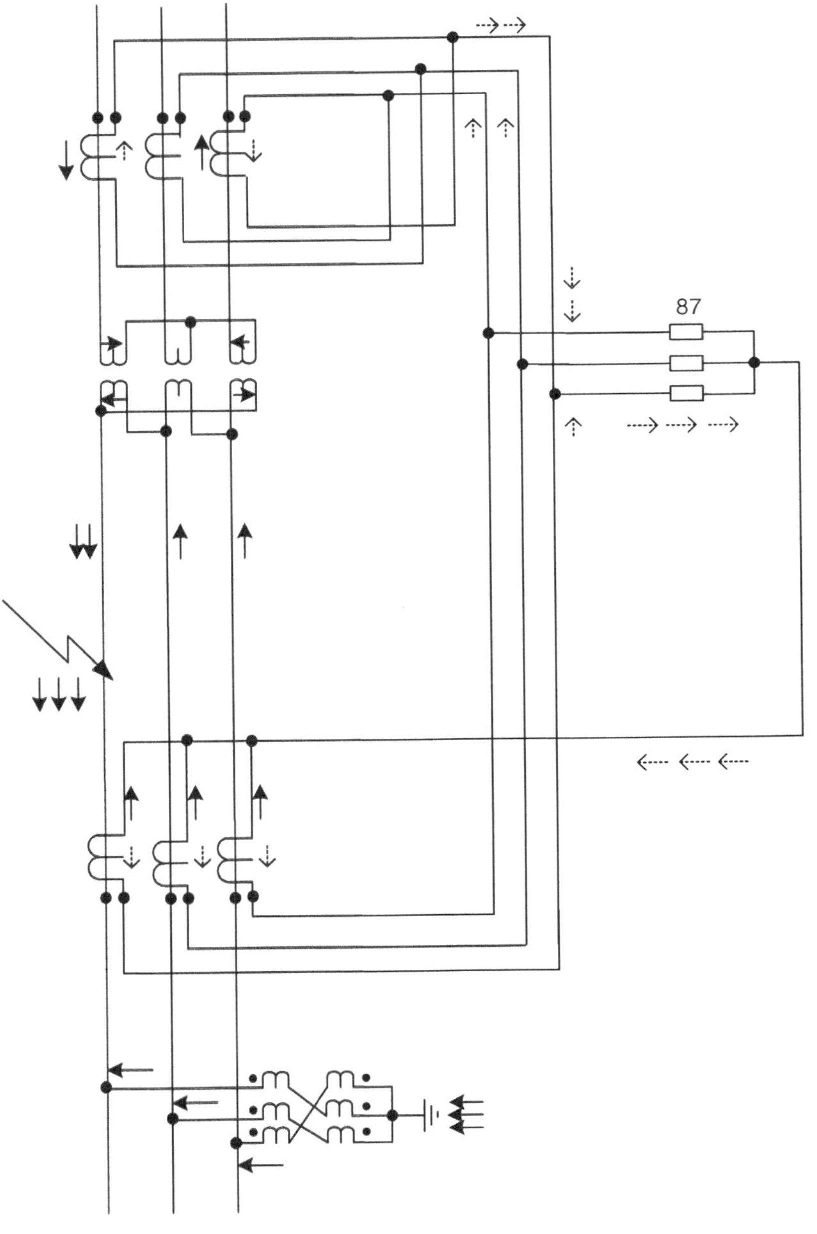

Figure 8.15 Zero-sequence current distribution for a wye–delta transformer with an out-of-zone grounding transformer and an internal ground fault.

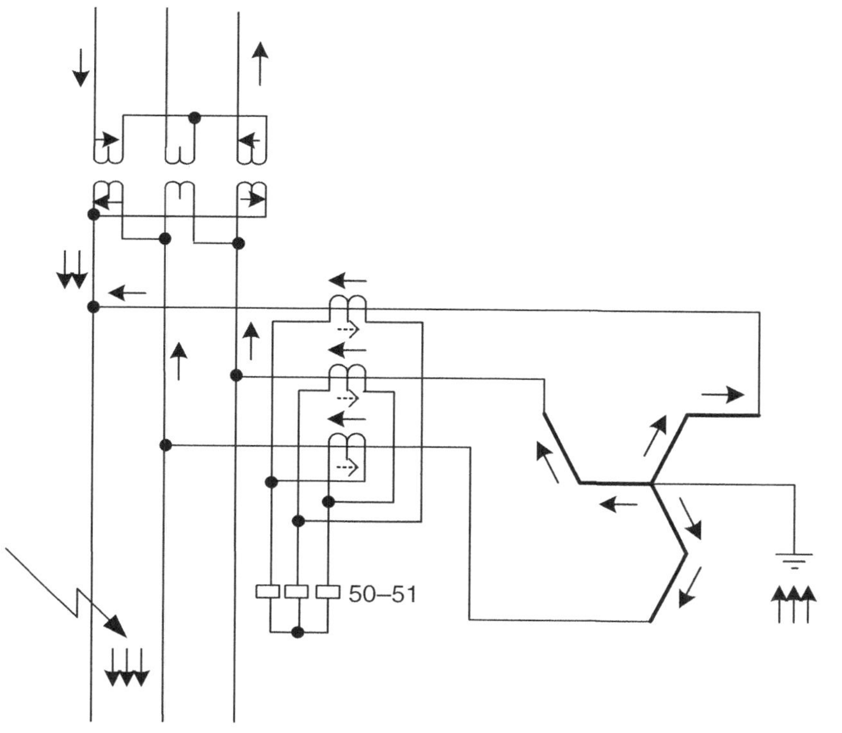

Figure 8.16 Phase protection of a zigzag grounding transformer.

adequate current for operation. Typically, utilities use two types of values for the neutral reactor in the transformer neutral to ground connection.

For three-wire distribution where normal unbalance current should not exist, the neutral reactor is typically rated around 7.5 Ω. This reactor will typically limit ground infeed per transformer high impedance grounded winding to 1000 A.

For four-wire distribution where normal unbalance current does exist, the neutral reactor is typically rated 1.5 Ω. This reactor will typically limit ground infeed per transformer low impedance grounded winding to 6000 A.

A separate RGF protection based on the high impedance principle is required where three-wire distribution type reactors are used that restrict the ground current to lower values. It should be noted that in neither case does the actual zero-sequence current flow through the differential relay for ground faults. Rather the positive and negative sequence components of the fault current are either high enough to operate the differential relay or not. Usually by the rule of thumb, unrestricted or partially RGF current results in sufficient positive and negative sequence fault current to operate the differential relay. When in doubt, it is best to provide RGF protection.

RGF protection, for example, would be standard primary protection for all power transformers when the neutral grounding reactor restricted ground infeed to a fault on the LV bus to 1000 A per transformer winding. For LV bus ground faults of 6000 A per winding, for example, RGF would typically not be applied. The primary reason for not using it is as shown in Figure 8.22, a CT is used in the neutral to ground connection of each secondary transformer winding. When RGF uses an electromechanical relay, this CT must ratio match the three LV winding CTs. Ratio matching

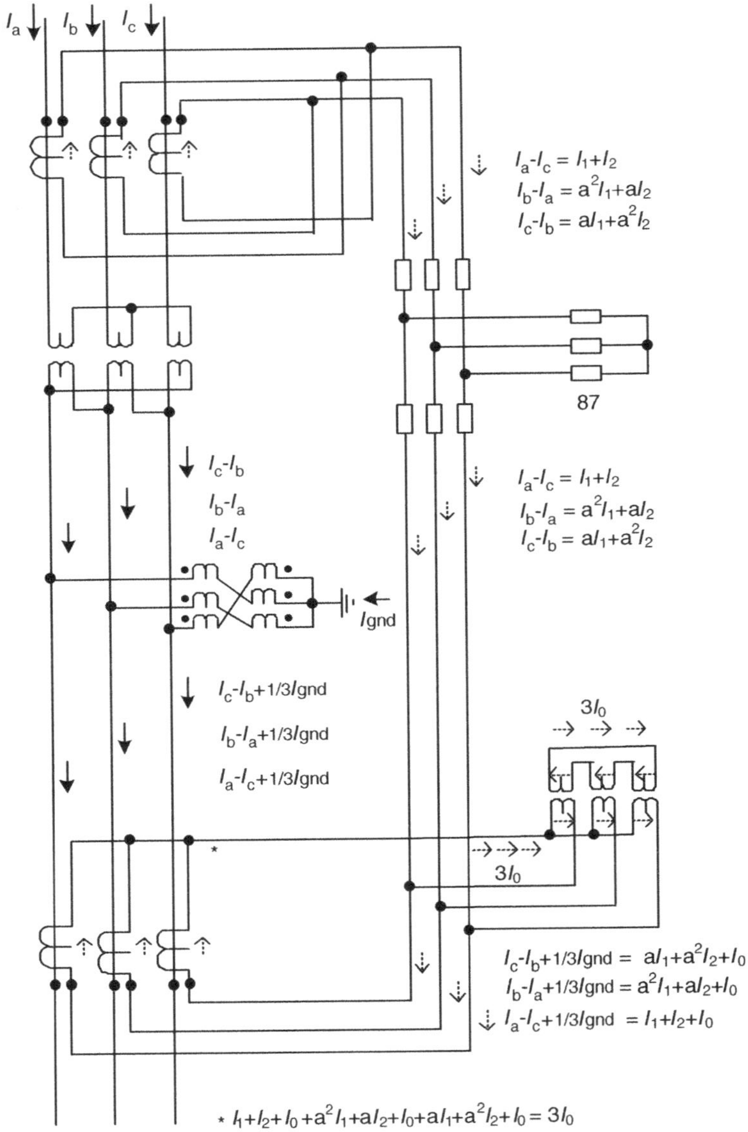

Figure 8.17 Positive, negative, and zero-sequence current flow in a differential circuit under normal unbalanced load.

auxiliary CTs is not permissible. The bushing CTs that transformers usually come with on the neutral are usually much lower rated than required. For applications where RGF is applied where an electromechanical relay is used, a difficult to install and costly outdoor type CT is required located between the transformer neutral and its ground connection. More recently with the advent of digital relays, this no longer was an issue as CT ratio matching is done internally in the relay. It is highly recommended to therefore always apply RGF when a digital relay is used as there is no additional costly CT required, and it provides much greater sensitivity for low-level and incipient ground faults.

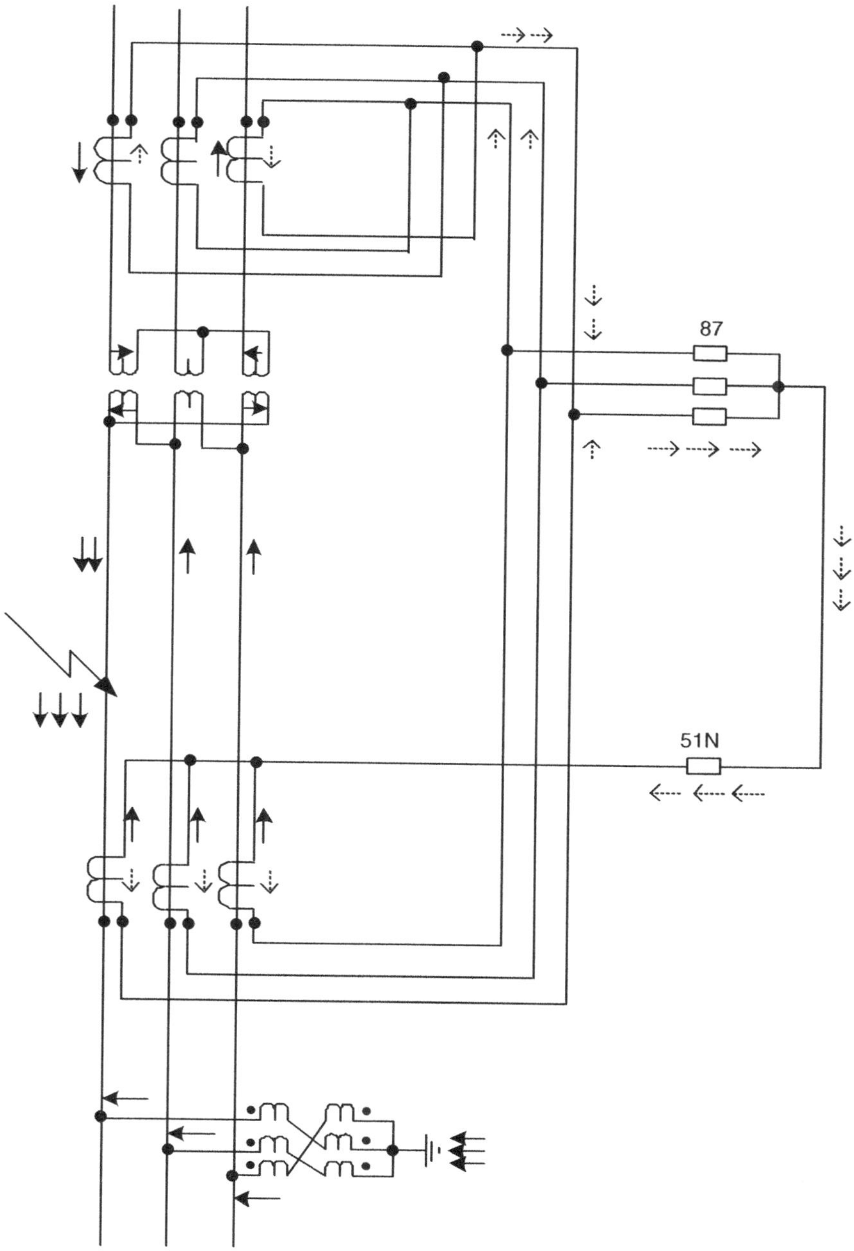

Figure 8.18 Application of ground overcurrent protection for an internal fault.

8.3.5.1 Theory of Operation

Refer to Figure 8.22 showing the CT connections and relay location for a typical application of RGF protection. A set of three CTs connected in parallel on the transformer LV winding side are connected in parallel with a CT connected in the transformer neutral to the ground. A sensitively set instantaneous overcurrent relay itself in series with a rheostat type resistance is connected in parallel with the CTs as illustrated.

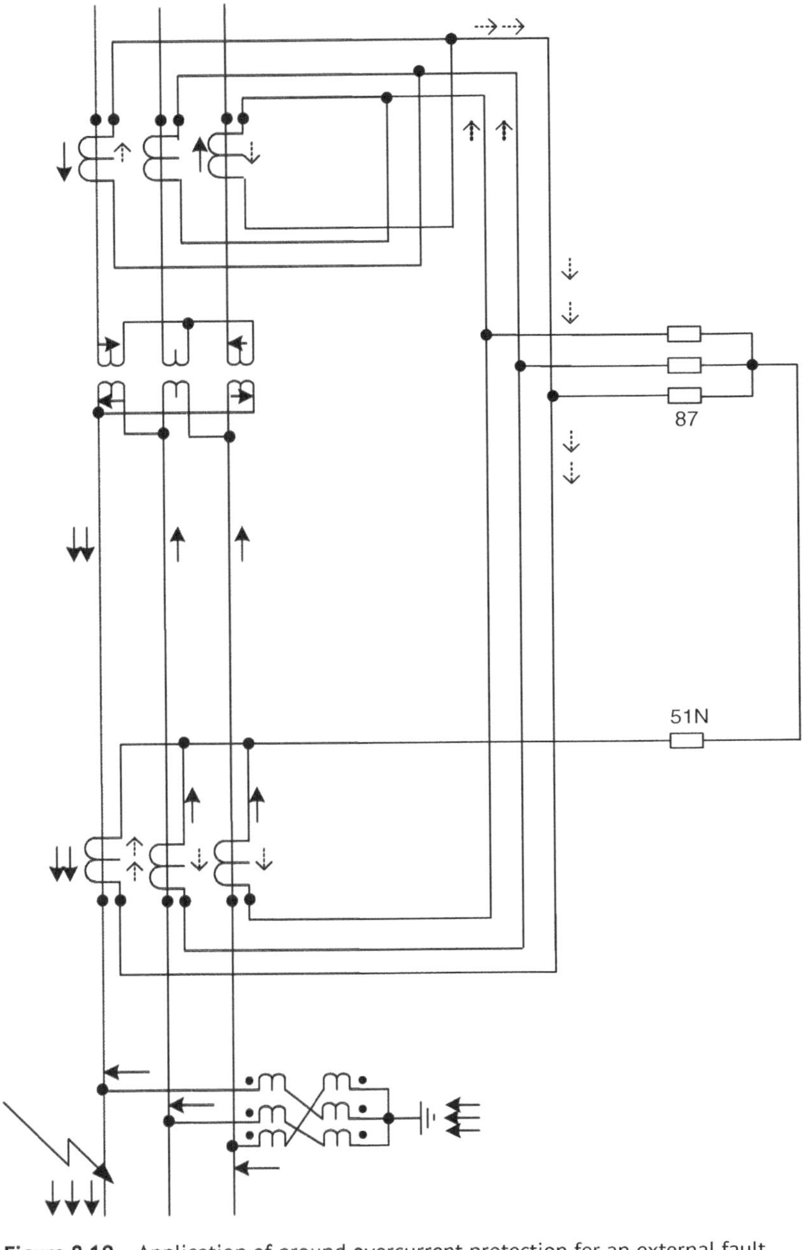

Figure 8.19 Application of ground overcurrent protection for an external fault.

The key to RGF protection is that the rheostat resistor connected in series with the overcurrent relay turns the resistor – relay combination in a sense into a voltage sensing device. With a fixed instantaneous overcurrent relay pick up, sufficient voltage must be applied across the resistor – relay nodes for the relay to measure sufficient current to operate. The real advantage to this circuit is that it is possible to calculate with certainty the value of the voltage applied across the resistor – relay nodes for an external three-phase fault for when one LV transformer CT completely

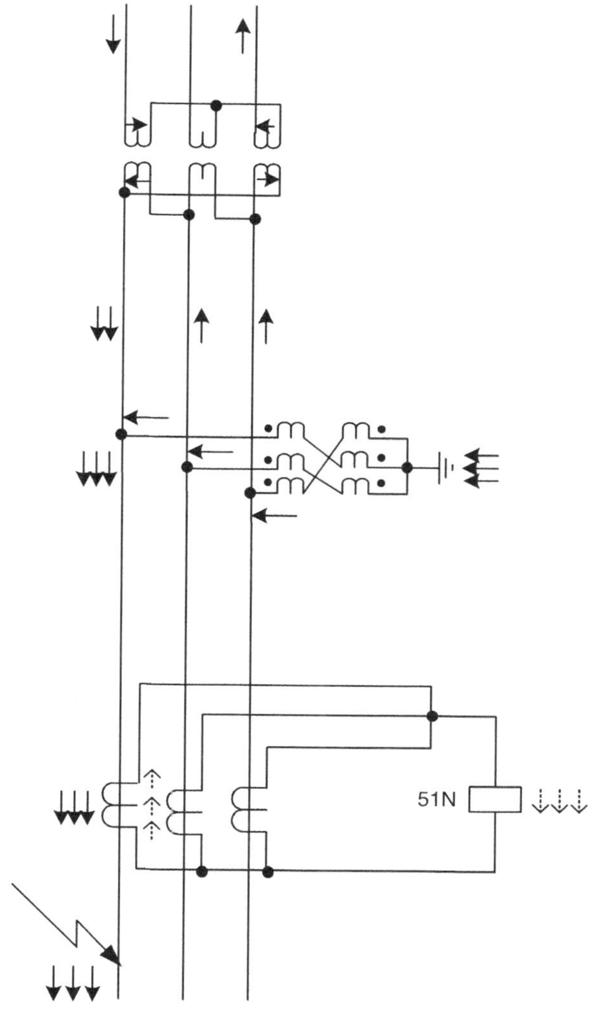

Figure 8.20 Location of grounding transformer with relay operation.

saturates producing no secondary current. When this happens, the fully saturated CT looks like a dead short. This is the worst-case for an external fault where the protection must not operate.

For an internal ground fault, the CT located in the transformer neutral to ground connection would also need to develop sufficient voltage across its secondary windings to operate the relay. However, in this case, the CT is chosen with an accuracy classification that ensures that sufficient CT secondary voltage exists to push current into the relay to operate the protection before the CT goes into severe saturation.

As previously discussed in the chapter on CTs, a very considerable build-up of flux with an asymmetrical fault current may take a CT into saturation, with the result that the dynamic exciting impedance is reduced and the exciting current greatly increases.

When a CT completely saturates, the shunt impedance of the saturated CT can be considered to have fallen to zero and is replaced on the equivalent diagram by a short circuit, thereby reducing the impedance of this CT to the value of the dc winding resistance.

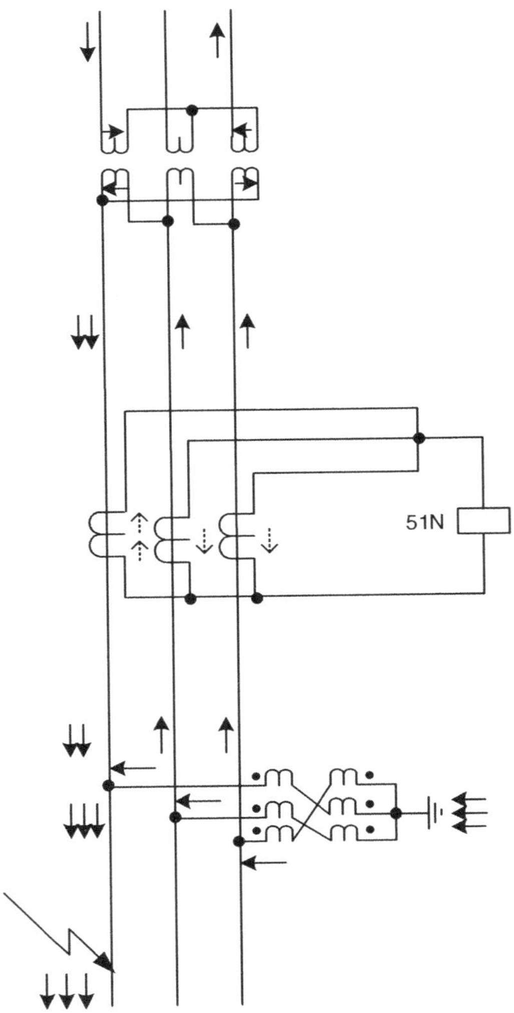

Figure 8.21 Location of grounding transformer without relay operation.

It is assumed that one CT of a balanced group becomes completely saturated, while the others remain on the linear portion of the excitation characteristic and require negligible exciting current.

The voltage developed at the relay connection point is given by the equation known as the stabilizing voltage $V_S = I_F\,(R_{ct} + 2R_L)$ as shown in Figure 8.23.

This equation shows that the relay pickup current can be reduced to a value below that of any given relay setting by a suitable choice of relay resistance. Also, the equation further shows that the maximum voltage which can be applied to the relay is given by this same equation. This value can be low compared to the CT knee point voltage so that a value of resistance can be applied to R_{st} which will ensure stability while still permitting sensitive operation for an internal fault. Resistance added to the natural resistance of the relay itself is known as stabilizing resistance or R_{st}. This value must be compared to the CT knee point voltage of the CT in the transformer neutral to ground connection that measures the zero-sequence current flowing from the location of ground fault back to the grounded neutral. Since the instantaneous overcurrent relay that measures this

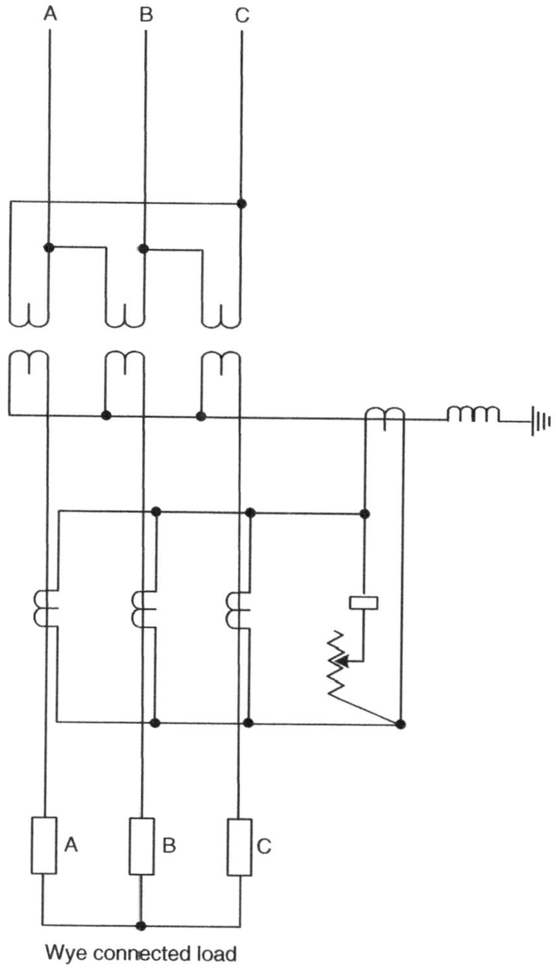

Figure 8.22 Restricted ground fault protection.

current is in series with the stabilizing resistor, this CT must generate a sufficient secondary voltage to overcome the stabilizing voltage.

The stabilizing voltage V_S must be compared to the knee point voltage of the transformer neutral to ground CT as shown in Figure 8.24.

Figure 8.25 shows the flow of zero-sequence currents for an internal transformer ground fault, and Figure 8.26 shows the flow of zero-sequence currents for an external transformer ground fault.

Conditions

LV Three-Phase Fault

CT1 fully saturates; therefore, secondary current can flow in it, in either direction.

CT connections are made in the relay building

CT2 and CT3 sum their respective fault currents in the relay building and push that secondary fault current through saturated CT1 (electrically just a resistance) and the loop leads between CT1 and the relay building.

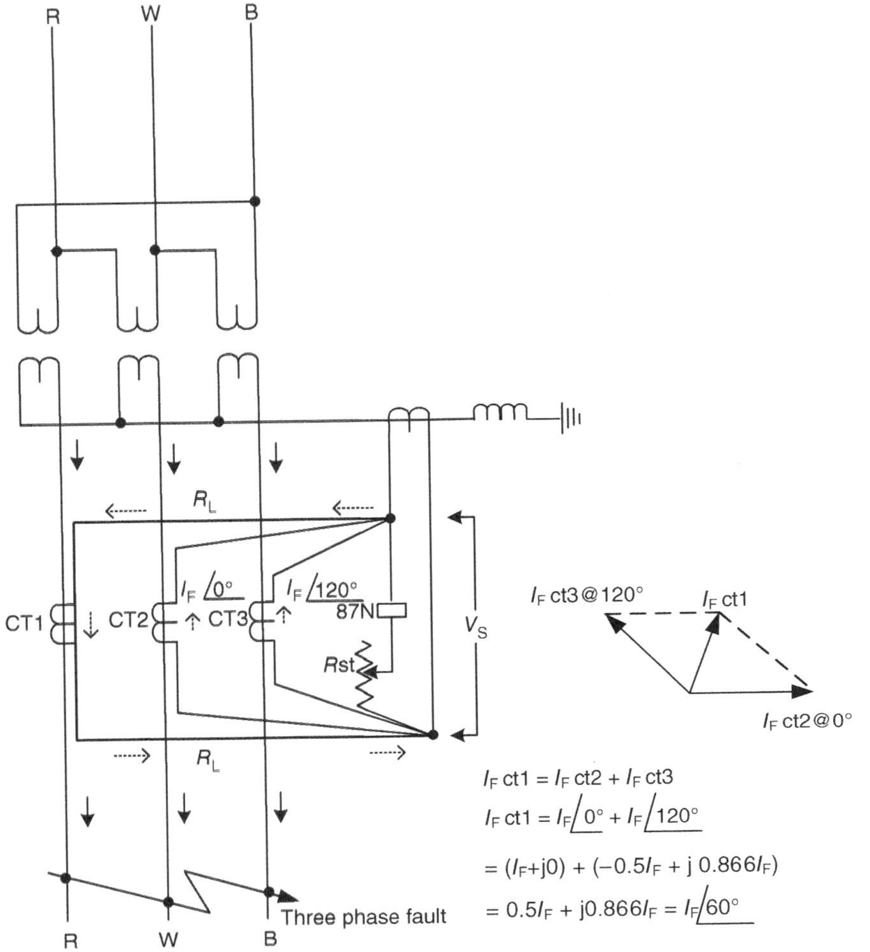

Figure 8.23 Restricted ground fault protection calculation.

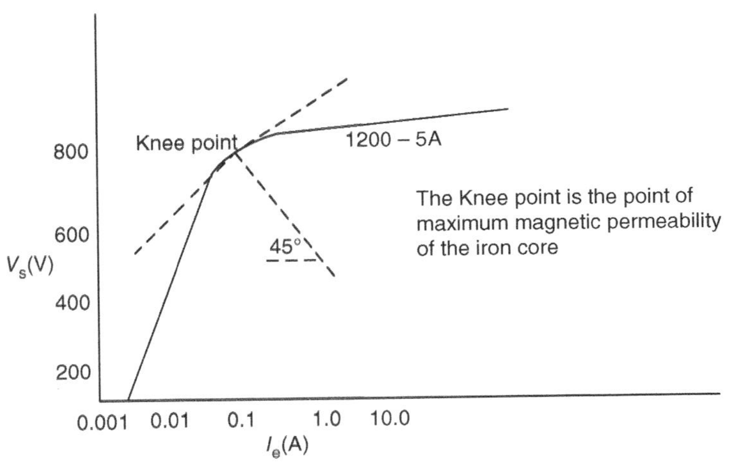

Figure 8.24 Typical CT excitation characteristic.

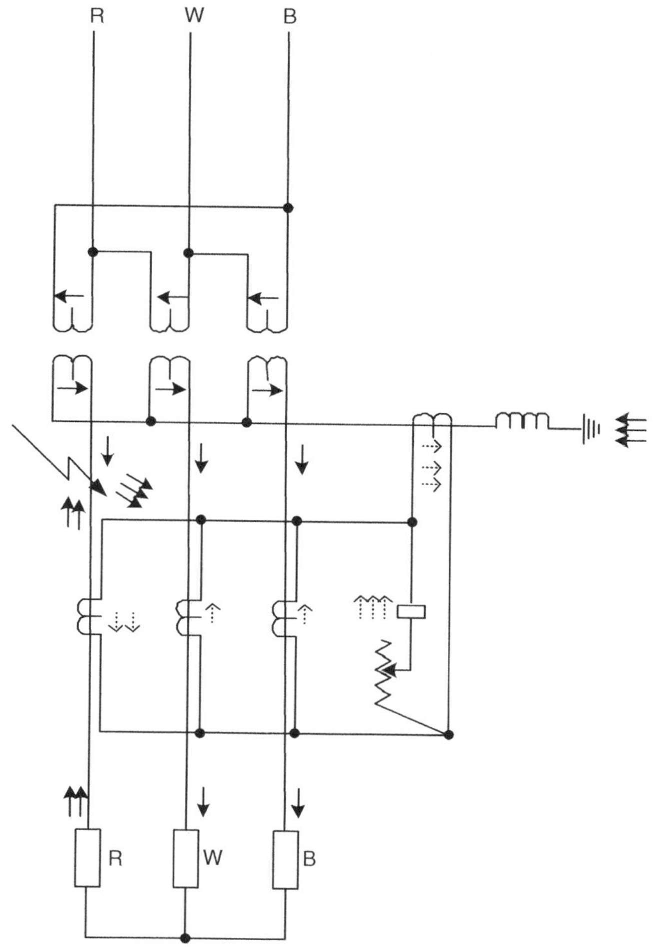

Figure 8.25 Restricted ground fault protection for an internal fault.

$$V_S = I_{Fct1}\,(R_{ct1} + 2R_L)$$

$$I_{PKP}\,(R_{ST} + R_{RLY}) \text{ must be} \gg V_S$$

RGF protection operates for ground faults within the confines of the protected zone bounded by the CTs as shown in Figure 8.25. It will remain stable for all faults outside of this zone as shown in Figure 8.26.

The gain in protection performance comes mainly from using an instantaneous relay with a sensitive setting. RGF protection is often applied even when the neutral is solidly grounded. Since fault current then remains at a high value even to the last turn of the winding, virtually complete coverage for ground faults is achieved.

RGF protection using digital relays is much more complex. Some digital relays use a digital overcurrent element in series with an external stabilizing resistor with the advantage over

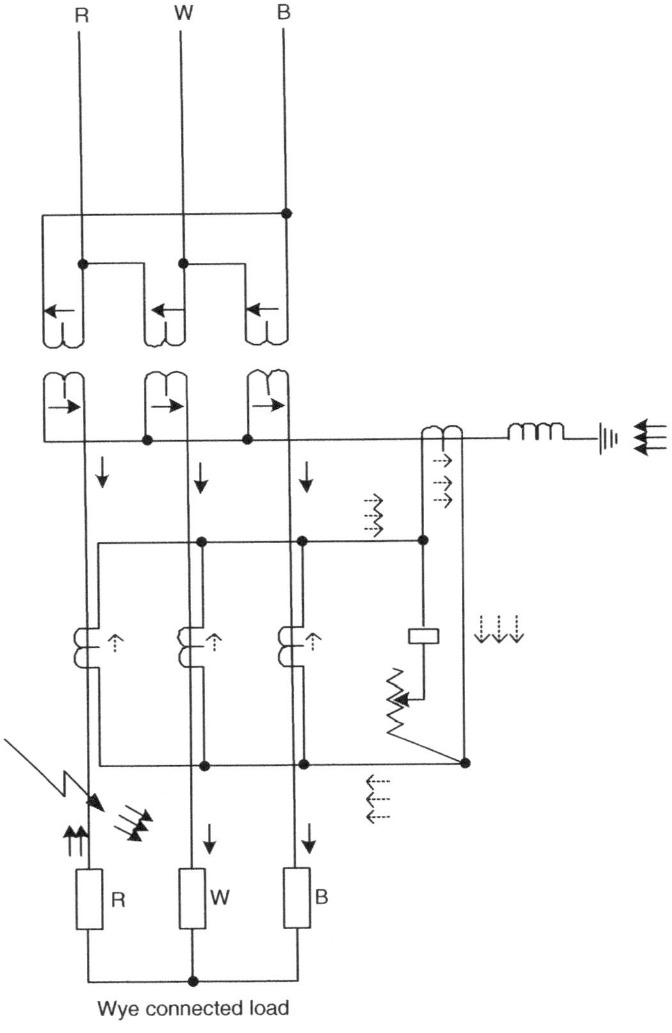

Figure 8.26 Restricted ground fault protection for an external fault.

electromechanical of being capable of compensating for differing CT ratios and sophisticated event capture. Other transformer digital relays use a true RGF protection using a sensitive overcurrent element and a proprietary sophisticated algorithm to maintain stability for out-of-zone faults. The advantage of this approach is that a stabilizing resistor is not required thereby simplifying relay mounting and location on a rack or panel.

8.3.6 Percent Differential Relay Settings

8.3.6.1 Current Transformer Mismatch and Under-Load Tap Changing

The two primary factors affecting the choice of percent slope are current transformers located on the primary and secondary of the transformer that comes close to but still does not exactly match the transformation ratio and transformer under-load tap changing.

These two factors introduce what is termed in this chapter as system unbalance. System unbalance means that there is by virtue of these two factors, a continuous amount of spill current flowing into the relay operate coil due to the overall unbalance. This spill current increases with through current such as load or external fault current in direct proportion to the system unbalance.

8.3.6.2 Percent Differential Relay Operation

The principle of percent differential relay operation is similar to differential protection using just simple overcurrent type relays except that the differential current required to operate this relay is a variable quantity, owing to the effect of a restraining coil as shown in Figure 8.27.

Refer to Figure 8.28 representing the basic construction and operation of a typical electromechanical percent differential relay. The relay depicted in Figure 8.28 has three coils wound around a magnetically permeable iron bar. The CT secondary leads are connected to restraint windings. The two restraint windings are then connected in parallel with each other and then in series with an operate coil. The other side of the operate coil is connected in parallel with the two non-spot sides of the CTs.

The operate coil has N number of turns, while each restraint winding has half the number of turns or $N/2$ turns. The operate winding is wound in the opposite direction to the two restraint windings. The induced magnetic flux ϕ_{operate} produced by the operate coil is in the opposite direction to the magnetic flux ϕ_{restrain} produced by the two restraint windings. A net magnetic flux in the operate direction will pull the plunger-type relay contact to the closed or trip direction.

The primary current is shown in the direction of supplying load where the CT secondary currents circulate around the two CTs. Any difference in value between them will flow through the operate coil.

The electrical circuit representation of the percent differential relay in Figure 8.28 is shown in Figure 8.29.

Referring to Figure 8.29, the differential current in the operating coil is proportional to $I_1 - I_2$ and the equivalent current in the restraining coil are proportional to $(I_1 + I_2)/2$ since the operating coil is connected to the midpoint of the restraining coil. If N is the number of turns on the operating restraining coil, the total ampere turns are $I_1N/2 + I_2N/2$ which is the same as if $(I_1 + I_2)/2$ were to flow through the entire coil.

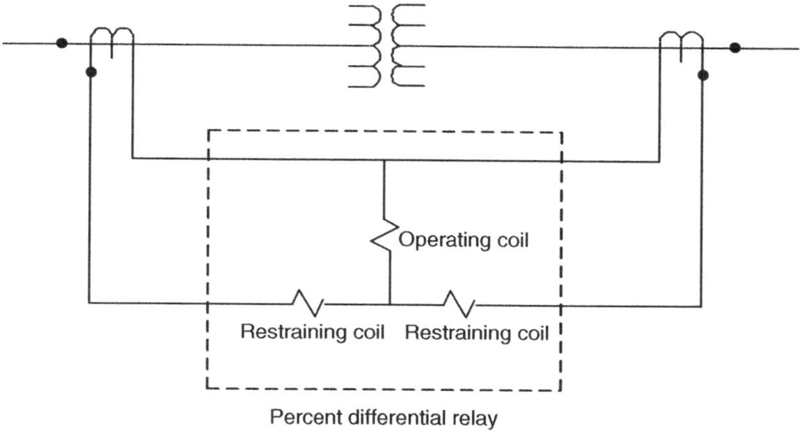

Figure 8.27 Percent differential relay.

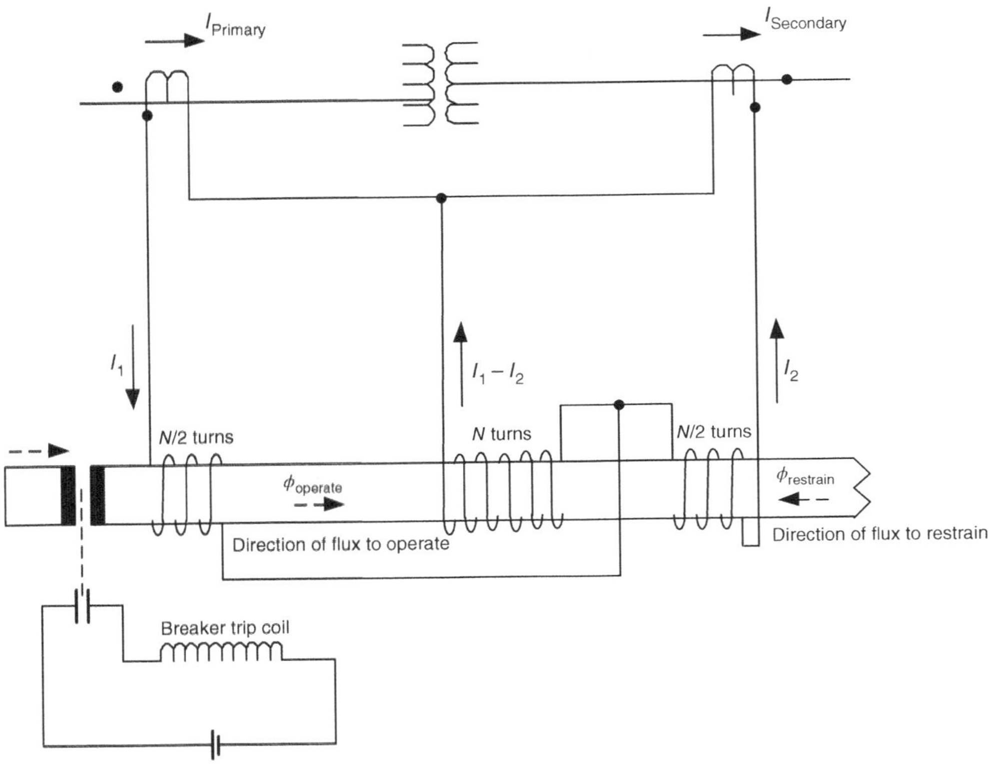

Figure 8.28 Percent differential relay basic construction and operation.

The percent slope shown in Figure 8.30 is derived in terms of competing for magnetic fluxes via the tendency to operate in ampere turns and the tendency to restrain in ampere turns. The percent slope is the tendency to operate in ampere turns divided by the tendency to restraint in ampere turns.

$$\text{Operate in ampere turns} = (l_1 - l_2)\,N$$

$$\text{Restraint in ampere turns} = \frac{l_1 N}{2} + \frac{l_2 N}{2}$$

$$\text{Slope in ampere turns} = \frac{(l_1 - l_2)\,N}{\frac{l_1 N}{2} + \frac{l_2 N}{2}}$$

$$= \frac{(l_1 - l_2)N}{\frac{(l_1 + l_2)N}{2}}$$

$$\%\text{Slope} = \frac{(l_1 - l_2)200}{(l_1 + l_2)}$$

The degree of percent slope is a function of the relative values of No for the operate coil and N_r the two restraint coils. When N_o and N_r are equal, the percent slope is fixed at 45%. When the number of turns N_o making up the operated coil is higher than the number of turns N_r making up the two restraint windings, the percent slope will be lower than 45%. When the opposite is true and there are more relative turns in the restraint windings, the percent slope will be higher than

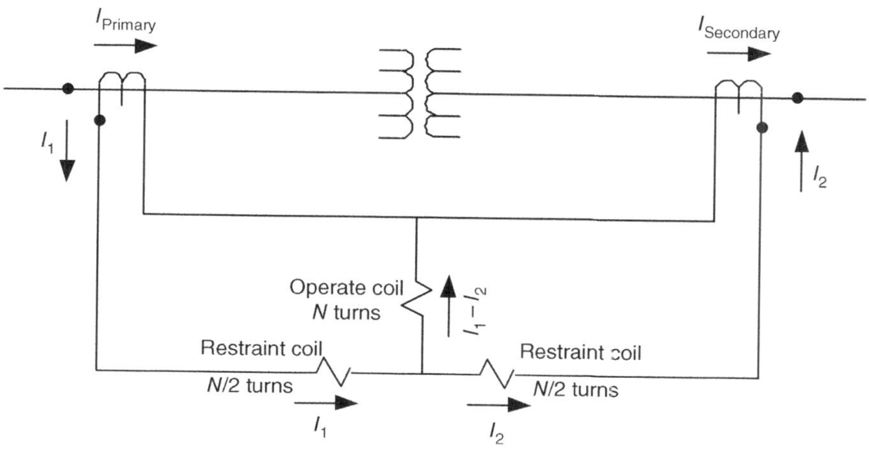

Figure 8.29 Electrical circuit representation of the percent differential relay.

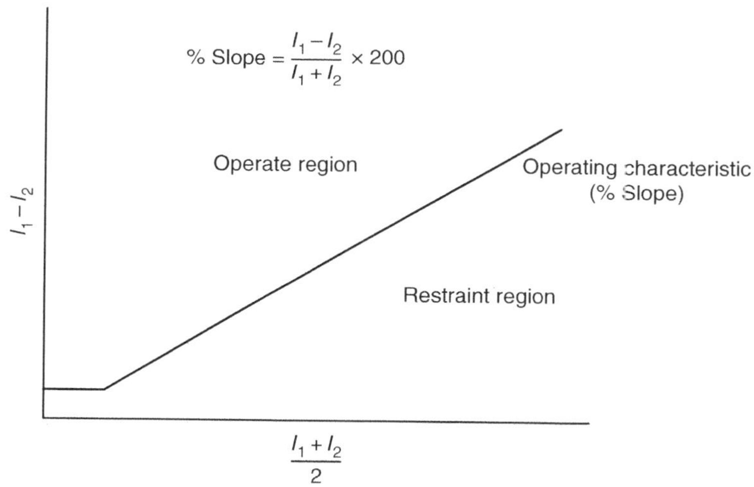

Figure 8.30 Operating characteristic for the percent differential relay.

45%. The percent slope is therefore adjustable by means of adding inter-turn taps to the operate and restraint windings.

All digital percent differential relays are programmed with algorithms that mimic this historical simple percent differential relay model. The definition of % slope is always the same regardless of relay type.

8.3.6.3 Maximum System Unbalance

Maximum system unbalance is the aggregate unbalance introduced by the various factors that cause spill current to flow in the operating coil of the percent differential relay per the same relationship that defines relay operation.

Refer to Figure 8.31 showing the two main factors causing spill current to flow. These are CT mismatches and transformer tap changing.

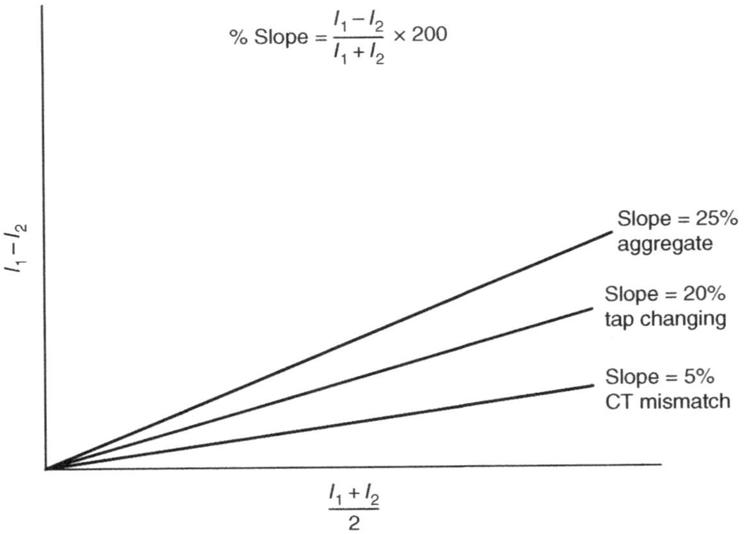

Figure 8.31 Maximum system unbalance.

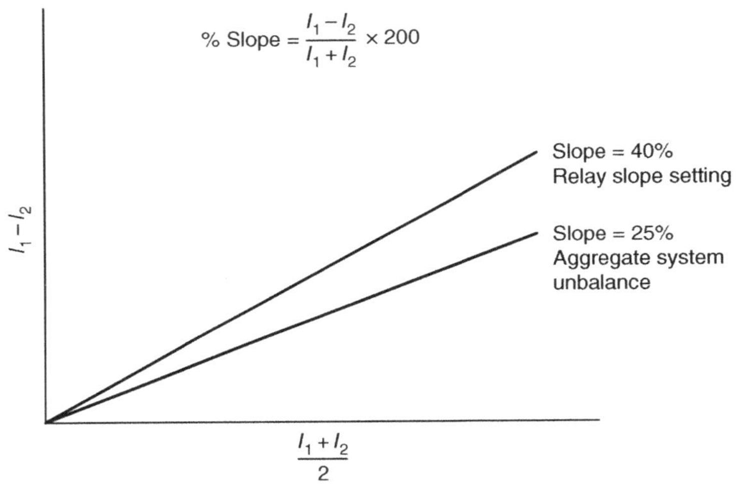

Figure 8.32 Comparison of maximum system unbalance and relay slope.

The method of calculating the required differential relay percent slope selection is to calculate the system unbalance for each of these two factors according to the same relationship that defines the transformer percent slope selection then take the aggregate of the two to obtain the maximum system unbalance in relay terms. The relay percent slope should then be selected to provide approximately 50% safety margin above this value as shown in Figure 8.32.

By comparing the maximum system unbalance plotted on the same graph as the relay percent slope setting, it can be assured that the relay will never operate incorrectly due to natural system unbalance at any value of through current.

The CT ratio mismatch is taken into account as well as any $\sqrt{3}$ factors introduced by CTs connected in delta. The effect of under-load tap changing is always taken as deviation from the

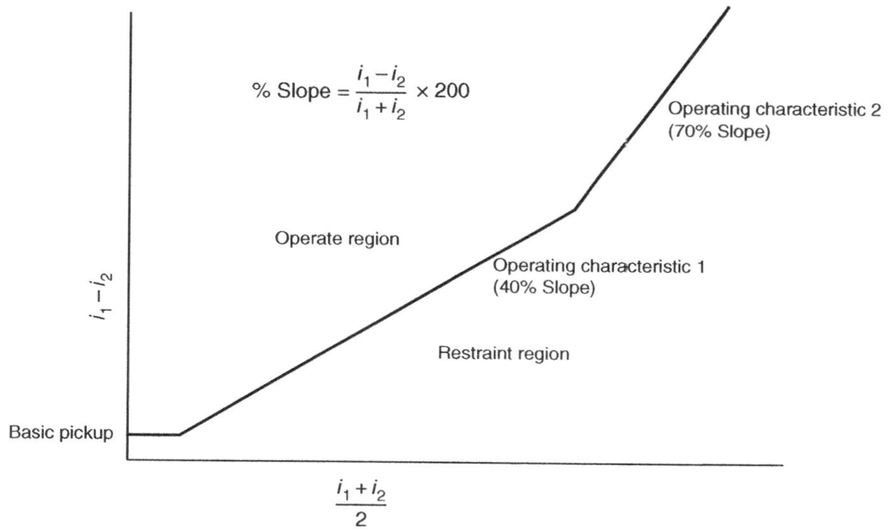

Figure 8.33 Double slope characteristic.

center tap. For accuracy, the percent transformer regulation should be used rather than voltage change from the center tap. It is the percent variation in current due to the voltage tap changing which affects the overall unbalance. This will become apparent in the percent slope calculation examples to follow (Figure 8.33).

8.3.6.4 Introduction of % Slope Under Fault Conditions

Regardless of % slope applied to a percent differential relay, it will on its own superimpose a steeper % slope that kicks in a value of through current that approximates high transformer fault current for an external fault. The intention of the very steep slope is to ensure that the overall differential protection remains secure even with severely saturating transformer LV winding CTs. These CTs usually on the bushings of LV breakers are of a lower rating than those on the transformer HV bushings and tend to go into saturation.

The advantage of this second slope characteristic is that it is less likely to operate incorrectly than a differentially connected overcurrent relay when a fault occurs external to the protected zone. Since the percentage-differential relay has a steeper rising pickup characteristic as the magnitude of through current increases, the relay is restrained against operating improperly.

8.3.6.5 Effect of Magnetizing Current Inrush on Differential Relays

How CTs are connected, and the way in which CT ratios and relay taps are chosen for differential relaying neglect the transformer exciting current component. This component causes the current to flow in the relay's operating coil, but it is so small under normal load conditions that the relay does not tend to operate.

The largest inrush and the greatest relay operating tendency occur when a transformer bank has been completely de-energized, and then, a circuit breaker is closed, thereby applying a voltage to the windings on one side with the windings on the other side still disconnected from the load.

The principle of 2nd harmonic restraint makes a differential relay self-desensitizing during the magnetizing current inrush period, but the relay is not desensitized if a fault should occur in the transformer during the magnetizing inrush period. Thus, the relay can distinguish the difference between magnetizing inrush current and fault current by the difference in wave shape.

Harmonic analysis of a typical magnetizing inrush current waveform:

Harmonic component	Amplitude in % of fundamental
2nd	63.0
3rd	26.8
4th	5.1
5th	4.1
6th	3.7
7th	2.4

The operating coil will receive from the differential current transformer only the fundamental component of the differential current, the harmonics being separated, rectified, and fed back into the restraining coil.

In many large cities, transformers are commonly supplied by underground cables. Underground cables have a high distributed capacitive effect which causes a discharge transient. The characteristics of this transient are very different from those of conventional magnetizing inrush currents. The discharge transients often cause the currents in all three phases to contain low-magnitude second harmonics. The reduced level of second harmonics has historically caused percent differential relays to lose restraint and misoperate.

8.3.6.6 Third Restraint Winding

Some substation transformers have double secondary windings. The transformer differential relays can either be optimized for connection of both secondary transformer windings CTs to be connected or the CTs may be connected in parallel as shown in Figure 8.34. At substations that supply radial

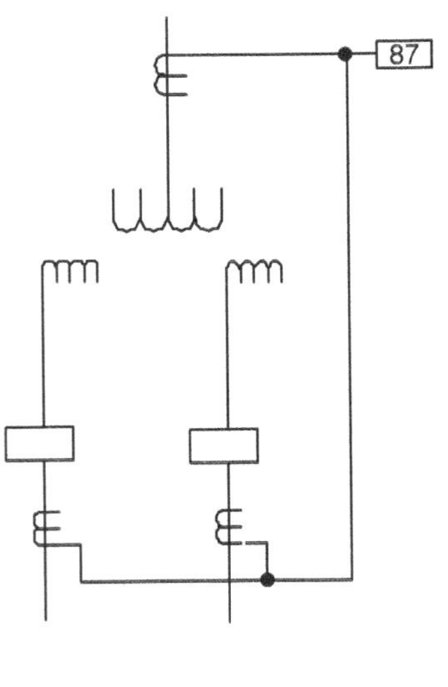

Figure 8.34 Zoning for a transformer with double secondary windings.

Figure 8.35 Percent differential relay misoperation.

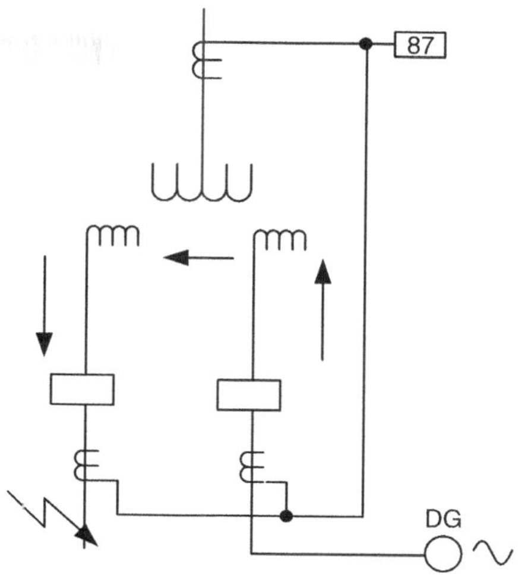

load as long as there is no local source of generation downstream of the transformer secondary windings, it is possible to connect the two transformer LV winding sets of CTs in parallel. The only method of adding a third restraint winding to an electromechanical percent differential relay was to add a specialized third relay. There are some applications where three-winding restraint is necessary at substations supplying large synchronous motors back-feeding onto LV bus faults. The problem with not supplying a third restraint winding is that faults on one bus supplied by the other especially when the transformer is off potential will result in a very sensitive overcurrent element measuring CT secondary currents that may have high harmonic content due to CT saturation and in the absence of restraint will trip as shown in Figure 8.35. Virtually, all digital transformer percent differential relays come with at least three CT inputs thereby eliminating the need to connect CTs in parallel.

8.3.6.7 Transformer Overload Protections for Double Transformer Contingency

Normally, transmission substation transformers are not protected against overload where each of the two power transformers is capable of carrying the total station load. However, there is a contingency termed double transformer contingency with certain substation configurations. Take for example a substation configuration as shown in Figure 8.36. When one LV bus is removed from service for maintenance in this example BU1 the entire station load is carried by two transformer secondary windings, one from each transformer. In this example, it is assumed that the feeder back-to-back switches are closed so that all feeders are supplied from bus BU2. The entire station load is supplied from the single winding of each of the two transformers. In the example shown in Figure 8.37, in the unlikely event that the line supplying T2 is faulted and trips even with automatic reclosure the single remaining transformer winding will be severely overloaded. The overload protection to cover this is known as double transformer contingency. The overload relays only monitor one phase current as the load is always balanced. The setting criteria typically used are to operate for load more than 1.75 times the transformer force cooled rating (FCR). The overload protection should not operate for genuine faults but to permit instantaneous and timed protections downstream to clear faults within their protective zones with proper coordination. In general, other similar substation configurations and operating conditions must be thoroughly analyzed.

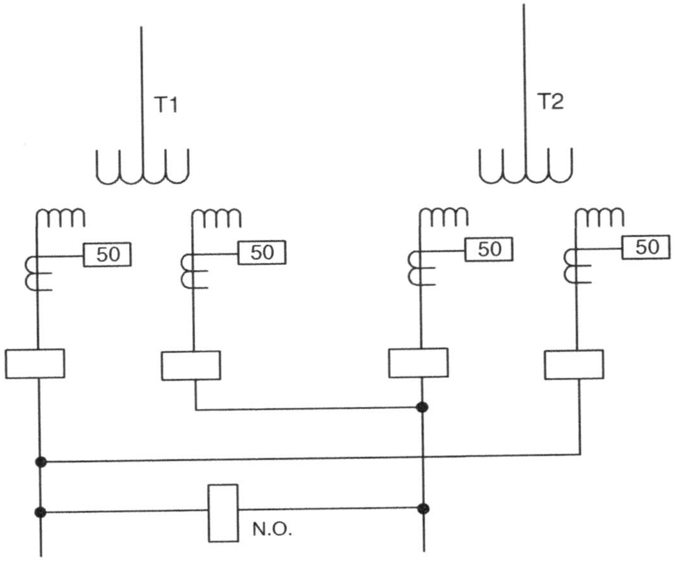

Figure 8.36 Substation configuration requiring overload type protection.

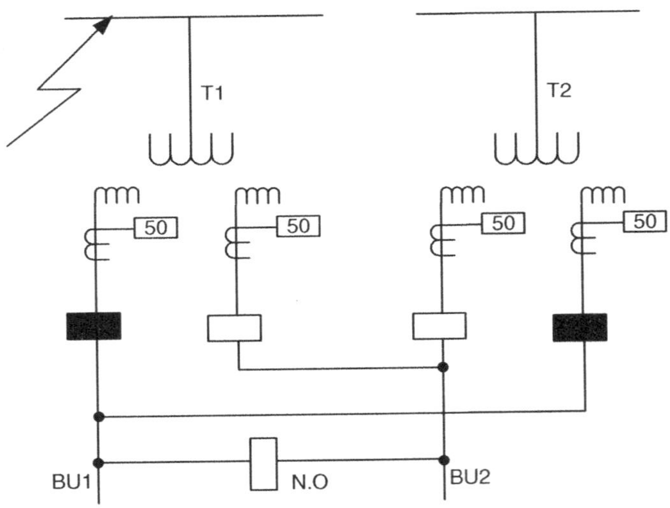

Figure 8.37 Transformer overload protection for double transformer contingency.

8.4 Percent Differential Protection Autotransformers

Of the four main factors influencing the design of differential protection for power transformers, two do not apply to autotransformers.

1. There is no phase shifting between the primary and secondary sides of an autotransformer.
2. The flow of zero-sequence current is not determined by different transformer configurations or whether an artificial ground source such as a grounding transformer is used and where it is electrically located.

Autotransformers do have one significant distinction compared to power transformers. They are always manufactured with a low voltage delta tertiary winding as an add-on. The delta tertiary may or may not be brought out to external bushings and even if it is it may not be used at all. In the case of the delta tertiary not being brought out to bushings, it is known as a buried tertiary. Buried tertiary windings need not be protected in any specific way as they are covered by the overall autotransformer differential protection.

The main reasons for the tertiary windings are as follows:

- To supply a station service load at a distribution level voltage.
- It offers a low impedance path for zero-sequence current to flow for a ground fault.
- It provides a closed path for third harmonic magnetizing currents to flow.
- It provides a polarizing current for directional overcurrent relays.

In most applications, the tertiary windings brought out to bushings are not connected to any type of load whether station service or any other distribution type. In this case, the practice is to ground one of the bushing terminals of an unloaded tertiary with lightning arrestors installed on the other two-phase bushings. The purpose of grounding is to allow protection to be installed to cover over-voltages induced from the high voltage side.

The flow of zero-sequence currents for external faults predictably flow through the operate coil of differential relays as long as zero-sequence current is not shunted away. This shunting is traditionally done by connecting the CTs on either side of the autotransformer in delta for electromechanical relays. Digital relays filter out the zero-sequence current components of the fault current by mathematical means within the relay itself.

Refer to Figure 8.38 showing the flow of zero-sequence current for an autotransformer internal ground fault. The entire zero-sequence current in the secondary CT circuit flows through the operate coil of the faulted phase differential relay coil.

Refer to Figure 8.39 showing the flow of zero-sequence current for an autotransformer external fault. The zero-sequence current in the secondary CT circuit is trapped in the delta CT.

Refer to Figure 8.40 where the tertiary windings are not buried but brought out to bushings. However, the tertiary is not connected to any load. The tertiary ground protection is implemented by connecting a non-directional instantaneous overcurrent relay fed by the secondary of a CT of the tertiary bushing whose terminal is grounded. Any single ground fault on either of the two ungrounded phases or an internal winding ground fault will be detected by this relay. On operation, the relay trips the autotransformer without any intentional time delay. This overcurrent relay will also sense internal ground faults too low to be measured by the differential protection.

When the tertiary winding is brought out and connected to a station service load paralleled with other sources, it is necessary to connect the CT leads from the tertiary to a third differential relay CT input for restraint.

As the tertiary windings are connected in delta, it is an ungrounded source. Ground faults cannot be detected by measuring ground as there is no path for the zero-sequence current to flow. However, zero-sequence voltage can be measured to detect a faulted single-phase. Immediate tripping of the autotransformer is not necessary but alarmed to give system operators time to redirect load before removing the autotransformer from service for repair.

Refer to Figures 8.41 through 8.43 showing the method of detecting ground faults on ungrounded (delta) systems. This is done by measuring zero-sequence voltage on the ungrounded system. With no ground fault, the measured zero-sequence voltage is essentially zero as shown in Figure 8.41. When one of the phases is grounded, the resultant zero-sequence voltage can be as high as three

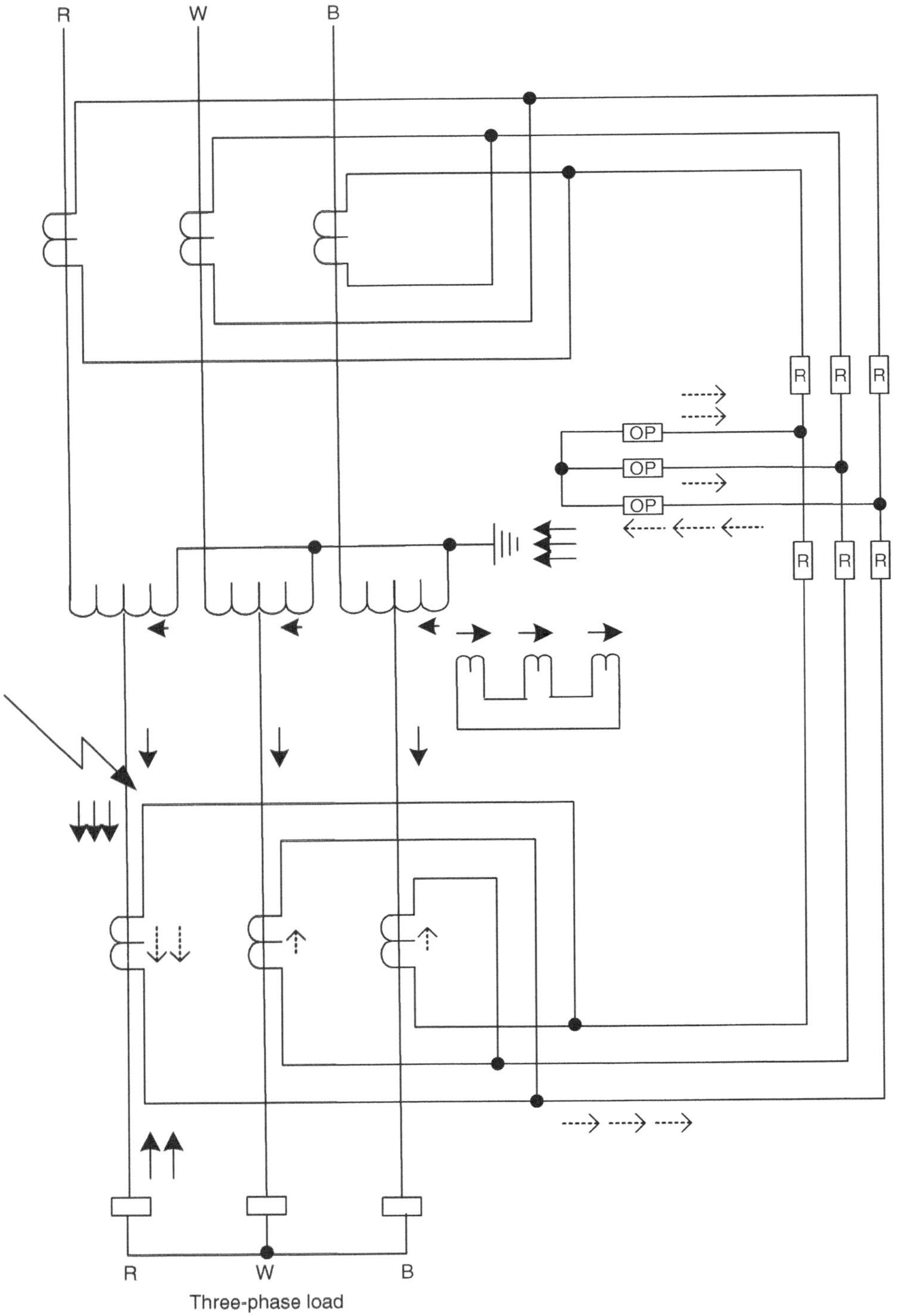

Figure 8.38 Flow of zero-sequence current for an autotransformer internal fault.

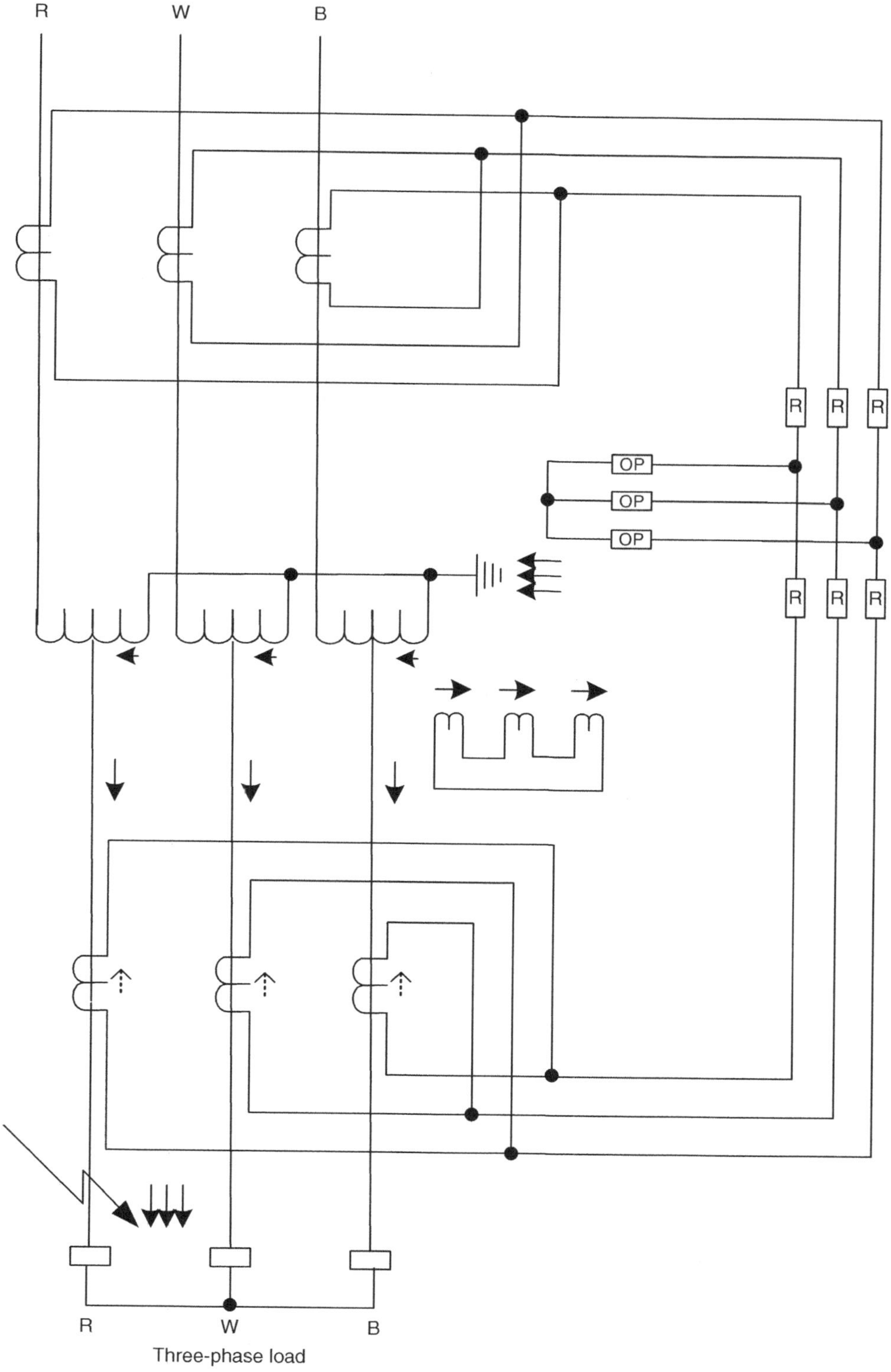

Figure 8.39 Flow of zero-sequence current for an autotransformer external fault.

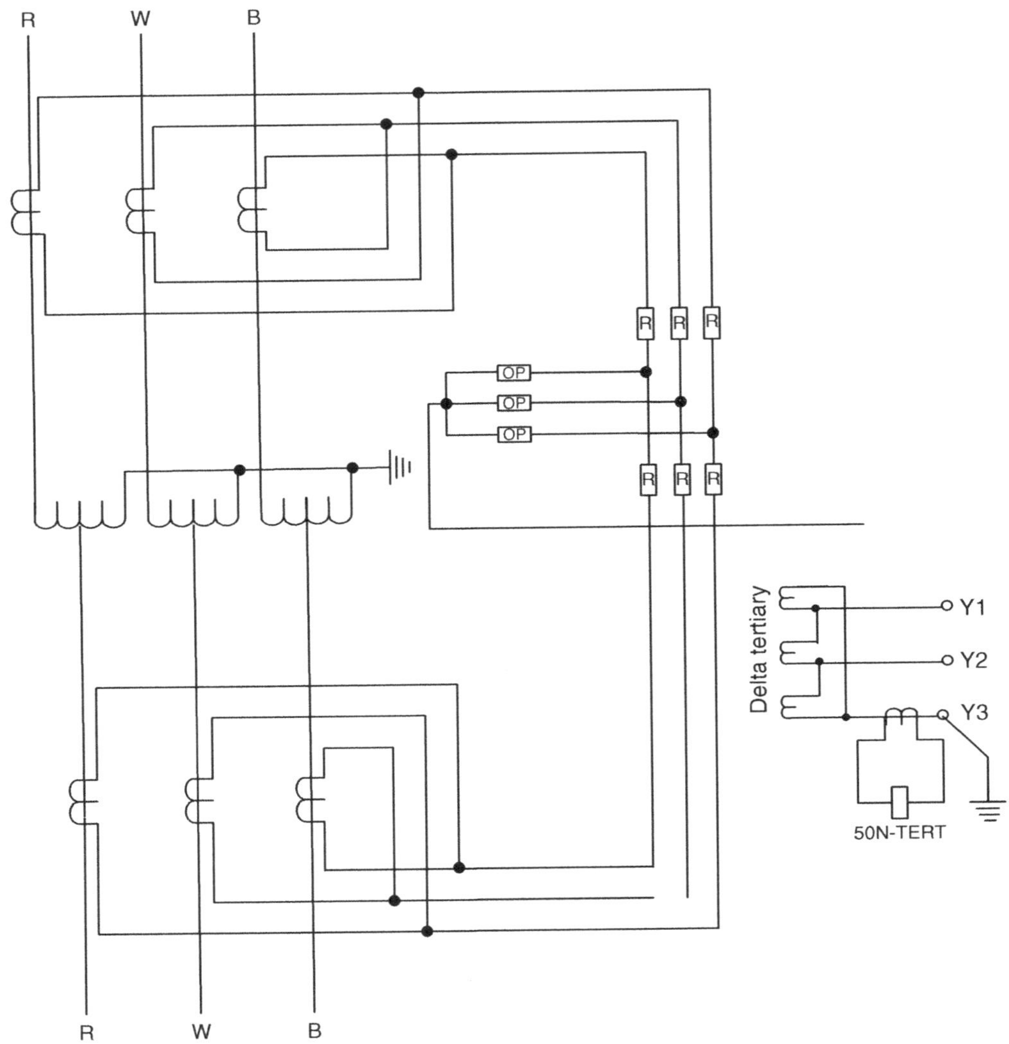

Figure 8.40 Differential connections for an unloaded tertiary winding.

times the pre-ground fault phase to imaginary neutral voltage as shown in Figure 8.42 and derived in Figure 8.43.

Three-phase voltages are taken from a voltage transformer supplying auxiliary VT's connected grounded wye–open delta. Prior to any of the phases being grounded, the three-phase voltages are referenced to an imaginary neutral at the center of the delta. The three-phase imaginary neutral voltages are not connected to each other. Therefore, the voltage measured across the open delta is nothing but the vector sum of these three voltages. This is the zero-sequence voltage $3V_0$ with the resultant being zero as shown in Figure 8.42.

The grounded wye connection of the auxiliary VT's makes it possible to short the faulted phase such that the voltage drops to zero on that phase as shown in Figure 8.43. When one phase voltage faults to the ground, the imaginary neutral shifts to the faulted phase. The other two-phase

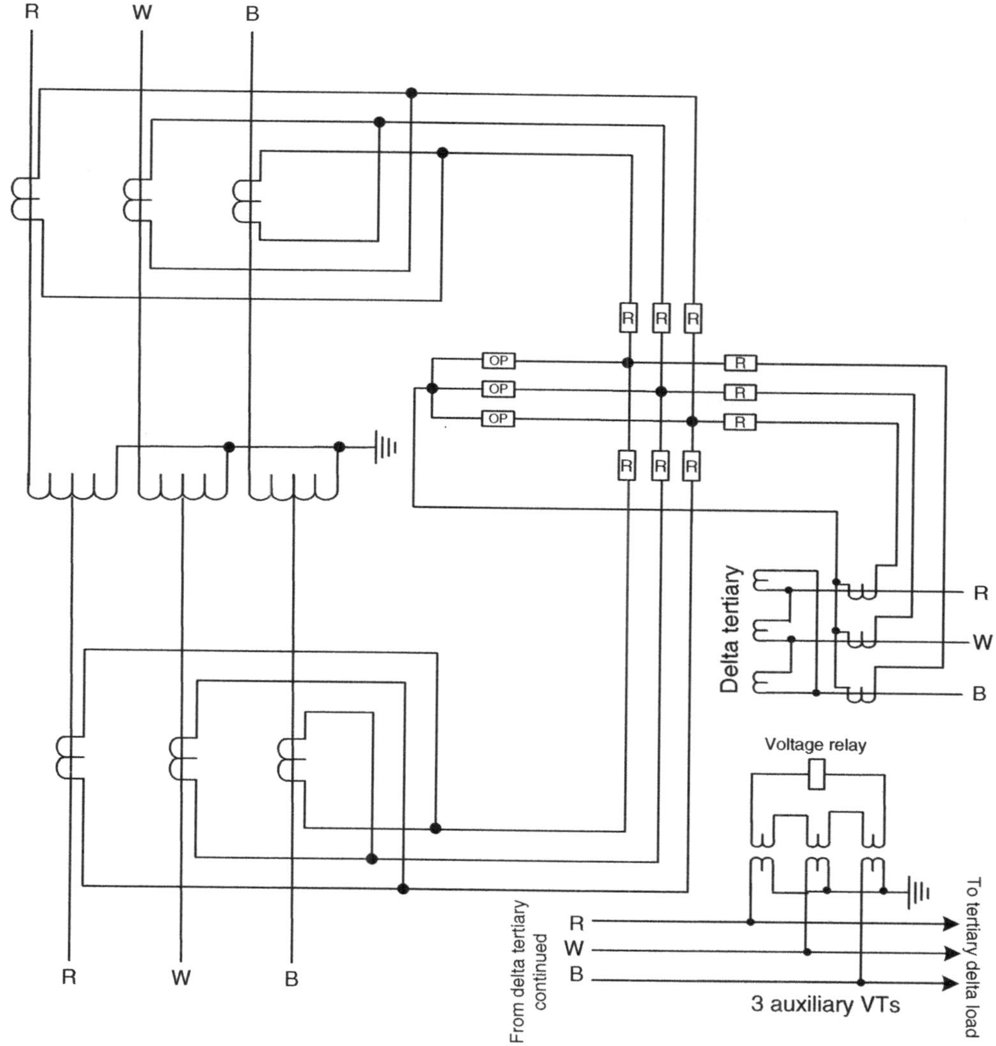

Figure 8.41 Differential connections for a loaded tertiary winding.

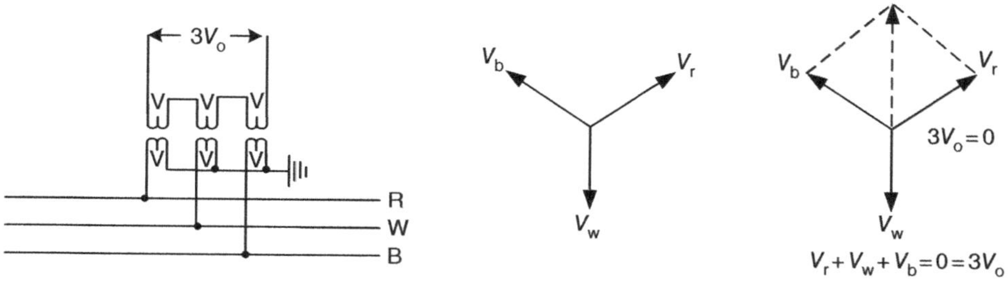

Figure 8.42 Ungrounded system with no phases faulted to ground.

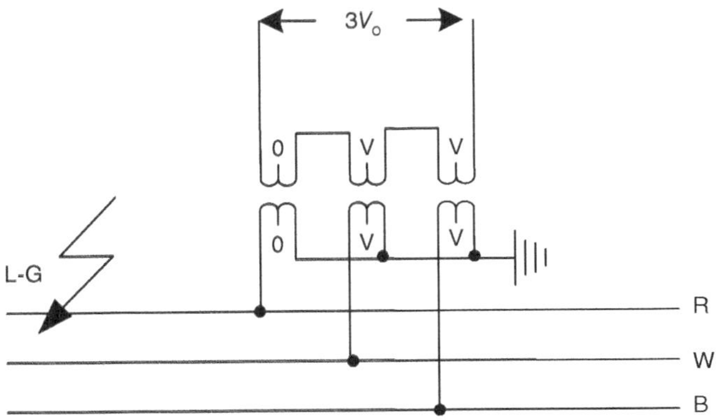

Figure 8.43 Single-phase ground fault on an ungrounded system.

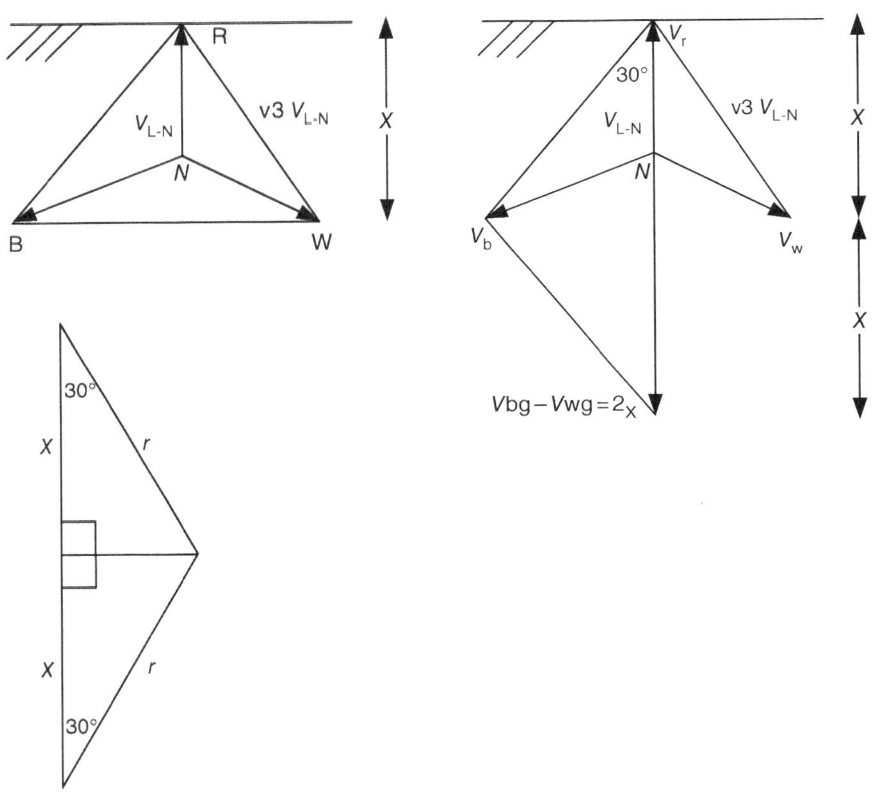

Figure 8.44 Derivation of 3V_0 in a delta system with a phase to ground fault.

imaginary neutral voltages are now connected at the point of fault. As a result, the voltage measured across the open corner delta is the vector difference between the two. This voltage is the zero-sequence voltage measured during a ground fault on a delta system. The other two voltages across the open delta now subtract from each other also after rising by a factor of $\sqrt{3}$. The result is three times the original phase to the imaginary neutral voltage measured as 3V_0 by the relay.

In an ungrounded system for a single-phase faulting to ground, there is no path for zero-sequence current to flow as zero-sequence impedance does not exist. Under healthy conditions, the three-phase voltages are referenced to an imaginary neutral located in the center of the delta. With one of the phase voltages grounded, the imaginary neutral shifts. Along with this shift, the other two-phase voltages increase by a factor of $\sqrt{3}$ as they take on the value of phase-to-phase voltage and are now connected to each other at the point of fault as shown in Figure 8.44.

$$\text{Cos } 30^{o} = \frac{X}{r}$$

$$X = \cos 30^{o} \times r \text{ and } \cos 30^{o} = \frac{\sqrt{3}}{2}$$

$$X = \frac{\sqrt{3}}{2} \times \sqrt{3}V_{L-N}$$

$$X = \frac{3}{2}V_{\text{L-N}}$$

$$2X = V_{wg} - V_{bg} = \frac{2 \times 3}{2}V_{\text{L-N}} = 3V_{\text{L-N}} = 3V_{o}$$

8.5 Transformer Percent Differential Setting Examples

All these setting examples are shown for an application where electromechanical percent differential protection relays are used. For modern applications where digital relays are used, all the principles employed for electromechanical are employed within the digital relays themselves with settings provided in menu form. While employing digital relays, it is easy to lose sight of what the various menu-driven settings mean. By becoming familiar with the electromechanical relay application, one gains a greater understanding and awareness of the digital relay setting process.

8.5.1 Power Transformer Percent Differential Protection Setting Example

In almost all power transformer applications, whether electromechanical or digital, there is a CT ratio mismatch whereby the given CT ratios are not a direct function of the transformation ratio of the transformer itself. The CT mismatch leads to a need to raise the percent differential slope. A raised slope reduces the sensitivity of the overall protection to operate for low-level and incipient transformer faults. For electromechanical relay applications, there is little that can be done but to live with the lowered sensitivity. However, for digital relays, there is a simple solution used by all relay manufacturers. The relay setting menu simply asks for the CT ratios and internally automatically compensates for them.

Another source of unbalance is on-load tap changing whereby the actual transformation ratio does not remain constant thereby resulting in various values of unbalance. Most modern digital transformer relays are provided with the ability to recognize which on-load tap is being used and compensate internally accordingly. However, utility practice is usually not to use this feature and to calculate the unbalance and appropriate slope setting similar to that done for electromechanical relays.

Almost all power transformers come with HV and LV windings not configured similarly. Some come with HV windings configured in wye, while the LV windings are configured delta and some the other way around with the HV in delta and the LV in wye.

Even the delta connection itself has variations affecting the phase relationships of currents from primary to secondary. The way this is done when using electromechanical relays is to configure the

CTs in delta and wye whereby if the transformer HV windings are connected in wye the HV CTs will be connected in delta and if the LV windings are delta the LV CTs will be connected in delta. The CT delta connection itself must be chosen such that the phase shifting of current from primary to secondary via the transformer configuration itself is corrected by the choice of CT configurations and ratios as $\sqrt{3}$ factors are introduced whenever transformers and CTs are connected in delta. When setting and configuring digital transformer relays, the phase relationship and $\sqrt{3}$ factors are taken care of when the transformer winding configuration is input into the setting menu. It is important to note for digital relays that for this reason, all CTs regardless of power transformer configuration are always connected in wye as all phase shifting and $\sqrt{3}$ factors are internally compensated for by digital means.

The determination of the maximum system unbalance is a two-step process. Refer to the following calculation template for the transformer configuration identified. Current transformers and auxiliary current transformers are chosen to take into account the need for phase-shifting compensation and any $\sqrt{3}$ that may be introduced. The primary and secondary currents are calculated at the transformer FCR. The voltages used to derive these currents are at the center tap of all transformer primary and secondary windings. There are three factors that contribute to overall unbalance. The three factors are:

a. unbalance due to difference in CT ratios;
b. unbalance due to CT errors;
c. unbalance to voltage regulation

The values of a, b, and c are added with the result being the maximum system unbalance for the given transformer configuration, CTs chosen, and the other specific parameters.

The % Slope transformer setting should be chosen to add a 50% margin.

The basic pickup current should be chosen with the understanding that the pickup should be higher than any spill current up to a transformer through current of 2.5 A secondary when the % slope begins. Two examples are given to illustrate the methodology.

TRANSFORMER PERCENT DIFFERENTIAL PROTECTION CALCULATION

TRANSFORMER CONFIGURATION: Delta-Wye-Wye

VOLTAGE RATING: 110 kV : 14.2 kV: 14.2 kV

MVA RATING: 45 / 75 MVA

HV TAPS:

137.5 kV
____ kV
____ kV
110 kV Centre Tap
____ kV
____ kV
91.7 kV

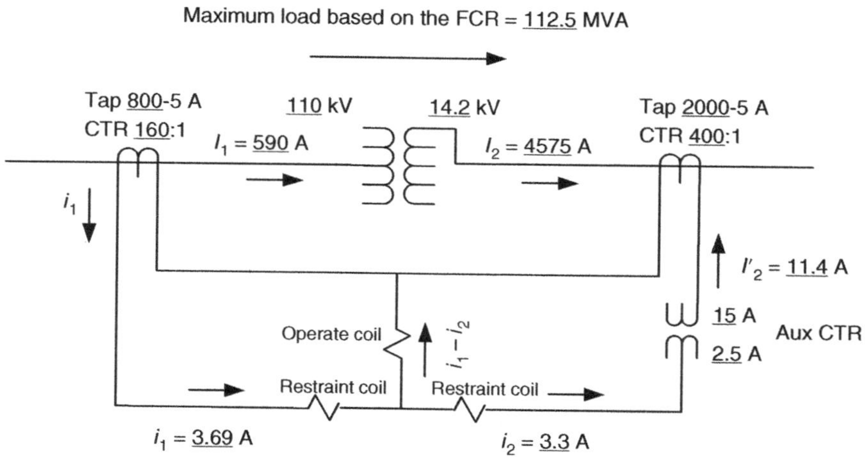

Transformer percent differential protection calculation
delta–wye transformer

a. Unbalance due to difference in CT ratios:
   $i_1 = 3.69$ A
   $i_2 = 3.3$ A
   $i_1 - i_2 = 0.39$ A
   $i_1 + i_2 = 6.99$ A
   $a = 2 \times 100\,(i_1 - i_2)/(i_1 + i_2) = 200 \times 0.39/6.99 = 11.2\%$
   $a = 11.2\%$
b. Unbalance due to CT errors:
   CT error = 2.5%
   Aux CT error = 2.5 %
   $b = 5.0\%$
c. Unbalance due to voltage regulation:
   HV Regulation = 20%
   LV Regulation = %
   $c = 20\%$
   Maximum System Unbalance:
   $a + b + c = 36.2\%$

Relay Settings			
Basic Pickup:	2.0 A sec	320 A pri	65 MVA
Slope:	50%		

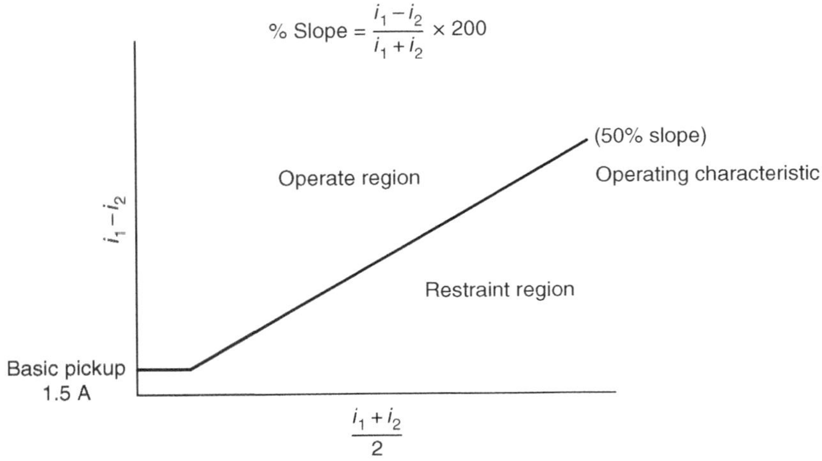

TRANSFORMER PERCENT DIFFERENTIAL PROTECTION CALCULATION

TRANSFORMER CONFIGURATION: Wye-Zig-Zag

VOLTAGE RATING: 215.5 kV : 28 kV: 28 kV

MVA RATING: 75 / 125 MVA

HV TAPS:

215.5 kV
_____ kV
_____ kV
175.5 kV Centre Tap
_____ kV
_____ kV
255.5 kV

Maximum load based on the FCR = 187.5 MVA

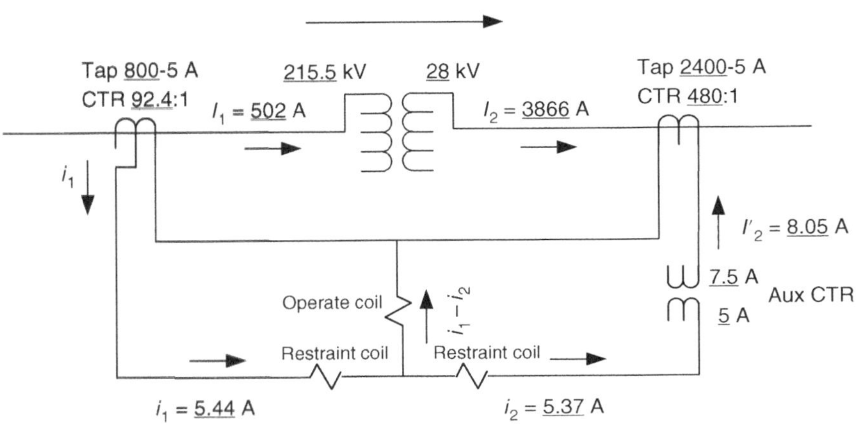

Transformer percent differential protection calculation
wye–zig-zag transformer

a. Unbalance due to difference in CT ratios:
   $i_1 = 5.44$ A
   $i_2 = 5.37$ A
   $i_1 - i_2 = 0.07$ A
   $i_1 + i_2 = 10.81$ A
   $a = 2 \times 100\,(i_1 - i_2)/(i_1 + i_2) = 200 \times 0.07/10.81 = 1.3\%$
   $a = 1.3\%$
b. Unbalance due to CT errors:
   CT error = 2.5%
   Aux CT error = 2.5%
   $b = 5.0\%$
c. Unbalance due to voltage regulation:
   HV Regulation = 20%
   LV Regulation = %
   $c = 20\%$
   Maximum System Unbalance:
   $a + b + c = 26.3\%$

Relay Settings			
Basic Pickup:	1.5 A sec	138 A pri	53 MVA
Slope:	40%		

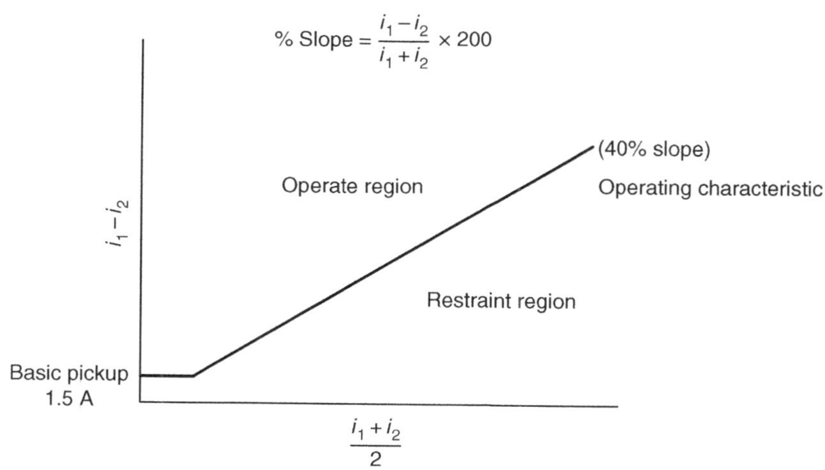

8.5.2 Autotransformer Percent Differential Protection Setting Example

The percent differential calculation to derive the percent slope and basic pickup current settings for a typical autotransformer is provided below and in the second example where a tertiary winding is brought out and needs to be zoned off.

The percent slope and basic pickup current settings must be calculated for the autotransformer high to low side voltage and high to tertiary independent of each other and the worst-case setting must be used as the percent differential relay slope restraint cannot be set independent of each other (in an electromechanical relay).

AUTOTRANSFORMER PERCENT DIFFERENTIAL PROTECTION CALCULATION

AUTOTRANSFORMER CONFIGURATION:

VOLTAGE RATING: 236.8 kV : 121 kV : 13.4 kV

MVA RATING: 150 / 250 MVA

HV TAPS:

236.8 kV
_____ kV
_____ kV
225 kV Centre Tap
_____ kV
_____ kV
214.5 kV

Maximum load based on the FCR = 500 MVA

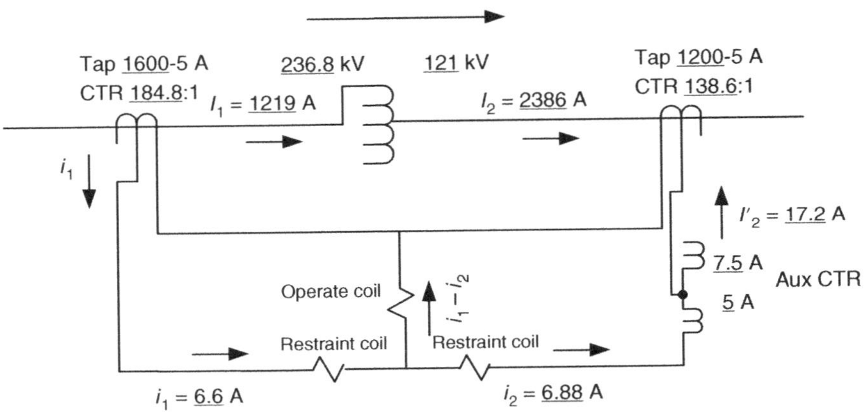

Autotransformer percent differential protection calculation

a. Unbalance due to difference in CT ratios:
   $i_1 = 6.6\text{ A}$
   $i_2 = 6.88\text{ A}$
   $i_1 - i_2 = 0.28\text{ A}$
   $i_1 + i_2 = 13.48\text{ A}$
   $a = 2 \times 100\,(i_1 - i_2)/(i_1 + i_2) = 200 \times 0.28/13.48 = 4.15\%$
   $a = 11.2\%$
b. Unbalance due to CT errors:
   CT error = 2.5%
   Aux CT error = %
   $b = 2.5\%$
c. Unbalance due to voltage regulation:
   HV Regulation = 5%
   LV Regulation = %
   $c = 5\%$
   Maximum System Unbalance:
   $a + b + c = 11.65\%$

Relay Settings

Basic Pickup:	1.0 A sec	185 A pri	70 MVA
Slope:	20%		

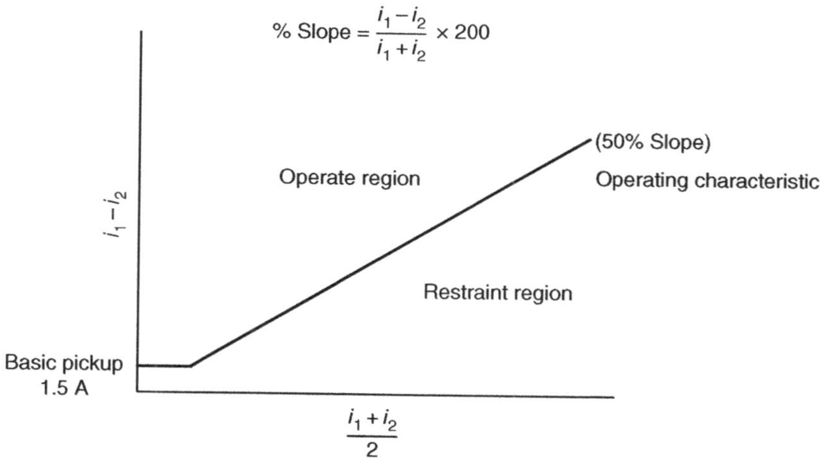

AUTOTRANSFORMER TERTIARY PERCENT DIFFERENTIAL PROTECTION CALCULATION

AUTOTRANSFORMER CONFIGURATION:

VOLTAGE RATING: 236.8 kV : 121 kV: 13.4 kV

MVA RATING: ____ / 50 MVA

HV TAPS:

236.8 kV
____ kV
____ kV
225 kV Centre Tap
____ kV
____ kV
214.5 kV

Maximum load based on the FCR = 50 MVA

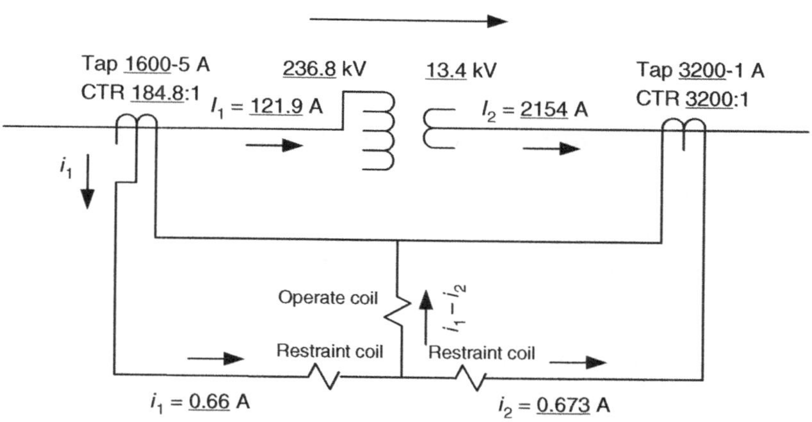

Autotransformer tertiary percent differential protection calculation

a. Unbalance due to difference in CT ratios:
   $i_1 = 0.66$ A
   $i_2 = 0.673$ A
   $i_1 - i_2 = 0.39$ A
   $i_1 + i_2 = 0.013$ A
   $a = 2 \times 100\,(i_1 - i_2)/(i_1 + i_2) = 200 \times 0.013/1.3 = 11.2\%$
   $a = 1.95\%$
b. Unbalance due to CT errors:
   CT error = 2.5%
   Aux CT error = %
   $b = 2.5\%$
c. Unbalance due to voltage regulation:
   HV Regulation = 5%
   LV Regulation = %
   $c = 5\%$
   Maximum System Unbalance:
   $a + b + c = 9.45\%$

Relay Settings

Basic Pickup:	1.0 A sec	185 A pri	70 MVA
Slope:	20%		

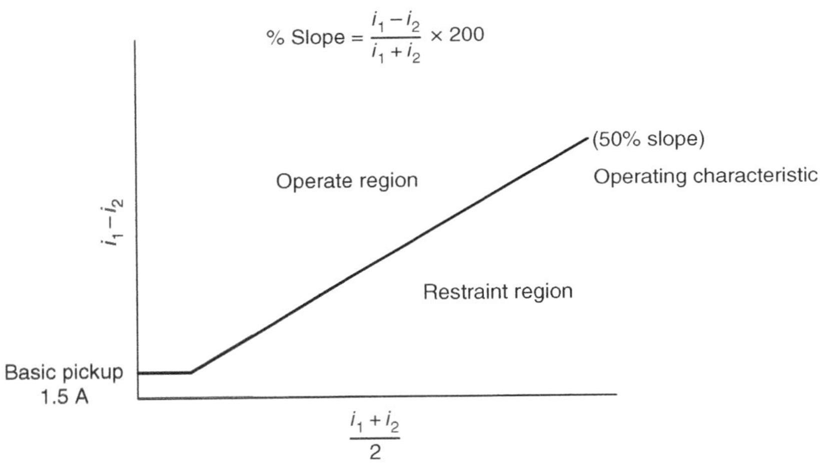

8.5.3 Power Transformer Restricted Ground Fault Setting Example

Example Given Conditions

LV Three-Phase Fault given as 400 MVA at 28 kV = 8250 A
LV Line-Ground Fault with a ground limiting reactor of 7.5 Ω given as 1750 A
CT ratio 400-5 A (80-1)
CT1 fully saturates
CT connections are made in the relay building
CT 2 and CT3 sum their respective fault currents in the relay building and push that secondary fault current through saturated CT1 and the loop leads between CT1 and the relay building.

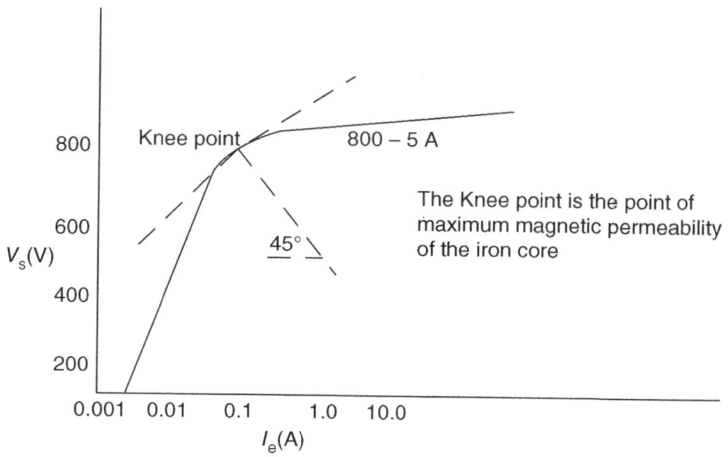

Figure 8.45 CT excitation characteristic in this example.

CT internal resistance = 0.15 Ω
CT cable resistance = 1.5 Ω

The voltage developed at the relay connection point is given by the equation known as the stabilizing voltage $V_S = I_F\,(R_{ct} + 2R_L)$

$$V_S = I_{Fct1}\,(R_{ct1} + 2R_L)\text{ and } V_S = I_{Fct1}\,(R_{ct1} + 2R_L)\,I_{PKP}\,(R_{ST} + R_{RLY}) \gg V_S$$

$$\text{Calculate } V_S = I_{Fct1}\,(R_{ct1} + 2R_L) = 8250/80\,(0.15 + 2 \times 1.5) = 325\text{ V}$$

Choose a relay instantaneous current pickup that will operate for the given ground fault of 1750 A with margin. Choose 0.25 A at 80:1 CT ratio the relay picks up at 437.5 A secondary which is 25% of the calculated RGF of 1750 A primary.

The value of the stabilizing resistor R_{st} whereby the relay will just operate at a pickup of 0.25 A is $R_{st} = 325\text{ V}/0.25\text{ A} = 1300\,\Omega$. Add an approximate 125% margin to the calculated R_{st} where it just operates for an external three-phase fault with a fully saturated CT is $1.25 \times 1300\,\Omega = 1625\,\Omega$.

For an internal line-ground fault for the overcurrent relay to operate a value of voltage applied across the relay = R_{st} node must be $1625\,\Omega \times 0.25\text{ A} = 406\text{ V}$.

This value must be compared to the CT knee point voltage of the CT in the transformer neutral to the ground connection that measures the zero-sequence current flowing from the location of ground fault back to the grounded neutral. Since the instantaneous overcurrent relay that measures this current is in series with the stabilizing resistor, this CT must generate a sufficient secondary voltage to overcome the stabilizing voltage (Figure 8.45).

A calculated value of voltage required to operate the relay with R_{st} of 1625 Ω is 406 V which is half the value of 800 V where this CT begins to saturate.

The protection will not operate for a maximum LV three-fault with a fully saturating LV CT and will operate for a low-level limited ground fault.

Reference

1 IEEE std C37.91-2021, Guide for Protecting Power Transformers.

9

Bus Protection

9.1 Introduction

Buses may be considered to be the most critical of all power system elements [1]. Many other power system elements connect or converge onto buses. Generators, transformers connect to buses while lines converge on to buses. The loss of a single bus affects whether the generation is bottled up, whether transmission between voltage levels is reduced or whether normal load transfers via transmission or distribution lines are prohibited. In many cases, depending on system architecture some or all of the above could be affected.

It is possible to classify buses into two basic voltage levels. High voltage is classified typically at many transmission systems at nominal 750, 500, 230, and 115 kV while low voltage is classified typically at many utilities at nominal 44, 27.6, and 13.8 kV.

Types of high voltage buses are air-insulated outdoor or gas-insulated within tubes using SF6 as an insulating medium. Types of low voltage buses are air-insulated outdoor or indoor metal-clad types.

High voltage buses are always protected with some form of differential protection. Low voltage buses are sometimes protected with differential protection or something similar and or overcurrent protection either as primary or as backup. The biggest problem facing differential protection has always been and still is the performance of current transformers (CTs) for of out-of-zone faults. Multiple infeeds saturating one CT to the exclusion of others and exceptionally high fault infeeds make it challenging to ensure bus protection security under all conditions.

Bus faults occur due to failure of switchgear insulation at in-door stations and due to animal contact or human error at outdoor stations. Most bus faults are phase-to-ground but on occasion, phase–phase and even three-phase faults do occur although rarely.

Most utility systems are solidly grounded with generator unit transformers solidly grounded on the high voltage (100 kV and above) (HV) system side. For that reason, HV buses are subjected to very high fault currents even for ground faults. At load-substations (referred to as substations for the rest of the chapter), this is not always so. Where ground faults are restricted, dedicated ground protections may be required otherwise low-level ground faults would go undetected.

9.2 Typical Bus Arrangements

The most applied architecture that defines the use of breakers at HV terminal stations is known as "breaker and half." The reason for this terminology is that two lines share three breakers or one line uses one and a half breakers so to speak. The alternative is to dedicate a single in-line breaker for

Power System Protection: Fundamentals and Applications, First Edition. John Ciufo and Aaron Cooperberg.

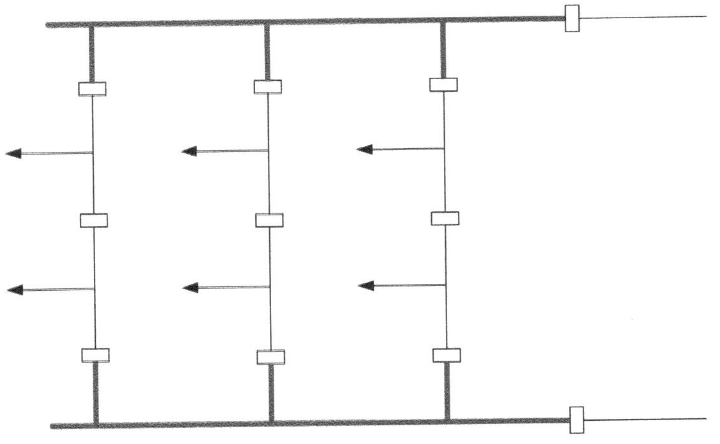

Figure 9.1 Buses at a typical terminal station.

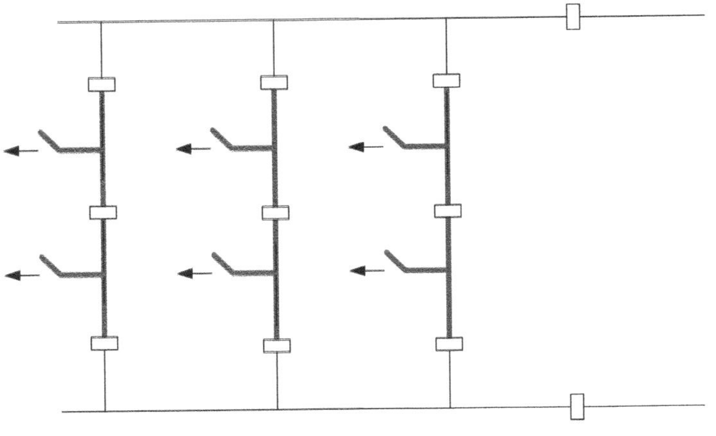

Figure 9.2 Stub-buses at a typical terminal station.

each terminating line. The advantage to the breaker and half architecture is to allow a single-line breaker to be maintained without the need to isolate the line from that line end. Refer to Figure 9.1 showing the buses at a typical terminal station built with breaker and half architecture. Also, refer to Figure 9.2 showing stub-buses on the diameters of breaker and half architecture at these same constructed stations.

Many transmission stations operating at lower system voltages such as 115 kV nominal are commonly designed and built to what is known as a ring bus arrangement. A ring bus makes very efficient use of breakers in a low-cost operating environment. Dedicated bus protections may or may not be required depending on whether there is local generation supplying the ring bus or not.

Refer to Figure 9.3 showing such a HV station with a typical ring bus architecture. The only buses in this arrangement requiring dedicated bus protection are those connecting to load transformers or generators. Ring buses with terminating lines are protected by line protection similar to stub-bus protection at breaker and half configured stations.

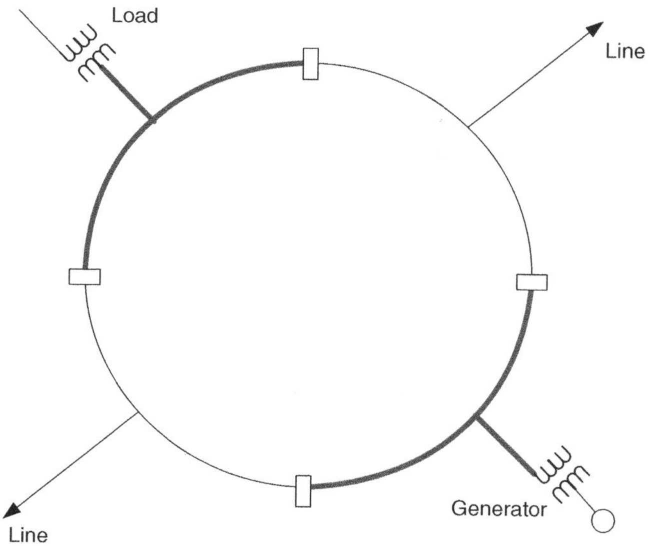

Figure 9.3 Typical ring bus.

Figure 9.4 illustrates the typical architecture for a low voltage (LV) substation. For this type of architecture, each of the two buses can be supplied from either one or two transformer LV windings depending on the station configuration.

9.3 Bus Protection Requirements

Buses are less prone to faults than any other power apparatus as buses are typically built with solid tubing mounted on robust pedestals. Except for strain buses where conductors are electromagnetically attracted to each other and could flash-over, solidly built buses are not affected by external faults. As opposed to transmission lines, they are little affected by the elements in general, including lightning.

A bus fault is a significant event as it involves extremely high fault currents and must be cleared in the shortest time. An uncleared bus fault in an indoor station could cause extensive damage resulting in fire and loss of life. Therefore, bus protections must be highly dependable. Since a bus fault involves the simultaneous loss of many line terminations, bus protections must also be highly secure.

High voltage buses are usually protected by dual-redundant high-speed A and B protections. Low voltage buses at substations are usually protected by single high-speed protection and an inverse time-overcurrent backup protection.

9.4 Methods of Protecting Buses

Primary bus protection must always be instantaneous due to the severity of the fault. It must also be secure from operating for external bus faults. If a bus protection operated for an external bus

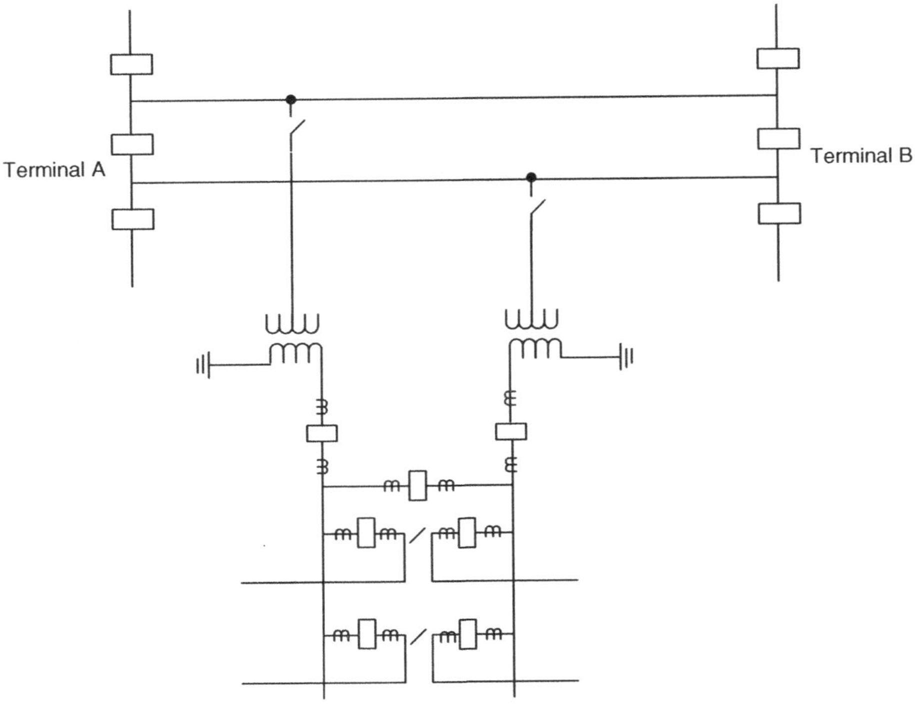

Figure 9.4 LV bus at a typical substation.

fault, depending on the station architecture, more than one bus could be lost. The simultaneous loss of two buses from service would place significant operating stress on the system.

The most effective method of protecting buses is by differential protection. Many utilities specify dual-redundant differential protections for their HV buses at terminal stations. There are two types of bus differential protection using electromechanical relays.

High impedance bus differential protection uses instantaneous overcurrent relays in series with a resistance in parallel with differentially connected CTs such that a significant voltage needs to be developed by the CTs to operate the relay. The required high voltage provides a stabilizing effect to the protection for external faults.

Low impedance bus differential protection uses low set inverse time-overcurrent relays in parallel with the differentially connected CTs. A high set instantaneous overcurrent relay element typically is also used operating independently of the inverse time-overcurrent relays.

Digital relays can be used for the application of low impedance bus differential whereby just an instantaneous overcurrent element is connected in parallel with the differentially connected CTs. The overcurrent element pickup is a function of through fault current similar to the percent differential protection of a transformer. Additional algorithms may still be necessary to ensure out-of-zone protection stability.

9.4.1 Differential Protection

There are two distinctly very different types of differential protection when electromechanical relays are used for protection. One type is known as "low impedance differential" and the other "high impedance differential."

Low impedance differential connects overcurrent relays in parallel with differentially connected CTs. These relays are usually standard off-the-shelf inverse time-overcurrent with an instantaneous element with both current operating elements connected in series and their tripping contacts in parallel. The term low impedance derives from the fact that overcurrent relays represent a low burden. The reason to install a low impedance bus differential is that it costs less and takes up less panel space than a high impedance differential.

High impedance differential also connects overcurrent relays in parallel with differentially connected CTs. These relays are always instantaneous overcurrent exclusively whose current operating elements are connected in series with a resistor known as a stabilizing resistance. The term high impedance derives from the fact that the overcurrent relays whose burden are low is connected in series with very high resistance. The huge advantage of a high impedance differential is that it can be calculated and predicted in advance what to set the protection at to ensure it never operates for external bus zone faults even with a fully saturated CT. Where bus protection security is paramount high impedance differential is typically used for all 230, 500 kV, and higher voltage buses.

With the advent of digital relays, neither low impedance nor high impedance methods are used. Low impedance differential mimicking the inverse time-overcurrent characteristic of an electromechanical relay but with digital means leaves the protection vulnerable to operation for external faults and would not be very secure. High impedance differential mimicking the voltage actuating instantaneous overcurrent element with a series resistor is essentially the same thing as the electromechanical method. There are manufacturers of digital bus protection relays that do just that. More recently available digital bus differential relays all use a common approach of percent differential such as used for transformers but with the added feature or algorithm that ensures that the relay remains secure for external faults.

9.4.1.1 Low Impedance Bus Differential Protection

All differential protections that use straight overcurrent relays with low burden fall into the category of low impedance differential. Either Induction disc inverse time-overcurrent relays or digital overcurrent relays optimized for this type of application can be used. In general, this type of differential protection can be made to be highly stable for applications up to 115 kV system typical short circuit levels when set properly.

CTs are located at all incoming and outgoing nodes of the protected bus. The CT secondary windings are then interconnected, and the coil of an overcurrent relay is connected across the CT secondary circuit. The CT polarities are critical to the correct operation of this type of connection. The effective CT ratios at either end of the differential connection must be the same for bus protection. Without the CT ratios being the same, there is a mismatch between secondary currents that would flow into the relay at the relay node. In an ideal situation, the secondary current representing the fault current would all circulate around the CT connection or all flow into the relay operate element at the relay node.

Refer to Figure 9.5 showing an external fault with the instantaneous primary current entering the CT causes a secondary excitation voltage resulting in instantaneous secondary current to leave X_1. At the same instant, the primary current is leaving the other CT causing a secondary excitation voltage resulting in the secondary current entering X_1. The secondary currents simply circulate around the CT connection with no current entering the relay operate element.

Refer to Figure 9.6 for an internal fault where the opposite is true. The instantaneous primary current entering the CT causes a secondary excitation voltage resulting in instantaneous secondary current to leave X_1. At the same instant, the primary current is also entering the other CT causing a

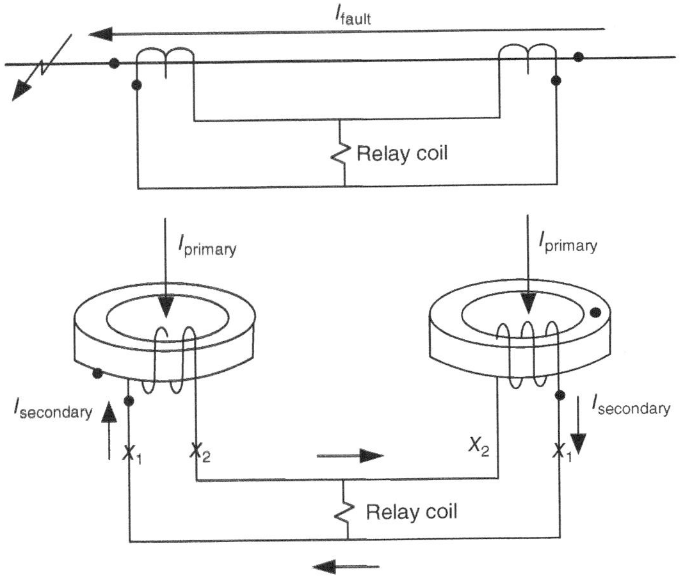

Figure 9.5 Simple two-CT differential bus protection – external fault.

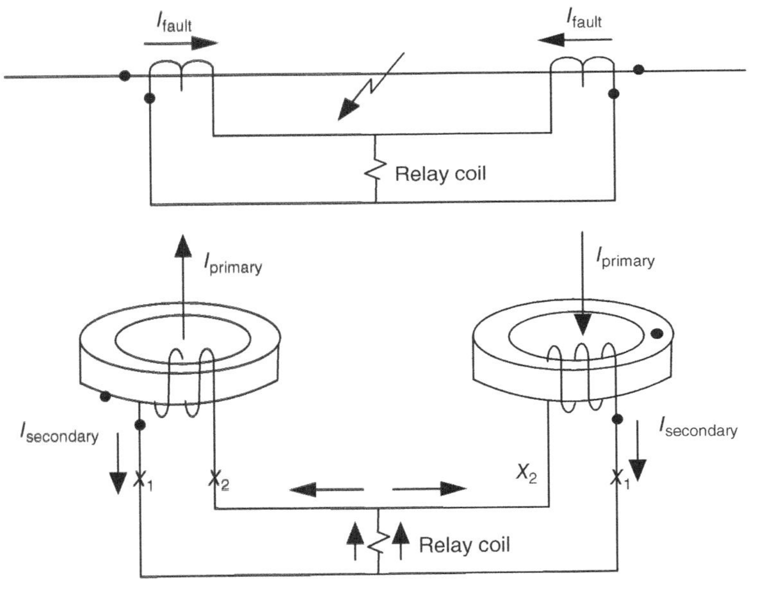

Figure 9.6 Simple two-CT differential bus protection – internal fault.

secondary excitation voltage resulting in the secondary current leaving X_1. The secondary currents meet at the relay node and both flow through the relay operating element.

An internal fault located anywhere between the two CTs causes current flows to the fault from both sides. The sum of the CT secondary currents will flow through the overcurrent relay operate element.

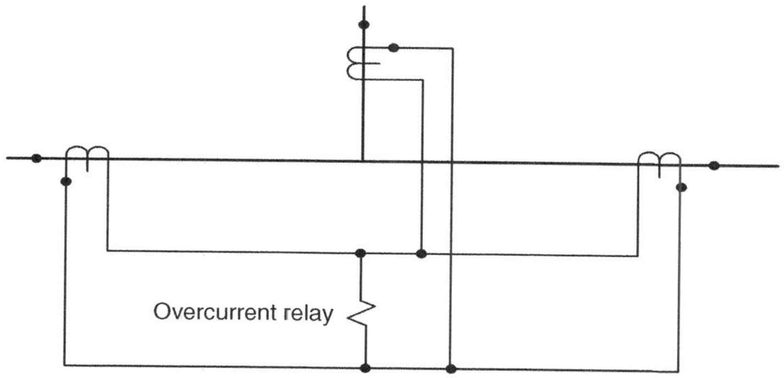

Figure 9.7 Differential bus protection.

It is not necessary that fault current flow from both sides to result in secondary current flow through the relay operate element. A flow on one side only, or even some current flowing out of one side while a larger current enters the other side, will cause a differential current.

The differential current will be equal to the vector difference between the currents entering and leaving the protected zone. Should the differential current exceed the relay pickup value, the relay will operate.

It is a simple step to extend the principle to a system element such as a bus having several connections as shown in Figure 9.7. It is merely necessary that all CTs have the same ratio and that they be connected so that the relay receives no current when the total current leaving the protected element is equal in magnitude and phase to the total current entering the element. In this case, the secondary exciting voltages of all the CTs are such that all the currents flow around the differential circuit and no current flows through the relay operate element.

The performance of CTs is a significant factor for bus differential protection. As shown in Figure 9.8, infeeds to a high magnitude bus fault can be through multiple sources. The one CT closest to the fault needs to transform the entire fault current while the others only need to transform a portion of it. The result of seeing the entire infeed is to possibly cause severe waveform distortion in that one CT. This issue is common for high voltage buses at terminal stations and low voltage buses at substations. There are two types of bus differential protection using electromechanical relays. One is known as low impedance bus differential and the other is

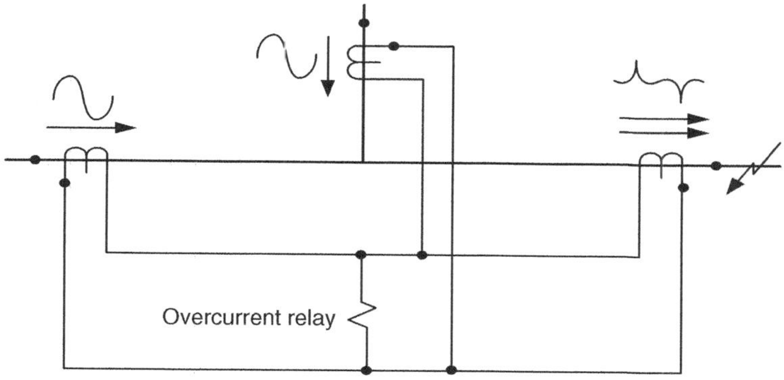

Figure 9.8 Waveform distortion for out-of-zone faults.

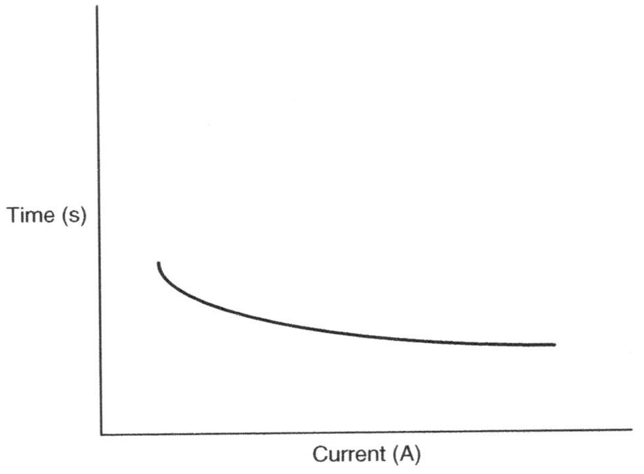

Figure 9.9 Short-time inverse characteristic.

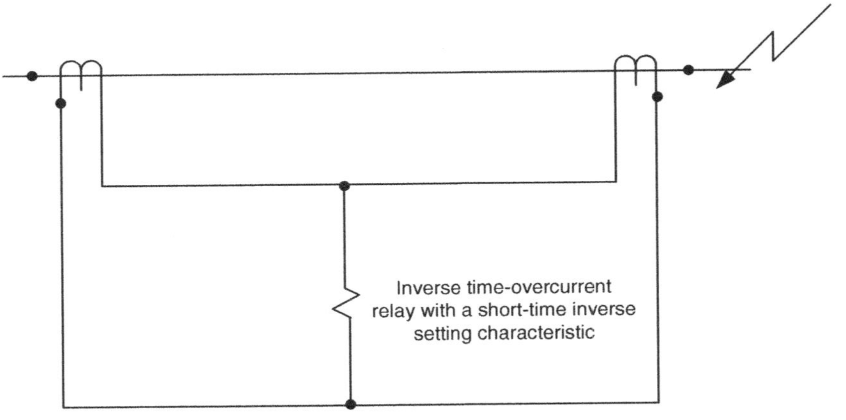

Figure 9.10 Low impedance differential protection.

high impedance bus differential protection. Digital relay is a third type using a different method of ensuring stability for out-of-zone bus faults.

The "short-time" inverse characteristic shown in Figures 9.9 and 9.10 provides superior performance for remaining stable for even severe out-of-zone faults.

The reason electromechanical relays were so stable was due to their fundamental operating property. Torque is applied to the disc when two sinusoidal fluxes out of phase by 90° are directed down into the disc. When CTs severely saturate, the secondary current loses its sinusoidal character. The additional harmonic components in distorted waveforms tend to interfere with torque production from power frequency components in a way that tends to retard disc rotation. With severely distorted waveforms, the two fluxes simply cannot react with one other in time. Without the ability to produce torque, the relay remains stable. The simple digital overcurrent relays convert the current via an analog to digital conversion and apply the digital representation of the current to the short-time inverse equation. Whereas the electromechanical relay provides natural filtering for severe waveform distortion and restrains, the digital relay does not restrain. Therefore, it tends to trip unless another add-on feature is applied.

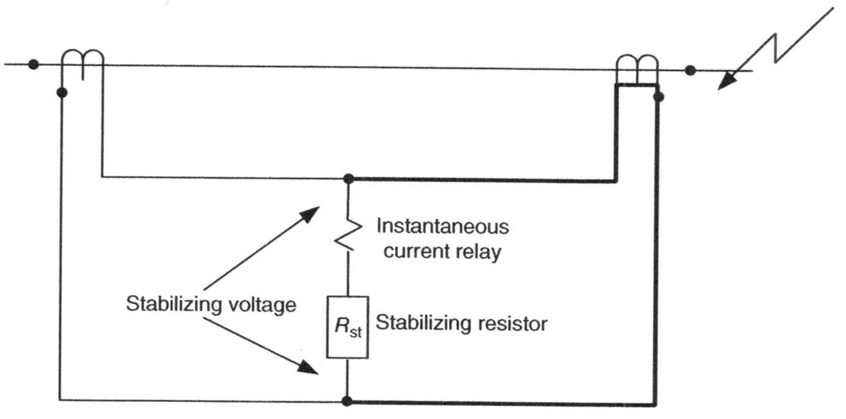

Figure 9.11 High impedance differential protection.

9.4.1.2 High Impedance Bus Differential Protection

Low impedance differential protection becomes unstable for applications at higher voltages usually at 230 kV and above. System short circuit levels at these voltages are higher resulting in a tendency of low impedance differential schemes to operate for out-of-bus zone faults. High impedance differential protection is generally specified at 230 kV and higher voltages.

Low impedance differential protection works on the principle that the relay burden in parallel with all the CTs is of low magnitude with stability calculated based on current. In high impedance differential protection schemes, the overall relay burden in parallel with the CTs is of high magnitude with stability calculated based on voltage. Refer to Figure 9.11 showing a typical high impedance differential connection.

A maximum three-phase external fault is considered along with the assumption that the CT closest to the fault needs to transform the entire infeed and goes completely into saturation as shown in Figure 9.12. A fully saturated CT looks electrically to the other CTs as a dead short as shown in Figure 9.11. The maximum voltage across the relay node is the sum of the secondary currents from all CTs (excluding the saturated CT) times the resistance of the CT leads and the internal resistance of the shorted CT. A stabilizing resistor is added such that the current seen by the relay will be just below its pickup setting. This protection is effective as the worst-case fully saturated CT is assumed in calculating the value of the stabilizing voltage and the value for the stabilizing resistor to ensure the relay does not operate for that value. The other option is a voltage measuring relay with a high impedance connected across the stabilizing resistor.

The voltage developed at the relay connection point is given by the equation known as the stabilizing voltage $V_S = I_F\,(R_{ct} + 2R_L)$ as shown in Figure 9.13.

This equation shows that the spill current can be reduced to a value below that of any given relay setting by a suitable choice of relay resistance. Also, the equation further shows that the maximum voltage which can be applied to the relay is given by this same equation. This value can be low compared to the CT knee point voltage so that a value of resistance can be applied to R_{st} which will ensure stability while still permitting the sensitive current operation. Resistance added to the natural resistance of the relay itself is known as Stabilizing Resistance or R_{st}. This value must be compared to the CT knee point voltage of the non-saturating CT(s) shown in Figure 9.14. The three-phase representation for this protection is shown Figure 9.15.

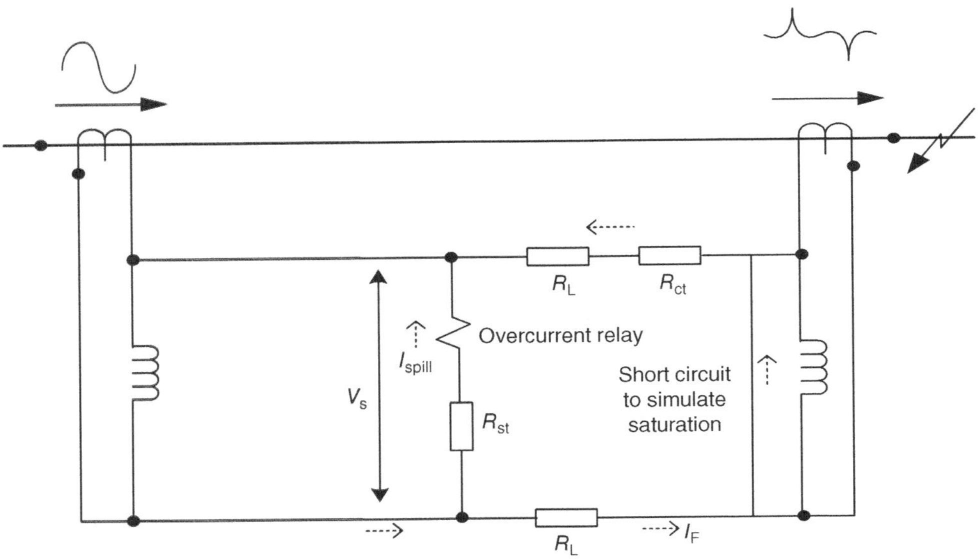

Figure 9.12 Equivalent differential circuit under transient conditions.

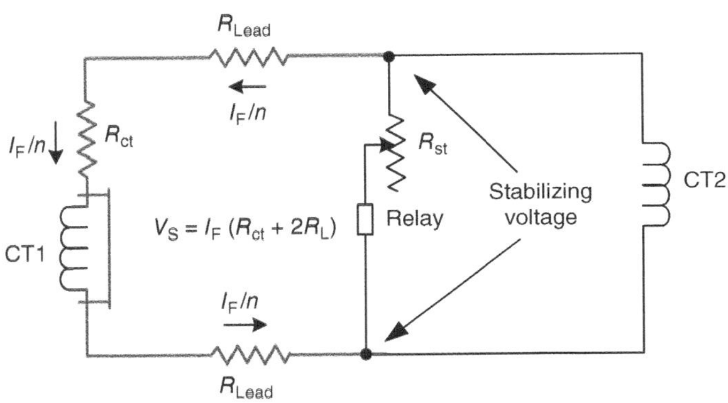

Figure 9.13 Calculation of stabilizing voltage V_s.

9.4.1.3 Differential Protection with Digital Relays

Digital bus relays all use similar methods to provide stability for external bus faults with severely saturating CTs. The same concept as percent differential relays for transformer protection is used but with an add-on feature. Algorithms are developed that allow for severely saturating CTs yet maintain security. There are two reasons why bus protection is such a challenge for digital relay manufacturers. Bus faults tend to be severe as the only impedance limiting the fault infeed is small source impedance. Buses with multiple nodes guarantee as a certainty that one CT will go into saturation, sometimes even heavy saturation, well in advance before the others. With the secondary current from the severely saturating CT being compared with the others differentially the relay tends to operate unless prevented from doing so by some other means. Added to these limitations is the fact that faulted buses must be removed from service at high speed. Digital relays do not possess natural filtering and therefore tend to operate under these conditions.

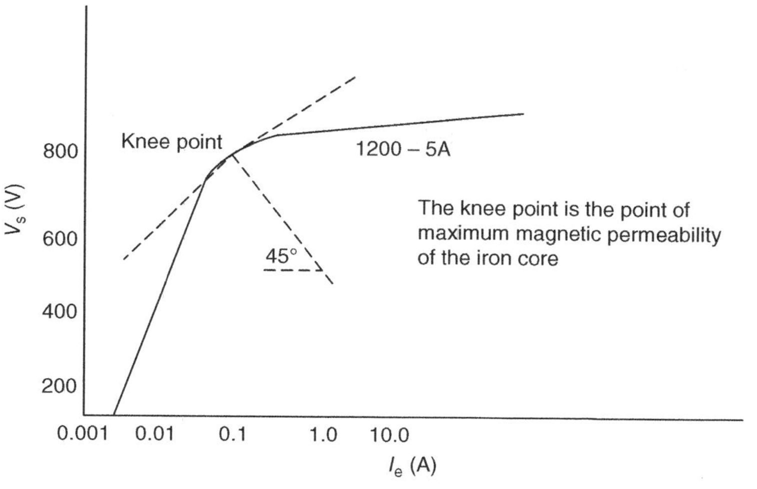

Figure 9.14 Typical CT excitation characteristic.

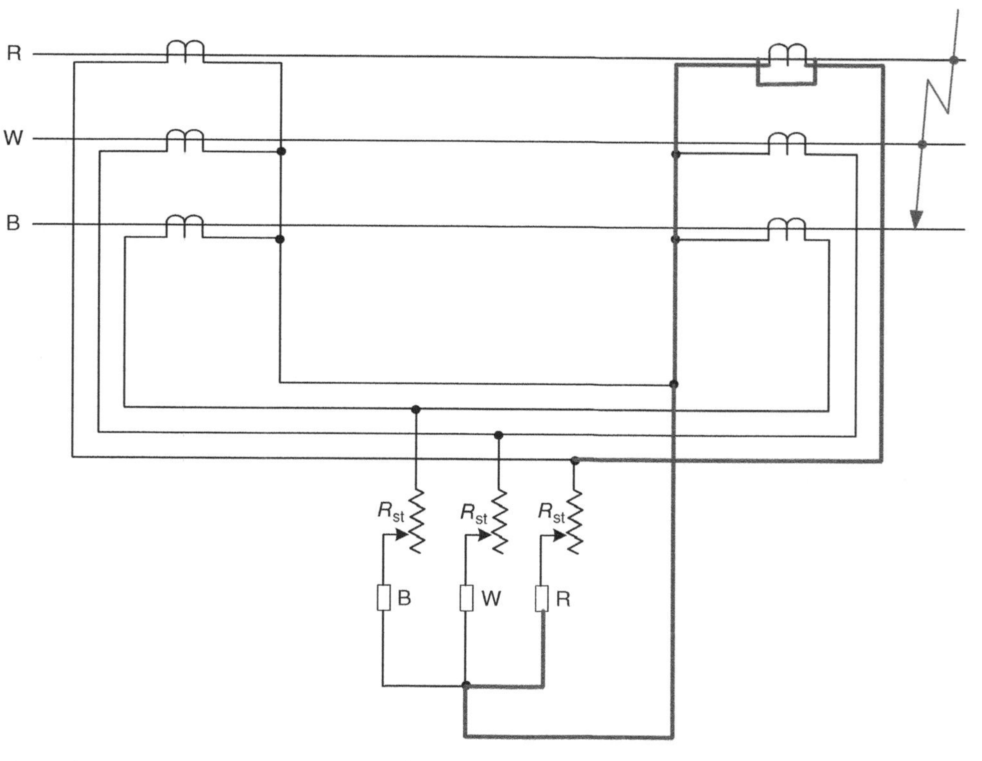

Figure 9.15 Three-phase representation of high impedance differential.

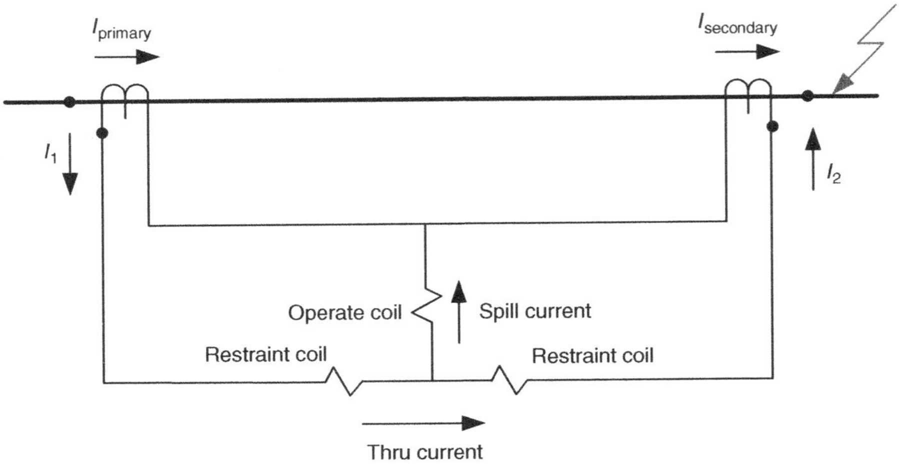

Figure 9.16 Simple two-node bus protected with percent differential.

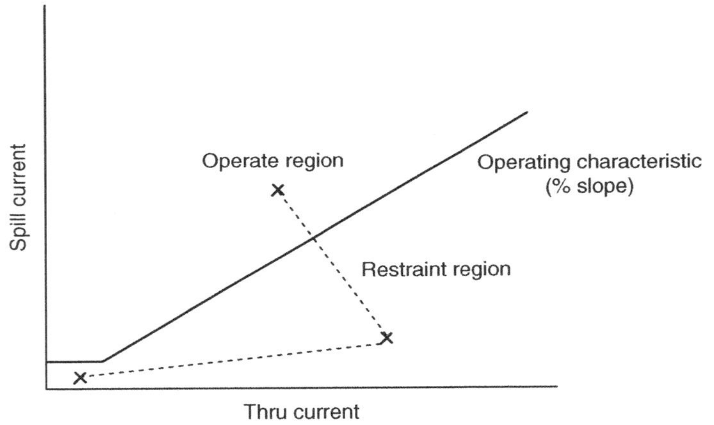

Figure 9.17 Trajectory of current with severely saturating CTs.

Manufacturers do not easily disclose the concepts underpinning their proprietary algorithms. However, a pattern has emerged as to what the algorithm looks like in concept as they are all based on the simple laws of physics and the fundamental limitations of distorted waveforms from severely saturating CTs being converted into binary data to be interpreted as being meaningful in a traditional differential calculation.

The approach taken is to create a percent differential calculation similar to that used for a transformer but with a unique algorithm to ensure out-of-zone protection stability. Refer to Figure 9.16 showing a simple two-node bus protected with a percent differential relay. Under normal operating conditions, when the CTs are not saturating, this protection is very dependable and secure. Refer to Figure 9.17 showing a typical trajectory of the current during fault conditions when the thru current dramatically increases with a severely saturating CT.

Consistently, this type of event captured by digital relays shows that the trajectory of current to correctly travel from a typical thru load location to a healthy restraint region location. Then, as one CT severely saturates, the current travels from the restraint region to the operate region. This happens as the secondary current begins to severely distort and no longer represent a replica of the primary. There is no natural filtering to retard relay operation as exists with the torque applied to an induction disc.

Referring to Figure 9.17 showing the added-on restraint region, the design is predicated on saturating CTs, even severely saturating CTs having a portion of the secondary waveform at the immediate onset of fault without significant distortion. The trajectory of current entering the added-on restraint region will reside in that region for a short time until severe CT saturation takes hold of the CT. When that happens, the trajectory of current will migrate to the operate region but not until it spent a finite amount of time in the added-on restraint region. It is the finite amount of time required spent in the restraint region that is used as a marker requiring restraint even as the trajectory goes on into the operate region. This time even for severely saturating CTs is typically one-quarter cycle.

With one-quarter good cycle required to ensure that the algorithm works to prevent out-of-zone operation, this may not always be possible. The high impedance bus differential scheme by comparison works to prevent out-of-zone operation no matter what the CTs do in terms of saturation. Before applying these digital bus differential relays, it is important to predict the time to severe saturation to determine a least a good quarter cycle exists. Without this information, it is not possible to apply a purely digital bus differential relay and it may be advisable to switch back to a high impedance bus differential scheme to ensure stability. A significant advantage a purely digital differential relay has over a high impedance differential relay is the ability to cater for CT ratio mismatch by a simple setting while programming the relay. Otherwise, it is virtually impossible to compensate for it by traditional means as auxiliary CTs tend to go into saturation much earlier than primary CTs. Also, auxiliary CTs could introduce magnified burdens by the square of the turn ratio of the auxiliary CTs themselves. It is for this reason that the use of auxiliary CTs is not recommended by the industry for bus differential.

9.4.1.4 Bus Differential Protection Zones

9.4.1.4.1 Terminal Stations

The bus differential protection covering high voltage buses at terminal stations is zoned off around the diameter breakers and the bus-tie breaker as shown in Figure 9.18.

Typically, buses between diameter breakers known as stub-buses are not covered by dedicated bus protection but by line protections. The line protection is zoned off around the two-diameter breaker sets of CTs defining the line zone. Should a three-phase fault occur on the stub-bus or on the line close-in to the station such that the voltage measured by the relay collapses to essentially zero the relay would not operate. However, the distance relays used for line protection have memory action that uses the pre-fault memorized voltage to polarize the relays to facilitate correct relay operation. For three-phase faults occurring without the possibility for memory action, for example when closing into three-phase maintenance grounds, another protection known as "switch onto fault" is used as shown in Figure 9.19. This protection is covered in detail in Chapter 14 on transmission lines.

Figure 9.20 shows typical overlapping protection zones found at terminal stations. The bus protections and line protections overlap the diameter breaker CTs that summate current entering and leaving the bus. The line protections are zoned off around the diameter breaker CTs that summate current entering and leaving the line.

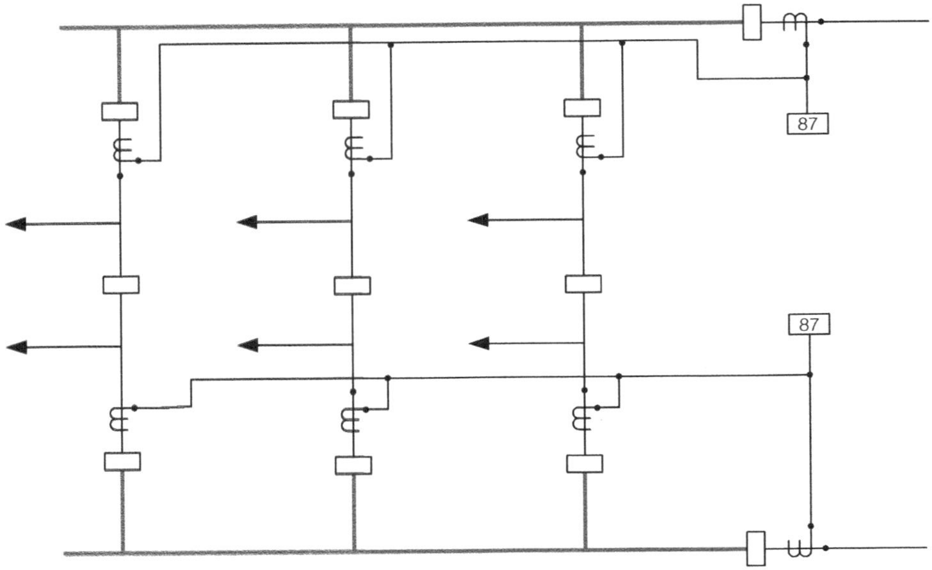

Figure 9.18 Bus protection zoning at HV terminal stations.

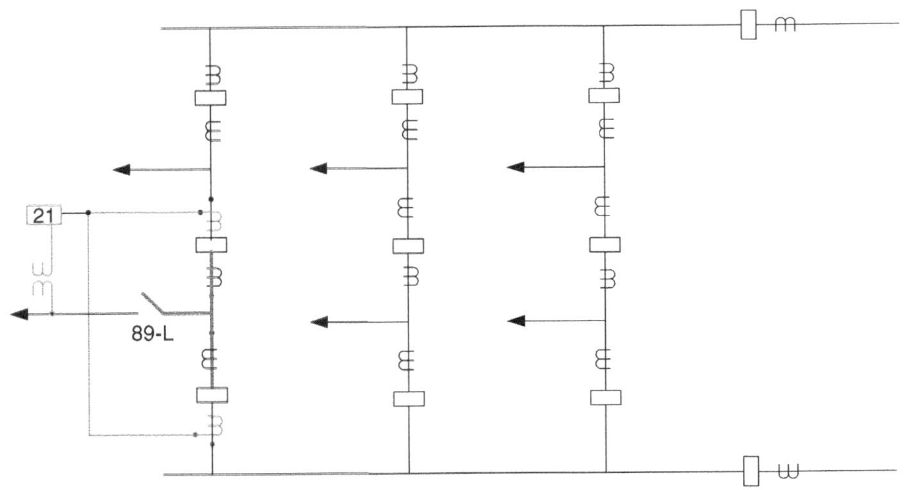

Figure 9.19 Stub-bus protected by switch onto fault protection.

9.4.1.4.2 Ring Buses

Many lower voltage system stations typically 115 kV are designed and built to what is known as a ring bus arrangement. A ring bus makes very efficient use of breakers in a low-cost operating environment. Dedicated bus protections may or may not be required depending on whether there is local generation supplying the ring bus or not.

When there is generation or local load ring, buses are covered by bus differential protection as shown in Figure 9.21. The bus differential protection is zoned off around breaker CTs where the breakers bound a section of the bus to either a local load transformer or generator. In either case, the

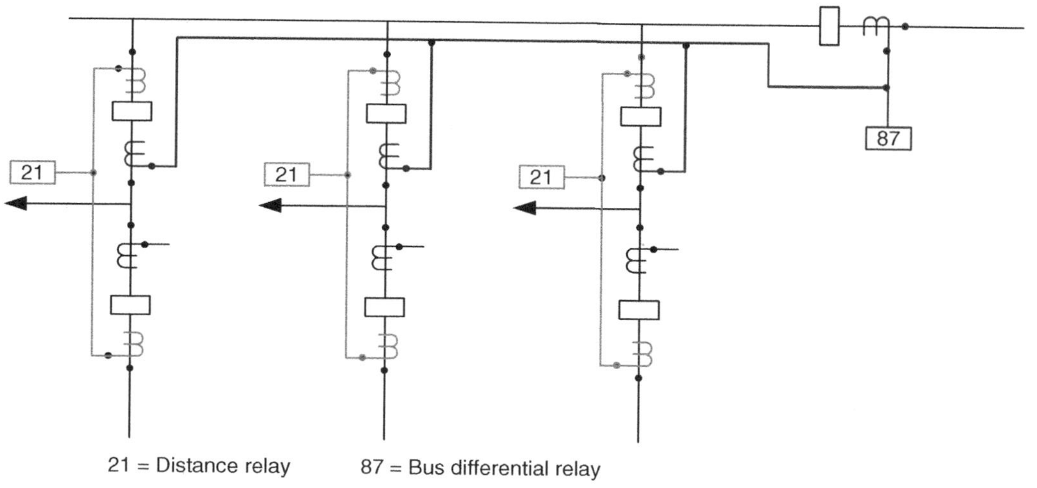

Figure 9.20 Typical zoning and overlapping zones at a terminal station.

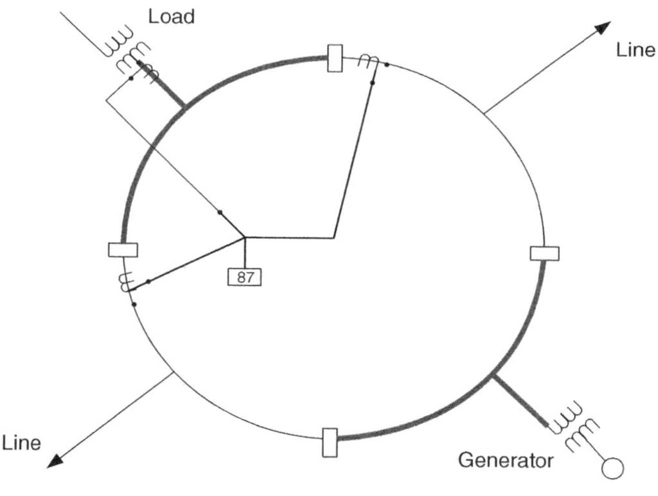

Figure 9.21 Ring bus protection zones.

CTs are paralleled with bushing CTs on either the load transformer or generator main output transformer. The other sections of the bus used to terminate lines are covered by the line protections. Memory voltage action and line test protection cover these sections of the bus similar to terminal station stub-buses.

9.4.1.4.3 Substation LV Buses

Figure 9.22 shows an LV bus zone at a typical substation. The LV bus differential protection is zoned off around the transformer LV breaker CT, the bus-tie breaker CT and the distribution feeder CTs in parallel. For a bus fault, all breakers are tripped by the differential protection. Many utilities choose to also trip the feeder breakers by the bus protection even when the feeders only supply radial loads with no possibility of back feed into a bus fault. The purpose for tripping the feeder breakers is to facilitate restoration of service. In general, the bus must be completely isolated

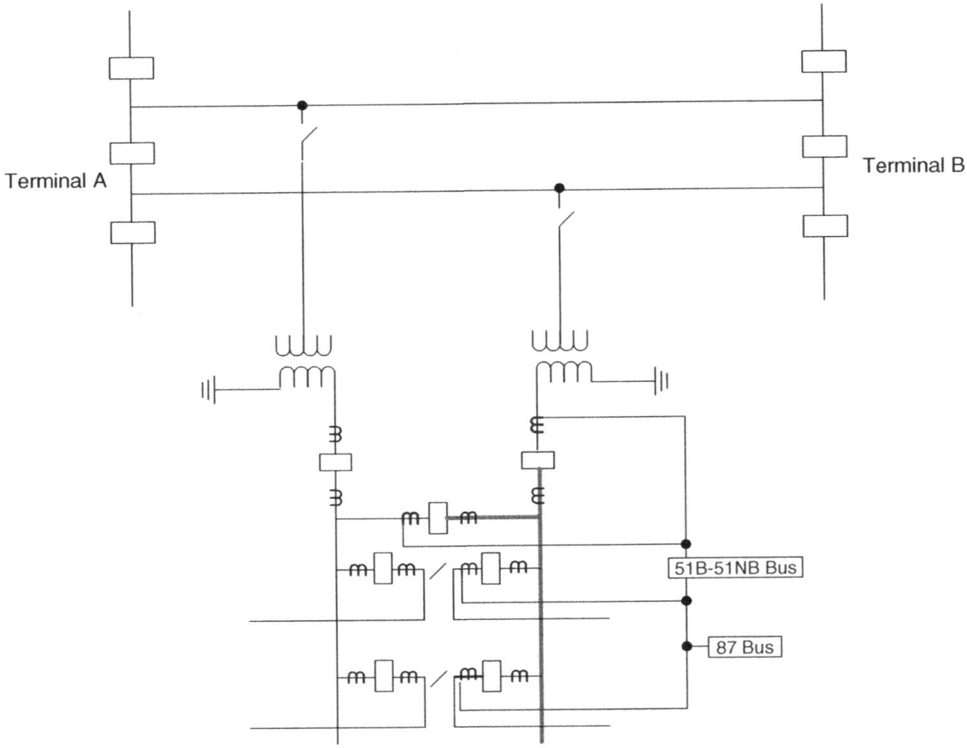

Figure 9.22 Typical substation LV bus protections.

prior to the system operators restoring it to service. If the feeder breakers were not automatically tripped by the bus protection, the system operator would need to do so either locally or via SCADA remotely.

Bus backup protection accepting current from the transformer LV breaker CTs and the bus-tie breaker CTs in what is known as "partial differential" protection is often used to supplement the bus differential protection. The bus backup protection is usually an inverse time-current characteristic timed to coordinate with feeder protections.

9.4.1.4.4 Substation Ground Source Locations

Where grounding transformers are electrically located has a significant effect as to whether the differential protection remains zone tight or trips for out-of-zone ground faults. Eight examples are given below of in-zone and out-of-zone ground faults under various types of zigzag grounding transformer zoning locations.

9.4.1.4.4.1 Zero-Sequence Current Distribution #1 In this example, the grounding transformer is located in the transformer zone. This location is mainly typical of substations where zigzag grounding transformers are used as artificial ground sources.

For an internal bus fault, as shown in Figure 9.23, zero-sequence current representing full ground current flows through the operating coil of the faulted differential phase relay to correctly operate the relay.

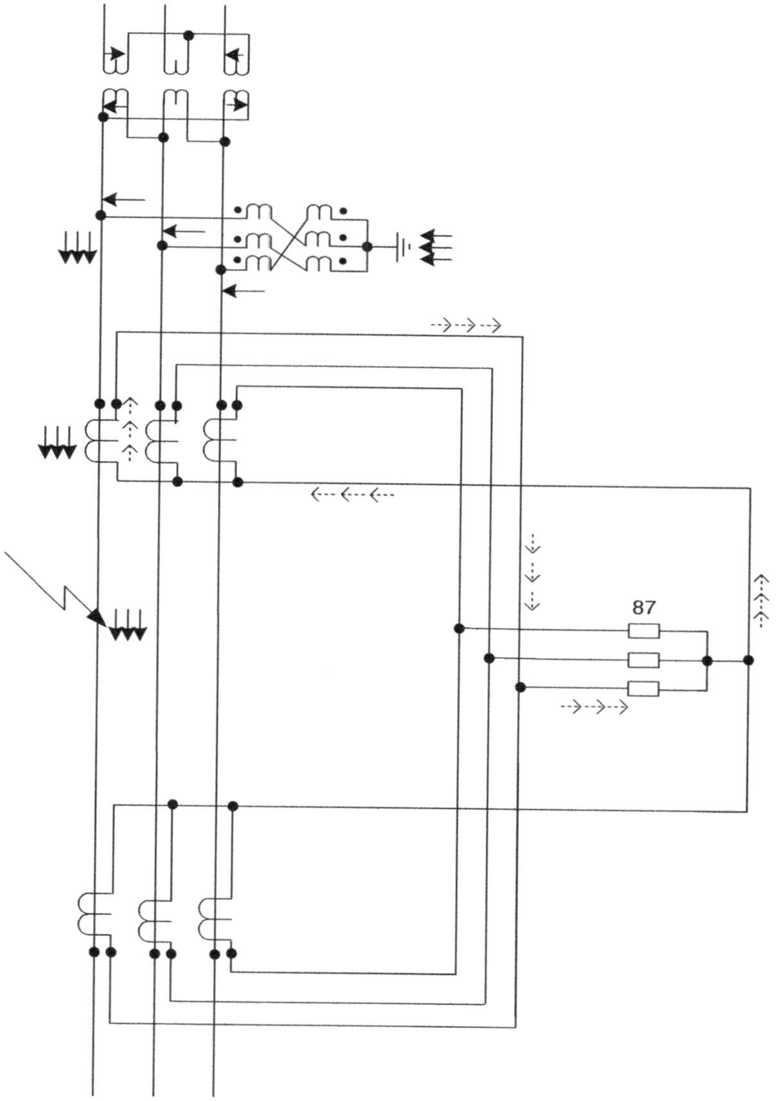

Figure 9.23 Zero-sequence current distribution Case #1.

9.4.1.4.4.2 Zero-Sequence Current Distribution #2 In this example, the grounding transformer is still located in the transformer zone.

For an external bus fault, as shown in Figure 9.24, zero-sequence current correctly flows around the differential circuit and not through the operate coil of the faulted differential phase relay.

9.4.1.4.4.3 Zero-Sequence Current Distribution #3 In this example, the grounding transformer is located in the bus zone.

For an internal bus fault, as shown in Figure 9.25, zero-sequence current flows through the operate coil of the faulted differential phase relay but also through the operate coils of the other two non-faulted differential phase relays. The relay operates to isolate the faulted bus. However,

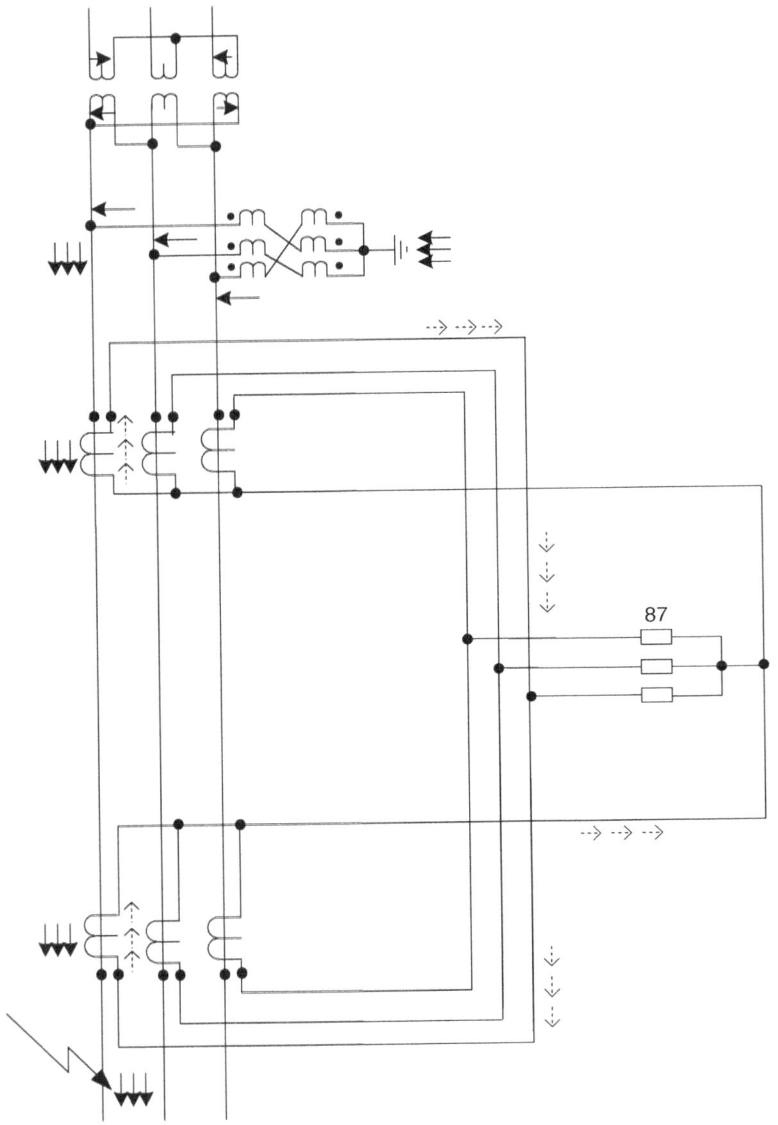

Figure 9.24 Zero-sequence current distribution Case #2.

only two-thirds of the total ground current flow through the faulted differential phase relay which reduces its sensitivity. Also, it is not possible to determine which phase was faulted by viewing targeting on the differential relay as all three phases operate.

9.4.1.4.4.4 Zero-Sequence Current Distribution #4 In this example, the grounding transformer is still located in the bus zone.

For an external bus fault, as shown in Figure 9.26, zero-sequence current will incorrectly flow into the operate coil of each faulted differential phase relay. The zero-sequence current seen by

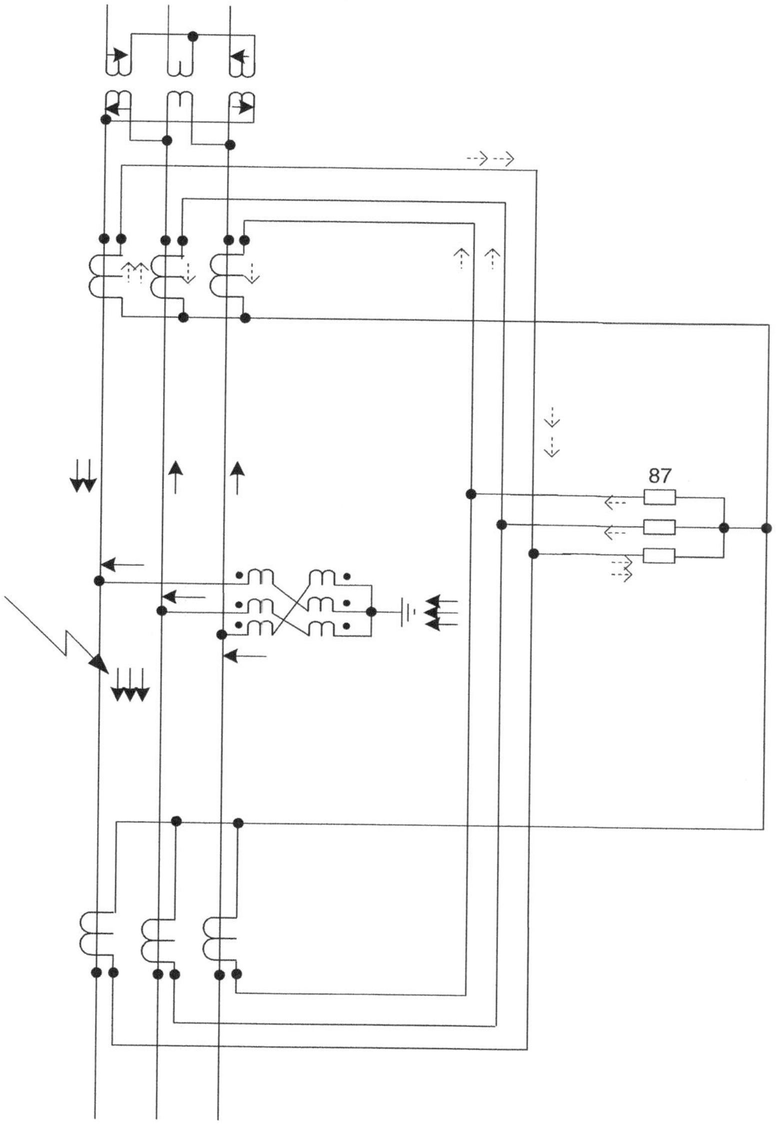

Figure 9.25 Zero-sequence current distribution Case #3.

each phase relay is one-third of the total ground-fault current. Nevertheless, it most likely would be sufficient to incorrectly operate all three differential phase relays.

9.4.1.4.4.5 Zero-Sequence Current Distribution #5 In this example, the grounding transformer is still located in the bus zone. However, the grounding transformer is now zoned by sets of CTs in parallel with the other differentially connected CTs. In effect, the grounding transformer is now zoned off to be outside of the bus zone.

For an internal bus fault, as shown in Figure 9.27, zero-sequence current representing the full ground current flows through the operate coil of the faulted differential phase relay to correctly operate the relay.

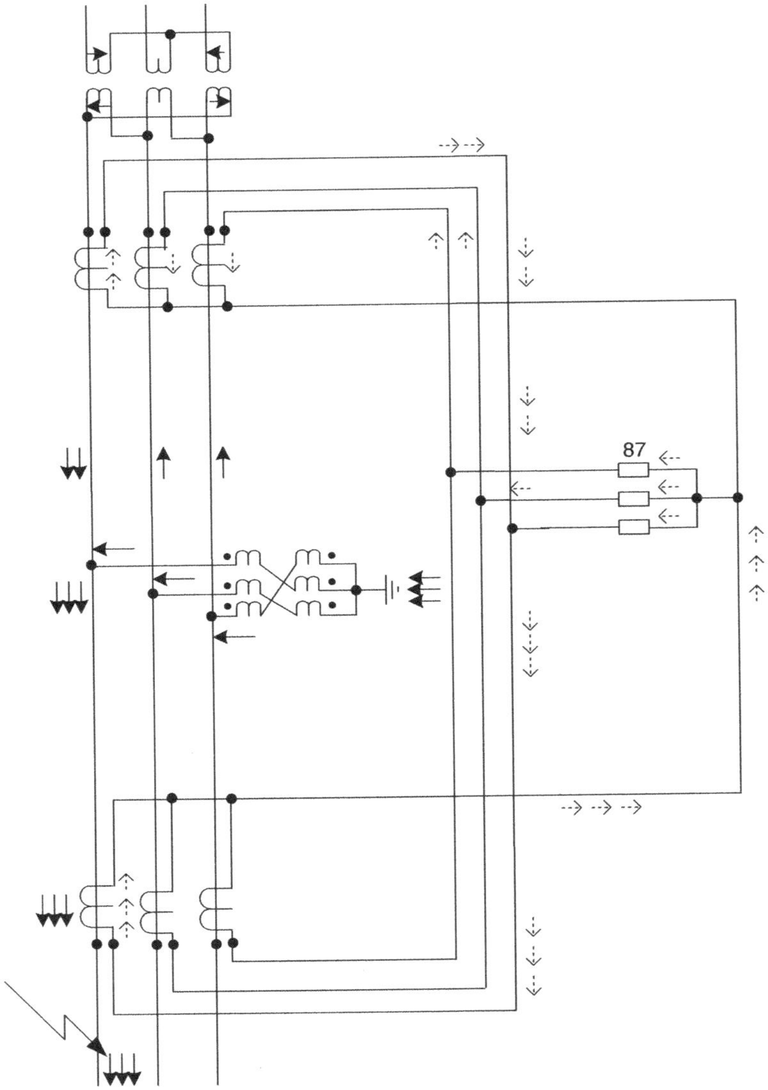

Figure 9.26 Zero-sequence current distribution Case #4.

9.4.1.4.4.6 Zero-Sequence Current Distribution #6 In this example, the grounding transformer is still located in the transformer zone. It is also still zoned off via CTs on the bus side of the grounding transformer.

For an external bus fault, as shown in Figure 9.28, zero-sequence current correctly flows around the differential circuit and not through the operate coil of the faulted differential phase relay.

9.4.1.4.4.7 Zero-Sequence Current Distribution #7 In this example, the grounding transformer is located in the next zone beyond the bus.

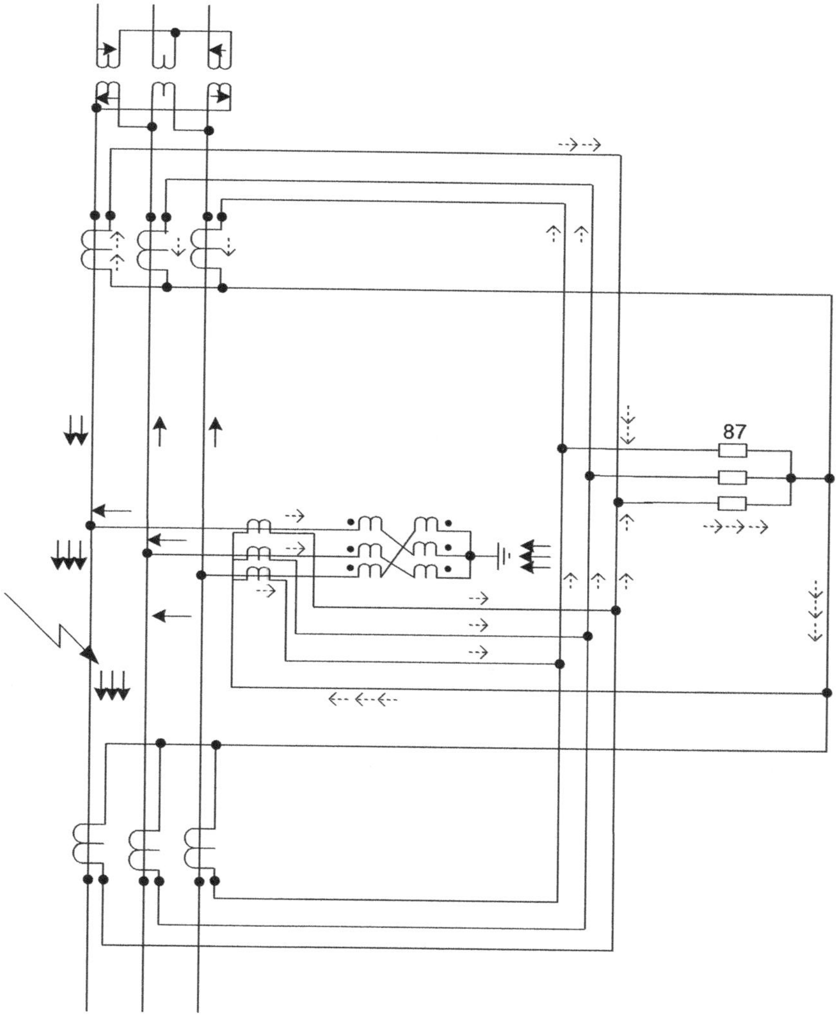

Figure 9.27 Zero-sequence current distribution Case #5.

For an internal bus fault, as shown in Figure 9.29, zero-sequence current representing full ground current flows through the operate coil of the faulted differential phase relay to correctly operate the relay.

9.4.1.4.4.8 Zero-Sequence Current Distribution #8 In this example, the grounding transformer is still located in the next zone beyond the bus.

For an external bus fault, as shown in Figure 9.30, zero-sequence current correctly flows around the differential circuit and not through the operate coil of the faulted differential phase relay.

9.4.1.4.4.9 Zero-Sequence Current Distribution #9 In this example, the grounding transformer is located in the transformer zone. This location is mainly typical of substations where zigzag grounding transformers are used as artificial ground sources.

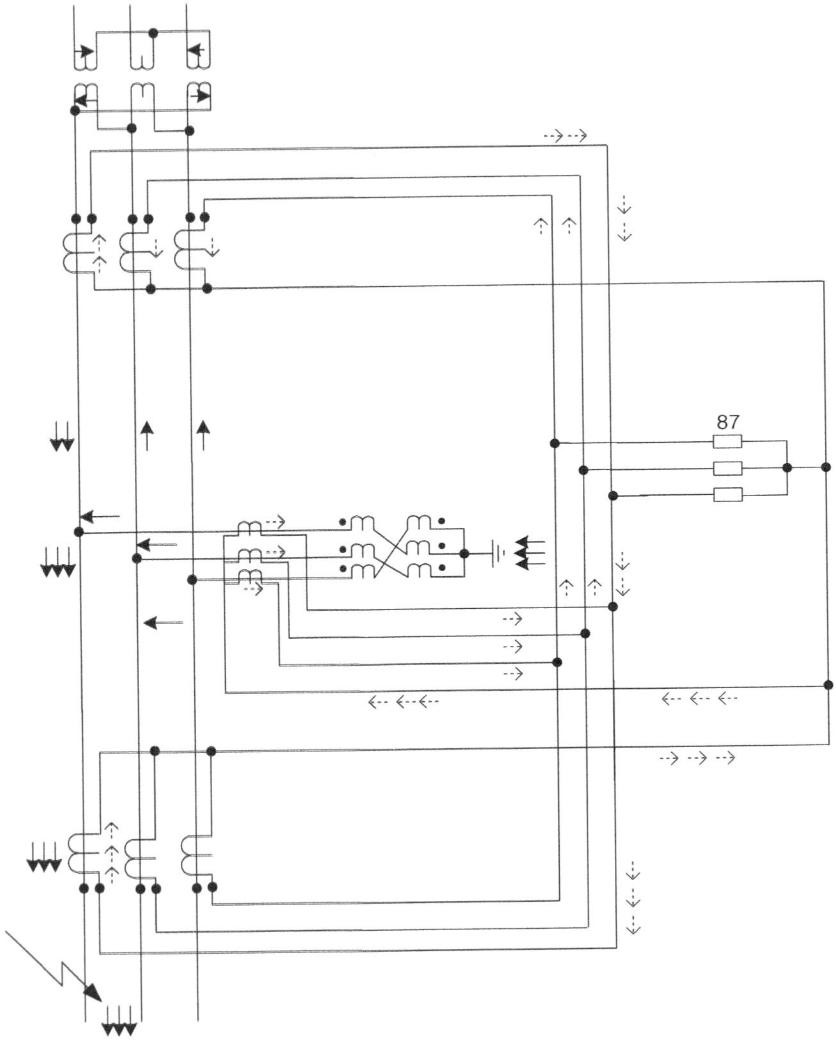

Figure 9.28 Zero-sequence current distribution Case #6.

Grounding transformers have different ratings depending on the purpose of the artificial ground in the first place. Utilities use grounding transformers for two distinctly different purposes.

One purpose is to provide a source of ground to facilitate naturally occurring unbalance while supplying four-wire distribution loads. In this case, the grounding transformer will be of the low impedance variety and will allow a hefty amount of ground-fault current to flow. It is meant to allow a significant amount of continuous ground current to flow.

Another purpose is to provide a source of ground to facilitate using overcurrent relays to detect ground faults. The purpose is not to facilitate the grounding of load and is typical of three-wire distribution loads. In this case, the grounding transformer will be of the high impedance variety and will allow a very little amount of ground-fault current to flow. It is not meant to allow anything but a token amount of continuous ground current to flow.

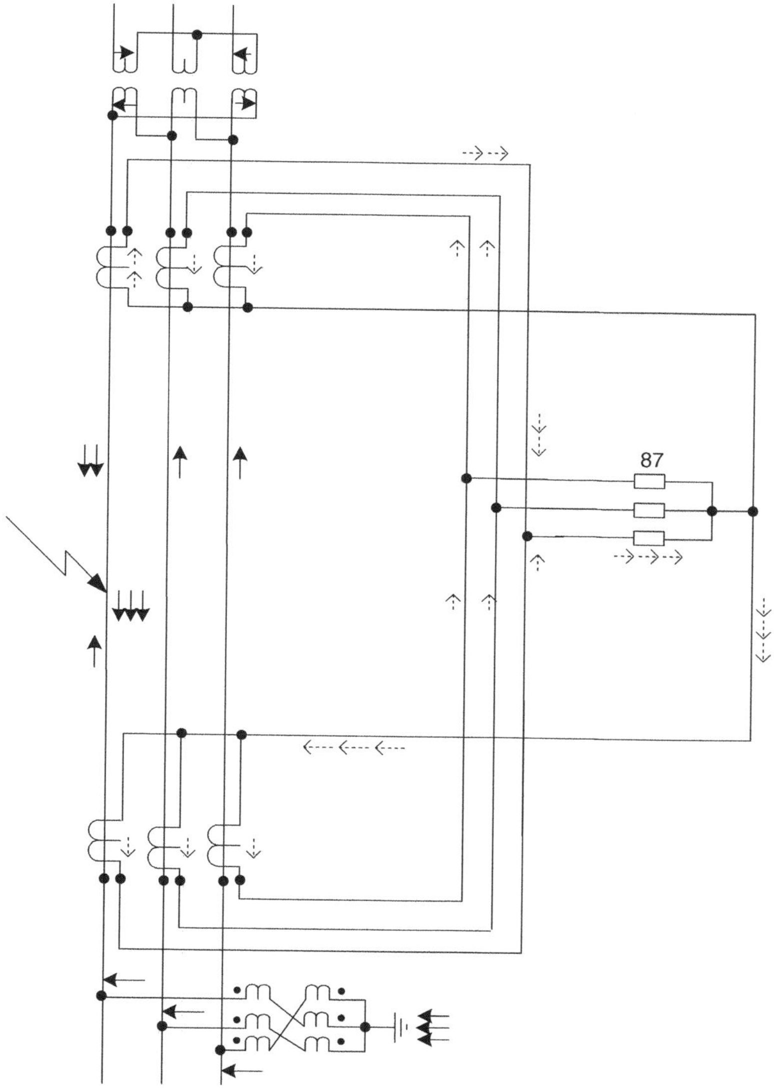

Figure 9.29 Zero-sequence current distribution Case #7.

For an internal bus fault, as shown in Figure 9.31, zero-sequence current representing full ground current flows through the operate coil of the faulted differential phase relay but also through the 87*N* ground relay. Given a situation where the phase differential relays are set high as would be the case if they were set to be immune to CT saturation they will not operate for a low-level ground-fault current. In this case, the dedicated ground relay 87*N* is connected as shown will operate. The ground relay can be set to be much more sensitive than the phase relays and will operate for even extremely low levels of ground-fault.

9.4.1.4.4.10 Zero-Sequence Current Distribution #10 In this example, instead of using a grounding transformer to introduce a source of ground current, a reactor located in the star point to ground connection of the Delta-Wye transformer is used as shown in Figure 9.32. The grounding reactor

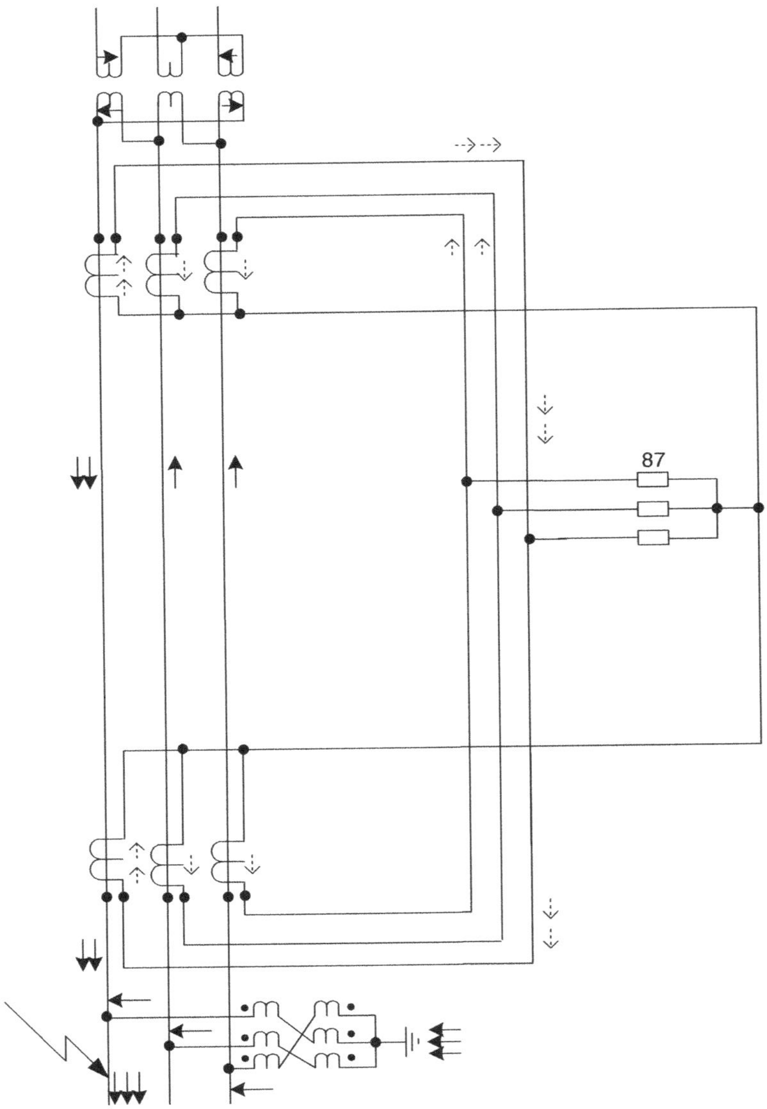

Figure 9.30 Zero-sequence current distribution Case #8.

can either be of the low impedance or the high impedance variety. Where the grounding reactor is used to introduce a limited amount of ground current to flow to allow detection of a ground fault it will be high impedance and will not have a meaningful continuous rating. Where the grounding reactor is used to introduce a large amount of ground current for a four-wire distribution system where unbalanced loads are being supplied it will be low impedance and will have a continuous rating matching the anticipated load unbalance.

9.4.2 Bus Blocking Protection

Digital relays are most suited for a bus protection scheme known as bus blocking. Bus blocking is mainly used at substations where a bus is protected by an overcurrent protection that is definite

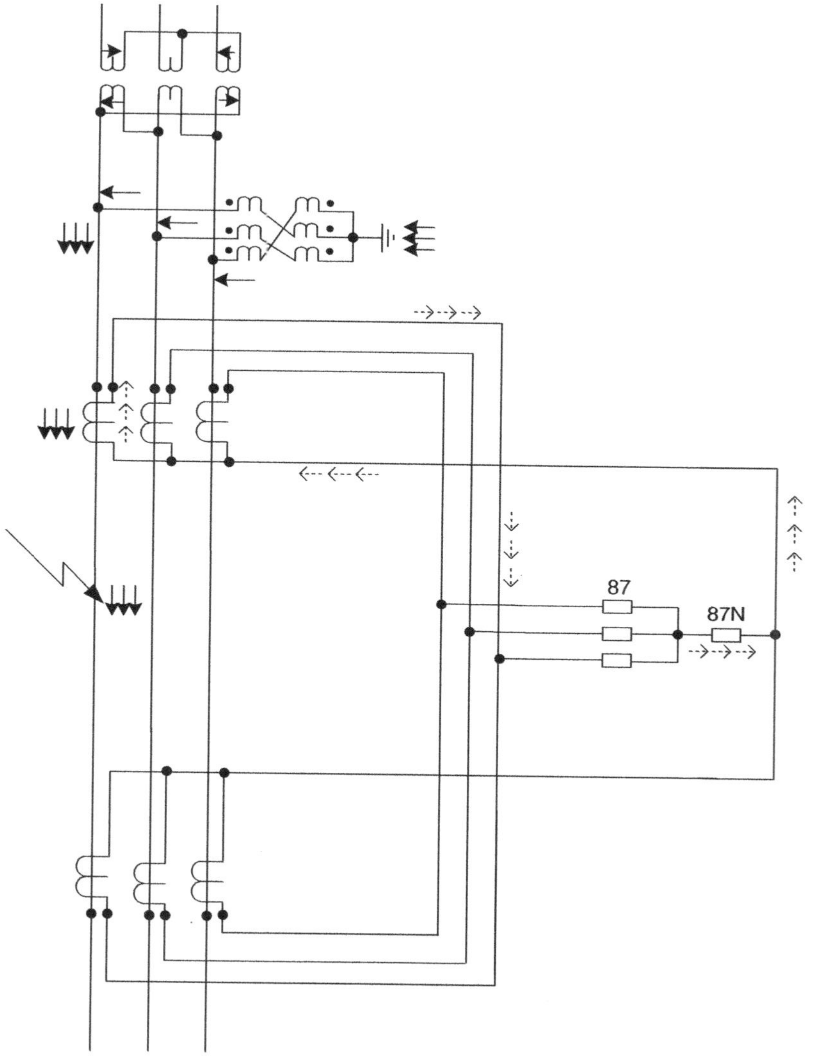

Figure 9.31 Zero-sequence current distribution Case #9.

time delayed waiting a short time for a block signal from a feeder protection. Low set instantaneous overcurrent relay contacts on all elements supplied by the bus, typically distribution feeders are connected in parallel and configured to block the bus overcurrent protection from operating for any faults on a feeder. In the absence of a block signal being received, it is assumed the recognized fault by the bus overcurrent protection is in fact a legitimate bus fault and not a fault on any of the distribution feeders egressing from it. When wiring blocking contacts directly into the digital input of bus relays, the time delay can be as short as one cycle thereby ensuring high-speed protection of the bus. The bus blocking bus protection scheme is illustrated in Figures 9.33 and 9.34.

Probably, the largest noticeable difference between bus differential and bus blocking protection is the absence of CT secondary connections on one side of the feeder breakers. Bus differential protection must connect in parallel all the CTs on the feeder egress side of the feeder breakers. With bus blocking protection, these CTs are not required and are shorted in the switchyard. Costly CT cabling

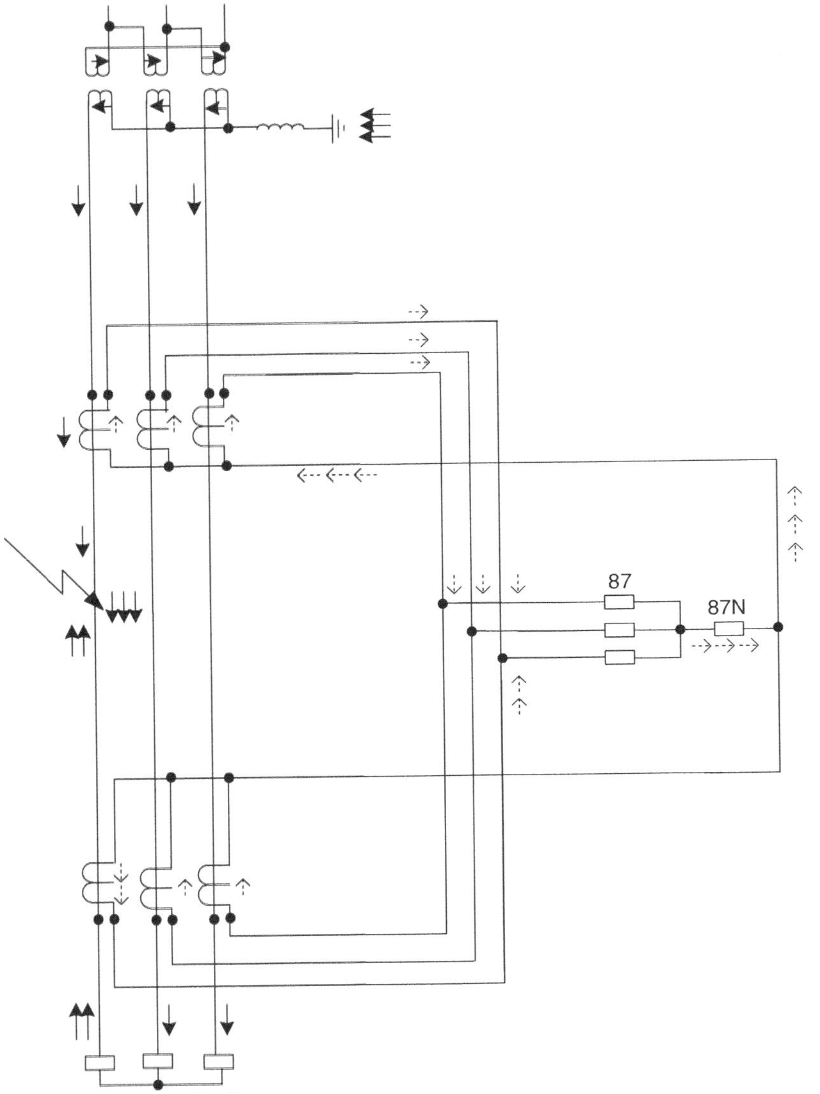

Figure 9.32 Zero-sequence current distribution Case #10.

is not required any longer for just under half the CTs in the switchyard as most switchyards contain anywhere from eight to twelve feeder breakers. Where LV capacitor banks are located tapped to the buses at substations, they are treated the same way as ordinary feeders as far as bus blocking is concerned. Their breakers only require one set of CTs to be cabled from the switchyard to the relay building. Bus differential protection is highly susceptible to the effects of CT saturation. CT lead burdens are a significant contributing factor to causing these CTs to saturate. For that reason, the CT cable lead resistances are kept arbitrarily low to ensure the overall burden does not exceed a predetermined value. This leads to, depending on the size of the switchyard, to some very hefty

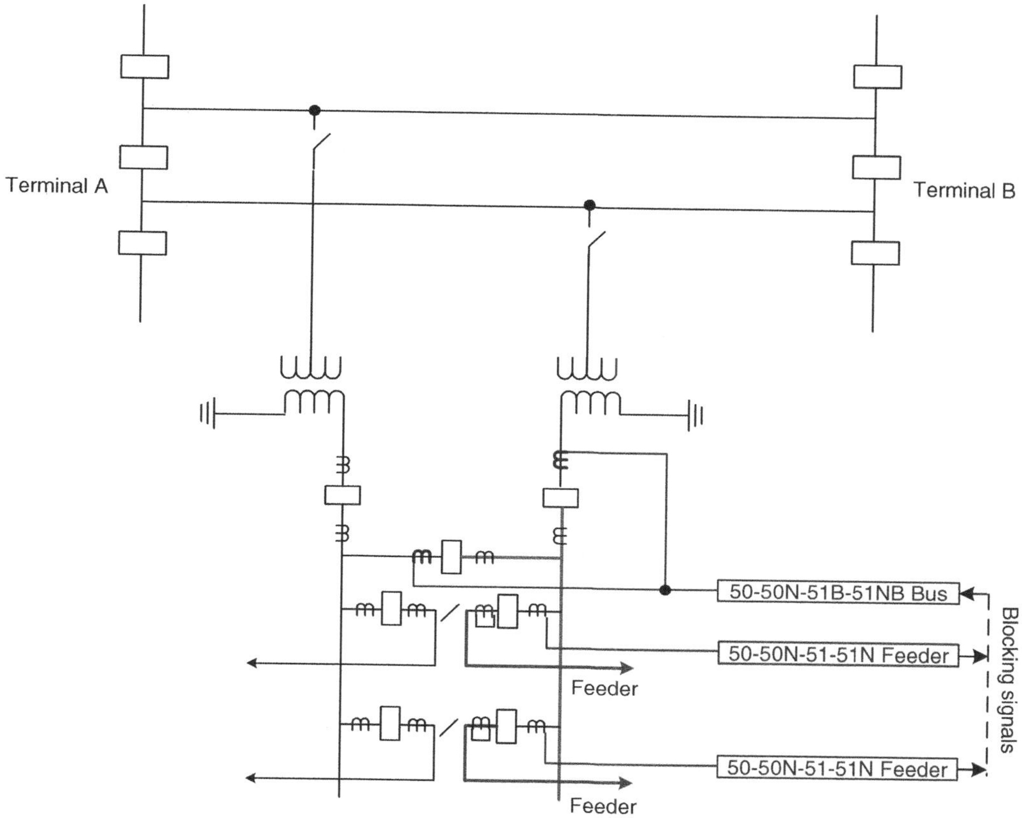

Figure 9.33 Station single-line diagram showing bus blocking protection.

cable sizes. It is not unusual to have CT lead cables specified at 8 AWG (American Wire Gauge) or even 6 AWG. Not only are there significant cost savings in terms of copper but there are other savings in terms of cable trenches and labor costs of installing such large unwieldy cables.

Figure 9.35 shows bus and feeder digital relays mounted on metal-clad switchgear. The switchgear compartments house 13.8 kV breakers designated M1 and M2 in this photograph. In the middle, is a compartment designated for protection and control equipment. The metal-clad switchgear itself is modern arc-proof making it safe to mount P&C equipment on it as shown. In this particular example, the bus protection relay is shown on the far right mounted on the door of the P&C equipment compartment. Two feeder protection relays are shown in the middle mounted on the door of the P&C equipment compartment. One relay protects the M1 feeder while the other relay protects the M2 feeder. The bus protection relay on the far right receives a bus blocking signal from either of the two M1 and M2 feeder protection relays in a bus blocking scheme. Since the feeders are radial, for a bus fault, only the bus protection operates, and after a short-time delay trips the bus breaker located in the breaker compartment shown on the far upper right. If the bus protection operates for a feeder fault, the respective feeder protection (either M1 or M20) relay will block the bus relay from tripping the bus breaker. The respective faulted feeder protection will of course trip its breaker.

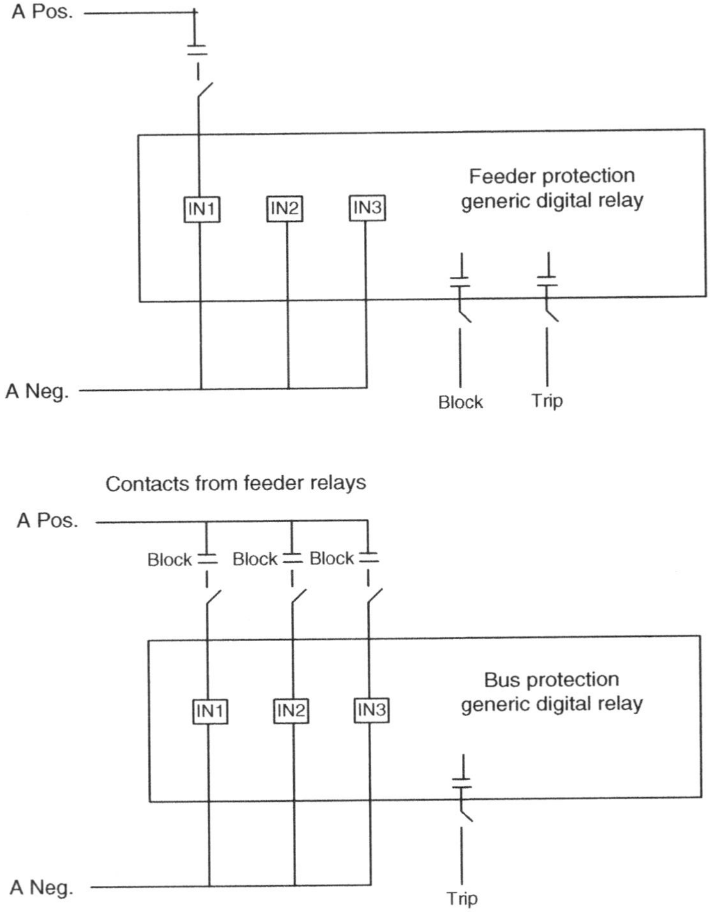

Figure 9.34 Bus blocking conceptual DC drawing.

9.5 Example High Impedance Differential Protection Setting

For a 13.8 kV bus given:

Maximum three-fault = 356 MVA
The maximum line to ground fault = 400 A (limited by a grounding resistor)
CT rating 2000-5A (400:1)
400 ft of 6 AWG cable from the switchyard to the relay building
50 ft of 12 AWG cable from the relay building CT rack panel to the protection relay panel
The internal resistance of the totally shorted CT 1.0 Ω
CT accuracy classification C200
Relay burden is given as 1 VA at chosen current pickup

The voltage developed at the relay connection point is given by the equation known as the stabilizing voltage $V_S = I_F\,(R_{ct} + 2R_L)$ as shown in Figures 9.36 and 9.37.

Figure 9.35 Bus and feeder protections mounted on metal-clad switchgear.

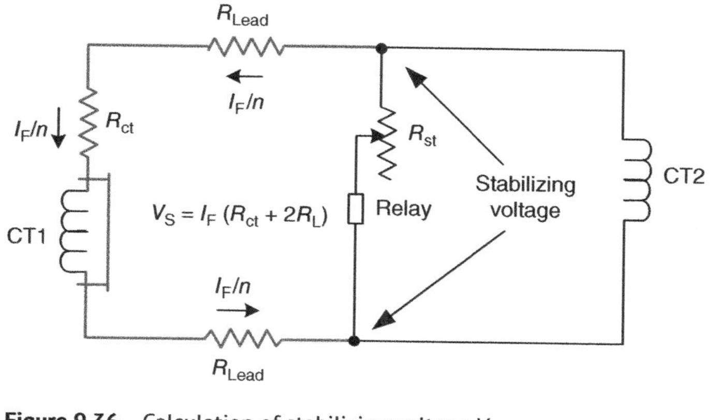

Figure 9.36 Calculation of stabilizing voltage V_s.

Calculate the maximum three-phase fault current I_F

$$I_F = 356\,\text{MVA}/\left(\sqrt{3} \times 13.8\,\text{kV}\right) = 15\,\text{kA}$$

$$I_F\ (\text{secondary}) = 15\,\text{kA}/400 = 37.5\,\text{A}$$

400 ft of 6 AWG cable at 0.4 Ω per 1000 ft is 0.16 Ω
50 ft of 12 AWG cable at 1.6 Ω per 1000 ft is 0.08 Ω

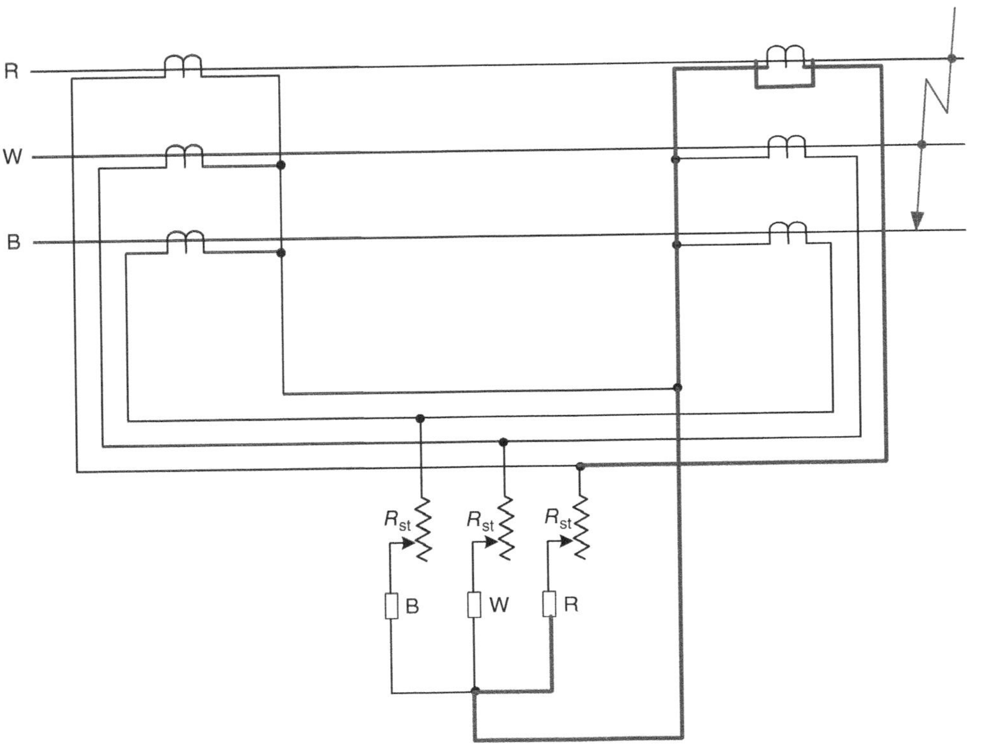

Figure 9.37 Three-phase representation of high impedance differential.

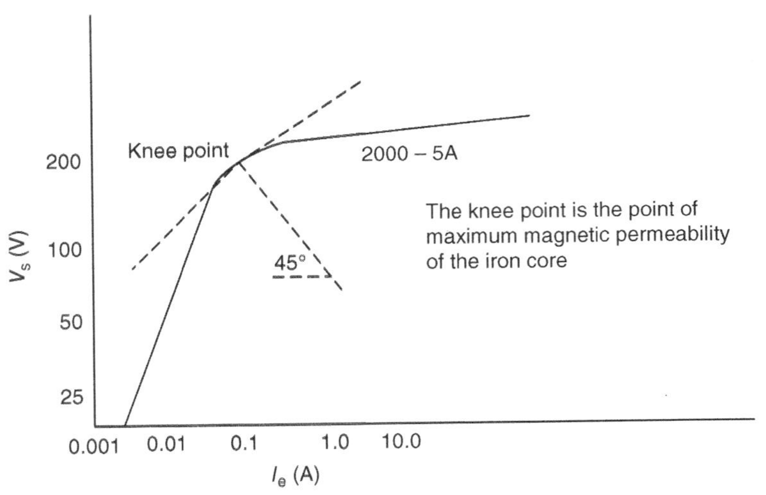

Figure 9.38 CT excitation characteristic representation C200.

The total loop cable resistance $R_L = 2 \times (0.16 + 0.08) = 0.48\,\Omega$

$$R_{CT} + R_L = 1.0 + 0.48 = 1.48\,\Omega$$

$$V_S = I_F\ (R_{CT} + R_L) = 37.5 \times 1.48 = 55.5\,V$$

Add a 10% margin for $V_S = 61\,V$

Arbitrarily choose a current pickup at approximately 30% of the maximum line to the ground fault of 400 A. Choose $I_{PKP} = 0.3\,A$

$$R_{st} = 61\,V/0.3\,A = 200\,\Omega$$

$$R_{burden} = 1\,VA/(0.3A)^2 = 11\,\Omega$$

$$R_{total} = 211\,\Omega$$

V_S to operate the relay must be greater than $211\,\Omega \times 0.3\,V = 63.3\,V$

The CT is classified as C200. Therefore, at relay operation, the CT is producing approximately one-third the amount of secondary voltage at saturation as shown in Figure 9.38. The protection is therefore sensitive and secure.

Reference

1 IEEE std C37.234-2009, Guide for Protective Relay Applications to Power System Busses.

10

Breaker Failure Protection and Automatic Reclosing

10.1 Introduction

This chapter deals with the two independent aspects of breakers, failure and reclosing. High voltage (HV) breakers are almost always provided with dedicated breaker failure protection. This protection caters specifically when a breaker is called upon to trip by a protection system and fails to adequately do so. HV breakers are usually also provided with a mechanism to automatically reclose after being tripped by protection. Of course, this is only applicable when the trip precipitating fault is deemed to be transient. A prime example of a transient fault is lightning striking an overhead transmission line where there is a legitimate reason to believe the initial cause of the fault is transient. The lightning will cause an arcing fault between phases or between phase(s) to ground. However, the arc will disappear upon line isolation with breakers having been opened with the source of magnetomotive force (MMF) no longer maintaining the arc. Where a fault is most likely to be permanent automatic reclosing is either not armed or is canceled. Furthermore, where a breaker has failed for any given fault, all other breakers tripped by protections to isolate that fault should have their automatic reclosing reset.

10.2 Breaker Failure General Background

Before the time, when modern tele-protection systems were available, dedicated breaker failure protections could not exist at stations where local breakers were hardwired to trip. Without the means to transfer trip remote breakers, fault isolation could not be achieved. Until that time, remote backup was used to isolate remote fault infeeds. The long-time Zone 3 elements used for remote backup themselves were difficult to set to ensure adequate security was achieved. Modern communication systems are now primarily used to isolate fault infeeds from remote sites for a local breaker failure using dedicated breaker failure protection.

The failure of a breaker is considered a first contingency, and it is not expected that the communication system fail along with it. High-speed line protection could then be easily achieved with either permissive or directional comparison blocking schemes. For this reason, breaker failure only requires one communication route.

10.2.1 Typical Breaker Failure Tripping Zones

When a breaker fails to clear a fault, breaker failure protection acts to isolate the fault by tripping adjacent zones. It does this by tripping the next zone breakers at the station by hard-wired tripping

Power System Protection: Fundamentals and Applications, First Edition. John Ciufo and Aaron Cooperberg.

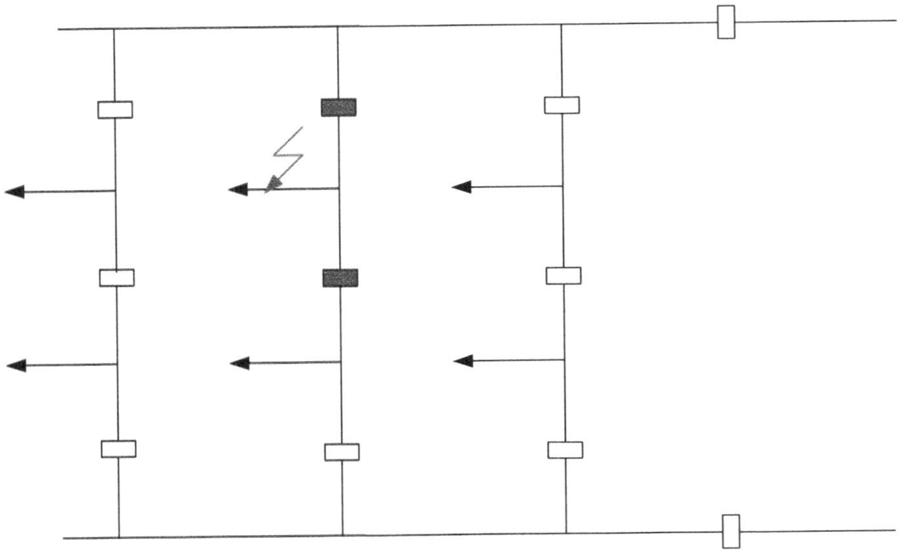

Figure 10.1 Typical terminal station with no breaker failure.

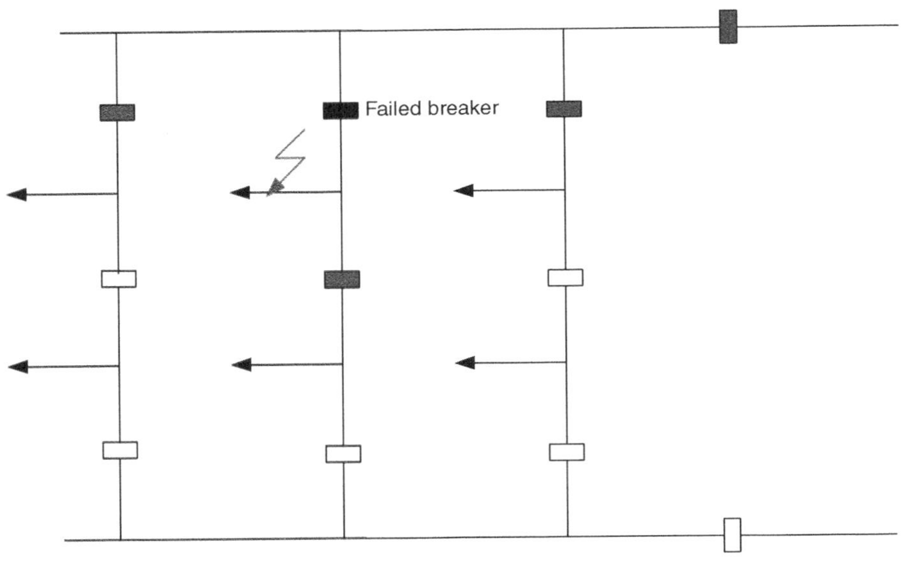

Figure 10.2 Typical terminal station with breaker failure.

and for breakers at remote stations by transfer tripping. Refer to Figure 10.1 showing a typical terminal station with healthy isolation of the faulted zone.

Compare this with Figure 10.2 showing the extended tripping of breakers in the adjacent zones called upon to trip thereby isolating the same fault but with one of the breakers having failed.

Refer to Figure 10.3 showing a breaker failure at a ring bus. In this example, communication facilities do not exist between the local ring bus station and the remote terminal station feeding into the uncleared fault. In this case, a dedicated breaker failure protection will work by tripping the local next zone by hardwired tripping of local breakers. The infeed from the remote terminal

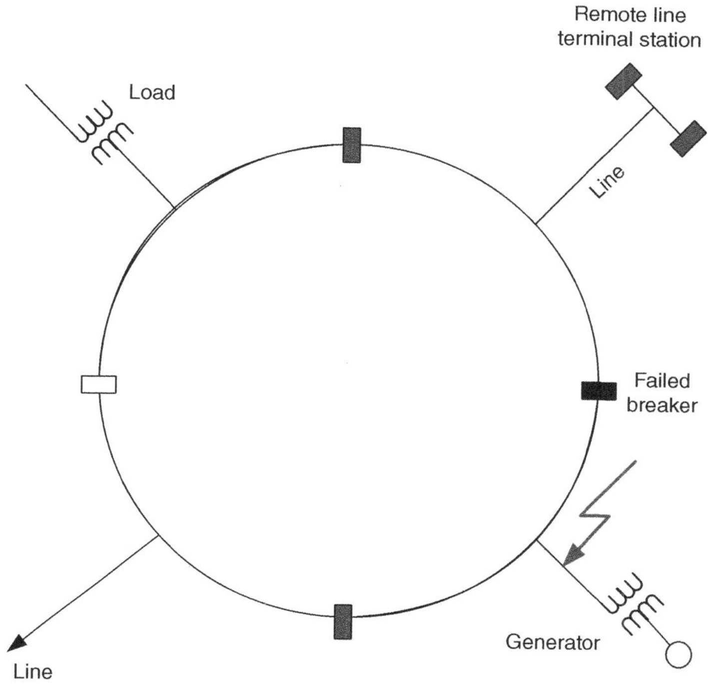

Figure 10.3 Ring bus with a failed breaker.

station at the far end of the line will be eliminated by the remote Zone 3 timed backup protection at that location.

Refer to Figure 10.4 showing a breaker failure condition at a substation for a low voltage (LV) bus fault. In this example, there does not exist an isolating device on the HV side of the transformer to isolate the infeed from the terminal stations. Communications in the form of either remote trip or transfer trip to the terminal station and then cascaded to the other terminal station are the only means of isolating this fault and failed breaker.

The breaker failure protection in this example will also open the motorized transformer HV disconnect. However, even though the faulted bus is isolated with the opening of the transformer disconnect, automatic reclosing of the terminal breakers is not permitted. Many utilities but not all have adopted the procedure that no automatic reclosing of any kind is permitted following a breaker failure. This is true even when the fault is then isolated from the breakers to be automatically reclosed. The reasoning is operators do not want to see any automatic reclosing of breakers anywhere once a single breaker has failed. They want total discretionary authority to close breakers themselves upon assessing it to be safe to do so.

10.2.2 Principles of Breaker Failure Protection

The significance of breaker failure protection cannot be overstated. Breaker failure protection is not required to operate very often but when required provides a very important function. There are three important principles fulfilled by breaker failure protection:

(a) When a breaker fails to isolate a fault, the fault stays on the system until breaker failure protection acts to isolate it by another means. As a consequence, the duration of uncleared fault is

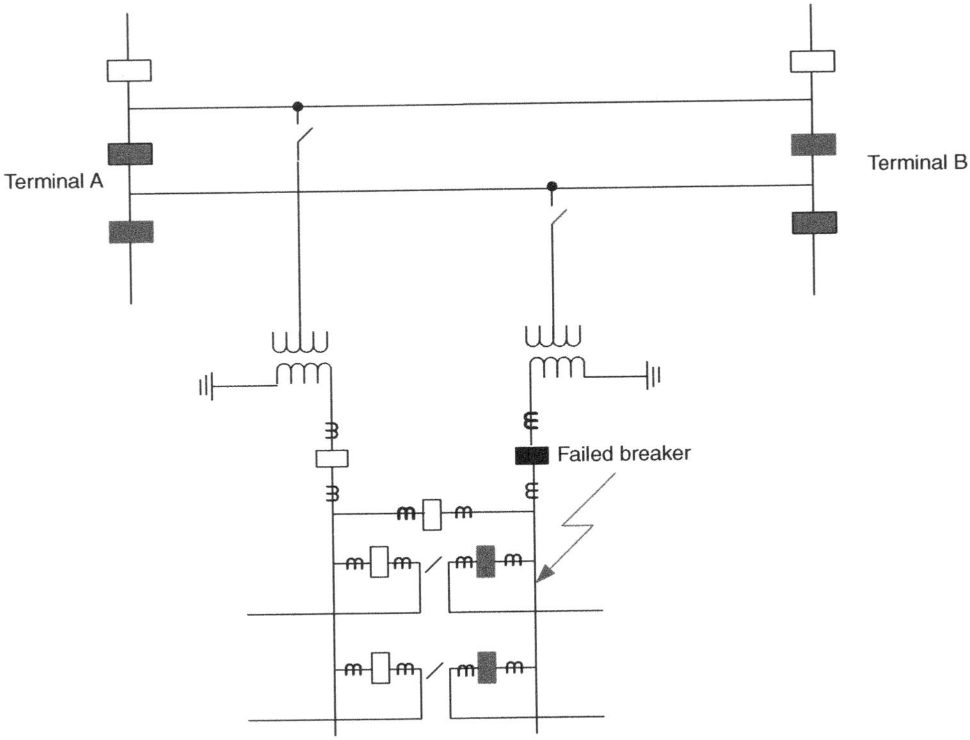

Figure 10.4 Faulted LV bus at a substation with a failed breaker.

the criteria used to maintain system stability. Breaker failure clearing times underpin the basic criteria in the design of the interconnected power system. It is therefore of the utmost importance that breaker failure clearing times be as fast as possible without jeopardizing security.

(b) When a breaker fails, the fault continues to persist until the next zone is isolated by the breaker failure protection. During this extra time to fault clearance, considerable damage to equipment can occur. Transformers are subjected to extraordinary internal magnetic forces and the enormous heating effect of sustained high currents resulting in the potential for fires to break out. One of the lesser understood phenomena of faults is the anatomy of an uncleared fault itself. Most faults are started by a transient phenomenon such as a lightning strike on an overhead transmission line. The air surrounding the stroke itself is ionized and allows a low resistance path for fault current to flow either to the ground or between phases. Upon the initial fault current or arc being established fault current will continue to flow until all sources of magneto motive force MMF is removed. Therefore, breaker failure protection is required regardless of whether the initial fault is transient or permanent.

(c) Whenever breaker failure protection operates, it trips a large number of breakers in adjacent zones. The inadvertent operation of breaker failure protection depending on system topology can lead to customer interruption and blackout. For this reason, it is also of the utmost importance to mitigate the possibility of operating breaker failure protection needlessly. A look at the number of breakers assigned to trip from breaker failure protection indicates that in almost all cases two independent and adjacent protection zones are affected. Refer to Figure 10.4 where the effect of this principle is readily apparent. Should a transformer LV breaker failure protection be inadvertently operated, the entire LV bus along with feeder breakers is tripped, as well as the HV line supplying the transformer. Without question, customer interruptions would occur.

10.2.3 Requirements for Breaker Failure Protection

There are several general requirements for breaker failure protection. Some of these requirements are applicable to all HV and LV applications while others are only applicable to HV applications.

10.2.4 Types of Initiation

Whenever a protection operates to trip breakers, simultaneously along with tripping those breakers, breaker failure protections associated with each breaker are initiated or armed. Whether breaker failure protection goes on to operate and trip next zone breakers depends exclusively upon whether the breaker is deemed to operate as expected or not.

For this reason, many utilities have adopted the principle not to use the same auxiliary relay or logic controller to trip and initiate its breaker failure. Should that one auxiliary relay or logic controller fail, the consequences could be substantial. This principle is valid regardless of breaker failure protection being initiated from two independent protections. The reason is that one of the two protections could be rendered ineffective for example being removed for service during maintenance. A fundamental tenet of breaker failure protection hence is it must always be available to be initiated reliably from at least one protection.

Many system operators are lenient when breaker failure protection itself is being maintained. Operators will provide protection maintenance staff a short time during clear weather conditions to maintain a breaker failure protection without needing to remove the affected breaker itself from service. This is provided the breaker is first test tripped, thus ensuring beyond a reasonable doubt its ability to trip and isolate a fault if called upon to do so during the short time required to do the maintenance.

10.2.5 Speed of Operation

All types of breaker failure protection have some form of time supervision. This means that following breaker failure initiation some predetermined time must elapse before the decision is made that the breaker failed and the need to trip the next zones. The speed of breaker failure protection is therefore very much dependent on the setting of these timers. Furthermore, even with some timers allowed to time out, additional criteria may need to be met.

10.2.6 Determination of Breaker Failure Condition

The normal trip time of a breaker is used to determine whether the breaker has in fact failed to open following a trip initiation. Many utilities and manufacturers of digital breaker failure protection relays use three parallel logic paths in the determination of a breaker failure condition for all HV breakers as shown in Figure 10.5.

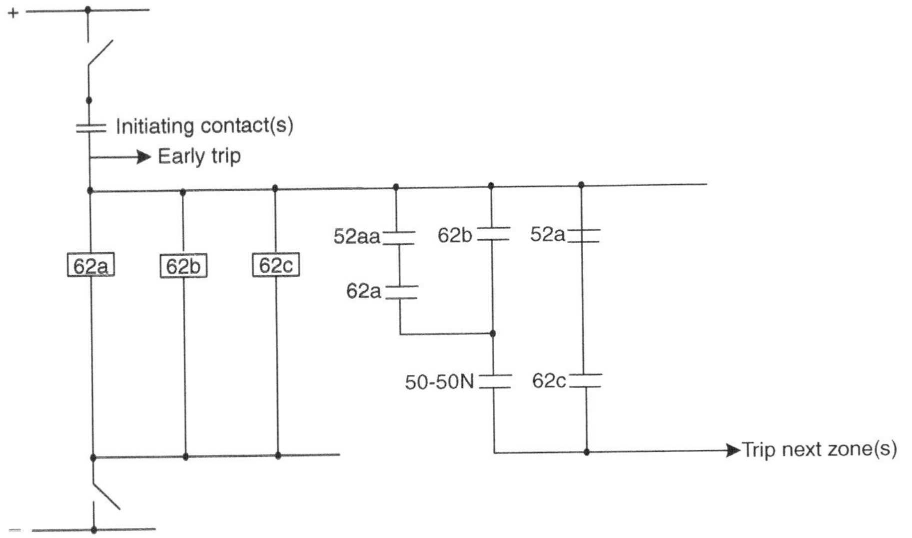

Figure 10.5 Conceptual representation of the three breaker failure logic paths.

10.2.6.1 62a Logic Path (Mechanical-Breaker Failure Detection)

The purpose of this logic path is to cover breaker mechanical failure in the shortest amount of time. The 62a timer in this logic path is supervised by what is known as a breaker advanced pallet switch with designation 52aa. This pallet switch changes state immediately upon the breaker mechanism beginning to move. Should the breaker mechanism show no sign of moving shortly after a trip initiation the breaker is deemed to have failed. The 62a logic path is also supervised by the current through the breaker to ensure a wide area breaker failure operation cannot take place for an inadvertent operation. In other words, if a for whatever reason no current is flowing through the breaker, a breaker failure condition is not declared regardless of the 52aa advanced pallet switch not changing state. It should be noted that for this logic path to be effective, the 52aa advanced pallet switch cannot be multiplied and is adjusted by station maintenance whenever the breaker is maintained. It can and does go out of alignment from time to time.

The 62a timer is set typically as 33 ms + 52aa auxiliary switch operating time (usually 1cycle) for a total of 50 ms following a trip initiation. Current through the breaker is very often set at some general predetermined setting typically 1000 A for both phase and ground supervising current elements.

10.2.6.2 62b Logic Path (Electrical Breaker Failure Detection)

The purpose of this logic path is to cover electrical failure regardless of whether the breaker contacts opened or not. The 62b timer in this logic path is supervised by the current through the breaker only. It is essential, that the current detecting elements, when using electromechanical overcurrent relays, have a very high pickup to drop out ratio. Typically, this requires specialized overcurrent relays where the relay's output contact drops out typically at approximately 97%. For example, should the relay be set to close output contact at 10 A it will open its output contact when the actuating current drops below 9.7 A. In the case of digital breaker failure relays, the overcurrent elements assert and de-assert at similar values if not better.

Typical settings for the 62b timer are 90 ms for two cycle breakers and 105 ms for three cycle breakers. Measured current through the breaker is typically the same standard-setting as for the 62a path which is 1000 A for both phase and ground supervising current elements.

Many utilities are governed by criteria that require their system to maintain stability with a permanent phase to ground fault on any transmission line, transformer, or bus section with delayed clearing due to breaker failure. The 62b path time delay is mostly used for this determination.

10.2.6.3 62c Logic Path (Low Magnitude Fault Breaker Failure Detection)

The purpose of this logic path is to cover a failed breaker regardless of fault current detection. It is meant to detect a mechanically failed breaker with a very low fault current flowing through the breaker. The 62c timer in this logic path is supervised by a regular breaker auxiliary 52a pallet switch which may be multiplied if insufficient spares exist.

The 62c timer is set typically at 500 ms for all types of HV breakers.

10.2.7 Additional Generic Features of Breaker Failure Protection

In addition to the three logic paths for breaker failure, there are two additional features meant to avoid the unnecessary operation of breaker failure protection mainly due to inadvertent action by local station staff usually during maintenance.

10.2.7.1 Early Trip Function

Whenever a breaker failure protection is initiated additional tripping of the same breaker is performed by an independent function known as an early trip. The early trip feature is an added level of security preventing a false breaker failure protection operation should any breaker failure initiating contacts be inadvertently closed during protection maintenance.

10.2.7.2 Breaker Test Switch Supervision

The 62c logic path is supervised by a contact of the breaker test (maintenance) switch such that inadvertent breaker failure operation is prevented by mistake during breaker maintenance. This feature is not needed for the 62a and 62b paths since they are both supervised by the current detecting relays or elements.

10.2.8 Circuit Operation of Breaker Failure

There are three distinct stages to breaker failure protection that take place in sequence. Firstly, there is an initiation by a protection or protections that trip the breaker. Secondly, there must be a clear determination that the breaker has failed mechanically or electrically or both. Thirdly, an energized breaker failure condition triggers a wholesale tripping of the surrounding zones required to isolate the uncleared fault. This may include hardwired tripping of local breakers and transfer/remote tripping of breakers at other locations.

When a breaker is tripped by a protection, the breaker failure protection dedicated to that breaker is simultaneously initiated. If any one of the three logic paths is satisfied as shown in Figure 10.5, tripping of the next protection zones begins.

10.2.9 Coordination with Other Protections

It is common utility practice for line protections to have a typical 400 ms unconditional tripping Zone 2 or Zone 3 characteristics to cover for the loss of communications. These Zone 2 or Zone 3 elements will reach into a next protection zone where a breaker may have failed. Referring to Figure 10.6 the 62a and 62b tripping paths must be faster than the 400 ms tripping elements to ensure they do not operate for a breaker failure but to leave it to the breaker failure protection to

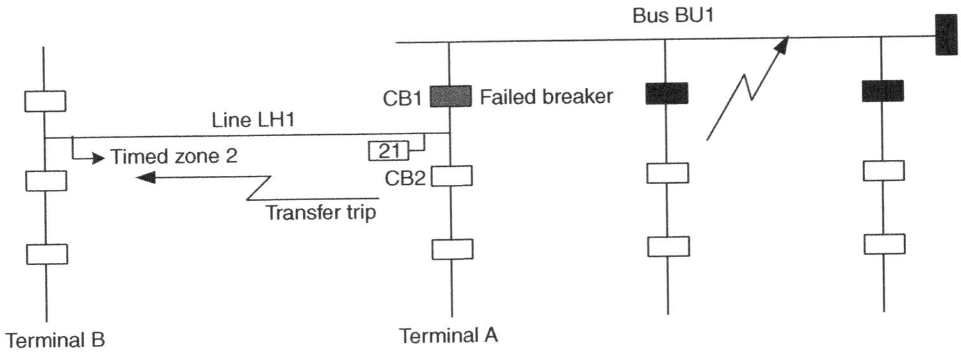

Figure 10.6 Breaker failure and need to coordinate with timed Zone 2/Zone 3.

isolate the failed breaker. The 62c timed at 500 ms will not coordinate with these other protections but this risk is considered acceptable by many utilities as any shorter time for 62c would reduce the overall security of breaker failure.

10.2.10 Non-Duplication of Breaker Failure

Usually, there is no requirement for duplicating breaker failure protection. Since breakers are adequately maintained to the same stringent maintenance requirements as the protections, a failed breaker is usually considered the first contingency. The protection philosophy adopted by the utility industry is that there is no requirement to cater for a double contingency failure of protection equipment. Consequently, it is considered unreasonable to expect breaker failure protection to also fail at the same time. By extension, the auxiliary station dc supply for breaker failure is typically only from a single "A" or main battery.

As every protection that trips a breaker also initiates its breaker failure protection, both "A" and "B" protections always initiate breaker failure. The actual tripping of breakers in the adjacent zones is also typically tripped via the "A" battery and "A" trip coils of those breakers. It is not necessary to consider two sets of outputs one "A" and the other "B" to trip the adjacent zones. A failure of either a single breaker trip coil or battery supply used to trip the breakers is considered double contingency to that of the failed breaker and double contingency failure of equipment is generally not catered for.

10.2.10.1 Non-Duplication of Breaker Failure Example

For simplicity to illustrate the concept of non-duplication of breaker failure, a simple design using auxiliary relays is shown. The concept shown would similarly apply when digital relays are used. Each utility has its method of doing a variation of this concept. Refer to Figure 10.7 showing a typical single "A" protection where a set of auxiliary relays are dedicated to a measuring relay or relays associated with a single protected element. The breakers are tripped via the "A" trip coil as this is an "A" protection. The contacts initiating breaker failure are wired to the respective breaker failure protections via inter-panel cabling where they are wetted by the "A" battery at that location.

Refer to Figure 10.8 showing a typical single "B" protection where a set of auxiliary relays are dedicated to a measuring relay or relays associated with a single protected element. The breakers are tripped via the "B" trip coil as this is a "B" protection. The contacts initiating breaker failure are wired to the respective breaker failure protections via inter-panel cabling where they are wetted by the "A" battery at that location.

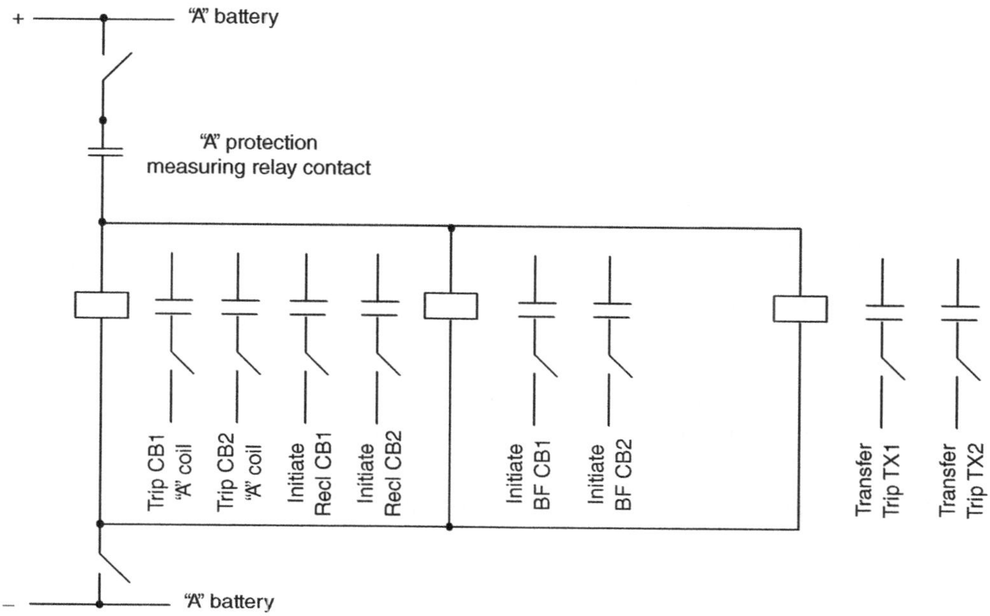

Figure 10.7 Discrete auxiliary relays dedicated to a single "A" protection.

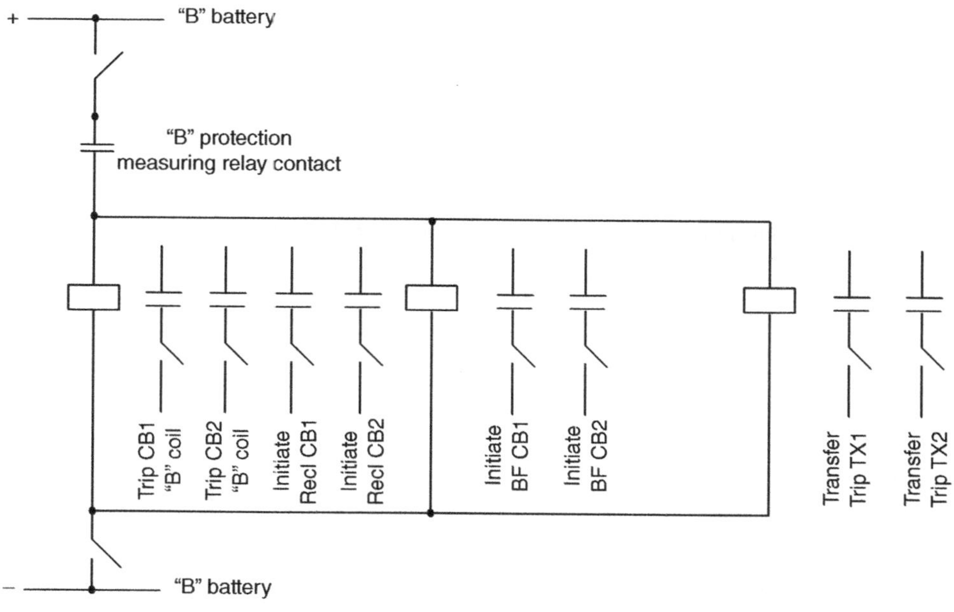

Figure 10.8 Discrete auxiliary relays dedicated to a single "B" protection.

10.2.10.2 Impact on Protection Separation

Breaker failure protection need not be duplicated for system reliability issues. Nevertheless, they require inputs from both the "A" protections and the "B" protections causing the two battery groups coming together. The close proximate of the two battery groups (a potential common mode failure) within the breaker failure protection may be an issue for various utilities to deal with.

In some electrical power system jurisdictions, the requirements for separation are becoming more stringent. Currently, more physical separation in addition to galvanic separation is suggested to improve protection system reliability against a single common mode of failure due to the two-station battery's close proximity, where two battery systems are required. The two-station batteries close proximity would be most pronounced within the breaker failure modules.

The subject of physical separation of protection systems is complex. There is a significant difference in whether: The equipment is grandfathered or new; protections that are essentially left untouched; or the work done to them is deemed to be inconsequential and therefore may be grandfathered. Each utility has guidelines that are to be adhered to.

10.2.11 Load-Substation Bus Fault and LV Breaker Failure

Breaker failure protection for load-substation LV breakers shown in Figure 10.9 usually only use the 62c path with the timer typically 300 ms. Timed pickup should the breaker pallet switch not change state indicates mechanical failure only. No provision is made at load-substations for the electrical failure of a breaker that otherwise behaved mechanically correctly.

Where an HV circuit switcher is used at a load-substation instead of a breaker, some utilities choose to mechanically open the associated disconnect switch simultaneously with tripping the circuit switcher interrupter. In this manner, should the interrupter fail, opening the disconnect still isolates the faulted transformer from the system. The disconnect would of course be destroyed while extinguishing the arc as it opens. The circuit switcher needs repair; however, the fault would be isolated despite the failure of the interrupter. A circuit switcher interrupter and disconnect are shown in the photo in Figure 10.10.

10.2.12 Pedestal Breaker and Free-Standing CT Frame Leakage Protection

Many utilities for breakers 500 kV or higher use pedestal type breakers and free-standing current transformers (CTs) independent of the breakers. The decision for this switchyard arrangement is based on economics as these breakers and CTs are much less expensive. The conventional switchyard arrangement is shown in Figure 10.11 where the CTs are built into the incoming and outgoing bushings of the breakers. This allows for the overlapping of protections around the

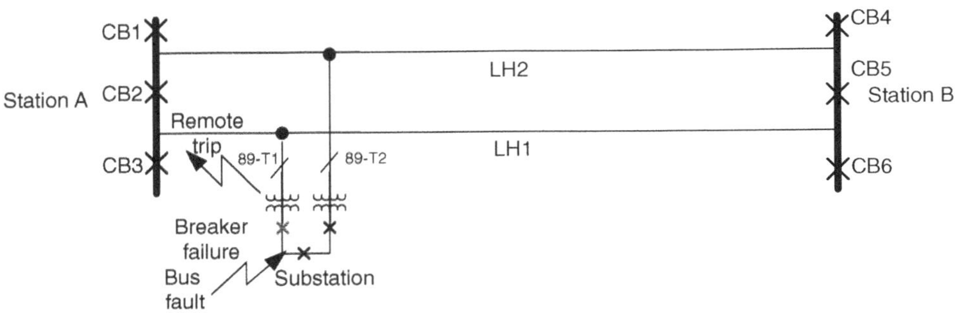

Figure 10.9 Transfer trip from a substation for a bus fault and breaker failure.

Figure 10.10 S&C Series 2000 115 kV circuit switcher and disconnect. Source: Photo Courtesy of S&C Electric Co.

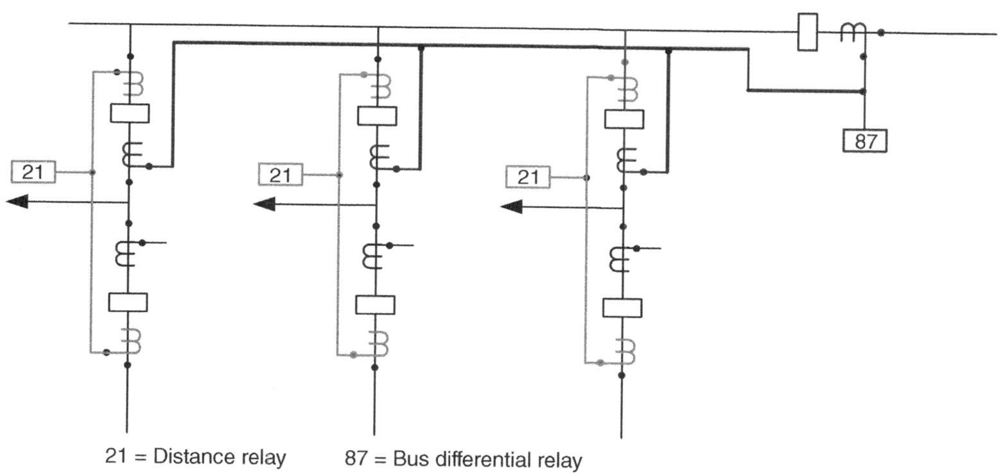

Figure 10.11 Switchyard with dead tank breakers and bushing CTs.

breakers. A fault between the CT's on the breaker was essentially a fault in the breaker itself. All protection zones bounded by the breaker would recognize the fault. All adjacent protection zones would be tripped thereby isolating the fault.

To reduce costs, live tank breakers mounted on pedestals are sometime used. Separate and independent CTs are also mounted on separate pedestals. The free-standing CTs come as a set of four with all four mounted on the same pedestal. Two are used for one protection zone supplying the "A" and "B" protections and the other two for the other protection zone supplying those "A" and "B" protections as partially shown in Figure 10.12.

Whereas live tank breakers and free-standing CTs all mounted on pedestals are economical they introduce unique protection challenges.

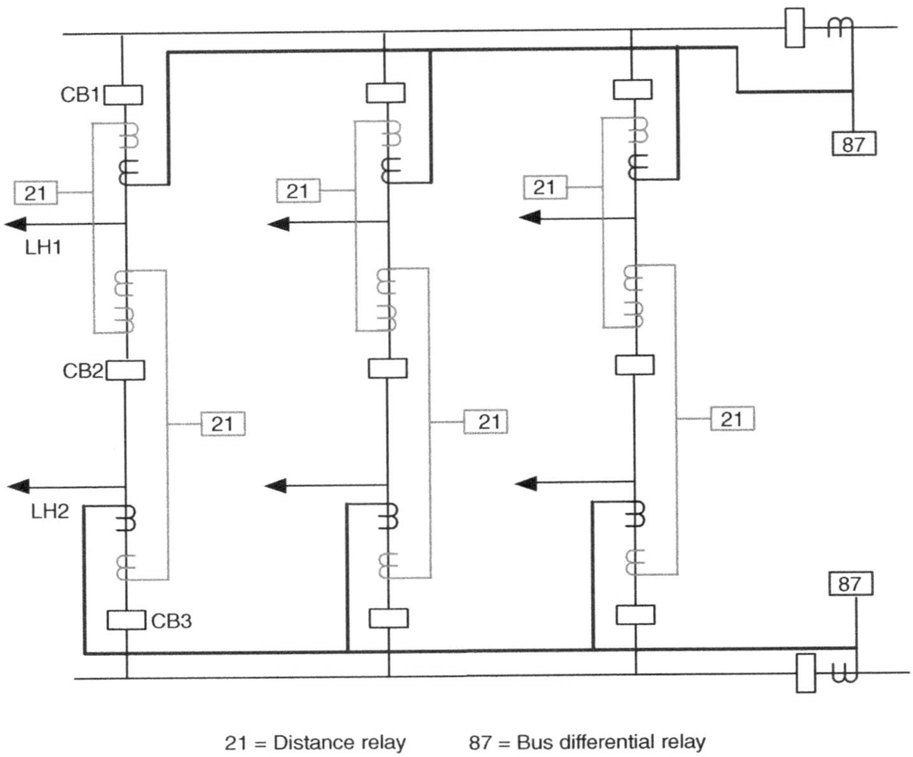

Figure 10.12 Switchyard with live tank breakers and free-standing CTs.

10.2.12.1 Protection Blind Spot

When CTs are located on only one side of a breaker protection, blind spots are created. Consider the fault shown in Figure 10.13 with a fault location as shown. For conventionally built switchyards, there is approximately one meter of exposed bus between the free-standing CT's and the live tank breaker. For a fault on this exposed bus, the 21-LH2 line protection would see the fault and trip breakers CB2 and CB3. However, tripping CB2 cannot eliminate the fault infeed from line LH1 and through breaker CB1.

However, when the LH2 line protection sees the fault and trips CB2, it also initiates CB2 breaker failure protection. Fault current will continue to flow through breaker CB1 which has not been tripped and also from the remote terminal from line LH1. These combined fault currents will continue to flow into the 50-50N-CB2 current sensing elements of CB2 failure protection. The fault current will continue to flow until the CB2 breaker failure 62b timing is complete typically 90 ms for 500 kV two cycle breakers. Upon 62b timing out, it will trip breaker CB1 as well as sending transfer trip signals to the remote terminal breakers of line LH1. With CB1 and line LH1 remote breakers tripped the fault is isolated.

10.2.12.2 Frame Leakage Protection

There are two fundamental issues with relying on breaker failure as the primary protection. The first issue is breaker failure is not duplicated leaving no backup to a breaker failure protection first contingency failure. The second issue is breaker failure is a delayed protection and therefore fault clearance is much slower compared to the other high-speed protections.

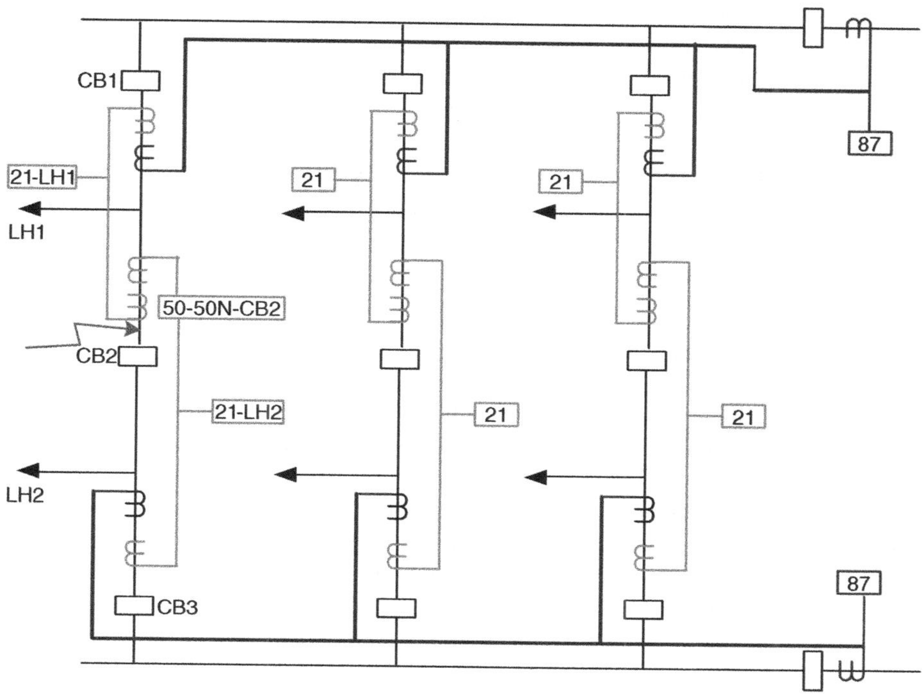

Figure 10.13 Location of fault that is a blind spot for conventional protections.

Most utilities address these two issues with a compromise solution. A reasonable postulation is that a three-phase fault on a 3-m-long piece of unexposed switchyard bus is not a credible contingency. Criteria for system stability assume a breaker failure along with a phase to ground fault as most faults fall into this category. It is not assumed that a breaker will fail for a three-phase fault which is extremely unlikely. In other words, it could be argued that the possibility of a three-phase fault on this section of bus is akin to a single contingency failure of protection equipment. Therefore, relying on breaker failure to clear this three-phase fault may be acceptable in terms of timing and system stability.

Since the vast majority of faults are phase to ground, a simple and elegant solution is to apply what is known as Frame Leakage Protection.

Frame leakage protection works when the switchgear and all supporting metal frameworks are mounted on ground isolating pedestals. A grounding strap is connected to the switchgear and associated supporting metal framing. The grounding strap is directed to pass through a window-type CT whose secondary leads are wired to an instantaneous overcurrent relay.

It is assumed any ground fault in the short section of the bus between the free-standing CTs, and the breaker will also involve a flashover to the grounded frame. Therefore, most if not all ground faults in the blind spot area will be covered not by the relatively slow clearing breaker failure protection via the 62b timing path but by a dedicated high-speed overcurrent-based protection. A ground fault in the blind spot itself not involving either the metallic frame of the breaker or CTs will not be covered with the frame leakage protection but must rely on the breaker failure protection just like for three-phase faults.

10.2.12.2.1 Frame Leakage Protection for Current Transformers

The most severe type of fault being covered by frame leakage protection is a fault within the CT frame itself which could disable all protections supplied by those CTs. Refer to Figure 10.14 showing a ground fault on the free-standing CT frame. This type of fault depending on where it is located within the frame could short the CT secondary leads thereby disabling the breaker failure protection 62b path that is reliant upon current from the supervising 50-50N overcurrent relays. In this situation, the only protections that can possibly clear the fault are the remote Zone 2 or Zone 3 backup protections. All the incoming lines at the station would trip after that definite time delay, quite certainly leading to serious system instability.

10.2.12.2.2 Frame Leakage Protection for Breakers

Refer to Figure 10.15 showing a ground fault in the breaker frame itself instead of in the CT frame as in the previous example. Here the breaker failure protection will work correctly to isolate the fault. Regardless, frame leakage protection also reduces the clearing time for this fault location provided the fault involves ground.

Frame leakage protection should be supervised by the breaker failure initiations to ensure the relay does not operate due to spurious currents which can flow through the ground strap on occasion. This protection has the same tripping functions as the breaker failure protection. For that reason, this protection may be considered a breaker failure protection with no intentional time delay.

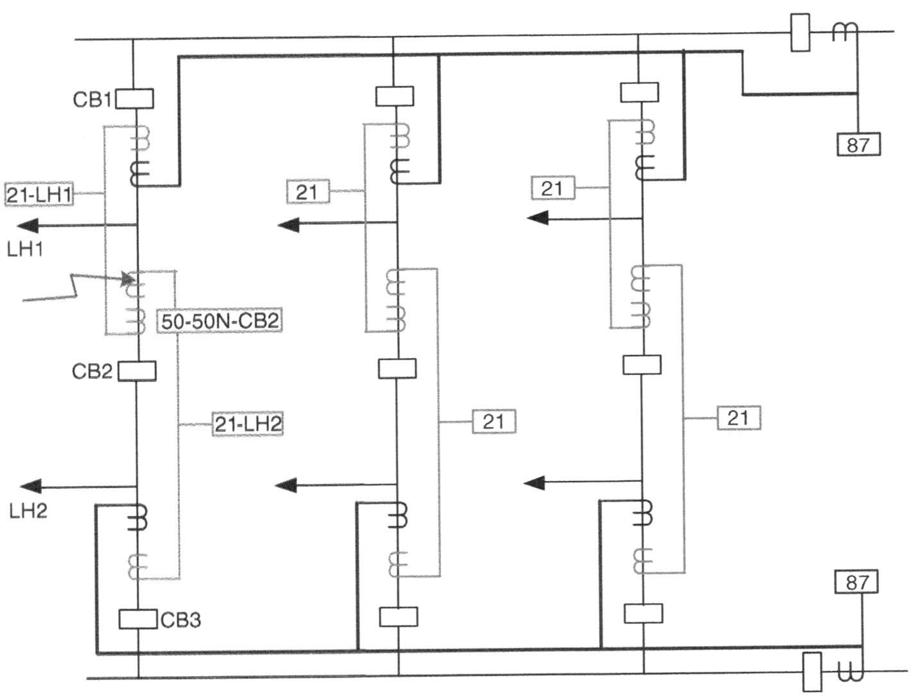

Figure 10.14 Location of fault in the free-standing CT.

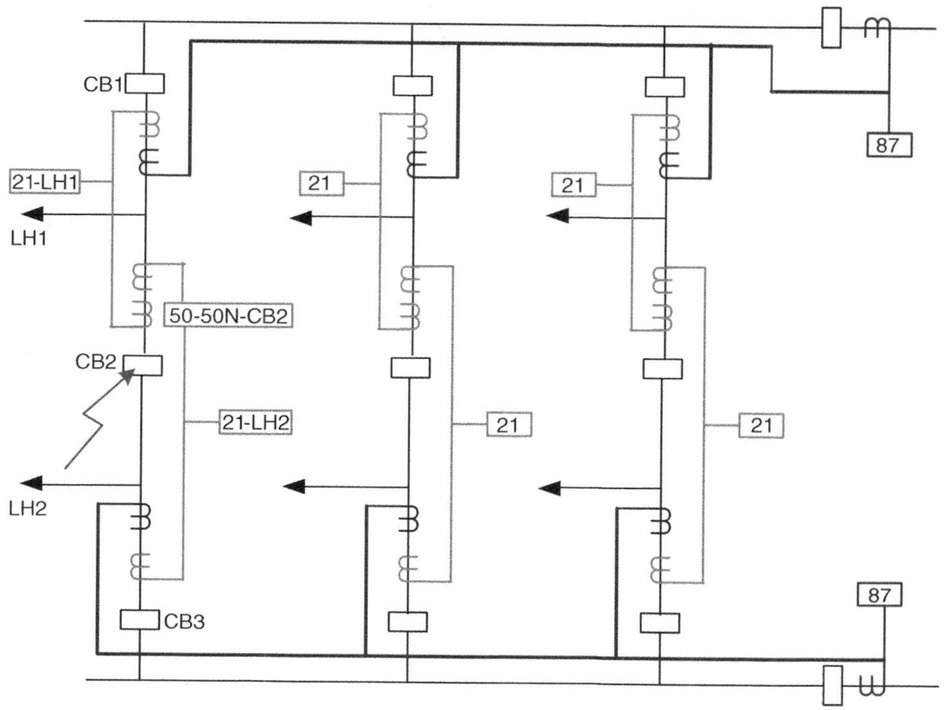

Figure 10.15 Location of fault in the breaker.

10.3 Breaker Automatic Reclosing General Background

The purpose for automatic reclosing of breakers following a protection operation is to restore equipment to service, provided the initial fault is categorized as transient. Some utilities do not rely on automatic reclosing as a method of maintaining system stability limits while other utilities do. When not relied on for system stability limits, its main function is to aid system operators in quickly re-establishing service. The function of quickly restoring service to customers without the need to rely on operator intervention depends on system topology. In general, regardless of system topology, it is always advisable to automatically reclose breakers following protection operations where possible rather than rely on operators to do the reclosing under stressful conditions.

Automatic reclosing can either be what is known as single-shot or multi-shot. It can also be single-pole or three-pole tripping and reclosing. Furthermore, it can be high-speed unsupervised or time-delayed with various forms of supervision.

10.3.1 Automatic Reclosing Timing

High-speed reclosing is to wait sufficiently long enough before reclosing until fault arc deionization has reasonably taken place. This time known as dead time is typically one cycle for 230 kV and one and a half cycles for 500 kV. Some utilities use high-speed reclosing as a means of maximizing load flow while others do not.

Normal reclosing time is chosen to allow system oscillations to decay to a level where reclosing into a fault that is permanent does not jeopardize system stability. The time to reclose based on these criteria can be as long as 5–10 seconds.

10.3.2 Reclosing Supervision

System operators usually have at their disposal five types of supervision requirements of which they choose one for each breaker in the system depending on the situation. In general, operators will look at specific system conditions that are met along with their chosen preferred sequence of breaker reclosing. Most breaker reclosing used by utilities for HV breakers typically allows for any one of the following supervision options or variations thereof to be chosen by system operators:

(a) **Voltage Presence + Time.** The breaker is automatically reclosed provided a minimum of 75% of nominal system voltage is detected on the line that the breaker is closing onto.
(b) **Undervoltage + Time.** The breaker is automatically reclosed provided a maximum of 25% of nominal system voltage is detected on the line that the breaker is closing onto.
(c) **Synchrocheck.** The breaker is automatically reclosed provided the systems on either side of the breaker are in synchronism for a set time, usually within 60° for two seconds.
(d) **Long Time.** The breaker is automatically reclosed unconditionally following a long- or significant-time delay, usually 10 seconds.
(e) **Short Time.** The breaker is automatically reclosed unconditionally following a short-time delay.
(f) **None.** The breaker is not automatically reclosed under any condition or amount of time.

10.3.3 Lockout

The lockout feature inherent in many automatic reclosing designs reasonably ensures that a breaker will not be automatically reclosed onto a permanent fault. It typically does this with a timer that prevents automatic reclosing unless the breaker has been in the closed position for at least a predetermined amount of time. The basis for lockout is that any trip and reclose followed by another trip within this time is probably either a permanent fault or one rapidly becoming permanent. A typical example is a broken or contaminated insulator whose insulation property is rapidly deteriorated. For lightning strokes on overhead transmission lines, it is assumed the same line will not be hit in rapid succession.

10.3.4 Long-Time Cancel

The long-time cancel feature inherent in many automatic reclosing designs ensures a breaker will not be automatically reclosed as long as a trip signal is still being applied to the same breaker and that previously armed reclosing will be reset after a predetermined amount of time. Therefore, for protections where reclosing is not wanted, a simple and effective method to cancel or reset automatic reclosing is to maintain the trip signal for more than this time.

10.3.5 Initiation and Cancellation of Reclosing

There are many situations where reclosing is initiated usually by default due to some sharing of communications or protection zone overlap. In all these situations, reclose cancellation can either be done through direct cancel reclose logic or by sealing in the initiating trip signal for longer than long-time cancel time.

11

Station Protection

11.1 Introduction

From the perspective of protection, each station has protection systems and schemes catering to the unique architecture and operating character of that station. In turn, that station as a whole must interact with other stations and the nearby network at all interconnecting voltage levels. A load substation has its own set of protections interacting with the supply lines and their terminals as well as the distribution lines and their load. This chapter, dedicated to stations, attempts to show the independence and the interdependence of the various protection systems at stations.

11.2 Types of Stations

Stations within the domain of most utilities can be broadly characterized as falling into one of two broad categories. The first station type is predominantly transmission switching with or without high voltage (HV) transformation via autotransformers. These stations are found at the 500, 230, 115 kV, and other voltage levels. Their purpose as a switching station is twofold. Transmission lines need to be sectionalized for load transfer to facilitate system operations and to optimize the application of protection schemes. They also are the sites where HV lines such as 500 kV are reduced to lower transmission voltages such as down to 230 kV or 115 kV which are still high but are more usable voltage levels. This is achieved via auto transformation. The second station type is the load substation where transmission voltages are transformed into sub-transmission voltage levels ready for distribution.

11.2.1 Switching Stations

Every transmission utility has switching stations also known as terminal stations where HV transmission lines are terminated and can be switched. Switching stations are mostly similar to the arrangement shown in Figure 11.1. There are typically two switchyards defined by location, e.g. the East Yard and the West Yard or the North Yard and the South Yard. Each yard has two buses with several diameters where lines egress the station in breaker and half terminal positions. The two switchyards are tied together with two bus-tie breakers. Sometimes there are three-breaker diameters and sometimes four-breaker diameters. Usually, the higher voltage level stations have four-breaker diameters. The advantage to breaker and half terminal positions whereby two lines share three breakers is that a single breaker can be removed from service for maintenance while still allowing the line to remain in service without affecting load transfers between terminal stations.

Power System Protection: Fundamentals and Applications, First Edition. John Ciufo and Aaron Cooperberg.

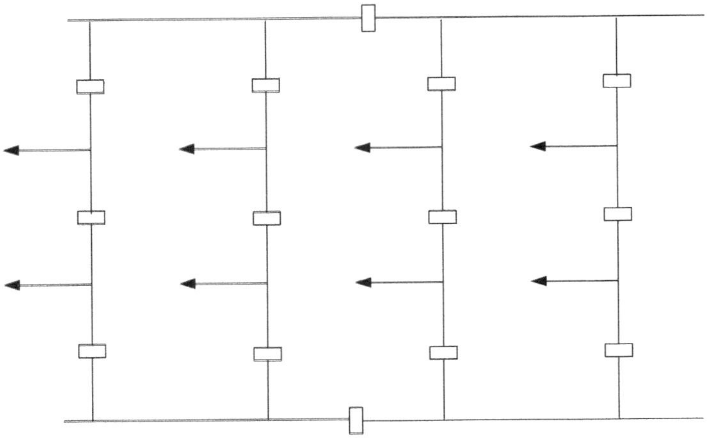

Figure 11.1 Typical terminal station arrangement.

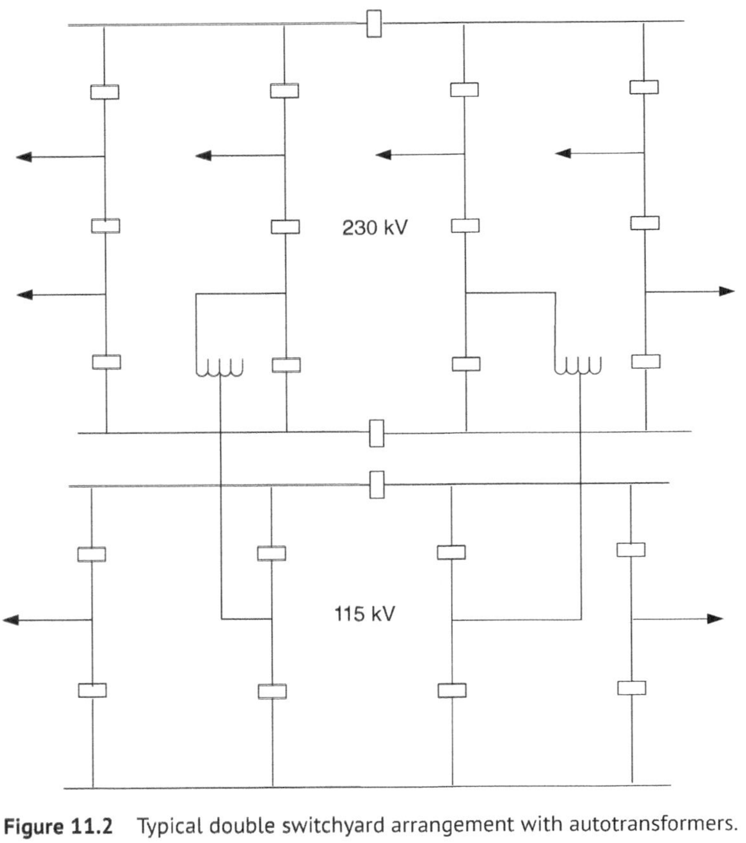

Figure 11.2 Typical double switchyard arrangement with autotransformers.

At locations where autotransformers are located at terminal stations to drop transmission voltage typically from either 500 to 230 kV or from 230 to 115 kV, there are two additional switchyards as shown in Figure 11.2.

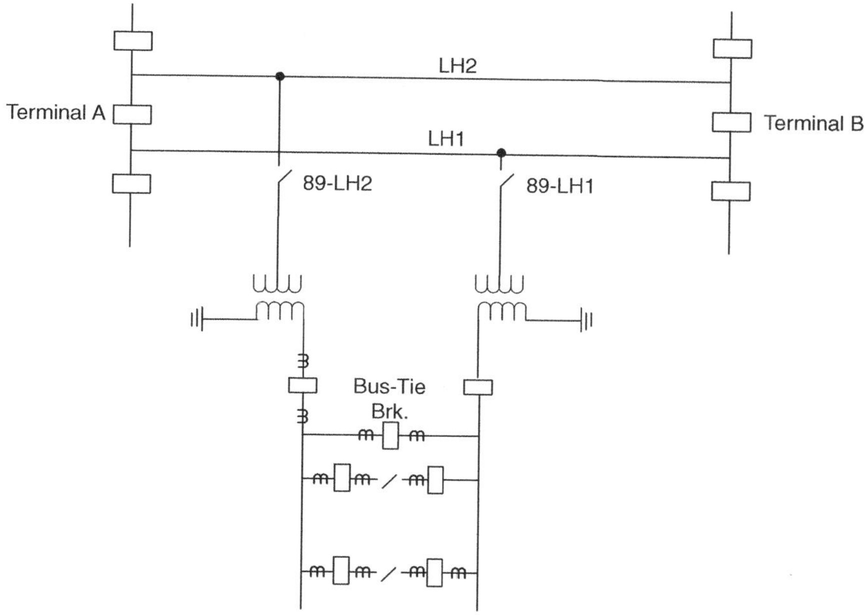

Figure 11.3 Single load substation tapped to two lines.

11.2.2 Load Substations

Most load substation architecture has commonality with some variation of the layout shown in Figure 11.3. In this load-substation architecture, the two transformer secondary windings can either be operated in parallel or not depending on the particular utility's philosophy of system architecture and operating principles. When paralleled, a normally closed low voltage (LV) bus-tie breaker makes the parallel. This architecture adds a high level of security to customer load. Each transformer would be rated in this case to handle up to half the station load under normal conditions and the entire station load with one transformer out-of-service. With a faulted line or transformer, the companion transformer tapped to the other line is capable of covering the entire station load without interruption of service to load customers.

11.3 Station and Protection Architecture

Stations whether terminal/switching or load-substation at various utilities, all have a similar look and feel about them unique to that utility. In the case of terminal/switching stations, there are two broad categories, the breaker and half arrangement which predominate, and the ring-type arrangement being less common.

11.3.1 Terminal/Switching Station

11.3.1.1 Breaker and Half Arrangement

In the classic switchyard breaker and half arrangement, the buses are protected by dedicated bus "A" and "B" protections. This is accomplished with either high impedance bus differential or digital relays optimized for bus protection. The bus differential protection current transformers (CT's) are zoned off around the diameter breakers and bus-tie breaker as shown in Figure 11.4

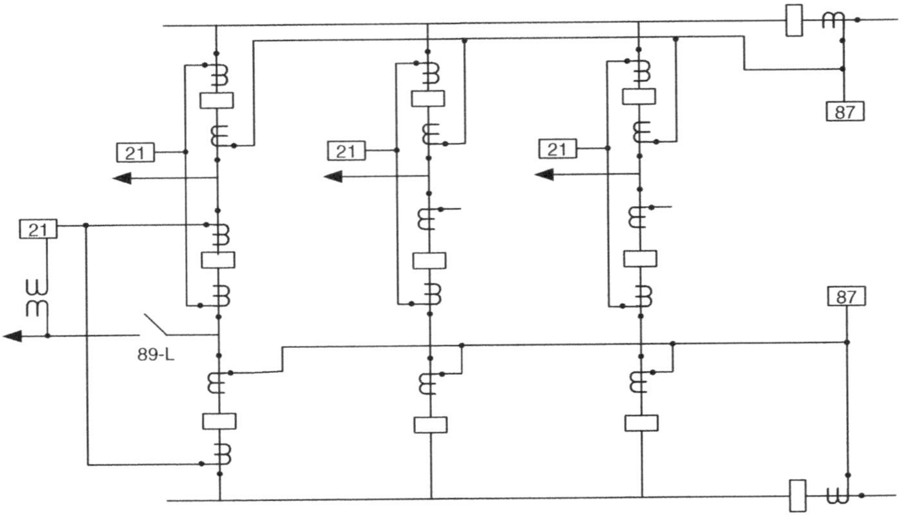

Figure 11.4 Protections for a typical breaker and half arrangement.

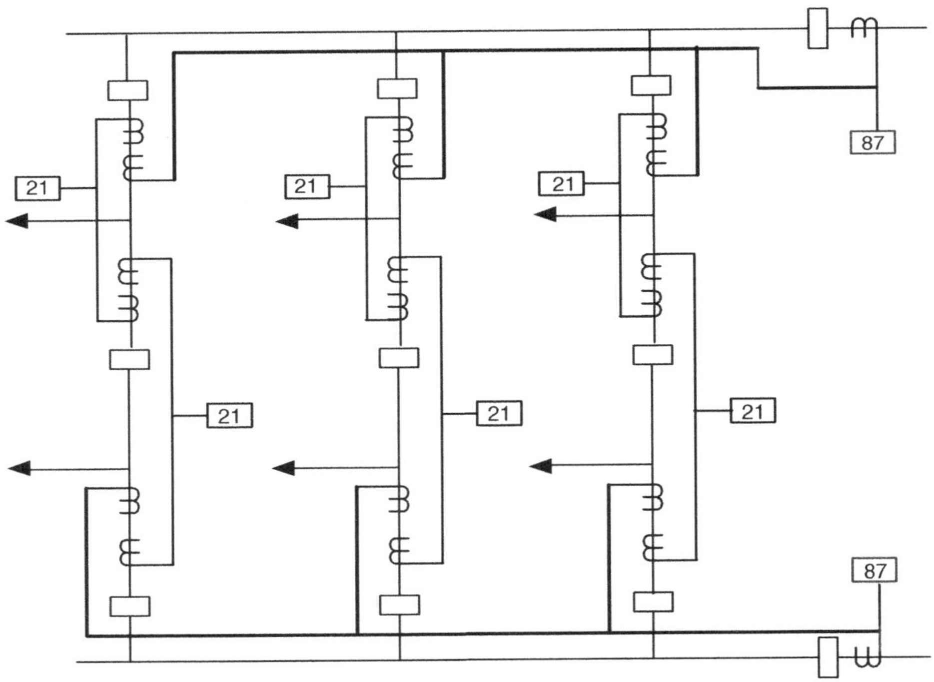

Figure 11.5 Protections for a typical higher voltage breaker and half arrangements.

The line protections are zoned off around the diameter breakers also as shown in Figure 11.4. The section of the bus bounded by those breakers is known as a stub-bus. The stub-bus is protected by the "A" and "B" line protection distance and switch onto fault protections.

Higher voltage switchyards such as 500 kV or higher also have breaker and half arrangements with similar protections. However, very often the protections are not zoned off around the breakers but around the CT's as shown in Figure 11.5. In this arrangement breaker failure and frame leakage

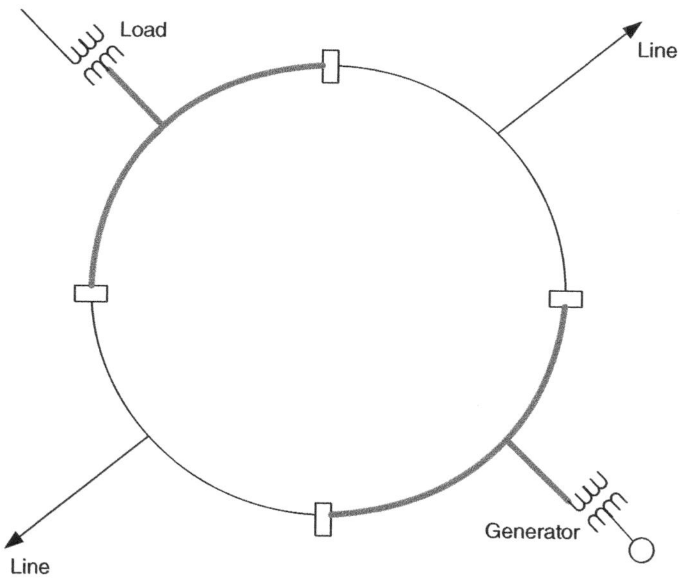

Figure 11.6 Typical ring bus.

protection cover, the small section of the bus between the breakers and the free-standing CT's as described in Chapter 10. Otherwise, the protections are similar to those for lower voltage breaker and half switchyard arrangements.

11.3.1.2 Ring Bus Arrangement

The ring bus arrangement is extensively used in more rural areas typically at lower voltages such as 115 kV systems. It is also becoming common practice to create collector stations, with a ring bus architecture, to connect distributed generation (DG) to the transmission network where line sectionalizing is required. The concept of a ring bus is similar to that of a breaker and half arrangement as shown in Figure 11.6. Each line is zoned off by two breakers. Also, either local load or generation is zoned off by breakers as well.

Typical protection for the ring bus arrangement is a dedicated bus differential protection zoned off using the ring bus breaker CTs and around the local load transformer CTs or local generation CTs. Where lines terminate onto the ring bus the bus is protected by the line protection similar to stub-bus protection at a switching station. Refer to Figure 11.7 showing typical protections covering the ring bus arrangement.

In some cases, the load or generation may be a considerable distance from the switchyard. In those situations, line current differential protection is an effective solution.

11.3.2 Load-Substation Typical Arrangement

The level of protections for load substations is dependent on a number of factors such as the power transformer MVA rating, the type of sub-transmission load whether three-wire or four-wire, type of switchgear, availability of communications, HV supply and voltage, operating mode and of course a balance of economics vs protection coverage.

11.3.2.1 Load Substation with Double Transformer Differential Protection

Substations with higher capacity transformers are usually protected with dual differential protection. The alternative to dual differential is single differential and timed overcurrent backup. The

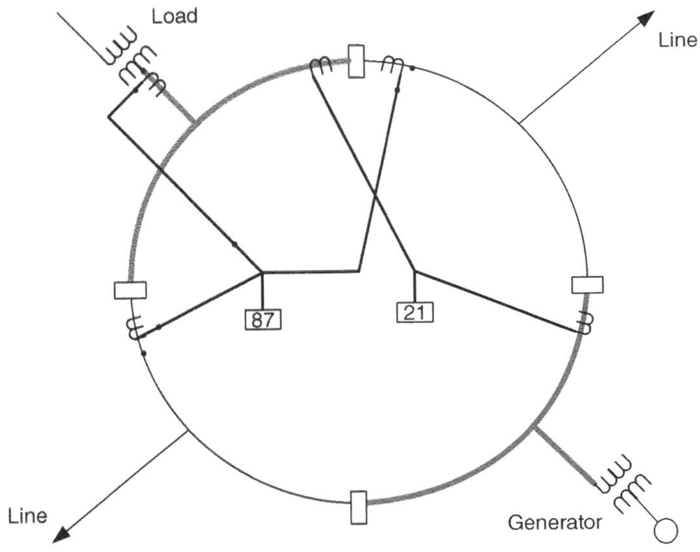

Figure 11.7 Protections for a typical ring bus.

purpose of dual differential is to satisfy operating who feel it necessary to keep transformers in service even while doing protection maintenance. Without dual differential instantaneous protections, the operating authority will take the transformer out of service rather than rely on slow clearing timed backup protection. Should the timed overcurrent protection be called upon to protect a faulted transformer it would most certainly result in a destroyed transformer or worse, a station fire (Figure 11.8).

11.3.2.2 Load Substation with Single Transformer Differential Protection

Substations with lower rated capacity transformers are usually protected with single differential protection. Backup protection is via phase and ground timed overcurrent. The advantage of this protection arrangement is to eliminate the need for dedicated LV breaker failure protection. The disadvantage is the inability of discriminating between faulted buses or transformers. As shown in Figure 11.9, a faulted bus results in equal infeeds through both transformer banks. A phase overcurrent relay on either bank will operate without the possibility of coordination. The same thing applies to ground faults. Overcurrent relays in the neutral to the ground connection of the transformers will measure the same zero sequence current and trip both banks.

The solution used by some utilities is to implement a two-step tripping. Upon measuring an LV fault, the protections associated with each of the transformers operate simultaneously to first trip the bus-tie breaker. This in a sense divides the station vertically in two. One side is faulted and the other is not. The relays on the faulted side proceed to isolate the fault whether by tripping the LV transformer breaker for a bus fault or by removing the transformer from service for a transformer fault (Figure 11.10).

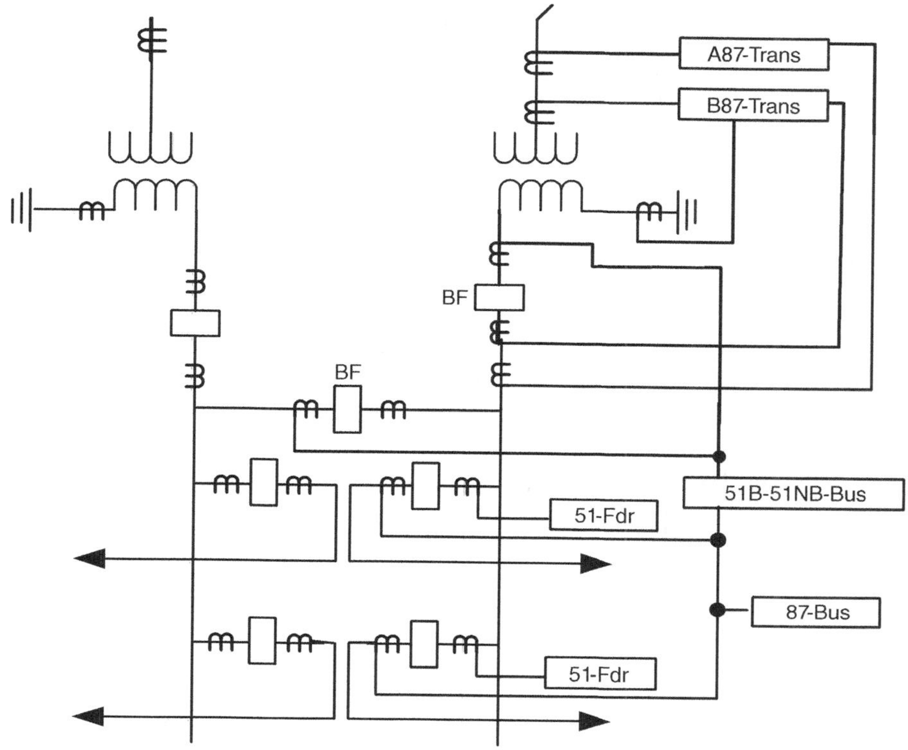

Figure 11.8 Dual-redundant protections for a load substation.

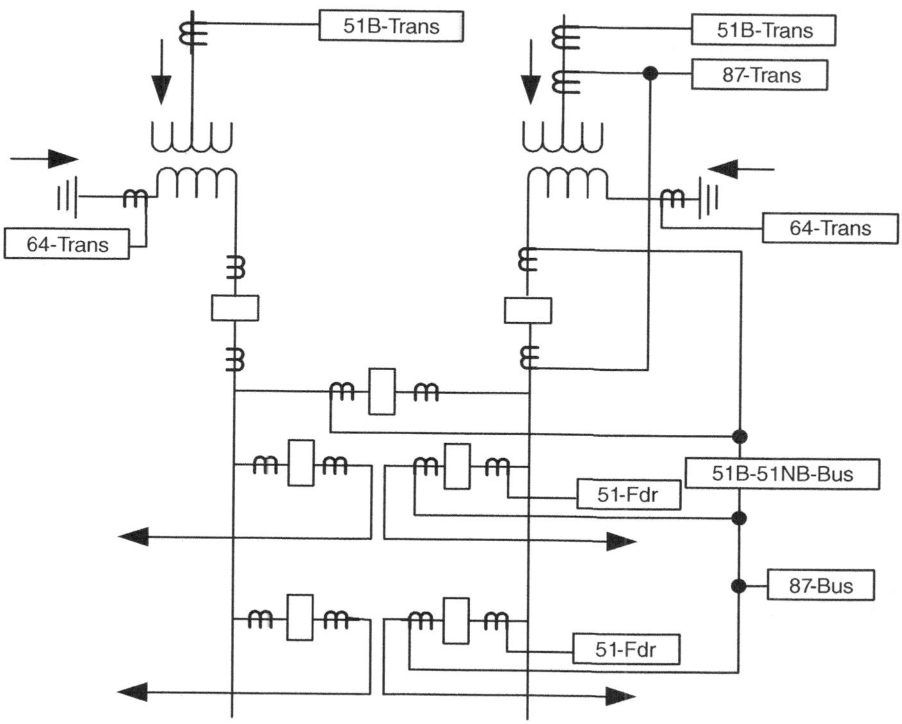

Figure 11.9 Single protections for a load substation.

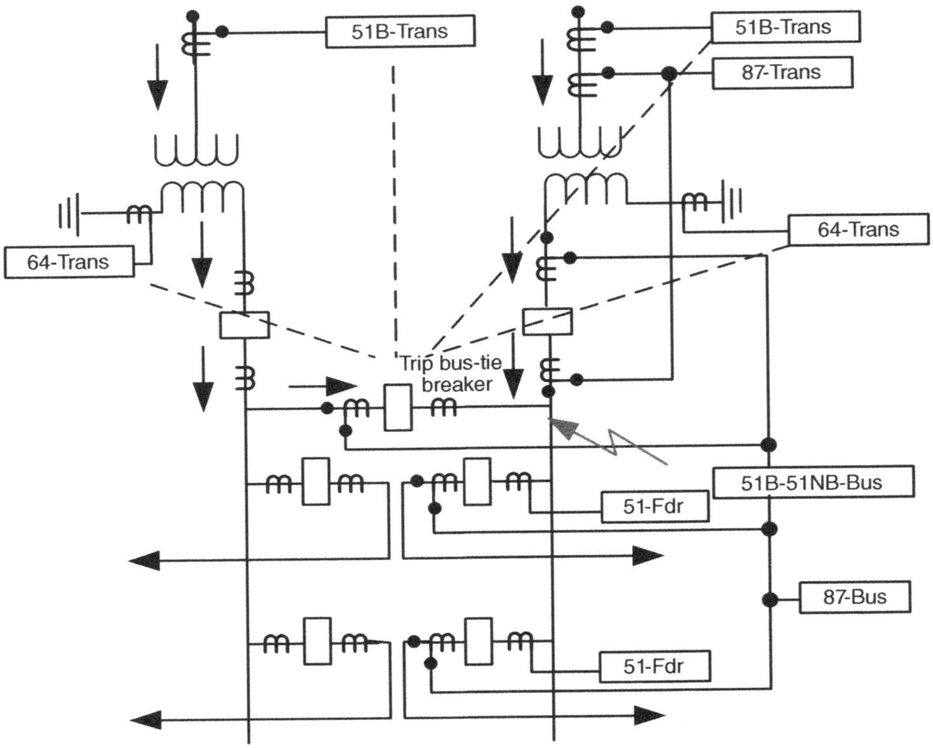

Figure 11.10 Two-step tripping scheme.

11.3.3 Fault Isolation via HV Isolating Devices

11.3.3.1 Normally Open Bus-Tie Breaker

When the LV bus-tie breaker is operated normally open and the distribution feeders are all radial supply an HV interrupting device, either an HV breaker or circuit switcher is usually used. This fulfills the need to isolate a faulted transformer at the load substation. Lower voltage load substations supplied at 115 kV could use HV circuit switchers with much lower installed cost instead of the more expensive HV breaker. In general, 230 kV short circuit levels in most situations tend to exceed the short circuit interrupting capability of the less expensive circuit switcher.

For a transmission line fault, since the LV bus-tie breaker is operated normally open and the station load is radial, there is no possibility for back feeding into the transmission line fault. Therefore, the load-substation HV isolating device need not be tripped for line faults.

11.3.3.2 Normally Closed Bus-Tie Breaker

When the bus-tie breaker is operated normally closed and or when the load is non-radial supply, the possibility exists for back feeding a transmission line fault. In this case, there exist one of two options. The first option is to install protections that would detect the load-substation back feeding into the line fault. This would involve for phase faults providing relays directioned toward the transmission line to detect phase faults on the line and trip the HV breaker or circuit switcher. For ground faults depending on the type of transformer configuration and grounding, there are various options. If the transformer is ungrounded on the HV side, a zero-sequence voltage measuring protection could be used supplied from an HV voltage source such as a CVT configured Wye-Ground: Open Delta. For a line fault, the zero-sequence voltage should rise to three times the system phase to neutral voltage provided the line protections have already tripped all the line breakers as shown in Figure 11.11.

However, if there exists other load substations supplied from the same line but their transformers are HV grounded, the zero-sequence voltage would be less than three times the system phase to neutral voltage. How much less depends on the location of those other stations relative to a worst case ground fault location. A study would be required to determine whether there is sufficient zero-sequence voltage to ensure line ground fault isolation at the load substation under all circumstances. If the transformer is configured Wye-Ground, a CT in the neutral to ground connection could be used to supply an overcurrent relay. The grounded transformer is a ground source and with sensitive low pickup settings an overcurrent relay should detect all line ground faults (Figure 11.12).

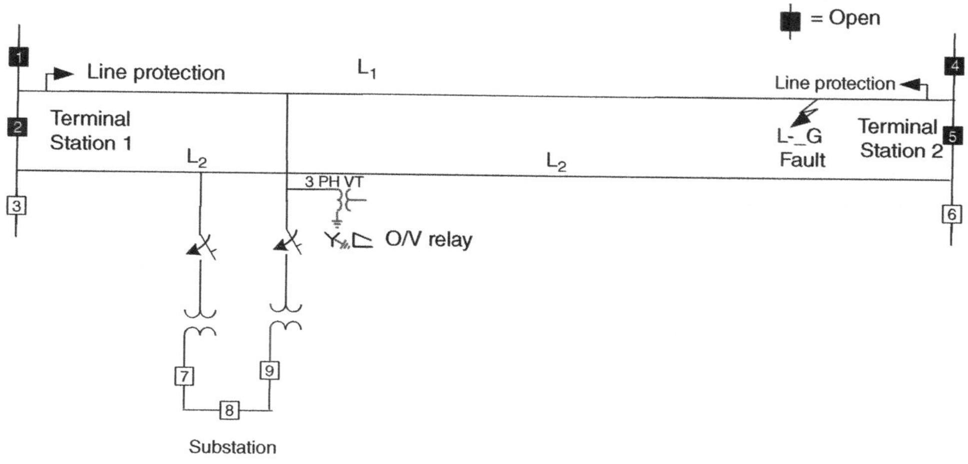

Figure 11.11 Measuring zero-sequence voltage for a line ground fault.

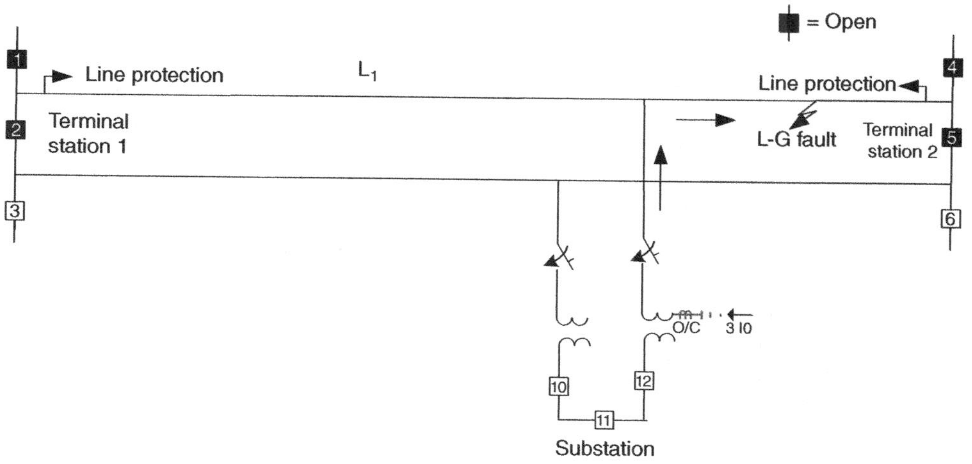

Figure 11.12 Measuring ground current for a line ground fault.

11.3.4 Fault Isolation via Tele-Protection

An alternative to using relays to detect line phase and ground faults is to rely on tele-protection.

For a transmission line fault, the protection at the switching station would send a transfer trip signal to trip either an HV isolating device if there is one or to trip the transformer LV breaker.

For a transformer fault, the same (two-way) tele-protection media would be used to send a transfer trip signal to one of the switching stations to trip the line breakers. There is no need to simultaneously send a transfer trip signal to the other line terminal as the transfer trip signal can be connected to cascade into the transfer trip tele-protection associated with the line protections. This in turn also trips other load substations supplied by the same line. This would also apply to tripping any DG connecting to the line as well. It should be noted that breaker and line disconnect pallet switch interlocks would be used to only trip the switching station breakers from transfer trip provided the switching station is feeding the line. Otherwise, that transfer trip signal would just be cascaded to trip the remote line terminal breakers along with any connecting other load substations and DG. The use of the interlocks ensures that the breakers at the switching station are not tripped needlessly.

The following is a possible method of isolation for illustrative purposes:

Refer to Figure 11.13 where three load substations are supplied from lines L_1 and L_2. There are various tele-protection paths that can be used to achieve isolation of the faulted transformer.

At Station A, the L_1 line protection for a line fault trips local breakers BR1, BR2 and transfer trips the remote line breakers BR4, BR5 at Station B. Line L_1 protection simultaneously sends transfer trip signals to Substations 1, 2, and 3 to trip breakers BRA, BRB, and BRC.

The transformer protection at Substation 1 can utilize the same transfer trip scheme economically by the transformer protection sending a transfer trip tele-protection signal to cascade into the line L_1 transfer trip tele-protection scheme. Similarly, for a transformer LV breaker failure operation, the same method of fault isolation would be adopted.

11.3.5 Station Tele-protection Subsystems

Individual station protection systems operate based on measurements gathered at the local relay location. However, for some protection schemes, this is not sufficient information to maintain selectivity. Moreover, for protection systems to isolate some faulted power system equipment, it needs

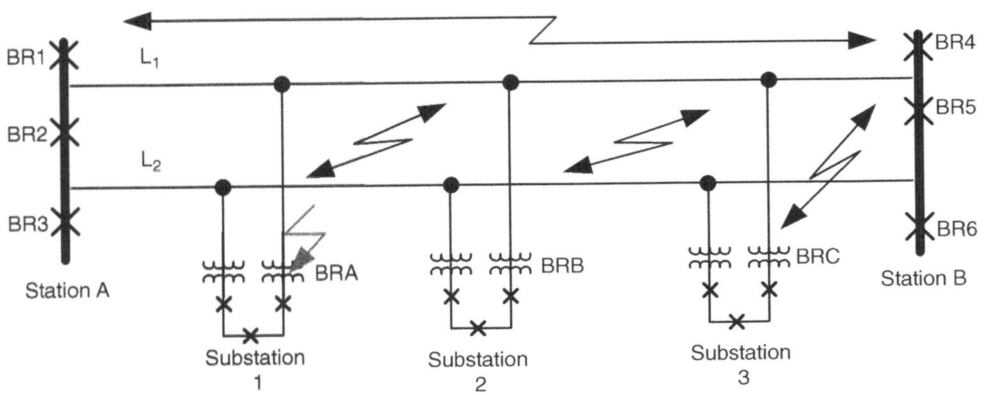

Figure 11.13 Fault isolation via communications example.

to trip the local breaker(s) and also the remote breaker(s). The remote breakers are geographically located at different stations and can be hundreds of kilometers/miles away. This is typical for line and breaker failure protections.

Tele-protection is the protection subsystem that allows sending and receiving local and remote station information. It represents typically high-speed, communication links between the line end terminals, or between stations – refer to Figure 11.14. Tele-protection subsystems utilize communications paths to send signals from the relaying systems located at one station to protection systems located at a remote station, and vice a versa, thereby, resulting in instantaneous (non-delayed) tripping for faults.

Tele-protection is critical in linking protection systems between stations, for the transfer tripping of remote power circuit breakers, to isolate faulted elements such as transmission lines, and for the transfer of sophisticated data between systems necessary for secure operation.

There are many methods available to achieve appropriate protection grade communication between protections systems located at various stations remote from each other.

Today, digital communications over a fiber optic cable are typically used. The fiber cable is installed between stations interfacing with access multiplexers (mux.) optimized for protections using T1 time division multiplexing (TDM). Direct fiber interfacing mux. to mux. can be done or the T1 signals could be uploaded onto a synchronous optical network (SONET). The fiber optic cable can be strung from transmission towers or distribution feeder poles or in urban areas underground from cable vaults to cable vault.

11.3.5.1 Tele-Protection Using Frequency Shift Keying

For load substations, many telephone companies lease dedicated communication facilities, optimized for protection systems interfacing between substations. The interface into the substation is preferably fiber otherwise, if metallic cable, there is an issue with ground potential rise, and it requires isolating transformers optimized for this purpose. Not all substations are located where telephone company's offer direct fiber interfacing. Telephone company leased communications are predominantly based on frequency shift keying (FSK) interfacing with such equipment located at the switching stations or substations. The analog frequency signals are converted at the telephone exchange stations also known as Central Offices in some jurisdictions to digital representations of the frequencies. When a frequency known as the guard signal is shifted to another specific frequency, a tripping output is generated at the remote end. In order to ensure high reliability, it is suggested that the communication FSK devices be programmed to provide a pattern of two levels of frequency shift from f_0 to f_1 to f_2 repeating itself over a short specified time (Figure 11.15).

11.3.5.2 Tele-Protection Using T1 (Digital)

The term T1 evolved from early telephone voice multiplexing and transmission facilities. T1 is currently used to describe almost any communications link operating at 1,544,000 bits/s. The main purpose of a T1 is to provide up to 24 channels of voice and data over a four-wire metallic circuit or fiber optic pair.

11.3.5.2.1 The Advantage of T1

The use of private T1 networks has grown rapidly. Several factors have contributed to this growth:

- Increase in voice and data traffic
- Communications cost reduction
- Quality and reliability of service

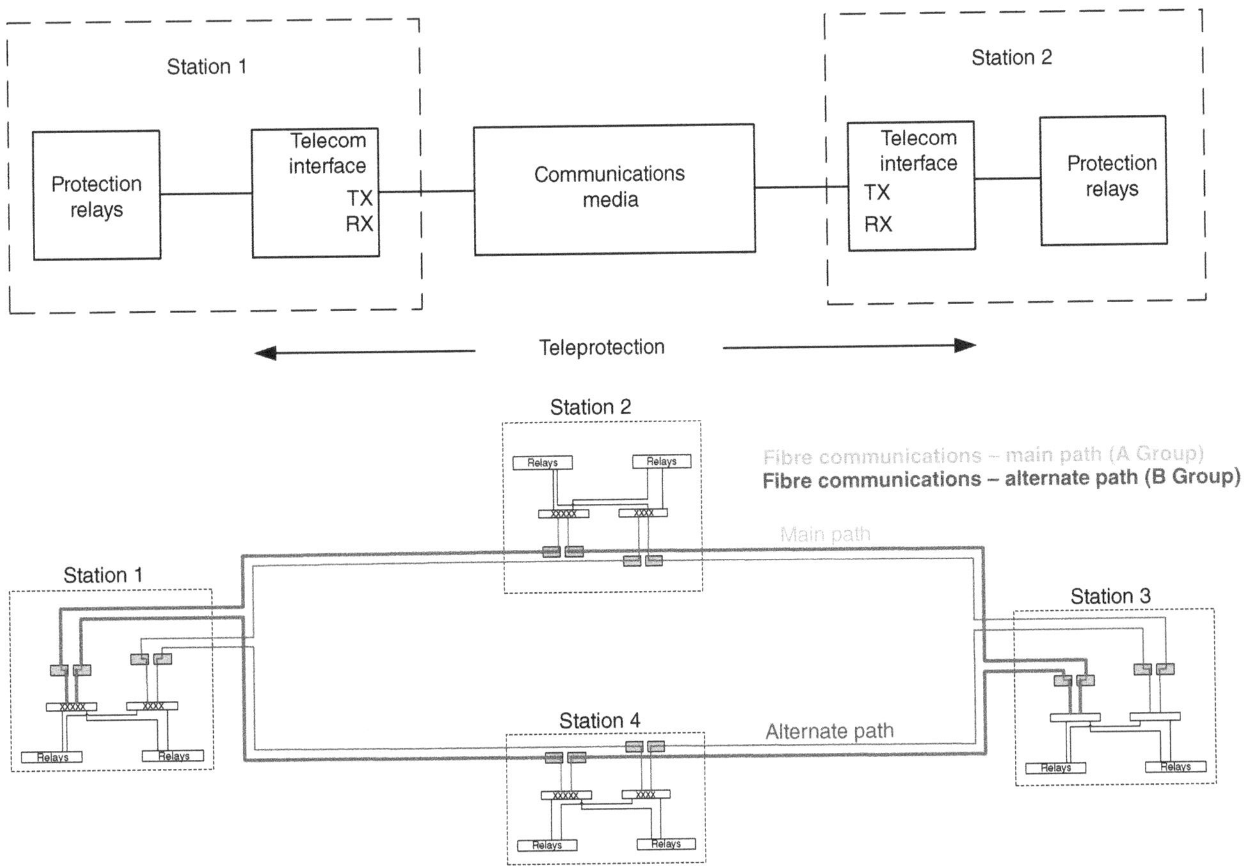

Figure 11.14 Illustration of stations tele-protection schemes.

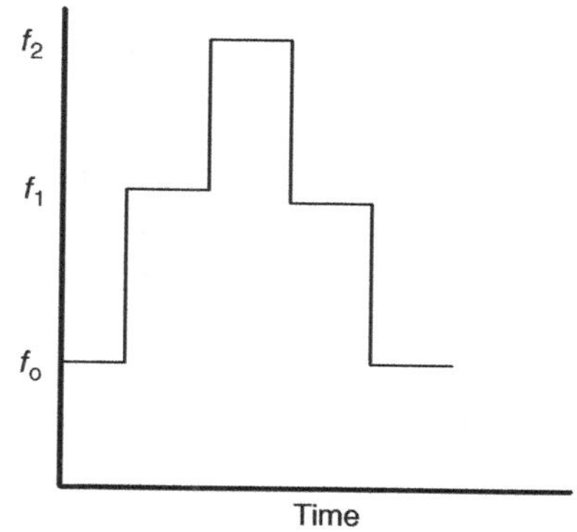

Figure 11.15 Example of FSK with a key pattern for security.

- Increased flexibility and control
- Simplification

11.3.5.2.2 Cost/Reliability

One of the benefits of using private T1 networks is the reduction of leased telephone service costs. The initial cost of T1 equipment and the high installation cost of fiber or the monthly cost of leased T1 lines is offset by the reduction in the number of dedicated telephone lines and monthly service charges. T1 systems may under many circumstances pay for themselves over time. The emphasis in power utility applications is primarily reliability and quality of service.

11.3.5.2.3 Simplification

T1 simplifies the task of networking different types of communication functions. Refer to Figure 11.16 illustrating what a typical utilities communication network might look like without T1.

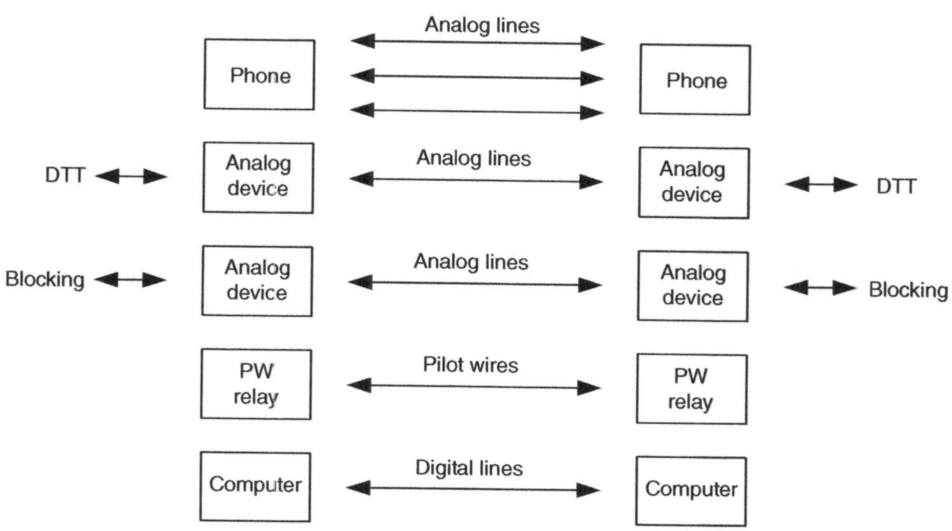

Figure 11.16 A communications network not using T1.

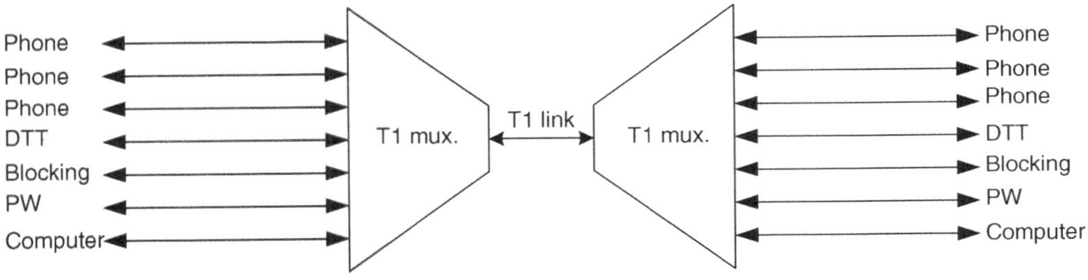

Figure 11.17 A communications network using T1.

Figure 11.17 depicts the same network with a T1 link installed. Utility grade T1 access multiplexers have different modules to support utility needs thereby making T1 a viable solution for all utility applications.

11.3.5.2.4 How T1 Works – Making Voice and Data Compatible

Many benefits of T1 are attributable to the fact that voice and data are transmitted over a single digital communication link. Since digital data consists of 1's and 0's, it is already compatible with T1's digital format. However, because voice signals are complex waveforms, they must be digitized to achieve compatibility with T1.

The most common method of digitizing analog voice signals is a technique called **pulse code modulation** (PCM). PCM is a sampling process that compresses a voice conversation into a 64kb/s standard rate known as **digital signal-level zero** or DS0.

PCM is a two-step process. In the first step, the incoming analog signal or tone is sampled 8000 times/s, a rate sufficient to adequately represent voice information. These sample values are then converted to pulses using a process known as pulse amplitude modulation (PAM) (Figure 11.18).

In the second step, the height of each pulse is assigned an equivalent eight-bit binary value. The resulting output is a digital representation of the pulse and by extension the sampled analog waveform or tone. The 64 kb/s DS0 rate is obtained by multiplying the number of samplings per second (8000) by the number of bits in each sample (8) (Figure 11.19).

$$\frac{8000\ \text{Samples}}{\text{Second}} \times \frac{8\ \text{Bits}}{\text{Sample}} = 64\ \text{kb/s}$$

11.3.5.2.5 Time Division Multiplexing

Once digitized, voice and/or data signals from many sources can be combined (i.e. multiplexed) and transmitted over a single T1 link. This process is made possible by a technique called TDM.

TDM divides the T1 link into 24 discrete 64 kb/s time slots. An identical number of DS0 signals (representing 24 separate voice and/or data channels) are assigned to each time slot for transmission from the link.

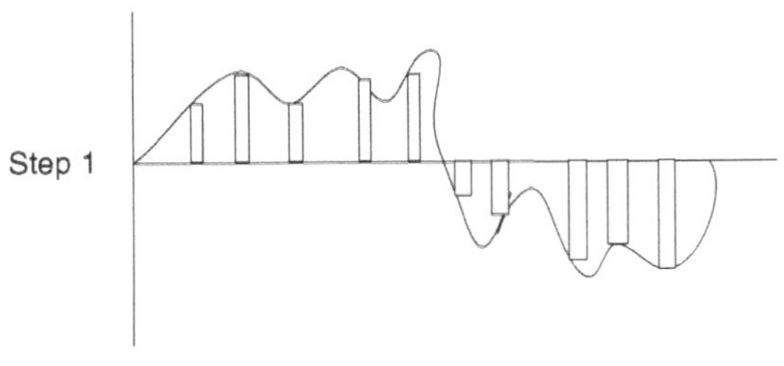

Figure 11.18 Pulse amplitude modulation.

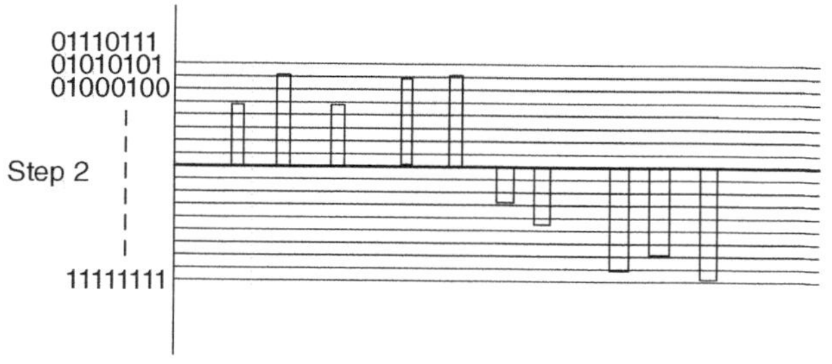

Figure 11.19 Assigning binary values to pulses.

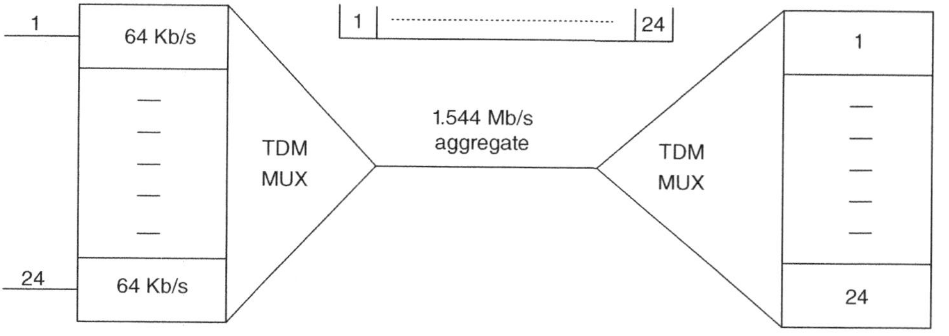

Figure 11.20 Time division multiplexing.

PCM and TDM explain the basic T1 rate of 1.544 Mb/s (Figure 11.20).

11.3.5.2.6 Understanding 1.544 Mb/s

In T1, the eight-bit digital samples created in the PCM step (for voice or tones only) are grouped into the 24 discrete DS0 time slots created by TDM. Each group of 24-time slots is called a T1 **frame**. Since each time slot contains eight bits, the number of information bits within each frame totals 192 (24 × 8). Additionally, a 193rd bit is added to mark the end of one frame and the beginning of the next. Appropriately enough, this added bit is called the **framing bit** (Figure 11.21).

Since the DS0 signals are sampled 8000 times/s, 8000 192-bit information frames are created during that period. Total: 1.536 Mb/s. At 8000 samples/s, framing bits are created at the rate of 8 kb/s. Result: A single 1.544 Mb/s signal known as *digital signal-level one* or DS1 (Tables 11.1 and 11.2).

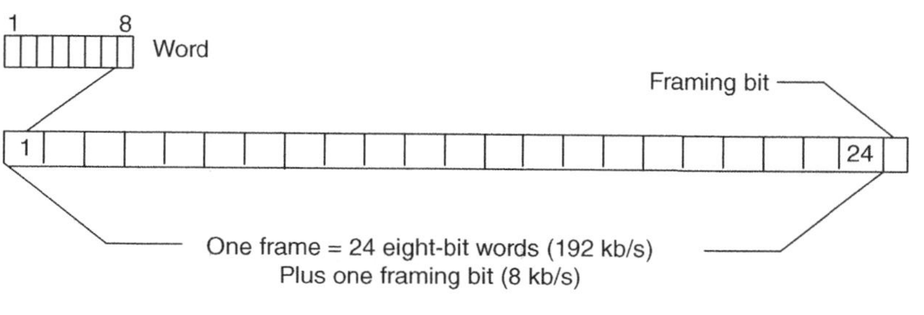

Figure 11.21 Illustration of a T1 frame.

Table 11.1 Calculating the 1.544 Mb/s T1 Rate.

Step	What happens	Calculation
1	The eight-bit digital samples created by PCM (for voice or tones only) are grouped into the 24 discrete time slots created by TDM. Each group of 24-time slots is called a T1 frame	$\frac{24 \text{ samples} \times 8 \text{ bits/sample}}{192 \text{ information bits/frame}}$
2	A framing bit is added to mark the end of one frame and the beginning of the next.	$\frac{192 \text{ information bits} + 1 \text{ framing bit}}{193 \text{ total bits/frame}}$
3	T1 frames are transmitted at the rate of 8000/s.	$\frac{8000 \text{ samples} \times 193 \text{ total bits}}{1{,}544{,}000 \text{ bits/s} \ (1.544 \text{ Mb/s})}$

Table 11.2 T1 process.

Process	What happens
Pulse code modulation (PCM)	1. Samples the incoming analog signals or tones 8000 times/s and converts them to pulses. 2. Assigns the height of each pulse with an equivalent eight-bit digital value. 3. Creates a 64 kb/s signal (8000 samples/s multiplied by eight bits).
Time division multiplexing (TDM)	Combines 24 DS0 signals to create a signal 1.544 Mb/s signal (DS1)

11.4 Station Switchgear Type

In general, there are two distinctly different types of switchgear. There is the outdoor type and the indoor type. Where land is not at a premium and there is little issue with esthetics, outdoor stations are built as they are always much less expensive. Where land is in short supply or esthetics is an issue indoor stations are built instead.

11.4.1 Terminal/Switching Station Outdoor Switchgear

Most HV terminal/switching stations are of the outdoor switchgear variety. The switchyard is spread out over a large area. In many stations, a large number of transmission lines terminate into a single terminal station. The switchyard itself is usually split vertically into two distinct areas bounded by the bus-tie breakers. It is not uncommon to find a load substation tapped to two of the local HV buses instead of on a line.

11.4.2 Terminal Station Indoor Switchgear

Less common but still popular is terminal/switching stations where switchgear is indoor. These stations use SF_6 indoor breakers and buses enclosed in SF_6 insulated conduit. From a protection perspective, exclusive of requiring added alarms for low SF_6 gas, there is little difference than with outdoor.

11.4.3 Load-Substation Outdoor Switchgear

Most load substations in rural areas are of the outdoor switchgear variety. Usually, load substations are tapped to lines at the transmission right-of-way where the aesthetics are not a major concern.

11.4.4 Load-Substation Indoor Switchgear

Load substations using indoor switchgear predominate in cities and other urban areas. At these locations, indoor switchgear is enclosed in a metal housing known as metal-clad. Metal-clad switchgear such as breakers and disconnects are located in their own individual metal-clad cell. These cells are installed in a metal-clad cell line-up such that the bus work travels through the length of the cell line-up. Where there are combined breaker cells owned by two independent utilities, one being transmission and the other being distribution, spacers are provided in the line-up as a demarcation point between the two.

For reasons of safety, all metal-clad switchgear manufactured before the 1980s made it unacceptable to mount protections and control devices on the switchgear. The main safety issue was the identified risk of an arc developing in live electrical parts in the cell such as a breaker or bus with the resultant rapidly expanding air having no place to go but to blow out the front door. Anyone standing in front of the door would be in harm's way. In arc-proof metal-clad, the arc-induced expanding air is directed away from the front to dedicated vents on top of the switchgear assembly. This provides a much greater safety margin for anyone who may be standing in front of the switchgear and prevents damage to protections mounted on the switchgear door and in the instrument compartment as shown in Figures 11.22 and 11.23.

Figure 11.22 Protection devices on arc-proof metal-clad cell doors.

Figure 11.23 Backside view of a metal-clad instrument compartment.

11.5 Sub-Transmission Types and Station Grounding

Two types of load are connected to sub-transmission feeders. There are single-phase loads and there are balanced three-phase loads. Where single-phase loads are supplied by sub-transmission feeders, the possibility exists for sometimes significant continuous neutral unbalance current to exist. For exclusive balanced three-phase loads, the continuous neutral unbalance current is essentially non-existent. In either case, protections must be designed in a manner consistent with the type of sub-transmission.

11.5.1 Exclusive Three-wire Sub-transmission

Exclusive three-wire sub-transmission is typically obtained from transformation using wye–wye transformers with a delta tertiary to circulate third harmonic currents and eliminate third harmonic voltages. This arrangement permits grounding the sub-transmission neutral at the station. It is not, however, brought out to the overhead lines. The main purpose is to stabilize the neutral point and provide a source of ground current for relaying during line to ground faults. Only balanced three-phase loads are permitted on the sub-transmission system.

Many distribution utilities do not transpose their sub-transmission feeders and do not strictly adhere to the geometric spacing between the three-phase conductors so that the phase impedances are not necessarily matched. Without a path for zero-sequence current to flow, the result of unbalanced phase conductor impedances is to skew the balancing of the phase voltages such that there is no longer a precise 120° angular displacement between them. A high impedance grounding source such as a zigzag grounding transformer accomplishes two things. It stabilizes the neutral point and provides a limited but usually sufficient ground current to flow for ground fault detection.

11.5.2 Exclusive Four-wire Sub-transmission

Exclusive four-wire sub-transmission grounding requirements are typically obtained from either delta-wye transformer with a low impedance grounding reactor, wye-zigzag transformers with a low impedance grounding reactor, or wye-delta transformers with a low impedance grounding transformer. In each case, the station neutral is brought out to the overhead lines, and unbalanced loads usually in the form of single-phase laterals are permitted on the sub-transmission system.

Four-wire sub-transmission is characterized by loads being a combination of three-phase as well as single-phase. Distribution station (DS) transformers are configured delta on the sub-transmission side thereby representing balanced three-phase load. Also, individual feeder phases can be separated from the three-phase trunk line. These are known as single-phase laterals. Single-phase laterals supply single-phase loads such as single distribution pole-mounted transformers connected phase to ground.

There is no way of guaranteeing whether single-phase loads are balanced across all three phases. There is going to be unbalanced current that continuously changes according to how well loads are balanced at any given moment. The diversity factor is a measure of how well this balancing takes place. With well-designed distribution systems, the maximum unbalance current never exceeds the continuous rating of the grounding transformer or grounding reactor. The zero-sequence current flowing back to the ground sources at the station is the sum of all the individual feeder unbalances.

Ground overcurrent relays such as for transformer, or bus backup at the substation that measure the total ground return must be set above the maximum expected continuous unbalance current which includes that of all feeders supplied from the substation.

11.5.3 Combined Three-Wire and Four-Wire Sub-Transmission

Sub-transmission that is a combination of three-wire and four-wire and is obtained from either delta-wye transformer with a low or high impedance reactor, wye-delta with a low or high impedance grounding transformer. Where the substation is designed to supply single-phase laterals, the station neutral is brought out, otherwise, it is not.

11.5.4 Station Grounding Bar

At load substations where the neutral is needed for sub-transmission, there exists a station grounding bar where the grounded neutrals of either power transformers or grounding transformers are connected. The station neutral wire for each four-wire sub-transmission distribution feeder is then tapped onto this same grounding bar. Ideally, the grounding bar should reside within the metal-clad line-up assembly of the metal-clad switchgear. For outdoor stations, the grounding bar is typically located at a convenient central location.

11.6 Master Ground

When a load substation has two transformers, many utilities operate the LV bus-tie breaker normally open. However, when a utility chooses to operate the LV bus-tie breaker normally closed,

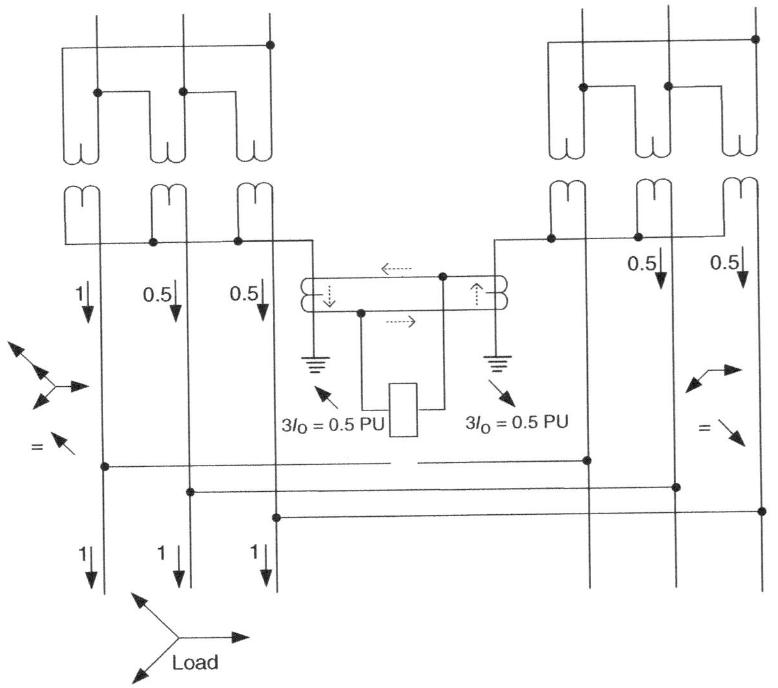

Figure 11.24 Master ground example with an open pole in the bus-tie breaker.

this leads to low level zero sequence currents to flow into the ground protections while closing the bus-tie breaker. During switching of the bus-tie breaker, it is likely to expect not all three poles to open and close in unison. The resultant temporary unbalance of phase current results in circulating neutral current. The switching operations where this is possible are:

When one pole of the bus-tie breaker is opened or closed compared to the other two poles
Switching load on of the buses while both transformers are in-service and the bus-tie breaker is closed.
When a feeder transfer switch is used to transfer load at the time the bus-tie breaker is open.

For all these situations, residual current will flow into and out of the grounded transformer neutrals. By differentially connecting a supervising overcurrent relay in the transformer neutrals as shown in Figure 11.24, a simple method of supervision is achieved. For a legitimate ground fault, the residual current will flow equally and in phase into the grounded neutrals of both transformers thereby operating the supervising overcurrent relay and supervise the ground protections at the station. The overcurrent supervision of ground protections at a -load substation is sometimes termed by utilities Master Ground.

The master ground is only necessary at substations supplying three-wire load as the feeder and capacitor bank ground overcurrent relays are set very sensitively below load current. At substations supplying four-wire load, the ground overcurrent relays are set above the unbalance current and master ground supervision is not necessary.

Referring to Figure 11.24 is an example of an open pole in the bus-tie breaker. The zero-sequence current which comprises the residual phase currents is opposite in phase on either side of the

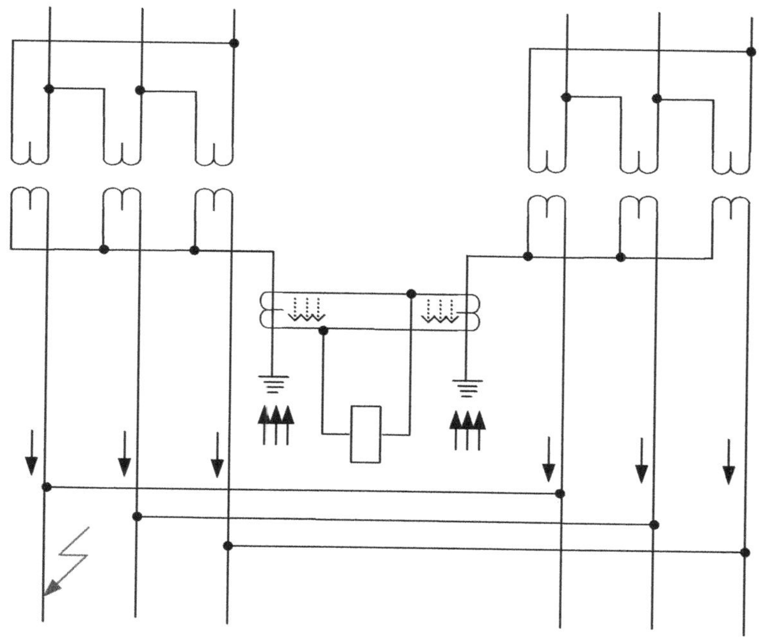

Figure 11.25 Master ground example with an external bus fault.

bus-tie breaker. These currents circulate around the differentially connected master ground relay circuit and do not operate the master ground relay that supervises the station ground protections.

Referring to Figure 11.25 is an example on an external bus fault. The zero-sequence currents flowing back to the transformer grounded neutrals are now in phase. The secondary CT currents summate to flow into the master ground relay that supervises the station ground protections.

12

Capacitor Bank Protection

12.1 Capacitor Banks

High voltage (HV) shunt capacitors are used on electric power networks at transmission and distribution levels. They are connected in parallel (shunt) to other power system elements, such as loads and lines.

The function of a shunt capacitor is to supply reactive power to the power network at the connection point. A shunt capacitor has the same effect as an overexcited synchronous generator or synchronous condenser.

12.2 Purpose for Shunt Capacitors on Power System Networks

Capacitor banks are found at substations for power factor (PF) correction and voltage control. In both cases, shunt capacitors affect reactive power consumption/availability in the power system.

In the case of PF, they deliver reactive power magnetizing type current for motors, transformers, etc. They do this by bringing the load current more in phase with the voltage. Ideally, the load current is completely in phase with the voltage. This would result in the least amount of I^2R losses and the lowest voltage drop from the sources of generation to the load centers. The load defines the amount of reactive power consumption. Shunt capacitors compensate for the low PF with additional reactive power sources to increase the PF.

Referring to Figure 12.1, assume a load of 1200 kVA, 0.7 PF.

$$\text{Cos}^1\, 0.7 = 45.6°$$

$$P = 1200 \times \text{Cos}\, 45.6° = 839.6\,\text{kW}$$

$$Q = 1200 \times \text{Sin}\, 45.6° = 857.4\,\text{kVAr}$$

If a 600 kVAr shunt capacitor, "C" is installed at the load bus:

$$\text{New } Q = 857.4 - 600 = 257.4\,\text{kVAr}$$

$$\text{New kVA} = \sqrt{(839.6^2 \times 257.5^2)} = 878\,\text{kVA}$$

$$\text{New power factor PF} = 839.6\,\text{KW}/878\,\text{kVA} = 0.96$$

Assuming a constant source voltage, this results in a percent lower current requirement from the source of:

$$\{1 - (I_2/I_1) \times 100\} = \{1 - (878\,\text{kVA}/1200\,\text{kVA}) \times 100\} = 26.8\%$$

Power System Protection: Fundamentals and Applications, First Edition. John Ciufo and Aaron Cooperberg.

$$E = I(R + jX) = RI\cos\Theta + jXI\sin\Theta$$

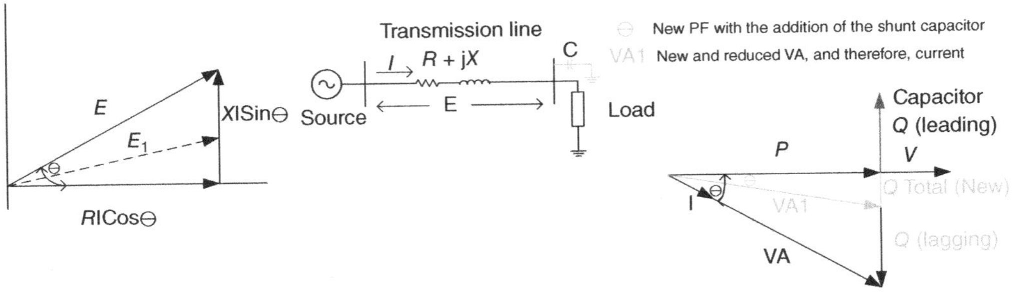

Figure 12.1 PF correction and transmission line voltage drop.

Shunt capacitors are also used for voltage support where the reactive power losses could be high enough during transmission/distribution that the local voltage is not sufficient enough to supply the load. This might lead to a voltage decrease; therefore, the system needs to supply reactive power locally at that bus to compensate for the reduction in voltage to be able to supply the local load at the appropriate voltage.

Shunt capacitors, properly sized and located, provide voltage regulation. By connecting a leading volt-Ampere reactive (VAr) loads near lagging VAr loads, the lagging VAr on a system basis is canceled, thereby increasing the voltage and, therefore, power capacity. Capacitors are typically sized for peak load requirements and removed from the system as the load drops. They are switched in and out of service typically by power system operators. Capacitors draw a specific leading current that generates a voltage rise through the reactive ohms of the system impedance, refer to Figure 12.1.

The voltage at the load bus is equal to the source voltage minus the line voltage drop. If the source voltage and the voltage drops are known, the voltage at any point may be determined. In Figure 12.1, the line drop equals E, and the ISinθ represents the VAr current. Connecting a shunt capacitor at the load bus would reduce the VAr current. This reduction in reactive current reduces the voltage drop, E1 in Figure 12.1, by reducing its reactive component by an amount equal to the reduction in VAr current times the reactance.

12.3 Capacitor Bank Construction

Capacitor banks are made up of individual capacitor units, that are in turn connected in a variety of series/parallel combinations.

12.3.1 Individual Capacitor Units

For all voltage level applications, the basic component of capacitor banks is the single encapsulated capacitor units, also historically referred to as capacitor cans. The internal capacitor units are connected in a series of parallel arrangements. A capacitor unit is the basic building block of a shunt capacitor assembly.

A capacitor unit is made up of multiple single capacitor elements connected in series or parallel arrangement as illustrated in Figure 12.2.

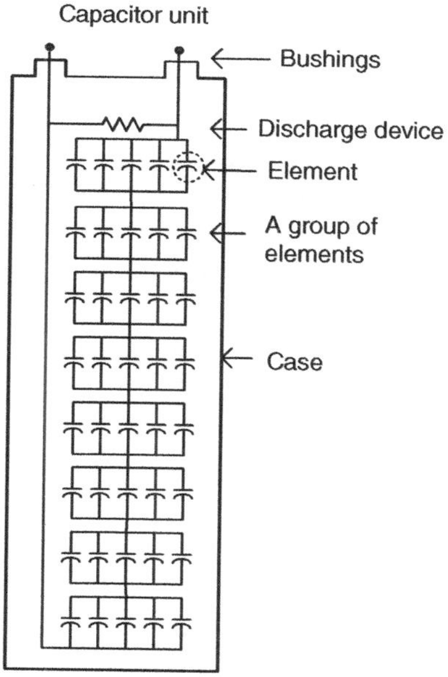

Figure 12.2 Illustration of a capacitor unit.

Typical capacitor units are available in the range of 240–25,000 V and 2.5–800 kVAr. A capacitor unit is encased in a metal case with external connection points via bushings.

A capacitor element predominant failure mode is a failure in the dielectric between capacitor plates resulting in a short circuit. That is why one needs to design protection schemes and operate shunt capacitors such that, they are within their voltage limits; they are voltage sensitive.

12.3.2 Basic Series – Parallel Arrangement

Refer to Figure 12.3 showing the basic series–parallel capacitor unit/can arrangement.

$$\text{Capacitive reactance}\, X_c = 1/\omega c \text{ and } \text{KVAR} = \text{kV}^2/X_c$$

When parallel capacitor cans are added to the capacitor bank, the effect is to lower the overall capacitive reactance, of the group, which in turn, increases the KVAR rating for a fixed voltage.

12.3.3 Capacitor Bank Configurations

Capacitor banks can be configured in several ways, such as wye grounded/ungrounded and double-wye grounded/ungrounded.

The three most popular capacitor bank configurations adopted by many utilities are the single-wye ungrounded bank, the single-wye grounded bank, and the double-wye ungrounded bank as shown in Figures 12.4–12.6.

12.3.4 Shunt Capacitor Grounding

The need for shunt capacitor installations, location, configuration, grounded or not grounded; and to be discussed shortly, fused or not used; are the accountability of the power system planning group.

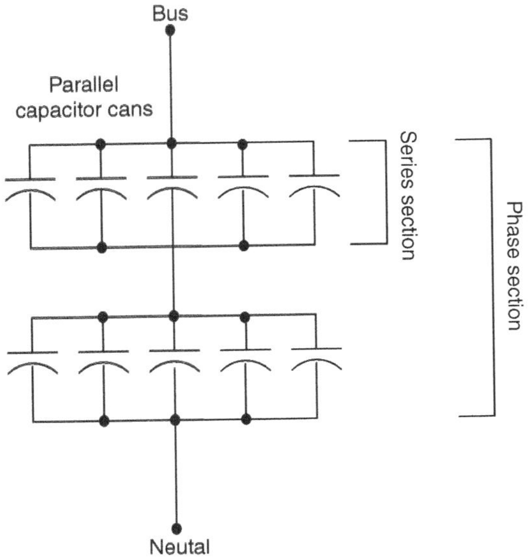

Figure 12.3 Basic series–parallel capacitor can arrangement.

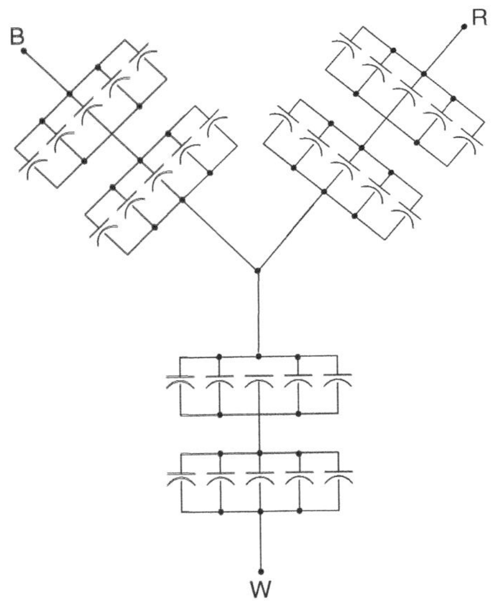

Figure 12.4 Single-wye ungrounded capacitor bank.

A protection practitioner is normally provided this information and is required to design the appropriate shunt capacitor protection for the configuration specified. Grounded or ungrounded banks will dictate the capacitor bank unbalance scheme to use and the associated number of and connections of the instrumentation transformers (current transformers [CTs] and potential transformers [PTs]).

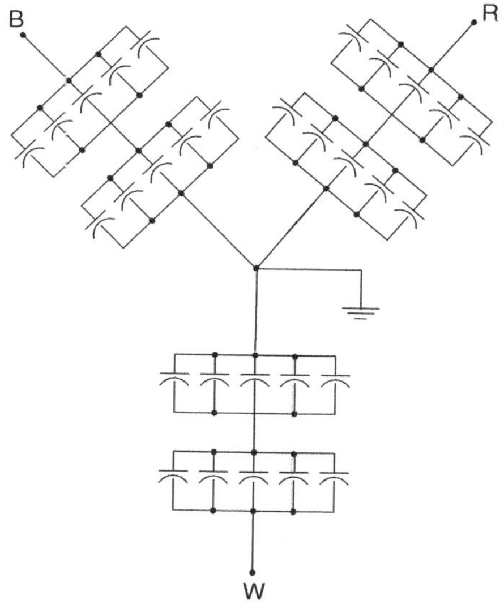

Figure 12.5 Single-wye grounded capacitor bank.

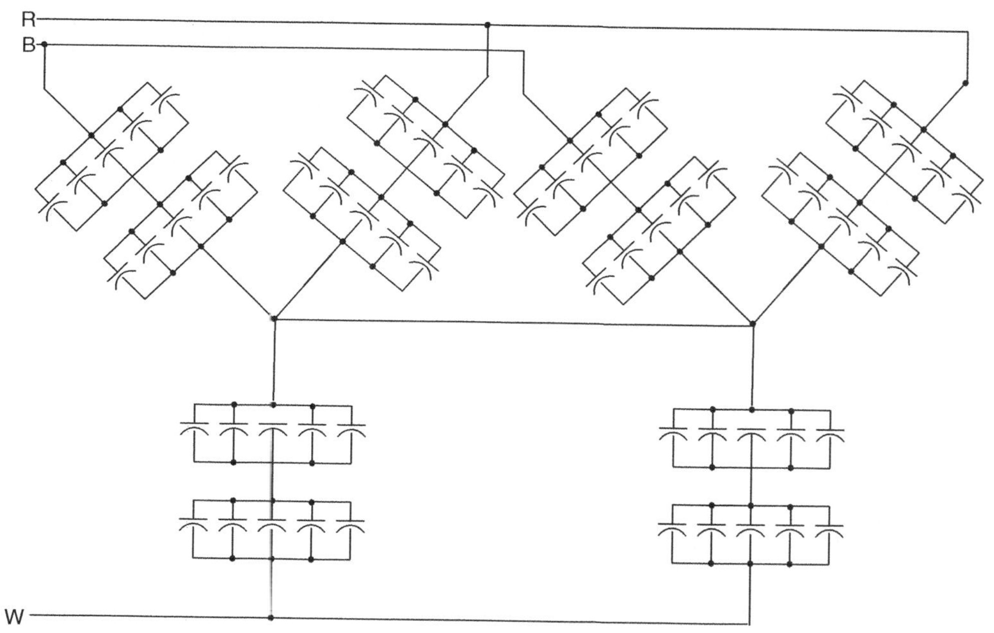

Figure 12.6 Double-wye ungrounded capacitor bank.

Most shunt capacitor bank applications are wye connected (single or double) and either ungrounded or grounded. The following are some key take-a-ways between grounded and ungrounded banks.

Grounding

- The capacitor bank neutral does not have to be insulated from the ground, and therefore, the cost should be less.
- Capacitor breaker recovery voltages are reduced; impacts breaker ratings.
- Provides a low impedance to ground for surges.

Ungrounded

- Ungrounded wye banks do not permit zero-sequence currents to flow.
- Does not provide a path for capacitor discharge currents for ground faults, which predominate; it does so for phase faults.
- The neutral should be insulated for full-line voltage.
- Capacitor breaker transient recovery voltage (TRV) requirement is higher and may require two breakers in series instead of a single breaker with a higher TRV rating to prevent restrike.

The use of grounded or ungrounded capacitor banks will depend upon a utility's system planning philosophy, compliance requirements, equipment ratings, costs, and risk.

In some applications, determined by system planning studies, limiting reactors are installed in series with the shunt capacitor bank. It may be installed upstream or downstream of the capacitor bank breaker(s). They may be required to limit high-amplitude, high-frequency current transients that may be caused by internal or external bank faults. These current transients can also be caused by shunt capacitor bank switching. The use of ungrounded banks eliminates very large outrush currents during system ground faults but outrush still occurs for phase faults. Series reactors are installed to limit outrush currents and can support breaker currents within ratings. Reactors will also limit inrush currents typically in the order of 10 times rated capacitor current and frequencies of 600 Hz. System Planning studies must be conducted to ensure that the introduction of the series reactor does not influence the breaker TRV rating thereby contributing to breaker restrike and therefore making the overall protection inadequate.

12.3.5 Capacitor Types by Fusing

Presently, there are four shunt capacitor bank types by the method of fusing:

(a) externally fused
(b) internally fused
(c) fuseless
(d) unfused

Historically, many shunt capacitor banks were externally fused. However, improvements in dielectric insulation have improved the integrity and decreased the failures of capacitor units. As a result, capacitor banks constructed from these types of capacitor units do not require any fusing at all.

Figure 12.7a depicts a fused arrangement in a normal state with no capacitor unit failures. In this case, the four series groups each equally share the total voltage of 24 V_{L-N} with the voltage drop across each being 6 V. Therefore, each capacitor unit has 6 V across it and 0.3 A through it, all within their respective capacitor unit limits.

Figure 12.7b represents the same fused arrangement with one blown fuse in Series Group 1 (C10). It should be noted that the overall capacitor unit phase current decreases; the series group (SG1) that contains the failed capacitor unit reactance increases, thereby increasing the voltage drop across the remaining nine parallel units in that group. The failure of one or two capacitor

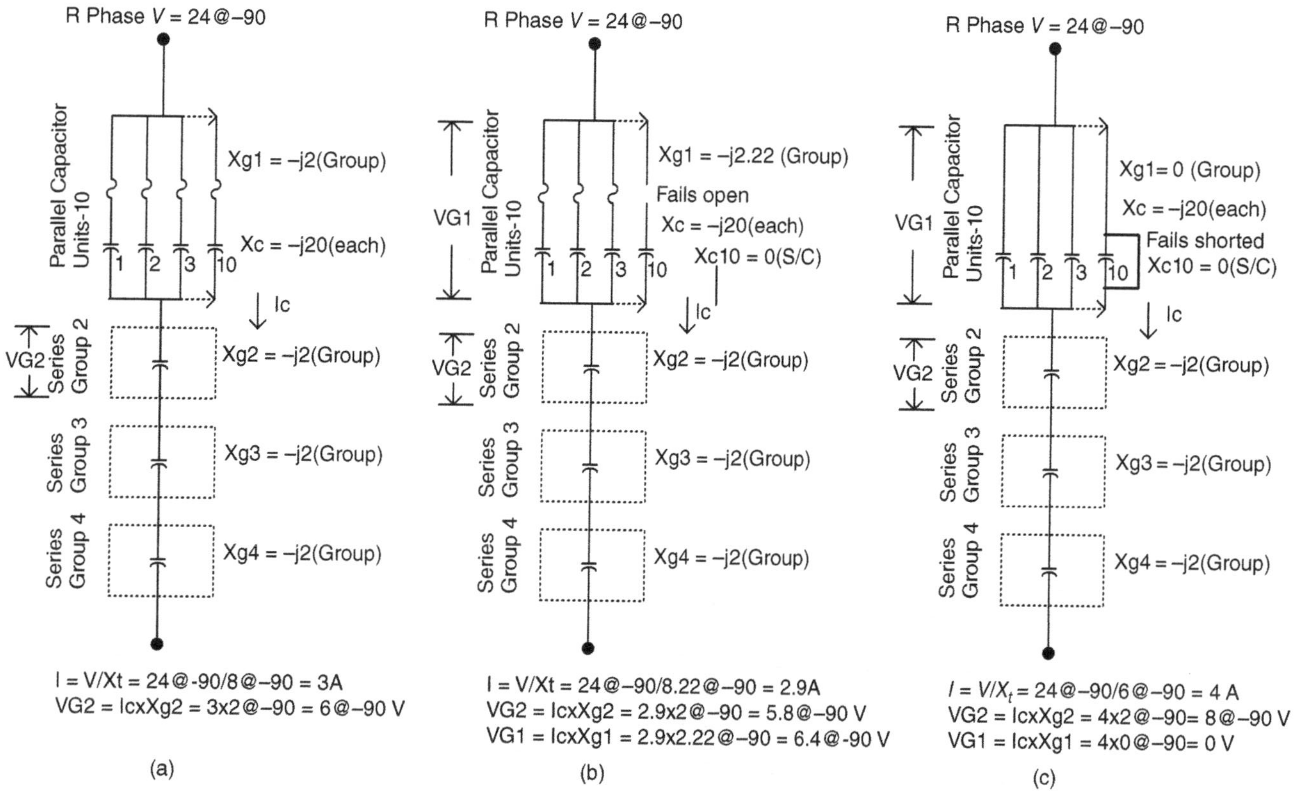

Figure 12.7 Fused and unfused capacitor bank with one unit failed (only one phase shown).

units may still contain the remaining capacitor unit's voltage within their limits. If not, then the capacitor bank unit must be removed from service.

Figure 12.7c represents an unfused bank with the same failed capacitor unit. In this case, the failed capacitor unit shorts out all of the parallel units in that series group, as there is no fuse protection to isolate the single failed unit. The phase current increases more substantially as well as the voltages across the capacitor units on the remaining serial groups.

12.3.5.1 Fused

The function of fuses for protection of the shunt capacitor elements and their location, external or internal to the capacitor unit is part of the design of shunt capacitor banks. It impacts the failure behavior of the capacitor elements and impacts the setting of the capacitor bank protection.

12.3.5.1.1 Externally Fused

Many of the legacy capacitor banks are externally fused types. A power fuse is installed in series with each capacitor unit. The fuse protects the capacitor bank by operating and quickly isolating the failed capacitor units within the bank refer to Figure 12.8.

A faulted capacitor unit will cause fault currents to flow between capacitor units in parallel in the same group, refer to Figure 12.7b. The remaining capacitor units in the bank stay in operation with an increased voltage across the units in the series group with the failed capacitor unit.

If a second unit fails, it causes an even greater voltage for the remaining units in the series group. Sequential faults within the same bank will cause unit ratings to be exceeded and develop bank unbalances; protections should initiate tripping of the bank.

Externally fused shunt capacitor units are typically configured using one or more series groups of parallel-connected capacitor elements per phase as shown in Figure 12.8. The unbalance reduces as the number of series groups is increased or as the number of capacitor units in parallel per series

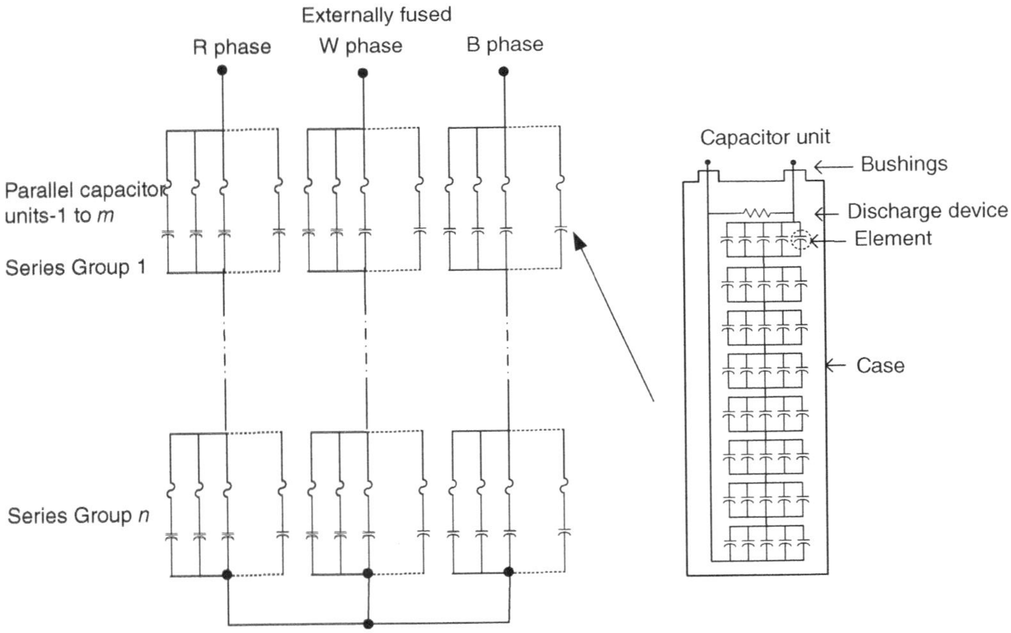

Figure 12.8 Externally fused capacitor bank configuration.

group is increased. However, the kVAr rating of the capacitor units may be smaller since a minimum number of parallel elements are needed to allow the shunt capacitor bank to stay in operation with one fuse or unit out.

12.3.5.1.2 Internally Fused

As the name implies, the fusing is provided within the capacitor unit. Each capacitor element within the unit is individually fused, refer to Figure 12.9.

The fuses are much smaller and are not explosive-type fuses as used for externally fused units. In this case, a capacitor element failure will only remove an element within the unit and not the whole unit. Internally within the unit, the series group of capacitors with the blown fuse will remain in service but will experience an increase in voltage.

Internally fused capacitor banks generally are configured with fewer capacitor units in parallel and more series groups of units. The capacitor units are typically large because the whole unit is not anticipated to break down.

12.3.5.2 No Fuses

12.3.5.2.1 Fuseless

The capacitor units for capacitor banks without fuses are the same as externally fused units. These types of capacitor units employs newer dielectric insulation technology resulting in lower failure rates. The capacitor bank is configured with several capacitor units in a series arranged in-between phase and neutral, as shown in Figure 12.10.

The protection is integral to the dielectric design. When a capacitor element fails in the unit, it shorts out the series group, refer to Figure 12.7c. Once the capacitor element breaks down, it welds, and the capacitor unit stays in operation.

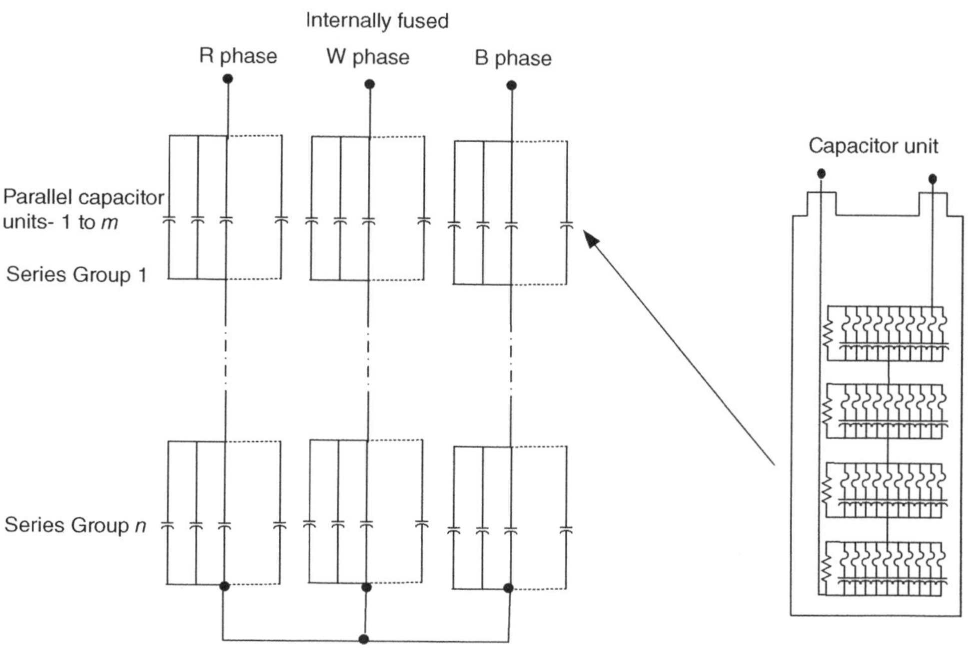

Figure 12.9 Internally fused capacitor bank configuration.

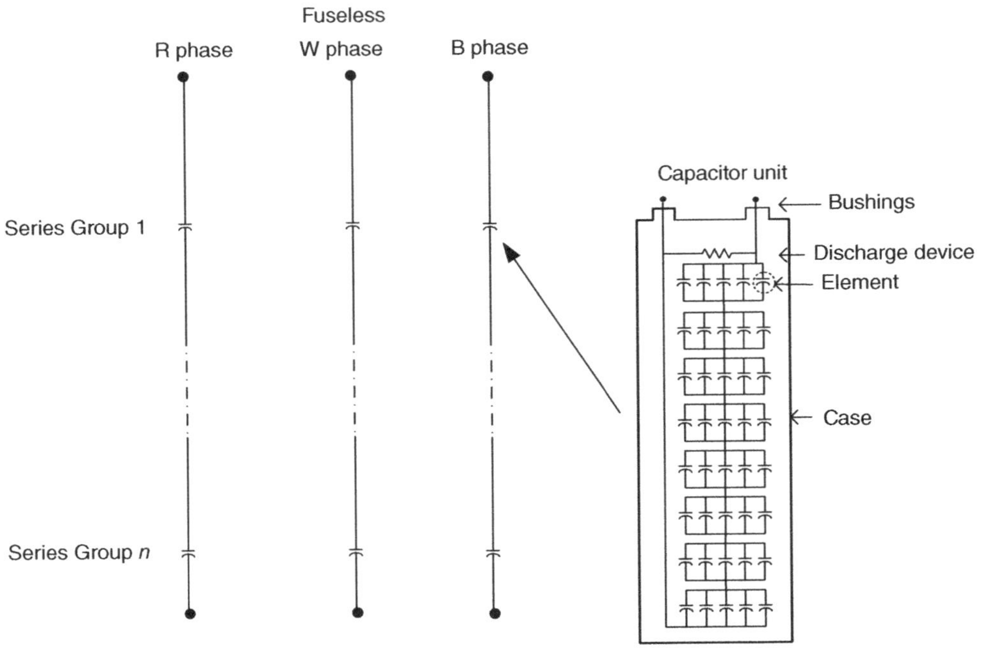

Figure 12.10 Fuseless.

The voltage across the failed capacitor element is then split among the leftover capacitor element groups that are connected in the series. As an example, assuming eight capacitor units in series, each unit containing 10 capacitor element groups ($E = 10$), this arrangement results in a total of 80 series groups. If one capacitor fails, the voltage on the remaining elements experiences an (80/79)–1 × 100 = 1.3% increase, and the bank remains in service. Typical configuration results in 10 or more serial connections so that the bank can remain operational with failures of more than one element.

Presently, fuseless shunt capacitor banks have typically replaced externally fused banks and are used for HV applications. The higher the system voltage, the more units connected in series. The fuseless banks have several series strings in parallel, per phase, to generate the MVAR capacity.

12.3.5.2.2 Unfused

The unfused capacitor bank configurations use similar capacitor unit designs as the fuseless. However, the unfused shunt capacitor bank configuration uses a series and parallel arrangement not just a series arrangement like the fuseless bank, refer to Figure 12.11. Typically, fuseless capacitor banks are used for distribution-based applications. This arrangement does not need as many capacitor units connected in parallel as a bank with external fuses.

12.3.6 Capacitor Bank Ratings

12.3.6.1 An Example Fused Capacitor Bank Rating

$M = 3$ parallel units per group
$N = 4$ series groups per phase
Unit rating: 9.96 kV, 300 kVAr

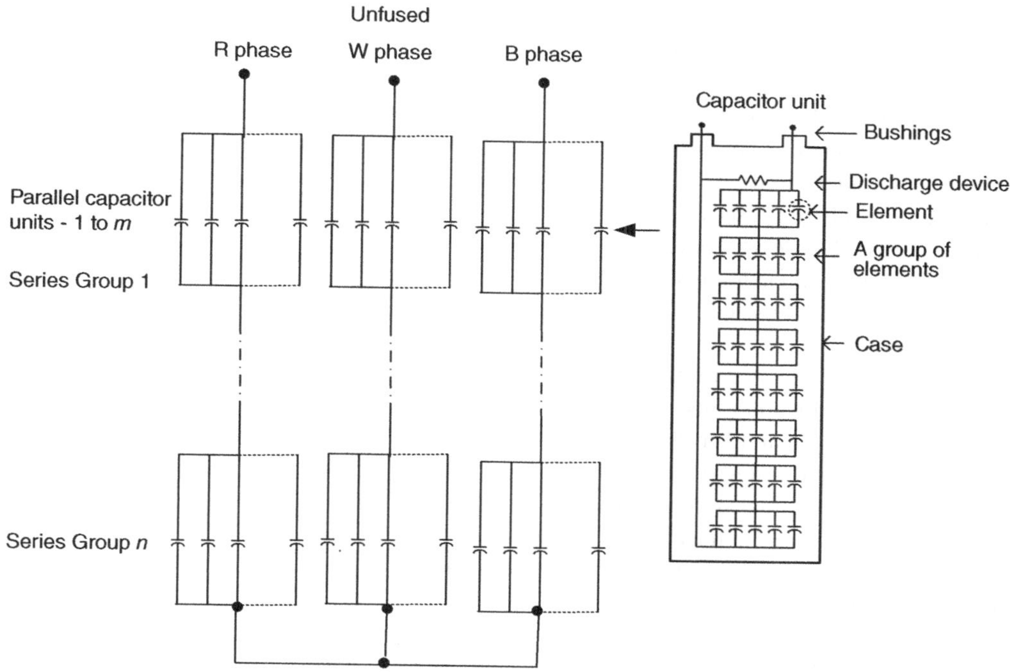

Figure 12.11 Unfused banks.

Bank Rating Calculation

Voltage rating: $\sqrt{3} \times kV_{LN} \times N$ series groups per phase $= \sqrt{3} \times 9.96\ kV \times 4 = 69\ kV$

MVA rating: #of phases × Unit rating $\times N \times M = 3 \times 300\ kVAr \times 4 \times 3 = 10.8\ MVAr$

12.3.6.2 An Example Fuseless Capacitor Bank Rating

$M = 4$ parallel strings per phase
N = 4 series units per string
Unit rating: $10.4\ kV_{L\text{-}N}$, 650 kVAr
System voltage: 69 kV

Bank Rating Calculation

Voltage rating: $\sqrt{3} \times kV_{LN} \times N$ series units per phase $= \sqrt{3} \times 10.4\ kV \times 4 = 72.05\ kV$

MVAr rating: #of phases × Unit rating $\times N \times M = 3 \times 650\ kVAr \times 4 \times 4 = 31.2\ MVAr$

MVAr rating at 69 kV: MVA Rating $\times (kV_{System}/kV_{Bank})^2 = 31.2\ MVAr$
$\times (69\ kV/72.05\ kV)^2 = 28.61\ MVAr$

12.3.7 Information Required Before Protection Settings Calculations

The following information is required to develop the settings for capacitor bank protection:

- Power System Information
  - o Nominal voltage (Φ-Φ and Φ-G)
  - o Maximum operating voltage (Φ-Φ and Φ-G)

- o Minimum and maximum three-phase fault current at the bus
- o Minimum and maximum single-phase to ground fault current at the bus
- Bank Arrangement
  - o Bank Type (Fused or Fuseless)
  - o Bank connection (Wye Grounded/Ungrounded, Double-Wye Grounded/Ungrounded, etc.)
  - o Units per Group (Fused)
  - o Series groups/Phase (Fused)
  - o Strings/Phase (Fuseless – For each side if double-wye)
  - o Series elements/Unit (Fuseless)
  - o Series units/Phase (Fuseless)
- Bank Ratings
  - o Rated voltage of bank and units (Φ-Φ and Φ-G)
  - o Rated kVAr of bank and units
  - o Rated current (Total and Per Phase)
  - o Current at nominal voltage (Total and Per Phase)
  - o Voltage/Unit at nominal system voltage
  - o Rated and the nominal voltage across each element (Fuseless)
  - o The impedance of each element and each series string (Fuseless)
  - o Number of Strings (Fuseless)
- Bank Protection
  - o Internal Utility Standard protection drawings, if they exist
  - o Protection one-line
  - o Relay modes/types that are going to be used
  - o Manufacturer's relay manuals
  - o Manufacturer's application guides
  - o CT and PT ratios

The configuration of the capacitor bank, fuses, or no fuses, size, grounding, etc., determines the types of relays and relaying schemes to be used.

12.3.8 General Shunt Capacitor Bank Protection Principles

Shunt capacitor bank protections typically protect against:

- Overvoltage across individual capacitor units (fused banks) or elements (fuseless banks) caused by failure of other units or elements.
- Overvoltage at the bank's terminals, caused by the System.
- High currents are caused by the failure of bank elements or units.
- High currents caused by faults in the bank, other than above

Protective relays and schemes are used to mitigate these conditions, and the following protections are typically applied:

- Internal or external fuses for protecting individual capacitor units (fused banks)
- Unit failure detection using overvoltage or neutral overcurrent, unbalance protection
- Bank system voltage detection and protection
- Bank system fault detection and protection – current

12.4 Capacitor Bank Protection

Capacitor banks should be maintained in-service when PF correction and voltage regulation are required. For that reason, capacitor banks should not be tripped off the system for when some individual capacitor cans have failed to leave the rest of the capacitor units still within voltage specification. This occurs in either fused or non-fused capacitor banks when one or more individual capacitor units are removed by fuse operation. Neither should the bank be tripped off because of inrush currents due to switching or because of outrush for external faults.

There are three basic types of protection for capacitor banks:

- Protection utilizing neutral unbalance to protect remaining capacitor units from excessive voltages when one or more capacitor units are removed from service – (50NU) or (59NU).
- Protection utilizing overvoltage for when the system operating voltage rises excessively from nominal – (59)
- Protection utilizing overcurrent when the capacitor bank experiences either phase or ground fault within the capacitor bank zone – (50/51/50 N/51 N)

As per IEEE Standard for Protection of Shunt Capacitor banks [1], capacitor units are required to be capable of continuous operation up to 110% of rated terminal RMS voltage including fundamental and harmonic voltages and up to 135% of rated RMS current, including fundamental and harmonic currents.

The units shall yield not less than 100% and not more than 115% of rated reactive power at rated voltage and frequency. The units shall be suitable for continuous operation at 135% of rated reactive power due to the combined effects of overvoltage, harmonics, and manufacturing tolerances. The capacitor bank configuration and design should ensure that these limitations of individual capacitor units are not exceeded under operating conditions (Figure 12.12).

12.4.1 Fused Capacitor Banks

12.4.1.1 Neutral Unbalance Protection

Capacitor bank manufacturers provide fuses for each individual capacitor can as part of the overall assembly. These fuses are either expulsion or current limiting type. Until recently, fuses were always external to the cans.

The fuses are rated to sense a failed capacitor can and isolate it fast enough to prevent rupture. At the same time, the fuses are rated to ride through normal transient inrush current without being damaged themselves.

These capacitor banks are designed to continue operating even with the loss of a limited number of fuses. The worst case is the blowing of more than one fuse in a series group. When a capacitor can is removed, the remaining capacitor cans in that series group represent a higher impedance since $X_c = 1/\omega c$, refer to Figure 12.7b.

The greater the number of capacitor cans removed within a series group, the higher will be the increase in impedance of that series section. Since most capacitor banks have many of these sections in series, the overall effect of increased impedance per section does not decrease the phase current in the same relative proportion. The slightly reduced current flowing through the much higher impedance results in an increased voltage across the remaining capacitor cans in that section.

A higher voltage across the remaining cans of up to 10% can be sustained for a reasonable amount of time. However, should the situation not be addressed the higher voltage would eventually lead to a cascading failure of the other capacitor cans in parallel leading to all the fuses in that series group

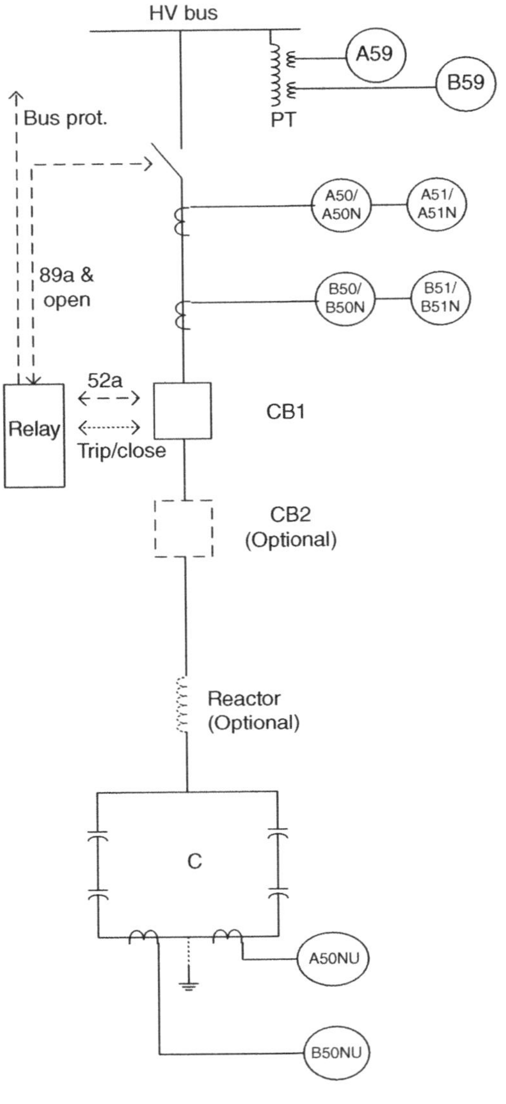

Figure 12.12 HV shunt capacitor protection one-line.

blowing. Following that, the remaining series sections would be subject to similar overvoltage and too would see their fuses blow.

Each of the configurations has its unique method of detecting blown capacitor can fuses typically, involving detecting and operating for unbalance current and or voltages. Reference should be made to the IEEE standard on HV Shunt Capacitor Banks [1].

12.4.1.1.1 Single-Wye Ungrounded Capacitor Bank

The single-wye ungrounded configured capacitor bank utilizes resistor potential devices connected in the neutral to the ground connection of the capacitor bank. A voltage relay measuring the zero-sequence voltage across the resistor potential device detects any impedance unbalance comprising the three phases of the capacitor bank.

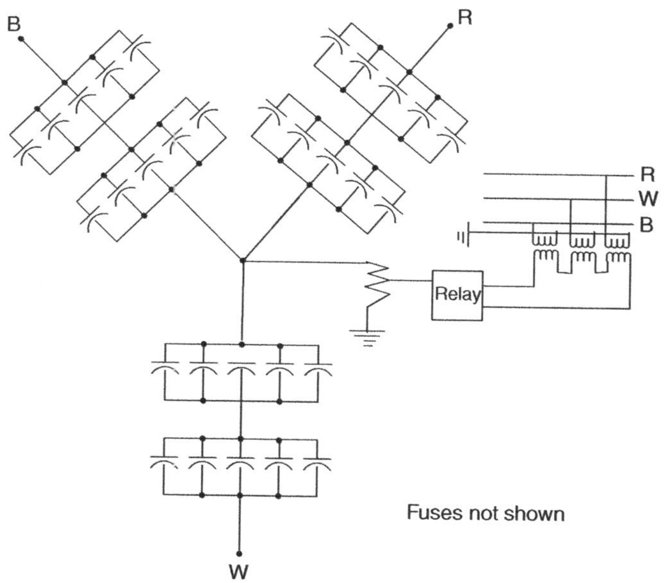

Figure 12.13 Neutral unbalance relay with neutral unbalance compensation (single-wye ungrounded capacitor bank).

Since the zero-sequence measured voltage may be due to power system voltage unbalance and not blown fuses, it is necessary to compensate for this phenomenon. Neutral unbalance compensation is the adopted method whereby the three-phase system voltage is obtained from a voltage source on the bus and is bucked off internally in the capacitor bank protection relay measuring the zero-sequence voltage across the resistor potential device as shown in Figure 12.13.

12.4.1.1.2 Double-Wye Ungrounded Capacitor Bank

A double-wye ungrounded configured capacitor bank protection measures the current flowing between the neutral star points of the wye sections. It does this by placing a 5-5 A CT in the solid connection between the two neutrals as shown in Figure 12.14.

12.4.1.1.3 Single-Wye Grounded Capacitor Bank

A single-wye grounded configured capacitor bank protection measures the current flowing from the neutral star point to the ground. It does this by placing a CT in the solid connection from the star point to the ground as shown in Figure 12.15.

12.4.2 All Capacitor Banks

12.4.2.1 Overcurrent Protection

Phase and ground timed inverse and instantaneous overcurrent relays are used to provide protection mainly for bus faults on the connecting bus to the capacitor bank outside of the bus differential zone as shown in Figure 12.16.

Another possible fault location is rack faults. A rack fault can be a short across stack links of different phases, a fault to ground from any one of the links in the stack, or arc-over a single series section. For rack faults, the current would be limited by the capacitive reactance of the series groups.

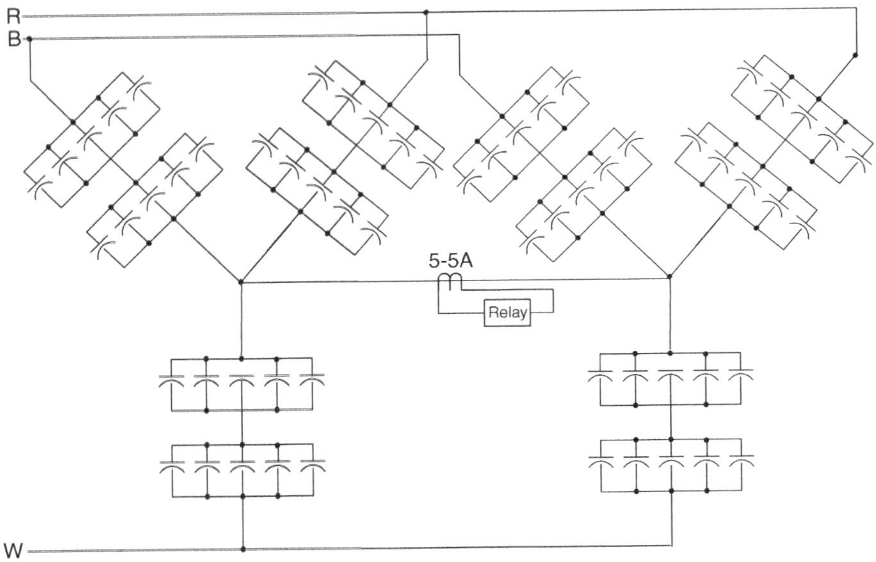

Figure 12.14 Neutral unbalance relay (double-wye ungrounded bank).

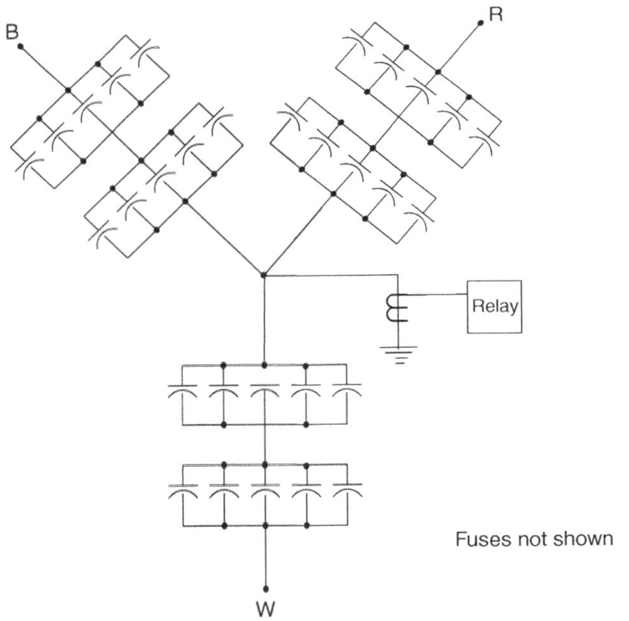

Figure 12.15 Neutral unbalance relay (single-wye ungrounded bank).

The use of inverse-time overcurrent characteristics allows for the overriding of transient inrush currents associated with capacitor banks.

The pickup setting of the phase instantaneous overcurrent protection should be set high enough to avoid a false trip due to high-frequency inrush currents, which can be in the order of three to four times rating. Present-day shunt capacitor protection relays include a 60 Hz band-pass filter to

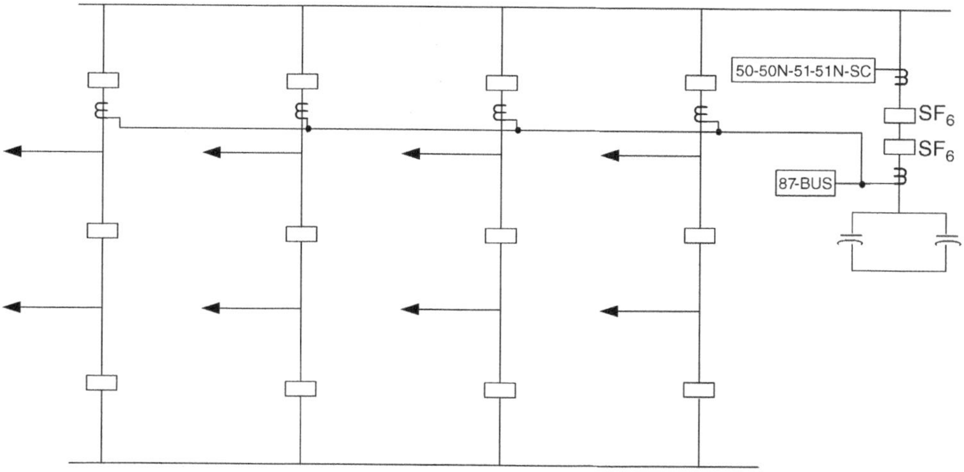

Figure 12.16 Phase and ground overcurrent capacitor bank protection.

mitigate in and outrush high-frequency currents. Conversely, it should be set less than the minimum three-phase bus fault with some margin.

The phase timed overcurrent protection should be set above the normal phase current with a margin to ride through transient currents. An extremely inverse-time characteristic is typically selected to allow transient current ride through. The time dial setting should be such to allow the transient ride through and to trip for minimum bus faults within the required clearing time.

Ground protection instantaneous protection should be above the bank rating and inrush with a margin and less than the minimum ground fault current. The ground timed overcurrent protection should be set to clear the minimum ground fault current with a margin within the required clearing time. The setting should be above any expected unbalance current.

12.4.2.2 Overvoltage Protection

System voltage that a shunt capacitor bank is connected to can, and does, vary based on power system equipment being switched on and off thus increasing source impedances. Power systems that operate at nominal voltages can typically operate at slightly higher maximum voltages; some examples are depicted below. The actual maximums can vary slightly from one utility to the next.

Nominal voltage (kV)	Maximum voltage (kV)
500	550
230	241
138	145
69	72
44	46

Capacitors are rated to continuous operation up to 110% of the bank rated RMS voltage. Therefore, capacitor banks can handle a sustained considerable overvoltage. However, capacitor banks are the apparatus contributing to the cause of some overvoltage. It is therefore prudent to trip off a capacitor bank automatically above a predetermined overvoltage. In other words, other elements of the power system can be vulnerable to damage from overvoltage than the capacitor bank the bank overvoltage protection is a system overvoltage protection.

The bank's overvoltage protection (59) can be set as follows.

Pickup setting is lower of:

- 103% of maximum system voltage
- 110% of capacitor's rated voltage

In some applications, two 59 protection elements are used, one for alarming set lower and the second that would trip the bank's breaker.

Typically, there is a time delay to prevent unwanted tripping.

12.5 Capacitor Bank Breakers

Many utilities for capacitor banks at switching stations use SF_6 breakers which are capable of handling capacitive fault currents. Standard breakers at many utility switching stations cannot handle these currents. Should the capacitor bank breaker fail, the breaker failure protection would not be capable of isolating the next zone by conventional means as the next zone breakers are not meant to trip capacitive fault currents. For that reason, some utilities choose to provide two identical SF_6 breakers in series at switching station installations. The capacitor bank protection, therefore, trip both series-connected breakers simultaneously thereby not requiring dedicated breaker failure protection as shown in Figure 12.17. Two breakers failing simultaneously is not a catered contingency.

Capacitor banks are not provided with automatic reclosing. Manual reclosing is usually blocked for typically five minutes following the opening of the breaker to ensure the capacitor bank has had sufficient time to discharge itself via discharging resistors.

12.6 Capacitor Bank Sample Settings

12.6.1 Double-Wye Ungrounded Configuration – Externally Fused

Given:

Bank rated kV = 28.8 kV

Bank rated MVAR = 23.40 MVAR

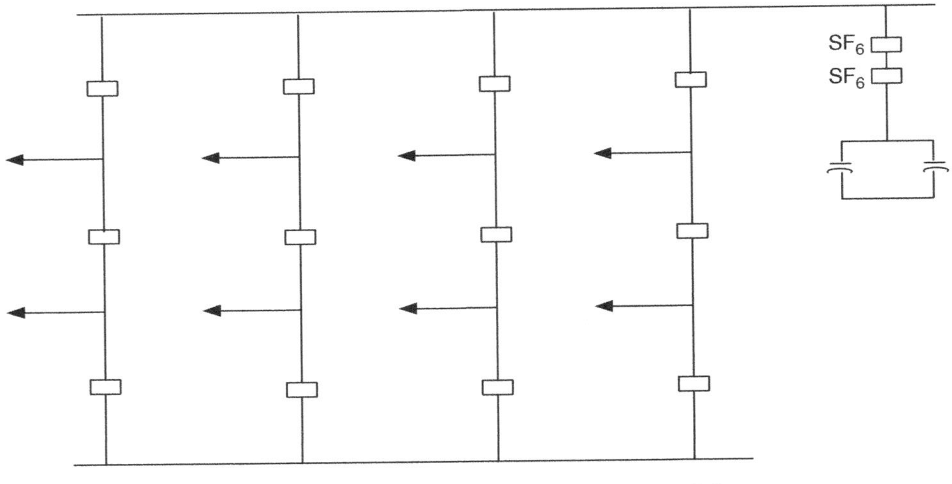

Figure 12.17 Two SF6 breakers in series instead of dedicated breaker failure.

Bank rated current = 469.1 A
Number of series sections per leg = 1
Number of parallel capacitor units = 13
Bank Configuration is one Double-Wye Ungrounded
Total number of capacitor cans = 78
Each capacitor can is rated 300 KVAR, 16.6 kV

Using the equation that governs the neutral unbalance current for various numbers of capacitor can units lost for a Double-Wye Ungrounded Capacitor Banks [1]

$$I_N = [I_{\text{PHASE}} \times (3F_1)]/[6S\,(P - 1F_1) + 5F_1]$$

where

P = Number of units in the group = 13
S = Number of series groups = 1
F_1 = Number of units removed due to an internal fault and blown series fuse

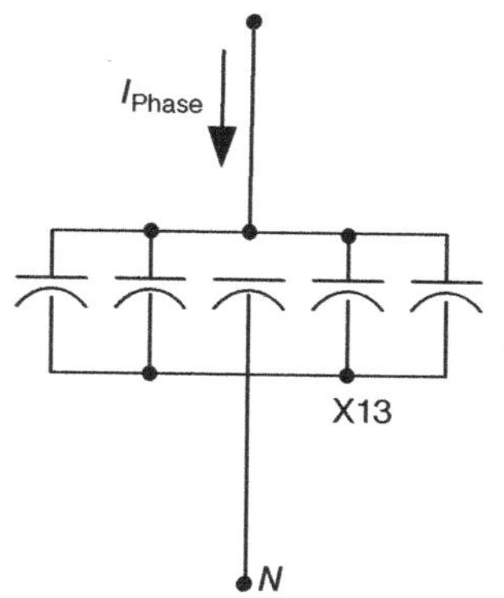

$$13 \times 300 \text{ KVAR} = 3.9 \text{ MVAR}$$

$$3.9 \text{ MVAR} \times 3 \text{ Phases} \times 2 \text{ Wye groups} = 3.9 \times 6 = 23.4 \text{ MVAR (the bank MVAR rating)}$$

$$16.64 \text{ kV} \times \sqrt{3} = 28.8 \text{ kV (the bank voltage rating)}$$

$$I_{\text{PHASE}} = 3.9 \text{ MVAR}/16.64 \text{ kV} = 234.4 \text{ A}$$

1. Blown fuse

   $$I_N = (234.4 \times 3 \times 1)/\{[(6 \times 1) \times (13 - 1)] + 5 \times 1\} = 9.13 \text{ A}$$

   Set to alarm system operators that one capacitor can is removed – a setting of 6 A pickup will suffice.

2. Blown fuses

   $$I_N = (234.4 \times 3 \times 2)/\{[(6 \times 1) \times (13 - 1)] + 5 \times 1\} = 18.51 \text{ A}$$

   Set to trip the entire bank as two capacitor cans are now removed – a setting of 12 A pickup will suffice.

Bus voltage (PU) % Rating		Number of blown fuse (Double-wye ungrounded)					
F		1	2	3	4	5	6
0.88	%OV	−10.86	−9.68	−8.48	−7.24	−5.97	−4.67
	I_N	3.61	7.32	24.76	35.89	42.40	47.62
0.9	%OV	−8.83	−7.63	−6.40	−5.14	−3.84	
	I_N	8.22	16.66	25.33	34.22	43.37	
0.92	%OV	−6.81	−5.58	−4.32	−3.03	−1.70	
	I_N	8.22	17.03	25.89	34.98	44.33	
0.94	%OV	−4.78	−3.53	−2.24	−0.92		
	I_N	8.59	17.40	26.45	35.75		
0.96	%OV	−2.75	−1.47	−0.16	1.19		
	I_N	8.77	17.77	27.01	36.51		
0.98	%OV	−0.73	0.58	1.92			
	I_N	8.95	18.14	27.58			
1	%OV	1.30	2.63	4.00			
	I_N	9.14	18.51	28.14			
1.02	%OV	3.32	4.68	6.08			
	I_N	9.32	18.88	28.70			
1.04	%OV	5.35	6.74				
	I_N	9.50	19.25				
1.06	%OV	7.38	8.79				
	I_N	9.68	19.62				
1.08	%OV	9.40	10.84				
	I_N	9.87	19.99				
1.1	%OV	11.43	12.89				
	I_N	10.05	20.36				

12.6.2 Single-Wye Ungrounded Configuration – Externally Fused

Given:
Bank rated kV = 28.8 kV
Bank rated MVAR = 23.40 MVAR
Number of series sections per leg = 2
Number of parallel capacitor units = 13
Bank configuration is one Single-Wye Ungrounded
Total number of capacitor cans = 78
Each capacitor can is rated 300 KVAR, 16.6 kV

Using the equation that governs the neutral unbalance current for various numbers of capacitor can units lost for a Single-Wye Ungrounded Capacitor Banks [1]

$$V_N = [V_{L\text{-}G} \times F_1)/[3S(P - F_1) + 2F_1]$$

where

P = Number of units in the group = 13
S = Number of series groups = 2
F_1 = Number of units removed due to an internal fault and blown series fuse

1. Blown fuse:

$$V_{N\text{-}G} = [16.64 \times 1]/\{[3 \times 2\,(13 - 1)] + (2 \times 1)\} = 225\ \text{V}$$

   Set to alarm system operators that one capacitor can is removed – a setting of 150 V pickup will suffice.

2. Blown fuses

$$V_{N\text{-}G} = [16.64 \times 2]/\{[3 \times 2\,(13 - 2)] + (2 \times 2)\} = 475\ \text{V}$$

   Set to trip the entire bank as two capacitor cans are now removed – a setting of 300 V pickup will suffice.

Bus voltage (PU) % Rating		**Number of blown fuse (Wye ungrounded)**					
F		**1**	**2**	**3**	**4**	**5**	**6**
0.88	%OV	−7.24	−1.94	4.00	10.71	18.34	27.11
	V_N	197.74	418.08	665.13	944.05	1261.45	1625.87
0.9	%OV	−5.14	0.29	6.36	13.23	21.03	30.00
	V_N	202.23	427.58	680.24	965.51	1290.12	1662.82
0.92	%OV	−3.03	2.51	8.73	15.74	23.72	
	V_N	206.73	437.08	695.36	986.96	1318.79	
0.94	%OV	−0.92	4.74	11.09	18.26	26.41	
	V_N	211.22	446.59	710.48	1008.42	1347.46	
0.96	%OV	1.19	6.97	13.45	20.77	29.10	
	V_N	215.72	456.09	725.59	1029.87	1376.12	
0.98	%OV	3.30	9.20	15.82	23.29	31.79	
	V_N	220.21	465.59	740.71	1051.33	1404.79	
1	%OV	5.41	11.43	18.18	25.81		
	V_N	224.71	475.09	755.83	1072.79		
1.02	%OV	7.51	13.66	20.55	28.32		
	V_N	229.20	484.59	770.94	1094.24		
1.04	%OV	9.62	15.89	22.91	30.84		
	V_N	233.69	494.09	786.06	1115.70		
1.06	%OV	11.73	18.11	25.27			
	V_N	238.19	503.60	801.18			
1.08	%OV	13.84	20.34	27.64			
	V_N	242.68	513.10	816.29			
1.1	%OV	15.95	22.57	30.00			
	V_N	247.18	522.60	831.41			

12.6.3 Wye Grounded Configuration – Fuseless

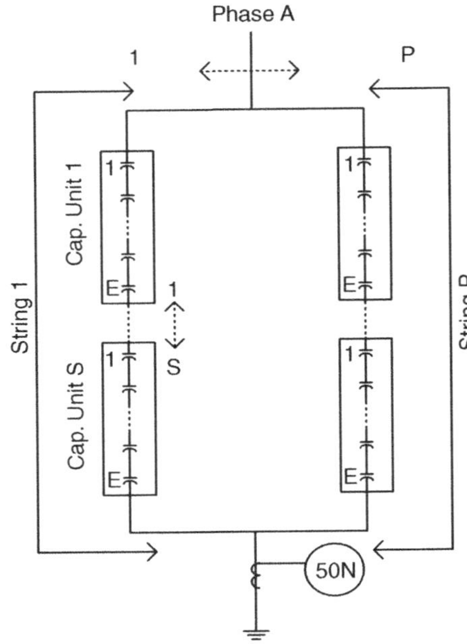

Given:

Capacitor Unit:

$V_{cap} = 20.8$ kV, 650 kVAr, E = 10 = number of series capacitors in one unit.

Z_{cap} unit $= 1000 \times (kV_{L\text{-}N})^2/kVAr$ unit $= 1000 \times 20.8^2/650 = 665.6\,\Omega$

Z_{cap} each within unit $= 665.6/10 = 66.56\,\Omega$

Nominal operating voltage = 69 kV

Capacitor Bank:

10 Strings, therefore P = 10

Two capacitor units in series per string, $S = 2$

Therefore, each string has $S \times E = 2 \times 10 = 20$ series capacitor elements: 10 strings,

$$P \times S \times E = 10 \times 2 \times 10 = 200 \text{ capacitor elements}$$

$$650 \text{ KVARs} \times 2 \times 10 = 13 \text{ MVAr per phase or } 13 \times 3 = 39 \text{ MVARs 3-PH rating}$$

$$\text{Capacitor bank voltage rating} = 1.732 \times 20.8 \text{ kV} \times 2 = 72.05 \text{ kV}_{(L\text{-}L)}$$

$$\text{Bank MVAr rating at 69 kV} = (69/72.05)^2 \times 39 = 35.76 \text{ MVAr}$$

$$\text{Phase current} = 35.76 \text{ MVAr} \times 1000/1.732 \times 69 \text{ kV} = 299.23 \text{ A}$$

The voltage across Non-failed Capacitor Elements – VH

It is assumed that all the element failures occur in one string. The voltage across the remaining non-failed elements can be expressed in terms of the rated voltage of one element.

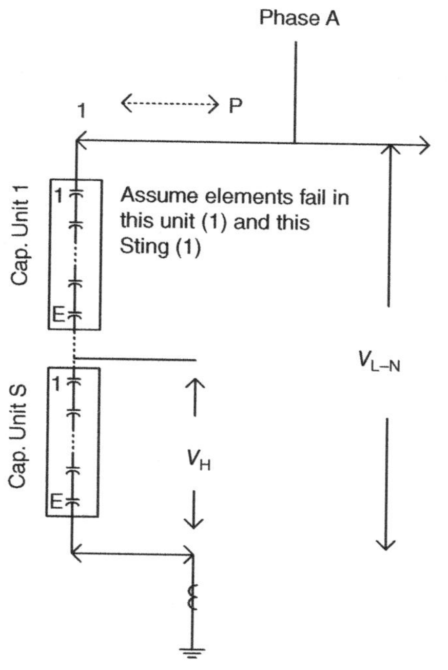

Applied voltage $= 69\ \text{kV}_{\text{L-L}} = 69/1.732 = 39.84\ \text{kV}_{\text{L-N}}$

Assume one of the 10 elements fails in the unit, N = 1:

$$V_H = 10/19 \times 39.8\ \text{kV} = 20.9473\ \text{kV} = E/(S_{\text{E-N}}) \times \text{V}_{\text{L-N}}$$

$$\%V_H = 100 \times V_H/V_{cap} = (20.97/20.8) \times 100 = 100.71$$

$$I_N = (1/665.6 \times 2 \times 19) \times 39.8 \times 1000 = 1.57\ \text{A} = (N/(Z\ \text{unit} \times S \times (S_{\text{E-N}}))$$

Bus voltage (PU) % Rating		**Number of failed elements (Wye grounded Fuseless)**					
F		**1**	**2**	**3**	**4**	**5**	**6**
0.88	%OV	88.71	93.64	99.15	105.34	112.36	120.39
	I_N	1.39	2.93	4.65	6.58	8.78	11.29
0.9	%OV	90.72	95.77	101.40	107.74	114.92	123.13
	I_N	1.42	2.99	4.75	6.73	8.98	11.54
0.92	%OV	92.74	97.89	103.65	110.13	117.47	125.86
	I_N	1.45	3.06	4.86	6.88	9.18	11.54
0.94	%OV	94.76	100.02	105.91	112.52	120.03	128.60
	I_N	1.48	3.13	4.96	7.03	9.38	11.54
0.96	%OV	96.77	102.15	108.16	114.92	122.58	131.34
	I_N	1.51	3.19	5.07	7.18	9.58	11.54

Bus voltage (PU) % Rating		Number of failed elements (Wye grounded Fuseless)					
F		1	2	3	4	5	6
0.98	%OV	98.79	104.28	110.41	117.31	125.13	134.07
	I_N	1.54	3.26	5.18	7.33	9.78	11.54
1	%OV	100.81	106.41	112.66	119.71	127.69	136.81
	I_N	1.58	3.33	5.28	7.48	9.98	12.83
1.02	%OV	102.82	108.53	114.92	122.10	130.24	139.54
	I_N	1.61	3.39	5.39	7.63	10.18	11.54
1.04	%OV	104.84	110.66	117.17	124.49	132.79	142.28
	I_N	1.64	3.46	5.49	7.78	10.37	11.54
1.06	%OV	106.85	112.79	119.42	126.89	135.35	145.02
	I_N	1.67	3.52	5.60	7.93	10.57	11.54
1.08	%OV	108.87	114.92	121.68	129.28	137.90	147.75
	I_N	1.70	3.59	5.70	8.08	10.77	11.54
1.1	%OV	110.89	117.05	123.93	131.68	140.46	150.49
	I_N	1.73	3.66	5.81	8.23	10.97	11.54

Unbalance Overvoltage Setting

Alarm after two capacitor elements fail
Trip after three capacitor elements fail; greater than 110% (112.66%)

Unbalance Overcurrent Setting

$$\text{In setting: } (5.28 + 3.33) = 4.3\text{ A primary}$$

$$\%V = (112.66 + 119.61)/2 = 109.6\%$$

Overvoltage 59 Settings

Trip setting: 1.03 × Maximum system voltage = 1.03 × 70 kV = 72.1 kV
Choose two levels of definite-time overvoltage protection.
Alarm with 10 seconds delay, at maximum system voltage = 70 kV primary
Trip with two seconds delay, at 103% of maximum system voltage = 72.1 kV (Figure 12.18)

12.6.4 Phase and Ground Overcurrent Protection

Given:

Capacitor Bank rating 250 kV, 410 MVAR
CT 3200-5 A (640:1)
Bus nominal voltage 230 kV
Phase fault on the 230 kV Bus 15,000 MVA

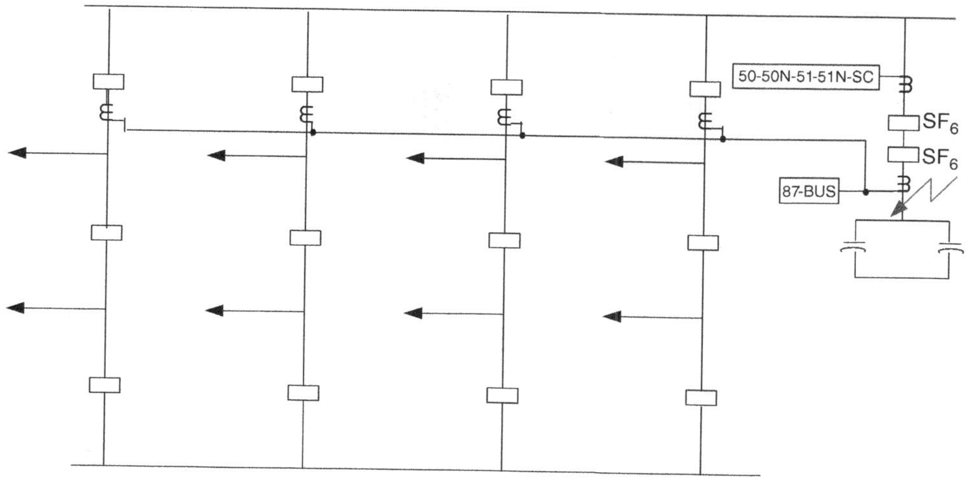

Figure 12.18 Phase and ground overcurrent protection.

Ground fault on the 230 kV Bus 64 kA
Capacitor bank load current = 410 MVAR/ ($\sqrt{3} \times 230$ kV) = 1.029 kA

50 Instantaneous Phase Overcurrent

Secondary	Primary kA	Primary MVA
12 A	7.68 kA	3060 MVA

Pickup approximately 7.5 times load current and 20% of bus phase fault

50N Instantaneous Ground Overcurrent

Secondary	Primary
12 A	7.68 kA

Pickup approximately 12% of bus ground fault, capacitor bank load is balanced

51 Timed Phase Overcurrent

Secondary	Primary kA	Primary MVA
2.0 A	1.28 kA	510 MVA

Curve Extremely Inverse

Time Dial – 3.0 (Figure 12.19)
3.5 seconds at 2 × Pickup
0.35 seconds at 10 × Pickup
Pickup approximately 125% of load current and 3.4% of bus phase fault
Fast clearing in 0.35 seconds at 34% of bus phase fault

51N Timed Ground Overcurrent

Secondary	Primary
2.0 A	1.28 kA

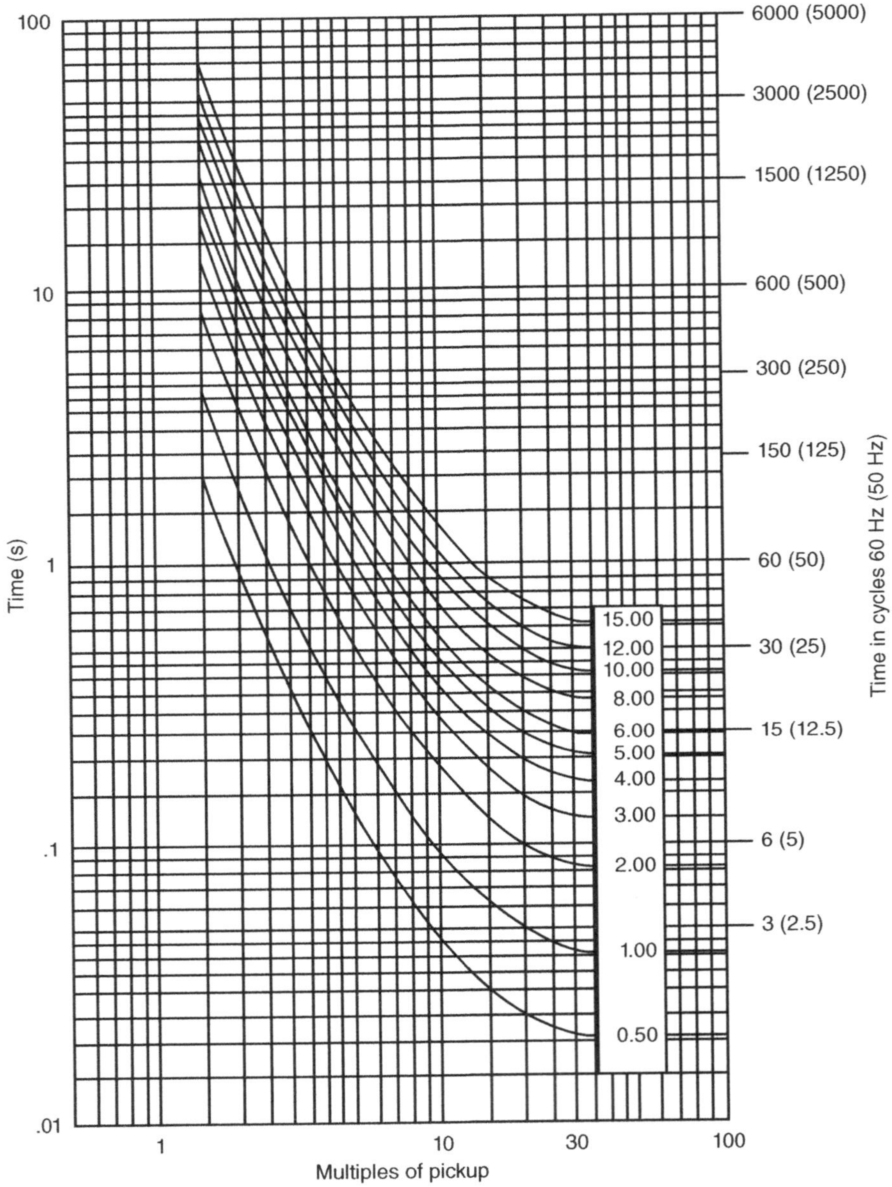

Figure 12.19 Standard extremely inverse-time overcurrent curves.

Curve Extremely Inverse

Time Dial – 3.0 (Figure 12.19)
3.5 seconds at 2 × Pickup
0.35 seconds at 10 × Pickup
Pickup approximately 2% of bus ground fault
Fast clearing in 0.35 seconds at 20% of bus ground fault

12.6.5 Overvoltage Protection

Given:

Bus nominal voltage 230 kV
Single-Phase VT 138 kV–69 V Ratio 2000:1

59 Overvoltage

Secondary	Primary (L-N)	Primary (L-L)
73.0 V	146 kV	252.9 kV

252.9 kV represents a 10% above nominal operating voltage of 230 kV.

Reference

1 IEEE Std. C37.99 – 2012, Guide for the Protection of Shunt Capacitor Banks.

13

Synchronous Generator Protection

13.1 Introduction

Generators represent one of the most important power system elements in an electric network. Generators are the energy source providers for electric power grids. They supply electric energy to the connected loads, via a very large network of transmission, sub-transmission, primary and secondary lines, and transformers that step up and down their operating voltages, respectively.

Generators are located in power generating plants that contain many other subsystems that all work together to convert one form of energy to electric energy. Generators are complex devices that consist of several subsystems that all need to operate in collaboration to supply electric energy to a power system network. Generators are specified, designed, and constructed for specific application sizes and prime movers. They represent significant investments and require a few years lead time to build. Generator protections are also complex and consist of many sub-protections for each of the generator's operating subsystems.

Power system networks typically consist of hundreds of geographically spread out and connected generators that operate in parallel. The number of operating generators at any given time is a function of the combined load demand and losses of the network. A generator's output capacity can either be rated in MVA or the lower MW value taking into account the expected load power factor. The total power of all the connected generators must equal at all times to the total sum of connected load and losses in terms of power.

Typically, a site that contains one or more generators is referred to as a generator plant or facility. In the last century, generator facilities were built containing several very larger rated generators, resulting in generator facilities with up to 5000 MW total capacities. In this century, as utilities moved away from being vertically integrated companies along with the opening of the power market, generator facilities were built that are now much smaller in capacity and more spread out. As a consequence, there are smaller generator facilities connected that are more geographically dispersed.

Large thermal units are typically rated 500–1100 MW while smaller hydraulic units are usually 50–100 MW. Most protections are equally applicable to either thermal or hydraulic types of generators. Nevertheless, some protections are uniquely applicable to either one or the other.

Most power system-connected generators are of the synchronous machine type. They represent approximately 80% of generator types that are currently used. Recently, a new type of electronic inverter-based generator technology is being adopted. These inverter-based systems are associated with sustainable generation such as wind and solar. This chapter covers typical protections of large thermal and hydraulic synchronous generators only.

Power System Protection: Fundamentals and Applications, First Edition. John Ciufo and Aaron Cooperberg.

Generator terminal voltage always requires a step up to higher voltage where it is transmitted. The step-up transformer in most utilities is almost always connected delta on the generator side and wye solidly grounded on the HV side. Therefore, most utility transmission networks are solidly grounded systems. It is fair to say that line to ground fault levels on solidly grounded transmission systems are approximately the same magnitude as three-phase fault for that reason.

13.2 General

13.2.1 Generator Basics and Functions

Generators supply power, real and reactive (*P&Q*), to the electric power network for customer load consumption, and can also absorb reactive power as required. To generate this power, several subsystems and components are required to work together to meet this overall objective. A protection practitioner must have a basic understanding of these subsystems when applying generation protection.

Figure 13.1 conceptually illustrates the fundamental components of a synchronous AC generator, depicting some salient generator subsystems.

Electric power is produced by coupling/connecting a prime mover, via a shaft, that rotates and converts mechanical energy, from a turbine, connected to the electric generator's rotor, to electrical energy. This transformation occurs via an intermediate form of magnetic energy using Faraday's and Lenz's laws.

Figure 13.2 conceptually illustrates some of the salient generator subsystem components.

13.2.1.1 Prime Mover

- A prime mover is directly connected to the synchronous generator's rotor. The rotor contains windings on the magnetic core material, and the rotor rotates on an axis/shaft within the generator stator housing.
- The prime mover is typically a turbine. There are steam-type turbines, and for hydraulic generators, hydro/water turbines. Turbines rotate the mechanical shafts that turn the generator's rotor.
- Steam turbines, which predominate, are rotated by pressurized steam that is in turn created by boilers that are heated by a fuel either fossil or nuclear-based.

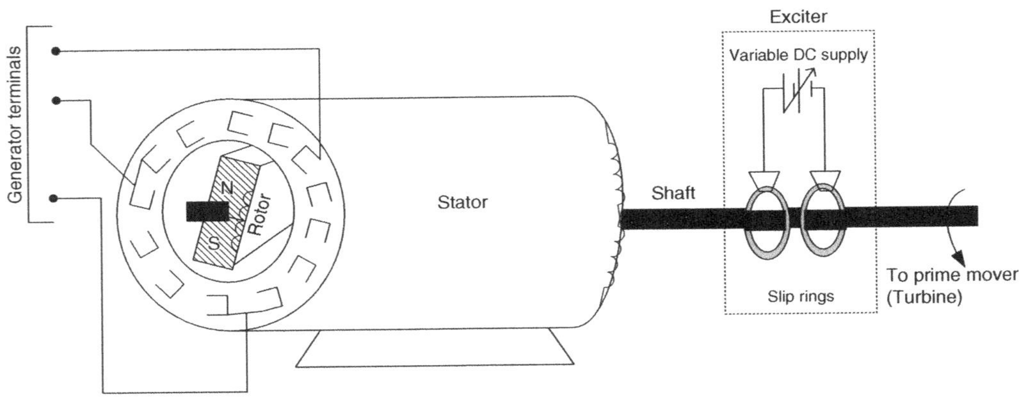

Figure 13.1 Generator conceptual illustration.

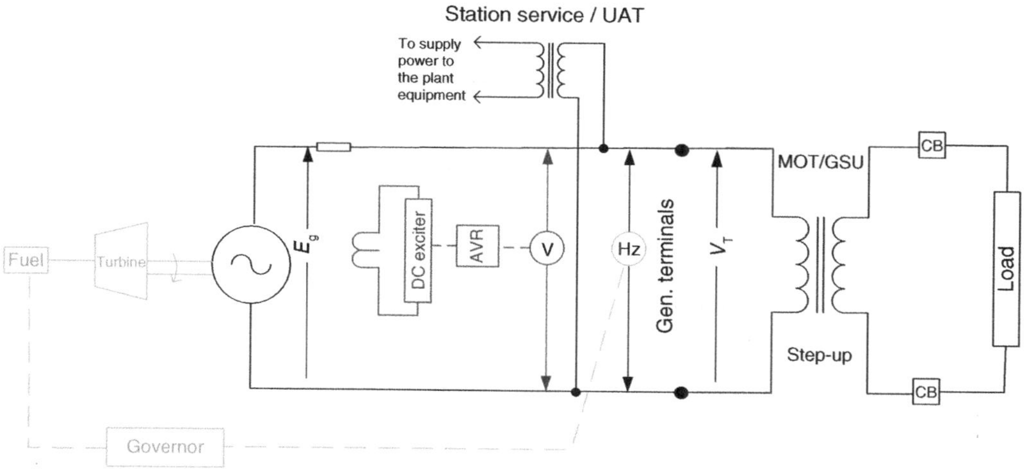

Figure 13.2 Generator subsystem component conceptual schematic.

- Currently, approximately 63% of the installed generators use fossil fuels such as gas, oil, and coal as fuel to create steam. Nuclear generators represent approximately 20% that also drive a series of steam turbines.
  The remaining 17% represent sustainable type generators, of which 7% are hydroelectric that use the natural flow of water to rotate the turbines.
  The remaining 10% of the sustainable generator types use primary energy such as wind, solar, and biomass; and transform them into electrical energy. Solar generation interface to the power network via electronic invertors. The solar panels transform solar energy into DC electricity and the inverter transforms it into AC electrical energy. Inverter-based devices are very different than synchronous generators.
- The fuel provides the energy that in turn rotates the turbine, and therefore, the connected shaft, which is connected to the generator's rotor.
- Generators convert the mechanical energy applied by the prime mover, into output power in watts, minus a small amount of mechanical and core losses. The prime mover rotates the rotor to create one of the two magnetic fields necessary to develop generator voltage in the stator windings.
- The protection practitioner should be aware that the prime mover or the shaft can get damaged if it rotates too fast or too slow and can also lead to vibrations.
- For synchronous generators, there is a direct relationship between shaft speed and power system frequency defined by: $F = n\ P/120$. Where F = frequency in Hertz, n = speed of the magnetic field, which is the speed of the rotor, in revolutions /minute, and P = number of stator poles.
  For example, a two-pole generator must rotate at 360 revolutions/m (RPM) to generate electricity at 60 Hz, which is the standard frequency in North America. Therefore, many generators measures and monitor frequency which is then converted into speed as part of their control system.
- Prime movers are rated by the manufacturers and are given power limits that must be adhered to, to prevent damage. The generator must operate within this rating. Furthermore, the prime mover limit should be coordinated with the generator's capability curve.

13.2.1.2 Rotor

- The rotor windings are connected to a separate DC source via slip rings or other means. The rotor develops a magnetic field that rotates at the speed of the shaft. This in turn generates and maintains 60 Hz (or 50 Hz) at the generator terminals.
- Functionally, the rotor is a rotating electromagnet inserted inside the generator's stator. The rotor's rotating magnetic field develops voltage at the stator winding terminals. It is the load connected to the generator that in turn leads to a counter magnetic field in the stator windings. Therefore, the greater the connected load the greater the amount of energy required to turn the rotor shaft.
- The rotor's magnetic field strength is a direct function of the amount of DC current, the number of coil turns, and the rotor's magnetically permeable core material.
- The rotor windings are connected via slip rings, or by some other means, to an external power source. The slip rings permit the DC current to flow through the windings while the shaft rotates.
- Generators convert the magnetic energy applied to the rotor into reactive power (Vars).
- The generator's internal and terminal voltage can change with the amount of DC current that is allowed to flow in the rotor's windings, more current more voltage/vars.

13.2.1.3 Stator

- The stator contains a set of stationary windings installed around the generator's housing. The windings are wound around magnetic core material referred to as poles.
- The number of poles of generators is defined by $F = n\,P/120$. Hydraulic turbines rotate at lower speeds and therefore are built with more poles.
- The stator coils are interconnected and are terminated and represent the generator's output terminals that are connected to the power system.
- It is important to note that generators provide the supply voltage to the power system, and dependent upon the amount of load connected, the load determines the current magnitude drawn from the generator.

  The magnitude of current at the generator terminal voltage determines the amount of power delivered to the power system. Each generator has a power capability curve provided by the manufacturer. The generator must be operated within this stipulated capability, otherwise the generator will be damaged.
- The stator current that flows, from the connected load, causes heat. If excessive, and beyond the stator coil's rating, it can cause insulation damage.

13.2.1.4 Governor Controls

- A power system network is dynamic. Loads are being added and removed as customer needs change. Power system equipment may fail or be taken out of service for maintenance.

  As such, generators need to dynamically change their operations to meet the instantaneous changing load demand. They do so, via some automatic closed-loop negative feedback control systems. One such system is the governor control.
- The heat/mechanical conversion subsystem via the fuel, turbine, and shaft is accountable for providing the power requirements of the generator to meet the combined load and losses.
- The generator develops a voltage when there is an interaction between the two rotor and stator magnetic fields.

  These fields in turn create magnetic torques. When no power flows, these two magnetic fields are in phase, but in the gap between the stator and rotor, as power flows, an angular difference between the two fields is developed referred to as Load Angle or Torque Angle. For clockwise shaft rotation, the rotor angle will be ahead of the stator's magnetic field. The maximum power transfer occurs when the torque angle reaches 90°.

- When a generator is first put into service with the generator breaker open (no load), the two fluxes are essential in phase, there is only a small torque angle. The frequency, which is a measure of the shaft speed, is maintained at 60 Hz.
  Upon the output breaker being closed, with load now being increased on the generator, the speed/frequency starts to decline and the torque angle starts to increase in proportion to the load. To maintain 60 Hz synchronism, more heat/mechanical energy conversion is required to increase the turbine/shaft speed to compensate to maintain 60 Hz. Therefore, the generator will need to increase the amount of fuel to increase the shaft speed.
- Monitoring speed and frequency and the adjustment of fuel to maintain 60 Hz is required continuously as the power system to which the generators are connected, is dynamically changing. The Operator's actions are generally not timely; therefore, a governor closed-loop control system is provided. The governor constantly monitors speed/frequency and makes fuel adjustments automatically. The governor also contains other related capabilities and limiters to protect that subsystem.

13.2.1.5 Voltage Control (Exciter and AVR)

- Generator voltage is induced in the stator, and the magnitude is based on the relationship of the speed (n) and the magnetic field between the rotor and stator (flux). The frequency and related speed are maintained close to 60 Hz under normal operating conditions via the governor control system.
  Therefore, the primary control for the generator voltage is the flux produced by the magnetic field between the rotor and the stator.
- The rotor's magnetic field strength is varied by the degree of DC current flowing in the rotor windings which then is controlled by the generator's excitation source.
- If the generator's terminal voltage (VT) decreases, an Operator would be required to increase the excitation source to permit a higher field current which results in a higher magnetic field thereby, increasing the generator voltage.
- Since the power system is dynamic, an automatic means is required to control the generator's voltage. Such a system is referred to as an automatic voltage regulator (AVR).
- The AVR constantly monitors the generator terminal voltage and controls the exciter's field current to make voltage adjustments. It uses a closed-loop control loop that measures actual generator terminal voltage against a set point and makes adjustments accordingly.
- The Exciter and AVR control system contain several other functions such as limiters and may include over and under excitation protections, and over-fluxing protections (V/Hz).
  Exciters can supply field current that is higher than normal operations. During a fault condition, normally the exciter forces the field (field-forcing) to increase current to increase reactive power. During the fault, the AVR requests this increase in excitation. However, there is a field current/heating limit that must be complied with to prevent damage. The limiter is designed to protect the field and interact with the AVR.
  Many AVRs include an over-flux limiter that monitors voltage–frequency (V/Hz) ratio, and when a ratio threshold is reached, it reduces excitation.
  Modern voltage control systems can also include, in addition to limiters, added protection functions such as over-excitation. These protection functions should be included when conducting a generator protection coordination study with other power system protections. Also, a protection engineer should coordinate with the generator plant's generator control engineer to ensure that generator and control protection/limiters that are in service, are coordinated.

13.2.1.6 Station Service/UAT (Unit Auxiliary Transformer)

- Generator facilities require a substantial amount of auxiliary equipment such as motors, controllers, valves, and switches to operate and maintain all of the generator's subsystems and functions.
- Most generating facilities have their own distribution network to power and operate these many auxiliary systems.
- The integrated functioning of these generator auxiliary systems is key to the reliable operation of the generator. As such, a generator facility has its own power supply via a generator station transformer also referred to as a station service transformer or Unit Auxiliary Transformer (UAT).
- The station service transformer is typically powered from the output of the generator. To ensure power integrity, an alternate power system feed is usually provided as a backup.

13.3 Generator/Unit Transformer Protections

Generators, unlike other power system equipment needing protections, require a large number of various protection types to adequately protect the generator. When the Main Unit Transformer is included in the generator zone as it is for many utility installations, the protections become even more elaborate. Further sophistication is introduced, depending on which protection sensing element has operated, determines which of the various generator shutdown modes are necessary.

In the past, with discrete electromechanical generator protection relays, there existed a unique combination of all the discrete measuring relays along with complex tripping schemes for the various shutdown modes. With the advent of digital relays, all of these protection measuring element types are found in one multi-functional generator protection relay. To accomplish the same thing, the software outputs of all the internal protection elements are assigned to specific outputs initiating the required shutdown modes. This is accomplished via the use of a simple to use software-driven menu system uniquely provided by the digital relay manufacturer.

Furthermore, to simplify the overall generator protection scheme, the current transformer (CT) and voltage transformer (VT) instrument transformer inputs are shared among the various protection element types thereby dramatically reducing cabling and wiring.

Regardless of whether the protections are electromechanical or digital the fundamental purpose for the protection and the setting policy stays consistent.

The following protection element types, each described later in detail, are typically used protections as included in IEEE Std C37.102 [1].

87G	Generator Differential
87T	Unit Transformer Differential
87O	Generator-Unit Transformer Overall Zone Differential
87SP	Generator Split-Phase Differential (hydraulic generators only)
59G	Generator Stator Ground
40	Generator Loss of Excitation
46	Generator Phase Unbalance
21/51V	Generator and System Phase Backup
51TG	Generator and System Ground Backup
81	Generator and System Underfrequency
24	Unit Transformer Over-Excitation
21–78	Generator Out-of-Step
32	Generator Reverse Power

Most protections initiate a complete shutdown of the generator. In addition to complete shutdown, some protections of hydraulic units additionally initiate a water deluge system to flood the generator to prevent fire.

The complete shutdown of a generator also known as a Class A Trip typically involves the following actions:

- Trip the generator breaker(s)
- Initiate breaker failure of the generator breaker(s)
- Trip the field breaker
- Begin mechanical system shutdown
- Initiate station service transfer

The separation from the HV system is referred to as a Class B Trip and involves the following actions:

- Trip the generator breaker(s)
- Initiate breaker failure of the generator breaker(s)

13.3.1 Generator Differential (87G)

Differential protection applied to generators is similar in concept to power transformers. Three CTs on the generator neutral side are connected in parallel with three CTs on the generator supply side. The currents in each are differentially compared with each other. There are, however, distinct differences between generators and power transformers. The CTs on either side of the generator usually have the same ratio and rating, do not have tap changing and require much smaller inrush proof characteristics. For these reasons, typical percent slope settings for power transformers are 40% while for generators it is 10%.

Differential protection is the only protection for phase–phase and three-phase faults in the generator stator windings. Stator ground faults may not be covered by the differential protection as generators are typically grounded through a high impedance.

A differential protection operation initiates a complete shutdown (also water deluge for hydraulic).

The connections for percent differential generator protection are shown in Figure 13.3.

13.3.2 Main Unit Transformer Differential (87T)

Differential protection applied to the Unit Transformer is similar in concept to power transformers. Three CTs on either side of the transformer are connected in parallel. The currents in each are differentially compared with each other. A distinct difference between unit and power transformers is that unit transformers do not have tap changing. For this reason, typical percent slope settings for power transformers are 40% while for unit transformers it is 10%; Figure 13.4.

A differential protection operation initiates a complete shutdown (also water deluge for hydraulic).

13.3.3 Generator-Transformer Overall Zone Protection (87O)

At many installations, the A group protections use dedicated generator and unit transformer differential protections while for the B group an overall differential protection zone is used for the sake of economy. The advantage of having at least one protection group being dedicated is to be able to determine which power system element faulted (Figure 13.5).

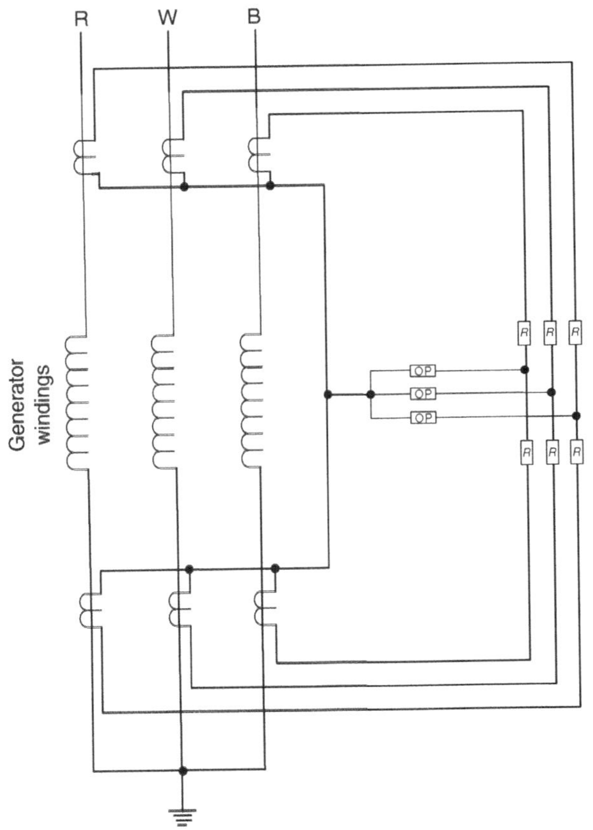

Figure 13.3 Percent differential connections.

13.3.4 Split-Phase Differential (87SP Hydraulic Only)

Differential protection cannot detect inter-turn faults on the same phase winding. When an inter-turn fault does occur and is not detected, substantial damage can happen to that part of the stator winding as it does take time for the fault to evolve into a phase–phase, or three-phase fault, when the differential protection would detect it. Hydraulic generators are uniquely designed with the stator winding split into two within the same phase. The larger number of poles and slower speed allows for this design.

Referring to Figure 13.6 a shorted turn will result in a voltage unbalance between the two split phases. The leads from each of the two split phases are passed through a window-type CT. When there are no inter-turn faults, the currents in each of the split phases are balanced. This is not the case when an inter-turn fault does exist in which case the difference between the two split phases is significant enough to operate an inverse-time overcurrent relay. The load current in each of the two spit phases is never entirely balanced. There is always some natural residual current needing to be taken into account when setting the relay. A typical natural unbalance at full load is typically 10–12 A. For this amount of natural unbalance, a current pickup would be 20–24 A using a moderately inverse characteristic. A split-phase differential protection operation initiates a complete shutdown with a water deluge.

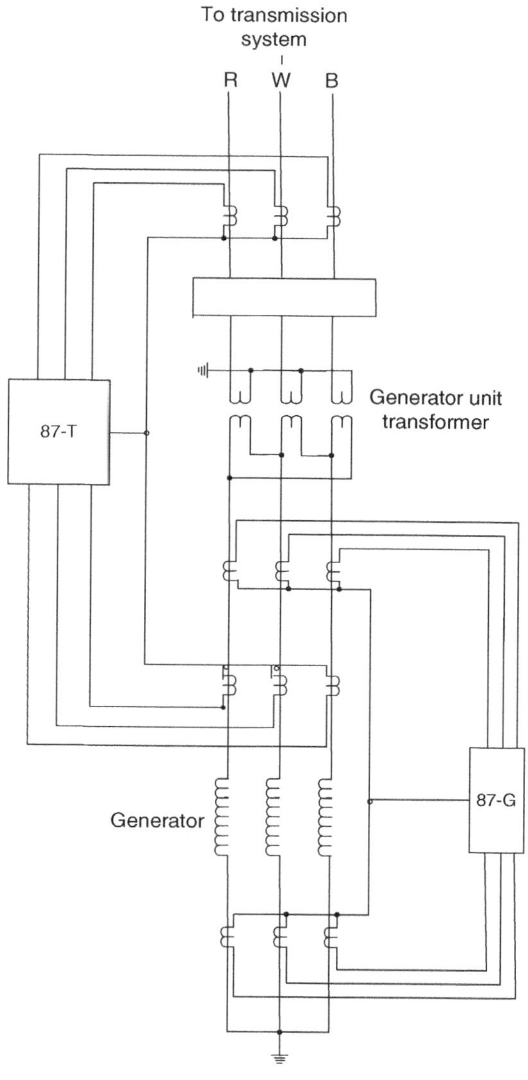

Figure 13.4 Independent generator and unit transformer differential zones.

13.3.5 Stator Ground (59G)

Stator ground faults are the most common faults in generators. This type of fault occurs due to a breakdown of insulation of the stator winding to ground through the core of the stator. The great danger from ground faults is the possibility of damage to the laminations of the stator core and to the stator winding due to the heat generated at the point of the fault.

Protection relays must trip the generator as soon as possible, tripping the main breaker, disconnecting the excitation field and tripping the prime mover. The stator ground function 59G is intended to detect a ground fault on the stator windings of a generator connected to a delta-connected winding on the generator step-up transformer.

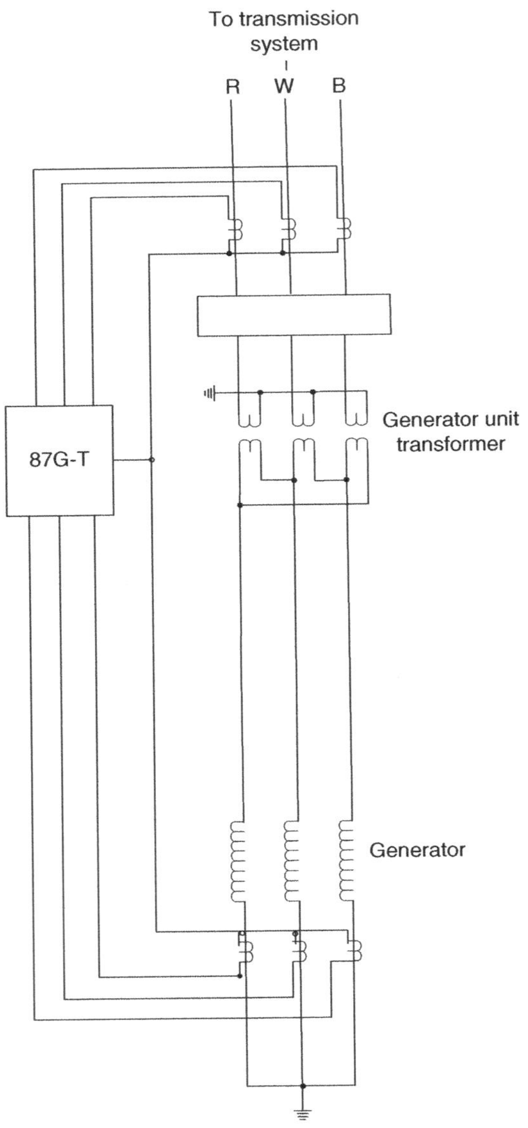

Figure 13.5 Combined generator and unit transformer zone.

Of the various methods of grounding generators, one of the more common methods is to use a distribution type transformer with a resistor connected across its secondary winding to measure the zero-sequence voltage as shown in Figure 13.7. The relay itself must be sensitive to a fundamental frequency only as third-harmonic and other higher harmonic voltages may exist at the generator neutral and need to be filtered out. The resistor is carefully chosen to limit any stator ground fault current to within 10–15 A.

The distribution transformer zero-sequence impedance is large compared to that of the generator. Therefore, for a ground fault at the generator terminals, the full phase to neutral voltage will be measured across the distribution transformer secondary winding. For a ground fault at the generator neutral, the zero-sequence voltage measured across the distribution transformer secondary winding is zero volts. The measured zero-sequence voltage is a maximum for a ground fault at the

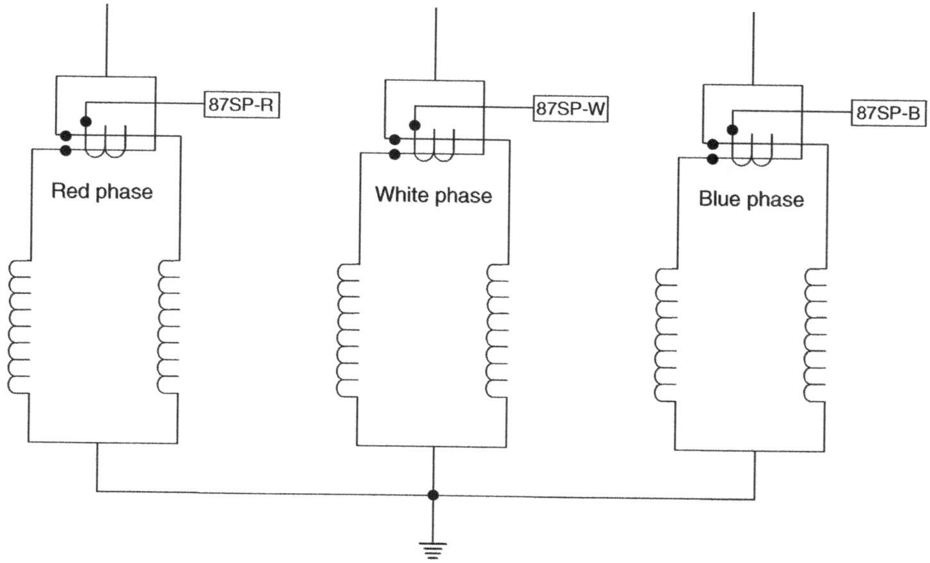

Figure 13.6 Split-phase protection.

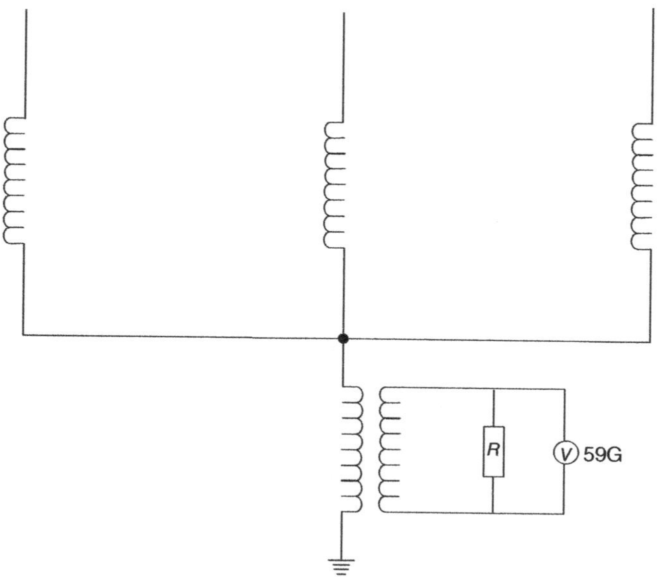

Figure 13.7 Stator ground fault protection.

terminals and decreases in magnitude as the fault location moves away from the terminals through the stator windings to the neutral.

Usually, a voltage relay connected across the distribution transformer secondary winding can be set sensitively enough to cover up to 95% of the stator winding for ground faults.

Modern digital generator protection relays also employ a means of protecting the remaining 5% of the stator winding for ground faults by measuring the signature harmonics present in the voltage impressed across the distribution transformer secondary winding in the presence of a ground fault in the stator close to the generator neutral IEEE Std.C37.102 [1]. Stator ground protection initiates a complete generator shutdown.

13.3.6 Loss of Excitation (40)

When a synchronous generator loses its excitation, it behaves like an induction generator. It will start to run above the synchronous speed as well. Asynchronous operation will cause AC currents to flow at twice the slip frequency in the rotor. This could result in significant overheating of the rotor leading to severe rotor damage. The generator will continue to supply power to the system behaving as an induction generator. However, excitation is now supplied from the system in the form of reactive power taken from the system.

The degree of generator slip frequency and power output operating as an induction generator is related to the initial generator loading. If a synchronous generator is operating at full load, upon loss of excitation it will speed up approximately by up to 5%. The reactive power drawn by the generator during this time could be as high as the generator MVA rating. The worst situation for the generator and the system is when the generator loses excitation while operating at full load.

The importation of reactive power from the system could, depending on the specific situation, cause system voltages to be depressed. The increased reactive power flow throughout the system could cause transmission line protections to operate thereby affecting overall system stability.

Loss of excitation can be the consequence of a short circuit or open circuit in the excitation circuit, a mal-operation of the AVR, and incorrect control of generators. The phenomenon of loss of excitation is detected as a change in apparent impedance seen at the generator terminals. Before the loss of excitation, the apparent impedance is in the first quadrant of the impedance diagram. Upon the loss of excitation, the trajectory of apparent impedance is such that it then enters the third and fourth quadrants. The loss of excitation protection should not operate for stable power swings whose apparent impedance seen at the generator terminals also passes through the third and fourth quadrants but much closer to the origin.

Loss of excitation is detected by a specially designed mho relay measuring voltage and current at the generator terminals. The relay is directioned toward the generator looking away from the system. The maximum torque angle or relay characteristic angle is at 90° while directioned in this manner. Furthermore, it is offset from the origin so that the measured impedance excludes stable power swings which always appear to be close to the origin. The maximum reach should cover the maximum apparent impedance looking into the generator upon loss of field (Figures 13.8 and 13.9).

The setting criteria for loss of excitation is the negative offset should exclude power swings near the origin. The maximum reach should ensure that when the field is lost, the operating point will come within the impedance characteristic.

The typical setting for loss of excitation for thermal generators is as follows:

Offset = $0.5\,X'_{\mathrm{d}}$ where X'_{d} is the generator's transient reactance
Diameter = X_{d} where X_{d} is the generator's synchronous reactance
Timing 1.0 second

The typical setting for loss of excitation for hydraulic generators is as follows:

Offset = $0.5\,X'_{\mathrm{d}}$ where X'_{d} is the generator's transient reactance
Diameter = $0.8\,X_{\mathrm{d}}$ where X_{d} is the generator's synchronous reactance
Timing 1.5 second

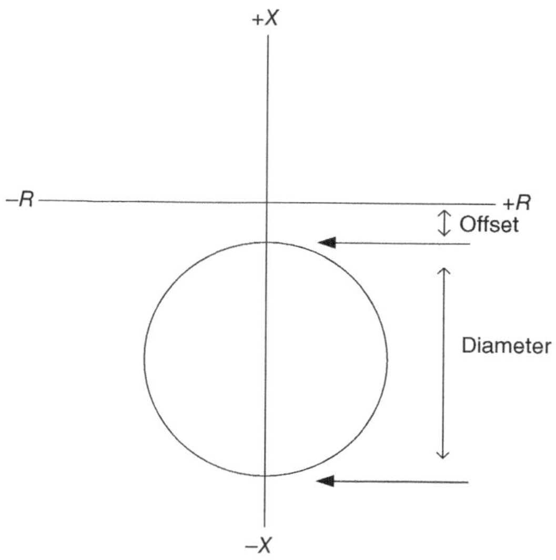

Figure 13.8 Setting criteria for loss of excitation protection.

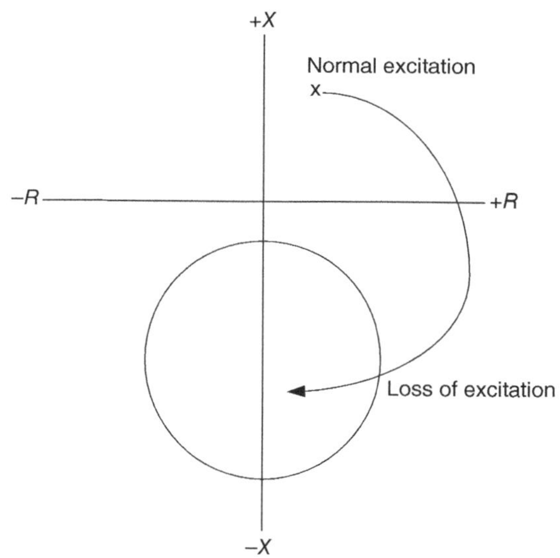

Figure 13.9 Trajectory of apparent impedance upon loss of excitation.

The reach is lower for hydraulic generators to prevent the generator from tripping when running under-excited in the condense mode something a hydraulic generator can do and a thermal generator cannot.

Loss of excitation operation initiates complete generator shutdown.

13.3.7 Phase Unbalance (46)

Generators can withstand a limited amount of unbalanced three-phase currents either continuously or for a short finite amount of time before being damaged. Many conditions can cause the three-phase currents to be not of the same magnitude. Examples of these are most obviously unbalanced system faults, unbalanced loads, open-phases in the switchyard, and even untransposed transmission lines. Any condition resulting in negative sequence current to flow in the stator windings can cause serious harm to the generator rotor. The presence of negative sequence currents in the stator windings induces AC currents to flow in the rotor. These induced AC currents flow along the surface of the rotor iron at double the system frequency. These currents will quickly cause rotor overheating seriously damaging the generator if left to operate under these conditions. The unbalanced three-phase stator currents could also cause severe mechanical vibration depending on the magnitude of the unbalance but of the two concerns overheating is by far the most serious.

The length of time T a generator can withstand the heating effect of the negative sequence currents is shown to be:

$$\int_0^T i_2^2 dt = K$$

where i_2 is the instantaneous negative sequence component of the current in the stator as a function of time.

The adopted convention by the industry is to express i_2 in per unit based on the generator full load rating. The factor K is a constant provided by the generator manufacturer and is a function of the capability of the particular generator to withstand the heating effect of the 120 Hz currents flowing and heating the rotor iron. Viewing i_2^2 as the average value of i_2^2 over the time interval T allows the heating effect to be expressed simply as:

$$i_2^2\ T = K$$

Protection is provided by an inverse-time overcurrent negative sequence sensing relay or digital relay characteristic which operates when $i_2^2\ T$ exceeds K where:

K – is the generator's negative sequence current heating constant
I_2 – is the generator's negative sequence current in per unit
T – is the time in seconds the generator is exposed to the negative sequence current

Exclusive of the negative sequence current K factor all generators have a pickup threshold for i_2 below which the relay will never operate. Therefore, to set the phase unbalance protection two items of information are required; the minimum i_2 pickup and the K factor both specific to the generator being protected.

A phase unbalance protection operation initiates a complete generator shutdown.

13.3.8 Under-Frequency (81)

Under-frequency protection is intended to prevent turbine blade damage, or loss of life due to the operation of the generator, which may be carrying a high load, during a low-frequency system condition. Under-frequency protection is generally set to protect the turbine rather than the generator. The turbine blades are subject to mechanical resonance at a certain speed below rated. The fatigue caused by operation at or near these speeds is severe and cumulative. It should be avoided.

Generators, when connected to the power system, are responsive to disturbances that occur on the network. Large steam turbines are designed to operate within a narrow band of speed such as 59.5 and 60.5 Hz. High frequency can occur as a result of load rejection caused by tripping of transmission line(s), or load. The turbine governor will control the speed and maintain frequency close to normal. For the possible contingency of the governor losing control, a turbine is fitted with an over-speed trip, which is set to operate at approximately 110%.

Low frequency can also occur as a result of system overloading. If the turbine-generator operates below 59.5 Hz, series vibration and consequent damage may occur. The turbine is permitted to operate at a low frequency only for very short periods of time.

Coordination between generator frequency protection, and the transmission system is necessary for off-nominal frequency events during which system frequency declines low enough, to initiate operation of the under-frequency load shedding (UFLS) scheme.

Sufficient frequency decline may occur to initiate UFLS operation as a result of tripping generators or tie lines or on weakly connected portions of interconnections.

Coordination is necessary to ensure that the UFLS scheme can operate to restore a balance between generation and load to recover and stabilize system frequency at a sustainable operating condition. Without coordination, generation may trip by operation of under-frequency protection to exacerbate the unbalance between load and generation resulting in tripping of more load than necessary, or in the worst case, resulting in system collapse if the resulting imbalance exceeds the design basis of the UFLS scheme.

Example 13.1

Generator: 24 kV, 960 MVA units

B87L (Low Set): Pickup frequency 56 Hz, timed at 0.35 seconds

VT: 200:1; 3-13.8 kV: 69 V

B81H (High Set): Pickup frequency 57.5 Hz, timed at 10.0 seconds (B81Ht)

B81A (Alarm): Pickup frequency 59.5 Hz

If the generator is connected to the HV system and under frequency conditions develop, the generator will be separated from the system but not shut down.

Assume this generator needs to coordinate according to regional frequency requirements depicted in Figure 13.10. The generator should not trip for frequencies above or on the curve.

As shown in Figure 13.10, the generator under-frequency protection function settings do coordinate with the generator and turbine capability and the regional requirements.

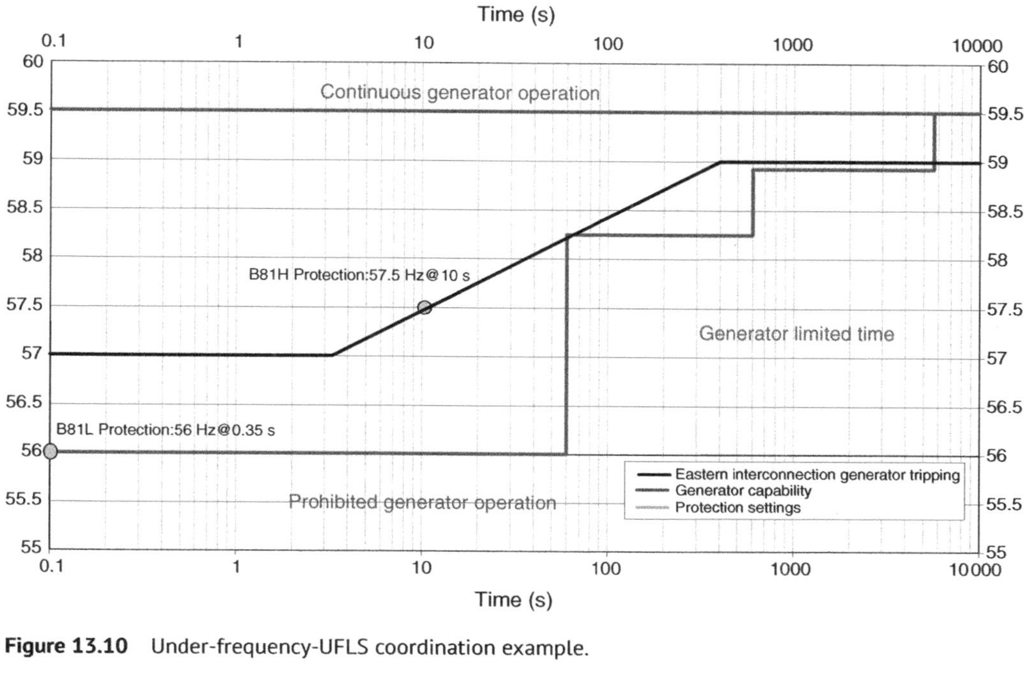

Figure 13.10 Under-frequency-UFLS coordination example.

13.3.9 Over-excitation 24

Over-excitation of generators and transformers occurs if the per unit Volts/Hertz exceeds the manufacturer's specifications and will result in equipment damage if it persists. Over-excitation will increase core losses and produce high core temperatures. The structural steel parts can also become flux paths with resulting dangerous temperature for the generator.

When a generator is first brought up to synchronous speed, the field may be applied before the terminal frequency voltage reaches 60 Hz. In this situation, the main output transformer (MOT) is subject to excessive over-excitation in the form of volts per hertz. Over-excitation can cause high core temperatures in the MOT.

Typically, a MOT is rated at 5% below the generator voltage rating. Therefore, the MOT excitation limit is reached when the generator is operating in as little as 5% above its voltage rating. When the permissible amount of excitation is exceeded, for even a short period of time, excessive damage could occur.

Regardless of this protection being mainly for the MOT and not the generator, it is associated with the generator as it is the generator excitation that needs controlling. Furthermore, the generator static exciter system will most likely have built-in over-excitation protection independent of anything external. This protection is a de facto backup to the generator static exciter system. It also provides backup to operating mistakes leading to an over-excitation condition.

The following are some settings for Volts/Hz over-excitation based on a nominal secondary relay voltage of 120 V:

Nominal Volts/Hz = 120 V/60 Hz = 2.0 V/Hz
24 setting 2.5 V/Hz timed five seconds which is 125% higher than nominal

Upon over-excitation protection operation, the excitation is removed but no complete shutdown is necessary.

13.3.10 Out-of-Step (21–78)

An electrical network operating in stable conditions is characterized by the energy balance equilibrium. Sudden occurrences in an electrical grid, such as load changes, short circuits, interruptions, and asynchronous closings which disrupt the equilibrium between generated power and load, are usually followed by oscillations known as power swings.

The power oscillations that happen could evolve toward another stable state of the electrical system, or if the oscillation is not stable, a loss of synchronism of one or more of the generators. If this happens, synchronous machines run with asynchronous speed, producing successive pole slips with load angles greater than 90° while the excitation is maintained. This state is characterized by strong oscillations of the reactive and active power, having negative effects on the generator and the network.

When a generator loses synchronism, the resulting high peak currents and off-frequency operation cause winding stresses, pulsating torques, and mechanical resonances that are potentially damaging to the generator and the turbine-generator shaft. Electrical system stability also will be affected because of the power oscillations and the difficulties for recovering the nominal voltage.

Generator out-of-step protection needs to coordinate with system stable power swings.

A generator may pole-slip (out-of-step or loss-of-synchronism) or falls out of synchronism with the power system for several reasons. The primary causes are prolonged clearance of a low-impedance fault on the power system, generator operation at a high load angle close to its stability limit, or partial or complete loss of excitation.

To properly apply out-of-step protection, stability studies are performed. Stability studies usually conducted by the regional operating authority evaluate a wide variety of system contingency conditions. Out-of-step protection is not applied unless stability studies indicate that it is needed.

The typical method for detecting out-of-step is to measure the change in apparent impedance as seen at the generator terminals. When a generator loses synchronism, the apparent impedance as viewed at the generator terminals varies as a function of the generator and system impedance, the system voltages, and the angular difference between the two.

Referring to Figure 13.11 is the classic Single Blinder Scheme used by many utilities to detect a generator falling out-of-step with the system. The apparent impedance must enter regions 1, 2, 3 in sequence and stay in each region for few milliseconds before going on to the next region to be seen as an unstable power swing resulting from the generator falling out of synchronism with the system.

Referring to Figure 13.12 is the plot of an unstable power swing from a typical computer simulation for a large thermal generator. Firstly, the apparent impedance enters into region 2 then leaves. Subsequently, the trajectory of apparent impedance enters into region 1 then region 2, and finally region 3. This trajectory of apparent impedance is a classic example of a generator losing synchronism with the system.

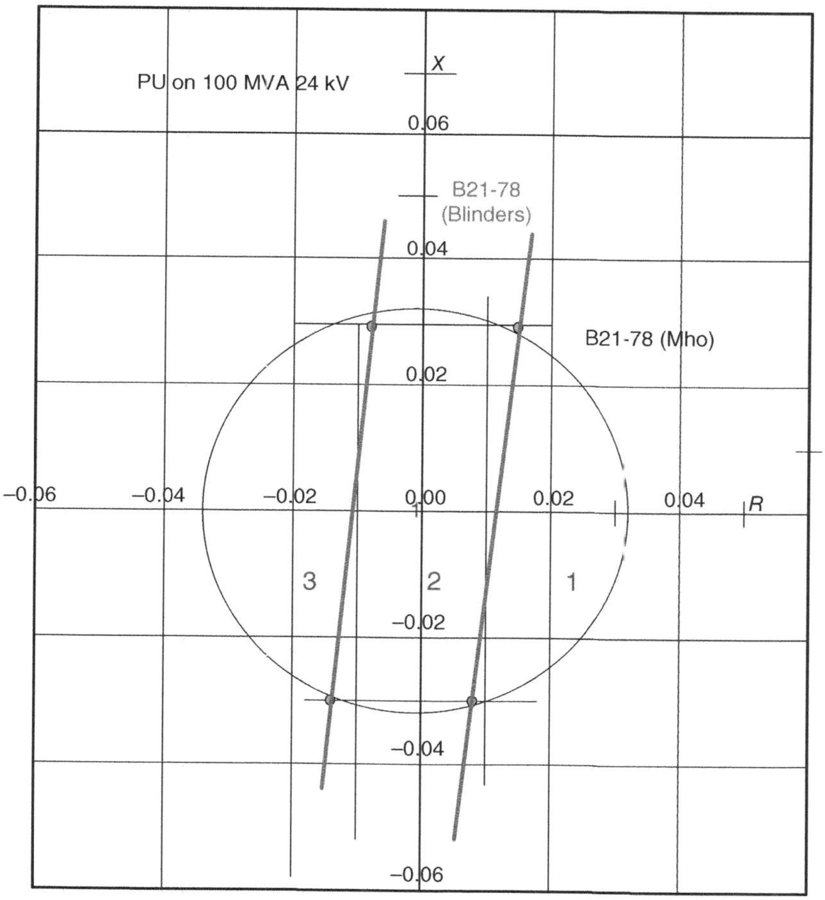

Figure 13.11 Single blinder scheme to detect out-of-step.

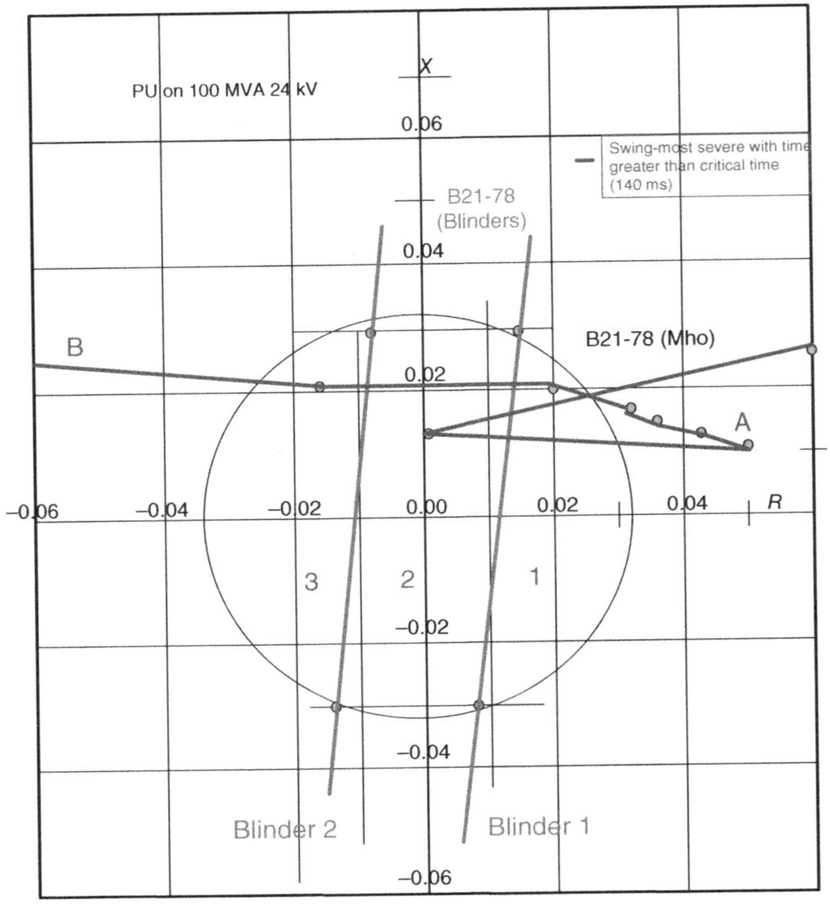

Figure 13.12 Example of an unstable power swing.

As the apparent impedance migrates from point A to point B during the system swing, the regions traversed cause blinder elements to operate and reset. As the swing progresses, Blinder 2 operates then Blinder 1 operates. As the swing progresses further, Blinder 2 resets. Timers and associated logic recognize the pickup and dropout of these relays taking into account the time spent in each region. If Blinder1 and Blinder2 remain picked up for a certain time period and Blinder 2 subsequently resets, an out-of-step condition is established and the protection operates to initiate full generator shutdown.

The significance of recognizing out-of-step in this manner is the swing impedance emerges from the side of the relay characteristic opposite to which it entered. This indicates a reversal of power flow as seen from the generator terminals intrinsic to an out-of-step condition.

The offset mho characteristic has a forward reach and a reverse reach:

The forward reach is set at two times the main output transformer reactance
The reverse reach is set at the generator transient reactance X'_d
The maximum torque angle (MTA) or relay characteristic angle (RCA) is 90°

The forward reach covers the HV transformer bus with sufficient margin to adequately define regions 1 and 3 in Figure 13.11.

The reverse reach covers the swing region into the generator's transient reactance with a margin that overlaps the loss of field relay's offset of $X'_d/2$.

Blinder 1 is set at 1.5 Ω with an angle of 85° (i.e. MTA = −5°). Blinder 2 is set with similar parameters to Blinder 1 but reversed 180° in the direction either externally by wiring convention or internally by a program in a digital relay.

The distance between blinders must be wide enough to give an unstable swing apparent impedance trajectory sufficient time to remain in each region thus ensuring an out-of-step condition is being detected.

13.3.11 Reverse Power (32)

There are possible operating conditions where the power direction can reverse for a generator, which is also referred to as "motoring." Reverse power or anti-motoring protection is normally applied in such situations where the prime-mover supply is removed from a generator supplying the network at synchronous speed with the field normally excited.

The power system will then drive the generator as a motor. A motoring condition may develop if a generator is connected improperly to the power system. This will happen if the generator circuit breaker is closed inadvertently at some speed less than synchronous speed. Typical situations are when the generator turning gear is slowing down to a standstill or has reached a standstill.

Motoring will cause adverse effects, particularly in the case of steam turbines. The basic phenomenon is that the rotation of the turbine rotor and the blades in a steam environment will cause windage losses. Windage losses are a function of rotor diameter, blades length, and directly proportional to the density of the enclosed steam. In any situation where no steam flow is present, harmful windage losses could occur.

The effect of the steam flow through a turbine is first to cause rotation of the turbine and second to carry away the heat of the turbine elements. In a motoring condition, the steam flow no longer exists, and consequently, the heat of the windage losses is not carried away. Because of this, parts of the turbine could be heated to an abnormal level during motoring. The maximum permissible time a turbine can sustain a motoring condition is normally available from the manufacture's data and is normally a function of rated speed.

A reverse power condition is an undesirable condition for generators. The power drawn by the generator during motoring is equal to the mechanical losses and they can be very low for large steam units (below 0.5% in some cases). Therefore, a reverse power function typically is set very sensitive to prevent mechanical damage on turbine blades, shaft, gear-box, etc.

13.3.12 Transmission System Backup

13.3.12.1 Phase Backup (21,51V)

Large thermal generators typically have dual redundant protections along with breaker failure protection of the synchronizing breaker. Furthermore, the transmission system being supplied also has dual redundant protections. For that reason, there is generally no need for phase backup protection for large generators. Smaller hydraulic units, on the other hand, are typically equipped with a phase backup protection to cover uncleared phase faults on the connecting network.

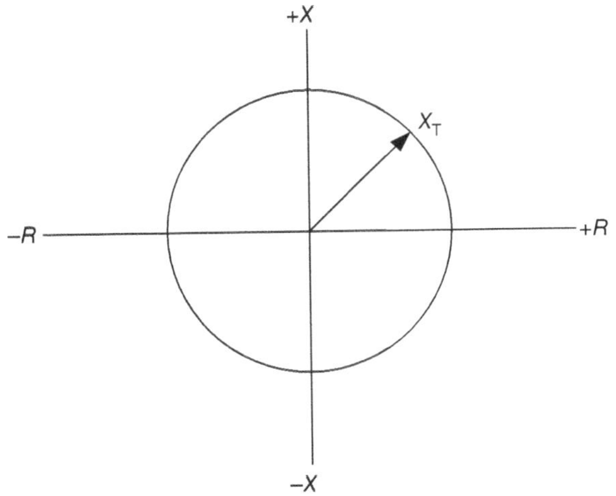

Figure 13.13 Phase backup reach setting.

There are two types of relays typically used for this purpose. The first type uses some form of impedance relay either a mho relay with a reverse offset or a simple impedance relay with the impedance characteristic encompassing the origin, as shown in Figure 13.13. They are both used with a definite time delay. These relays obtain currents from current transformers in the neutral ends of the generator phase windings and voltage from the generator terminals as shown in Figure 13.14. Both the impedance and the mho with reverse offset provide backup for faults internal to the generator while it's connected to the system. Both provide backup or last resort tipping of the generator for external transmission system phase faults.

The impedance reach setting looking into the transmission system is usually set to cover at a minimum the main output transformer impedance with adequate margin to cover close-in transmission phase faults and trips typically 1.5 seconds. This protection upon operating initiates a complete generator shutdown.

Inverse-time overcurrent relays can also be used for phase backup provided they have added on voltage restraint. The pickup setting of an inverse-time overcurrent relay needs to be at least 1.5 times and preferably 2 times the generator maximum load. The fault current will drop after 0.5 seconds as the generator enters synchronous reactance mode. This would result in the fault and load current being similar values making it virtually impossible while just measuring current to distinguish between the two. Therefore, a setting sensitive enough to see a fault at the generator terminals with such a long-time delay would always trip on load unless something can be done to prevent it from doing so.

A solution is to supervise the inverse-time overcurrent relay with voltage restraint. Under normal load conditions, the voltage at the generator terminals stays constant and the relay is blocked from operating. For a fault at its terminals, or even much further away from the generator terminals, the voltage collapses and the overcurrent backup would be an effective backup protection.

13.3.12.2 Ground Backup (51TG)

Main unit transformers are always solidly grounded on the wye transmission side for grounded transmission systems. An effective ground backup protection is to apply instantaneous overcurrent relays with a definite time delay connecting to a CT in the neutral to ground connection of the transformer as shown in Figure 13.14.

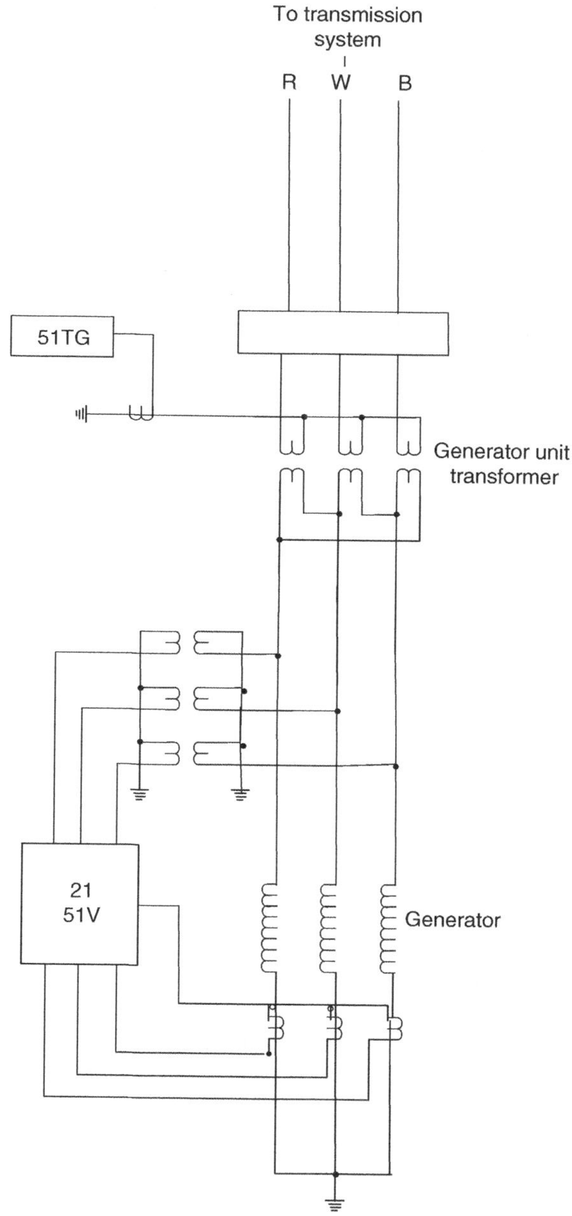

Figure 13.14 Phase and ground backup system protections.

13.4 Current Transformers

Typical generator output voltages are 13.8 and 24 kV. This presents a problem in terms of CT ratio ratings. CT ratios must closely match the full load current of the equipment to guarantee the CT does not overheat under normal loading conditions. Typically, CTs are sized so that the secondary CT current does not exceed 5 A under normal loading conditions. Furthermore, this usually also guarantees the CT secondary current does not exceed 100 A under maximum fault conditions.

For those reasons, large thermal or nuclear generators rated at 24 kV and typically 600–700 MVA with full load current approximately 15,000 A comes with generator bushing CT's rated 16,000-5 A.

13.5 Generator Protection Sample Settings

Following are two sample generator protection settings. The first sample settings are typical for a large hydraulic generator. The second sample settings are typical for a large thermal generator.

13.5.1 Sample Setting 1 – Large Hydraulic Generators

Large hydraulic generators are rated typically anywhere between 50 and 100 MW. This example is for a typical large hydraulic generator rated 110 MVA, 13.8 kV.

13.5.1.1 Differential (87)

CT ratings for generators are chosen such that 5 A flows in its secondary winding at full load current. Also, identical CT ratings are used on either side of the generator.

$$\text{Full load current} = 110\,\text{MVA}/(\sqrt{3} \times 13.8\,\text{kV}) = 4600\,\text{A}$$

The generator comes equipped with 5000-5 A rated bushing CTs on the generator neutral side. The generator breaker comes equipped with 5000-5 A bushing CTs to match.

Since there is no CT ratio mismatch or underload tap changing as is present with power transformers, any spill current if present at all is due to CT excitation current differences between the two sets of CTs which is negligible.

Minimum pickup: 0.5 A secondary, 500 A primary
Percent slope: 10%

13.5.1.2 Split-Phase Differential (87SP)

Split-phase protection unique to hydraulic generators only detects inter-turn faults on the same phase winding. It does this by taking leads from each of the two split phases and passing them through a window-type CT. Without an inter-turn fault, the split phases are close to being balanced. With an inter-turn fault developing significant secondary CT current develops operating overcurrent relay elements. Typically, there are two overcurrent elements used. One is instantaneous and is set with a high pickup to detect catastrophic inter-turn failures while the other is set with a sensitive low pickup and timed with an inverse-time characteristic to detect low-level incipient inter-turn faults.

Before setting this protection, the generator must be placed in-service and field measurements taken of the residual current from each of the split-phase CTs that normally occur as split-phases rarely have identical impedances.

Given: from field measurements for a typical hydraulic generator: Normal operating phase unbalance current is no greater than 0.3 A secondary, 12 A primary with a CT ratio of 40:1.

Instantaneous high-set pickup	3.6 A secondary, 144 A primary
Timed low-set pickup	0.6 A secondary, 24 A primary
Timing	0.3 seconds at 10 × pickup

13.5.1.3 Stator Ground (59G)

In this example, the digital generator protection measures the signature harmonics present in the voltage impressed across the distribution transformer secondary winding used to ground the generator neutral to ground connection to limit fault currents. This method of protection covers all ground faults in the stator, even those close to the generator neutral.

Setting criteria: 59G 100% stator ground protection is generally set for approximately 50% of the minimum third-harmonic voltage observed during various loading conditions. The protection also has what is known as an undervoltage inhibit generally set equal to 80–90% of nominal voltage at the generator terminals.

Third-harmonic undervoltage pickup is field settable after measurements are taken. A typical example setting would be:

Third-harmonic undervoltage pickup	1.2 V secondary, 72 V primary VT ratio of 60:1
Undervoltage Inhibit pickup	11.7 kV 85% of nominal

13.5.1.4 Loss of Excitation (40)

Loss of excitation is detected by a specially designed mho relay measuring voltage and current at the generator terminals. The relay is directioned toward the generator looking away from the system. The maximum torque angle or relay characteristic angle is at 90° while directioned in this manner. Furthermore, it is offset from the origin so that the measured impedance excludes stable power swings which always appear to be close to the origin. The maximum reach should cover the maximum apparent impedance looking into the generator upon loss of field (Figure 13.15).

Given:	CT rating 5000-5 A, 1000:1 VT rating 13.8 kV-115 V, 120:1 Impedance ratio 120/1000 = 0.12:1

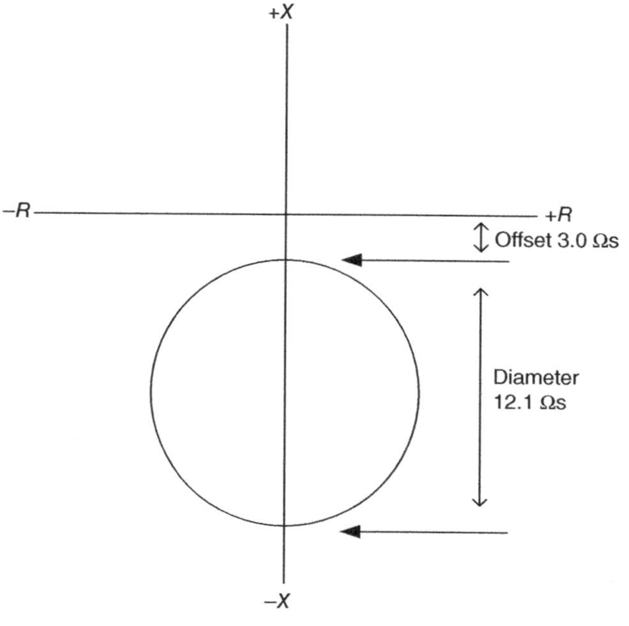

Figure 13.15 Loss of excitation setting example impedance diagram.

Given: $X_d = 1.31$ PU, $X'_d = 0.417$ PU

where: PU (per unit) values are based on 13.8 kV, 100 MVA.

Diameter: 12.1 Ω secondary, 1.45 Ω primary, 0.838 PU

where: Diameter = $(0.8\, X_d) - (X'_d/2)$.

Offset: −3.0 Ω secondary, −0.36 Ω primary, −0.208 PU

where: Offset = $0.5\, X'_d$.

13.5.1.5 Phase Unbalance (46)

The adopted convention by the industry is to express i_2 in per unit based on the generator full load rating. The factor K is a constant provided by the generator manufacturer and is a function of the capability of the particular generator to withstand the heating effect of the 120 Hz currents flowing and heating the rotor iron. Viewing i_2^2 as the average value of i_2^2 over the time interval T allows the heating effect to be expressed simply as $i_2^2\, T = K$.

Exclusive of the negative sequence current K factor all generators have a pickup threshold for i_2 below which the relay will never operate. Therefore, to set the phase unbalance protection two items of information are required; the minimum i_2 pickup and the K factor both specific to the generator being protected.

Given: CT rating 5000 − 5 A, 1000: 1

where:

4600 A equals full load current (110 MVA, 13.8 kV)
K rating provided by the generator manufacturer is 40
The maximum continuous I_2 recommended by the manufacturer is 7–10%
Negative Sequence Current Alarm 7% of nominal 4.6 A secondary with 0.5 second time delay
Negative Sequence Current pickup 10% of nominal 4.6 A secondary

13.5.2 Sample Setting 2 – Large Thermal Generators

Large thermal generators are rated typically anywhere between 500 and 1100 MW. This example is for a typical large thermal generator rated 635 MVA, 24 kV (540 MW at 0.85 PF)

13.5.2.1 Differential (87)

CT ratings for generators are chosen such that 5 A flows in its secondary winding at full load current. Also, identical CT ratings are used on either side of the generator.

$$\text{Full load current} = 635\ \text{MVA}/(\sqrt{3} \times 24\ \text{kV}) = 15{,}276\ \text{A}$$

The generator comes equipped with16,000-5 A rated bushing CTs on the generator neutral side. The generator breaker comes equipped with 16,000-5 A bushing CTs to match.

Since there is no CT ratio mismatch or underload tap changing as is present with power transformers, any spill current if present at all is due to CT excitation current differences between the two sets of CTs which is negligible.

Minimum pickup	0.5 A secondary, 1.6 kA primary
Percent slope	10%

13.5.2.2 Stator Ground (59G)

Of the various methods of grounding generators, one of the more common methods is to use a distribution type transformer with a resistor connected across its secondary winding to limit the primary ground-fault current for a ground faulted stator winding as shown in Figure 13.16.

Calculate the value of the line to ground fault limited by the 0.765 Ω resistor:

(1) Reflect the resistor to the primary of the distribution transformer
Transformation ratio 22 kV/240 V = 92:1
The reflected impedance varies inversely as the transformation ratio squared
$92^2 = 8464{:}1$
$0.765 \times 8464 = 6475\,\Omega$

$$\text{Base } \Omega \text{ on a 22 kV base} = 22\text{ kV}^2/100\text{ MVA} = 4.84\,\Omega$$

$$\text{Reflected impedance in per unit} = 6475/4.84 = 1338\text{ PU}$$

(2) Calculate $3I_0$ based on the calculated value above

$$3I_0 = 3/13{,}338 = 0.0022\text{ PU}$$

$$\text{Base Amp} = 100\text{ MVA}/\sqrt{3} \times 22\text{ kV} = 2624\text{ A}$$

$$3I_0 = 0.0022 \times 2624 = 5.8\text{ A}$$

The 0.764 Ω resistor limits the line to ground fault current on the output terminal of the generator to a mere 5.8 A using the distribution transformer in the neutral to ground connection.

Usually, a voltage relay connected across the distribution transformer secondary winding can be set sensitively enough to cover up to 95% of the stator winding for ground faults.

The distribution transformer zero sequence impedance is large compared to that of the generator. Therefore, for a ground fault at the generator terminals, the full phase to neutral voltage will be measured across the distribution transformer secondary winding. For a ground fault at the generator neutral, the zero-sequence voltage measured across the distribution transformer

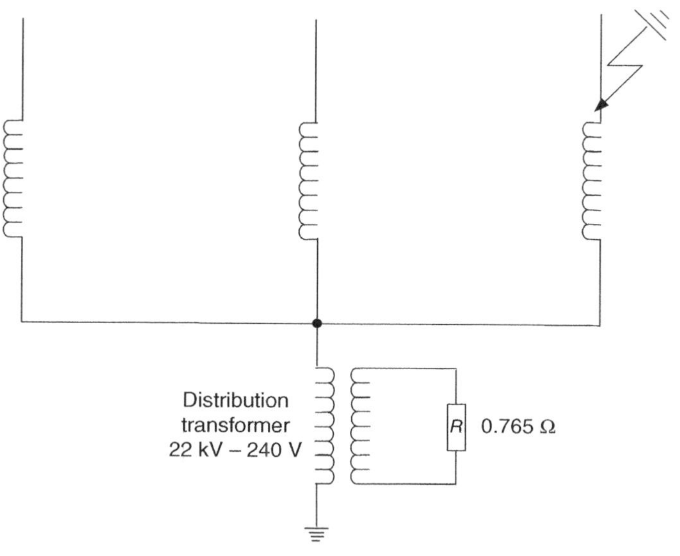

Figure 13.16 Stator ground limiting resistor.

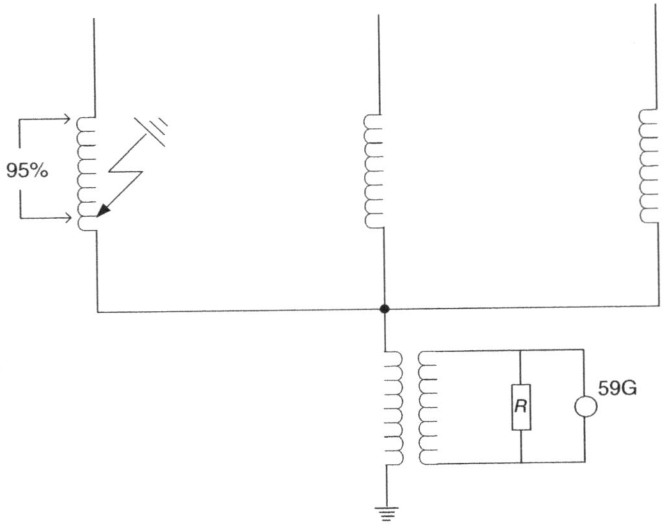

Figure 13.17 Calculation of the maximum stator ground fault coverage.

secondary winding is zero volts. The measured zero-sequence voltage is a maximum for a ground fault at the terminals and decreases in magnitude as the fault location moves away from the terminals through the stator windings to the neutral (Figure 13.17).

The generator is rated 24 kV L-L = 24 kV/$\sqrt{3}$ = 13.8 kV L-N.

Choosing a pickup voltage for 3 V_0 measured across the resistor of 10 V.

Consider the stator windings as providing a physical linear voltage relationship calculate what primary voltage would cause the relay to just pickup (Figure 13.18):

$$10\,\text{V} \times 22\,\text{kV}/240\,\text{V} = 0.917\,\text{kV}$$

Percentage of stator winding covered:

$[1 - (0.917\,\text{kV}/13.8\,\text{kV})] \times 100 = 93.4\%$ a definite time delay of one second is usually applied.

The voltage relay should be set at 10 V providing ground fault coverage to 93.4% of the stator winding. For 100% stator ground fault coverage, measuring just zero-sequence voltage across the resistor is not achievable.

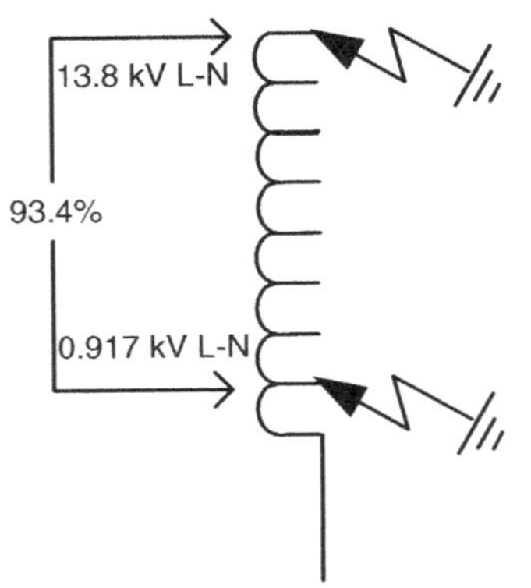

Figure 13.18 Percentage of stator winding protected for ground faults.

Modern digital generator protection relays also employ a means of protecting the remaining 5–7% of a stator winding for ground faults by measuring the signature harmonics present in the voltage impressed across the distribution transformer secondary winding in the presence of a ground fault in the stator close to the generator neutral.

13.5.2.3 Loss of Excitation (40)

Loss of excitation is detected by a specially designed mho relay measuring voltage and current at the generator terminals. The relay is directioned toward the generator looking away from the system. The maximum torque angle or relay characteristic angle is at 90° while directioned in this manner. Furthermore, it is offset from the origin so that the measured impedance excludes stable power swings which always appear to be close to the origin. The maximum reach should cover the maximum apparent impedance looking into the generator upon loss of field.

Given:	CT rating 16,000-5 A, 3200:1
	VT rating 13.8 kV-69 V, 200:1
	Impedance ratio 200/3200 = 0.0625:1

Given: $X_d = 0.331$ PU 1.91 Ω, $X'_d = 0.063$ PU = 0.363 Ω

where: PU (per unit) values are based on 24 kV, 100 MVA.

Diameter: 30.6 Ω secondary, 1.91 Ω primary, 0.332 PU

where: Diameter = $(0.8\,X_d) - (X'_d/2)$.

Offset: − 3.0 Ω secondary, −0.187 Ω primary, −0.0325 PU

where: Offset = $0.5X'_d$.

Usually, a definite time delay of one second is applied (Figure 13.19).

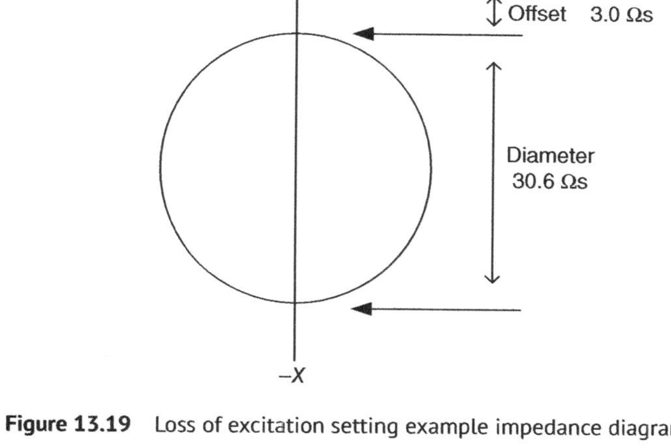

Figure 13.19 Loss of excitation setting example impedance diagram.

13.5.2.4 Over-Excitation (24)

Given: VT rating 13.8 kV – 69 V, 200 : 1

Nominal Volts/Hz = 120 V/60 Hz = 2.0 V/Hz

Typical Setting = 2.36 V/Hz timed 5s which is 118% higher than generator rated voltage

13.5.2.5 Out-of-Step (21–78)

Given:	CT rating 16,000-5 A, 3200:1
	VT rating 13.8 kV-69 V, 200:1
	Impedance ratio 200/3200=0.0625:1

Given: $X'_d = 0.063$ PU, 24 kV Unit transformer $X_T = 0.024$ PU

where: PU (per unit) values are based on 24 kV, 100 MVA.

The mho characteristic has two components:

The long reach is the forward reach at the RCA = $2X_T$ where X_T = the transformer reactance.

The short reach is the reverse reach at the RCA = X'_d (generator transient reactance)

The long reach should cover the HV transformer bus with sufficient margin to adequately define regions 1 and 3 in Figure 13.20.

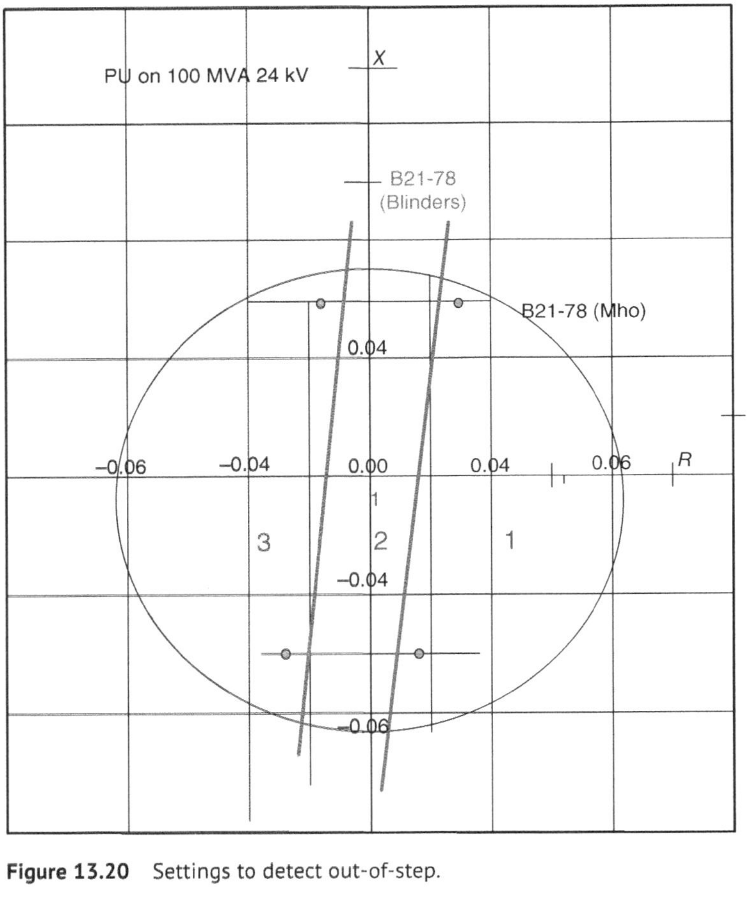

Figure 13.20 Settings to detect out-of-step.

The short reach covers the swing region into the generator transient reactance with a margin that overlaps the loss of field relay offset of $X'_d/2$.

Setting 21 (mho Characteristic)

Long Reach:
RCA = 90°
Forward reach at RCA = 4.37 Ω secondary, 0.273 Ω primary, 0.047 PU
Short Reach:
RCA = 270°
Reverse reach at RCA = 5.8 Ω secondary, 0.353 Ω primary, 0.063 PU

Setting 78 (Blinder Characteristics)

RH Blinder: 1.5 Ω secondary at −5°, 0.094 Ω primary 0.0163 PU
LH Blinder: 1.5 Ω secondary at 175°, 0.094 Ω primary 0.0163 PU

Two parallel straight line impedance characteristics intersecting at 85° to the *R* axis.

13.5.2.6 Phase Unbalance (46)

A minimum of 3% continuous I_2 must be considered always present due to the mutual induction effects in a transmission network where the transmission lines are not geometrically spaced and are not transposed. A typical setting for $K = I_2^2$ t of 7.5% is typical for large thermal generators.

13.6 Generator Control and Protection Systems Coordination

13.6.1 Synchronous Generator Phasor Diagrams/Capability

Generators, like most power system equipment, are designed to operate within specific thresholds and manufacturer ratings. These ratings define a set of operating limits and are intended to protect the generator from damage caused by excessive operations beyond their tested limits.

Most North American generators operate at 60 Hz, and in most other countries in the world, they operate at 50 Hz. On larger generators, the terminal voltages can vary; some typical terminal voltages are in the range of 13.8–25 kV. A generator's terminal voltage is normally step-up to the power system voltage that it connects to, via a generator transformer. Typically, for larger generator units, the system isolation breaker is located between the generator transformer and the system bus. On smaller units, the breaker may be located between the generator and the generator transformer.

A generator's voltage depends on the flux; the higher voltages require higher flux densities. The flux is limited by the field current. Higher-rated generators require more winding insulation, and the generator must operate within the insulation breakdown limits.

The main limits of a generator are the shaft torque/prime mover and the heating of the stator and rotor windings (thermal ratings). The prime mover is rated by the manufacturer and can be provided as a Watt rating. It is typically rated higher than the generator power rating, and Operations need to operate within its limits.

A manufacturer typically provides, as a minimum, a generator's power rating, in MVA at a maximum power factor, its rated terminal voltage and frequency, its Impedance, and the generator's power capability curve.

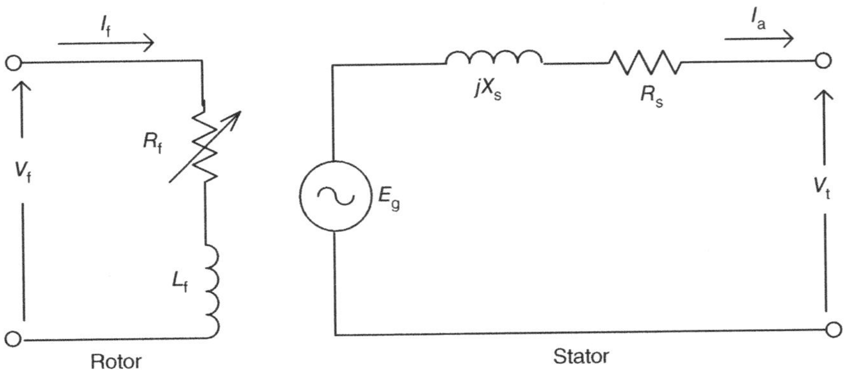

Figure 13.21 A generator's equivalent single-phase circuit.

The following is a simple generator's single-phase equivalent circuit (Figure 13.21):

The above equivalent circuit depicts a reactance jX_s, synchronous reactance (X_d), that represents the armature reaction voltage and the stator winding inductance. R_s represent the winding resistance.

The rotor field circuit is supplied with DC power; R_f represents the field current adjustment. The terminal voltage $V_t = E_g - (R_s - jX_s)I_a$. The value of R_s is much less than jX_s, and therefore; can be neglected to simplify the generator phasor diagrams.

13.6.1.1 Capability Curve

A generator's capability curve represents the generator's power limits. It is a plot of complex power $S = P + jQ$. It depicts the stator and rotor heating limits. External limits can be drawing on this plot to access operational suitability. This curve defines a boundary within which the generator can safely operate.

Figure 13.22 is a phasor diagram that has converted the generator voltages to apparent power by multiplying the voltage vectors by $3V_t/X_s$. If one rotates these phasors such that the X-axis represents power (P) and the Y-axis represents Vars (Q), it results in the following generator plot shown in Figure 13.23.

Generator capability plots are used by protection practitioners similar to how they use transformer damage curves. Some generator protection functions should be plotted and compared to the

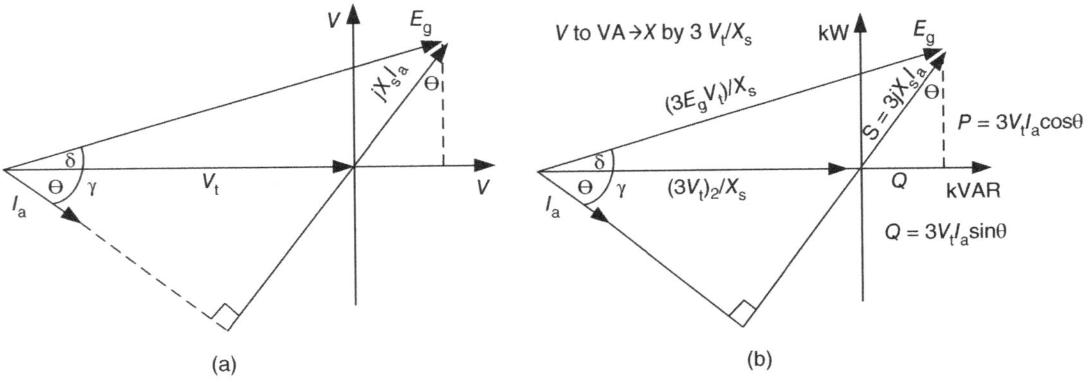

Figure 13.22 (a) Generator voltage phasor diagram. (b) Generator power phasor diagram.

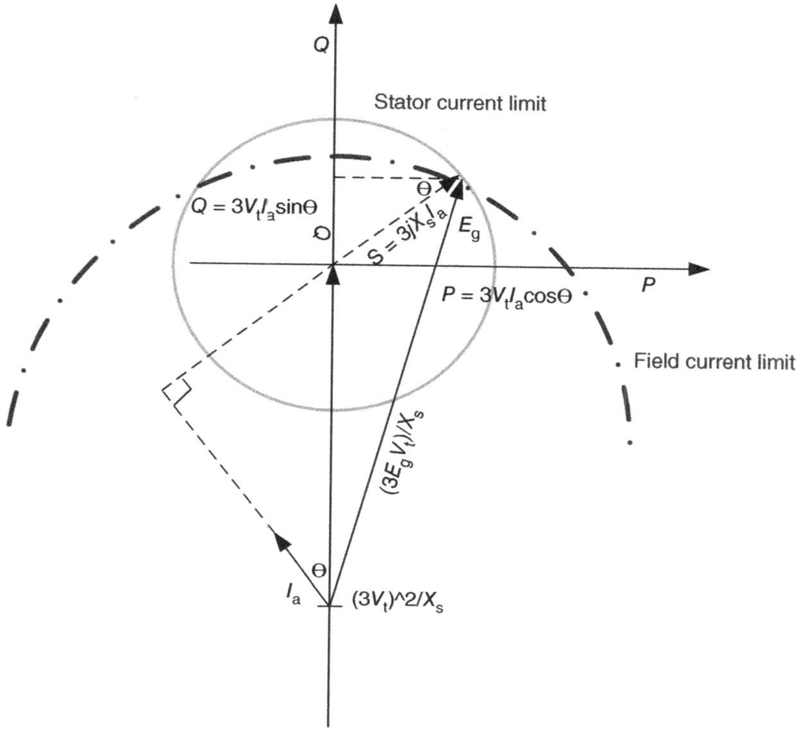

Figure 13.23 A generator's power capability plot.

capability curve to ensure settings that prevent generator plant damage, and to ensure coordination among different generator protection functions and power system protection.

Every generator has its specific capability curve and it is a function of its design. The above generator plots are intended to provide an understanding of the development of a generator capability curve based on theoretical ratings. The actual generator capability curve deviates from such a plot because of the design. A protection practitioner should use the actual generator capability curve provided by the manufacturer which is normally derived by tests.

Figure 13.24 shows an example of a typical generator capability curve:

Thermal heating is one of the most important limits for a generator. As depicted in the above generator capability curve, there are three generator heating/thermal limits, the field current limit, the stator current limit, and the stator end region heating limit.

The stator current limit is the maximum stator current that the generator can continuously carry without exceeding this heating limit. On the capability curve, this limit represents a circle with a center at (0, 0) and radius equaling the maximum rated generator's MVA. It should be noted that a generator's prime mover/shaft also has a power limit and the generator needs to operate within all of these limits.

IEEE C37.102 [1] provides short time stator/armature winding thermal capability that can be exceeded for a short period of time, during an emergency operation conditions as follows:

Time (seconds)	10	30	60	120
Armature current	218%	150%	127%	115%

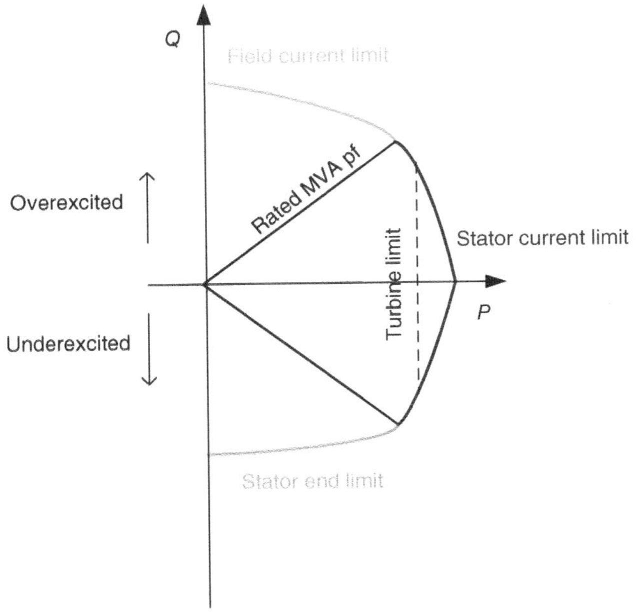

Figure 13.24 A typical synchronous generator capability curve. Source: Provided by a Manufacturer.

where 100% current is the rated current of the generator.

Similarly, the field current limit is the maximum field current that the generator can continuously carry without exceeding the heating limit. It is represented on a generator capability *P*-*Q* plot, as a circle with a center at $-V_t^2/X_s$ or $-3V_t(l-n)/X_s$ per phase; and a radius of $-(E_gV_t)/X_s$ or $-[3E_g(l-n)V_t\,(l=n)]/X_s$ on a per-phase basis.

IEEE C37.102 [1] states that the continuous field current can be exceeded for a short time as follows:

Time (seconds)	10	30	60	120
Field current (percent)	209%	146%	125%	113%

Some generator's stator end leakage flux will cause eddy currents which thereby cause heating of the stator ends. The field current can also contribute to this heating. This limit is a function of the type of generator, turns, design, etc. Some generators may not have this limit, and others may have magnetic shields installed. Therefore, a protection practitioner must use the actual generator capability curve provided by the manufacturer.

Example 13.2 ***Generator Capability Curve Example***

Given: a 234 MVA, 18 kV, 0.85 PF Generator with an $X_s = 2.94\,\Omega$.

$$V_t = 18\text{ kV L–L or }10.4\text{ kV L–N}, S = 234\text{ MVA}@31.8°$$

The maximum steady-state stator current $= S/(3V_t\text{ L-N}) = (234)\times 1000/(3\times 10.4) = 7500$ A. This defines the stator current limit and circle (Figure 13.25).

Center of the Field Current limit circle (E_g):

$$Q = -(3V_L - N^2)/X_s = -(3\times 10.4\times 1000)^2/2.94 = 110.4\text{ MVar}$$

$$E_g = V_t\,\text{L}-\text{N} + jX_sI_a = 10.4\times 1000@0 + j2.94\,(7500@-31.8) = 28.9\text{ kV}@40$$

$$\text{Radius} = (3E_gV_t)/X_s = (3\times 28.9\times 10.4)/2.94 = 306\text{ MVar}$$

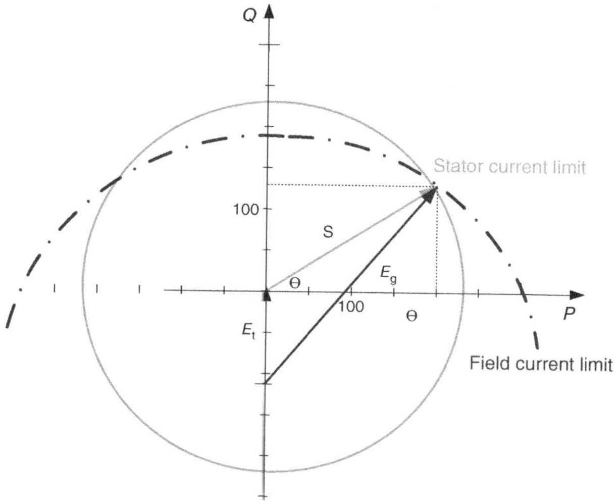

Figure 13.25 Example generator's capability plot.

13.6.2 Steady-State Stability Limit (SSSL)

A generator, upon being connected to the power system, will be synchronized to the system at 60 or 50 Hz and have the same voltage as the connecting system bus. It will provide its share of the total power system load via a network of transmission lines to loads.

A generator can provide power safely; only when it is operated within its capability. If there are sudden power system changes such as a loss of one or more transmission lines, the load requirements from that generator will change to support the loss. There is a maximum steady-state load limit that the generator can supply without losing synchronism; this is referred to as the generator's steady-state stability limit (SSSL).

The generator SSSL is the ability of the generator to adjust to load changes and to operate within a limit so to maintain synchronism.

This limit is a function of the generator voltage, impedance of the generator, the generator transformer (MOT/GSU [generator step-up transformer]), and the power system impedance that the generator is connected to.

Per IEEE C37.102 [1], SSSL can be plotted on a P-Q plot using the following:

$$\text{Center offset} = (\tfrac{1}{2}\,\text{kV}^2\text{L} - \text{L})\,(1/X_s - 1/X_d)$$

$$\text{Radius} = (\tfrac{1}{2}\,\text{kV}^2\text{L} - \text{L})(1/X_s + 1/X_d)$$

$$X_s = X_{tg} + (X_{min}),\ \text{units in primary ohms}$$

It is a circle on a P-Q plot. All impedances should be converted to the same MVA base, usually the generator's.

13.6.2.1 On a *RX* Plot

$$\text{Radius} = (X_d + X_s)/2$$

$$\text{Center offset} = (X_d - X_s)/2$$

Impedances are typically plotted in secondary ohms.

Example 13.3 ***Steady-State Stability (SSSL) Example***

For the same generator used in the above example (Figure 13.26):

On 234 MVA and 18 kV Base

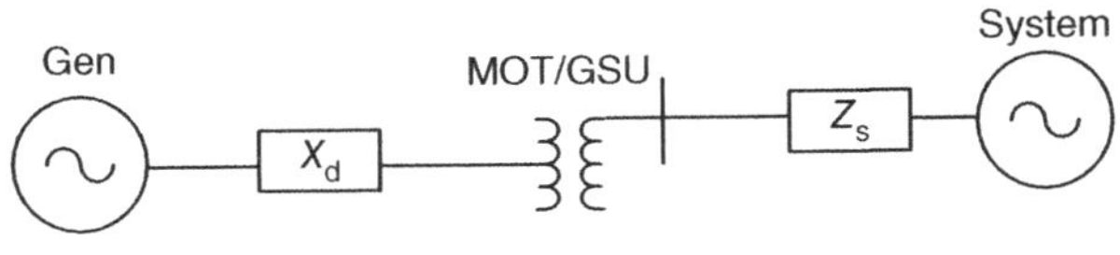

Figure 13.26 Example case one-line.

Given: GSU $X_t = 0.148$ PU
System $X_s = 0.18$ PU
Gen $X_d = 2.12$ PU

$$\text{Center offset} = (\tfrac{1}{2}\,\text{kV}^2\text{L-L})\,(1/X_s - 1/X_d)$$
$$= \tfrac{1}{2} * 1^2 * [1/(0.148 + 0.18) - 1/2.12]$$
$$= 2.77\ \text{PU}$$
$$= 2.77 * 234\ \text{MVA} = 650\ \text{MVar}$$

$$\text{Radius} = (\tfrac{1}{2}\,\text{kV}^2\text{L-L})\,(1/X_d - 1/X_s)$$
$$= \tfrac{1}{2} * 1^2 * [1/(0.148 + 0.18) + 1/2.12]$$
$$= 3.25\ \text{PU}$$
$$= 3.25 * 234\ \text{MVA} = 760\ \text{MVar}$$

From Figure 13.27, it can be seen graphically, that the steady-state stability curve coordinates with the generator's capability.

13.6.3 Over-excitation V/Hz Protection and Control System Coordination

There are no coordination issues for system faults for this protection function, and also, there are no coordination issues related to loadability for this function. However, as discussed previously, a generator AVR/excitation control system also provides over-excitation (V/Hz) protection. The control system may provide a limiter and also a control V/Hz protection. The protection practitioner should coordinate the external V/Hz protection with the control system-enabled protections, refer to Figure 13.28 below as an example.

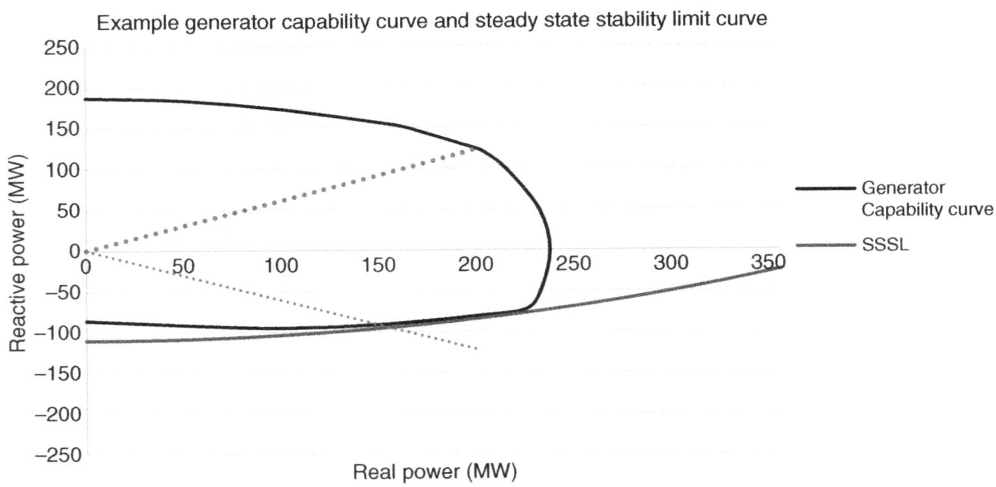

Figure 13.27 Example generator capability curve using actual overexcitation limiter (OEL) data including the SSSL.

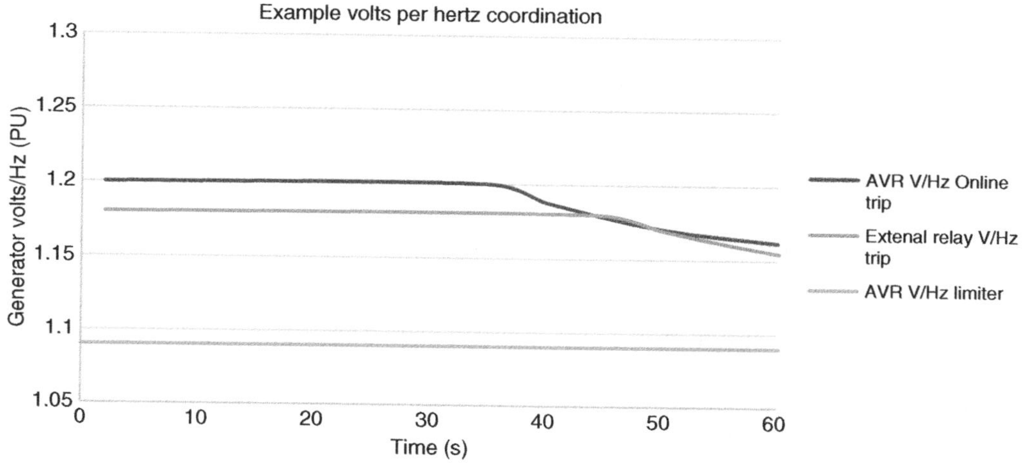

Figure 13.28 Example volts per hertz coordination.

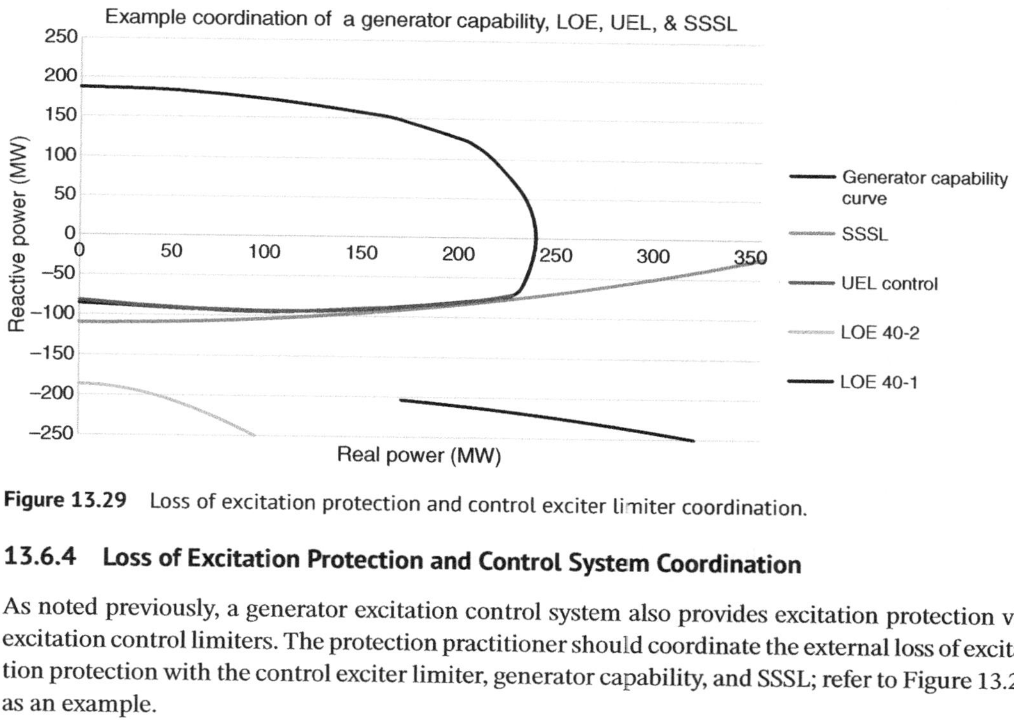

Figure 13.29 Loss of excitation protection and control exciter limiter coordination.

13.6.4 Loss of Excitation Protection and Control System Coordination

As noted previously, a generator excitation control system also provides excitation protection via excitation control limiters. The protection practitioner should coordinate the external loss of excitation protection with the control exciter limiter, generator capability, and SSSL; refer to Figure 13.29 as an example.

13.7 General Generator Tripping Requirements

- Typical protection systems only need to trip their associated breakers to isolate the faulted protected element from the power system network. Generator facilities are more complex, and tripping only the breakers that isolate the generator from the power system is not sufficient. They will also need to trip one or more of the generator's subsystems.

- A protection practitioner should consult with the generator's plant staff and develop a plan to determine what each of the individual generator protections need to trip.
- The prime mover should always be included in the trip decision. For example, in response to a winding failure, it should initiate a trip to the main breaker to isolate the generator from the system, and a trip to shut down the prime mover to prevent damage.
- If there is a problem with the prime mover that requires tripping the generator, then a trip signal should be sent to the main breaker.
- When a generator is tripped from the power system, it should also de-energize the excitation system by tripping the field breaker.
- The generator facilities' auxiliary equipment is powered from a generator service transformer/UAT connected directly to the generator terminals. The service transformer unit bus normally has an alternative feed. In case of an internal generator fault, the main breaker and perhaps the station service breaker may have to be tripped to completely isolate the generator.
- Generator auxiliary equipment failures can also occur which is mechanical in nature: bearing problems, vibration, cooling system problems, high winding temperatures, prime mover failures, etc. This equipment is continuously monitored, and in some cases, tripping may be initiated. Generator digital protection relays generally come equipped with what is known as digital inputs initiated by DC voltage. Many of these other systems that need to shut down the generator identical to the protection elements could do so via these digital inputs. Besides the large advantage in reducing cabling is the secondary advantage of the digital generator relay capturing the trip initiating event. These secondary events can be parsed along with the protection elements in time and captured as a Sequence of Events (SER).
- Generator protection systems that trip the generator and/or its subsystems should also alarm to alert an Operator as to the cause of the trip. In addition, a generator facility deploys non-trip type of alarms that also assist the Operator in determining appropriate actions.

13.8 Breaker Failure Initiation

As a general rule, all protection elements that trip a breaker also simultaneously initiate its breaker failure protection. However, for generators, not all protection elements result in sufficient current detection through the breaker to indicate the breaker has in fact failed to trip.

There may be faults or abnormal operating conditions such as stator ground, over-excitation, phase unbalance, underfrequency, and reverse power that do not produce sufficient current to pick up the breaker failure protection current detectors.

In this case, the breaker failure protection relies exclusively on the breaker pallet switches not changing state and timing out. Therefore, when using single pole generator breakers that operate independently, all three breaker pallet switches need to be paralleled. IEEE C37.102 [1].

Reference

1 IEEE Std. C37.102 - 2006, IEEE Guide for AC Generator Protection, IEEE Power Engineering Society and Sponsored by the Power System Relaying Committee.

14

Transmission Line Protection

14.1 General

Interconnected transmission systems typically consist of hundreds of transmission lines transmitting electrical power between generators and load centers. They represent the foundation of the power system. The majority of transmission line construction is of overhead type and therefore is easily susceptible to various transient and permanent faults. These faults can lead to damage to the line itself and can cause power system instability. It is of the utmost importance that protective relaying systems be capable of clearing all faults within the designed operating speed and have a high degree of dependability and security.

There are various methods of protecting transmission lines from simple overcurrent relays to impedance relays with and without the use of communications, to the most sophisticated line differential protections utilizing fiber optic communication links. The application of simple overcurrent relays is limited in scope which lead to the development of relays that operate based on measured line impedance to a fault. To be applied effectively and to decrease tripping time, impedance relays require a communication link between the impedance relays located at all of the line terminals.

An estimated 90% of all existing and very high percentage of new line protections use impedance relays also known as distance relays. The remainder mostly use line differential relays.

Protections that rely on communications are known as tele-protection or pilot-type protection. The various types of line protections utilizing all these protection methods including communications are covered in this chapter. The format of the chapter is to describe why simple and inexpensive overcurrent relays are not suitable for most transmission line networks. It then continues with an emphasis on impedance relays followed by line differential relays utilizing fiber optic communication links.

14.2 Basic Line Protection Requirements

The basic requirements for line protections are that they be reliable. This requirement normally, dictates the need for at least, two independent protection systems to clear any line fault. These two independent systems can either be two local similar systems working in parallel or be one local and the other remote with a time delay as the remote protection backs up the local protection. Either way, the failure of one system should not allow a fault to go uncleared.

Power System Protection: Fundamentals and Applications, First Edition. John Ciufo and Aaron Cooperberg.

Some of the essential criteria associated with transmission line protection are outlined below:

- Each one of the two alternate protective schemes must recognize and clear any line fault without the overall system being affected.
- One of the two schemes, and preferably both, must recognize any line fault fast enough to allow a backup system for a failed breaker to still be fast enough to clear a fault as to not cause a major power system disturbance.
- A protective scheme that will see faults beyond the area it is to protect must be selective in time taking into account failed breaker clearing of the area it is overreaching.
- Neither scheme must limit the load-carrying ability of the system (normal or emergency).
- Neither scheme must operate for stable system swings which would not cause instability.
- Either or both alternate schemes may require communication facilities to operate as intended.
- These communication facilities may be a carrier, microwave, or digital SONET, or other, depending on the performance/requirements of the scheme.
- The protective schemes must be so designed that a loss of communication or misoperation of any one communication facility will not result in tripping of multiple lines between stations for a fault in the next zone.
- The schemes must also be designed so that if both alternate schemes use communication facilities, the loss of any one facility will not prevent both schemes from operating instantaneously for any line fault.

To satisfy the first requirement, distance relay line protections generally will have overreaching zones that are set to operate for faults beyond the protected line. The overreaching zones are coordinated to respond by permission, or by not being blocked or by time delay as will be described later in the section on line protection schemes.

Establishing settings in accordance with the above requirements, in general, requires information on the characteristics of the protected line, i.e. positive sequence line impedance and zero-sequence impedance. Also, the maximum apparent impedance (ZMA) on three-phase faults (TPF) and single-line ground faults (SLG) may be necessary. The apparent impedances as "seen" by a distance relay element are required for certain line configurations and are obtained from fault studies on the power system that represents the protected line and an appropriate network of surrounding lines. The apparent impedance generally depends on the line configuration (e.g. tapped, multi-terminal), and the source impedance behind the relay terminal. Hence, so that apparent impedances be accurate, it is important that the power system model used in the fault studies be current.

14.3 Impedance Relays and Why Not Just Overcurrent Relays

Overcurrent relays are simple protective devices that restrain for low-level current typical of load while operating for higher-level current typical of a fault. They are suitable for radial lines and for lines where fault currents are predictably constant along the entire line. However, they are not generally suitable for lines in a transmission network. The reason being that in networked transmission lines, generators, and other transmission lines constantly go in and out of service, thereby, altering the level of fault current along the protected transmission line itself. In other words, fault current levels vary over time. This variation of measured electrical quantities will impact the application of protections in two-terminal and multi-terminal transmission lines. The alternative to overcurrent relays are impedance relays where line coverage known as relay reach is predictable, deterministic,

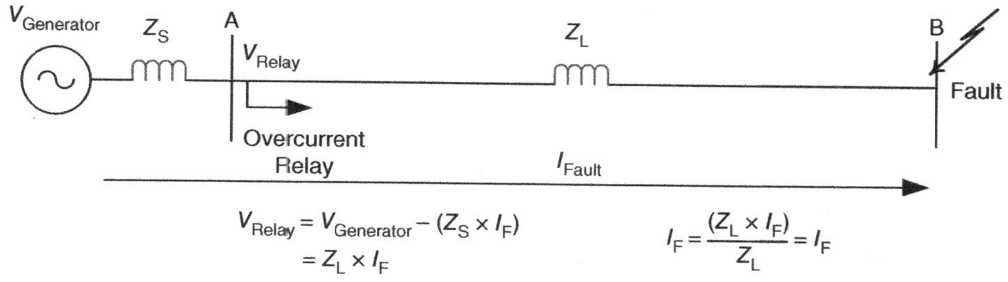

Figure 14.1 Fault detection using a simple overcurrent relay.

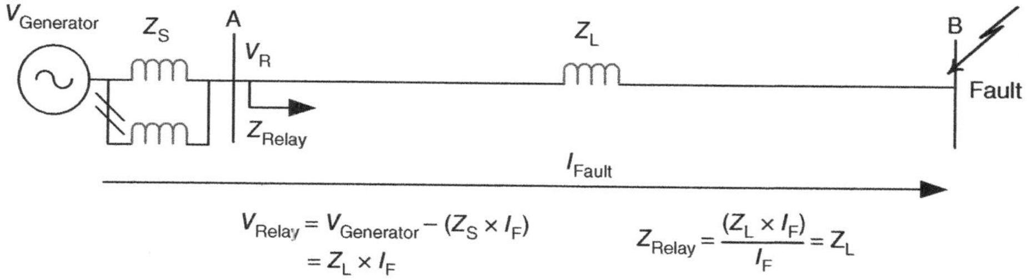

Figure 14.2 Fault detection using an impedance relay.

and as long as the line impedance is not altered by restringing conductors, it is also timeless and independent of changes to the overall system.

The source impedance (Z_S), refer to Figure 14.1, is the Thevenin equivalent impedance behind the bus where the protection relays are located. In an ideal network, where the source impedance is constant, overcurrent relays could be used whenever there is a need for an underreaching Zone 1. The fault current is simply $I_F = V/Z_L$ with Z_s not affecting the set reach of the relay, as shown in Figure 14.1. In reality, however, Z_s is not constant and changes with the number of lines in service at any given time that makes up the Thevenin equivalent network impedance, as shown in Figure 14.2. For that reason, it is not feasible to use overcurrent relays for most lines in an interconnected network where the source impedance varies and is not predictable.

Distance relays measure the impedance between the relay location and the fault, and if this impedance falls within the relay operating characteristic, which it is set to reach, it will operate to trip. The impedance to the fault is the ratio between the voltage (IZ_L) and the current (I) supplied to the relay. In other words, for a fault along the transmission line, the voltage at the relaying location is the IZ_L voltage drop of the line from the relaying location to the point of fault. Therefore, the voltage to the current ratio for this fault will be $V/I = Z_L$. For a fault inside the line section, $V/I < Z_L$ and conversely, outside the line section $V/I > Z_L$. Therefore, a measuring relay designed to operate if it measures a ratio $<Z_L$ and restrains for ratios $>Z_L$ provide a very predictable form of protection independent of source impedance consideration.

The impedance of a line is generally uniform in terms of its ohms per kilometer. The relay that measures the impedance to the fault essentially measures the distance to the fault and hence the name distance relay. The per unit impedance value of transmission lines, whether positive sequence, negative sequence, or zero sequence, is reasonably constant and determined by its geometry namely physical arrangement of conductors over ground and physical properties of the conductors and the soil below the line.

For example, a typical 230 kV line is designed and constructed to yield a measured positive sequence impedance of approximately 0.5 Ω/km. A typical 230 kV line that is 100 km long has a positive sequence impedance therefore of 50 Ω.

14.3.1 Source Impedance

Lines and or generation could be either placed into service or removed from service to alter the value of source impedance. The maximized source impedance is the term described for a source impedance of the lowest value that provides maximum fault current. The minimized source impedance is the opposite term that describes a source impedance of the highest value that provides minimum fault current. When one does a fault analysis on a modern interconnected system, the maximized source impedance is calculated for all possible lines and generators in service. This value can be arbitrarily doubled to provide a reasonable assumption of the minimized source impedance, when such information is not available.

Impedance relays in most cases eliminate the need to calculate the value of source impedance. Impedance relays are therefore optimized for use on two-ended lines where there are no additional sources of generation feeding a fault not measured by the local impedance relay. For more complex systems where there are multiple sources of generation, the source impedances become relevant.

Figure 14.3 shows a typical 230 kV terminal station. In this example, each of the diameter breakers is rated for a short circuit of 63 kA.

Since one breaker may be open by configuration prior to a fault, it must be assumed based on the 63 kA breaker rating the maximum fault infeed to a fault on the line just outside the station is the same 63 kA value. It is easy to convert this fault value to a per unit impedance that is the source impedance Z_s for a fault on the line as shown.

$$\text{Fault MVA} = \sqrt{3} \times 220\ \text{kV} \times 63\ \text{kA} = 24{,}000\ \text{MVA}$$

$$\text{Per unit source impedance } Z_s = 100\ \text{MVA}/24{,}000\ \text{MVA} = 0.004\ \text{PU}$$

An impedance of 0.004 PU is the maximized source impedance $Z_{S_{max}}$ at the station.

Consider the pre-fault configuration in Figure 14.4 for the same station. In this configuration, one entire diameter is removed from service along with one of the bus-tie breakers. Any infeed to a

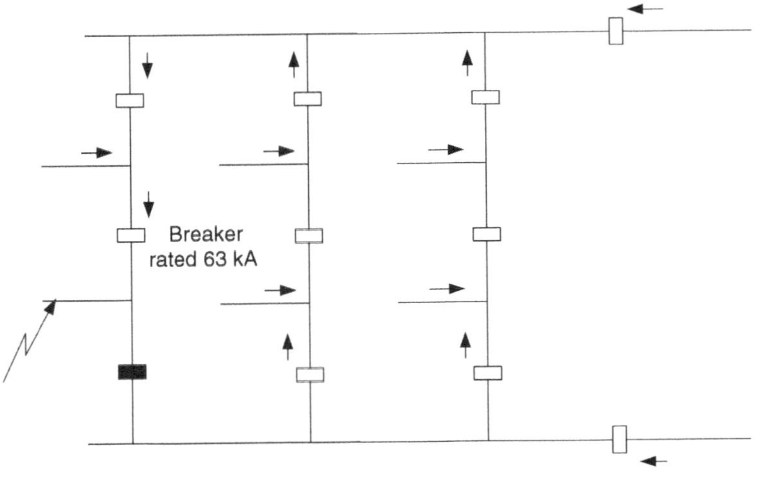

Figure 14.3 Typical 230 kV terminal station.

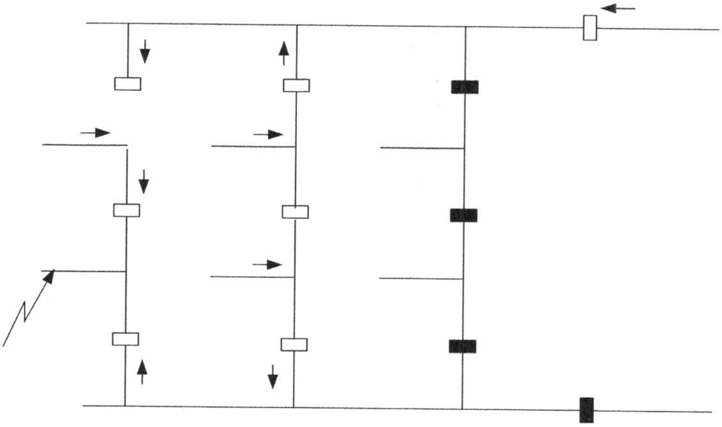

Figure 14.4 Terminal station with half the infeeds to a faulted line.

fault on the same line is essentially half of what it was previously as the source is now minimized. The accepted method at many utilities to calculate the minimized source impedance is to arbitrarily double the value of maximized source impedance.

An impedance of 0.008 PU is the minimized source impedance $Z_{S_{min}}$ at the station.

For a phase fault on the line one kilometer from the station calculated for maximum and minimum pre-fault conditions is as follows:

The impedance of a 230 kV line is 0.001 PU/km.
A line fault with maximized source is 100 MVA/(0.001 + 0.004) = 20,000 MVA or 52 kA
A line fault with minimized source is 100 MVA/(0.001 + 0.008) = 11,100 MVA or 29 kA

Figure 14.5 shows the calculation of impedance measured at the relay location when the source is maximized, i.e. all lines in service. Figure 14.6 shows the same calculation of impedance measured

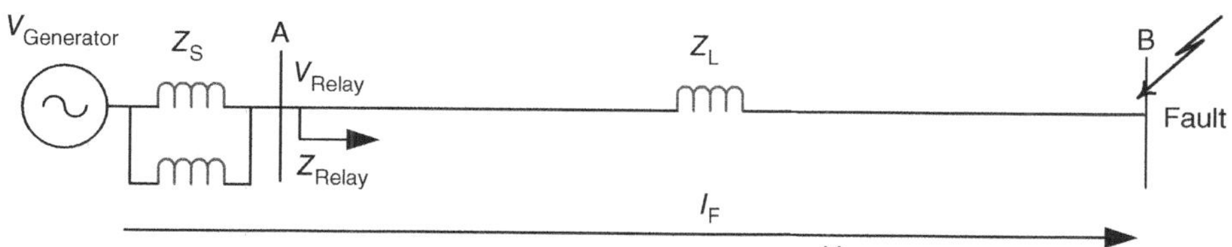

$$V_{Relay} = V_{Generator} - (Z_S \times I_F) = Z_L \times I_F$$

$$Z_{Relay} = \frac{V_{Relay}}{I_F} = Z_L$$

$$Z_{Relay} = \frac{(Z_L \times I_F)}{I_F} = Z_L$$

$Z_S = 0.004$ PU
$Z_L = 0.001$ PU
$V_{Generator} = 1.0$ PU
PU Current = 100 MVA/(√3 × 220 kV) = 262 A
I_F = 52 kA = 52 kA/262 A = 198 PU
$V_{Relay} = 1.0 - (0.004 \times 198) = 0.208$ PU
$Z_{Relay} = 0.208/198 = 0.001$ PU $= Z_L$

Figure 14.5 Maximized source impedance.

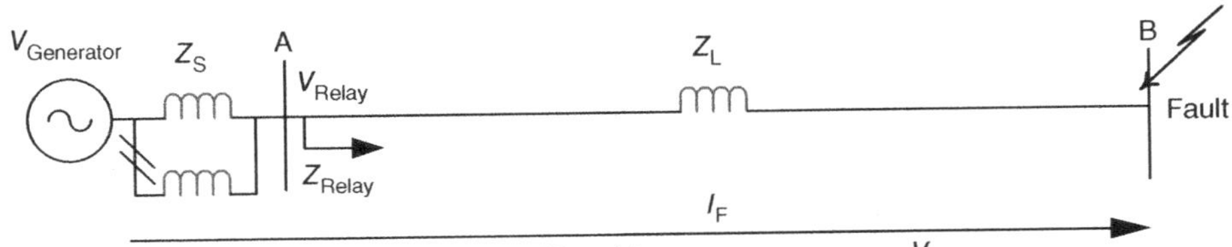

$$V_{Relay} = V_{Generator} - (Z_S \times I_F)$$
$$= Z_L \times I_F$$

$$Z_{Relay} = \frac{V_{Relay}}{I_F} = Z_L$$

$$Z_{Relay} = \frac{(Z_L \times I_F)}{I_F} = Z_L$$

$Z_S = 0.008$ PU
$Z_L = 0.001$ PU
$V_{Generator} = 1.0$ PU
PU Current = 100 MVA/($\sqrt{3} \times 220$ kV) = 262 A
$I_F = 29$ kA = 29 kA/262 A = 111 PU
$V_{Relay} = 1.0 - (0.008 \times 111) = 0.112$ PU
$Z_{Relay} = 0.112/111 = 0.001$ PU $= Z_L$

Figure 14.6 Minimized source impedance.

at the relay location when the source is minimized, i.e. half the lines out of service. In either case, the impedance measured by the relay is just the line impedance.

An overcurrent relay would be significantly affected by the deviation in source impedance (almost by a factor of 2 in this example) while a distance relay is not.

Note, that the generator voltage is maintained at 1.0 PU even under fault conditions. This is true for large interconnected systems with large amounts of generation.

14.4 Distance Relay Response to Fault Types

14.4.1 Phase Fault Response

When impedance relays were developed around 1928, the phase relays that were used to detect phase to phase and three-phase faults, measured phase to phase voltage and single-phase current. For example, a red phase relay would by convention, measure $V_R - V_W$ phase to phase voltage, and I_R phase current. As shown in Figure 14.7, this type of relay connection leaves the relay with two different reaches depending on whether the phase fault is phase–phase or three-phase. A relay calibrated to be set for a certain line impedance reach for three-phase faults would underreach by a factor of 0.866 for a phase–phase fault. For a relay calibrated to be set for a certain line, impedance reach for a phase–phase fault would overreach by a factor of 1.15 for a three-phase fault.

Some manufacturers chose to set calibration according to phase–phase faults while others preferred to calibrate for three-phase faults. Both approaches are acceptable provided that the application practitioner recognizes what type of fault the relay was calibrated for. As will be covered later, a Zone 1 distance relay is set to cover 80% of the line, and the Zone 2 relay is set at 125% of the line. Therefore, there is a material difference as to whether the relay was a Zone 1 underreaching relay or a Zone 2 overreaching relay. If it was a Zone 1 underreaching relay, the relay practitioner would not mind that the relay under reach by a factor of 0.866 for some other types of faults. Conversely, if it was a Zone 2 overreaching relay, the relay practitioner would not mind that the relay overreach by a factor of 1.15 for some other types of faults. However, it was of the utmost importance for the relay practitioner to be aware of the type of calibration and its related limitations prior to setting

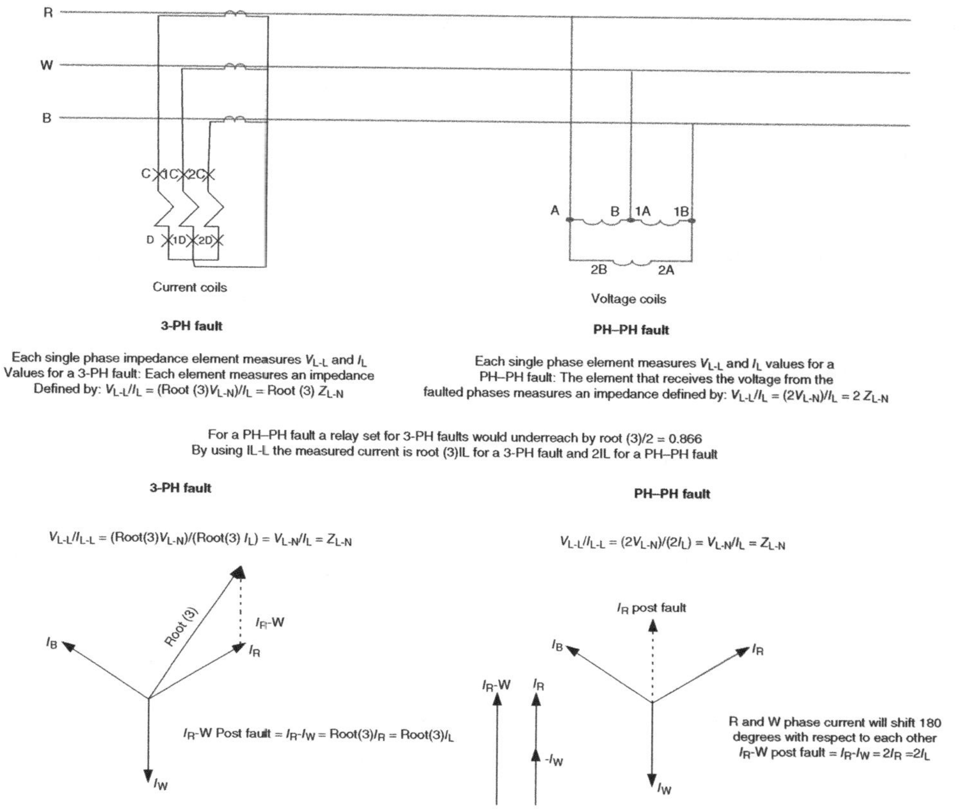

Figure 14.7 Two fault detecting techniques used for impedance relays.

these relays for Zone 1 and Zone 2. This limitation added complexity while applying impedance relays at that time.

With the objective of measuring the same distance or impedance regardless of fault type, the industry recognized around 1950 that a distance relay supplied with the voltage between the two faulted phases and the vectorial difference of their currents could achieve this. For example, for an R-W phase fault, the relay measures their currents, in this case, the relay measures $V_{\text{R-W}}/(I_{\text{R}}-I_{\text{W}}) = Z_1$, the impedance of the line between the relay and the point of fault. For a three-phase fault, the relay measures current as $\sqrt{3}\,I_{\text{L}}$ and measures voltage as $\sqrt{3}\,V_{\text{L-N}}$, the $\sqrt{3}$ multipliers cancel.

For a phase-to-phase fault, the two-phase currents that are normally out of phase by 120° shift to be 180° out of phase. Shown in Figure 14.8 is a red phase to white phase fault where I_{r} and I_{w}

Figure 14.8 Direction of currents in a phase-to-phase fault.

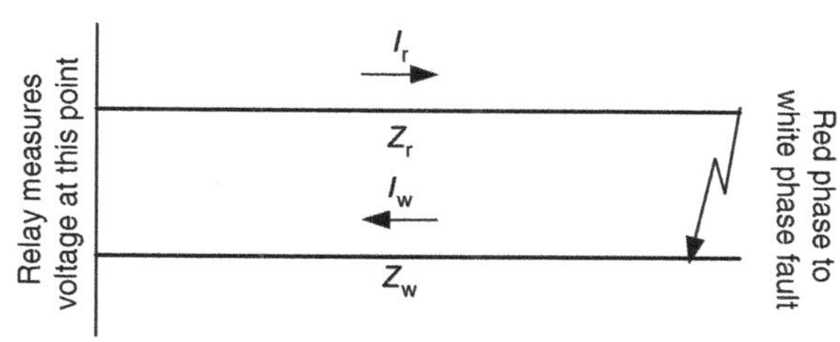

are equal in magnitude but out of phase by 180°. Since the relay measures $I_r - I_w$, the resultant is $2 \times I_r$. The voltage measured by the relay is the voltage drop across both phase impedances which is $(I_r \times Z_r) + (I_w \times Z_w)$ or two times each phase voltage drop. The relay measures current as $2\, I_L$ and measures voltage as $2\, V_{L\text{-}N}$, the two multipliers cancel.

Thus, the relay will interpret the measured impedance similarly whether to a three-phase fault or to a phase-to-phase fault.

14.4.2 Ground Fault Response

As it's advantageous for relay practitioners to set the reach of phase relays simply based on the straight positive sequence line impedance, so too it's desirable to do the same for ground distance relays. For a line to ground fault, the fault current flows through the line to the fault and then returns to the source terminal via two ground paths. The fault current will flow along the metallic return paths, such as sky wires or distribution feeder neutral conductors and through the ground itself. The sky wire impedance is a fixed value but the ground resistivity is not constant and changes with the season and weather conditions as well as where the fault is on the line. The ground resistivity would be very different in winter compared to a wet summer.

The method used to allow relay practitioners to set ground distance relays based on straight positive sequence line impedance is to supply the relay with phase to neutral voltage and phase current plus a pre-determined proportion of the residual current. The proportion of residual current, known as Residual Compensation, is a function of the protected line's positive, negative, and zero-sequence impedance.

Consider a single line-to-ground bolted fault, refer to Figure 14.9.

Z_1, Z_2, and Z_0 are the Positive, Negative, and Zero-Sequence Line Impedance
I_1, I_2, and I_0 are the Positive, Negative, and Zero-Sequence Line Currents
V_{PN} is the Phase to Neutral Voltage measured at the Relay Point
I_P is the phase fault Current

$$I_P = I_1 + I_2 + I_0$$

$$V_{PN} = I_1 Z_1 + I_2 Z_2 + I_0 Z_0$$

For transmission lines, $Z_1 = Z_2$

$$V_{PN} = Z_1\,[I_1 + I_2 + CI_0] \text{ where } C = Z_0/Z_1$$

$$V_{PN} = Z_1\,[I_1 + I_2 + I_0 - I_0 + CI_0]$$

$$V_{PN} = Z_1\,[I_P + (C-1) \times I_0]$$

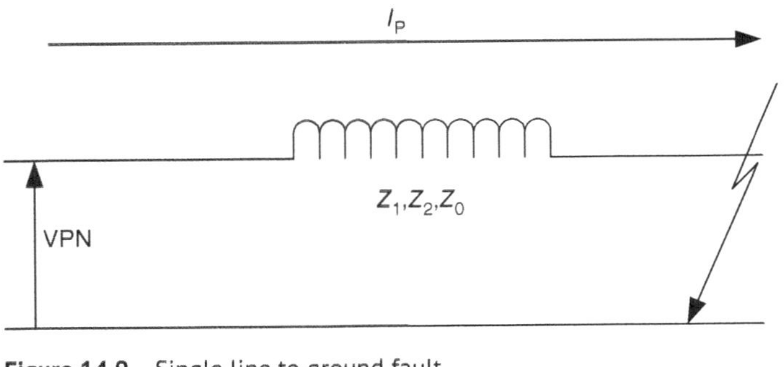

Figure 14.9 Single line to ground fault.

Since $C = \dfrac{Z_0}{Z_1}$ and $3I_o = I_{\text{Residual}}$

$$(C-1)I_0 = \left[\frac{Z_0}{Z_1} - 1\right] \times \frac{I_{\text{Residual}}}{3}$$

$$(C-1)I_0 = \left[\frac{Z_0 - Z_1}{Z_1} \times \frac{1}{3}\right] I_{\text{Residual}}$$

$$(C-1)I_0 = \left[\frac{Z_0 - Z_1}{3Z_1}\right] I_{\text{Residual}}$$

Solve for Z_1:

$$Z_1 = \frac{V_{\text{PN}}}{I_{\text{p}} + (C-1) \times I_0}$$

$$Z_1 = \frac{V_{\text{PN}}}{I_{\text{p}} + k\, I_{\text{Residual}}} \quad \text{where } k = \frac{Z_0 - Z_1}{3Z_1}$$

Thus, a relay that is supplied with a line to neutral voltage and line current plus a proportion of the residual current will measure and can be set for the straight positive sequence impedance of the line. For electromechanical relays, the proportion of zero-sequence current that is added to the phase current is done using a tapped primary winding of a standard auxiliary current transformer as shown in Figure 14.10.

The combination of phase current measured by the relay is the vectorial sum of the phase current and a fraction of zero-sequence residual current. The multiplying factor k does have magnitude as well as angle. There is no method of including the angular value of k with electromechanical relays. However, digital relays have no such restriction and the angular value of k is computed by the relay and a more precise method of zero-sequence compensation is achievable as shown in Example 14.1 for a typical digital distance relay.

Zero-Sequence Compensation Factor Settings

Zone 1. Zero-Sequence Comp. Factor 1 Magnitude: (0–4 unitless) = **0.318**
Zone 1. Zero-Sequence Comp. Factor 1 Angle: (±180°) = **−3.8**
Zones 2, 3, and 4. Zero-Sequence Comp. Factor 2 Mag.: (0–4 unitless) = **0.318**
Zones 2, 3, and 4. Zero-Sequence Comp. Factor 2 Angle: (±180°) = **−3.8**

$$Z_1 = 0.000344 + \text{j}0.00707 \text{ PU} = 0.0070784\underline{/87.2^\circ}\text{PU}$$

$$Z_o = 0.001116 + \text{j}0.013781 \text{ PU}$$

$$Z_o\text{–}Z_1 = 0.000772 + \text{j}0.006771 \text{ PU} = 0.006755\underline{/83.4^\circ}\text{PU}$$

$$k = (Z_o - Z1)/3Z_1 = 0.006755/83.4^\circ/(3 \times 0.0070784\underline{/87.2^\circ}\text{PU})$$

$$k = 0.318\underline{/-3.8^\circ}$$

Example 14.1 ***Typical Zero-Sequence Compensation Setting for a Digital Relay***
The above example is for a shorter 7 km line where the ratio of zero-sequence impedance to positive sequence impedance is approximately 2:1. However, for most long lines typical of most 230 kV networks, the ratio of zero-sequence impedance to positive sequence impedance is 3:1.

For most lines found in transmission networks except for very short lines: $Z_0 = 3Z_1$.

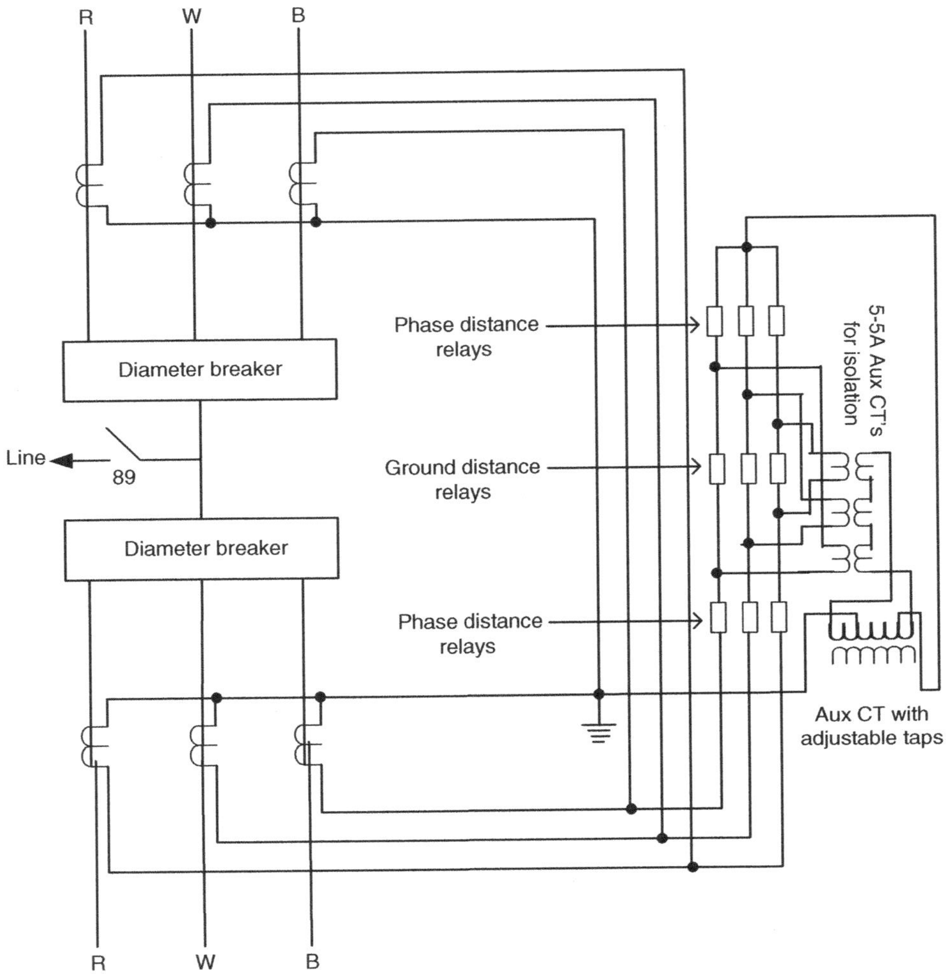

Figure 14.10 Zero-sequence compensation with electromechanical relays.

Therefore

$$k = \frac{3Z_1 - Z_1}{3Z_1}$$

$$= \frac{(3-1) \times Z_1}{3Z_1}$$

$$= \frac{2}{3} = 66\% \text{ Compensation}$$

Distance relays were first used in 1928 and from that time until about 1950 relay practitioners would set ground distance relays for the positive sequence line impedance multiplied by factor k plus some margin. Since 1950, which defines the modern era of distance relay design, a proportion of the zero-sequence ground return current according to factor k is added to the phase current of the faulted phase. With the advent of digital relays magnitude and phase angle are included in the calculation. In either case a more precise method of ensuring that the ground return impedance was adaptively represented regardless of the season.

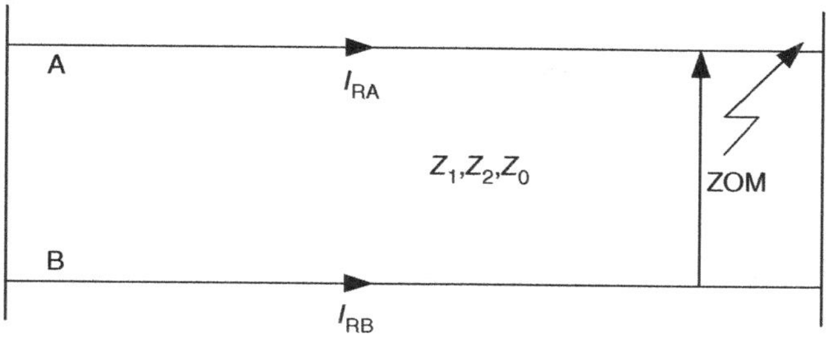

Figure 14.11 Effects of mutual impedance between two parallel lines.

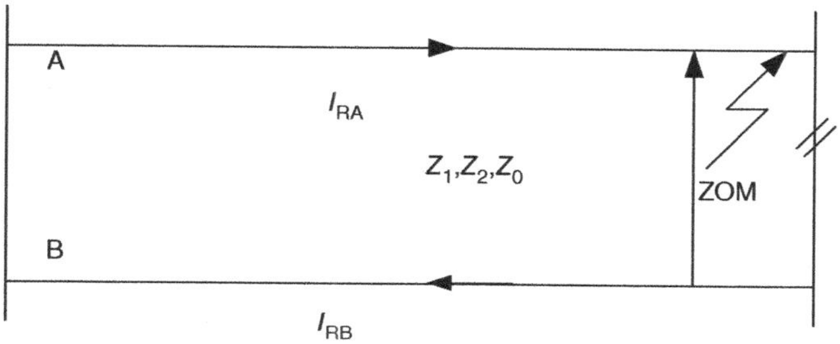

Figure 14.12 Effects of mutual impedance with remote breakers open.

The above analysis neglects the effect of mutual impedance between two parallel lines, refer to Figure 14.11.

The voltage seen by a ground distance relay for a fault on line A is $V_{\text{R-N}}$ as follows:

$$V_{\text{R-N}} = I_{1A}Z_1 + I_{2A}Z_2 + I_{0A}Z_0 + I_{0B}Z_{0M}$$

The voltage seen by the relay is either larger or smaller depending on the direction of the zero-sequence current in line B. This factor will cause the ground distance relay to underreach when the zero-sequence ground current is flowing in the same direction in both lines which is the normal case when remote breakers are all closed. It will also cause it to overreach when the currents in the two lines are flowing in opposite directions which happens when remote end breakers are open. This effect is shown in Figure 14.12.

The voltage seen by a ground distance relay for a fault on line A is $V_{\text{R-N}}$ as follows:

$$V_{\text{R-N}} = I_{1A}Z_1 + I_{2A}Z_2 + I_{0A}Z_0 + (-I_{0B}Z_{0M})$$

This overall phenomenon can contribute significantly to the underreaching effect as the fault currents are equal in both the faulted line and the healthy line. However, the overreaching effect is less pronounced as the fault current in the healthy line has to travel through a much larger amount of line impedance. Simple reach adjustments are used to accommodate for this except for when generation is directly tapped to lines.

14.5 Apparent Impedance

When a relay measures some impedance other than the impedance of the equipment intended, what is being measured is the apparent impedance. To cover the intended protected equipment, an impedance relay must be set based not only on the equipment impedance but on the apparent

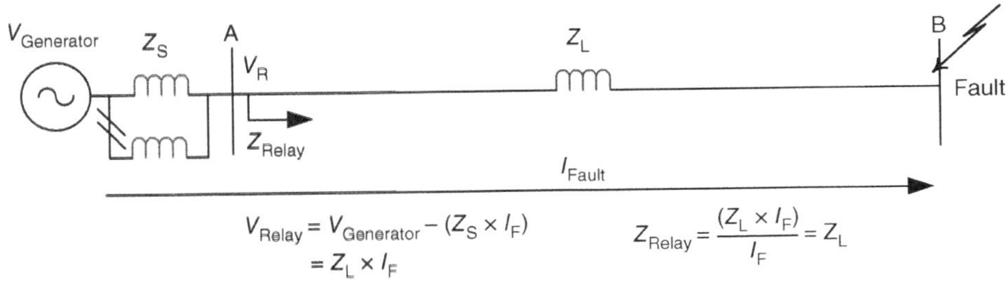

Figure 14.13 Distance relay impedance measurement in a simple system.

impedance. Almost always the apparent impedance is larger than would have been measured if the apparent impedance did not exist.

For a fault on a transmission line, a distance relay will measure impedance equal to the line positive sequence impedance, provided there are no other sources of fault current between the line terminal at which the relay is located and the fault. The distance relay measures impedance by comparing the voltage drop between its location and the fault with the current at the relay as shown in Figure 14.13.

Impedance relays in most cases eliminate the need to calculate the value of source impedance (Z_s). Impedance relays are therefore optimized for use on two-ended lines where there are no other additional sources of generation feeding a fault current not measured by the local impedance relay. For more complex systems where there are multiple sources of fault current, the source impedances become relevant. This is the basis for the phenomenon known as apparent impedance. When setting impedance relays, it is important to take into account the effect of apparent impedance.

14.5.1 Example of Apparent Impedance

Refer to Figure 14.14, the actual line impedance from the relay terminal (Terminal A) to the fault is not always the impedance seen by the relay. This is due to the third line terminal (Terminal C) tapped (T point) to a line being an additional source of current for a line fault. Current will be supplied to a fault that occurs on the line section beyond the tap of Terminal C through both Terminal A and Terminal C. The voltage drop resulting from the input of fault current from each of these sources into the common section of the line will be seen by the distance relay at Terminal A.

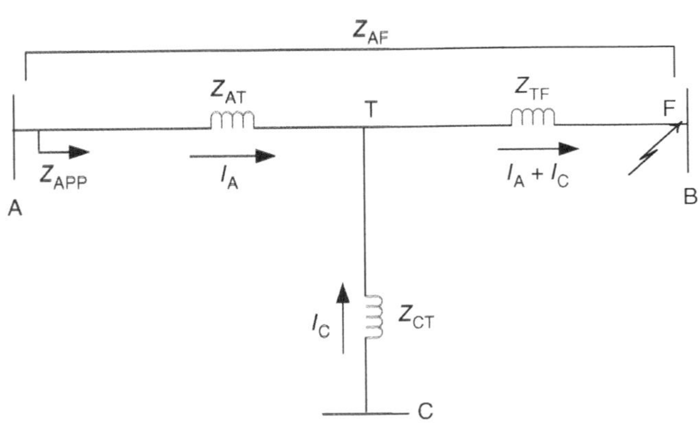

Figure 14.14 Effect of infeed from the third terminal on a three-terminal line.

Since the current input from Terminal C is not applied to the relay at Terminal A, the impedance seen by this relay is higher than the actual impedance from Terminal A to the fault. The relay will underreach, that is for a given relay setting, the relay does not cover the same length of line it would have if the additional current source were not present.

14.5.2 Derivation of Apparent Impedance

Referring to Figure 14.14

The voltage at Terminal A (relay location) to fault F, with Terminal C open:

$$V_A = V_{AT} + V_{TF} = I_A Z_{AT} + I_A Z_{TF} = I_A(Z_{AT} + Z_{TF}) = I_A Z_{AF}$$

Impedance as measured from Terminal A:

$$Z_{AF} = V_A/I_A = (I_A Z_{AF})/I_A = Z_{AF}$$

Z_{AF} = the actual line impedance that appears at the distance relay terminal

The voltage at Terminal A (relay location) to fault F, with Terminal C closed:

$$V'_A = V_{AT} + V_{TF} = I_A Z_{AT} + (I_A + I_C) Z_{TF}$$

Impedance as measured from Terminal A:

$$Z_{APP} = V'_A/I_A = (I_A Z_{AT})/I_A + (I_A + I_C) Z_{TF}/I_A = Z_{AT} + Z_{TF} + (I_C/I_A) Z_{TF}$$

$$Z_{APP} = Z_{AF} + (I_C/I_A) Z_{TF}$$

Z_{app} = the apparent line impedance that appears at the distance relay terminal $\frac{I_C}{I_A}$ is defined as the infeed factor for Terminal A which is the ratio of tapped infeed current to relay location infeed current for a relay located at Terminal A.

The effect of the fault infeed I_C from Terminal C increases the measured impedance at Terminal A which is larger than the actual line impedance. Since the measured impedance is larger than the actual line impedance, it is known as the apparent impedance.

Whenever a distance relay measures more impedance to a fault than the actual line impedance, it will not operate even though it should. This tendency to not operate is known as underreaching.

The underreaching tendency at Terminal A with infeed from Terminal C is a function of the ratio, $\frac{I_C}{I_A}$.

The apparent impedance for a fault at the same location but viewed from Terminal C is as follows:

$$Z_{APP} = Z_{CT} + Z_F + (I_A/I_C)/Z_{TF} = Z_{CF} + (I_A/I_C) Z_{TF}$$

14.5.3 Apparent Impedance Effect on Distance Relay Settings

To overcome this effect, the relay setting has to be calculated in terms of the maximum impedance as measured by the distance relay located at the line terminal. The required setting providing complete coverage of the line can be much larger than the setting necessary without the three-terminal configuration.

For apparent impedance to have meaning, it must be calculated for the highest or worst-case value of apparent impedance. Setting a distance relay for the maximum apparent impedance with sufficient margin guarantees the relay will always see the intended fault. Maximum apparent impedance is also referred to in some utilities as ZMA (Impedance Maximum Apparent). For the purpose of abbreviation, that's what it will be referred to further in this book.

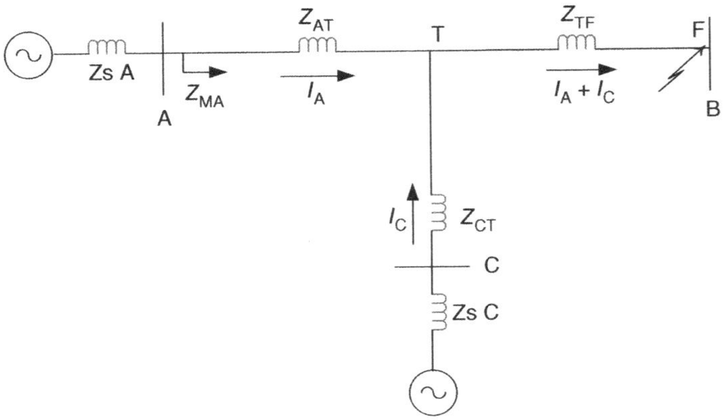

Figure 14.15 Source impedances and effect on apparent impedance.

Whenever apparent impedance may exist, distance relay reaches are based on approximated ZMA. The ZMA is not a true value but an approximated value by arbitrarily maximizing and minimizing the various source impedances.

Source impedances are maximized or minimized to generate the maximum infeed effect. The longest T length determines the fault location. As an example, for Terminal A, with a fault at Terminal B, assuming Z_{TB} is larger than Z_{CT}, the source impedance at A should be maximum (minimum system fault infeed), and at Terminal C, the source impedance should be minimum (maximum system fault infeed). This will result in the largest infeed factor (Figure 14.15).

A similar conclusion may be arrived at when considering a phase-to-ground fault provided the Z_{L0}/Z_{L1} ratio for each branch of the protected line is the same. The infeed effect for phase-to-ground faults is very much a function of the system grounding and needs to be determined by conducting system fault studies for the specific application.

14.5.4 Maximum Apparent Impedance Calculation Example

For a phase fault at Terminal C, find the value of the Maximum Apparent Impedance ZMA as seen by a distance relay located at Terminal A as shown in Figure 14.16.

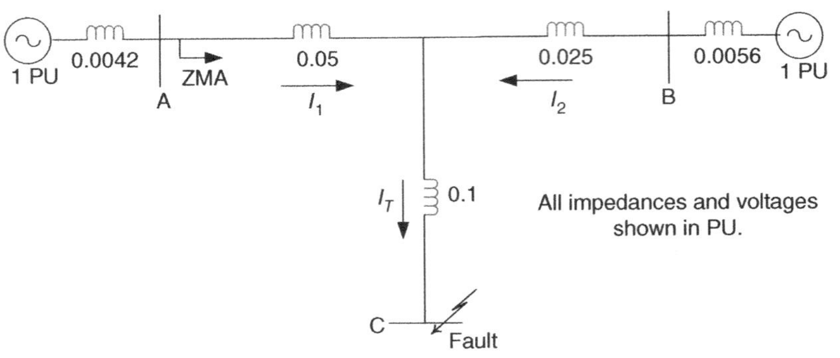

Figure 14.16 Example three-terminal line with one terminal faulted.

$Z_{SA} = 0.0042$ PU and $Z_{SB} = 0.0056$ PU need to minimize Z_{SA} and maximize Z_{SB} to minimize Z_{SA} multiply it by 2 which is the same as taking half the lines feeding the bus at Terminal A out of service. Z_{SA} (minimized = 0.0084 PU)

$$Z_T = 0.1 + [(0.0084 + 0.05)/(0.0056 + 0.025)]$$

$$= 0.1 + [0.0584/0.0306]$$

$$= 0.1 + 0.02 = 0.12 \text{ PU}$$

$$I_T = 1.0/0.12 = 8.33 \text{ PU}$$

$$I_1 = [8.33(0.0056 + 0.025)]/[(0.0056 + 0.025) + (0.0084 + 0.05)]$$

$$= 8.33 \times 0.0306/0.089 = 2.86 \text{ PU}$$

$$V_A = 1.0 - V_{ZSA} = 1 - (2.86 \times 0.0084) = 1.0 - 0.024 = 0.976 \text{ PU}$$

$$Z_{MA} = V_A/I_1 = 0.976/2.86 = 0.341 \text{ PU (2.3} \times \text{the line impedance Terminal A to C)}$$

14.5.5 Apparent Impedance and Paralleled Conductors

This type of line arrangement is occasionally utilized to reduce losses on heavily loaded transmission interfaces in the power system. In this type of conductor arrangement, the line for each phase consists of twinned conductors that are paralleled at the terminals only as shown in Figure 14.17. Twinned conductors can present difficulties in setting distance relays in a direct underreaching transfer trip/permissive overreaching transfer trip (DUTT/POTT) scheme. Scheme types are described in Section 14.7.2.

The Zone 2 ground reach is set at 125% of maximum apparent impedance that in this case can be exceptionally large. With the twinned conductor arrangement, the maximum apparent impedance occurs with one of the conductors broken and on the ground near the local terminal. The impedance seen by the relay at the local terminal is magnified by the infeed from the remote terminal (similar to the infeed effect on three-terminal lines). In some cases, the maximum apparent impedance may be too large for the distance element to accommodate. The required reach could be reduced with a tie loop at the midpoint of the line as shown in Figure 14.18. Where the apparent impedance is significantly large even with a mid-point tie, another two ties may be added on either side half-way between the center tie and each terminal. The application of line current differential protection is a good alternative to the impedance measuring protections as that technology does not involve the measurement of impedance or apparent impedance.

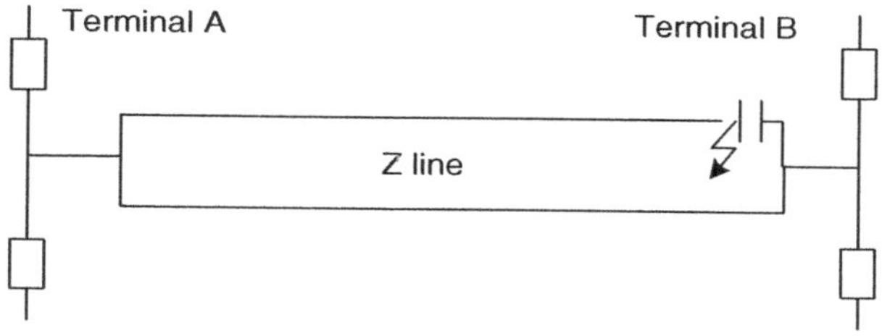

Figure 14.17 Paralleled conductors (only one phase shown).

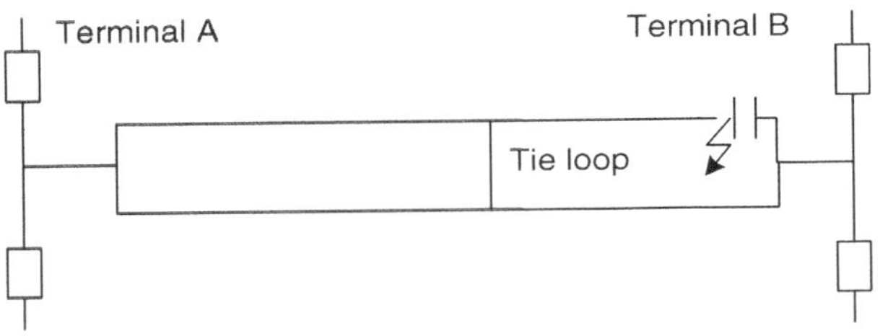

Figure 14.18 Paralleled conductors with a mid-point tie loop.

14.5.6 Load substations and Apparent Impedances

There are four important issues all relating to apparent impedances involving load substations. The first is how the presence of tapped load substations affects apparent impedance seen by the line protections. The second is how a line backup protection at a load substation is affected by apparent impedance when the load substation is operated with the LV bus-tie breaker closed as is the situation at some utilities. The third is how a long Zone 2 reach due to apparent impedance on the line may create a situation where that Zone 2 could possibly reach into the LV side of other tapped load substations. The fourth is load substations supplied at the end of long line taps creating large apparent impedances.

14.5.6.1 Presence of Tapped Load substations

Refer to Figure 14.19 showing the effect of a tapped load substation on the apparent impedance seen by the line protection at the terminal station. In this example, the line protection at Terminal A contributes most of the short circuit current to a phase fault on the line. Nevertheless, some short circuit current will back feed through the load substation thereby creating apparent impedance at the terminal station. Since the line impedance is much smaller than the transformer impedances at the load substation, the apparent impedance is relatively very small.

The short circuit current splits according to Kirchhoff's current law. A good proven rule of thumb is to add 2.5% per load substation to the actual line impedance to cater for this back feeding and apparent effect. The closer the load substation is located to the terminal station where the measurement takes place, the higher the apparent impedance effect. The opposite is also true, the closer the load substation to the remote terminal the less the apparent impedance effect. Where there are multiple load substations tapped to the line, each one adds a percent increase to the overall apparent impedance. The apparent impedance can be derived by computer fault programs or as an estimation, can be calculated using this simple rule of thumb.

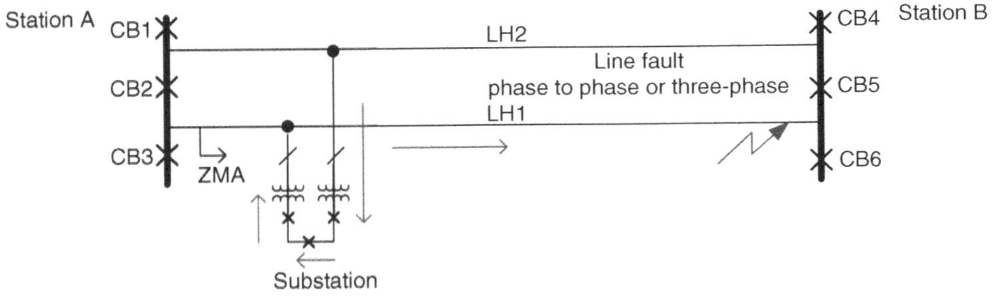

Figure 14.19 Effect of a tapped load substation on line protection.

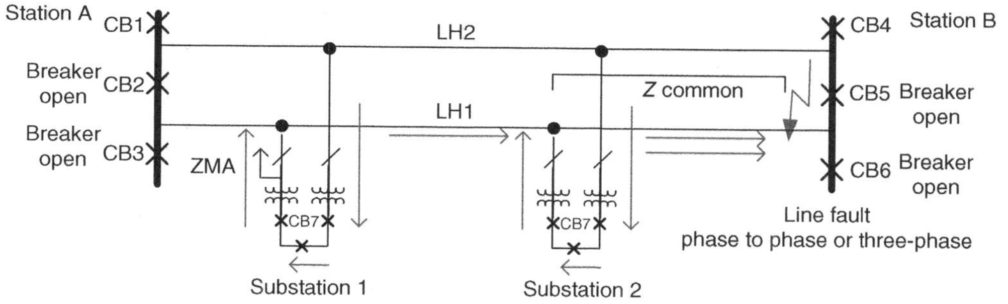

Figure 14.20 Line phase backup protection using distance relays.

14.5.6.2 Line Backup Protection at Load substations

When distance relays are used for phase line backup protection at load substations where the bus-tie breaker is operated normally closed, the apparent impedance must be taken into account. Depending on the number of other load substations tapped to the same line, the apparent impedance can be significant and only exists following the opening of the line breakers.

Referring to Figure 14.20, the impedance seen by the relay at load substation 1 must contend with the voltage drop it measures not relating to its own back feed but relating to the back feed from load substation 2. The voltage drops across the common impedance shown will cause a significant increase in measured impedance seen from load substation 1.

A common practice is to arbitrarily assume all the tapped load substations are located at the load-substation farthest from the fault, in this case, load substation 1. Multiply the line impedance from that farthest load substation to the fault by the number of load substations tapped to the line. In this example, the maximum apparent impedance (ZMA) is no greater than $2 \times Z_{\text{Line}}$ from load substation 1 to the fault at Station B.

14.5.6.3 Large Zone 2 Reaches Seeing into the Load Substation LV Side

Large Zone 2 reaches on occasion see into the LV side of load substations. When this occurs, the load substation usually sends a blocking signal to the terminal station line protections to not operate.

Referring to Figure 14.21, a relatively large ZMA at Station A for a fault at Station C causes a large Zone 2 reach. This large Zone 2 reach would see through the load-substation transformer to an LV fault. A typical large load-substation transformer rated at 100 MVA is approximately 0.2 PU which is similar in magnitude to about 200 km of 230 kV line at 0.001 PU/km. A long 200 km line with some

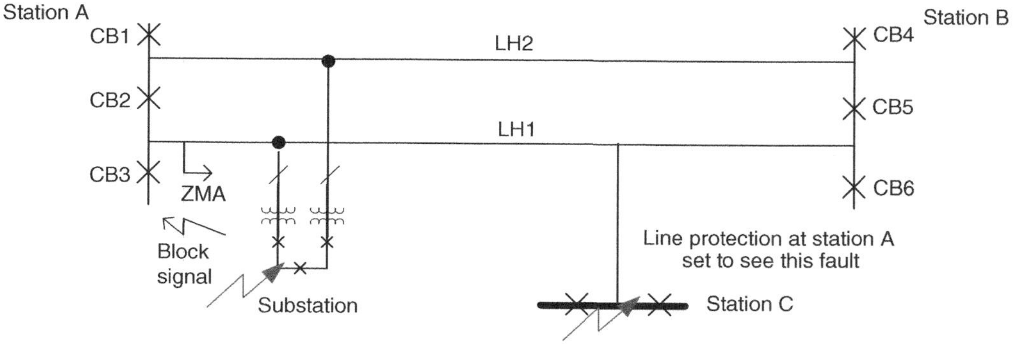

Figure 14.21 Line protection Zone 2 reaches into a load-substation LV side.

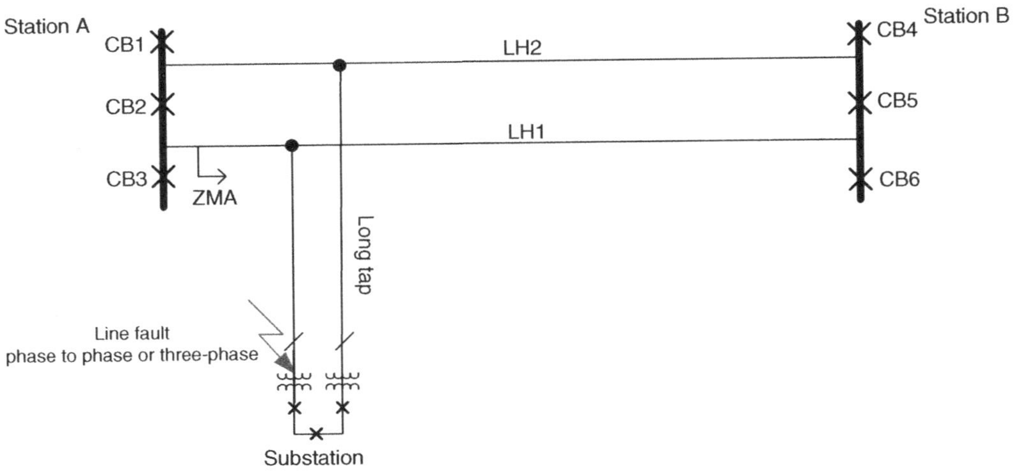

Figure 14.22 Fault at the end of a load-substation long tap.

apparent impedance is therefore vulnerable to having its zone 2 seeing into the load-substation LV side.

14.5.6.4 Load Substations Supplied by Long Taps

Referring to Figure 14.22, where a load substation is supplied at the end of a long tap that is faulted as shown, the ZMA at the terminal station could be significantly higher than the line impedance. In this case, the Zone 2 reach must be set based on this large apparent impedance. There is however a problem created that could lead to this reach seeing through the load-substation transformer. The Zone 2 reach underreaches the load-substation transformer LV with infeed from Station B. Should this infeed disappear for when the line breakers at Station B are operationally open, the Zone 2 no longer underreaches. In other words, the Zone 2 reach becomes even much longer as the reach reducing effect of the infeed from Station B is no longer there. A solution would be to use blocking signals generated by the LV protections at the load-substation to block the Station A LH1 Zone 2 from operating for this condition.

14.6 Redundancy/Backup

14.6.1 Need for Protection Backup

Protection systems can and do fail from time to time. As it is of the utmost importance that all power system faults be isolated by the protections, it is imperative that all protection systems be backed up. A failure of protection equipment is a contingency that must be catered for, and therefore, redundant or backup protection is provided. It is not expected, an industry accepted contingency, which the primary and backup protection would ever fail for the same fault condition. That is to say, double contingency failure is not reasonable, and need not be generally catered for.

14.6.2 Remote and Local Protection Backup

There are two possible methods to provide acceptable backup to primary protection. One is known as remote backup, and the other is known as local backup. Both are effective in isolating a fault should the primary protection fail for whatever reason.

Remote backup is defined as providing backup up to local primary protection with one that is located at a remote location or locations feeding into the local fault. This or these backup protections would be time coordinated to allow the local primary protection a chance to operate.

Local backup refers to the duplication of the primary protections themselves, such that if one fails the other is still functioning to carry out the necessary fault isolation process. Since each of these systems is functionally identical, they are seen as being redundant to each other when there is no equipment failure. It is for that reason that local backup is also known as dual redundancy.

Historically, most transmission utilities in North America subscribed to the concept of remote backup until the 9 November 1965, blackout shut down half of North America due to remote backup protections operating without discrimination. Following that event, many utilities turned to the application of dual redundancy or local backup.

The key incentive to initially use remote backup is that not only are the local protections backed up but also the breakers so that no discrete breaker failure protection is necessary. Breaker failure protections by definition cannot work without elaborate and expensive communication systems between terminals of transmission stations. With the removal of remote backup protections, breaker failure protections along with communications must be installed. Also, with the removal of remote backup protections, local backup is the only method available to provide protection backup.

The key difference between these two backup approaches is therefore the willingness to spend money on a sophisticated communication system that itself requires dual redundancy.

14.6.2.1 Remote Backup

For remote backup, all stations in the network typically have single high-speed protection for each power system apparatus protected locally. The backup protections are located remotely and are timed to allow the remote high-speed protection to operate first. The local protection equipment typically is described as follows:

- Only one set of current transformers required
- Single protection per protected apparatus
- Single station battery
- Single trip coils for circuit breakers

14.6.2.2 Local Backup

For local backup, the two groups of protections are self-contained and independent of each other. Each group is capable of providing complete high-speed protection for all classes of line faults. Separation of the two groups of protections is achieved by:

- Separate current transformers for each protection group
- Dual protections mounted separately
- Separate batteries or DC sources, for each protection group
- Routing of wiring for a reasonable separation between the two groups
- Dual trip coils for circuit breakers

One may postulate an A group protection could fail but never that the A group protection fails along with the failure of the B group protection simultaneously for the same fault. Breaker failure is not duplicated since one would not expect breaker failure protection apparatus to fail simultaneously with the actual breaker failing. This adheres to the basic underlying premise that double contingency failure of protection equipment is not anticipated as a credible possibility, and is therefore, not catered for in general.

14.7 Tele-Protection (Also Known as Pilot-Protection)

Remote backup does not require communications since the remotely located backup protections are time coordinated with the local primary protections. Local backup does require communications to trip remote breakers and to provide logic to the schemes employed.

It is the introduction of high-speed and reliable communications that allows for the use of local backup and the deployment of dedicated breaker failure protections.

14.7.1 Tele-protection Architecture

To achieve dual redundancy matching the protections, geographically diverse main and alternate communication routes via microwave in a ring-type topology were established by many utilities after the 1965 blackout. Many utilities adopted multiple frequency shift tones for reliability. Six tones for reliability were typically multiplexed onto microwave communications. Line protections used these tones for transfer trip, permission, and blocking signal logic. Local breaker failure protection very often used the same tones to transfer trip remote breakers to isolate a local breaker failure condition.

Microwave communications are now mainly phased out being replaced with fiber optic-based digital communication systems. Regardless of communication medium, the fundamentals of geographically diverse main and alternate communication routes in a ring-type topology must be maintained.

14.7.2 Protection Scheme Types for Local Backup

Remote backup does not require any specific protection scheme type to work correctly. This is valid provided that the backup protection elements usually in the form of what is known as a remote timed Zone 3 are set to cover all apparent impedance conditions.

For local backup to work correctly, various protection scheme types have been developed and employed by utilities in one form or another.

14.7.2.1 Direct Underreaching Transfer Trip (DUTT)

Direct underreaching transfer trip is an established scheme type as part of an overall scheme solution. It was not meant to be used alone but used in conjunction with other scheme types as it did not cover the operating contingency of line end open. This scheme is meant to be duplicated with one in the A and one in the B protection group.

14.7.2.1.1 Underlying Principles

The direct underreaching transfer trip scheme uses the Zone 1 distance elements usually set at approximately 80% of the line for the phase elements and 75% of the line for the ground elements. Some utilities set it as high as 85% for the phase elements.

- Once asserted, the Zone 1 elements usually trip locally, initiate breaker failure and automatic reclosing, and transfer trip via communication channels to the remote end. Receipt of a transfer trip receive signal from the remote end will also initiate a local trip, breaker failure, and automatic reclosing – refer to Figure 14.23.
- This scheme requires only one Zone 1 operation at a single end to trip both ends of the protected line. Fault detection by a Zone 1 at either end will transfer trip the other end. This scheme

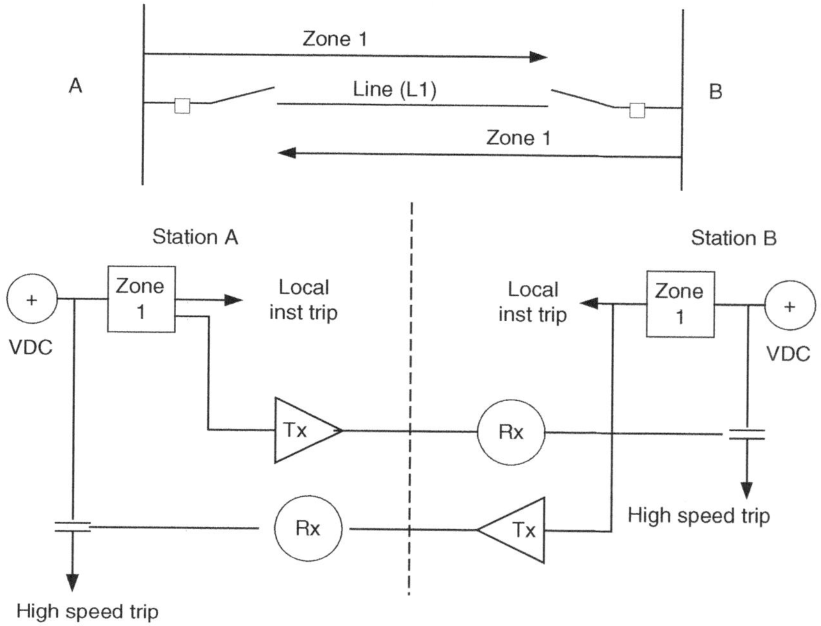

Figure 14.23 Direct underreaching transfer trip.

covers the entire line dependent upon having reliable communication channels and sufficient Zone 1 overlap. This scheme will not detect faults beyond the Zone 1 reach upon loss of total communication channels, or if the remote breakers or the remote line disconnect happens to be open.

14.7.2.1.2 Zone Overlap

Underreaching for the Direct Underreaching Transfer Trip scheme is achieved by setting the Zone 1 elements (phase and ground) at both ends of a line to reach about 75–80% of the line section from the terminal, see Figures 14.24a,b.

When a fault occurs outside the Zone 1 of one terminal, transfer trip is received from the other terminal. In this case, the time of fault clearance includes communication channel time. For the correct functioning of this scheme, there should be an adequate overlap of the zones from each terminal. A 20% overlap has been considered to be the requirement and under no circumstances should the Zone 1 element reach past the remote terminal.

Normally, faults on any part of the line are within the reach of at least one Zone 1 element. With direct underreaching transfer tripping, Zone 1 alone provides fast clearance for faults anywhere along the line between the two terminals. However, if the line disconnect at one terminal is open, or the fault infeed is low at one terminal, the underreaching scheme will not cover all fault locations on the line. This is covered either by the Permissive Overreaching Transfer Trip or the Directional Comparison Transfer Trip schemes. This scheme also serves as alternate instantaneous protection, with separate elements and communication channels.

14.7.2.1.3 Timing

Direct underreaching transfer trip is the fastest line protection for most faults. High-speed tripping, delayed only by the relay response time, is achieved for faults on 60% of the line where there is

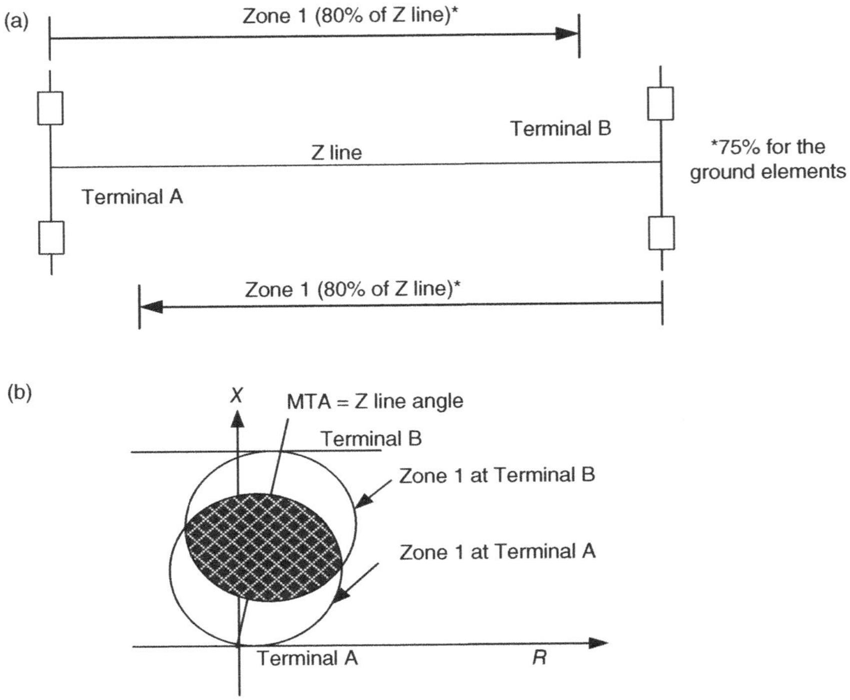

Figure 14.24 Zone 1 coverage from the two terminals of line. (a) Zone 1 coverage on a single line. (b) Zone 1 coverage – MHO elements on a *RX* plot.

overlap. For faults at the far end of the line, where there is no overlap, the local protection is still high speed while the remote end must wait an additional half cycle when using digital communications to receive transfer trip.

14.7.2.1.4 Line End Open

If it was not possible to operate two-ended lines with one end open, the direct underreaching transfer trip scheme with duplicated A and B protections and Main and Alternate routes would be all that would be necessary. No other scheme type would be required. However, since this is not the case, it is simply not possible to cover the last 20% of the line when the remote end is open. It is for this reason that the other following scheme types have been developed.

14.7.2.2 Permissive Overreaching Transfer Trip (POTT)

The permissive overreaching transfer trip scheme is established as one of the standard schemes which are part of an overall scheme solution. It is not meant to be used alone but typically in conjunction with direct underreaching transfer trip. This scheme is meant to be either for the A protection group or the B protection group or both for local backup.

14.7.2.2.1 Underlying Principles

Zone 2 overreaching elements key permissive signals to the remote end, refer to Figure 14.25.

- The Zone 2 overreaching elements wait for a permissive signal to be received from the remote end.

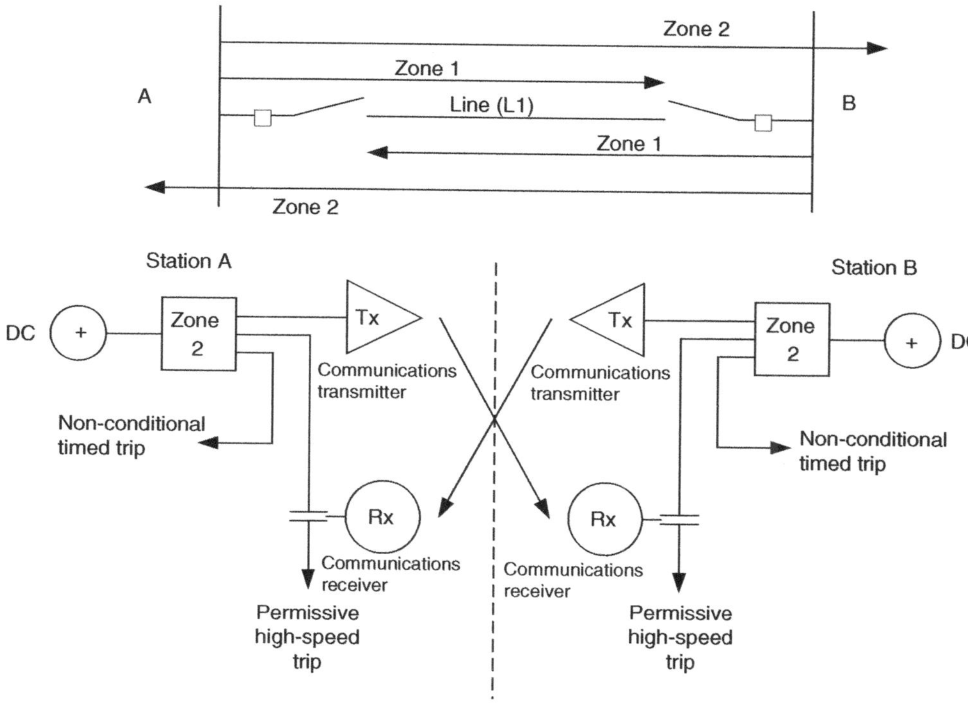

Figure 14.25 A permissive overreaching scheme.

- Upon receipt of a permissive signal from the remote end, these elements are permitted to trip locally without any time delay. In addition, Zone 2 elements initiate breaker failure and automatic reclosing of those breakers and send transfer trip.
- A select number of transmission utilities have a practice for the Zone 2 overreaching elements to trip the local end unconditionally after an extended time delay, up to typically 400 ms, even without receipt of permission. However, no transfer trip or automatic reclosing of local breakers is initiated. This practice is intended to provide protection upon loss of total communication channels.

14.7.2.2.2 Zone Overlap

The overreaching Zone 2 phase and ground elements are typically set to 125% of the line's positive sequence impedance or of the maximum apparent impedance seen by the distance measuring element at the terminals, see Figures 14.26a,b. A fault anywhere along the protected line, the Zone 2 elements at both ends of the line will pick up. Each of them then sends a permissive signal to each other. On receipt of the permissive signal, the local relay at each terminal trip their respective breakers to clear the fault.

If a fault occurs beyond the line terminal, e.g. beyond Terminal A as illustrated in Figure 14.27, it may be seen by the Zone 2 distance element at Terminal B which then sends a permissive signal to Terminal A. However, because the Zone 2 element at Terminal A does not see this fault and hence is not picked up, no tripping will occur at Terminal A.

Also, since no permissive signal is sent from Terminal A, no tripping will occur at Terminal B either. For the correct functioning of this scheme, the reach settings of the Zone 2 ground elements

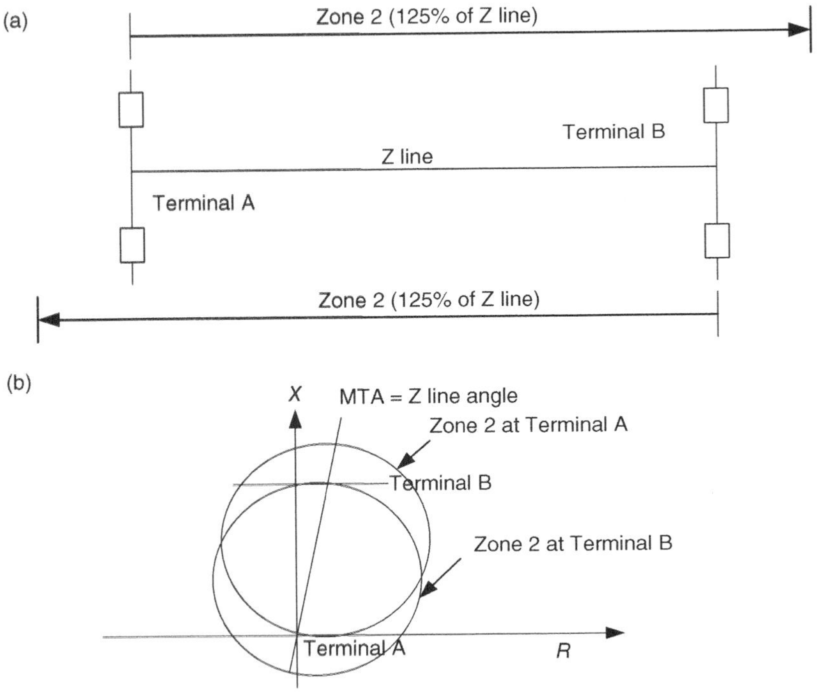

Figure 14.26 Zone 2 coverage from the two terminals of line. (a) Zone 2 coverage on a single line. (b) Zone 2 coverage – MHO elements on a *RX* plot.

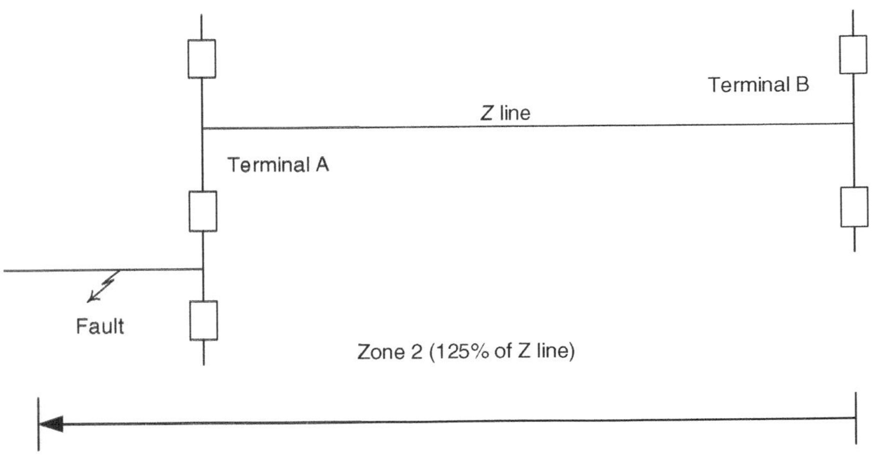

Figure 14.27 Zone 2 sees some faults in the next line section.

must be selected with consideration for the underreaching effects of mutually induced impedance on double circuit lines.

14.7.2.2.3 Timing

Permissive overreaching requires that local tripping only take place upon the receipt of permission from the remote end. Communication time to key a permissive signal at the remote end and for it to be received at the local end is typically a half cycle for digital communications. The time from fault inception to total clearing including relay and breaker operation is approximately 100 ms for three-cycle breakers and is therefore considered high-speed protection.

The Zone 2 elements usually have a timed backup function, set by many utilities at typically 400 ms as an unconditional trip. This covers the loss of communication for a fault on the last 20% of the line beyond the local Zone 1 reach.

14.7.2.2.4 Line End Open

In the permissive scheme, there is a requirement to have a special provision for when the line is open at one terminal. In this situation, with the fault near the open end and out of Zone 1 reach, the Zone 2 elements operate and send a permissive signal to the open end. On receipt of the permissive signal, the line end open logic in turn sends a permissive signal back to the other end (even though the Zone 2 element did not operate). Instantaneous tripping is achieved. This concept, also called "permissive echo," is shown in Figure 14.28a.

The advantage of this scheme is that high-speed tripping is achieved for all faults within the protected zone. For that reason, the term accelerating channel is also used to describe the permissive channels used in the echo scheme. The disadvantage is that reliable high-speed communication is required for the proper functioning of the scheme.

14.7.2.2.5 Ground Overcurrent Delay Timer

The directional ground overcurrent element, when used with the permissive scheme, needs to be time delayed, typically in the order of 120 ms, to prevent operation of the parallel healthy line directional overcurrent elements due to current reversals as shown in Figure 14.28b and 14.28c.

Prior to any breaker tripping, the infeed to the fault from Terminal A is via both lines Line 1 and Line 2 as both ground directional relays see the same fault. Both these relays at Terminal A key permission to Terminal B.

Breakers BRK2B and BRK3B will trip by local protections due to receipt of permission. Should the breakers at Terminal B trip first before those at Terminal A, the previous infeed to the fault via Line 2 will reverse. This could likely happen as the fault infeed for a close-in fault at Terminal B is substantially higher thereby resulting in faster relay response.

With the current reversal on healthy Line 2, the ground directional overcurrent relay at Terminal B will operate. Since permission was previously being received at Terminal B prior to the current reversal, there is a window of time when the ground directional overcurrent relay tripping contact is closed and yet the permissive receive contact has still not reset itself. Due to this possibility, a delay timer at Terminal B must be timed to delay local tripping until the permissive signal is no longer received from Terminal A on Line 2.

Without taking into account the ground directional overcurrent initial contact closing time at Terminal A, the time delay required at Terminal B is the total resetting time 58 ms of the permissive signal sent from Terminal A as follows:

40 ms (Ground directional overcurrent relay slowest drop out time)
18 ms (Tele-protection equipment opens local permissive receive contact)
58 ms

However, offsetting this total time at Terminal B is the time initially required by the ground directional overcurrent relay at Terminal A to close its contacts at the outset of the fault:

58 ms
−15 ms
43 ms

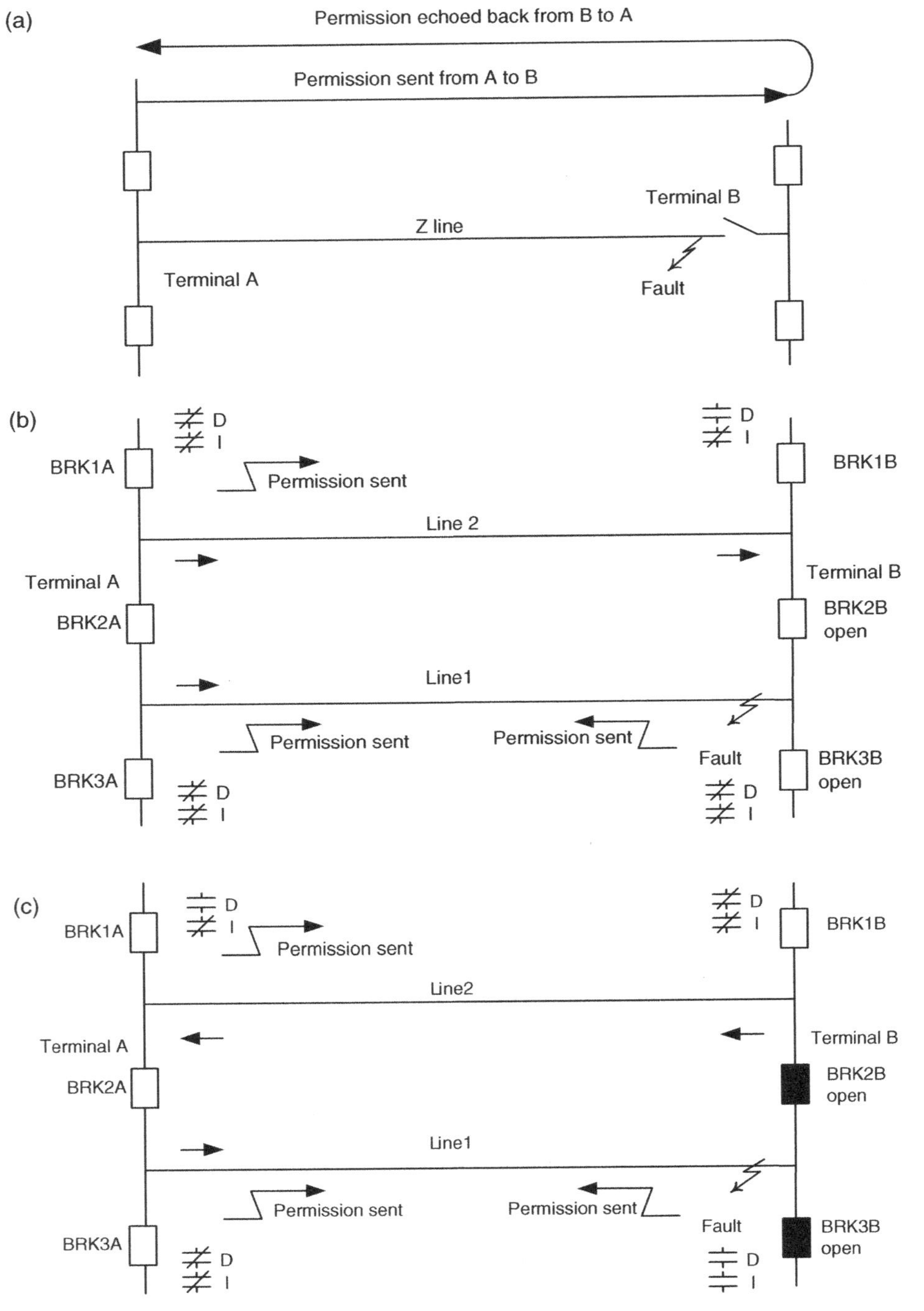

D is the directional contact of the relay
I is the overcurrent contact of the relay

Figure 14.28 (a) Illustration of the eco concept with remote disconnect open. (b) Directional ground overcurrent before current reversal. (c) Directional ground overcurrent following current reversal.

Add another one cycle to cater for the breakers at Terminal B possibly being three cycle breakers while at Terminal A being four cycle breakers. The total required time delay is therefore 60 ms (43 ms + 17 ms).

A setting of 120 ms provides a 2:1 margin of safety.

14.7.2.3 Directional Comparison Blocking Transfer Trip (DCBTT)

The directional comparison blocking scheme is established as one of the standard schemes as part of an overall scheme solution, refer to Figure 14.29. It is not meant to be used alone but typically in conjunction with direct underreaching transfer trip. This scheme is meant to be either for the A protection group or the B protection group or both when used as local backup.

14.7.2.3.1 Underlying Principles

The Zone 2 overreaching elements are time delayed typically two or three cycles to wait for a blocking signal to be received from the remote end. If a blocking signal is received, within the wait time, the Zone 2 elements will not trip locally, or transfer trip to the remote end.

- Should no blocking signal be received after the short wait time two or three cycles, the Zone 2 elements will trip locally. In addition, Zone 2 elements initiate breaker failure and automatic reclosing of the local breakers and send transfer trip.
- A select number of transmission utilities have a practice for the Zone 2 overreaching elements to trip the local end unconditionally after an extended time delay, typically 400 ms, even without receipt of permission. However, no transfer trip or automatic reclosing of local breakers is initiated. This practice is intended to provide protection upon loss of total communication channels.
- The reverse-looking Zone 3 elements must be set to cover more line impedance, with margin, than that seen by the forward-looking Zone 2 relays at the remote end (Figure 14.29).

14.7.2.3.2 Zone Overlap

The overreaching Zone 2 phase and ground elements are typically set to 125% of the line's positive sequence impedance or of the maximum apparent impedance seen by the distance measuring element at the terminals, see Figures 14.26a,b. On a fault anywhere along the protected line, the Zone 2 elements at both ends of the line will pick up. Each of them then sends a permissive signal to each other. On receipt of the permissive signal, the local relay at each terminal trip their respective breakers to clear the fault.

The overreaching Zone 2 phase and ground elements are typically set to 125% of the line's positive sequence impedance or of the maximum apparent impedance seen by the distance measuring element at the terminals The Directional Comparison blocking is achieved by the Zone 2 elements together with Zone 3 elements that are set to see faults in the reverse direction. The zones at each of the terminals of a two-terminal line are as shown in Figure 14.29.

On external faults, the reverse-looking Zone 3 operates and sends a blocking signal to the remote end to block tripping by the Zone 2 elements. The Zone 3 element is set with sufficient margin to see all faults behind the relay terminal that are seen by the overreaching Zone 2 element at the remote terminal. This ensures that a blocking signal is sent on all external faults for which the Zone 2 element at the remote terminal would operate.

For coordination, the tripping by the Zone 2 element has to be delayed long enough to receive a blocking signal. For example, consider a fault at X in Figure 14.30 (a). This fault may be seen

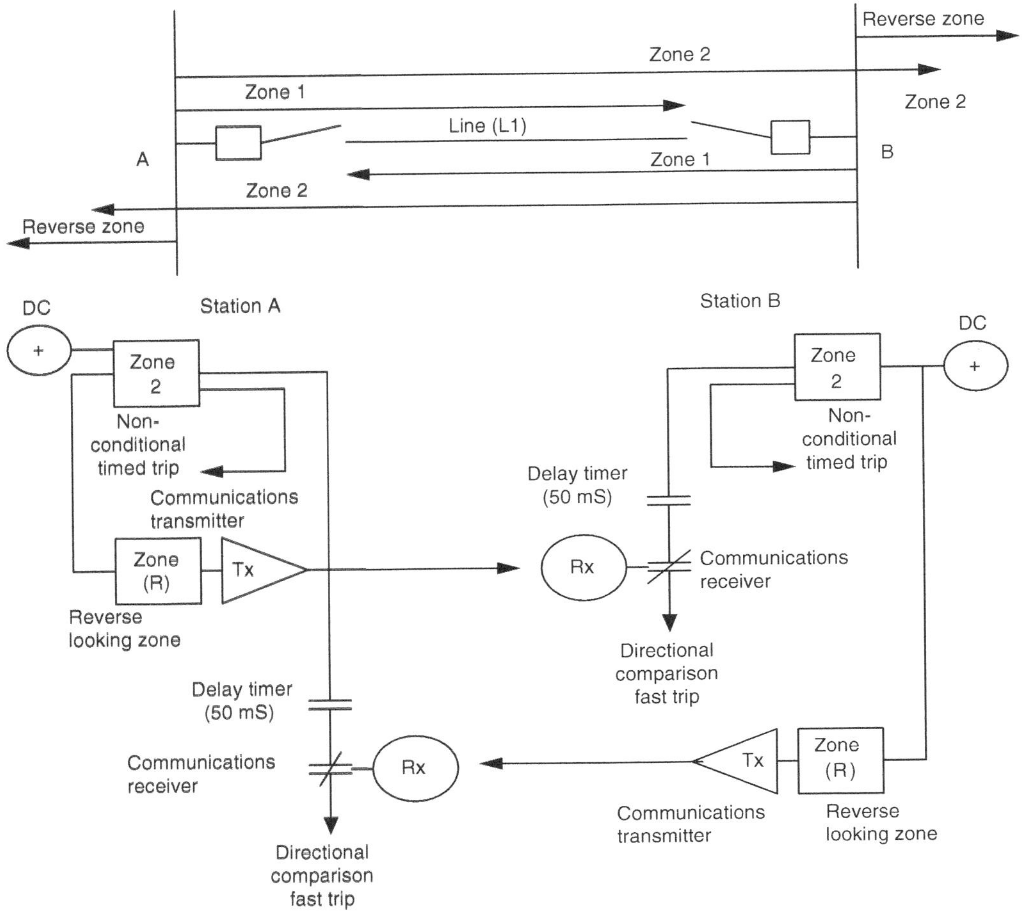

Figure 14.29 A directional comparison blocking scheme.

by Zone 2 at Terminal A and the reverse-looking Zone 3 at Terminal B. A blocking signal is sent to Terminal A to block tripping. If the fault occurs at Y, the Zone 2 distance elements at both Terminal A and Terminal B will operate. No blocking signal is sent and both terminals will trip after the set time delay.

14.7.2.3.3 Timing

Directional Comparison Blocking requires that local tripping only take place following a suitably sufficient time delay to receive a blocking signal(s). When using digital communications, the time delay could be as short as one cycle for half cycle communication time giving a 2 to 1 margin of safety. However, a more common two cycle time delay is more typical.

The time from fault inception to total clearing, including relay and breaker operation, is approximately 115 ms for three-cycle breakers which is 20 ms or approximately one cycle slower than a comparable permissive overreaching operation.

The Zone 2 elements usually have a timed backup function, set by many utilities at typically 400 ms as an unconditional trip. This covers the loss of communication for a fault on the last 20% of the line beyond the local Zone 1 reach.

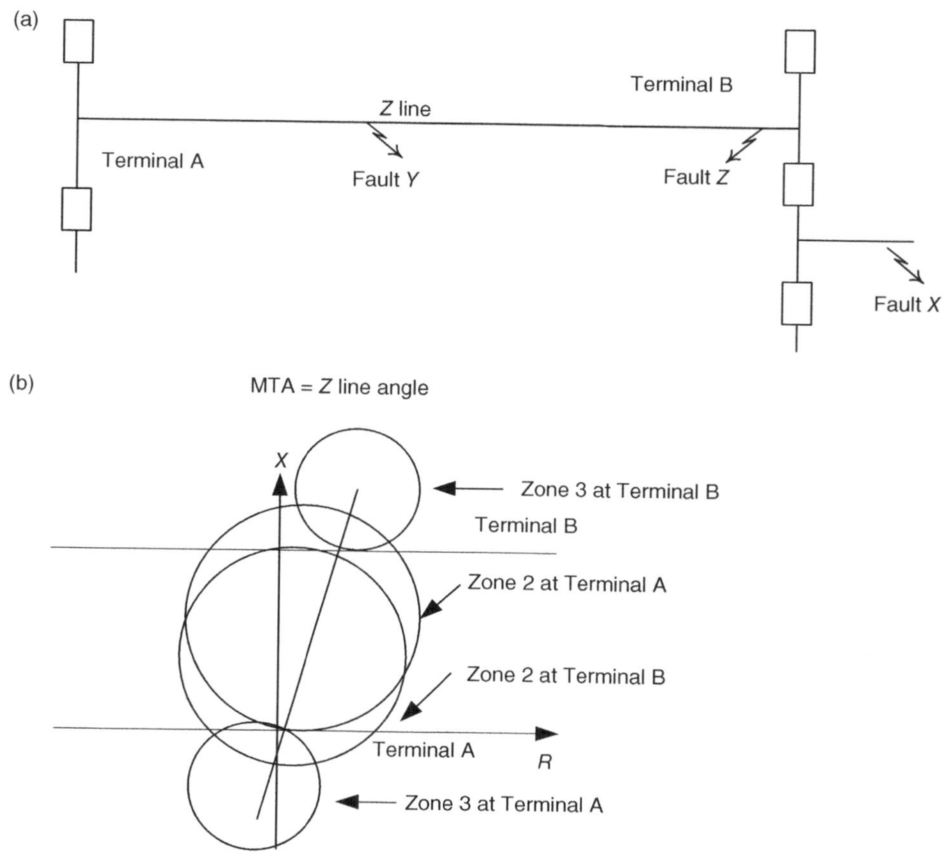

Figure 14.30 Illustration of the DCB scheme. (a) Fault locations on a single line diagram. (b) Illustration of the DCB scheme.

14.7.2.3.4 Line End Open

The directional comparison blocking scheme is ideally suited for fault clearance with a line end open operating condition. Local Zone 2 elements trip unconditionally after an intentional time delay waiting for a block signal which of course will not be sent. With line end open, it is not possible for both the local Zone 2 and remote reverse looking Zone 3 to see the same fault.

The advantage of this scheme type is that communication is not required for the proper functioning of the scheme. The disadvantage of this scheme is that timed delayed tripping is required for faults beyond the local Zone 1 or on the last 20%-line segment for a fault at Z Figure 14.30 (a).

14.8 General Implications

14.8.1 Apparent Impedance Implications

Zone 2 reaches at the line terminals are very unlikely to see through transformers at tapped load substations in the majority of cases. Thus, it is rare for blocking signals to be required and happen rarely only when apparent impedances are exceptionally large due to some other factor such as long line taps to load substations, exceptionally long lines, or large discrepancies between maximum and minimum source impedances.

14.8.2 Line Protection Zone 2 Reach Implications

It does happen on occasion that Zone 2 reaches based on large apparent impedance can reach through transformers into the LV side of load substations. Typical transformer impedance at large load substations is in the order of 0.2 PU which represents approximately 200 km of 230 kV line. Since Zone 2 is set at 125% of apparent impedance, a 175 km line with a significant but not unusual apparent impedance adding 50% to the required reach will certainly see-through locally tapped load-substation transformers. In this situation, the typical solution to guarantee the line protection does not operate for distribution system faults is to provide blocking signals from the load-substation LV protections to block the line Zone 2 from tripping the line.

An interesting situation can arise where the A and B group Zone 2 line protections see into a load-substation transformer but not through to the LV side. The A and B group protections must cover enough of the transformer to where the fault levels are no longer high enough if uncleared to cause power system instability. This is because load-substation protections are not usually dual redundant as there is only one station battery to operate transformer protections and to trip breakers either locally or remotely via transfer trip. What if a fault beyond the reach of the Zone 2 line protections no longer reaches into the transformer windings until the fault levels drop off to below bulk power system (BPS) levels and the load-substation protections are not dual redundant as required for all BPS system elements? In this case, either the line protection Zone 2 reach will need to be increased with the issues previously described or the transformer protections at the load substation will need to be designed and maintained to BPS protection criteria.

Note that the blocking signals do not need to be designed and operated to BPS protection criteria even though they are part of the BPS line protections since without the receipt of blocking the line protection will just trip unconditionally. Where regional operating authorities are not comfortable with line protections tripping indiscriminately for LV distribution faults with only one piece of protection equipment failing then dual redundant blocking signals should be implemented.

14.8.3 Communication Implications

In general, most A and B group line protections that are serviced by a standalone communication system use the permissive overreaching transfer trip scheme. This is true for most lines where the highest speed clearance times are necessary.

Power line carrier (PLC) instead of standalone communications is popular where high-speed clearance is not a priority. PLC communication means the lines themselves are the communication media for protections. Since PLC relies on the lines for a permissive signal to be received it is obvious that a fault affecting the line such as a downed conductor or tower would preclude permissive signals being received. Therefore, the use of a permissive scheme in both groups is limited to those applications where standalone communications exist. Otherwise, it is common practice to adopt directional comparison blocking in at least one protection group whenever PLC is used.

14.9 Peripheral Requirements of Distance Protection

Distance relays do have limitations and requirements that must be addressed when applying them. This section addresses the peripheral requirements of distance protection.

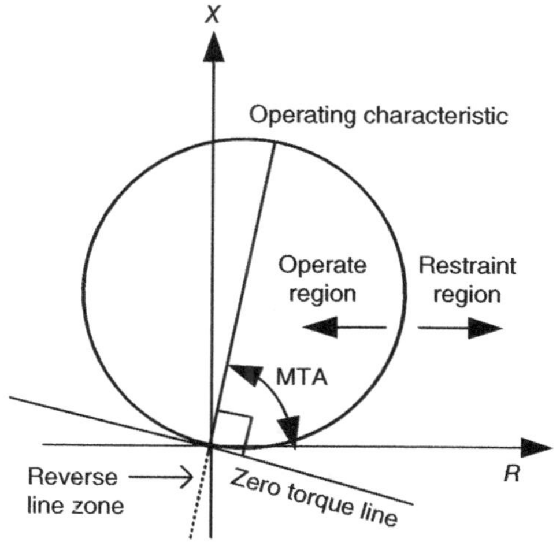

Figure 14.31 MHO relay operating characteristic.

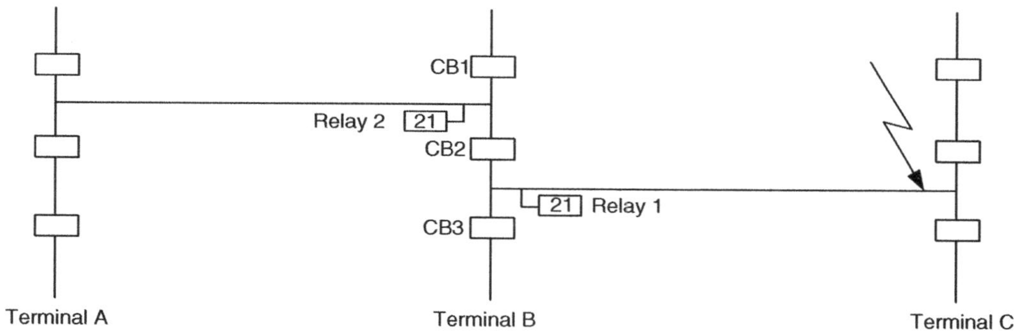

Figure 14.32 Typical line terminations and directioning of relays.

14.9.1 Distance Relay Response to Three-Phase Faults

The vast majority of distance relay applications use mho relays whose characteristic passes through the origin. As shown in Figure 14.31 the zero-torque line that defines the area of restraint for faults in the reverse direction passes through the origin along with the operating characteristic. The need for the characteristic and zero-torque line passing through the origin is to ensure that the relay is exclusively directioned toward the protected line as shown in Figure 14.32 where Relay 1 should operate and not Relay 2 for the fault location. Relay operation for faults in the forward direction and not for faults in the reverse direction or behind the relay's intended line zone of protection is essential for protection schemes such as permissive overreaching or directional comparison blocking to function.

Refer to Figure 14.33 showing a section of a typical terminal station. The architecture that defines the use of breakers is known as "breaker and half." The reason for this terminology is that two lines share three breakers or one line uses one and a half breakers so to speak. The advantage to the breaker and half architecture is to allow a single line breaker to be maintained without the need to isolate the line from that end.

The section of the bus bounded by the two diameter breakers is known as a stub-bus. The line protection is zoned off around the two-diameter breaker sets of CTs defining the line zone. Should a three-phase fault occur on the stub-bus or on the line close-in to the station such that the voltage

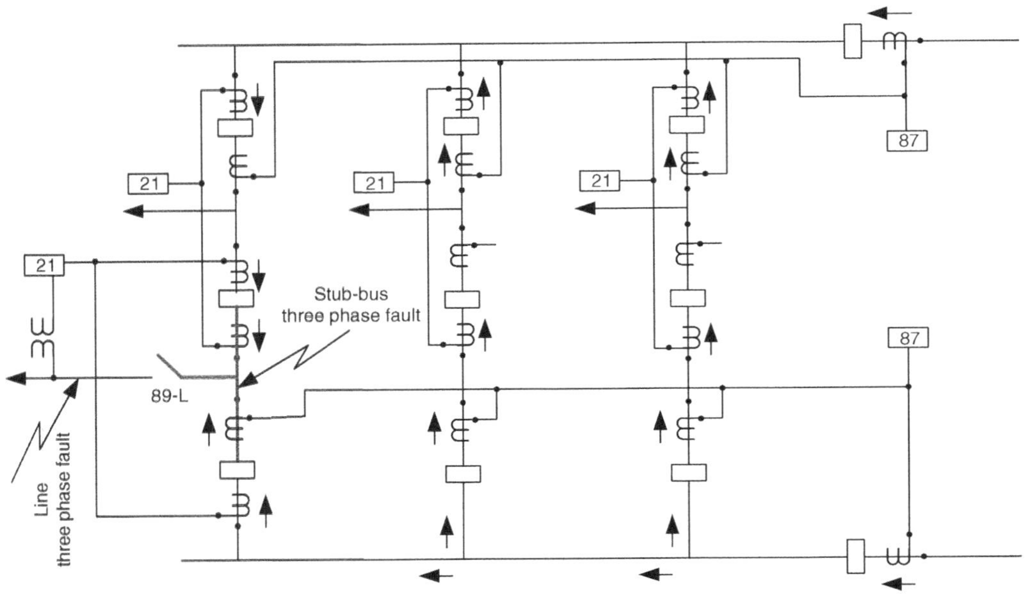

Figure 14.33 Typical line terminations and a close-in fault.

measured by the relay collapses to essentially zero the relay would not operate. For a mho relay, that is not offset, the origin on the R–X plane is defined by the location the relay measures the voltage, in this case, zero.

14.9.2 Memory Action

The purpose of memory voltage is to ensure that reliable polarization is available in the case of a balanced three-phase fault. In a three-phase fault when all of the voltages collapse, it becomes necessary to use the memorized pre-fault voltages. Digital relays are equipped with memory voltage. The directional element typically needs at least a minimum level of the normal measured polarizing voltage for correct operation (e.g. 10%). When the measured voltage is below this level, the device uses the memory voltage (e.g. for close-in short circuits). This memory voltage supports two important aspects of the distance protection function.

- Allows the distance function to operate for close-in three-phase faults in front of the protection location.
- Prevents the distance function from operating for close-in three-phase faults behind the protection location.

In a digital relay, the protection function itself decides whether to use measured or stored voltage or a mixture of these voltages. The memory time is mostly variable. It typically starts with two cycles but can be extended up to 20 cycles when there has been no earlier trip decision. After the maximum memory time has elapsed, the directional decision is maintained until the drop-off of fault detection.

14.9.3 Stub-bus and Switch Onto Fault Protection

Refer again to Figure 14.33 where the voltage source to the mho relay is located on the line side of the line disconnect switch 89-L. This is the typical location for the voltage source to a line relay at many utilities due to switchyard space requirements. For balanced three-phase faults when the line disconnect is just closed or when it's open for some time and there is a fault on the stub-bus, distance relays would not work as they do not operate on current only.

The stub-bus is not protected by any other relay except the distance mho relay. The mho distance relay does not protect under these very legitimate operating conditions. When the 89-L line switch is open, the mho relays whether phase or ground would not operate for any type of fault on the stub-bus. Also, when the 89-L line switch is first closed by system operators when the remote end happens to be open, there is always a chance that maintenance crews left the three-phase grounds on the line without informing system operators that they are still there. Since the mho relays did not have voltage for any length of time prior to the closing of the 89-L switch, they will not respond and hence there would be no protection for close-in faults either. It should be recognized that the memory action inherent in mho relay design would be useless in this situation as the voltage memory is only good for a very short length of time while this operating condition could exist for days or even weeks.

For electromechanical relays, protection is achieved by an independent high-speed three-phase overcurrent relay where the three overcurrent output contacts are connected in parallel such that an overcurrent on any one phase provides an output. The logic has two paths that lead to local tripping. One path monitors the status of the 89-L line switch such that the stub-bus protection is invoked when its position is open. The other path monitors the voltage on the line CVT, and when it is absent for more than two seconds, the switch onto fault (SOTF) protection is invoked. In either case, an overcurrent above a settable threshold on any one single phase would result in a local trip provided any of the line test logic as described is met. In digital relays, this logic is either created internally by settable logic identical to the one used for electromechanical relays or comes as native logic usually known as SOTF to be enabled during the setting process (Figure 14.34).

14.9.4 Overcurrent Supervision

Distance relays whether electromechanical or digital as a general rule need to have their impedance elements supervised such that those elements do not trip until a minimum amount of fault current

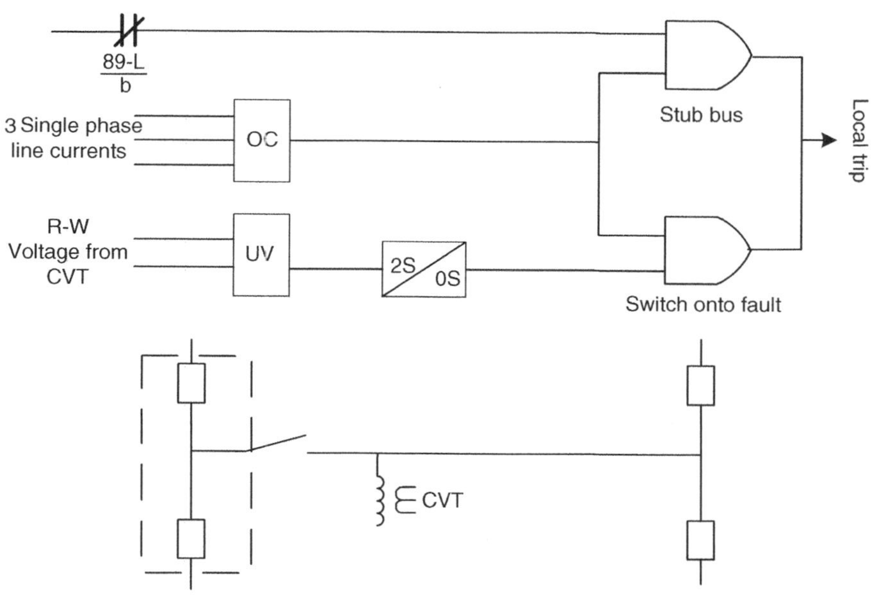

Figure 14.34 Line test logic.

is detected. The setting criteria are to set it above line charging current and below minimum remote-end fault current.

14.9.4.1 Supervising Current Elements

Two independent operating conditions necessitate the application of overcurrent supervision.

The first operating condition is reserved exclusively for electromechanical phase and ground distance relays where poly-phase compensator type relays are being used. In particular, it must be the ones that provide a single zone of phase protection for all three phases. This type of relay provides instantaneous tripping for all combinations of phase-to-phase faults, two-phase-to-ground faults, and three-phase faults. Overcurrent supervision is necessary otherwise these relays tend to misoperate due to the following condition. As shown in Figure 14.35 where line L1 is de-energized with 89-L1 open at Terminal A and CB4, CB5 open at Terminal B. Line L2 can induce some voltage onto line L1. Since many transmission utilities do not transpose their lines, the induced voltage may contain negative sequence components that are capable of causing these relay phase to phase elements to operate. The compensator-type relays are particularly susceptible to this phenomenon which has nothing to do with contact drift as is commonly thought.

The second operating condition occurs upon line de-energization where the discharge of stored energy in the line causes not just electromechanical relays but also digital distance relays to operate. The discharge of stored energy into the relay affects all distance elements whether phase or ground.

Instantaneous overcurrent relays that operate within one cycle above the overcurrent threshold setting and possess a high pickup to dropout ratio in the order of 0.98 are used for this use purpose. Modern digital distance relays possess built-in phase and ground overcurrent supervision elements that are user-settable. Digital relays by the nature of their design and also many non-compensator type electromechanical relays are not vulnerable to misoperate under the first condition described above but all of them are vulnerable under the second condition when the line is de-energized.

Typical line charging currents are dependent on the line construction which is mainly standard in most utilities for each voltage level. Line charging currents are directly proportional to line length.

14.9.5 Reclosing Coordination with Line End Open – Permissive Echo Timer

In the permissive scheme, there is a requirement to have a special provision for the case in which the line is open at one terminal. In this situation, with the fault near the open end and out of Zone 1 reach, the Zone 2 elements operate and send a permissive signal to the open end. On receipt of

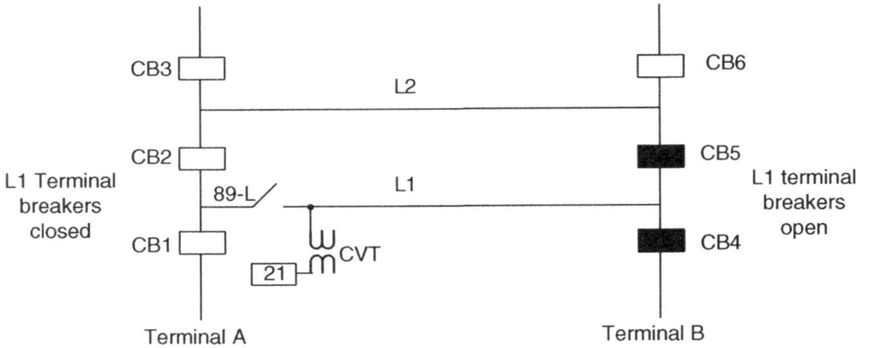

Figure 14.35 Operating conditions that cause misoperation of compensator type relays.

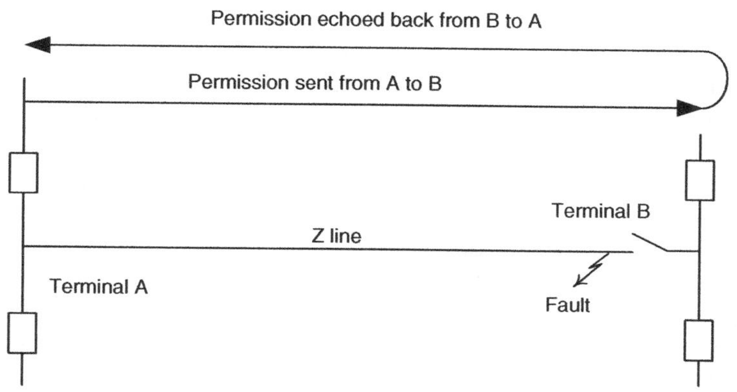

Figure 14.36 Illustration of the echo concept with remote disconnect open.

the permissive signal, the line end open logic in turn, sends a permissive signal back to the other end regardless of whether the Zone 2 operates at that location. Instantaneous tripping is achieved. This concept, also called "permissive echo," is shown in Figure 14.36.

A short-time delay of standard 0.5 seconds is necessary before the line end open logic enables for when one of the line breakers at the remote end is open by pre-fault configuration and a fault is in the next line zone. Without a time delay, as soon as the single breaker is open and line end open logic is thereby enabled, the local Zone 2 will receive an echoed permissive signal and also trip. This is applicable to breaker and half switchyard configurations.

When system operators choose the shortest time for automatic reclosing, it cannot be made to compete with the 0.5 second time to enable permissive echo logic. Ideally, operators should not set the short time reclose selection for less than 1.0 seconds. For an internal fault beyond the local Zone 1 reach. Should the line end open logic enable longer than the reclosing time, the local breakers will reclose onto the fault that is beyond Zone 1. Zone 2 will need to have its permissive signal echoed back immediately.

Automatic reclosing in conjunction with a permissive scheme must therefore not be allowed to be less than one second providing sufficient margin for the permissive echo logic to enable.

14.9.6 Potential Sources

Distance relays are designed to accept three-phase-neutral voltages of 69.3 V RMS (120 V RMS phase–phase) or 66.4 V RMS (115 V RMS phase–phase). In installations of 115 kV and below, the potential source is usually a magnetic voltage transformer in each phase. In installations of 230 kV and above, the potential source is usually a capacitive voltage transformer (CVT). The magnetic or CVTs provide appropriate primary to secondary voltage ratios (at any common primary voltage level) to achieve the required 66.4 V or 69.3 V RMS secondary voltage to the relay.

14.9.6.1 Transient Response

CVTs represent a very good source of voltage to distance relays when used with electromechanical relays. With the advent of digital relays, this may no longer be the case. Digital relays are susceptible to misoperate due to CVT transients resulting from the higher operating speed and sensitivity of digital relays.

Line faults in systems with a large Source Impedance Ratio (SIR) cause very depressed phase voltages where the CVT output does not closely follow its input voltage due to the internal CVT energy storage elements. As these elements take time to change their stored energy, they introduce a transient to the CVT output following a large input voltage change. These CVT transients reduce the fundamental component of the fault voltage. The decrease in the fundamental voltage results in a decrease in the measured impedance by the relay that computes a $Z = V/I$. When the measured voltage is lower than it should be based on the voltage drop from the relay to the point of fault, the computed impedance Z_{relay} is lower than the actual line impedance Z_{line} to the fault. When this happens, the relay will tend to have a larger reach than the intended reach setting. Underreaching Zone 1 elements could easily overreach into remote bus or line zones due to this phenomenon. There are a number of other factors that negatively influence the degree of CVT transient response besides high SIR. The coupling capacitor values used in the manufacture of the CVT, the design of the ferro resonance suppression circuit whether active or passive are examples. In general, utility experience shows that high SIR on their system does lead to transient overreach sufficient to cause Zone 1 tripping for faults on remote buses.

SIR is a measure of the minimum voltage on a line-end fault and is defined as the ratio of the source impedance behind the relay bus to the line impedance. In terms of impedances $SIR = Z_{source}/Z_{line.}$ Typical source impedances are low relative to line impedances. As a general rule, in large networks interconnected with a significant generation where there are low source impedance values relative to line impedances it is not unusual for SIR values to be as low as 0.1 to 0.5. At these low values of SIR, there is little concern for transient overreach. For lines supplied by an autotransformer as shown in Figure 14.37, the high source impedance represented by the autotransformer results in a significantly large SIR. It has been observed that the value of SIR for these lines could be as high as 30. At an SIR value of 30, there is potential for significant transient overreach.

Refer to Figure 14.38, showing the pre-fault voltage dropped across the source impedance Z_s and the line Z_L. The voltage drop is divided between the two impedances by the location of the relay such that the relay only measures the voltage drop across the line $Z_L \times I_F$.

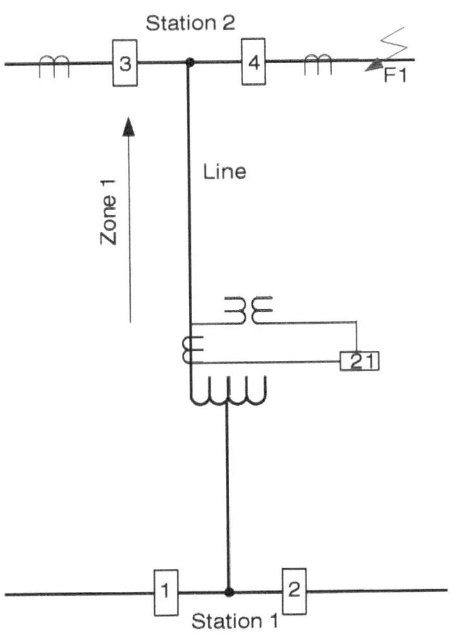

Figure 14.37 Line supplied by an autotransformer.

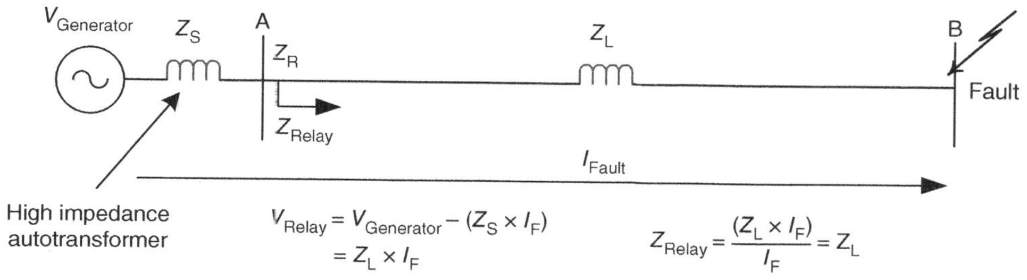

Figure 14.38 Example of high source impedance ratio.

The solution adopted by many transmission utilities for lines supplied by an autotransformer is to intentionally definite time delay the Zone 1 distance elements by 1.5 cycles to ride through the CVT transient. Other solutions would be to install digital relays with native transient overreach algorithms. These algorithms depending on the relay manufacturer either internally add an intentional time delay to the Zone 1 elements or pull back the Zone 1 reach whenever a CVT transient response is sensed. This is usually done via a set point of voltage deviation from system nominal. Another more elegant solution is to install line current differential relays instead of distance as they do not measure voltage.

14.9.7 Loss of Voltage to Distance Relays

Distance measuring elements tend to operate on the loss of one or more voltage signals to the relay, which could be the result of blown fuses in the circuit supplying the voltage signals. Many distance relays include a feature that detects a loss of voltage condition. A loss of voltage can be programmed to either inhibit the operation of the distance elements and/or to just provide an alarm.

On loss of all three voltages, a loss of voltage condition exists if the three voltages are below a given threshold and the current levels are not above the supervising current levels of the distance elements. The current supervision elements would typically be set above maximum anticipated line load and below minimum calculated fault current for a three-phase fault.

A common technique to detect the loss of one or two voltage signals is to monitor the negative and zero-sequence voltages and currents. Loss of voltage condition is detected if unbalance voltage exists without the presence of unbalance currents as measured zero sequence or negative sequence currents.

In most cases, the loss of voltage feature can be selected to alarm only. This is the established procedure in some utilities. The reason not to block the relay operation for a loss of voltage is that the distance elements could be blocked on some unbalanced faults in certain situations, depending on the settings that need to be used on the unbalance voltage and current detectors.

14.9.8 Self-Monitoring Relays

Up until the age of digital relays, there was no way of knowing whether a distance relay lost its voltage source. On the front of every electromechanical distance relay panel was a set of three fuses and a dummy fuse for the neutral. Each of the phase fuses had a dedicated pilot light showing the presence of potential to the relay. This was a visual indication that was only sufficient at manned stations at best.

Even more fundamental was the lack of truly knowing whether an electromechanical relay was functional altogether after it was maintained and placed back into service.

It was the lack of an intrinsic assurance of relay health that led many regional operating authorities to adopt a strict two-year maintenance cycle for distance relay systems.

One of the many features of digital relays is the self-monitoring or self-diagnosing feature whereby the relay checks automatically for the health of its vital signs and alarms should any of them fail the test. Both relay failure alarm and loss of voltage alarm are now monitored by supervisory control and data acquisition (SCADA). These new alarms assuring a healthy relay led, in many jurisdictions, to the relaxing of maintenance cycles from two to four then to six-year cycles, or longer.

14.10 Tele-Protection (Pilot-Protection) A Historical Perspective

A new approach using high-speed duplicated protections along with a new geographically diverse main and alternate path microwave system for tele protections was proposed for application by many utilities following the North American blackout of 1965. Utilities began installing new protection systems designed to the concept of local backup. This included transfer tripping of remote breakers for a local breaker failure.

Reliability was achieved by complete duplication with two groups of line protections, identical in function for each line. Security was improved by the elimination of the timed backup third zone. Local breaker backup was improved by using a new high-speed breaker failure protection dedicated to all breakers.

The installation of new high voltage line protections was made possible due to the installation of a high-speed and reliable communication backbone. To achieve duplication that matched the protections, geographically diverse main and alternate communication routes at that time usually via microwave in a ring-type topology were established. Line protection would use these communication systems for transfer trip and for either permission or blocking signal logic. Local breaker failure protection made use of these communication systems for transfer trip of remote breakers to isolate a local breaker failure condition.

With microwave and PLC communications becoming available between terminal stations, transfer tripping of the remote line terminal breaker for a local breaker failure became possible. It was the introduction of high-speed and reliable communications that allowed for the use of local backup and the deployment of dedicated breaker failure protections along with duplicated line protections.

Presently, the same fundamentals are still applicable and adhered to for the deployment of fiber optic digital-based tele-protection systems.

14.11 Tele-Protection via Power Line Carrier

The reliability of power line conductors for the transmission of information is very high and therefore, this medium is successful for the transmission of high-frequency signals for protections. Since PLC systems are susceptible to line noise, frequency-shift modulation is standard. Frequency-shift PLC is used in protection applications for transfer trip, permissive, and blocking functions. At many utilities using PLC for transfer trip applications, diversity is somewhat achieved by transmitting simultaneous tripping commands via frequency-shifting the PLC and via frequency-shifting audio tones on Single Side Band.

The carrier tones mostly used in PLC applications are radio frequency tones 40–200 kHz and 415–490 kHz. These carrier tones are propagated over the metallic transmission line conductors. Transmission line towers and conductors are structurally solid. The conductors are generously spaced providing a reliable path for PLC signals. The radio frequency tones employed are sufficiently high to be isolated from the 60 Hz power system. This includes noise generated from the power lines and attenuation of the signal.

A PLC system consists of three distinct parts:

(1) The carrier terminal equipment, consisting of transmitters, receivers, and associated components.
(2) The line coupling equipment, which provides a means of connecting the carrier equipment to the high voltage line.
(3) The high voltage line, which must provide a suitable path for transmission of the high-frequency signals between terminals.

Carrier systems are coupled to the HV lines in one of two methods, either on a phase-ground or on a phase–phase method.

Phase-ground coupling involves the transmission of the signal on one phase and returns over a variety of paths depending upon ground resistance and the power line configuration. This mode of coupling results in higher attenuation and is less secure in the event of a fault on the coupled line. However, since only one phase is coupled, it tends to be less expensive

With phase–phase coupling, there is a return path over the other phase of the same circuit or phase of an adjacent circuit when inter-circuit phase–phase coupling is used. This method provides lower losses, lower attenuation, and greater security against loss of signal due to line faults. Therefore, phase–phase or inter-circuit phase–phase coupling is common at many utilities. Approximately 90% of all line faults are phase-ground, and therefore, the phase–phase coupling provides optimum reliability and security.

14.11.1 Protection Architecture with PLC

Since it was always possible for the towers themselves to come down coincident with a fault such as during hurricanes or tornadoes, it is common practice to use permission in one protection group for high speed and directional blocking in the other group to guarantee operation should a tower come down.

The problem with this scheme arrangement is that remote terminal protections are exposed to over-tripping for faults in a next line zone whenever local blocking type protection is removed from service for maintenance. Ideally, both protection groups should be directional comparison blocking with cross blocking across A and B groups.

14.11.1.1 Applications

The characteristics considered when applying PLC are line attenuation, impedance, and noise. The overall losses in carrier signal strength influence the transmitted power and receiver sensitivities. A cause for PLC system failure is the noise generated on the power line. Noise is generated from electric discharge (corona), switching, lightning, and line faults. To prevent these failures, the receivers are optimized functionally to prevent operating above a predetermined noise to signal ratio.

The permissive overreaching scheme requires a permissive signal to be transmitted to the remote end across a faulted line. In the presence of high noise levels, it may not always be possible to receive a permissive signal. Failure to receive the permissive signal would hinder high-speed tripping.

14.12 Synchronous Optical Network (SONET)

SONET refers to a fiber optic transmission system for high-speed digital traffic. It is an intelligent system that provides advanced network management and a standard optical interface. For tele protection, many utilities use SONET as a backbone to aggregate T1 (DS1) signaling. It does this by using time-division multiplexing (TDM) to send multiple data streams simultaneously.

14.12.1 Fiber Installation

Utilities started to install SONET systems over fiber by installing fiber either as Optical Ground Wire (OPGW) or All Dielectric Self Supporting (ADSS).

14.12.2 Multiplexer (MUX) Equipment

Many utilities are replacing legacy tone equipment with new digital communications based on T1 multiplexers. T1 is a term that evolved from early telephone voice multiplexing and transmission facilities.

Communication manufacturers offer utility-grade T1 multiplexers (T1 MUX) optimized for protection applications. The T1 MUX can accept many DS0 modules that can be inserted in the various slots in the MUX shelf. Besides the typical voice frequency, the MUX allows for a plug-in transfer trip module. Transfer trip modules consist of two optically isolated inputs and two output contacts per module. One input can be designated for transfer trip and the other for either permission or blocking. Conversely, one output contact can be designated for transfer trip and the other for either permission or blocking. Refer to Chapter 2 Figure 2.9 showing a typical T1 MUX rack used for line protection.

14.12.3 Definition of T1

The term T1 evolved from early telephone voice multiplexing and transmission facilities. T1 is currently used to describe almost any communications link operating at 1,544,000 bits/s. The main purpose of a T1 is to provide up to 24 channels of voice and data over a 4-wire metallic circuit or fiber optic pair. Refer to Chapter 11 for a detailed description of T1.

14.13 Three-Terminal Lines

Three-terminal lines are used for bulk power transfer and to supply loads from three power sources. It is essential that the protections at the three terminals operate for faults in the protected line zone and not outside the line zone. Protections are similar to that of two-terminal lines except with more sophisticated techniques to ensure selectivity.

Very often, an existing two-terminal line is converted to a three-terminal line as part of a program to reinforce the power system, and more recently, by the tap connection of distributed generators.

Multi-terminal line configurations increase the complexity of line protections based on distance relays. The complexity is caused by the existence of apparent impedances. These apparent impedances can be significantly large necessitating special protection scheme types to cater for them.

14.13.1 Scheme Type and Setting Considerations

14.13.1.1 Permissive Overreaching Scheme

With a permissive overreaching scheme being used for a three-terminal line, the distance relays at each terminal must be set to cover the maximum apparent impedance. Referring to Figure 14.39,

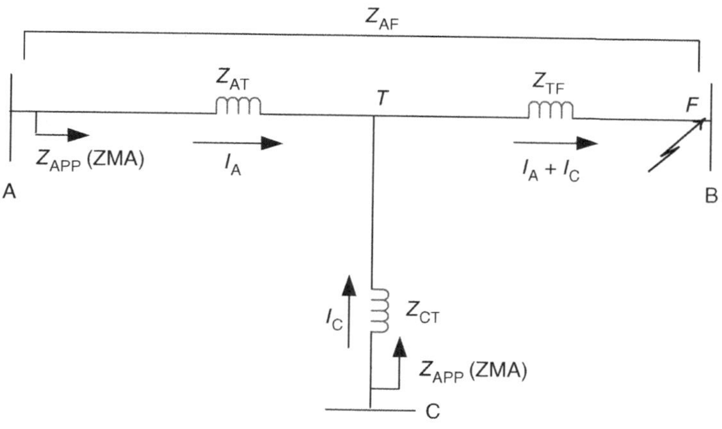

Figure 14.39 Three terminal lines with permissive overreaching.

for a fault at Terminal B as shown, for a permissive overreaching scheme to be successful both Terminal A and Terminal C must see the fault and key permission to each other respectively. To guarantee this, Terminal A and Terminal C protections must be set for the maximum apparent impedance (ZMA).

Very often a permissive overreaching scheme is simply not workable for a three-terminal line. Each terminal must be set to cover the maximum apparent impedance. However, this could lead to some very large Zone 2 settings. Electromechanical relays tend to lose directional stability in this circumstance. Correct directioning of these relays could not be assured by design when the ratio of Zone 2 to Zone 1 is larger than eight times. Digital relays are not affected the same way as they use different methods of obtaining directioning. Nevertheless, very large Zone 2 reaches are susceptible to loadability issues. In many cases, even blinding or using load encroachment techniques are not possible.

14.13.1.2 Directional Comparison Blocking Scheme

A permissive overreaching scheme must be set at all terminals for the maximum apparent impedance. Not so for directional comparison blocking schemes.

The apparent impedance at Terminal A

$$Z_{\mathrm{APP}} = Z_{\mathrm{AF}} + (I_{\mathrm{C}}/I_{\mathrm{A}})\, Z_{\mathrm{TF}}$$

The apparent impedance at Terminal C

$$Z_{\mathrm{APP}} = Z_{\mathrm{CF}} + (I_{\mathrm{A}}/I_{\mathrm{C}})\, Z_{\mathrm{TF}}$$

Assume that I_{A} and I_{C} are equal which means that the source impedances behind the buses at Terminal A and B are of equal value. Then set the Zone 2 relays at each terminals according to $I_{\mathrm{A}} = I_{\mathrm{C}}$. Then set each terminal for the following calculated apparent impedances:

Apparent impedance setting at Terminal A

$$\begin{aligned} Z_{\mathrm{APP}} &= Z_{\mathrm{AF}} + Z_{\mathrm{TF}} \\ &= Z_{\mathrm{AT}} + Z_{\mathrm{TF}} + Z_{\mathrm{TF}} \\ &= Z_{\mathrm{AT}} + 2Z_{\mathrm{TF}} \end{aligned}$$

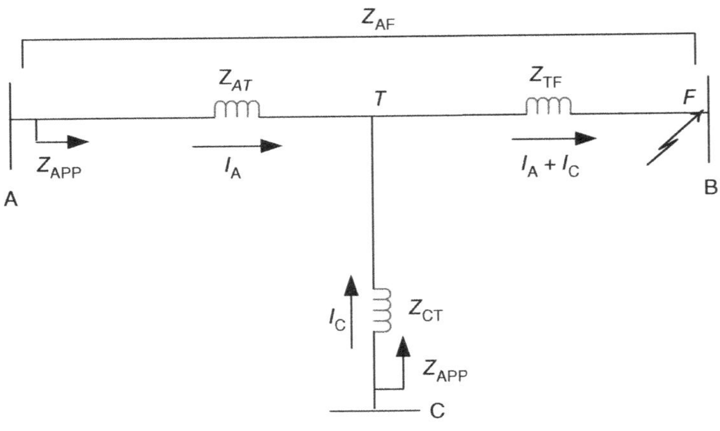

Figure 14.40 Three terminal lines with directional comparison blocking.

Apparent impedance setting at Terminal C

$$
\begin{aligned}
Z_{APP} &= Z_{CF} + Z_{TF} \\
&= Z_{CT} + Z_{TF} + Z_{TF} \\
&= Z_{CT} + 2Z_{TF}
\end{aligned}
$$

By setting each terminal Zone 2 at the line impedance to the tap then two times the line impedance to Terminal B with a 125% margin it is a certainty that at least one of the terminals either A or C Zone 2 will see the fault at Terminal B.

With equal infeeds, both Terminal A and C Zone 2 reaches see the fault at Terminal B. When the source impedance behind the bus at Terminal A is of a lower value than that of Terminal C, Terminal A will trip. When the source impedance behind the bus at Terminal C is of a lower value than that of Terminal A, Terminal C will trip.

Upon one terminal tripping, the other terminal trips sequentially as the apparent impedance are eliminated. In a permissive scheme, both terminals must see the fault for the scheme to operate correctly. In a directional comparison blocking scheme, only one terminal must see the fault. For sequential tripping to be successful, directional comparison blocking schemes must be applied at all terminals (Figure 14.40).

14.13.1.3 Weak End Infeeds and Current Reversals

Referring to Figure 14.41 for some system conditions, the current I_C at Terminal M can also be in the opposite direction where I_C is negative. The out-feed into Terminal C causes the apparent impedance at Terminal A to be less than the line impedance thereby resulting in an overreaching effect in the distance relay at Terminal A.

Where current reversals at one terminal can happen or where there is a chance for weak infeed at one terminal, a permissive scheme cannot be used. When distance relays are used, a permissive scheme must be converted to directional comparison blocking. The setting approach of one time the line impedance to the tap point and twice the line impedance from the tap point to the remote end guarantees that all terminals will eventually trip sequentially after a short intentional

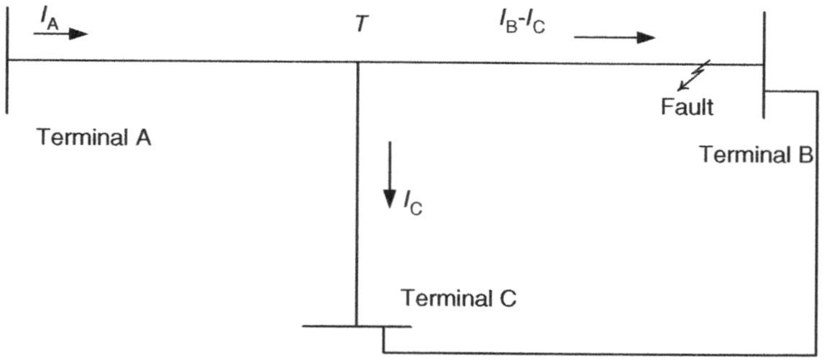

Figure 14.41 Outfeed on a three terminal line.

time delay, typically two cycles or less depending on the type of communications, waiting for a block signal.

14.14 Distributed Generation

Historically, generation would always supply a major bus at a terminal station. However, with new market rules, that is no longer the case, as many jurisdictions and utilities are opting for allowing generators, in many cases large ones, to tap directly into long transmission lines at multiple locations.

Multi-tapping of generation to transmission lines requires a trade-off of planning, economics, and protection complexities and can lead to compromises in reliability. The complexity of protecting these line configurations increases significantly from the two-terminal line with no tapped generation. Protecting multi-terminal lines in general has and continues to be a challenge for protection practitioners.

14.14.1 Traditional Protection Schemes

The traditional protection schemes placed into operation at many utilities beginning in 1968 following the 9 November 1965, wide-area blackout were very reliable. The introduction of multi-tapped generation into the transmission system transforms many of the two-terminal lines into multi-terminal lines. As such, there is some loss of protection integrity due to the reduction of Zone 1 reaches and increasing Zone 2 reaches with increased susceptibility to power swings. These protection schemes are no longer timeless and require attention as system topology evolves. This was not the case with the existing protection schemes that were fundamentally unaffected by changes to the power system when based exclusively on a fixed line impedance.

Protection complexities are introduced by the multi-tapping of generation. These complexities must be considered when evaluating high-voltage transmission plans that include multi-tapped generation. Past experiences leading to wide-area cascading blackouts have indicated that multi-terminal lines, in general, are more susceptible to outages than other line configurations, and are usually, the first to misoperate.

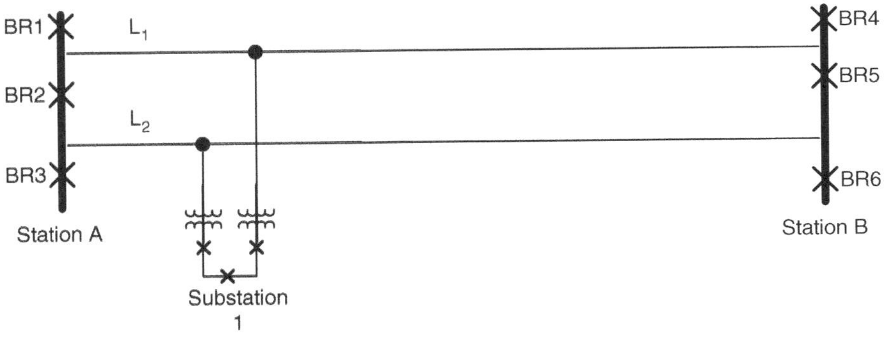

Figure 14.42 Transmission lines with no tapped generation.

14.14.2 Conventional Generation vs. Tapped Generation

14.14.2.1 Conventional Generation

Refer to Figure 14.42 showing two ended transmission lines supplied by two buses one at station A and the other at station B. Each of the buses is supplied by the interconnected grid including multiple generators behind it. The transmission line protections are designed and operate independently of the generation and are unaffected by it. The load substations with only load connecting to it are tapped to the transmission lines. There are no sources of generation on the transmission lines themselves.

14.14.2.2 Tapped Generation

Refer to Figure 14.43 showing two ended transmission lines supplied by two buses one at station A and the other at station B. Each of the buses is supplied by the interconnected grid including multiple generators behind it. However, there are also other sources of generation tapping directly to the transmission lines. The transmission line protections cannot exclude these sources of generation. The protections must be designed and operated to include the tapped generation that directly affects it.

Multi-tap generation is the term uses to describe when multiple sources of generation are tapping directly to the transmission line.

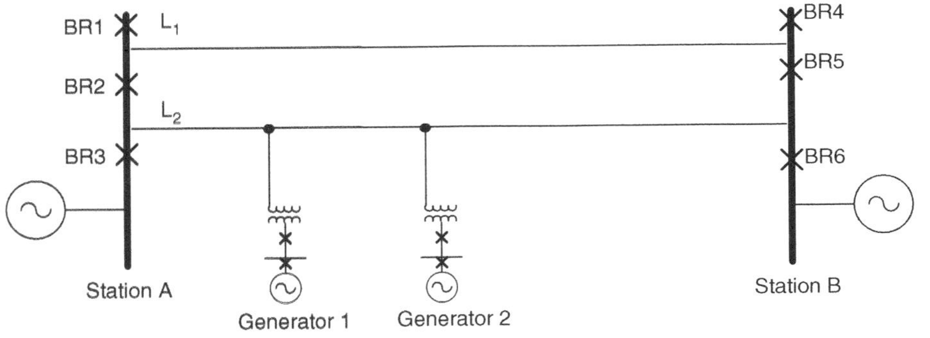

Figure 14.43 Transmission lines with tapped generation.

14.14.3 Impact on the Traditional Protection Schemes

Multi-tapped generation imposed on transmission lines required a material change in protection philosophy. The actual protection schemes to be used, and the associated settings would no longer be standard but would-be application dependent. It must take into consideration the specific topology of the transmission network, the protection scheme and associated settings used. Yet it must be adequate to meet the necessary clearing times and the reliability and security of the power system.

The protection complexities due to the multi-tapping of generation must be considered when evaluating high-voltage transmission plans that include multi-tapped generation. Past experiences leading to wide-area cascading blackouts have indicated that multi-terminal lines, in general, are more susceptible to outages than other line configurations, and are usually, the first to misoperate.

The introduction of multi-tapped generation into the transmission system transforms many of the two-terminal lines into multi-terminal lines. As such, there is some loss of protection integrity due to the reduction of Zone 1 reaches, and increasing Zone 2 reaches with increased susceptibility to power swings.

Take for example a typical 230 kV line whose impedance is typically 0.001 PU/km based on 100 MVA and base system voltages. Wind farm transformer impedances are in the order of typically 0.05 PU based on 100 MVA and base system voltages. This means that a wind farm transformer represents the equivalent of 50 km of 230 kV line. For any Zone 1, the reach of line 50 km or greater the Zone 1 could see into the generator-side of the wind farm as shown in Figure 14.44. It is unacceptable for the line to trip for generator sided faults at the wind farm. Zone 1 protections cannot be time delayed and blocked. The Zone 1 reaches therefore are reduced.

These protection schemes will no longer be timeless and will require attention periodically as system topology evolves. This was not the case with the existing protection schemes that were fundamentally unaffected by changes to the power system.

The following describes and documents the protection impact of connecting multiple generators to many transmission networks that allow for the introduction of tapped generation.

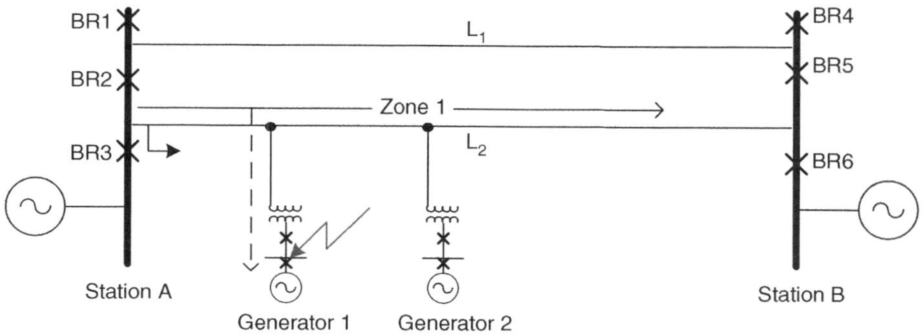

Figure 14.44 Zone 1 reach seeing into the generator side of a wind farm.

14.14.3.1 Permissive Overreaching Scheme

The Permissive Overreaching Scheme has always been used in conjunction with Direct Underreaching Zone 1 relays that overlap the transmission line. Any Zone 1 operation would trip locally and transfer trip the remote end via high-speed communications. Thus, the entire transmission line is protected via high-speed protections with communications. Faults on three quarters of the line are cleared from the local end instantaneously and unconditionally without communications altogether.

For faults beyond the reach of Zone 1 and in particular, for line-end-open, the permissive overreaching scheme will clear faults via communications with no intentional time delay.

The effect of tapped and multi-tapped generation is to force a reduction in Zone 1 reach and to force an increase in Zone 2 reach.

Zone 1 reaches at either end of the transmission line no longer typically overlap each other. For tapped generation close to one of the line terminals, the Zone 1 reach could dependent on apparent impedance become very short.

The Zone 2 reaches will become very large to cater for apparent impedances since in a permissive overreaching scheme the reaches must be set based on maximum apparent impedance.

Since the Zone 2 relays cannot be allowed to see into any of the tapped generation plants, it must be intentionally time delayed to wait for a blocking signal sent from all the tapped generation facility protections.

14.14.3.2 Directional Comparison Blocking Scheme

The Directional Comparison Blocking Scheme has always been used in conjunction with Direct Underreaching Zone 1 relays that overlap the transmission line. Any Zone 1 operation would trip locally and transfer trip the remote end via high-speed communications. Thus, the entire transmission line is protected via high-speed protections with communications. Faults on three quarters of the line are cleared from the local end instantaneously and unconditionally without communications altogether.

For faults beyond the reach of Zone 1 and in particular, for line-end-open, the directional comparison blocking scheme will clear faults without communications however with a short typically two cycle intentional time delay to wait for blocking signals to be received.

The effect of tapped and multi-tapped generation is to force a reduction in Zone 1 reach and to force an increase in Zone 2 reach.

Zone 1 reaches at either end of the transmission line no longer overlaps each other. For tapped generation close to one of the line terminals, the Zone 1 reach could dependent on apparent impedance become very short.

The Zone 2 reaches will still increase but not as much as in permissive schemes since it does not always have to cater for maximum apparent impedance as in a permissive overreaching scheme.

Since the Zone 2 relays cannot be allowed to see into any of the tapped generation plants, it must be intentionally time delayed to wait for a blocking signal sent from all the tapped generation facility protections.

14.14.3.3 Line Differential Scheme

Line differential schemes only compare currents without the need to measure voltages and are therefore immune to apparent impedance issues. However, they must compare the currents via either directly connected fiber or a SONET communication system. Present line differential technology is readily available for two and three-terminal transmission lines and less so for more than three terminals. There are five-terminal line differential protection relays available on the market.

The Permissive overreaching and Directional Comparison Blocking schemes have an enormous advantage over line differential since they can easily work with the existing communication systems between line terminals along with leased communication facilities to the multi-tapped generators. Line Differential relays are not designed to operate over leased communication facilities such as these. It is cost-prohibitive to install fiber to the generators with main and alternate separate routing complying with regional operating authority BPS requirements. There are also technical issues with timing delays which limit applications to only shorter line lengths. Therefore, line differential is typically not used for multi-terminal generator connections to transmission.

However, where the tapped generation is limited, it could be used along with a blocking signal from the tapped generation in case of faults internal to the generation facility. The stipulation is that the differential pickup sensitivity must be set above the outfeed component of the tapped generation to an external line fault and yet be below any minimum line faults with sufficient margin.

For two-terminal lines with tapped load substations and some tapped generation, the line current differential will not see faults through the much higher transformer impedance at the large load substations which are typically 0.2 PU. representing 200 km of 230 kV line. Therefore, there would be no need to send block signals from the load substations. However, the differential pickup sensitivity must be set above the load-substation maximum load.

14.14.3.4 Issues with Either Permissive or Blocking Schemes

When Zone 1 reaches from both of the terminals overlap, there is little difference between these two scheme types except for line end open. For faults at the far-end of the line which is open, the local protection is delayed by twice the communication time catering for permissive echo, for SONET two cycles, and for a blocking scheme delayed by two cycles. However, with both terminals operating, Zone 1 at least at one end would operate instantaneously and transmit a trip signal to the other terminal in high speed.

However, with multi-terminal generation, the Zone 1 reaches very unlikely overlap at best and at worst are very short while taking the underreaching effect of apparent impedance into account. This means that a timed delayed Zone 2 trip would take place for even close-in faults to the terminals. The worst case would be a breaker failure for a close-in fault in the adjacent line zone. The time delay for this scenario could be up to 250ms (depending on several local factors), and it is this protection total operating time that is used by the regional operating authorities reviewing the impact of protection on BPS stability.

Of concern, whether employing permissive overreaching or directional comparison is the outfeed effect that occurs when there is a concentration of generation near one of the line terminals. When the equivalent combined impedances of the generators are equal or lower than the source impedance, the apparent impedance could become impossibly large. In the worst case, fault currents could reverse and will leave the closest terminal instead of entering it. Both of these cases will severely affect the protection scheme and must be avoided. Line differential would not, however, be affected by the current reversal.

14.14.3.5 Preferred Scheme to Cater for Multi-Tapped Generation

Line protections based on permissive overreaching schemes would require conversion to directional comparison blocking. Thus, the line protection owned by the generator at the point of common coupling (PCC) will need to transmit blocking signals to each of the line protections at the terminals. It would not be necessary to transmit a block signal to other generator line protections as those protections could be time delayed to coordinate with adjacent high-speed line protections.

14.14.3.6 Converting from Permissive to Blocking Schemes

The procedure to convert from permissive overreaching to a directional comparison blocking scheme is relatively straightforward where there are digital relays and no electromechanical relays. Where the relays are still electromechanical, they are usually replaced with digital relays.

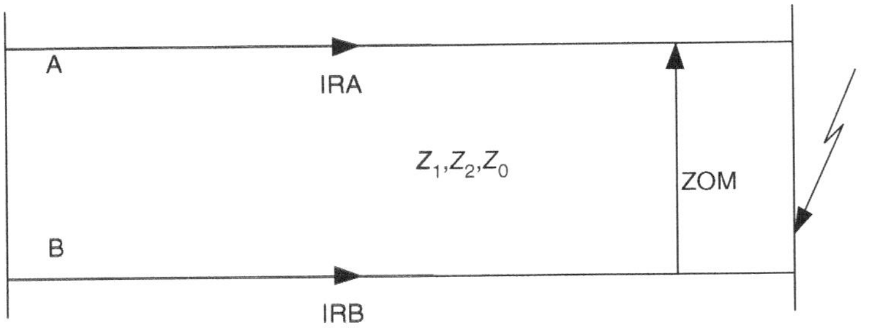

Figure 14.45 Effect of mutual impedance on Zone 2 reaches.

14.14.3.7 Ground Distance Protection Settings

Until the advent of tapped generation, the effect of mutual impedance between two parallel lines could for the most part be neglected while setting Zone 2 ground impedance reaches, refer to Figure 14.45.

The voltage seen by a ground distance relay for a fault on line A is $V_{\text{R-N}}$ as follows:

$$V_{\text{R-N}} = I_{1\text{A}}Z_1 + I_{2\text{A}}Z_2 + I_{0\text{A}}Z_0 + I_{0\text{B}}Z_{0\text{M}}$$

The voltage seen by the relay is either larger or smaller depending on the direction of the zero-sequence current in line B. This factor will cause the ground distance relay to underreach when the ground current is flowing in the same direction in both lines which is the normal case when remote breakers are all closed. The Zone 2 ground at the local end will underreach the fault due to the mutual effect.

The 125% margin built into the Zone 2 reach may or may not cover this underreaching effect for faults close to the end of the line near the next terminal. For those fault locations, the local Zone 1 at the remote end operates and sends transfer trip to the local end thus ensuring fault clearance even with severely underreaching distance ground relays due to mutual impedance.

When the line end is open with a fault close to the end of the line beyond the reach of the local Zone 1, the local Zone 2 relay must operate. Since the fault current can only flow along the faulted line and not on the adjacent line when the remote end is open, there is no mutual impedance to cause any type of underreaching. The local Zone 2 relays will certainly operate with the line end open. Philosophically, the overall protection scheme worked based on overlapping Zone 1 reaches with transfer trip always guaranteed to clear faults anywhere on the line with both terminals closed. With the remote end open, the local Zone 2 with permissive echo will always clear a remote end fault as mutual impedance does not exist when the remote end is open.

This philosophical basis for this line protection scheme is no longer valid. With Zone 1 reaches pulled back, the Zone 1 overlap is no longer a requirement. In essence, the entire line is only covered by Zone 2 with communication usually now in a directional comparison blocking scheme. It is essential that the Zone 2 relays be set for apparent impedance that includes the mutual impedance effect.

14.14.3.8 Load substations

In general, load substations do not affect line protections at the terminals of transmission lines in the same way that tapped generation does. The transformer impedance of a typical load substation is approximately three times larger than for a tapped generator transformer. This means that Zone 2 reaches at the terminals are very unlikely to see through the load-substation transformer to the LV distribution. Thus, it is rare for blocking signals to be required and happen rarely when apparent impedances are largely due to some other factor.

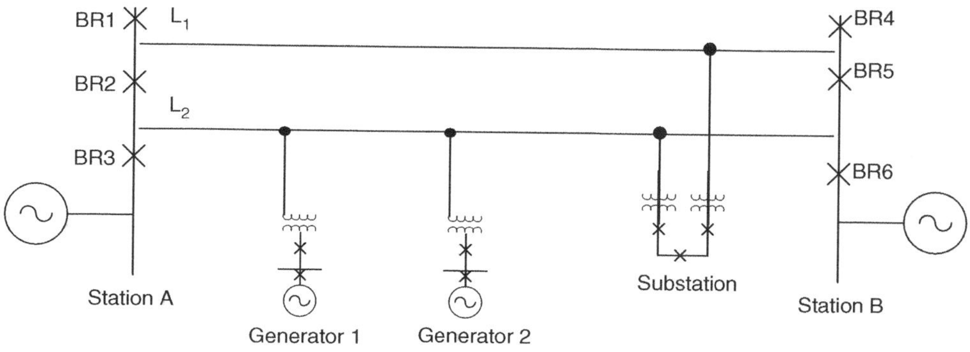

Figure 14.46 Effect of tapped generators on Zone 2 seeing the LV side.

14.14.3.8.1 Blocking Signals Required from Load substations

Refer to Figure 14.46 where there are a combination of multi-tapped generators and load substations on transmission lines. The Zone 2 reaches will have to cater for large apparent impedances created by the multi-tapped generators. Under normal circumstances, the Zone 2 reaches would not see the load-substation transformers but since the Zone 2 reaches are now set much longer, they may now see through the load-substation transformers and would need to be blocked.

14.14.3.8.2 Zone 1 Reach

From the utility perspective, there is no need to achieve overlapping Zone 1 reaches as the Zone 2 reaches in a blocking scheme protects equipment adequately.

There are redundant Zone 2 relays in two separated and independent protection groups at each line terminal. Zone 2 relays trip unconditionally after typically two cycles in the absence of any block signals. Each of the A and B protection groups are both served by separate and independent Main and Alternate communication facilities.

Regional operating authorities covering BPS facilities, in general, stipulate that Zone 1 reach with no intentional time delay cover a minimum amount of line distance typically 20 km seen from each line terminal where the fault level is the highest.

Since typical 230 kV line impedance is in the order of 0.001 PU/km, even covering an absolute minimum of 20 km of the line represents an impedance of 0.02 PU for Zone 1 reach. The smallest transformer impedance on system base for tapped distributed generation is 0.05 PU. Zone 1 set for 20 km of line will therefore never see through the transformers at the new distributed generation directly tapped.

14.14.3.8.3 Communications with Multi-Tapped Generation

Tapped generators must communicate with transmission line protections for transfer trip and blocking signals among other reasons. In most cases, more than one line terminal must be communicated simultaneously.

Referring to Figure 14.47, the generators must communicate with terminal station A and terminal station B for transfer trip, breaker failure, blocking, and possibly distributed generation end open (DGEO) to block automatic reclose without Synchrocheck.

It is usually acceptable for the generators to just communicate with station A. Then, station A will cascade these signals to and from station B via the utility's communication network. In effect, this would treat the utility like a black box that only needs to communicate with one location.

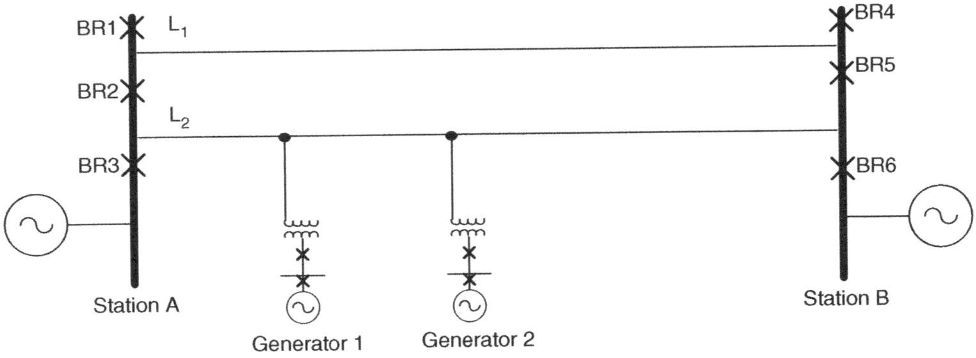

Figure 14.47 Distributed generators tapped directly to the transmission line.

14.14.3.8.4 Effects on Automatic Reclosing

When generation is connected to a transmission line, a DGEO signal is required from the generator station to the line terminal station where reclosure is enabled. This signal indicates that generation is isolated from the line and supervised reclosure of the line is allowed to proceed.

If the generator end open (GEO) signal fails and is not received, reclosure would not be allowed to proceed at the station where the running breaker reclosure is set to undervoltage plus time, refer to Figure 14.48. To block possible reclosure at other terminal stations and prevent possible damage to the generator, modifications are required since normally provision is included for the generator to communicate with just one terminal station. Reclosure should be delayed for up to 0.5 seconds to allow the GEO signal to be received and initiate reclosure. If the delay times out, indicating the generator is still connected, transfer trip will be transmitted to the terminal stations, thereby canceling reclosure, provided the protection system is designed for a trip signal to always override a close signal.

Another possible solution is to arrange reclosure settings as follows. The running breaker would be set at undervoltage plus time. All other breakers connected to the line would be set to Synchrocheck. At stations where the Synchrocheck function is performed by electromechanical relays, these would probably require replacement due to their unreliable performance with age. If this option is used, then the tapped load substations must also have Synchrocheck reclosing which therefore must be enabled.

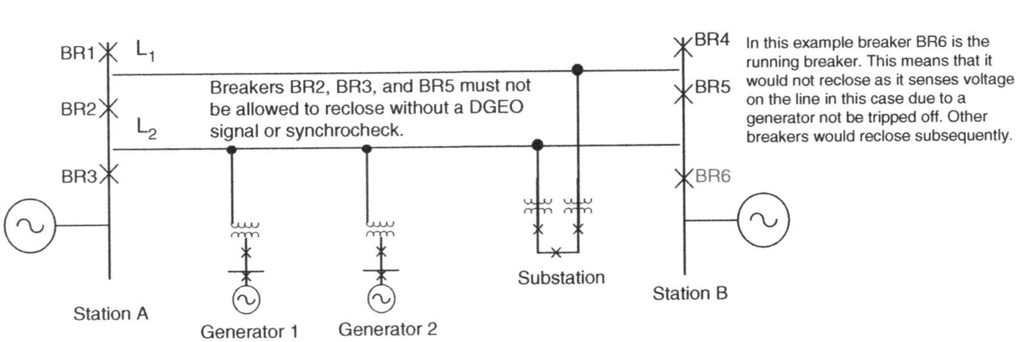

Figure 14.48 Automatic reclosing of terminal breakers.

14.15 Distance Relay Response to Resistive Faults

14.15.1 Background to Resistive Faults

Until now, it was assumed that a fault was a perfect short circuit for the development and illustration of protection principles. In reality, this is not the case as faults are not perfect short circuits and generally involve impedance at the fault itself that limits the short circuit current. Fault impedance can be made up of several things such as ionized air in the electric arc, tower grounding, tree branches, dry asphalt, concrete, wet grass to name a few. All of this phenomenon lead to fault resistance that has a significant effect on distance relays.

14.15.1.1 Resistive Faults

An electric arc is a non-linear phenomenon; however, it can simply be considered as a resistance that is dependent upon fault current and length.

The effective grounding impedance of towers is mainly resistive, its inductive part is greater when ground wires are used and its value is usually not dependent upon fault current.

The impedance of objects in the path of fault current is usually mainly resistive. Its value can drastically range from zero to an extremely high value; therefore, fault impedance can be described as an unpredictable quantity.

Line protections require minimum values of measured voltage and current to operate properly. Most line protections will not operate for a high impedance arcing fault such as a broken conductor falling on dry asphalt (1500 Ω) where the fault current is well below normal load currents. The industry is presently developing and researching other detection technologies (fault signatures, inference engines, etc.) to detect such high impedance broken conductor-type faults. However, this new technology is still very primitive and is not being used to trip but possibly to alarm only as it tends to be low in security.

This section is intended to provide a basic understanding of fault resistance and how it affects line protections. There is a need to set distance relays such that they do not overreach or underreach due to the combination of the arcing component and the remote end infeed.

Note, that this is a very important issue specifically for short transmission lines due to the fault resistance representing a substantial amount of the line impedance.

14.15.1.2 Distance Relays and Fault Resistance Coverage

Distance relays are mainly used to protect high voltage transmission lines. In the past, MHO distance relays were self-polarized. Modern MHO distance relays use cross-polarization, memory action, or a combination of both. The self-polarized relays have a disadvantage as it provides limited coverage for fault resistance.

The characteristic of an MHO impedance element as shown in Figure 14.49 when plotted on an RX diagram is a circle whose circumference passes through the origin. The equation that describes the reach at various line angles is as follows:

$$\text{Reach at any angle } \theta = \text{Reach at the MTA Cos (MTA} - \theta)$$

The impedance element is inherently directional and it will operate only for faults in the forward direction along line A–B. The impedance reach of the relay is adjusted by setting Zn, which happens to be the diameter of the circle. The angle is known as the maximum torque angle (MTA) in electromechanical relays is the angle of displacement of the diameter from the *R* axis. The MTA angle is also known as the relay characteristic angle (RCA) in digital relays.

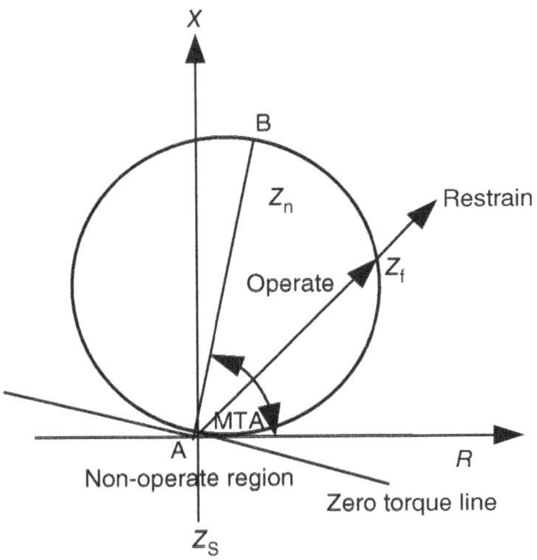

Figure 14.49 MHO relay characteristic.

The difference between electromechanical and digital is quite evident here. The torque applied to the induction cup of an electromechanical relay is a function of angle which is the maximum value at the MTA. A digital relay on the other hand will either operate or restrain depending on whether the fault is in the forward direction or the reversed direction as a function of the RCA. A zero-torque line can be drawn plus or minus 90° from the diameter of the circle. The relay operates for values of fault impedance Z_f everywhere within the Operate region of the circle.

The impedance reach varies with fault angle. As the line to be protected is made up of resistance and inductance, its fault angle will be dependent upon the relative values of R and X at the system frequency. For an arcing fault, or a ground fault involving additional resistance, such as a tower footing or resistance of fault through vegetation, the value of the resistance component of fault impedance will increase to change the impedance angle.

In the past, it was common practice to set the MTA less than the line angle so that a reasonable amount of resistance could be introduced without causing underreaching. In general, relay manufacturers did not recommend greater than 15° between the line impedance angle and the MTA.

In the example of Figure 14.50, the reach A-C at the MTA is the diameter of the circular characteristic and hence the longest reach from the origin. The line impedance angle is greater than the MTA whose reach A-B = Reach at the MTA (Cos θ – MTA).

If the line angle is 85° typical for 230 kV lines and an MTA of 75° typical of typical electromechanical relays is used, the reach is extended by a factor of 1 – (1/Cos 10°) = 0.015.

Therefore, by choosing an MTA lower than the line impedance angle by 10°, there is an extended reach of 1.5% to cater for fault resistance B-C.

Modern digital relays are not manufactured with fixed MTAs but have Characteristic Relay Angles set based on applied line impedance angle. The relay menu requests the line impedance upon set up and chooses this angle as the RCA. Therefore, the method previously described to provide some additional resistive coverage cannot be adopted while using digital relays.

Since the RCA is the same as the line impedance angle, another approach is to add resistive coverage by extending the reach of the relay at the line angle by a factor of 1/Cos (85°–75°) = 0.015. Therefore, the reach at the line angle should be arbitrarily increased by 1.5% to obtain the required reach when including some fault resistance as shown in Figure 14.51

Figure 14.50 Added resistive reach with MTA lower than line angle.

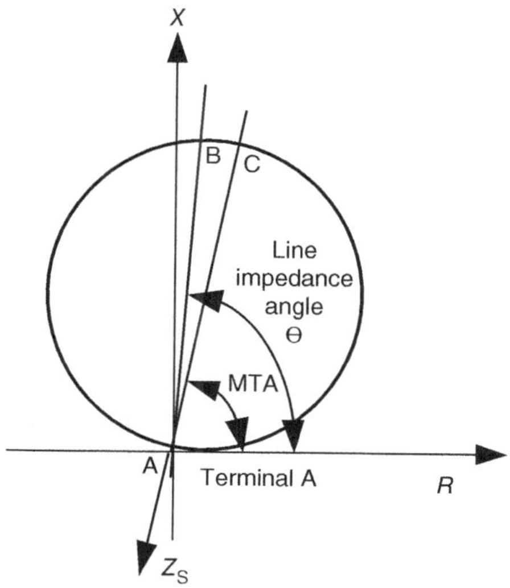

Figure 14.51 Reach extension to cover limited fault resistance.

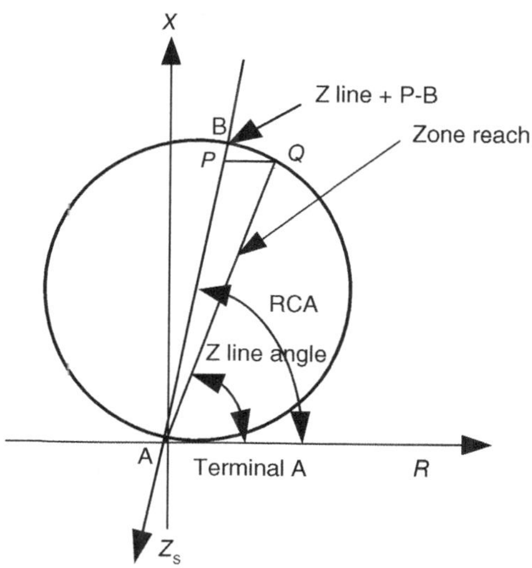

14.15.1.2.1 Fault Resistance

Faults on overhead transmission lines are usually accompanied by fault resistance. Phase–phase faults are often the result of high winds swinging the phase conductors close enough to flash or arc over. The resistance of the arc is dependent on the current in the arc and the length of the arc and can be approximated by using the following formula proposed by C. Russell Mason [1] and converted into metric IEEE std. C37.113-1999 [5] for phase–phase faults greater than 1000 A

$$R_{ARC} = (1804 \times L)/I\ \Omega$$

where

L = arc length in meters
I = arc current in amperes (RMS)

In the case of phase–phase faults, the arc length is largely defined by conductor spacing, with some variations introduced by conductor swinging and wind.

Phase-ground faults are usually the result of flashover of insulators started by lightning or by reduced insulation due to contaminants or defects. In addition to the arc resistance, the total fault resistance would include the tower resistance, the tower footing resistance, and the ground wire resistance. It is for this reason that distance relay MHO characteristics should not be exclusively relied on to detect high resistance ground faults. For Zone 1 applications, where it is essential that the relay not operate for faults beyond the next line terminal, directional ground overcurrent relays can be used but need to be set to cover very little line length. MHO relays can be set to cover much more line length. For Zone 2, where it is essential that the relays operate for all types of in-line faults regardless of fault resistance, sensitive ground directional overcurrent relays should be used. Refer to Section 14.7.2.2.5 where the ground directional overcurrent relays for Zone 2 need to be time delayed typically up to 120 seconds. It is for this reason that with the advent of digital relays it is advisable to invoke for Zone 2 the ground MHO elements for instantaneous tripping when there is little ground resistance and operating in parallel timed delayed ground directional overcurrent relays to cover the line when there is high fault resistance.

14.15.1.2.2 Mason – IEEE Formula

Due of the many variables, it is difficult to represent fault resistance with any degree of certainty. Small arc resistance values of 1 Ω or 2 Ω could present significant errors that can lead to erroneous distance relay operation, especially for short lines.

The shorter the line the more significant the fault resistance is. Take for example a 230 kV line that is 50 km long with a phase–phase fault.

Industry-standard is that a typical value for an arcing fault is 1 Ω or 2 Ω. These values are reasonably consistent with the calculated data in Table 14.1 for a phase–phase fault.

Referring to Table 14.2, assuming a value of 2 Ω as the arc resistance on a 50 km long line; the overall impedance including the arc is $4.42 + j24.2 = 25/80°$ Ω as opposed to a line impedance of $2.42 + j24.2 = 24/84°$ Ω

Now take the same 2 Ω as the arc resistance on a 10 km long line; the overall impedance including the arc is $2.48 + j4.84 = 5.4/63°$ Ω as opposed to a line impedance of $0.484 + j4.84 = 4.9/84°$ Ω

The change in reach and more significantly the change in reach angle is significant at 10 km but insignificant at 50 km.

Table 14.1 Phase-to-phase fault Arc resistance.

Arc length (m)	Arc current (A)	System voltage (230 kV)	Arc resistance (Ω)
7	1,000	230	12.63
7	2,000	230	6.31
7	5,000	230	2.52
7	10,000	230	1.26
13	1,000	500	23.4
13	2,000	500	11.7
13	5,000	500	4.69
13	10,000	500	2.34

Table 14.2 Typical 230 kV line impedances.

Line length (km)	Line impedance (no Arc) (Ω)	Line impedance (with 2 Ω Arc) (Ω)
10	484 + j4.84 = 4.9/84°	2.48 + j4.84 = 5.4/63°
50	2.42 + j24.2 = 24/84°	4.42 + j24.2 = 25/80°
100	4.84 + j48.4 = 49/84°	6.84 + j48.4 = 50/82°

Figure 14.52 10 km 230 kV line.

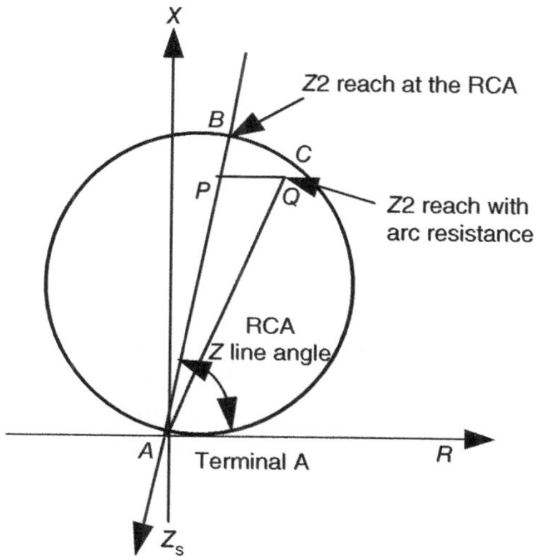

Referring to Figure 14.52 10 km 230 kV Line

A-P = 4.9 Ω
Add 25% margin for Zone 2 reach
A-B = 4.9 × 1.25 = 6.125 Ω
A-C = 6.125 Cos 21° = 5.72 Ω
A-Q = 5.4 Ω
C-Q = 5.72 − 5.4 = 0.32 Ω
% Margin = 0.32/5.4 × 100% = 5.9%

The introduction of a 2 Ω arcing fault reduces the % margin from 25% to less than 6% for Zone 2. The introduction of a realistic value for an arcing resistance between faulted phases leaves the Zone 2 underreaching by approximately 19%.

Referring to Figure 14.53 50 km 230 kV Line

A-P = 24 Ω
Add 25% margin for Zone 2 reach
A-B = 24 × 1.25 = 30 Ω
A-C = 30 Cos 4° = 29.9 Ω
A-Q = 25 Ω
C-Q = 29.9–25 = 4.9 Ω
% Margin = 4.9/25 × 100% = 19.6%

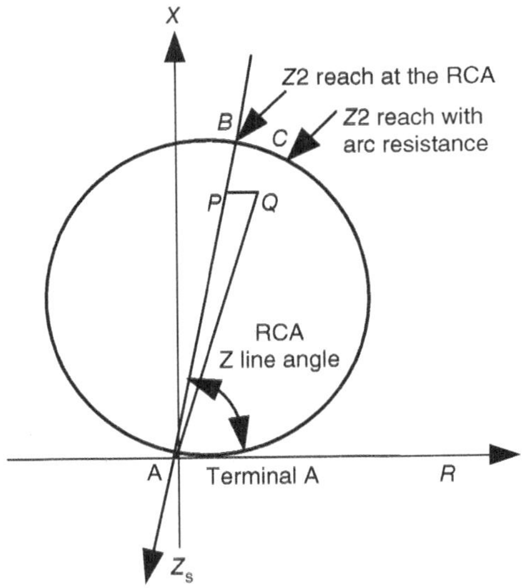

Figure 14.53 50 km 230 kV line

The introduction of a 2 Ω arcing fault reduces the % margin from 25% to just less than 20% for Zone 2. The introduction of a realistic value for an arcing resistance between faulted phases leaves the Zone 2 underreaching by approximately 5.5%.

The underreaching effects of fault resistance calculated above similarly apply to the Zone 1 reach.

For a Zone 1 reach of 80% of line impedance, the underreaching effect of the arcing resistance for a 10 km line reduces the overall reach to just 65%.

For a Zone 1 reach of 80% of line impedance, the underreaching effect of the arcing resistance for a 50 km line reduces the overall reach to 75%.

14.15.1.2.3 Close-in Resistance Faults

The shorter the line the more significant the fault resistance is. The reach in this example is almost along the *R* axis where the reach is short.

MHO elements that do not have inherent expanding characteristics have limited resistive reach on close-in faults as shown in Figure 14.54. The MHO characteristic representing a self-polarized

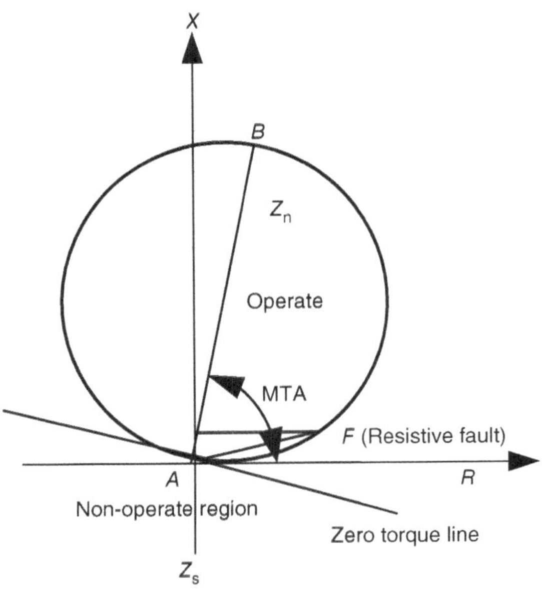

Figure 14.54 Example of a short line and a close-in resistive fault.

MHO element, whose characteristic passes through the origin on the R–X diagram, has very little relay reach along the resistive axis.

14.15.2 Distance Relay Response to Resistive Faults

14.15.2.1 Expanding MHO Characteristics

It is the use of sound phase polarization and memory of stored voltages that allows the MHO characteristic to expand upon indication of fault that leads to much greater close-in resistive fault coverage.

14.15.2.1.1 Polarizing Methods

The dependability of a relay's functional response to a specific fault is related to its polarizing quantity. Distance elements may be voltage or current polarized (or both) or by zero or negative sequence quantities. Different polarizing techniques may be applied depending on the manufacturer, the type of impedance characteristic used, or the type of fault. The most commonly used techniques are as follows:

- Positive sequence voltages.
- Cross-polarization.
- Self-polarization.
- Zero-sequence voltage polarization.
- Zero-sequence current polarization.
- Negative sequence voltage polarization.
- Memory voltage polarization.

When a fault occurs, the polarizing quantity should be stable and last long enough to guarantee that the protection element remains picked up until the fault is cleared.

14.15.2.1.2 Polarizing Quantity

There are several various polarizing quantities developed for phase and ground MHO distance functions. The following are some of the more commonly used:

- Self-polarized (V_a for Phase A function, V_{ab} for the Phase AB function, etc.)
- Positive sequence voltage (V_{a1} for Phase A, V_{ab1} for Phase AB function, etc.)
- Quadrature voltage (V_{bc} shifted leading 90° for Phase A, etc.)

An MHO characteristic other than being self-polarized is referred to as cross-polarized. All MHO characteristics require voltage to operate. For a fault at the relay location, the voltage will be small and a self-polarized MHO characteristic may not operate at all. A cross-polarized MHO characteristic will except for a three-phase fault. For a three-phase close-in fault, all three voltages will be small such that even a cross-polarized MHO characteristic will not respond. To deal with this apparent shortcoming of cross-polarized MHO characteristics, memory is added to the polarizing circuits.

14.15.2.1.3 Influence of Polarization on Expanding MHO Characteristic

The result of memory action is to produce a dynamic (time-varying) response from the function that is different from the steady-state response. This results in a dynamic characteristic shown in Figure 14.55 indicating an MHO Characteristic B, an expanded characteristic about the reach point, has significantly more resistive coverage. This is the influence of polarization in the implementation of the MHO element. The polarization is typically accomplished with memory voltage and with voltage from healthy phases. Many relays use the positive sequence voltage for polarization. The expansion is below the origin and is a function of the source impedance behind the relay terminal so the resistive reach may vary to a large degree.

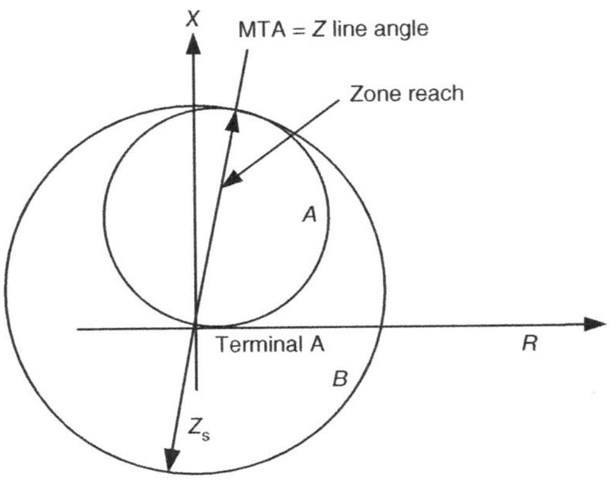

Figure 14.55 (a) Self polarized and (b) cross or memory polarized MHO elements.

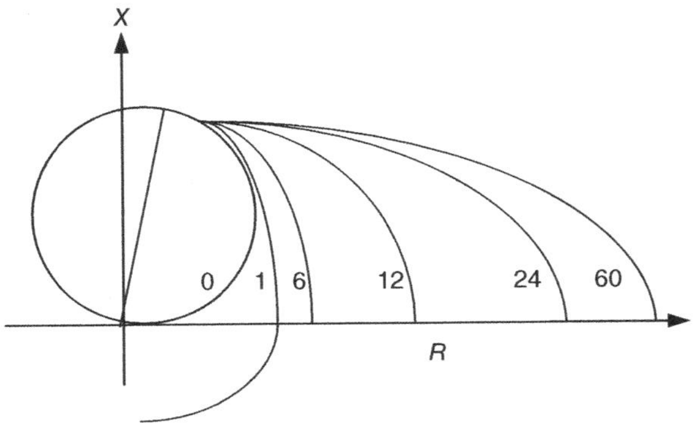

Figure 14.56 Resistive expansion of a cross-polarized MHO with increasing SIR.

14.15.2.1.4 Influence of SIR on Expanding MHO Characteristic

In some installations with a strong source behind the relay terminal, the expansion could be negligible such that the MHO characteristic will resemble that of the self-polarized MHO Characteristic A in Figure 14.56. A strong source is characterized by a low Source Impedance Ratio (SIR), where SIR = $Z_{\text{Source}}/Z_{\text{Line}}$. In most utility systems, source impedances (>230 kV) are typically in the order of 10% of line impedance which means that a common SIR is about 10.

It should be noted, that the apparent extension into the negative reactance quadrants does not imply that there would be operation for reverse faults. With cross-polarization/memory, the relay characteristics expand to encompass the origin for forward faults only. For reverse faults, the effect is to exclude the origin of the impedance diagram, thereby ensuring proper directional response for close-up forward or reverse faults.

14.16 Power System Considerations

Line protection systems must detect and clear all-in zone faults without needlessly restricting maximum load transfer capability for the protected line wherever possible.

Notwithstanding the previous statement, when load current flows through transmission lines at a given voltage, the load being supplied via the transmission line can be seen by distance relay line protections as an impedance. It is necessary that the measured load impedance not encroach upon the MHO relay impedance characteristic under any circumstance.

14.16.1 Loadability During Normal System Conditions

Refer to Figure 14.57 showing a Zone 2 MHO distance relay which is set with a reach of 0.2035 PU @82.4°. It must be set to protect for all line fault types on the line including apparent impedances.

Typical inductive load on power systems is in the range of 25°–30°. An industry-accepted angle of 30° representing most power system loads having a Power Factor of 0.87 is used. Refer to Figure 14.57, based on a required setting of 0.2035 PU and an MTA angle of 82.4°, the maximum load transfer capability for this reach setting is as follows:

$$\begin{aligned}\text{Point A on Figure 14.57} &= 0.2035\ \text{Cos}\ (82.4°-30°)\ \text{PU}\\ &= 0.124\ \text{PU}\\ &= 0.124\ \text{PU} \times 484\ \text{Base Ohms (on 220 kV, 100 MVA)}\\ &= 60\,\Omega\ \text{Primary}\\ &= 60/5 = 12\,\Omega\ \text{Secondary (with a } Z \text{ Ratio} = 5)\end{aligned}$$

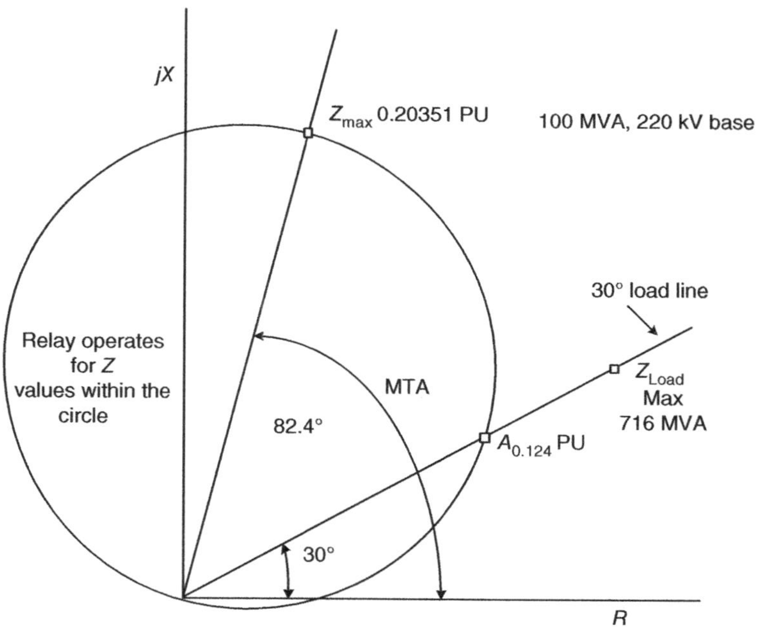

Figure 14.57 Typical loadability issue.

A load of (100 MVA/0.124 PU) = 806 MVA represents the load value that would cause this relay and setting to marginally operate. To prevent misoperation, this line should be restricted to the maximum load transfer of less than 806 MVA with a sufficient margin.

14.16.2 Loadability During Extreme System Conditions

In the previous loadability example, the system voltage is presumed to be maintained at 1.0 PU. However, during the 14 August 2003, North American blackout, the system voltage dropped well below 1.0 PU. The main cause for the voltage drop was the shaking off of a large number of generators during the event. Until this blackout, it was always assumed that even during a significant transmission event that the system voltage would be maintained at or close to 1.0 PU. Therefore, when protection practitioners plotted load on an R–X diagram, the load and MHO characteristic were always plotted in PU impedance based on 220 kV, 100 MVA or 500 kV, 100 MVA. The effect of reduced system voltage is to bring the load point closer to the relay for the same amount of load current. Refer to Figure 14.58 showing the change in load impedance seen by a relay with varying generator voltages while the load transfer across the line remains constant.

With a load current of 1.0 PU and Generator voltage of 1.0 PU

$$V_{\text{Relay}} = (1.0\text{–}0.05)\ \text{PU} = 0.95\ \text{PU}$$

$$Z_{\text{Relay}} = 0.95\ \text{PU}$$

with a load current of 1.0 PU and Generator voltage of 0.85 PU

$$V_{\text{Relay}} = (0.85\text{–}0.05)\ \text{PU} = 0.80\ \text{PU}$$

$$Z_{\text{Relay}} = 0.8\ \text{PU}$$

Since the relay measures an impedance of 0.8 PU instead of 0.95 PU, the relay reach will be extended by approximately 16% due to the reduction in system voltage.

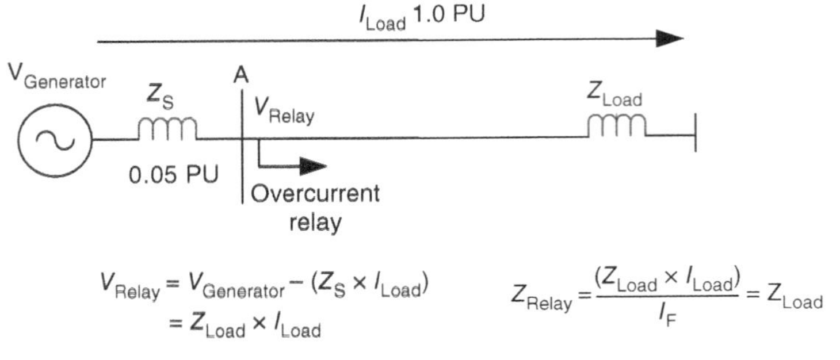

Figure 14.58 Load impedance seen by the relay with varying generator voltages.

14.16.3 New Loadability Criteria in North America

A significant conclusion of the 14 August 2003, blackout in the Northeast region of North America was protective relaying should not preclude operator action during extreme system emergencies. The conclusion was the system operator should be allowed up to 15 minutes subsequent to an extreme contingency in which emergency actions including load shedding could be performed. To meet this requirement, a thermal rating recommendation was established. A load of 150% of the transmission line's highest seasonal ampere circuit rating that most closely approximates a 4-hour rating was set as the maximum loading criteria for this purpose.

Two parameters were used in the subsequent North American Electric Reliability Corporation (NERC) Blackout Recommendation. Firstly, the system voltage that dropped across the entire region on August 14 needed to be addressed. For load encroachment purposes, it could no longer be a given that system voltage would remain at 1 PU. A system voltage of 0.85PU was chosen. A power factor of 0.87 representing a power factor angle of 30° current lagging voltage was also chosen. Just like the thermal rating, the voltage value of 0.85 was an observed value when the system was in an extreme condition but not in a cascading mode. Following the 14 August 2003, blackout, NERC initiated a loadability review and mitigation program that led to setting changes all utilities in the interconnected transmission system under their jurisdiction, must comply with.

Figure 14.59 shows where the closest load point is placed with respect to the MHO characteristic previously catered for by many utilities for load encroachment and where it is now under the new NERC PRC-023-4 standard [2]. Previously, many utilities adopted a 20% margin based on the reach from the origin to the relay characteristic at 30°. The new NERC standard stipulates a much higher margin based on the new criteria. The example in Figure 14.59 does not include the effect

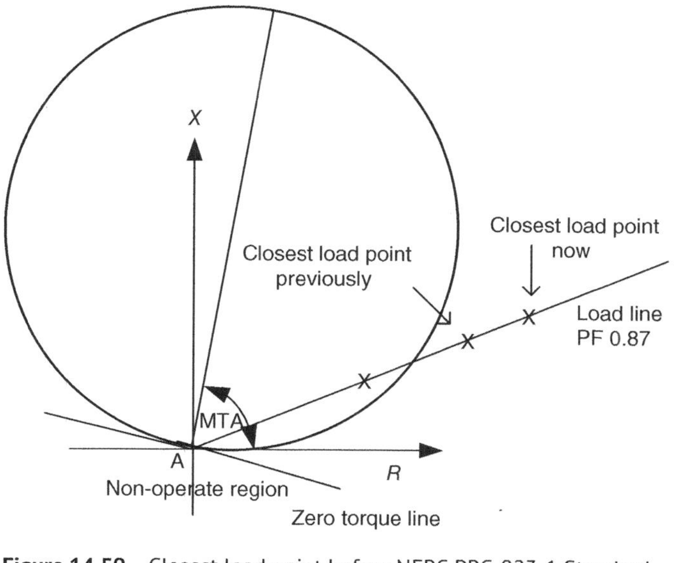

Figure 14.59 Closest load point before NERC PRC-023-1 Standard.

of lowering the system voltage to 0.85 PU from 1.0 PU or the actual definition of maximum line loading which when all taken into account makes it significantly more conservative. Where these criteria cannot be met the relays must be "blinded" to not operate for the given load and system conditions.

14.16.3.1 Various Methods of Blinding Relays to Load

Refer to Figures 14.60–14.62 showing various techniques used to blind relays to load, (block relays from operating on load). Any one of these techniques is an acceptable method of ensuring distance relay reaches at line angle are not affected by line loadability under any foreseeable circumstance [3].

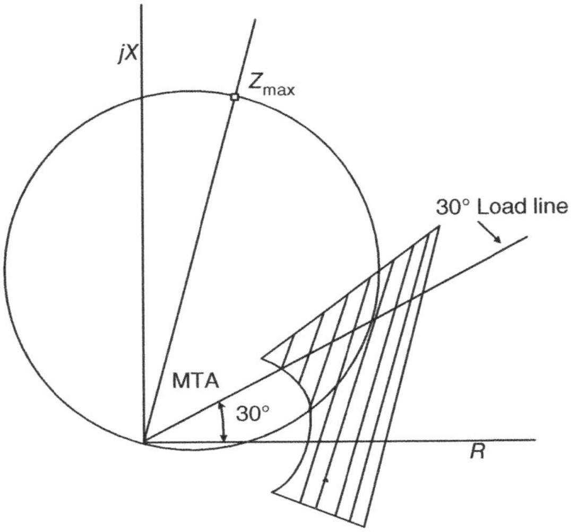

Figure 14.60 Load encroachment by Schweitzer Engineering Laboratories™.

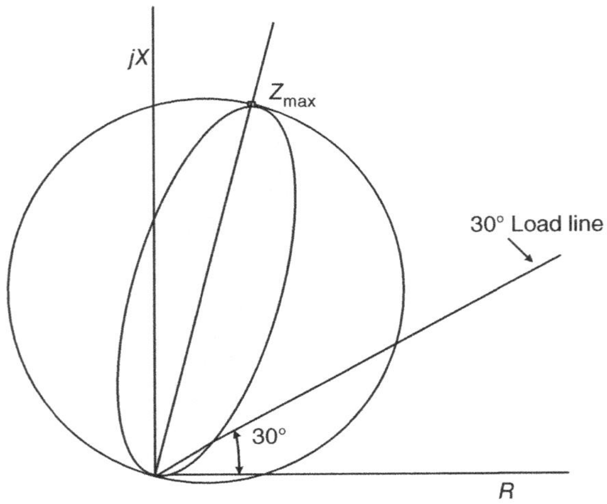

Figure 14.61 Lens characteristic solid state and digital relays.

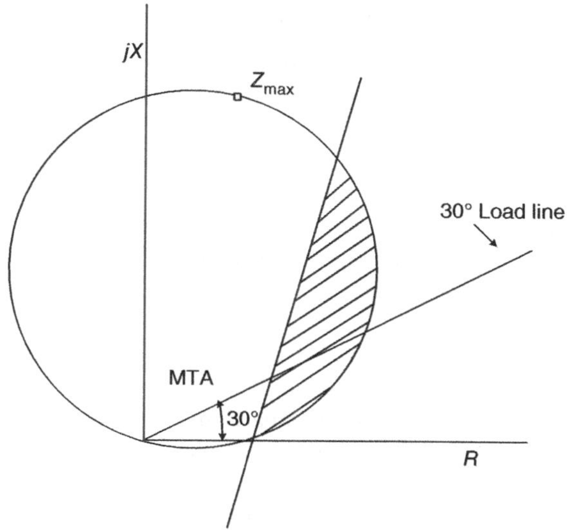

Figure 14.62 Traditional electromechanical relay blinder.

14.17 Line Current Differential Protection

14.17.1 Introduction

The current differential principle was initially used and is still being used, as electromechanical pilot-wire protection. In an electromechanical pilot-wire relay, the three-phase currents are combined into one composite signal that is transmitted over traditional metallic wires known as pilots. The composite signals are compared differentially with respect to one another in each relay located at all ends of the line.

With present-day advancements in digital communications and microprocessor-based protection relays, high-speed reliable current differential protection systems are now available. The scheme performs a differential comparison (Kirchhoff's current law) on a per-phase basis and communicates using several types of communication media. SONET/fiber communication systems, that multiplex 64 kb/s digital channels, or by direct fiber connection provide a reliable and high-speed communication system for line current differential protection.

The current differential principle is suited to protect two-ended or three-ended lines and need not contend with problems associated with voltage, loading, or power swings.

Line current differential relays are mostly supervised with distance elements. Some utilities assign a dual function to the distance elements and allow them to trip unconditionally after a predetermined time delay, typically 400 ms. The purpose for these timed distance elements is to back up the communication systems used for line current differential relays. Where line loading or system swings place these elements at risk, they can be supervised with a loss of communications alarm point before they can trip.

14.17.2 Functional Description

The overall functionality of line current differential is similar in concept to analog line differential schemes using electromechanical pilot wire relays. All differential schemes are based on Kirchhoff's current law at a system level. The main difference between traditional pilot wire relays and the new digital line differential relays is that digital relays utilize digital computations of analog signals. This signal is used to exchange data via digital communications media between terminals for current comparison. Each relay, located at every terminal, continuously takes digital samples of currents and their phase angles at the local line terminal and makes tripping decisions based on a comparison of the local values with samples received from the remote relays over the communication channels.

At each local terminal, a set of current samples for all three phases and neutral is used to compute what is known as phaselets. These phaselets are sent in data packets every one-half cycle to the remote terminal(s) for comparison. Each phaselet has real and imaginary components that are required to represent a complete phasor. This technique allows information for all three phasors and neutral to be sent over a 64 kb/s DSO communication channel. The reconstructed phasors, as received from the remote terminal(s), are compared. Any calculation of differential vs. restraint current in excess of the setting will operate the relay. In addition to tripping locally, any one or more relays making decisions to trip locally will also send a transfer trip signal to the remote relay(s) over the line differential communications channel. Transfer tripping caters for weak infeeds or even current reversal at one or more terminals.

An extremely important aspect of line current differential protection is the need to maintain the relays at all line terminals in constant synchronization with respect to the communications data exchanged between units. Each relay needs to know the precise time each digitized data packet was sent and received.

Line differential relays achieve synchronization by continuously adjusting the internal clock of all relays during initial startup and subsequently during operation to allow the sampling time for each data packet to be successfully compared.

Digital communications adopted by most utilities for interfacing with the SONET system incorporates the IEEE C37.94 [4] fiber optic interface standard that permits direct connection with SONET. Direct fiber is also possible including dual redundant direct fiber for short distances between stations or from one switchyard to another at the same station.

14.17.3 Lines with Tapped Load Substations

Line current differential protection is best applied to clean two or three-ended lines. Where lines are tapped with load substations the application of this form of protection becomes more complex. It may be possible for the line current differential protection to operate for out-of-zone faults due to the uncompensated infeed from tapped load substations where the LV bus-tie breakers are operated normally closed as shown in Figure 14.63. In most cases, the infeed is relatively low for phase faults not involving ground and can be accommodated via differential current setting sensitivity.

However, where the HV neutral of the load-substation transformers are grounded they may represent a significant ground source that could render the application of line differential protection impossible as shown in Figure 14.64.

Zero-sequence compensation native to many line differential relays as a settable feature accommodates zero-sequence out-flow to an external fault from other sources such as from a grounded primary neutral of a tapped load-substation transformer. This method can be used effectively to deal with ground sources not bounded by the differential zone.

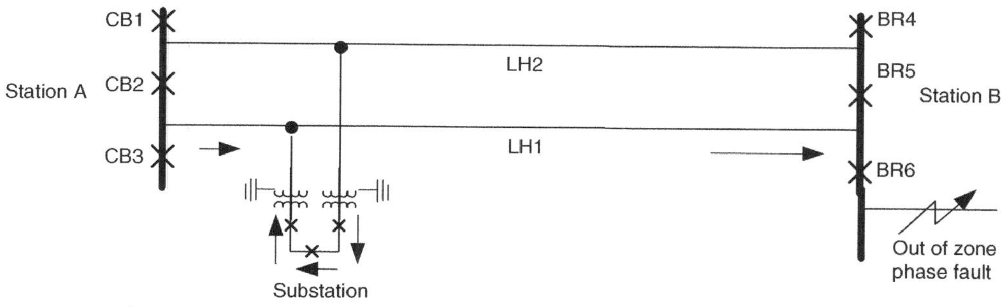

Figure 14.63 Out-of-zone phase fault with a tapped load substation.

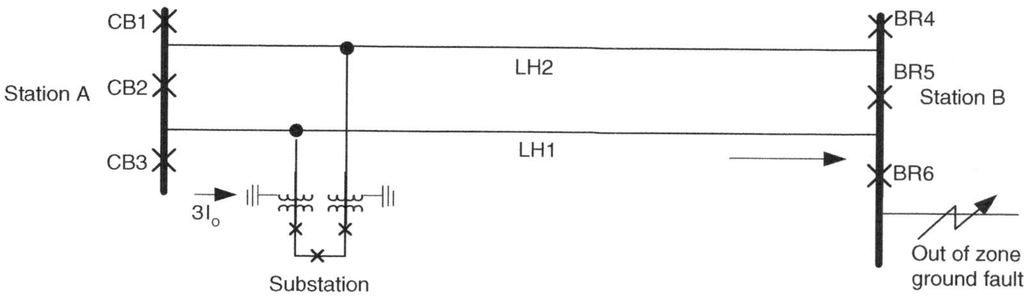

Figure 14.64 Out-of-zone ground fault with a tapped load substation.

In general, where there is more than one load substation tapped to the line it may become increasingly difficult to apply line current differential protection. It may not be possible to set the main differential element pickup above the total connected load-substation load plus margin with adequate sensitivity to operate correctly for low-level internal line faults.

To allow the use of line current differential on lines tapped with load substations, independent phase and ground MHO distance elements supervising the line differential relays may be used to block the differential elements. In this case, the distance elements are set not to reach into the LV side of load substations.

The minimum pickup of the differential current element must be higher than the aggregate maximum emergency loading of the tapped load substations with sufficient margin. The supervising distance elements must be set to cover at least 125% of the maximum apparent impedance of the line being protected while still not reaching through transformer winding impedances to the LV side at load substations. A measure of how successful a particular application is in meeting this competing criterion will determine whether it is acceptable to employ line current differential for tapped lines.

For two-ended lines with one tapped load substation, there is a possible option of applying three ended line current differential to include a relay at the load substation. This may only be possible if the load substation has a set of three HV CVTs on the transformer primary for the supervising distance relays. Also, appropriate communications must be available to the load substation. A DS1 communications link to the load substation would work. However, this may not be a viable option for system planners who must be cognizant of future planned load substations tapped to the same line where this option would no longer be possible.

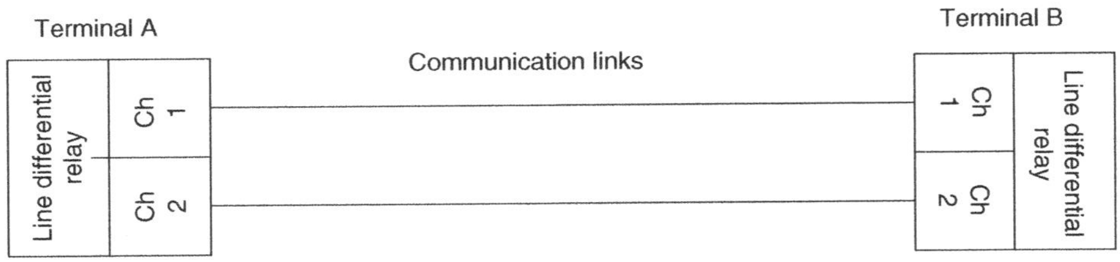

Figure 14.65 Communication links for a two-ended scheme.

Regardless of the interface, redundant communications over physically different paths are usually required by utilities. The function of the two channels differs depending on whether a two or three-ended scheme is applied.

Distance supervision is usually used in all applications even where there are no tapped load substations to prevent misoperation when the line is energized when line differential relays tend to trip on line in-rush current due to capacitance to ground.

14.17.4 Two-Ended Scheme

Referring to Figure 14.65, each line current differential relay comes with two communication ports. Channel 1 of each relay is the main channel and Channel 2 is the alternate channel. The relays will transmit differential data over both channels simultaneously. Channel 1 is applied over the main diverse route and T1 access multiplexer. Channel 2 is applied over the alternate diverse route and T1 access multiplexer. Thus, any single loss of either the main or alternate route does not affect the functioning of either the A or B group line differential protection.

Each of the relays independently makes a local tripping decision based on a comparison of the local and remote waveform samples. When local tripping occurs, a backup transfer trip signal is also sent to the remote relay. This transfer trip feature is also used to support breaker failure and remote trip/transfer trip cascading to/from tapped load substations.

14.17.5 Three-Ended Scheme

Referring to Figure 14.66, in a three-ended scheme, the use of the dual communication channels is different from the two-ended schemes. Each relay connects to the two remote units using both channels. Channel 1 of any given relay connects to Channel 2 of the next relay in a clockwise direction, forming a ring between all three relays. Under normal operation, with all communication routes in service, each relay receives two remote current waveform samples and makes its local tripping decision based on a comparison with its local acquired samples. A local trip decision also causes transfer trip to be sent to the two remote terminals.

If, however, one of the three bidirectional communication paths is interrupted, two of the three relays will not be able to exchange current samples with each other. Under this condition, the one remaining relay will still be able to receive remote samples from the other two and is still capable of making a local tripping decision and sending transfer trip which becomes the primary method of tripping the remote terminals. A single communication channel failure changes the system configuration from Master–Master–Master to Slave–Master–Slave where the Master is the relay that has both channels live. Transfer trip is also used for breaker failure and load-substation cascading tripping.

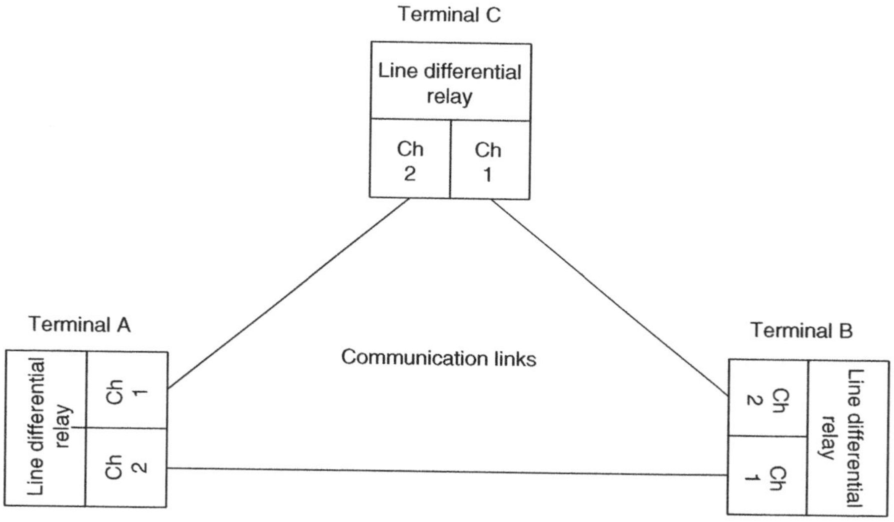

Figure 14.66 – Communication links for a three-ended scheme.

It should be noted that unless the routing of Channel 1 and Channel 2 interfaces with a different T1 access multiplexer, a single T1 access multiplexer outage will cause the differential elements of all three relays to be blocked due to the loss of communication on two of the three communication paths.

14.17.6 Clock Synchronization

A very important difference exists between line current differential using digital communications compared with older analogue circulating current systems (pilot wire). Due to the digital sampling and communication techniques, there is a critical need to maintain the internal clocks within the same protection zone in constant synchronization. The instantaneous digital snapshots of currents entering and leaving the protected zone must be compared in real time. Where the two-line differential relays are direct fiber connected, this is not a problem. However, it does become a problem where the communications linking the two relays adds latency as when multiplexer interfaces are used. The communication latency once established does not change or drift. Therefore, the method used by these relays to compensate for latency is upon initial set-up these relays use a Ping-Pong synchronization algorithm that measures and memorizes the latency and takes it into account while comparing local and remote terminal currents.

Newer versions of these relays allow for external IRIG-B time synchronization at all terminals. However, caution should be applied as these relays may be blocked on the loss of IRIG-B and likely would not revert to the original independent Ping-Pong synchronization algorithm.

14.17.7 Settings

14.17.7.1 Line Current Differential Element Settings

The differential element must be set with adequate margin to see the worst-case remote fault with the remote terminal(s) open (single end infeed). Also, when tapped load substations are connected to the HV line but not included in the line differential zone, the differential element must be set above the total maximum emergency load of the connected load substations assuming single end infeed.

14.17.7.2 Distance Element Settings

These elements supervise the current differential element and provide timed distance backup functions similar to traditional line protections. These elements are set for the maximum apparent impedance with margin. The mutual impedance of parallel conductors must also be taken into account.

14.17.7.3 Phase and Ground Distance Settings

Zone 2 MHO timed backup elements are set for the maximum apparent impedances with margin and timed typically at 400ms.

14.17.7.4 Distance Elements Settings – Tapped Load-substation Stations

The distance elements in this case are used to supervise the main differential element without intentional time delay. Therefore, they must be set at maximum apparent impedance with margin and also block short of the LV of all load substations. This is usually not a problem. However, when load substations are located close to one terminal of a long line or in three-terminal lines, it may not be possible to satisfy both of these criteria simultaneously.

14.17.7.5 Three-Terminal Lines

Setting distance elements are complicated for three-terminal lines with or without load substations due to the high apparent impedances. Weak end infeed from one terminal under certain power system conditions make things worse. For this reason, when the logic for line current differential relays is established for three-terminal lines, the supervision signals of the internal distance elements are cascaded around the loop. Functionally, any terminal with any supervision element picked up is capable of enabling the line current differential elements at both remote terminals.

14.17.8 System Considerations

14.17.8.1 Weak End Infeeds and Current Reversals

Section 14.13.1.3 describes the difficulties in applying traditional distance relaying schemes for systems that possess weak end infeeds or even current reversals, (Figures 14.67).

An option to using distance relays is line current differential whereby the currents are converted to digital signals at each terminal and compared with respect to each other via SONET. The advantage to line current differential is that regardless of whether fault currents are entering or leaving

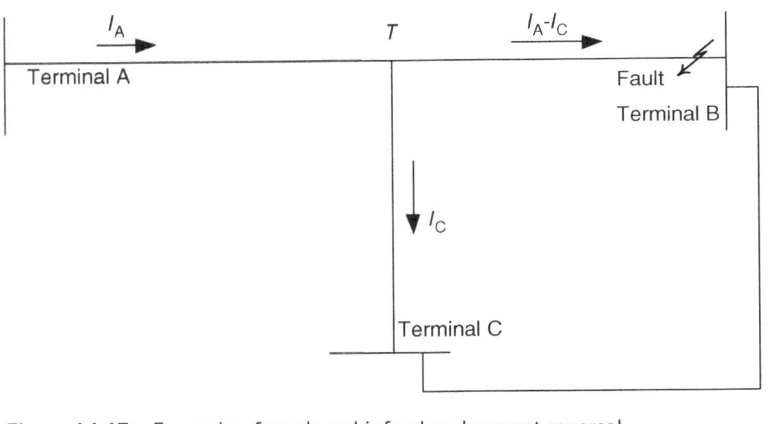

Figure 14.67 Example of weak end infeed and current reversal.

the protected zone, as long as Kirchhoff's current law is respected, correct protection operation is assured without intentional time delay.

14.17.8.2 Line Loading and System Swings

The possibly large Zone 2 settings due to large apparent impedances could affect the loadability of multi-terminal lines. The load impedance could encroach into the MHO characteristic of the phase distance element. This could also occur under dynamic, stable power swing conditions, where the swing locus could enter the MHO characteristic. To avoid operation of the distance element under these situations, the tripping characteristic would need to be shaped, or in other words, the distance element characteristic needs to be modified to increase the loadability of the transmission line.

Three terminal lines that use permissive overreaching are at greatest risk to line loading and system swing issues. To guarantee tripping at all terminals, the Zone 2 relays at each terminal must be set to reach the calculated maximum apparent impedance. These Zone 2 reaches can be very large and are highly vulnerable to line loading and system swings.

The alternatives to permissive schemes with their large Zone 2 reaches is to either use directional comparison blocking with reaches set for sequential tripping or use line current differential protection.

14.17.8.3 Tapped Generation

Permissive overreaching and Directional Comparison Blocking schemes have an enormous advantage over line differential as they can easily work with existing communication systems between line terminals along with leased communication facilities from telephone companies to the multi-tapped generators. Line differential cannot be applied on leased telephone facilities. It is cost-prohibitive to install fiber to the generators with main and alternate separate routing complying with regional operating authority requirements for BPS facilities. There are also technical issues with timing delays which limit applications to only shorter line lengths. Therefore, line differential has little appeal for multi-terminal generator connections to transmission.

However, where the tapped generation is limited to a small fixed amount of generation it could be used along with a blocking signal from the tapped generation. The stipulation is that the differential pickup sensitivity must be set above the outfeed component of the tapped generation to an external line fault and yet be below any minimum line faults with sufficient margin.

For two-terminal lines with tapped load substations and some tapped generation, line current differential will not see faults through the relatively high transformer impedance at the load substations. Therefore, there would be no need to send block signals from the load substations. However, the differential pickup sensitivity must be set above the load-substation maximum load.

14.18 Pilot Wire Protection

14.18.1 Introduction

Pilot wire relaying may be used to provide complete phase and ground protection of short transmission lines up to approximately 16 km in length. A communication circuit usually in the form of a shielded telephone cable is used to compare system conditions at each line terminal. Pilot wire relaying is a simple straightforward method of providing high-speed protection for short transmission lines. The pilot wire relays, being self-contained, require no additional communication equipment to transmit and receive information.

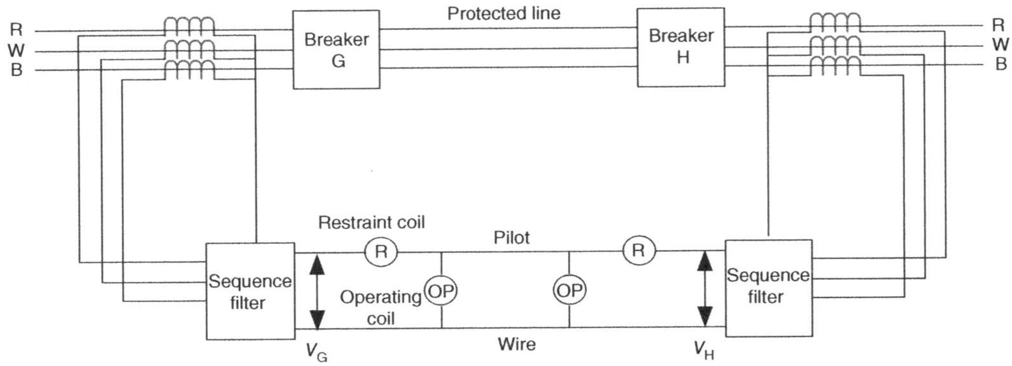

Figure 14.68 Typical two-ended pilot wire scheme.

14.18.2 Theory of Operation

Pilot wire relays use a composite filter to convert the three-phase currents at each terminal into a single-phase voltage. These single-phase voltages are compared with respect to each other to determine whether the fault is on the protected line or external.

For a through fault, the relative polarities of filter output are such as to cause current to circulate around the pilot wires as well as through the restraint coils of those relays. For an internal fault, the relative polarities of filter output reverse, with most of the current forced to flow through the operating coils.

This current will also flow through the restraining coils, and however, the tendency to operate is much greater than to restrain and tripping will occur, (Figures 14.68).

All relays in either a two- or three-terminal line arrangement can operate with heavy infeed at only one terminal thereby providing simultaneous tripping of all the terminals even if the other terminals have weak or no infeed at all. Furthermore, for a three-ended line, high-speed tripping is possible even when a small amount of current flows out of one terminal for an internal fault near another terminal with heavy infeeds from the other terminals.

14.19 Power System Considerations

The purpose of a line protection system is to detect faults on the protected line, and to trip the appropriate breakers to isolate the faulted line from the power system. When the protection system succeeds in doing this, it protects the power system by disconnecting only those facilities necessary to clear the fault.

The objective of the fault clearance system is to clear the fault fast enough so that the power system is most likely to recover and remain stable to prevent cascading outages and or blackouts. There are various scheme types for distance relays working with communication each with its unique fault clearing times. This section examines those basic scheme types and assesses fault clearing times for each of them.

14.19.1 Fault Clearing Time Criticality

Protection clearing time is the total time it takes to clear/isolate/remove a faulted transmission line from service. This is achieved by the determination of an abnormal condition by the protection system and the rapid removal of the abnormal condition. This is accomplished by removing the energized power system element from service via opening all associated circuit breakers that provide an electrical source of energy into the faulted line.

One of the most important objectives of any high voltage line protection scheme is to clear a fault fast enough to minimize equipment and environmental damage, maintain the stability of the generators, and minimize shock to customer loads and to guard human life.

A transmission line fault is very detrimental to any transmission network as it causes an enormous amount of energy to be delivered by all of the connected generators in the region of fault as well as neighboring inter-connected generators outside of the region. Each connected generator has a defined capability and can only be operated beyond this capability for very short periods. Faults not cleared within a critical clearing time defined by the specifics of the power system will cause the generators to lose synchronization and start tripping, thereby, losing system integrity. Such events cause major power system disturbances triggering an investigation by regulators, reliability organizations, and compliance organizations.

If a line protection system fails to isolate a fault within the power system required times to maintain reliability, equipment will get damaged, and fires could start possibly leading to environmental spills and will lead to major disturbances and blackouts. Most transmission power systems cannot sustain prolonged faults – in the North East of North America typically they need to be cleared in less than 175–275 ms.

14.19.2 Fault Clearing Performance Categories

Bulk power system (BPS) line protections need to meet clearing time requirements where applicable for the following power system performance categories:

- Normal Clearing Times
- Breaker Failure Clearing Times

14.19.2.1 Normal Clearing Times

Normal clearing time does not take into consideration protection system failure and assumes that the protection system is fully functional and will operate as designed and intended. Normal clearing time for the Protection System is based on the time in which each protection system component is expected and designed to operate.

A pilot (communication assisted) protected line, typical of all 500 kV and most 230 kV lines, has a protection system that is designed to provide instantaneous operation for all faults on the line. The Normal clearing time for this example might be seven cycles (two cycles for relay time, two cycles for aux. relays and telecom, and three cycles for breaker time).

A non-pilot (no communication) protected line, typical of many 115 kV lines, uses step distance protection. Faults at the end of the line would be cleared by time-delayed relaying and the normal clearing time for this fault might be 22 cycles (20 cycles for relay time and two cycles for the breaker).

14.19.2.2 Breaker Failure Clearing Times

Breaker Failure clearing time considers that the protection system is fully functional and operates as designed and intended. It does take into account a breaker needed to isolate the fault did not interrupt the fault current and remained closed or stuck.

Typical Breaker Failure clearing time is 12 cycles leading to a setting for the breaker failure timer at eight cycles (two cycles for relay time, eight cycles for breaker failure timer, and two cycles for breaker tripping).

14.19.3 Protection Planning

Transmission utilities need to assure clearing time requirements are met for the fault clearing performance categories listed above.

It is the accountability of the regional operating authority to determine critical clearing times for system performance contingencies. The critical clearing time is the single most important information protection planning requires to define the appropriate protection scheme for any given power system elements, i.e. is high-speed tripping required or not, such decision reflects in the design and operational costs.

The electric system performance varies based on the combined probability of an electric system event occurring and the performance/level of the protection system under consideration. Some reliability performance categories assume that the protection system operates normally. Other categories consider breaker failure and some delayed clearing times for protection system failure.

14.19.4 Fault Clearing Components

Power system protections are an assembly of electric components. The assembly of components can be referred to as modules. A line protection system consists of several interconnected modules for various scheme types. The basic protection modules are the measuring relay(s), trip auxiliaries, breaker trip module, tele-protection, and breaker operation. Typical protection operating times are listed in Table 14.3.

Table 14.3 Typical protection operating times.

Protection component	Assumed operating time (ms)
Measuring relays	
Distance	25
Relay auxiliary systems	
Trip aux. relays	4
Breaker trip modules (E/M)	4
Breaker trip modules (IED)	6
Tele-protection	
Microwave analog	25
PLC-SSB	25
SONET	15
Breaker trip	
115 kV	83
230 kV	50
500 kV	33

14.20 Line Setting Application Example

These examples will demonstrate methods of creating and applying settings for typical line protection relays. Since all new relays being installed are digital, the examples will exclusively focus on them.

For the most part, the application of digital relays closely follows or mimics those of the simpler electromechanical relays but with the added external auxiliary logic included. However, it is important to bear in mind that whereas there is always a one-to-one relationship between a settable element and an electromechanical relay there are multiple settable elements in digital relays.

In these examples, line impedances and other values such as CT and VT ratios are arbitrarily chosen but realistic. There are many different methods used for tele-protections, for example, relays directly transmitting and receiving simultaneous communication signals via fiber. In this example for illustrative purposes and to describe the methodology in simple terms, architecture using access multiplexers (MUX) hard-wired to the digital relays is used.

14.20.1 Two-Terminal Line Setting Example

Scheme type	Permissive overreaching transfer trip (POTT)
Positive sequence impedance	0.000344 + j0.00707 PU 0.00708/87.2° PU
Zero-sequence impedance	0.001116 + j0.01378 PU 0.01383/85.4° PU
Current source	3, (3200)/2400/2000/1600/1000 – 5 A on each breaker
CT ratio	640:1
Voltage source	3 × 152/(138) kV – 69/69 V Y_g:Y_g/Y_g CVTs
VT ratio	2000:1
Z ratio	2000/640 = 3.125:1
Bases	220 kV, 100 MVA

14.20.1.1 Setting Zone 1

The underreaching Zone 1 elements are set for 80% of the Z_1 (positive sequence impedance) of the line for the phase elements and 75% of the Z_1 of the line for the ground elements as shown in Figures 14.69, 14.70, and 14.71.

14.20.1.1.1 Setting of Relay Elements

A21P1	A – A group, 21 – Distance, P – Phase, 1 – Zone 1
A21G1	A – A group, 21 – Distance, G – Ground, 1 – Zone 1

A21P1 reach at line angle	0.88 Ω secondary	2.75 Ω primary	0.0057 PU
A21G1 reach at line angle	0.825 Ω secondary	2.58 Ω primary	0.0053 PU

14.20.1.2 Setting Zone 2

Referring to Figure 14.72, the two-terminal lines have no tapped load substations and consequently no apparent impedance. The overreaching Zone 2 elements are set for 125% of the Z_1 (positive sequence impedance) of the line for the phase and ground elements, as shown in Figures 14.73, 14.74, and 14.75.

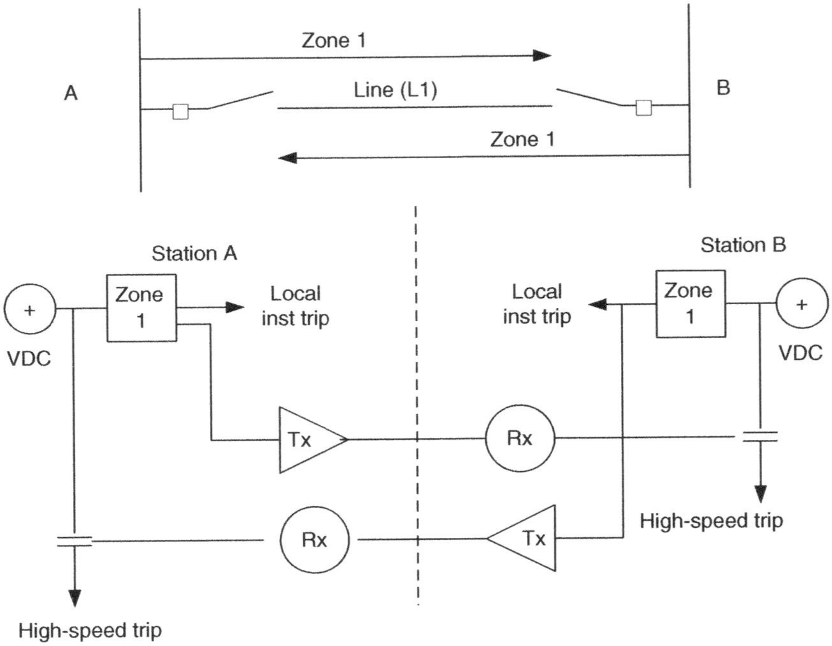

Figure 14.69 Underreaching Zone 1 logic.

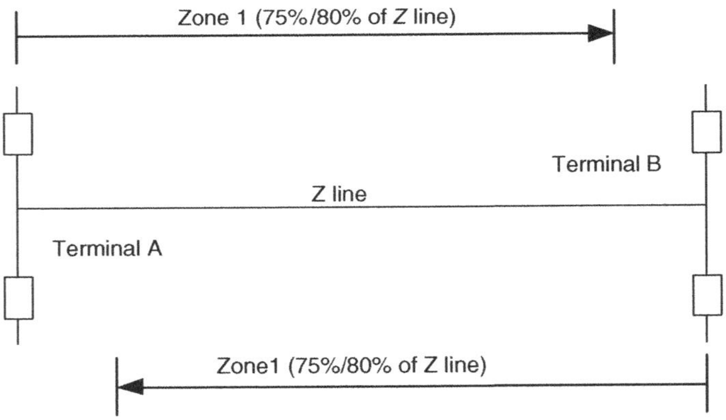

Figure 14.70 Underreaching Zone 1 reaches.

14.20.1.2.1 Setting of Relay Elements

A21P2	A – A group, 21 – Distance, P – Phase, 2 – Zone 2
A21G2	A – A group, 21 – Distance, G – Ground, 2 – Zone 2

A21P2 reach at line angle	1.375 Ω secondary	4.3 Ω primary	0.0089 PU
A21G2 reach at line angle	1.375 Ω secondary	4.3 Ω primary	0.0089 PU

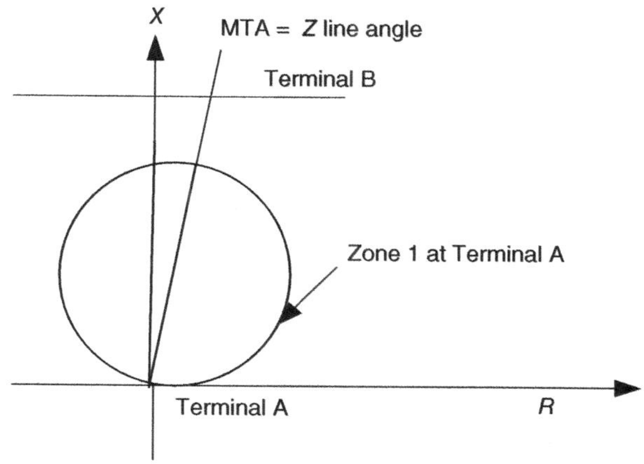

Figure 14.71 *R–X* plot of Zone 1 distance elements.

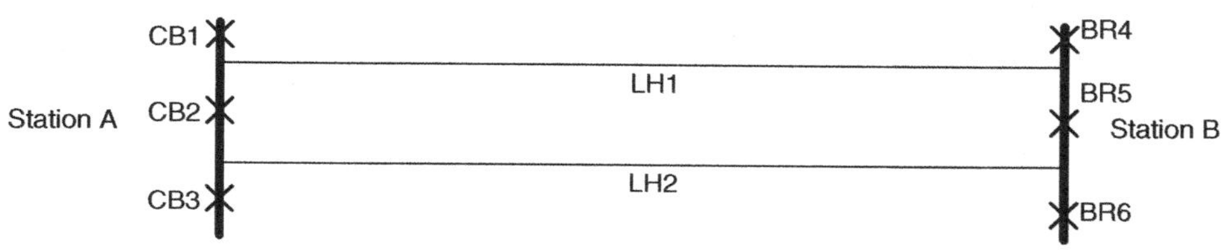

Figure 14.72 Clean two-terminal line.

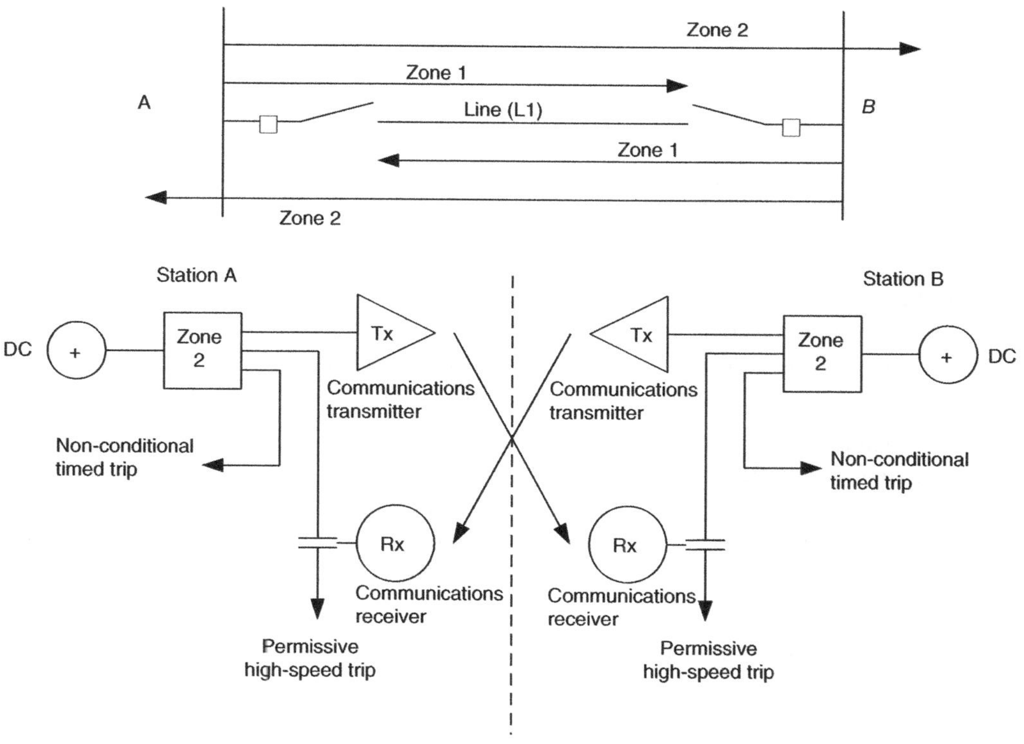

Figure 14.73 Overreaching Zone 2 logic.

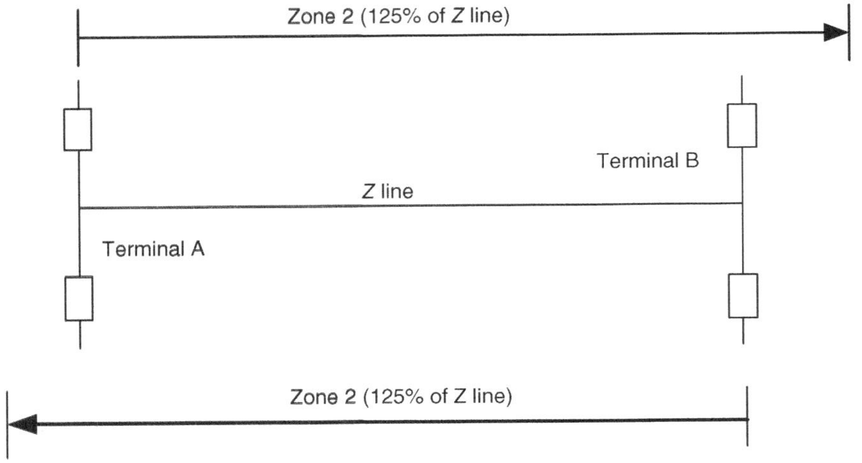

Figure 14.74 Overreaching Zone 2 reaches.

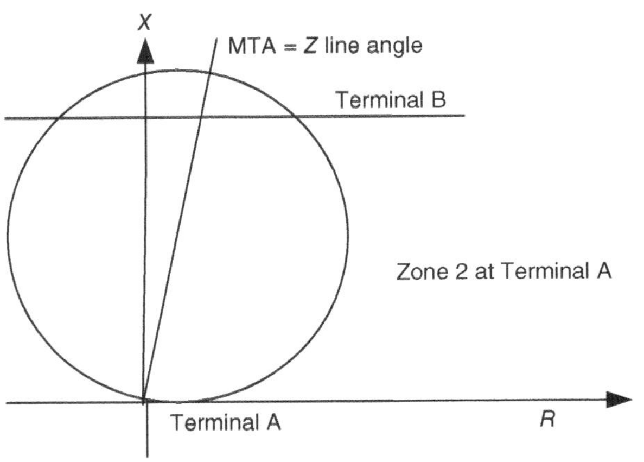

Figure 14.75 *R–X* plot of Zone 2 distance elements.

14.20.1.3 Peripheral Settings

14.20.1.3.1 Overcurrent Supervision

Overcurrent elements are used to supervise the phase and ground distance elements. The supervision is necessary as distance elements are known to misoperate under two conditions:

- If the protected line is de-energized, the parallel line can induce come voltage on the protected line. If the line is not transposed, the induced voltage, may contain negative sequence components, is capable of causing phase to phase distance elements to operate.
- When the line is de-energized, the discharge of stored energy in the line in the form of a capacitive charging current may cause the distance element to operate.

Typical setting criteria are to set the phase and ground overcurrent supervision at least twice the line charging current and no more than half the minimum remote end fault.

Refer to Section 14.9.4 for a full description.

A50	Phase to phase overcurrent supervision
A50N	Ground overcurrent supervision

A50	1.0 A secondary	0.64 kA primary (typical setting value)
A50N	1.0 A secondary	0.64 kA primary (typical setting value)

14.20.1.3.2 Switch Onto Fault

Switch onto is the function of protecting the stub bus in a breaker and a half scheme where the potential source is located on the line side of the line disconnect. For close-in line faults, even for those on the stub bus, when the line is in service, are covered by distance relays via what is known as voltage memory action.

However, for faults when the line disconnect is just closed or when it is open for some time and there is a fault on the bus stub, distance relays would not work as they do not operate on current only.

In SOTF protection, typically a phase overcurrent element is enabled by an undervoltage element when the voltage has been absent typically for more than 2 sec. Refer to Section 14.9.3 for a full description.

A50LT	Switch onto fault overcurrent
A27LT	Switch onto undervoltage

A50LT	10.0 A secondary	6.4 kA primary (typical setting value)
A27LT	20.0 V secondary	40.0 kV primary (typical setting value)

14.20.1.3.3 Line End Open – Permissive Echo Timer

In the permissive scheme, there is a requirement to have a special provision for the case in which the line is open at one terminal. In this situation, with the fault near the open end and out of Zone 1 reach, the Zone 2 elements operate and send a permissive signal to the open end. On receipt of the permissive signal, the line end open logic in turn sends a permissive signal back to the other end (although the Zone 2 element did not operate). Instantaneous tripping is achieved. Refer to Section 14.7.2.2.4 for a full description.

A33T	Permissive echo coordinating timer	
A33T	30.0 cycles	0.5 seconds (typical setting value)

14.20.1.3.4 Ground Directional Overcurrent Timers

When a directional ground overcurrent element is used with the permissive scheme, it needs to be time delayed typically 120 ms to prevent operation of the parallel healthy line directional overcurrent elements due to current reversals. Refer to Section 14.7.2.2.5 for a full description.

14.20.1.3.5 Zero-Sequence Compensation

Zero-Sequence Compensation factor $k = (Z_o - Z_1)/3Z_1$.

Refer to Section 14.4.2 for a full description.

$$Z_1 = 0.000344 + j0.00707\ \text{PU} = 0.0070784\underline{/87.2^\circ}\text{PU}$$

$$Z_o = 0.001116 + j0.013781\ \text{PU}$$

$$Z_o - Z_1 = 0.000772 + j0.006771\ \text{PU} = 0.006755\underline{/83.4^\circ}\text{PU}$$

$$k = (Z_o - Z_1)/3Z_1 = 0.006755/83.4^\circ/(3 \times 0.0070784\underline{/87.2^\circ}\text{PU})$$

$$k = 0.318\underline{/-3.8^\circ}$$

14.20.2 Long-Tapped Transformer Line Setting Example

Refer to Figure 14.76.

Scheme type	Directional Comp Blocking Transfer Trip (DCBTT)
Pos. seq. impedance A-T	0.000344 + j0.00707 PU 0.00708/87.2° PU
Pos. seq. impedance B-T	0.000344 + j0.00707 PU 0.00708/87.2° PU
Pos. seq. impedance A-B	0.000688 + j0.01414 PU 0.01416/87.2° PU
Pos. seq. impedance T-F	0.0002 + j0.005 PU 0.005/87.7° PU
Pos. seq. impedance T1	j0.25 PU
Current source	3, (3200)/2400/2000/1600/1000 - 5A on each breaker
CT ratio	640:1
Voltage source	3 × 152/(138) kV - 69/69 V Y_g:Y_g/Y_g CVTs
VT ratio	2000:1
Z ratio	2000/640 = 3.125:1
Bases	220 kV, 100 MVA

The apparent impedance at Terminal A

$$Z_{APP} = Z_{AF} + (I_B/I_A)\,Z_{TF}$$

The apparent impedance at Terminal B

$$Z_{APP} = Z_{BF} + (I_A/I_B)\,Z_{TF}$$

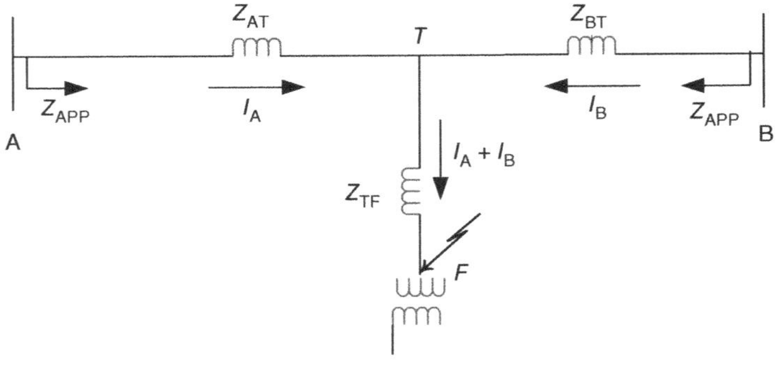

Figure 14.76 Long line tap protection with directional comparison blocking.

Assume that I_A and I_B are equal which means that the source impedances behind the buses at Terminal A and B are of equal value. Then set the Zone 2 relays at each terminal according to $I_A = I_B$. Then set each terminal for the following calculated apparent impedances.

Apparent impedance setting at Terminal A

$$
\begin{aligned}
Z_{APP} &= Z_{AF} + Z_{TF} \\
&= Z_{AT} + Z_{TF} + Z_{TF} \\
&= Z_{AT} + 2Z_{TF}
\end{aligned}
$$

Apparent impedance setting at Terminal B

$$
\begin{aligned}
Z_{APP} &= Z_{BF} + Z_{TF} \\
&= Z_{T} + Z_{TF} + Z_{TF} \\
&= Z_{BT} + 2Z_{TF}
\end{aligned}
$$

Set each terminal Zone 2 at the line impedance to the tap then two times the line impedance to the fault at the load-substation transformer HV winding with a 125% margin. Thus, it is a certainty that at least one of the terminals either A or B Zone 2 will see the fault at the load-substation transformer HV winding.

With equal infeeds, both Terminal A and B Zone 2 reaches see the fault at the load-substation transformer HV. When the source impedance behind the bus at Terminal A is of a lower value than that of Terminal B, Terminal A will trip. When the source impedance behind the bus at Terminal B is a lower value than that of Terminal A, Terminal B will trip.

Upon one terminal tripping, the other terminal trips sequentially as the apparent impedance is completely eliminated. In a permissive scheme, both terminals must see the fault for the scheme to operate correctly. In a directional comparison blocking scheme, only one terminal must see the fault. For sequential tripping to be successful, directional comparison blocking schemes must be applied at all terminals.

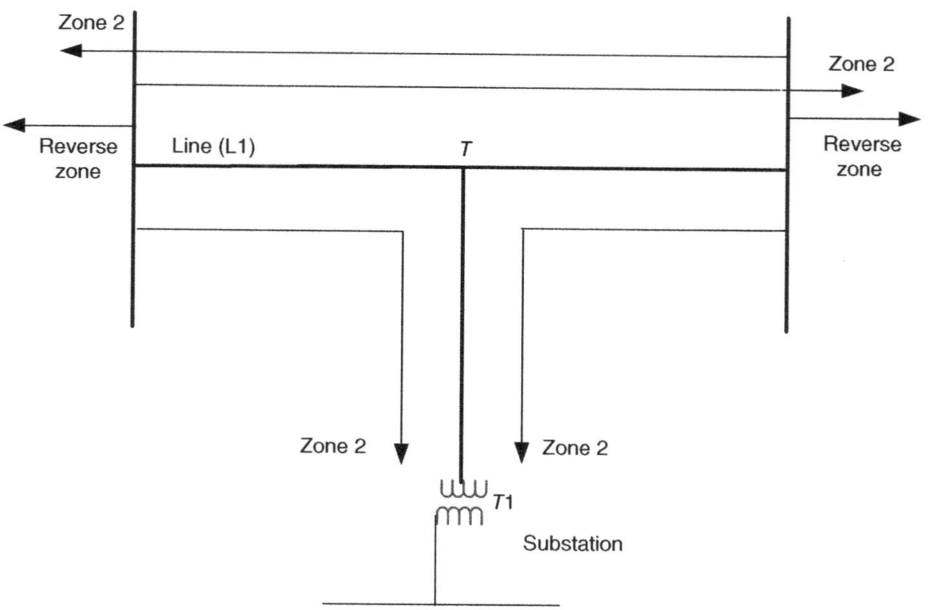

Figure 14.77 Zone 2 reaches.

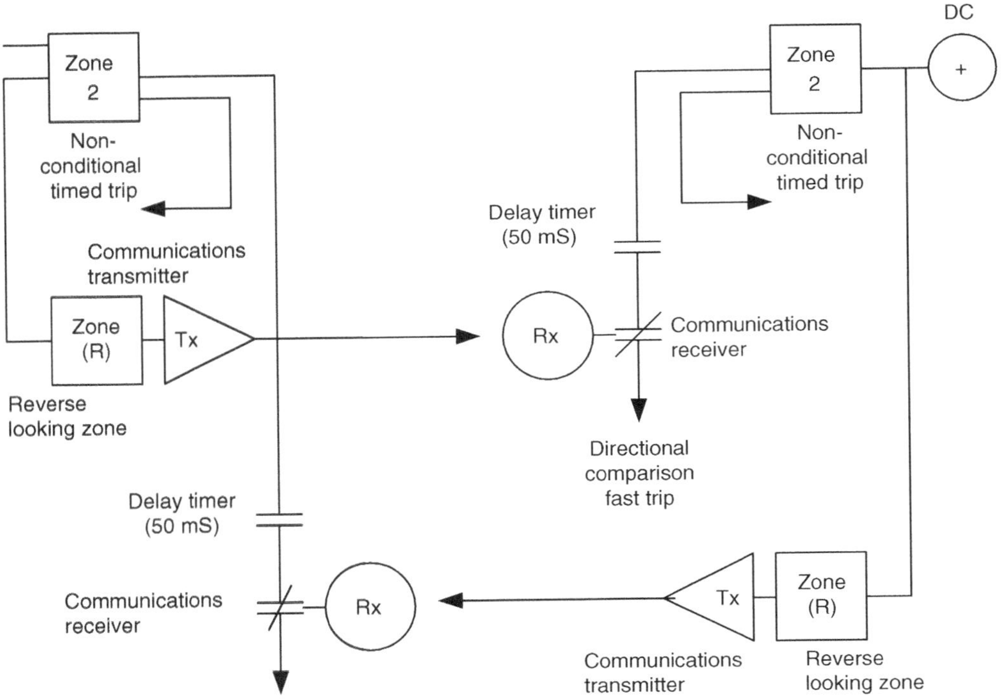

Figure 14.78 Overreaching Zone 2 logic.

It is essential that the Zone 2 settings cover the entire line impedance Z_{AB} with a 125% margin. The calculated Zone 2 settings must be checked that this is the case as the line tap to the load substation may be shorter than the entire line from Terminal A to Terminal B.

Furthermore, since no reverse-looking blocking signal is sent to the two terminals A and Terminal B, the Zone 2 protections must not see through the load-substation transformer LV side. This check must be done when no apparent impedance exists as when the load substation is only supplied from one terminal only. The means the Zone 2 reach must block short of the load-substation transformer LV bus. This is done by comparing the transformer winding impedance plus total line impedances to the transformer with the Zone 2 reach and doing this for both Zone 2 settings at Terminal A and at Terminal B (Figures 14.77 and 14.78).

14.20.2.1 Setting Zone 1

The underreaching Zone 1 elements are set for 80% of the Z_1 (positive sequence impedance) of the line from station A to station B for the phase elements and for 75% of the Z_1 of the line from station A to station B for the ground elements as shown in Figure 14.79.

14.20.2.1.1 Setting of Relay Elements

A21P1	A – A group, 21 – Distance, P – Phase, 1 – Zone 1
A21G1	A – A group, 21 – Distance, G – Ground, 1 – Zone 1

A21P1 reach at line angle	0.88 Ω secondary	2.75 Ω primary	0.01133 PU
A21G1 reach at line angle	0.825 Ω secondary	2.58 Ω primary	0.01062 PU

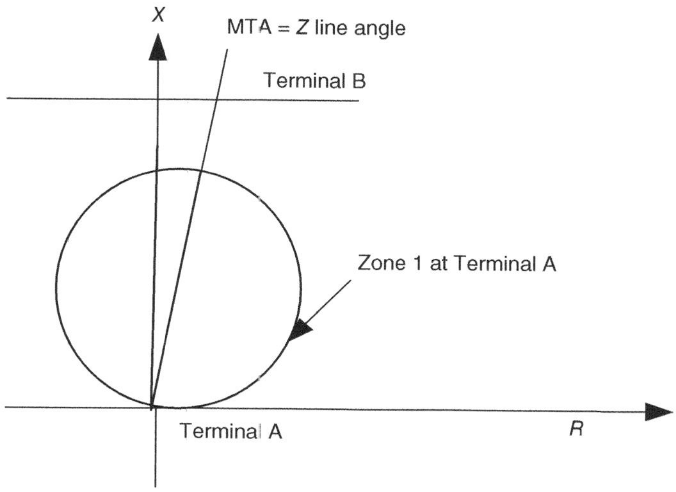

Figure 14.79 *R–X* plot of Zone 1 distance elements.

14.20.2.2 Setting Zone 2

Referring to Figure 14.77, the two-terminal line has a tapped load substation with a large tap to transformer T1 and consequently has an apparent impedance. The overreaching Zone 2 elements at station A are set for 125% of ZAT + 2ZTF and at station B for 125% of ZBT + 2ZTF for the phase and ground elements. At these settings, the Zone 2 reaches covers 150% of the line Z_1 (positive sequence impedance) and reach approximately into 8% of Transformer T1 when supplied from only either station A or station B and less when supplied by both stations when there is an underreaching apparent impedance affect. Also, with these reaches the long tap line supplying T1 at the load substation is always protected as either the station A Zone 2 will see the fault and then the station B Zone 2 will trip sequentially or the other way around Station B will trip first then Station A sequentially. These settings only work when the logic is directional comparison blocking (Figures 14.80 and 14.81).

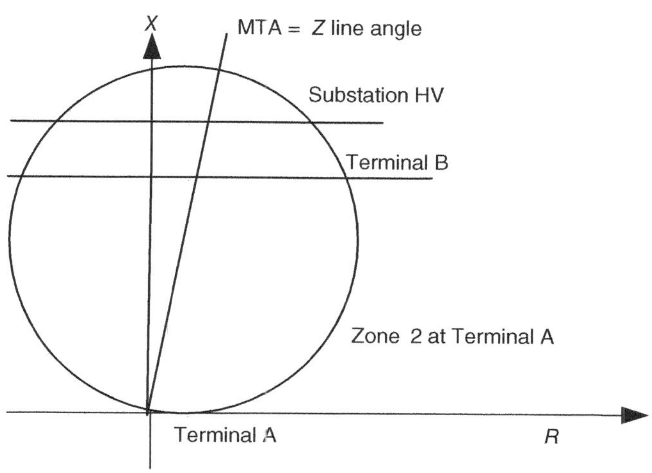

Figure 14.80 *R–X* plot of Zone 2 distance elements.

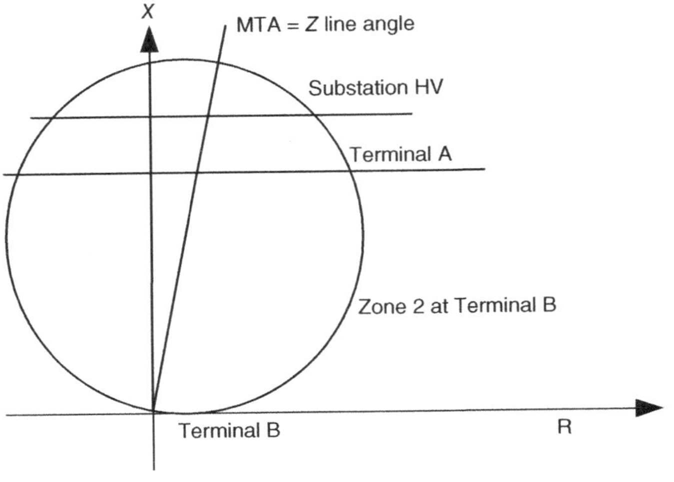

Figure 14.81 *R–X* plot of Zone 2 distance elements.

14.20.2.3 Setting Zone 3

On external faults, the reverse-looking Zone 3 operates and sends a blocking signal to the remote end to block tripping by the Zone 2 elements. The Zone 3 element is set to see all faults behind the relay terminal that are seen by the overreaching Zone 2 element at the remote terminal. This ensures that a blocking signal is sent on all external faults for which the Zone 2 element at the remote terminal would operate (Figure 14.82).

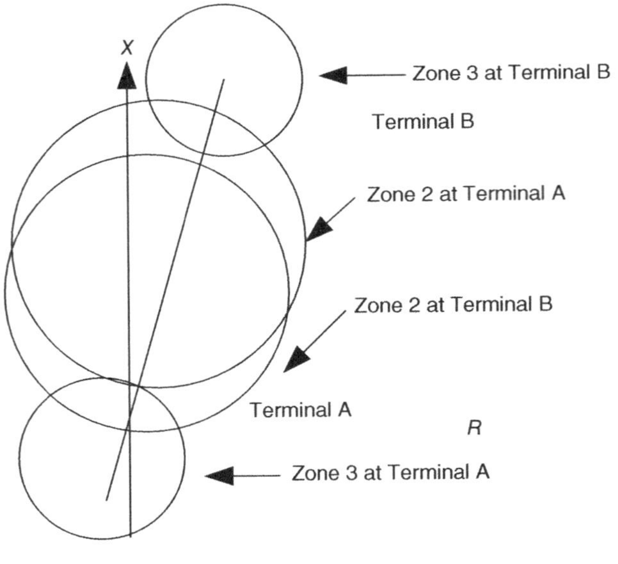

Figure 14.82 *R–X* plot of Zone 3 distance elements.

14.20.2.4 Setting of Relay Elements

Station A

A21P2	A – A group, 21 – Distance, P – Phase, 2 – Zone 2
A21G2	A – A group, 21 – Distance, G – Ground, 2 – Zone 2

A21P2 reach at line angle	3.31 Ω secondary	10.3 Ω primary	0.02135 PU
A21G2 reach at line angle	3.31 Ω secondary	10.3 Ω primary	0.02135 PU

Station B

A21P2	A – A group, 21 – Distance, P – Phase, 2 – Zone 2
A21G2	A – A group, 21 – Distance, G – Ground, 2 – Zone 2

A21P2 reach at line angle	3.31 Ω secondary	10.3 Ω primary	0.02135 PU
A21G2 reach at line angle	3.31 Ω secondary	10.3 Ω primary	0.02135 PU

Station A

A21P3	A – A group, 21 – Distance, P – Phase, 2 – Zone 3 Reverse Blocking
A21G3	A – A group, 21 – Distance, G – Ground, 2 – Zone 3 Reverse Blocking

A21P3 reach at line angle	3.31 Ω secondary	10.3 Ω primary	0.02135 PU
A21G3 reach at line angle	3.31 Ω secondary	10.3 Ω primary	0.02135 PU

Station B

A21P3	A – A group, 21 – Distance, P – Phase, 2 – Zone 3 Reverse Blocking
A21G3	A – A group, 21 – Distance, G – Ground, 2 – Zone 3 Reverse Blocking

A21P3 reach at line angle	3.31 Ω secondary	10.3 Ω primary	0.02135 PU
A21G3 reach at line angle	3.31 Ω secondary	10.3 Ω primary	0.02135 PU

References

1 C. Russell Mason, *The Art and Science of Protective Relaying*. Schenectady, New York: John Wiley & Sons, Inc. 1956, General Electric Company.

2 NERC Standard PRC-023-4, Transmission Relay Loadability.

3 North American Electric Reliability Corporation NERC paper titled Increase Line Loadability by Enabling Load Encroachment Functions of Digital Relays, System Protection and Control Task Force, NERC Panning Committee, December 7, 2005.

4 IEEE Std C37.94, IEEE Standard for N Times 64 Kilobit Per Second Optical Fiber Interfaces Between Tele-protection and Multiplexer Equipment, defines the rules to interconnect tele-protection and multiplexer devices of power utility companies.
5 IEEE Std. C37.113-1999, IEEE Guide for the Protection of Transmission Lines.

15

Subtransmission/Distribution Feeder Protection

15.1 Subtransmission/Distribution Characteristics

The nature and architecture of subtransmission and distribution differ from transmission. The voltage levels are much lower, and its main focus is on customers and loads as opposed to transmission where the main focus is power transfer.

The overall reliability of protections fundamentally differs from transmission as failures do not have the same consequences.

15.2 Definitions/Characteristics

Almost all electric utility customers obtain their electrical services from a network of subtransmission, primary, and secondary feeders operating at a variety of voltages. These subnetworks are categorically referred to as distribution, and they provide electric services to load customers. The distribution network is electrically supplied from the transmission network via many vastly geographically located load stations, refer to Figure 15.1 below. Distribution assets greatly outnumber transmission assets by factors of 100–1000 or more. In North America, it is estimated that there are 40–50 million distribution transformers, which represents only one of many distribution-type assets. This is due mainly to the need to connect and reach out to all residential, commercial, and industrial customers – the last mile.

In general, assets operating at less than 50 kV are classified as distribution assets; however, some utilities also include assets operating at 69 kV and below. Distribution network assets are further classified as subtransmission, distribution, and secondary equipment.

15.2.1 Distribution Network Feeder Definitions

In transmission networks, transmission lines represent the power system elements that transport electric energy from station to station. In distribution networks, the equivalent power system elements are generally referred to as distribution feeders. Feeders are further classified into subtransmission, primary, and secondary feeders.

Subtransmission feeders are defined as the lines emanating from the low-voltage (LV) side of transmission connected transformer stations that feed (energizes) the primaries of the step-down distribution stations (DS) and/or pole/pad mount utilization step-down transformers, Figure 15.1. They represent the bulk feeder power transfer lines, and some typical operating voltages are 69/44/27.6/13.8 kV.

Power System Protection: Fundamentals and Applications, First Edition. John Ciufo and Aaron Cooperberg.

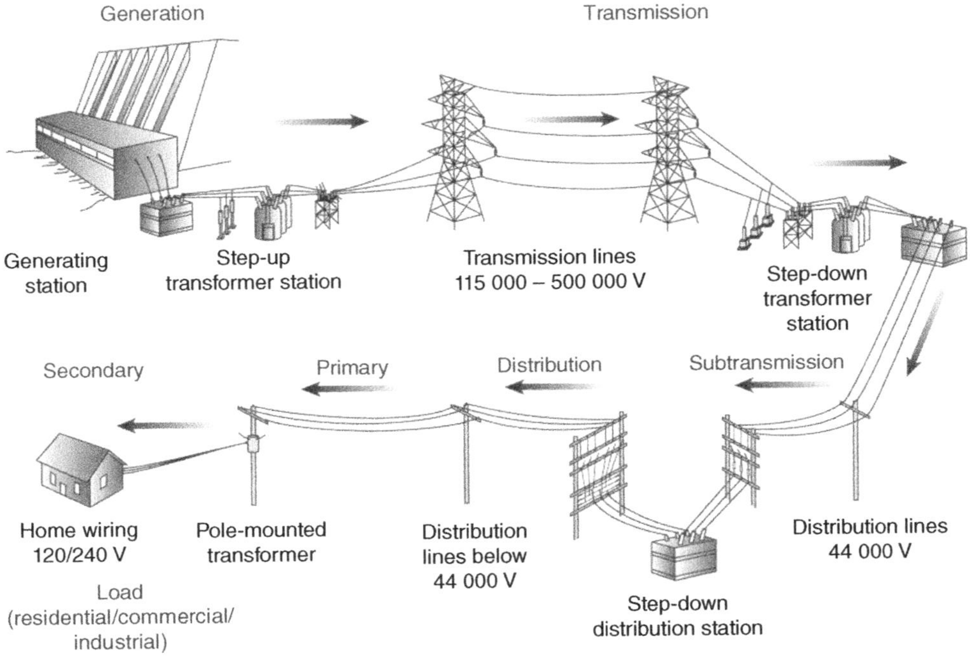

Figure 15.1 Typical electrical power system.

Primary feeders are lines that supply from the secondary of the step-down distribution transformer stations to the primary of the pole/pad mount distribution-to-utilization transformers. Some typical operating voltages are 2.4/4.16 kV, 4.8/8.32 kV, 7.2/12.48 kV, and 14.4/24.9 kV, although some distribution networks may operate at different voltages. Feeders are predominately constructed as overhead assemblies on pole-type structures, as well as underground and submarine cable construction. Feeder networks provide three-phase and single-phase construction to cater for three-phase and single-phase loads, respectively.

Secondary feeders are lines that supply load at utilization voltages. Some typical operating voltages used in North America are 120/240 and 600/340 V.

15.2.2 Feeder Characteristics

- The main purpose of a distribution feeder is to supply electrical power to utility customers at economical voltage levels. Feeders are the interface between the bulk power system (BPS) and the customers.
- The objective of any electrical utility is to provide a reliable supply of electrical power at an economic cost. To this end, the simplest and least expensive method of distribution is a radial or loop system using predominately overhead lines and some underground cables. Many legacy distribution networks are characterized as being mainly radial system with limited backup supply from an alternate circuit. Component failures require immediate repair and/or replacement to restore service. Radial feeders are lines that supply power from one source only. Faults that occur on these feeders are supplied from the source end with negligible or no fault current from other taps. Feeders with multiple sources are commonly referred to as ring circuits.

- More recently, distribution networks are required to connect tapped distributed generators (DGs) on their radial feeders, as such, converting them to multi-source feeders requiring different protection schemes than legacy simple overcurrent schemes.
- Typically, underground cables cost three-five times more to install than overhead lines. Therefore, overhead lines are predominately used.
- Subtransmission feeders are mostly three-phase and can be either 3-wire or 4-wire. For a 3-wire feeder, the tapped loads consist of DSs which are typically three-phase transformers with ungrounded primaries. For 4-wire feeders, the tapped loads may be the same as the 3-wire feeder, with additional single-phase connected loads.
- Operating voltage levels are selected based on such factors as maximum expected loads, distances, losses, and acceptable voltage drops of typically ±6%. In general, costs increase with voltage as the equipment is more expensive and clearances become larger. Standard voltage levels have changed several times since the inception of electric power. Refer to Table 15.1 below, for some typical North American distribution operating voltages.
- Connections to individual customers are economically made by using taps from the main feeder, and this results in a feeder having multiple taps – Refer to Figure 15.2. Taps can be laterals or a trunk. A lateral is a line that directly supplies customers but does not branch again before it ends so it does not supply customers on other line sections. A trunk section supplies customers directly but also carries power through to other customers on other line sections that branch off the trunk section. Trunk sections are usually constructed along major roads. Most laterals are single-phase.
- Feeders interface to the transmission network and supply the customer power at the utilization level. They represent a significant number of power system assets for utilities; therefore, the protection used for feeders should be economical. Due to the large volume of applications, one cannot afford to use expensive protection devices and schemes. Feeder protection is designed based on the most economical solution that achieves the company's operating philosophy and objectives, e.g. philosophy on outages to the entire feeder or outages to a portion of a feeder.
- Distribution systems have predominantly, due to costs, used overcurrent type protections such as overcurrent relays, fuses, and reclosers.
- Protective relays and associated breakers are used for feeders emanating from the transformer stations – subtransmission feeders. Predominately overcurrent relays are used based on costs and simplicity.
- Fuses are used protecting individual taps/laterals points and sometimes on longer feeders for sectionalizing. Fuses are also used to protect the primaries of the step-down DS transformers. DSs are tapped directly from the trunk feeders and are used to step the voltage from subtransmission levels to distribution levels. The DS transformer secondaries typically feed three primary feeders each protected by a set of reclosers. The distribution feeders are further protected downstream by fuses and some reclosers.

Table 15.1 Typical North American distribution voltages.

Utilization	120/240 (1-phase,3-Wire)	120/208 4-Wire	347/600 4-Wire	600 3-Wire		
Primary	2,400/4,160 4-Wire	4,800/8,320 4-Wire	7,200/12,500 4-Wire	8,000/13,800 4-Wire	14,400/24,900 4-Wire	16,000/27,600 4-Wire
Subtransmission	13,800 4-Wire	27,600 4-Wire	27,600 3-Wire	44,000 3-Wire		

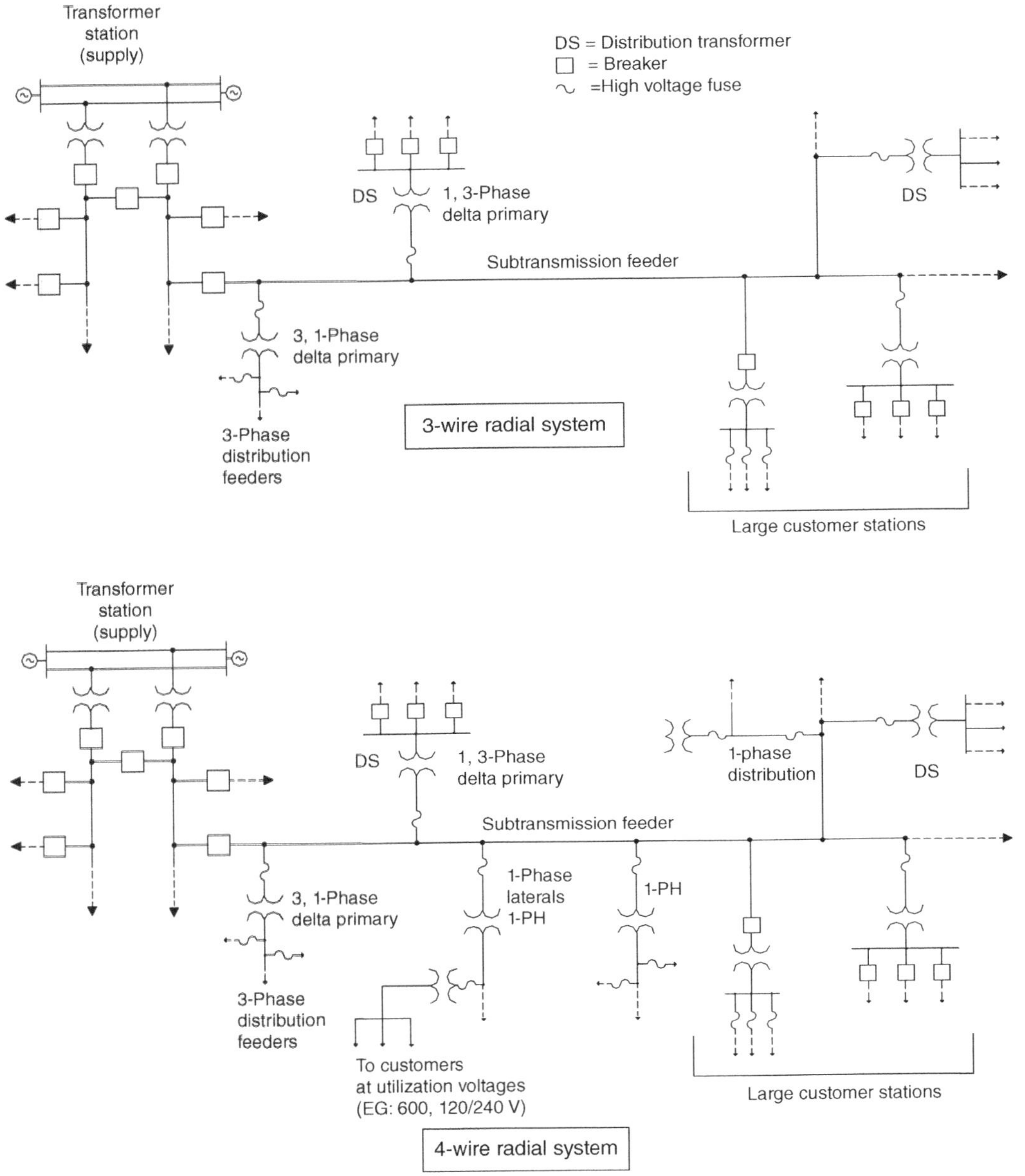

Figure 15.2 Some typical radial distribution feeders.

- More recently, due to the proliferation of feeder-connected DGs, such as wind and solar, distance relays are being used in place of overcurrent relays. Distance "mho" relays provide directional sensing and immunity to variations in source impedance and better control of the protection reach. The reach of distance relays refers to an electrical distance at which fault detection and clearing are possible. For overcurrent devices, however, reach is a highly variable quantity, and the fault detection zone of these devices will expand and contract with variations in fault types, environmental and system operating conditions.

- Feeder protections are normally categorized as open-zone types, that is, they operate when the measured quantity exceeds the preset threshold regardless of the system's conditions. The correctness of the operation of open-zone protection heavily relies on detailed system studies and fault level calculations. Setting and designing an open-zone protection system are usually a trade-off between security and dependability.
- Feeders are normally planned and designed with one or more series-connected protection detection devices. A protection detection zone starts at the location of the protection device and ends at the start of the next sequential serial protection device.
- The fault clearing zone of a protection detection device is a function of the "reach" setting, and it may or may not extend into multiple protection detection zones.

15.3 Distribution Feeder Protection Devices

The purpose of feeder protection, as with other forms of protection, is to detect faults on the feeder and isolate the faulty feeder/section from the distribution power system. By doing so, it protects the healthy system by disconnecting only those facilities necessary to clear the fault thereby, maintaining continuity of service to others where possible. It also mitigates damage to equipment and injury to people nearby.

This section is intended to provide an overview of typical distribution protection devices used in the industry.

15.3.1 Protection Devices

Refer to Chapter 4 where a more detailed description of overcurrent and distance protection relay types and their operation are covered.

15.3.1.1 Overcurrent Relays

15.3.1.1.1 Why Overcurrent Relays for Subtransmission Feeder Protection

One of the main objectives of any protection system is to design it economically and as simple as possible. To this end based on the large volume of distribution equipment, overcurrent relays are normally used for feeder subtransmission protection.

More recently, distance mho type relays are also being used for DG applications; however, it should be noted that the majority of the existing feeders employee overcurrent relays and represent the predominant protection type in the industry at large.

Overcurrent relays have been and still are the simplest and most economical type of protective relay available, making them universally used for feeder protections where low cost is an important factor.

The application of overcurrent relays is more difficult than other types of relaying. This is mainly due to their dependence on current, which is affected by variations in short-circuit current caused by changes in system operation and configuration. Overcurrent relaying should be reviewed periodically and especially when the system configuration changes (Figures 15.2 and 15.3).

15.3.1.1.2 General Overcurrent Types and Characteristics

As the name implies, these relays are designed to operate at a settable predetermined amount of current flow. There are two forms of overcurrent relays: the instantaneous and time delay types. The instantaneous is designed to operate with no intentional time delay when the current is greater

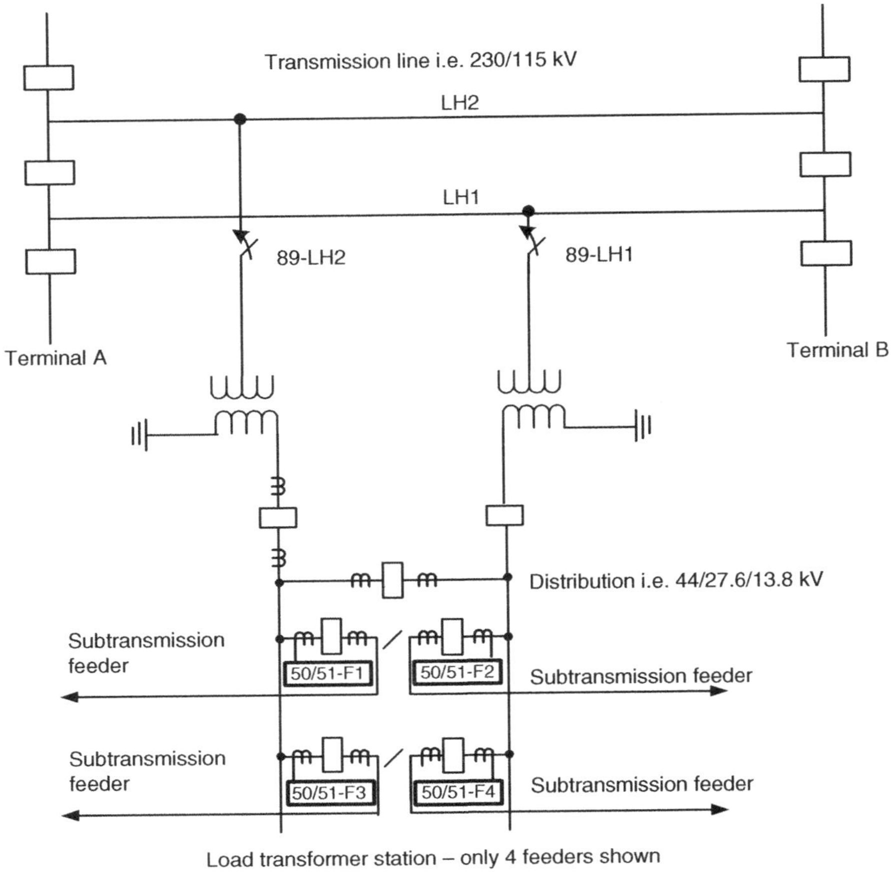

Figure 15.3 Typical load transformer station and subtransmission feeder arrangement.

than the relay setting. The operating time of the instantaneous element is a function of the relay's design and can vary from 16 to 100 ms.

The timed-overcurrent type of relay has an operating characteristic so that its operating time varies as the inverse of the current flowing through the relay, refer to Figure 15.4. Normal inverse, very inverse, extremely inverse, long time, and definite time, time-overcurrent characteristics are available. In addition, each inverse type has a time multiplier setting which results in a family of curves for each inverse type which a user can select from for a particular application, refer to Figure 15.4.

Most modern-day microprocessor/digital overcurrent relays are packaged as such, that they have both instantaneous and timed elements, with all the time characteristics available to be used.

15.3.1.1.3 Discrimination with Overcurrent Relays

Overcurrent relays are inherently non-selective in that they pick up on overcurrent conditions not only to their assigned protected element but also in adjacent equipment.

Overcurrent relays are made to be selective by using a magnitude of current (pickup or sensitivity), time, or a combination of both. Directional relays can also be used in conjunction with overcurrent relays to achieve selectivity. The application of overcurrent relays is generally more

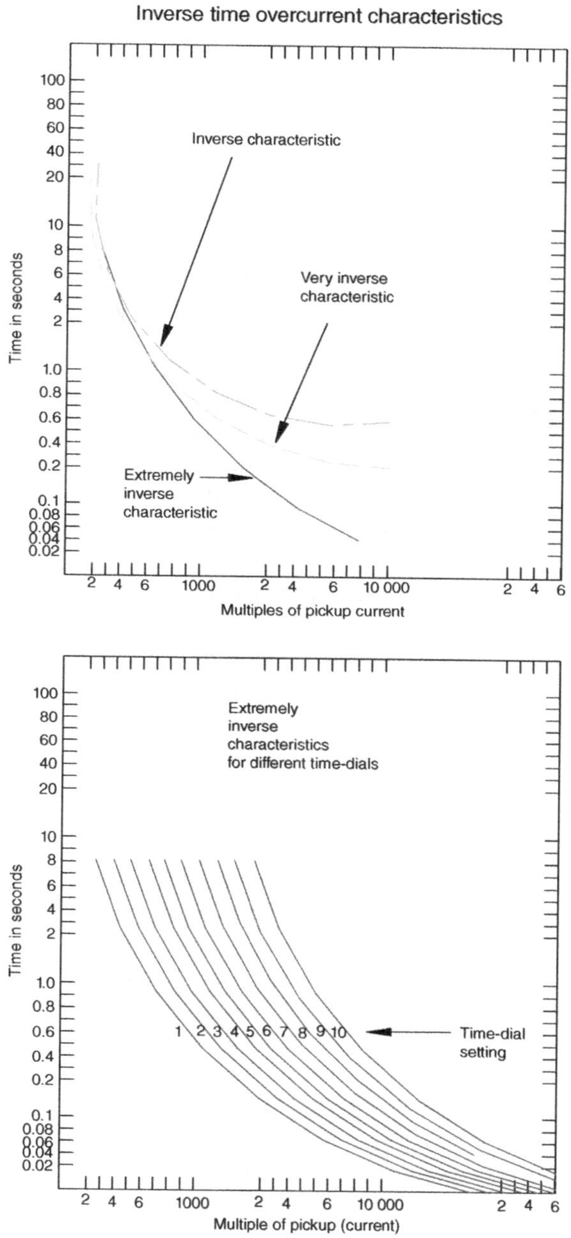

Figure 15.4 (a) Inverse overcurrent characteristics. (b) Time dials.

difficult and requires their setting to be reviewed more frequently. This is due to their dependence on the fault current magnitude, which changes as a function of system operations and configurations (source impedance).

15.3.1.1.3.1 Discrimination by Current Overcurrent relays, discriminating by current, rely upon the fact that the fault current varies as a function of the location of the fault along the feeder. Variation is due to differences in the impedance at the various fault locations. The relays are set to operate at suitable values such that the relay closest to the fault operates.

Protection practitioners should take into account when using current discrimination, the fact that the fault current on either side of a circuit breaker will be approximately the same.

The setting of a relay up to the isolation device must be such that it will detect a fault along the feeder to its end, terminating at the isolation device. The fault level on either side of the isolation device will be the same, and correct discrimination with a relay on the other side of an isolation device for faults within close proximity to the isolation device, will be nearly impossible.

The power system is continuously changing to accommodate load variations as a function of load diversity and seasonal changes. As a consequence, the source impedances and with it the fault levels change. A maximum/minimum ratio of 2:1 can be experienced. Relays set for maximum conditions may not operate under minimum conditions.

15.3.1.1.3.2 Discrimination by Time (Definite) If fault values along the feeder are fairly constant, then an overcurrent relay cannot discriminate by the current. Time can be used for these cases as an alternative. The overcurrent relay when it picks up is given a fixed time delay before it operates to close its output contact. The setting is such that the overcurrent relays farthest away from the source have the shortest time delay. Time discrimination, however, inherently causes the overcurrent relay at the source to have the longest time delay, where it normally experiences the highest fault level. The thermal rating of the current measuring element should not be exceeded; it is normally rated for a maximum current and time duration.

15.3.1.1.3.3 Discrimination by Time and Current (Inverse) The independent use of either current or definite time imposes application limitations. To overcome these limitations, the inverse time overcurrent relay was developed. The time of operation is inversely proportional to the current; therefore, the higher the current the faster the relay operates. The operating characteristic is a function of both time and current.

15.3.1.1.4 Overcurrent Relay Setting Adjustments

There is only one current pickup setting for instantaneous relays, and there are two settings for inverse time overcurrent relays: the current pickup setting and the time multiplier or time dial.

The current setting is adjustable by utilizing "plugs" or by dials or in more modern relays by the configuration software. Various setting ranges are available and are normally based on a percentage of CT secondary rating. As an example, for a 5 A rated CT, a 50–100% setting range would result in a current pickup setting range of 2.5–5 A; for a 1 A rated CT, it would be 0.5–1 A. As depicted in Figure 15.4b, the inverse time characteristics are normally drawn on log/log graph paper with the "Y" axis scaled for the time in seconds and the "X" axis scaled for current in multiples of the pickup setting. This is an industry standard, which allows the application to any relay regardless of setting range and nominal rating.

As an example, to obtain the relay operate time for the following inverse overcurrent relay:

Assume CT ratio 500/5
Relay setting range 50–100%
Relay setting range in amps $(0.5\text{–}1) \times 5 = 2.5\text{–}5$ A
Primary fault current equals 5000 A
Relay setting equals $100\% = 5$ A
Secondary fault current $= 5000/500 \times 5 = 50$ A
Multiple of pickup $= 50/5 = 10$
Assume time dial of 1.0
From the extremely inverse curve, the relay will operate in 0.8 second (approximate).

For the most effective use of an inverse time relay, the pickup should be chosen so that the relay operates on the most inverse part of its time curve over the intended range of current operation.

The resetting time of a relay should also be considered when calculating protection system operating times and coordination. A relay will reset, contact changes state from operation to normal, when the current is reduced to a percent of the current setting; typical values are 85–90% of pickup. Electromechanical relays require a definite time to reset, which is a function of their design. Digital relays provide the option of mimicking the reset time of electromechanical relays or can be set for instantaneous reset. This may have implications for coordinating electromechanical and digital relays.

The time multiplier setting or time dial setting results in a multiplying factor being applied to the relay operate time. The multiplying factor for electromechanical relays is close to but not exact given it's a mechanical setting. As an example, at 10 times pickup, the relay operates at three seconds with time setting, e.g. 1.0, for a time setting of 0.5; it will operate in 1.5 seconds. Also, refer to Chapter 4, Section 4.3.2.4.

15.3.1.1.5 Overcurrent Relay Selectivity (Coordination)

Overcurrent relay protection schemes, like most other relay types except for the unit types, will inherently extend their protection coverage to adjacent zones. Moreover, by design, protection systems overlap. Therefore, the settings of the overcurrent relays must be such that they do not operate for faults in this overlapping zone except as backup if designed to do so, until the primary relays have operated. This process of setting is called selectivity or more commonly referred to as relay coordination. For feeders, phase and ground coordination must exist between primary, and backup overcurrent relays, fuses, reclosers, etc.

The objective of coordination is to set the protections to operate as fast as possible for faults in the primary zone and operate slower for faults in the backup area. Settings should be below the expected minimum fault currents for which they should operate, but not operate under normal operating currents such as load. For some feeders, a sufficient margin does not exist between these two currents; therefore, coordination is not possible and compromises will have to be made and or distance, pilot, etc. relaying may have to be used.

15.3.1.1.5.1 Coordination Time Interval For correct coordination, it is necessary to have a time interval between the operations of two adjacent relays. This time is referred to as coordination time interval (CTI) and is illustrated in Figure 15.5. The time interval used for coordination is based on the following:

(1) **The circuit breaker fault interrupting time**. This time is typically eight cycles for feeder breakers.
(2) **The overshoot time or over-travel of the relay**. The operation of the relay continues for a short time after the relay is de-energized until the stored energy is dissipated. This time is minimized as much as possible by design but must be considered. Typical values are 30–60 ms and are a major contributor to the relay's reset time.
(3) **Margin for errors**. All measuring devices such as relays and CTs are subjected to some degree of error. A safety margin of 100 ms is normally added to account for general errors.

CTIs normally used are in the range of 0.3–0.5 seconds; with 0.4 seconds being typical and depends on the particular application.

Figure 15.5 shows the application of time overcurrent relays to a radial feeder and the total tripping time characteristics for faults at any location along the circuit. Note that an increase in the minimum tripping time for Breaker A for faults closer to Bus A and that the inverse characteristic helps reduce this increase. Also, note the CTI between the two adjacent relays for coordination.

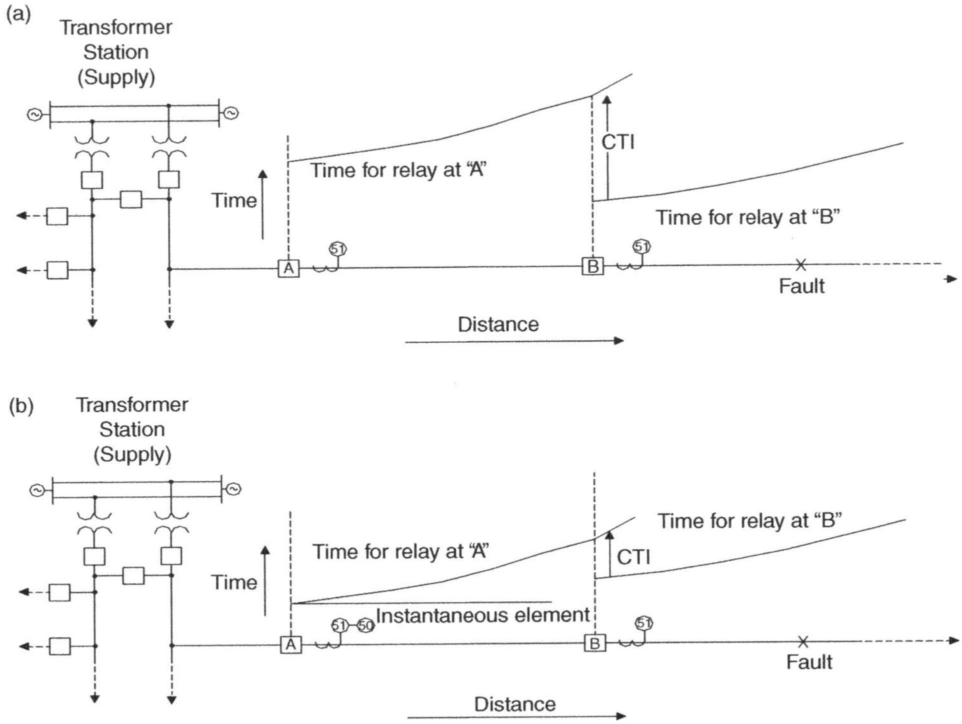

Figure 15.5 Operating time for inverse overcurrent relays.

Where inverse relays must be mutually selective, it is a good practice to use relays with the same degree of being inverse; otherwise, obtaining selectivity may be difficult.

15.3.1.1.6 Instantaneous Overcurrent

Instantaneous overcurrent relays are used in feeder protections to address the inherently long tripping times for close-in faults. Long tripping times are required to achieve feeder coordination with downstream devices. Instantaneous overcurrent relays are used in the primary relaying to supplement the inverse time relaying. The zone of protection is established entirely by setting and is set to "block" short of any remote faults. This is achieved by calculating the maximum remote end fault level and setting the instantaneous relay higher with sufficient margin to ensure it never operates for a remote end fault. By doing this, close-in faults that are close to the relay location where the fault levels can be very high are cleared instantaneously without definite time delay shown in Figure 15.5(b).

A reduction in the magnitude of fault current shortens the distance over which the instantaneous relay operates. However, it is not of general concern since the objective is fast tripping under maximum fault value conditions – close-in faults. Instantaneous tripping is possible only if sufficient fault value differences exist between close-in and remote faults.

15.3.1.1.7 Type of Inverse Overcurrent Relay for Feeder Protection

Timed overcurrent relays are available with different degrees of being inverse. The more inverse the characteristic, the faster is the operating time for larger magnitude faults, and hence, the duration of the fault is decreased.

The long operating time of the relay at high transient values of load current makes the relay suitable for coordinating with fuses and for the protection of feeders which are subject to transient currents on switching – hot and cold load pickup.

15.3.1.1.8 CT Connections

A minimum of three overcurrent relays and three CTs are required to detect all possible faults in a three-phase system as any inter-phase fault must affect at least one of the relays. Two of the relays are connected in the phases (50-A and 50-C) and the third in the residual (50N) – zero-sequence, current that exist when a current flows to ground. Ground faults, which are by far the most frequent type of faults, will be detected by the associated phase, phase overcurrent relay, e.g. A-G fault, will be seen by the A phase relay (50-A) and the residual relay (50N). A residual overcurrent relay allows for more sensitive protection. However, it must be set above the normal unbalance current associated with single-phase loads on 4-wire systems.

15.3.1.2 Distance Relays

Due to the recent phenomena of connecting distributed generation (DG) onto feeders, overcurrent relays are starting to be replaced by distance relays. DGs tapped to feeders and the resulting protection impact including the reason for distance relays will be discussed in Section 15.7. Another reason for the migration of overcurrent relays to distance relays is based on integrated multifunctional microprocessor-based technology, thereby reducing overall cost of installation. Moreover, compared to overcurrent relays, distance relays are inherently directional, less susceptible to source impedance variations, and have higher loadability limits, but they do require a voltage source (PT).

Distance relays compare voltage and current and operate when the ratio is below a prescribed value. Under fault conditions, the ratio of the voltage to current applied to the relay is used to calculate apparent impedance that is compared to a threshold setting for trip decisions. Multifunctional microprocessor-based distribution relays provide overcurrent as well as distance elements all in one device making them suitable for detecting faults based on known impedance for the zone that it is protecting.

The characteristics of distance relays are described by an impedance R-X diagram, where R is resistance on the horizontal axis and X is reactance on the vertical axis. The origin is depicted by the relay location and the operating area is usually in the first quadrant. When the ratio of the system voltage and current falls within a circular zone on the diagram, the relay operates. The mho relay is a commonly used distance type relay that is available in three-phase and single-phase versions – refer to Figure 15.6.

Distance relays may be set to overreach the load impedance and, in some cases, the maximum load impedance could fall within the operating zone. For such cases, blinders/load encroachment type functions may be used to restrict relay operation in the load area, allowing longer line reaches. Another possibility is to send a blocking signal from relays located at the load to block its operation upon receipt of that signal for faults at the load and beyond.

With the use of modern microprocessor-based relays, unique impedance characteristics can be created such as a "peanut" shape or "quadrilateral" shape. In some instances, these shaped characteristics can help to avoid the use of blinders by providing more specialized impedance characteristics.

The MTA (maximum torque angle) of an electromechanical relay or RCA (relay characteristic angle of a digital relay, see Figure 15.6, indicates the angle of maximum sensitivity for the relay and is usually set at or near the angle of the line being protected (the angle being the angle derived from the complex impedance of the line). This too can be used to advantage as the relay impedance

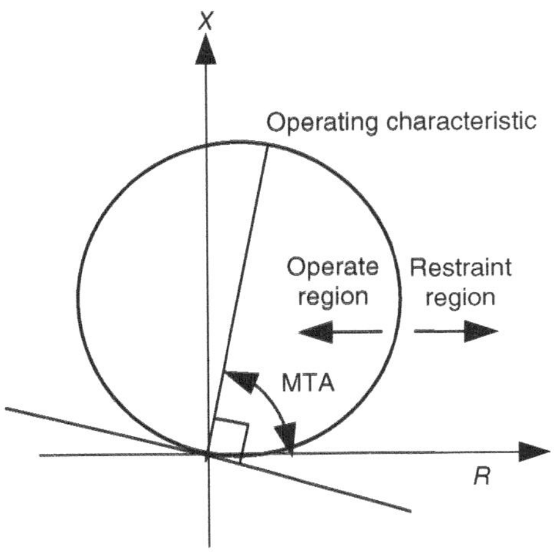

Figure 15.6 Operating characteristic of a Mho relay on an *R*-*X* diagram.

reach at other angles than the MTA will have shorter reach thereby making it less susceptible to operating on load.

15.3.1.2.1 Distance Relay Selectivity

There are two distinct methods of achieving protection selectivity using distance relays for feeders. One method is to provide overlapping zones with timed coordination. Another method is to provide overlapping zones with communication and some form of sophisticated logic scheme. The latter is generally not used for distribution systems mainly due to costs; however, new distribution smart grid protection schemes are being investigated and are starting to be deployed, which do require the use of communications.

Distance relay applications for feeders traditionally have zones defined by feeder line length or distance from the relay location. Either timed coordination in what is known as stepped time coordination may be applied or; in more rare cases, smart grid/DG applications, communication-based schemes can be used to achieve selectivity. The former case generally satisfies distribution fault clearing times.

15.3.1.2.1.1 Stepped Distance Scheme Step distance protection is generally used for non-pilot applications; an example of such a scheme is illustrated in Figure 15.7.

In this stepped distance scheme example, three protection zones are used for Terminal A:

Zone 1. Set with no intentional time delay and covers 80% of the line so as not to overreach terminal B (selectivity).

Zone 2. Set to cover 100% of L_1 plus at least 25% of L_2, covering faults for the section between 80% and 100% of L_1, with a time delay selected to coordinate with Terminal B protection systems.

Zone 3. This zone is typically applied as overall backup protection for single protection failures at Terminal B (breaker failure, battery, etc.) It is set to cover 100% $(L_1 + L_2)$ plus at least 25% of L_3 and is time coordinated with protection systems at Terminals B and C.

Figure 15.8 shows an example of a communication-based feeder protection scheme. In this example, a communication link is provided between the feeder protection and a downstream recloser protection device. A blocking scheme is used for the instantaneous protection whereby the feeder relay is intentionally timed delayed to wait 50 ms to receive a block signal for faults

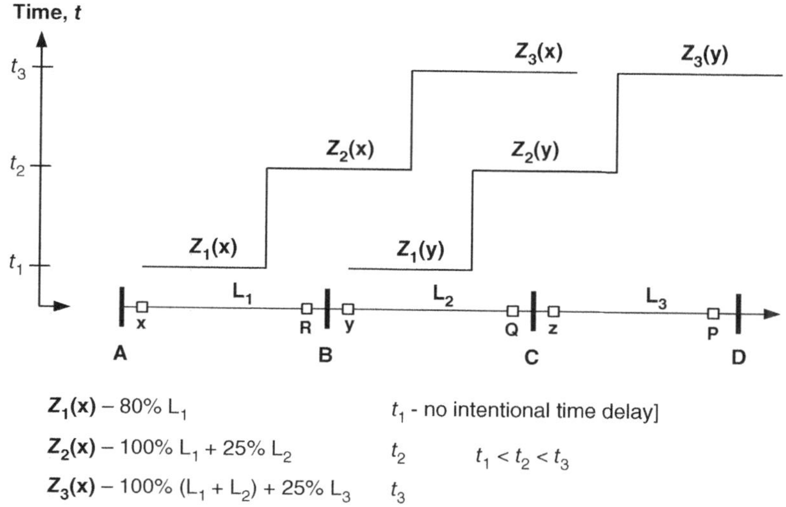

Figure 15.7 An example of a stepped distance scheme for Terminal A.

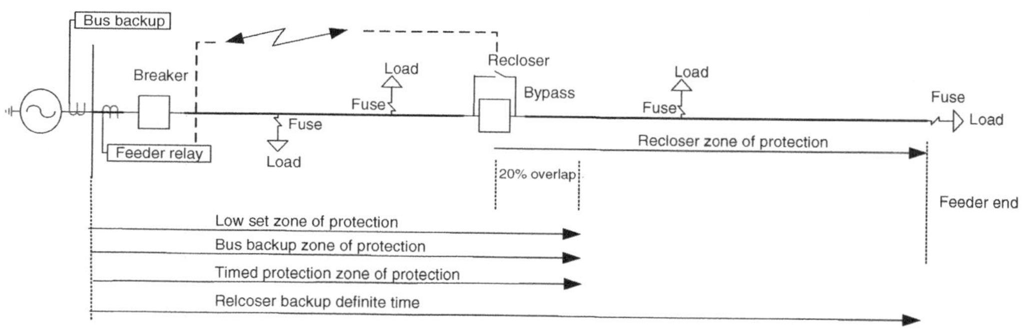

Figure 15.8 An example of a communication-based scheme.

downstream from the recloser if it does not receive a block signal, it initiates a trip. The major improvement is that all faults between the feeder protection and the recloser are cleared instantaneously, therefore, reducing the time for any sustained voltage variations.

15.3.1.3 Fuses

15.3.1.3.1 General

15.3.1.3.1.1 What Is a Fuse? In simple terms, it is an overcurrent protection device. Traditional fuses contain an opening element that melts or burns open. It opens once a certain designed quantity of current that passes through it is reached. It is a series-connected device, and it can be viewed as a low valued series-connected resistor that is designed to melt and then opens, to provide overcurrent protection.

15.3.1.3.1.2 What is the Purpose of a Fuse? Fuses are protection devices. All protection devices are designed to operate and isolate, by opening the circuit at their location. They operate for specific magnitudes and duration of fault currents. A fuse's basic functions are to limit current, isolate the failed equipment, and minimize damage to the protected equipment.

Fuses are designed and rated to operate for specific predetermined values and, therefore, must be correctly sized for the applications. Fuses are classified as overcurrent protection devices, but the destruction of the fuse element differentiates fuses from overcurrent devices, such as overcurrent relays and reclosers.

Fuses are the most common type of protective device applied to electric distribution systems because:

- It is a simple form of overcurrent protection.
- It is the most economical (device and construction cost) protection available.
- They require no maintenance.
- A fuse operation is inherently automatic, unlike circuit breakers that require relay systems for fault detection and trip initiation.
- Fuses do not require DC batteries.
- Fuses provide both fault detection and isolation.

Some disadvantages of fuses are as follows:

- Once a fuse operates, it requires more time to restore the circuit as field crews need to be dispatched to replace the fuse.
- The above impacts distribution reliability performance.
- The current-time characteristic of a fuse cannot always be coordinated with that of the protected apparatus.
- Fuse applications are limited by short-circuit fault levels.

15.3.1.3.2 High Voltage Fuses

Many varieties of fuses are available today. Fuses are available and are rated to operate from fractions of 1 A to many thousands of amps. Therefore, there are thousands of different fuses available. They are differentiated by DC and AC and from electronic to high voltage applications, each available with a family of operating characteristics. However, regardless, they are functionally an overcurrent protection device.

Fuses can be grouped into two general categories: high-voltage (HV) fuses designed for operation at voltage above 1000 V, and LV fuses operating up through 600 V AC. HV fuses are governed in North America by American National Standards Institute (ANSI)/IEEE. LV fuses standards are established by ANSI, National Electrical Manufacturers Association (NEMA), and Underwriters' Laboratories (UL).

HV fuses are used in distribution systems, and they are the primary type of fuse that is used for feeder protection.

15.3.1.3.2.1 Where Are HV Distribution Fuses Used? HV fuses are extensively used in distribution electric power systems.

They are used as overcurrent protection for the following types of applications:

- to protect distribution transformers and their secondary distribution systems
- they are connected on the primary side of the DS transformers
- used for sectionalizing and are installed at the tap points for laterals, branches, and also can be used at mid-points for long sections of distribution feeders, although reclosers are generally used for this type of application
- to protect distribution capacitor banks.

For HV fusing, there are numerous classifications that are based on electrical operating characteristics, mounting options, and operating environment. The HV fuses, by definition (IEEE) [1], are for use in AC systems (50 and 60 Hz) with rated voltages exceeding 1.0 kV.

15.3.1.3.2.1.1 HV Distribution Fuse Types There are two distinct types of HV fuses that are used in power system distribution applications, non-limiting and current-limiting fuses (CLFs). The main difference between them is the method used to manage the extinguishing of the arc and the fault and the speed of operation.

Non-limiting, also referred to as expulsion type fuses, and limiting type are both used for distribution protection; however, expulsion fuses predominate in distribution. Current limited type fuses are used for specific applications and, in general, in series with expulsion type fuses where fault currents are high, which could lead to damaging the protected equipment. Some newer CLFs are available that combine both elements into one fuse assembly.

The primary function of an HV fuse is to respond to a fault current and to open the electric circuit, thereby isolating the protected power system element from the fault condition. Fuses vary in how much fault current they can interrupt, and the proper fuse must be selected for the application. It is the behavior of the fault current that causes fuses to be classified as current-limiting or non-current limiting. The most commonly used non-CLFs are expulsion fuses; therefore, these two terms are used interchangeably.

Another possible function of an HV fuse is to respond to excessive overload currents. However, not all fuses are designed to operate for overloads. In some cases, two series fuses are required, one to respond to faults and the second for overload protection. CLFs are typically designed to operate for high fault current operation but not generally at lower currents; however, recent technology provides many new CLF ratings.

It should be noted that current IEEE [1] fuse standards classify HV fuses as Class A and Class B. The distinction is based on where in the distribution system the fuse is likely to be suitable for use. Class B fuses are applied in closer proximity to supplying stations where the fault levels are typically much higher. Class B expulsion fuses generally have higher maximum interrupting current and voltage ratings than Class A expulsion fuses.

Historically, Class A fuses were known as distribution fuses and Class B as power fuses. Class A or distribution type fuses are typically used for lower-rated distribution equipment and are located external to DSs – further from a strong source resulting in lower fault levels. Class B or power fuses are typically used within DSs, switchgear, and applications where fault values are much higher. Power fuses are expulsion fuses characterized by higher voltage rating, high continuous current rating, and the use of an interrupting medium.

Expulsion fuses that may be of cutout type or the solid material type operate differently from the CLFs. The solid material power fuse type has the inner lining of the tube coated with acid. Expulsion fuses are zero waiting devices. Power fuses normally use a mechanical means of separating the terminals after the fusible element melts. The arc is lengthened and forced into a chamber where the gap is deionized quickly by gases generated from the chamber lining due to the heat of the arc. The deionization leads to the extinction of the arc at the first or second current zero.

Expulsion fuses using non-damageable silver elements permit a very low fusing ratio, thus providing more sensitive protection for transformer secondary faults. The fusing ratio is defined as the ratio of the fuse link ampere rating to the transformer full-load current rating. In other words, if the fuse is a 5E type and if the transformer full load current is 2.62 A, then the fusing ratio is: $5/2.62 = 1.9$

Regardless of the classifications, all fuses provide the same overcurrent protection function. It is extremely important to select and use a fuse with the ratings appropriate for the intended use and application. One must take into account such ratings as voltage, short-circuit interruption, continuous current, required time to operate, the operational environment, etc.

15.3.1.3.2.1.1.1 Expulsion (Non-Limiting)

Fuse Cutout Type – Typically Class A The construction and operation of a Class A type expulsion fuse are different than a Class B expulsion fuse. They are used in primary electrical distribution systems to protect overhead feeder lines taps and equipment such as distribution transformers from faults and overloads. A fault will cause the fuse to melt, disconnecting the equipment from the fault condition. Some come with a fuse cutout being a combination of a fuse and a knife-type switch. The cutout can also be opened manually by standing on the ground and using a long insulating hot stick.

A cutout consists of three major parts. The body, it looks like an open "C" shape that supports the "fuse holder," and an insulator that isolates the conductors from the support, and the "fuse holder" contains the interchangeable fuse element and also acts as a knife switch.

When the fuse blows, the fuse holder will drop open, disengaging the knife switch and allowing it to hang from a hinge assembly. This operation provides a visible indication that the fuse has operated. It also provides visible assurance that the downstream circuit is electrically isolated. The fuse link is the replaceable portion of the assembly that melts/operates under high current.

Once a fuse operates, a crew is dispatched and they remove the fuse holder and replace the burnt-out fuse element with a new one; and the fuse holder is re-installed and closed in with the hot stick, refer to Figure 15.9a below.

Distribution cutouts are predominantly used for the protection of small distribution transformers, capacitor banks, and branch lines. Distribution cutouts are available with voltage ratings for all the common primary distribution system voltages, continuous current ratings as high as 200 A, and interrupting ratings as high as 20 kA RMS asymmetrical. The open-type distribution cutout is the most common packaging option for expulsion fuses.

Expulsion Power Fuses General – Class B A Class B, or power fuse, tends to be a more complex and elaborate design than a cutout type fuse. Some use a renewable fuse unit that can be replaced

(a)

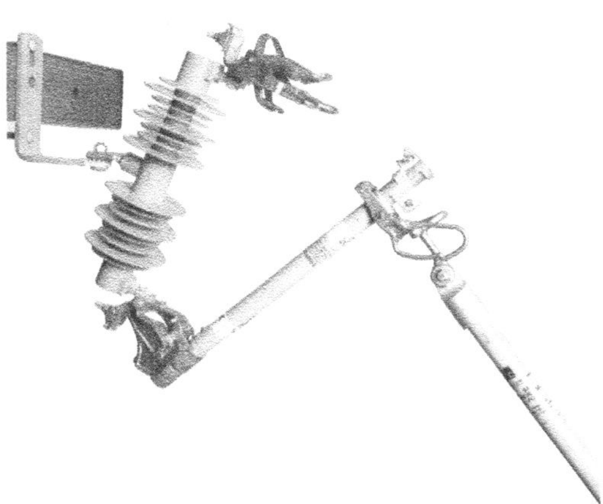

(b)

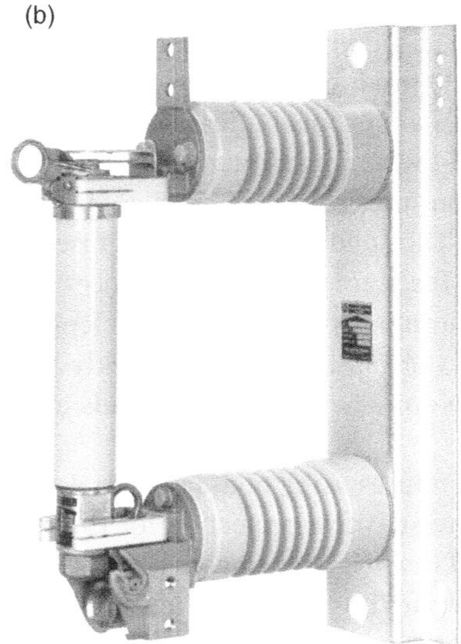

Figure 15.9 (a) S&C SMD-20 14.4 kV overhead – pole top fuse. Source: Photo Courtesy of S&C Electric Co. (b) SM-4 power fuse (outdoor distribution 4.16-34.5 kV). Source: Photo Courtesy of S&C Electric Co.

after an operation. Not all power expulsion fuses drop to an open position after an operation like a fuse cutout. Many Class B fuses use acid as part of their interruption mechanism.

This type of expulsion fuses vents out the expulsion materials internally produced by the action of arc distinguishing and current interruption. In simple terms, the fuse when it operates produces an explosion-type event during the venting.

Power expulsion fuses are characterized by higher interrupting capabilities, higher continuous current ratings, and higher voltage and basic impulse level (BIL) ratings (150 for 27.6 kV) than fuse cutouts. Power expulsion fuses find application for the protection of substation transformers with voltages up to 115 kV. They are also employed in metal-enclosed switchgear of 46 kV or less and in various configurations for the protection of the overhead and underground feeder and branch circuits.

15.3.1.3.2.1.1.2 Current-Limiting Fuses (CLF) The main difference between expulsion and current limiting is the method used for extinguishing the arc established after the fusible elements melt and the speed of operation. Functionally, an expulsion (non-limiting) fuse can limit the time exposure to a fault, but a CLF can limit the time and the magnitude of the fault. An expulsion fuse limits the duration of a fault (time–current curve) and allows the full fault magnitude to flow until interruption occurs, normally at the 1 or 2 times current waveform zero-crossing.

CLFs, as the name implies, limit the current by "chopping" the prospective current before it reaches its peak value, thereby limiting the let-through current – refer to Figure 15.10a below. The chopping action functions, in effect, by inserting high resistance into the circuit in the first half-cycle after the short circuit. The prospective current is the current that would have flowed due to the fault and in the absence of a fuse. When a high current fault occurs, heat is generated so rapidly that almost none of it can spread.

CLFs are used for applications where one is required to limit the maximum instantaneous current – peak let-through current. CLFs reduce the peak current of the available fault current to a value less than would occur without it, refer to Figure 15.10a. CLFs can be used as standalone devices or in series as backups. They are also available as an integrated device as depicted in Figure 15.10b.

15.3.1.3.2.1.1.3 Electronic Fuses Electronic-type high voltage fuses are also available. Electronic fuses combine the sum of the features and benefits of power fuses and protective relays, refer to Figure 15.10c.

They consist of two components, an electronic control unit and an interrupting unit. The electronic control provides a family of settable time–current characteristics functions and operates to initiate the tripping of the interrupting unit. The control unit can offer instantaneous and time delay tripping functions.

The interrupting unit interrupts the current when the overcurrent occurs as per the setting threshold settings of the electronic control. These two components are contained within one mountable electronic fuse. The interrupting unit melts and burns back the fusible element and once operated the interrupting module needs to be replaced.

It should be noted that the type of HV fuses that are used, are normally based on a utility's protection philosophy and standards. One company may choose only to use expulsion type fuses, where another may use a combination of expulsion and current limiting to achieve their protection and reliability goals.

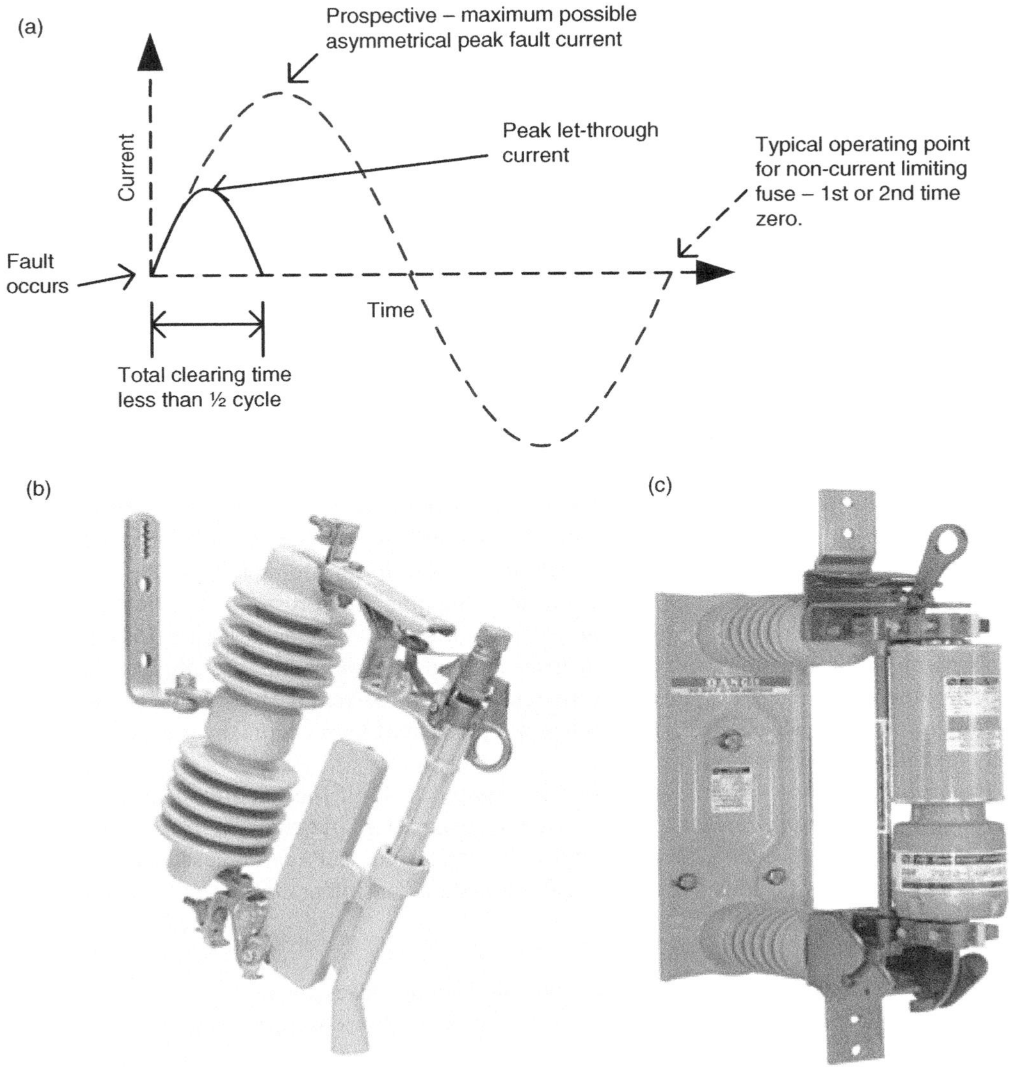

Figure 15.10 (a) Current-limiting fuse illustration of operation. (b) S&C current-limiting backup and expulsion combination fuse. Source: Photo Courtesy of S&C Electric Co. (c) S&C Fault Fiter® electronic power fuse. Source: Photo Courtesy of S&C Electric Co.

15.3.1.3.3 Fuse Operating Characteristic

The total clearing time of a fuse is the total time required to melt the fusible link and to extinguish the arc created after the fuse melts, at a specified value of short circuit current, plus a manufacturing tolerance. A typical fuse curve is shown in Figure 15.11.

Fuse time–current values are plotted as minimum melting time–current characteristic curves and maximum or total clearing time–current curves.

15.3.1.3.3.1 Time Current Characteristic (TCC) Curves Time–current characteristics (TCC) for overcurrent protective devices define the operating time at various currents. TCC curves are

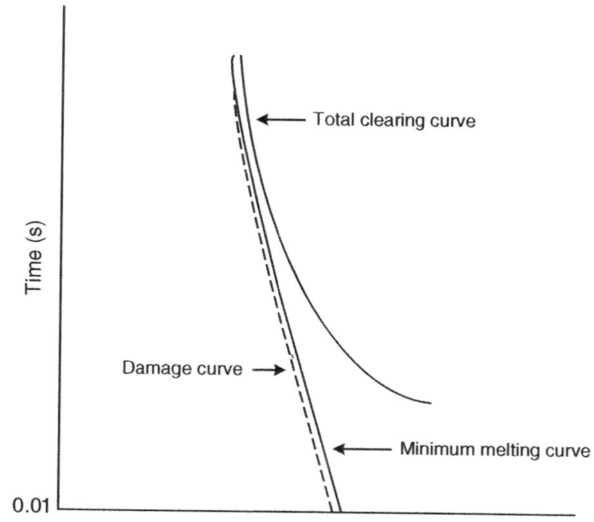

Figure 15.11 A fuse's minimum melting and total clearing time–current curves.

plotted on standard log–log graph paper, and multiple curves are plotted to permit comparisons and coordination of different curves. For a specified short circuit current, coordination may be determined by comparing properly adjusted, same voltage base, time–current characteristic curves.

15.3.1.3.3.1.1 Minimum Melting Time–Current Characteristics The melting time of a fuse is the time from the commencement of a current that will cause the melting of the fuse to the instant that arcing begins. The minimum melting current is the smallest current that will melt the element of the fuse in any specified time. The minimum melting time–current curve of a fuse is defined by tests that determine the minimum time required to operate a fuse at a specified current. The tests are performed with no preload on the fuse and with an ambient temperature of between 20 and 30 °C.

The time and current values are plotted on log–log paper with RMS symmetrical current on the horizontal axis and time in seconds on the vertical axis. The time range for the curve is 0.01–300 or 600 seconds for expulsion fuses and up to 1000 seconds for CLFs.

15.3.1.3.3.1.2 Total Clearing Time–Current Characteristics The total clearing time for a fuse is the time from the beginning of an overcurrent to the interruption of the circuit at a specified voltage. It is the sum of the average melting time, plus manufacturing tolerance, plus arcing time. The arcing time is the period from the start of the melting of the fuse element up to the final interruption. The total clearing characteristic curve defines the relationship between the overcurrent and the total clearing time. This curve is plotted on a log–log scale in a similar manner as the minimum melting curve.

The total clearing curves for expulsion fuses become asymptotic with a horizontal line at about 0.0133 seconds or 0.8 cycles. This indicates the lower limit of the arcing time curve. Even though higher currents cause shorter melting times, arcing times in expulsion fuses cannot be reduced below a lower limiting value. This is because, regardless of the current, the expulsion fuse always continues to arc until a current zero is reached.

15.3.1.3.3.1.3 Speed Ratio A method to describe the shape of the minimum melting curve is to state its speed ratio. The speed ratio is the ratio between the 0.1-second minimum melting current and either the 300- or 600-second melting current of the fuse, depending on the fuse rating. Comparison of the speed ratio for different types of fuses gives an indication of the relative speed of the fuses when operating on the same overcurrent; the lower the speed ratio, the faster the fuse operation. The K-type fuse-link with a speed ratio of about 7 operates more quickly at high currents than the T-type fuse with a speed ratio of about 12. Low-speed ratio (fast) fuses interrupt faults more quickly and coordinate well with inverse relays. Slower fuses withstand higher transients and inrush currents

15.3.1.3.3.1.4 Allowance Factors Manufacturers' published time–current characteristic curves are representative of the fuse link operating characteristics at a specified ambient temperature (i.e. 25 °C) and no preloading. Certain manufacturing tolerances are incorporated in these published curves, and additional allowances have to be made for other various operating conditions. Since the ambient temperature may be either below or above 25 °C and preloading may vary anywhere from 0% to 100% of the nominal current rating, the published TCC curves are normally adjusted either up or down depending on the magnitude of each variable.

TCC curve adjustments are typically necessary to predict under operating conditions the anticipated damage and clearing times necessary for coordination with other devices and protection of system components under fault conditions. Therefore, the following allowances should be considered:

(i) **Damageability allowance**. Damageability is the susceptibility of a fuse link to a permanent change in its minimum melt time–current characteristic curve, refer to Figure 15.11.
Damageability allowance is the difference between the published minimum melt time–current characteristic and a curve indicating values of time and current below which the fuse link cannot be damaged.
This value can be approximated to be 6% of the time axis value, and published minimum melt time–current curves can be adjusted accordingly.

(ii) **Preloading allowance**. Preloading allowance takes into consideration that a fuse element being a thermal device absorbs heat as the load current through the link increases from 0% to 100% of its nominal rating.
Since preloading may vary significantly over a period of time, a preload shift based on a design loading of 75% of the nominal rating of the fuse-link can be used as an approximation. One should use more accurate determination of preloading conditions, if available.

(iii) **Ambient temperature allowance**. The minimum melt time–current characteristic curves are based on tests starting at no initial load and with an ambient temperature of 25 °C. A change in this temperature above or below 25 °C would affect the minimum melting curve slightly, and the representative curve should reflect this consideration.

15.3.1.3.3.1.5 Application of Allowance Factors to Fuse TCC Curves A general rule of thumb, to account for the effects of ambient temperature and preloading, a 75% factor can be used [2]. The 75% factor is applied to the minimum melting times to account for ambient temperature and preloading effects. More precise methods for determining the required adjustment of the minimum melting curve are also possible. For simplicity, however, the adjustment for ambient temperature and preloading will generally be referred to as a 75% shift in the melting curve.

On a time–current graph, an adjusted minimum melting curve for a protected link is created by shifting the published minimum melting curve downward to points that are 75% in time of the published curve. Coordination using the shifted curve will ensure that the maximum clearing time of the protecting link does not exceed 75% of the minimum melting time of the protected link.

15.3.1.3.4 Fuse Application Considerations

The selection of a fuse should take into consideration the following:

15.3.1.3.4.1 Voltage Rating The fuse should have a maximum voltage rating equal to or exceeding the maximum RMS voltage at which it is designed to operate: line-to-line for phase applications and line-to-ground for single phase. The rated maximum voltage of a fuse is designed by the manufacturer. Some HV rated fuse voltages are listed below in Table 15.2.

Typically, the specified voltage rating for a fuse is related to its capabilities through the standard interruption and dielectric tests. Power fuses are not "voltage critical" in that power fuses can be applied at any system operating voltage lower than the voltage rating of the fuse. This is because the expulsion fuses do not produce overvoltage as the arc is interrupted at its natural current zero.

CLFs, on the other hand, chop the current to an artificial current zero causing the development of overvoltage. It is, therefore, prudent to limit the use of CLFs to systems whose operating voltages are in the same class as the voltage rating of the fuses to avoid undue voltage stress on the system.

15.3.1.3.4.2 Rated Continuous Current The rated continuous current of a distribution fuse is generally the maximum RMS current that it can carry continuously, at a specific ambient temperature, without exceeding standard temperature-rise limitations for the fuse components. Typically, melting of fusible elements in distribution fuses will not occur until the applied current is about 200% of the rated value.

The letter "E" in the ampere rating of a power fuse indicates that the melting time–current characteristic conforms to the electrical interchangeability requirement. The requirement is that the current element with ratings 100 A or below shall melt in 300 seconds at an RMS current within the range of 200–240% of the continuous current rating of the fuse unit. At currents above 200–240%, the manufacturer will show the characteristics' published TCC since the current element is a specific feature of each manufacturer.

Similarly, the letter "C" identifies the CLFs. The requirement is that the current responsive element shall melt in 1000 seconds at an RMS current within the range of 170–240% of the continuous current rating of the fuse unit.

Table 15.2 Some HV fuse voltages ratings [3].

System voltage kV (ANSI)	
Nominal	**Maximum**
2.4	2.8
4.8	5.1
7.2	7.8
14.4	15.0
34.5	38.0
46	48.3

The ampere rating and speed characteristic of the fuse link on the transformer primary side must be selected keeping the following considerations in view:

1. It should provide protection against damaging through-fault currents and should coordinate with the transformer damage curve.
2. It should coordinate with secondary side fuses as well as other upstream protective devices.
3. It should permit the transformer to be loaded by the continuous normal maximum load as well as emergency peak loads.
4. It should withstand the combined magnetizing in-rush and load pickup current after a short time service interruption. A current having a value of 12 times the primary full-load current for a duration of 0.1 seconds can be used as an approximation for an equivalent to the combined magnetizing and load in-rush current heating effect in absence of specific data.

15.3.1.3.4.3 Short-Circuit Interrupting Rating The symmetrical short-circuit interrupting rating of the fuse should equal or exceed the maximum RMS short-circuit current at the fuse location. To consider the asymmetrical interrupting rating, the value of the X/R ratio of the system must be taken into account. Typically, fuses are designed to interrupt full asymmetrical current based on a system of X/R ratio of 14. One should refer to the fuse data specific to the fuse being considered for the application. The multiplying factor to obtain the RMS asymmetrical interrupting current from the symmetrical interrupting current varies from 1.56 to 1.6. For the case when the system X/R is higher than 15 (as for a system at a Generating Station), the fuse symmetrical interrupting rating should be de-rated.

15.3.1.3.4.4 Primary Fuses Current for Secondary Faults – Delta/Y Grd. Bank The most important consideration when selecting a primary fuse is its ability to operate for all types of secondary-side faults. The degree of protection, relative to the damage curve of the transformer, provided by the primary fuse should be verified for the level of fault current and type of fault that produces the lowest ratio of per-unit primary-side line current to the per-unit transformer winding currents. Refer to Figure 15.12 for the cases of a secondary L-G fault. For these cases, one or more of the primary fuses will be exposed to a lower level of current than the windings and the primary fuse must be selected to operate fast enough to avoid damage to the transformer windings.

Figure 15.12 shows the magnitudes of fault current seen by the primary fuse under various secondary fault conditions for a Delta/Y Grd.-connected transformer. It is seen that for phase-to-phase and phase-to-phase-to-ground faults, that the primary per unit fuse line currents is proportionately different than the secondary per unit line currents.

It can be observed from Figure 15.12 that the ratio of per-unit primary line current to the per-unit transformer winding current is lowest for a phase-to-ground secondary fault. The ratio is 0.58:1. Accordingly, to ensure proper coordination, it will be necessary to shift the basic transformer through-fault current characteristic curve to the left (in terms of current) by the ratio of 0.58:1. The primary fuse TCC and the secondary fuse TCC in terms of PU line current will be then directly comparable to the transformer damage curve.

15.3.1.4 Automatic Reclosers (AR)

An automatic circuit recloser is a self-contained overcurrent protection device. It is similar in function to a fuse in that it senses, operates, and isolates faults; however, in addition and unlike a fuse, it can operate more than once, typically up to four operations. It is a device that is settable to operate through several trip/close operations and, therefore, is more akin to an integrated relay and breaker combined device.

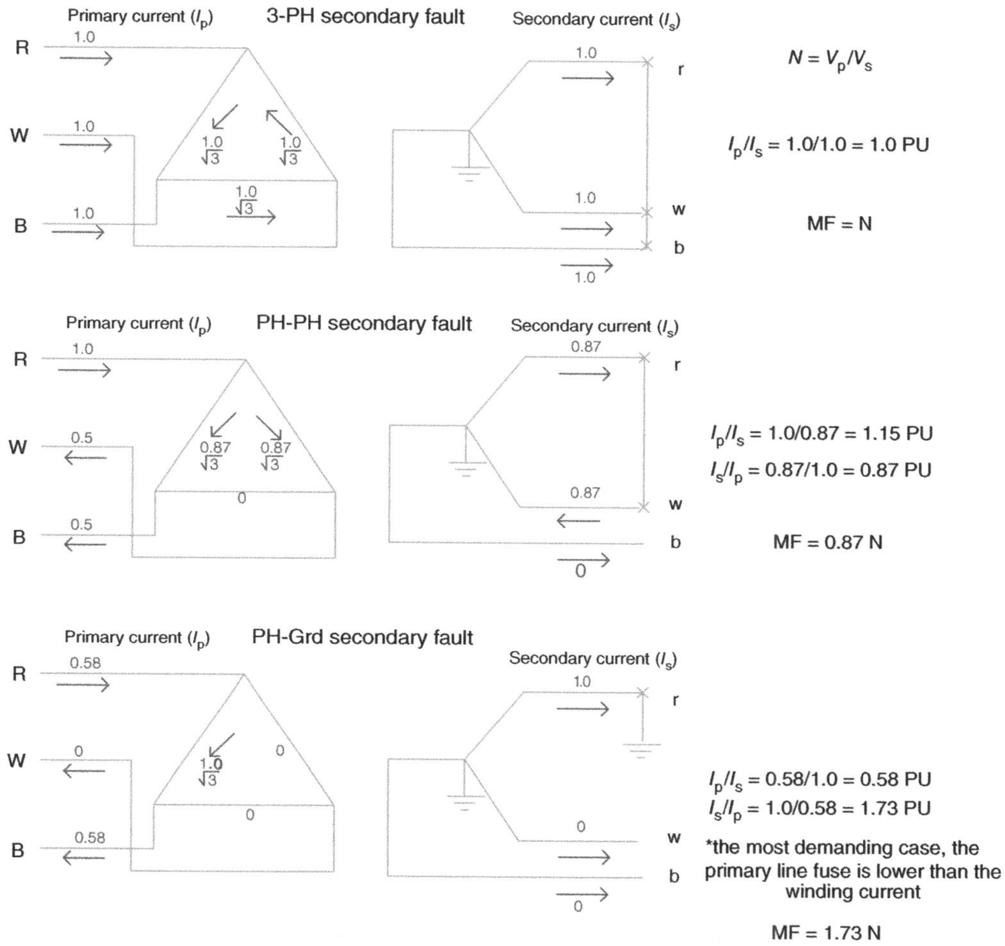

Figure 15.12 Primary fuse currents for secondary faults on Delta/Y-Grd. transformers.

An automatic recloser (AR) senses current, and upon detection of an overcurrent, operates, based on time–current curves, to open the device and then recloses automatically to re-energize the feeder. If the fault is permanent, the recloser will lock open after a preset number of operations and, by doing so, isolates the faulted section from the main part of the feeder.

Approximately 80% of the faults on overhead systems are transient in nature, and these fault types are maintained through an arc. If the feeder is de-energized for a short period of time and then restored, the fault current is interrupted and the arc is extinguished. An AR uses these phenomena by first interrupting the fault and then "reclosing" after a brief time to allow the arc to extinguish. This sequence of operations reduces customer power outages, for transient type faults, which represents the majority of the fault types experienced on distribution systems (Figure 15.13).

Automatic reclosers are designed for application primarily on overhead radial distribution systems. They improve continuity of service to distribution feeders for transient faults such as lightning or conductor clashing due to wind, ice or tree limb, etc.

ARs quickly open, after detecting an initial fault, before subsequent close open operations. It does this so that downstream fuses do not melt, thereby causing a permanent outage. After a delay,

Figure 15.13 An example 1-PH recloser – S&C – TripSaver® II cutout-mounted recloser. Source: Photo Courtesy of S&C Electric Co.

typically one second, the automatic recloser closes and re-energizes the feeder. If the fault still remains at the time it closes, the automatic recloser will reopen. It does so using a time–current characteristic operating characteristic selected by the protection practitioner, refer to Figure 15.14 below.

The automatic recloser continues to operate in this close–open sequence for up to four operations if selected. After a settable number of operations (1–4), the automatic recloser opens (locks out) as the fault is considered permanent. The automatic reclosers must then be manually closed by an operator after the event is investigated and corrected.

In the event of live feeder work being required, there is an integral safety feature that permits the automatic reclosing to trip/open once only. Automatic reclosers used in the distribution systems are designed according to IEEE Specifications [4, 5]. These standards cover recloser ratings, testing, and construction procedures. The reader is encouraged to read these standards for more detailed information on automatic reclosers.

15.3.1.4.1 Configurations and Interrupters

Reclosers are distinguished by several means including the:

- Physical construction (pole, pad-mounted, substation)
- Interrupter technologies (oil, vacuum)
- Internal insulation technologies (oil, SF_6)
- Control technologies (hydraulic, electronic)
- The sensing technology (series trip, shunt trip)
- The number of phases

Automatic reclosers are available from several manufactures and can be specified as single-phase or three-phase units, dependent upon the application. Some ratings for single-phase and three-phase reclosers are given in Tables 15.3 and 15.4 below.

Table 15.3 Some typical North American ratings for single-phase hydraulic control reclosers [6].

Nominal voltage (kV)	Maximum continuous current rating (A)	Interrupting rating (RMS sym. A at maximum voltage)	Interrupting medium	Control
2.4–14.4	50	1,250	Oil	Hydraulic
	100	2,000	Oil	Hydraulic
	200	2,000	Vacuum	Hydraulic
	280	4,000	Oil	Hydraulic
	280	6,000	Vacuum	Hydraulic
	560	10,000	Oil	Hydraulic
24.9	100	2,500	Oil	Hydraulic
	280	4,000	Oil	Hydraulic
24.9–34.5	560	8,000	Oil	Hydraulic

Table 15.4 Some typical ratings for three-phase reclosers [7].

Nominal voltage (kV)	Maximum continuous current rating (A)	Interrupting rating (RMS sym A at maximum voltage)	Interrupting medium	Control
2.4–14.4	100	2,000	Oil	Hydraulic
	200	2,000	Vacuum	Hydraulic
	400	6,000	Oil	Electronic
	560	10,000	Oil	Hydraulic
	560	10,000	Oil	Electronic
	560	12,000	Vacuum	Hydraulic
	560	12,000	Vacuum	Electronic
	560	16,000	Vacuum	Electronic
	800	18,000	Vacuum	Electronic
24.9	560	12,000	Vacuum	Hydraulic
	560	12,000	Vacuum	Electronic
34.5	400	6,000	Oil	Hydraulic
	560	8,000	Oil	Hydraulic
	560	12,000	Vacuum	Electronic
	560	16,000	Vacuum	Electronic
	560	12,000	Vacuum	Hydraulic
	560	8,000	Oil	Electronic

Automatic reclosers contain interrupters that physically trip and close and must do so under fault conditions. The interrupters can be oil-type or vacuum-type and can use oil, vacuum, or SF_6 as insulation mediums.

Oil-type reclosers have been used quite extensively in the past. They use oil for insulation, interruption, and smaller rated units, the hydraulic control. They are lower in cost but require more maintenance; they are used for low fault level applications. Oil-type reclosers are typically provided with hydraulic controls. Vacuum type reclosers are applicable for high-level fault applications and require less maintenance than oil-type.

15.3.1.4.2 Recloser Controls

An automatic recloser is a self-contained unit that not only contains an interrupter, but also a controller that directs the interrupter to operate, including what operating characteristics and how many times to operate. The control is the recloser's intelligence unit much like a protective relay is to a protection system. It senses the current, and once it has reached the designated value, it will initiate the actuation of the recloser (trip/close). There are two basic types of recloser controls, hydraulic or electronic.

15.3.1.4.2.1 Hydraulic Control Hydraulically controlled reclosers are predominantly legacy devices; however there are hundreds of thousands of such devices still installed on distribution systems. A hydraulic controlled recloser connects a solenoid in series (trip coil) with the feeder circuit that it is protecting. The pickup value is determined by the trip coil, (e.g. 25/50/70/100/140/225/280 A).

Typically, the trip coil size is based on the minimum line-end fault within the recloser's zone of protection with margin. Trip coils have predetermined values that must be ordered from the manufacturer. The pickup value is typically twice the trip coil rating and is load limiting with usually a short-time rating. The fault current on the feeder will cause the current to flow through the solenoid, and once the current has reached it pickup value, will in turn, open a trip contact in the recloser.

The hydraulic controls typically provide at least two inverse-time type curves, a fast curve (A) and a slow curve (B), refer to Figure 15.14 and the Recloser Operation section below. The operating time is a function of the selected curves and the number of operations.

15.3.1.4.2.2 Microprocessor Controlled Reclosers Modern reclosers are manufactured with electronic/microprocessor-based controls similar to microprocessor-based protection relays. As such, they offer many more functions than legacy controllers. Reclosers are available with their own integral electronic controller from the manufacturer, or an external electronic controller can be used from a compatible third-party vendor.

Electronic controls sample and convert AC measured quantities and digitizes the waveforms. They use the measured AC quantities as inputs and the integral control logic and protection functions to determine the trip/close actions. Electronic controls, in addition to providing recloser functions, offer many more protection characteristics and features. Electronic controllers' settings are programmable, normally via a computer interface, and they also provide several communication ports for local and remote access. They can provide monitoring, and some also can provide Supervisory Control and Data Acquisition (SCADA)-type control actions.

Since electronic controllers have communication ports, more advanced protection schemes can be deployed by the use of network communications and protocols such as peer-to-peer. These advanced schemes can quickly isolate and restore power. This is based on the communication of operating information of each of the installed reclosers, to determine the fault location, and from that information, determine the appropriate operating condition.

15.3.1.4.3 Recloser Operation

Reclosers operate in a sequence of one and up to four trip/reclose operations. The first operation is "fast" to quickly de-energize the fault, followed by a re-close action approximately 0.6–1 seconds later to allow the extinguishing of the arc. The number of subsequent trip/re-close operations is dependent upon the application and a utility's protection philosophy.

Each recloser manufacturer provides the protection practitioner with a family of time–current curves to choose from. Typically, fast and slower curves are available to choose from as well as the number of operations (1–4), typically referred to in the industry as "shots." The time–current curves represent the total time to clear for values of current within the rating of the recloser. The recloser operates on one time–current curve at a time. It typically starts by using the fast curve and then according to the setting, transfers operating to one of the slower curves. After a preset number of trip/reclose operations, the recloser locks its contacts in the open position on the assumption that the fault is permanent. The recloser must then be manually closed.

Time–current curves fall into one of two basic categories, definite and inverse.

A definite time–current curve characteristic describes the total time from fault initiation to fault clearing regardless of fault current magnitude. Response time may be instantaneous or subjected to a predetermined, intentional time delay. For example, curve "A" or fast curve has no intentional time delay, shown in Figure 15.14 below.

An inverse time–current curve characteristic is when the clearing time varies depending on the fault current magnitude. There may be different time delays. For example, curve "C" is more delayed than curve "B," as shown in Figure 15.14 below.

Recloser controllers offer several time–current curves, each curve offering different possible operating times based on the application needs. It should be noted that the total clearing time of the recloser is the published time–current curve time selected plus the interrupting times for the operation of the recloser. Electronic recloser interrupting time is in the range of 2.5–4 cycles, and for hydraulic reclosers, it is in the range of 1.8–2.4 cycles; one should refer to the vendor's specific product manual.

Recloser operating sequence can be set to open two, three, or four times before locking out. Or it can have one trip to lockout. The recloser may be modified to provide all fast operations, all delayed

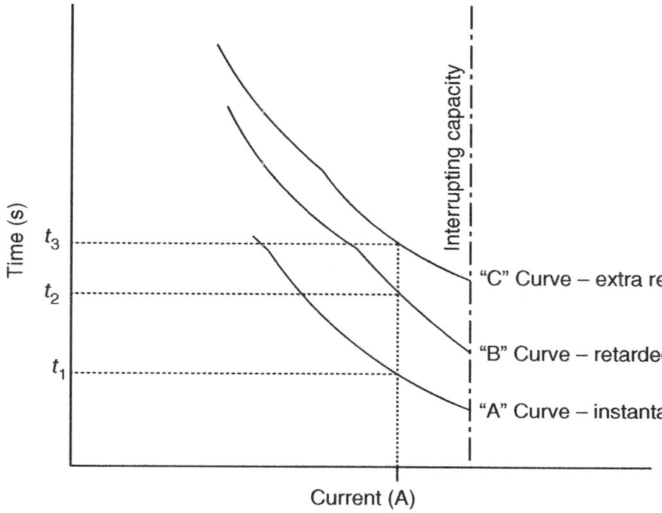

Figure 15.14 Typical recloser operational curves.

operations, or any combination of fast followed by delayed operations. In all cases, however, fast operations, if any, occur first followed by delayed operations, up to the total selected number of operations to lockout.

After each tripping operation, the recloser remains open for approximately 1–1.5 seconds. This is called the reclosing interval. A typical operating sequence may be described as "2A 2B". This means that the recloser operates twice on its "A" curve, twice on its delayed "B" curve, and then to lockout. Since the "A" curve operations always occurs first, this sequence may also be described as "B22" meaning two fast "A" curve operations and two delayed operations utilizing the "B" delayed curve.

A sequence of "1A2C" (or C12) means one fast operation followed by two extra-delayed "C" curve operations. Four fast operations could be expressed as "4A" (or "A40") and four delayed operations as "0A4B" (or "B04").

In the application of a reclosing overcurrent protective device, the sequencing of its mode of operation determines the degree of coordination that can be achieved with adjacent protective devices. For a primary feeder example, one can use one fast followed by three retarded (time-delayed) operations typically denoted as A+3B or B13 that are performed before the lockout of the device. On subtransmission feeder reclosers, they tend to be set for 1A1B operations.

15.3.1.4.4 Recloser Selection/Ratings

A recloser should satisfy the following criteria:

(a) The maximum phase-to-phase voltage should not exceed the design voltage of the recloser. In the case of a single-phase recloser protecting a single-phase tap from a three-phase multi-ground system, the maximum phase-to-neutral voltage should not exceed the design voltage of the recloser.
(b) The continuous current rating of the recloser should be equal to or greater than the maximum load current through the recloser.
(c) The interrupting, rating of the recloser should be equal to or greater than the maximum available fault current at the recloser location.
(d) The minimum trip current selected should permit the recloser to sense fault current throughout the desired protection zone.
(e) Time–current curves and operating sequences selected shall allow coordination with other protective devices on both sides of the recloser.

15.4 Protection Coordination Principles

This section is intended to provide some basic protection coordination principles for the above-described protection devices primarily used at the distribution level.

15.4.1 Functions of Distribution/Overcurrent Protection

Distribution protections, which are mainly overcurrent types, perform several of critical functions to meet the objectives of distribution planning.

(a) **Isolate permanent faults.** Protection systems are employed to isolate permanent faults from the unfaulted portion of the distribution system. The isolation of permanent faults limits the number of customers affected by a fault. Reliability is a function of the fault frequency, the number of customers affected, and the service restoration time. In distribution protection applications, this function is related to protection selectivity.

(b) **Minimize number of permanent outages.** A function of a distribution protection system is to quickly de-energize temporary faults. As distribution systems experience mostly transient or temporary faults, the ability of the protection system to discriminate between temporary and permanent faults will produce improvements in service continuity. Temporary faults can degrade into permanent faults if they are allowed to persist. Quick fault clearing can prevent this fault from evolving and provide an opportunity for the source of the fault to clear on it own. Rapid fault clearing and fast circuit re-energization will help with this objective. The most common method of achieving fast circuit re-energization is through the use of automatic reclosing protective devices.

(c) **Minimize service restoration time.** A function of the protection system is to minimize the time required to localize and sectionalize permanent faults. The goal is to optimize the overall design of the protection system and the specific devices used, such that all customers outside the permanently faulted line section can be restored to service at the earliest possible time. Advanced and sophisticated protection and control schemes are available which can perform this function automatically, through the use of automatically controlled and coordinated sectionalizing switches/devices. However, this type of system is generally more expensive. The system design has an important influence on the ability to minimize restoration time since the location and number of protective devices help to localize faults. In the event of a fault, the affected customers should be geographically localized, and customer trouble calls will provide clear indications of the general area of the problem. The other influence is the specific selection of protective devices. Many protective devices offer a physical indication of operation, which further assists field crews in localizing the fault location.

(d) **Mitigate equipment damage.** Protection systems function to mitigate damage in unfaulted equipment. Fault current withstand capabilities are specified for numerous types of system components, including bare conductors, cables, and switches. Fault withstand capabilities are defined in terms of the current–time combinations which are allowable. For overhead conductors, the damage conditions relate to the loss in conductor mechanical strength. For underground cables, the relevant damage condition is the extent of degradation of the cable insulation. For distribution transformers, the damage is measured in terms of the loss of life due to transformer insulation degradation.

15.4.2 Coordination General

Distribution protection, among other functions, is applied to isolate faulted equipment and to provide sectionalizing at strategic locations on the feeder circuit. Sectionalizing points are usually established at the substation, at various points along the feeder, in-branch and sub-branch lines tapped off from the feeder, and on the primary side of all distribution transformers. Overcurrent protection is a science in terms of fault current calculations, equipment specification, and coordination; however, there is a craft component to it and subjected to utility distribution operational philosophies.

The concept of protection selectivity refers to the capability of the protection system to detect faults on a power system and as a result, initiate the opening of circuits to isolate the faulty part of the system. It is designed, as much as possible, to select and isolate only the faulty part of the system leaving all other parts in normal operation. The protection must be discriminative, and this can be achieved by current, or by time, or by a combination of current and time, for when overcurrent protection applies, or by distance/ohmic reach for distance type relays. The coordination of distribution protective devices is primarily accomplished, for overcurrent devices, with a set of time–current characteristic curves that define the operating characteristics for each device.

Coordination of protective devices involves the selection of appropriately rated devices that will provide selective operation during transient and permanent faults. The operation of any single overcurrent protective device should not have any substantial or permanent effect on the operation or condition of adjacent protective devices. This concept forms the basis of overcurrent protective coordination.

Refer to Figure 15.15 for the given fault location, protection device PR1 referred to as the "Protecting Device," (downstream from PR2) should operate to lockout before PR2, the Protected "Device," (upstream from PR1) to maintain service to upstream customers connected to Section 2. The term downstream is used to describe the "load" side of a protection device, and upstream refers to the "source" side of a protection device.

For example, if PR1 is a fuse and PR2 is a recloser, and the fuse link ratings and recloser are properly applied based upon the magnitude of the calculated available fault current; the fuse should not blow or even be damaged by a transient fault located beyond it. The recloser should open the circuit one, or more times depending upon the sequence of recloser operation, without the fuse being damaged. On a permanent fault, the fuse should be allowed to operate and open the faulted circuit before the recloser reaches the second last operation. This selective operation (coordination) permits continued service to customers that are connected outside of the faulted Section 1 and allow transient faults to be cleared without a fuse operation.

15.4.2.1 Fuse – Fuse Coordination

Fuses are located at the tap point of single-phase laterals and sometimes in main distribution trunks; however, if the laterals or branches are considered long, then additional fuses are also located mid-point, resulting in two series-connected fuses. These fuses are strategically located and applied to isolate faulted equipment and to provide sectionalizing.

Fuse-to-fuse coordination is accomplished through the use of published fuse minimum-melting and total-clearing TCC curves. Coordination is achieved between two series-connected expulsion fuse links with the total clearing time–current curve of the protecting fuse link, plotted to maximum values, is compared with the minimum melting values of the protected link.

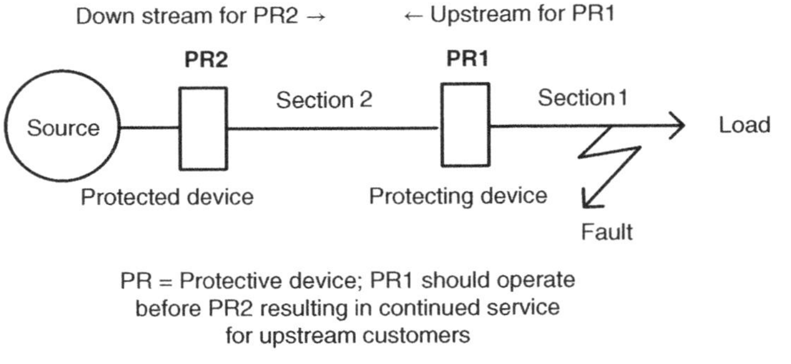

Figure 15.15 Basic distribution coordination principles.

Refer to Figure 15.15 and assume PR1 and PR2 are both fuses; to coordinate two fuses connected in series, the damage time–current characteristic curve of the protected fuse (PR2) must be greater than the total clearing time of the protecting fuse for all fault current values in the protective zone of the protecting fuse – refer to Figure 15.16 below.

It is common practice to adjust the minimum melting curve downwards (PR2) to ensure that the protecting link will clear before the protected link is damaged.

A shift in the melting curve of 75% on the time axis is typical. This shift provides a margin for such operating variables as preheating due to load and ambient temperature. All expulsion fuses must wait for the first current zero, followed by element melting to interrupt their total clearing time for any fault current and cannot be less than about 0.8 cycles. This is indicated by the total clearing curve, for expulsion type fuses, being a constant 0.013 for all high fault currents.

Therefore, any two expulsion fuses will only coordinate up to a certain maximum fault current at which point the let-through current of the protecting fuse will operate the protected fuse – refer to point "I" in Figure 15.16 below. This limit of coordination is indicated by the intersection of the total clearing curve of the protecting link and the shifted minimum melting curve of the protected link. Two expulsion fuses will only operate selectively when the available fault current is below the coordination limit.

15.4.2.2 Fuse – Automatic Recloser (AR)

A typical application for a fuse to recloser coordination is at DS for transformer protection as depicted below in Figure 15.17.

The use of an AR protecting device with a single-shot fuse eliminates unnecessary outages that occur with the use of fuses only. The AR should be set to trip for a fault before the fuse can blow and then recloses the circuit. If the fault is temporary, the AR will recognize the fault on the circuit and trip to clear it before the operation of any fuse, thus restoring normal conditions.

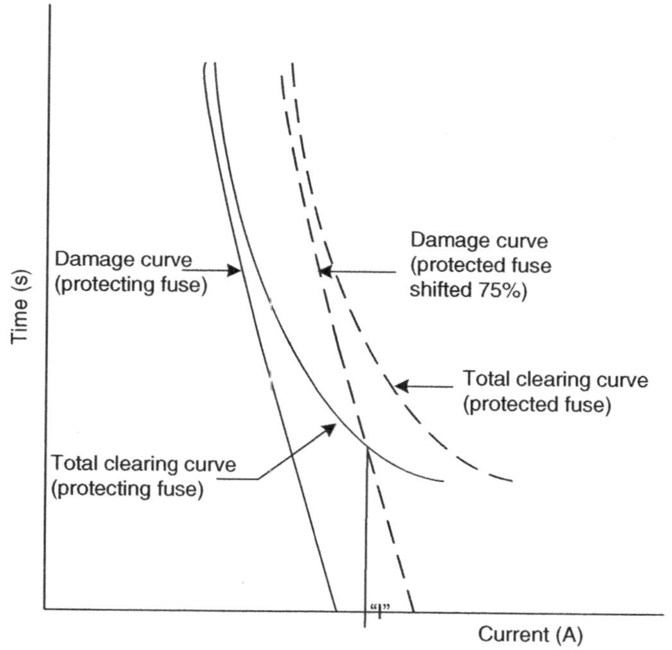

Figure 15.16 Fuse to fuse coordination.

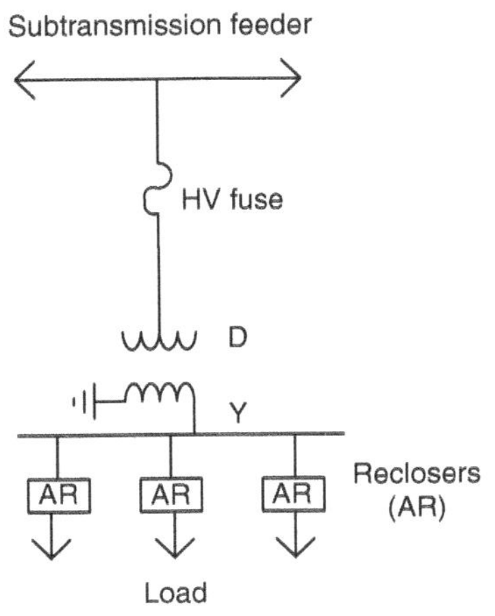

Figure 15.17 A typical DS fuse to recloser overcurrent protection scheme.

For a permanent fault beyond the AR, it will operate once, twice, or three times, depending on the operational sequence setting, in an attempt to clear the fault, and then lock-out. The permanent fault will then have to be cleared by the fuse. The rationale being to allow the AR, with additional operations, to clear the fault by "self-clearing," thereby, maintaining services to customers connected on the source side of the reclosers.

In this case, the transformer primary-side fuse is located upstream of the automatic circuit recloser. The goal is that the fuse not melt before the recloser operates to lockout in response to a permanent fault. The maximum fault current value up to which the fuse and the recloser will coordinate is generally the lower of:

1. The maximum interrupting capacity of the recloser or fuse, or
2. The intersection of the minimum melting curve of the fuse and the maximum equivalent operating curve of the recloser (i.e. the "lockout" curve).

Legacy coordination principles used a conservative coordination method where cooling of the fuse during reclosing time intervals was ignored. One simply summed the heating effect, or heat input, of each recloser operation. That is, the lockout curve of the recloser was developed by summing the total clearing times for the proper number of fast and slow operations, at various current levels, refer to Figure 15.18 below.

Note, that coordination is achieved up to the value of I_1 in Figure 15.18.

Current practice requires that the heating and cooling of the fuse during reclosing operations be taken into consideration when coordinating such overcurrent devices. At the onset of the fault, the fuse is heated by the fault current and then partially cools when the recloser opens and heats to a higher temperature with the next reclose cycle. Further heating and cooling occur during subsequent operations. To cater for this fuse heating/cooling effect, the industry, via tests, has established a set of "K" factors as provided below in Table 15.5.

The recloser should complete its entire operating sequence without causing damage to the upstream fuse. The set of "K" factors has been developed to multiply the time values of the delayed curves (B, C, etc.) of the recloser to compensate for the cumulative heating. The recloser's adjusted delayed curve should be below the shifted minimum melting curve of the fuse for all fault currents available at the recloser location. The intersection of the adjusted recloser delayed curve and the shifted minimum melting curve of the source side fuse indicates the maximum coordination current of these two devices – refer to Figure 15.19 below.

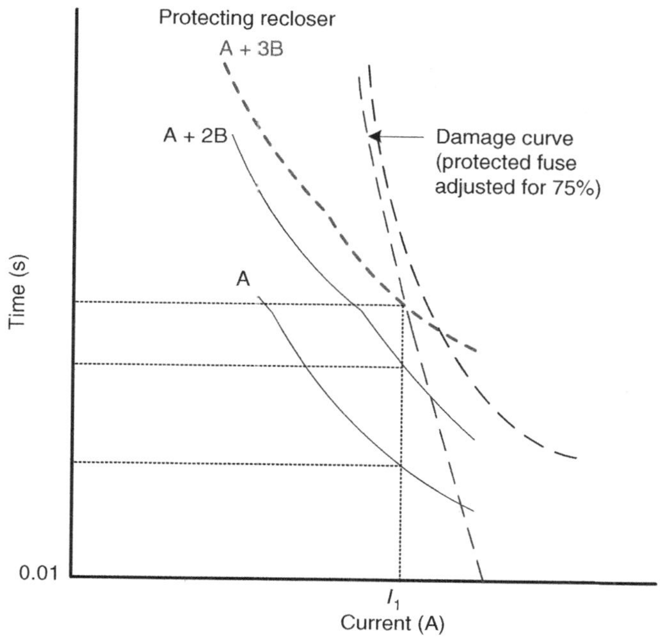

Figure 15.18 Legacy fuse-recloser coordination.

Table 15.5 "k" Factors for delayed curve of load-side reclosure [8].

Reclosing interval (cycles)	Two fast-two delayed	One fast-three delayed	Four delayed
25	2.7	3.2	3.7
30	2.6	3.1	3.5
60	2.1	2.5	2.7
90	1.85	2.1	2.2
120	1.7	1.8	1.9
240	1.4	1.4	1.45
600	1.35	1.35	1.35

Accounting for fuse cooling results in an increase in a fault coordination margin (I_2).

As discussed in Section 15.3.2.1.1.4, with an unsymmetrical transformer connection, such as the Delta–Wye Grd., the ratio of primary to secondary fault current will be different, depending on the type of fault – refer to Figure 15.12. As such, the fuse TCC must be shifted normally, to the secondary of the DS transformer for coordination.

For example, assume a DS transformer voltage ratio of $N = 46/12.47$ kV $= 3.7$.

For the N ratio of 3.7, the multiplying factors are, refer to Figure 15.20 below:

Three-Phase: Fuse curved moved to the right by $N = 3.7$ (shifted to the LV).
Phase-to-Phase: Fuse curve moved to the right by $0.87 \times N = 0.87 \times 3.7 = 3.2$ factor.
Phase-to-ground: Fuse curve moved to the right by $1.73 \times N = 1.73 \times 3.7 = 6.4$ factor.

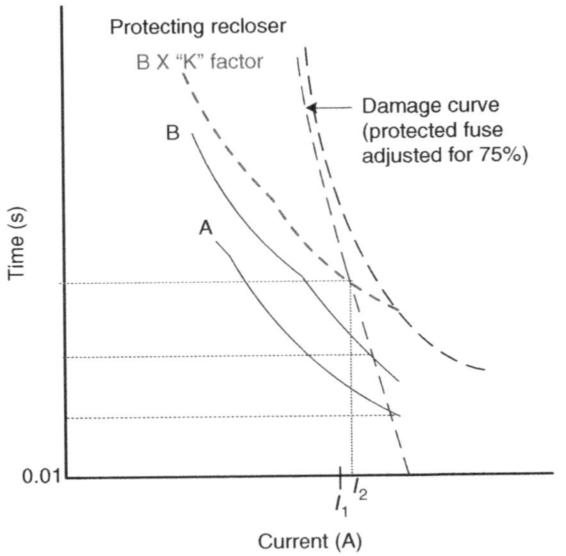

Figure 15.19 Fuse-recloser coordination.

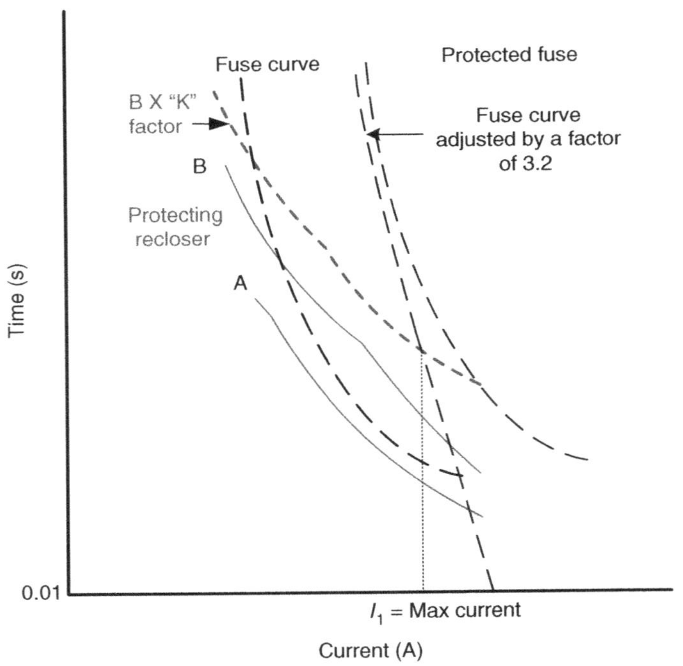

Figure 15.20 Fuse adjusted recloser coordination.

From a coordination perspective, the most demanding cases are when the per-unit primary line current is greater than the per-unit secondary current. Since the phase-to-phase fault factor of 3.2 will result in the tightest coordination that should be used as the limiting factor. Any other factor would result in a larger shift and allow more coordination space between the recloser and the fuse curves.

15.4.2.3 Primary Fuse – Transformer Damage Curve

Transformers, like most equipment, have continuous and short-term operating ratings that if exceeded, will be damaged and or have their expected lives shortened. One of the functions of a transformer's primary fuse is that it must protect the transformer against damage from mechanical and thermal stresses resulting from a secondary side fault that is not promptly interrupted.

The total clearing curve of the fuse selected for transformer protection should coordinate with the transformer thermal damage curves. A fuse should operate to clear for secondary faults before the magnitude and duration of the overcurrent exceed the short-time loading limits recommended by the transformer manufacturer.

In the absence of specific information applicable to an individual transformer, the primary fuse should be selected in accordance with recognized guidelines for the maximum permissible transformer short time loading limits. Through-fault protection curves are presented by IEEE [9, 10] for different transformer categories. Most DS transformers are Category II or III. For Category II transformers (up to 5 MVA), the recommended duration limits for through faults of various magnitudes are based upon the two curves provided in Figure 15.21. The choice of the curve depends on the expected fault frequency. The standard was developed with the recognition that transformer damage is cumulative and that the number of through faults will be different for different transformer applications. Frequent fault incidence is considered to be more than 10 in a transformer lifetime, while infrequent faults occur less than 10 times in the transformer life.

The curve on the left (frequent faults) reflects both thermal and cumulative mechanical damage considerations. It is dependent on the impedance of the transformer for fault current above 70% of the maximum possible through fault and is based on the I^2t of the worst-case mechanical duty (i.e. the maximum fault current for two seconds). The curve on the right for infrequent faults represents thermal damage considerations only and is not dependent on the impedance of the transformer. The upper solid portions of the curves are based on multipliers of the full load current.

The location and function of a station protective device, as well as fault frequency, should be considered when deciding which transformer damage curve should be coordinated with the curves for that device. For example, the secondary feeder recloser, which may see numerous downstream faults, should be coordinated with the frequent fault incidence curve. The primary fuse, on the other hand, would operate on relatively rare faults between the transformer and feeder protection and should be coordinated with the infrequent fault incidence damage.

An important function of the DS transformer primary fuse is to protect the transformer against damage from mechanical and thermal stresses resulting from a secondary fault that is not promptly interrupted. A properly selected primary fuse should operate to clear such a fault before the magnitude and duration of the overcurrent exceed the short-time loading limits recommended by the transformer manufacturer. The degree of transformer protection provided by the primary fuse should be checked for the level of fault current and fault type producing the most demanding conditions for each application. That is, for those for which the ratio of primary side line currents to the transformer winding currents is the lowest. For such cases, one or more of the primary fuses will be exposed to a lower level of current than the windings and the primary fuse should operate fast enough to avoid damage.

Example 15.1 ***Through Fault Damage Curve Calculation***

Assume a 5 MVA DS transformer subjected to frequent faults, $Z = 7\%$.

Transformer damage curve would include both the thermal and mechanical portion (dog leg):
Per Figure 15.21; $k = I^2\,t$; $k = (1/Z\ \text{PU})^2 \times 2$ for mechanical 70–100 and max through current.

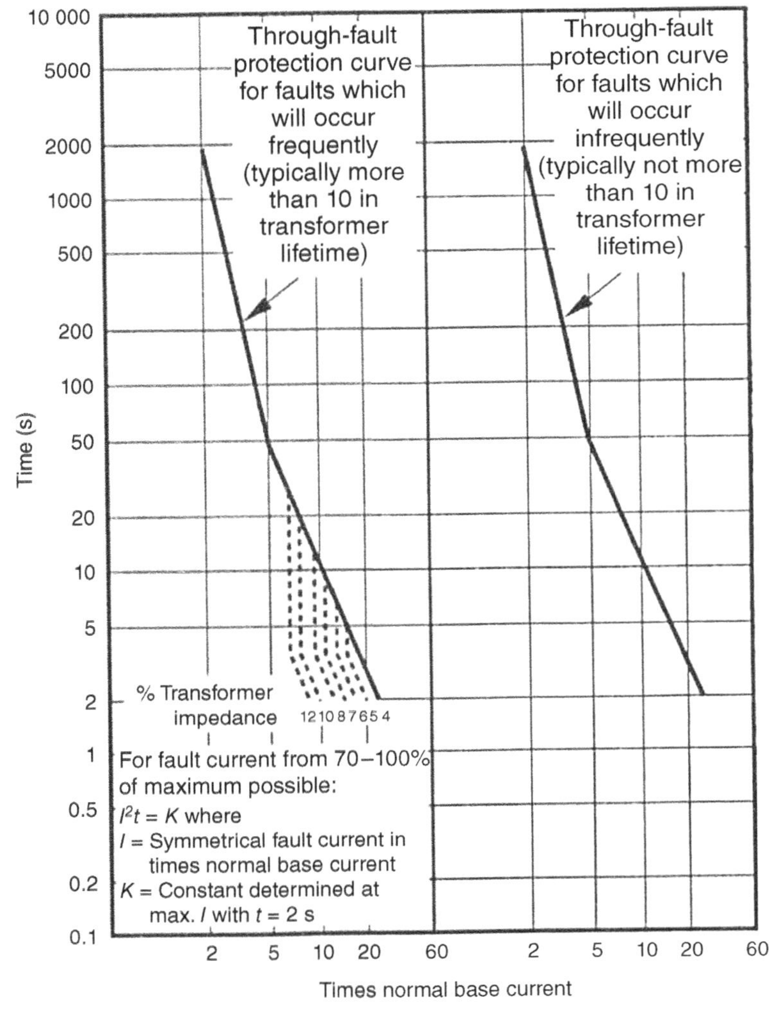

Figure 15.21 IEEE C37.91 Category ll Transformer Damage Curves [11].

Therefore:
At two seconds, $I = 1/Z$ PU $= 1/0.07 = 14.3$ PU
Calculate the thermal energy: $I^2\ t = 14.3^2\ t = 409@2$ seconds
$t@70\%$ of $14.3 = 0.7 \times 14.3 = 10$ PU
$I = 10$ PU find $t = 409/10^2 = 4.09$ seconds

Dog leg points are: (14.3, 2), (10, 4.1) [IPU, t seconds]

It should be noted that the transformer damage curve is based on, as per IEEE C57.109, winding currents not on line currents. There is only one damage curve defined by a set of points based on time and PU base current or rated current, ampere turn balance (MVA on the primary equals MVA on the secondary). As an example, with a damage curve point of five seconds rating at 18 PU and a transformer turns ratio of 2.0. On the primary it will result in two seconds at $18 \times 1/2 = 8$ A, on the secondary it is two seconds at $18 \times 2/1 = 16$ A. It is important to note that transformer winding

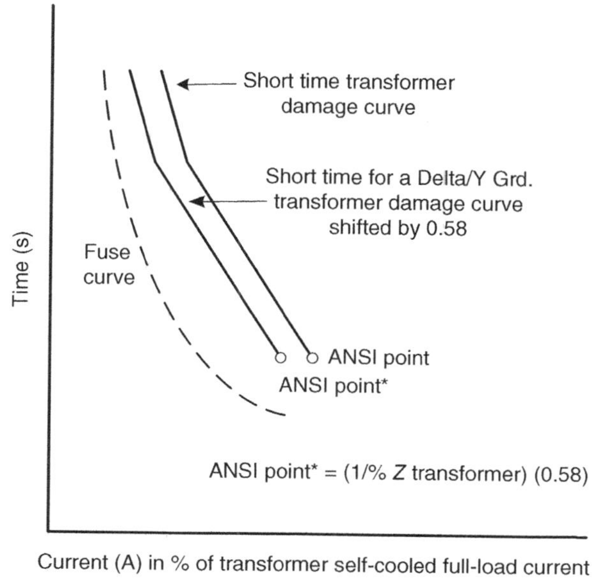

Figure 15.22 Transformer short time damage curves.

currents and line currents are only the same for wye connections and not for delta connections. For these cases, one need to shift the damage curves appropriately.

Refer to Figure 15.12 for a Delta/Y Grd. transformer, a phase-to-ground secondary fault produces the most demanding conditions for this type of transformer configuration, since the per-unit primary current is less than the per-unit winding currents, 0.58/1.0 = 0.58. To ensure proper transformer protection for this case, it is necessary to "shift" the basic transformer short-time characteristic curve to the left (in terms of current) by the ratio of the per-unit primary side line current to the per-unit transformer winding current. The shifted transformer short time characteristics curve will then be in terms of the primary side line current and, as such, will be directly comparable with the total clearing TCC of the transformer primary fuse, refer to Figure 15.22.

15.5 Feeder Energization

When a feeder is first energized, there occurs a short-duration inrush of current. Typically unloaded transformer inrush is 12 times primary full load for an approximate duration of 100 ms.

This inrush of current is caused by:

(1) The magnetizing inrush currents of transformers, motors, and the like;
(2) Motor starting currents.

A feeder and its associated protection must be able to account for this inrush of current. It must be able to withstand both magnetizing inrush and inrush currents that occur due to carrying the load as a consequence of a momentary (recloser) or longer loss of supply.

The inrush current due to magnetizing currents and momentary load disruptions is referred to as "hot-load pickup" current. If the disruption is longer than 20–30 minutes, this is referred to as "cold load pickup" current. Cold load pickup currents are higher, and it can make re-energizing the feeder difficult without causing the protective relays to operate. As a significant portion of a distribution feeder's load includes intermittent loads such as air conditioners, electric heaters, and refrigerators, these loads cycle on and off at different intervals; thus under normal conditions, only a portion is on at any given time creating load diversity on the feeder. However, this load diversity is lost after a scheduled or forced outage; most of these loads will be switched on. Motors, transformers, heaters will draw significant start-up currents. This can cause a significant surge in load current resulting in the operation of overcurrent relays. This condition is known as cold load pickup.

The magnitude of the current is related to the amount of load diversity, but it is difficult to determine – outage time and the cycling of on/off of load during that period. Inrush currents due to magnetizing and raising temperatures of heaters are very high, but of such a small duration, that it would not affect the overcurrent relay operation. Cold load pickup can be in the order of 4–6 times load for 1–2 seconds and reduce to 2 times in the order of several hundred seconds.

Motor starting can cause these high currents to remain for such duration as to initiate the pickup of some of the overcurrent protection devices and cause a misoperation.

Overcurrent protection systems should offer secure operation. It should be secure from a false operation, in that it does not cause de-energization of circuits due to load unbalances, inrush currents, cold load pickup, and other transient or steady-state conditions which will not be harmful to the system components. Typical, feeder current relay settings are in the range of 3–4 times the maximum load current; these settings may be too low to prevent pickup on cold load pickup. Increasing the feeder relay setting will affect the feeder fault coverage or restrict settings with upstream devices. The extremely inverse overcurrent relay provides a satisfactory solution to this problem. The relay can ride over the cold load inrush and, therefore, prevent a misoperation. The inverse type of overcurrent relay can also be used to solve this problem, but the extremely inverse can also clear high magnitude faults faster.

Some microprocessor-based feeder relays offer dedicated cold load pickup logic, allowing for a dynamically changing pickup setting; therefore, the relay can operate with a higher pickup for a given time-period during energization and then return to the normal modes after the preset timer expires, without compromising the protection performance, refer to Figure 15.23 below for an example.

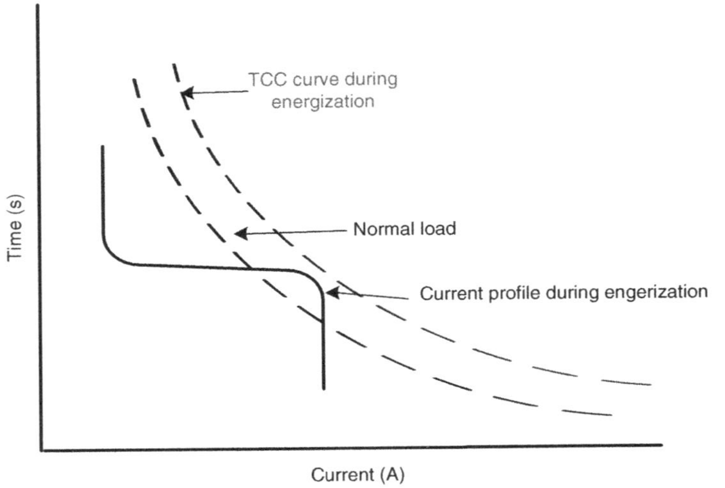

Figure 15.23 An example of dynamic TCC curve [12].

Table 15.6 Typical transformer short time loading [13].

Time (s)	Load (PU)	Condition
8 × 20 μs	2,000–10,000 A peak	Lightning current
0.01	25	Magnetizing inrush
0.02	7 · KVA · PIV/KV[2a)]	Lightning saturation
0.10	12	Inrush and hot-load pickup
1.0	10	Motor start-up
1.00	6[b)]	Cold load pickup (CLPU)
10.0	6	Motor start-up
10.0	3[b)]	CLPU
100	2[b)]	CLPU and short-time overload
100	3	Motor start-up
300	2[b)]	CLPU
600	2[b)]	CLPU
1000	1.5	Long-time overload

a) Where
KVA = transformer kVA
PIV = peak impulse voltage level
kV = transformer primary voltage in kV.
b) CLPU factors are multiplied by the pre-outage load, which in many cases is assumed to be 1 PU of the transformer rated load.

Some typical transformer short-time loading data are provided in Table 15.6.

Motor start-up surge magnitudes depend on the amount of motor load on the transformer. The time–current values in Table 15.6 are for motor loads which represent 75–100% of the transformer rating. These values can be scaled for transformers with smaller motor loads or neglected if there is insignificant motor loading.

15.6 Subtransmission Feeder Protection

15.6.1 General

Subtransmission feeders are predominantly radial type feeders, a single source with only one direction for current flow. The protections and associated settings discussed below apply to radial type feeders unless otherwise stated.

Most legacy feeders are radial and use electromechanical protection devices; however, the distribution systems are evolving. Recently, electric utilities are obliged to connect feeder DGs, also referred to as distributed energy resources (DERs), which has transformed some radial feeders into multi-sourced feeders. For such cases, the radial protections described below are maintained but will need to be modified to suit; such protection modifications will be described in Section 15.7.

Furthermore, electric utilities are embarking on advanced distribution system (ADS) projects or also referred to as Smart Distribution Systems. These systems deploy microprocessor-based protection devices, communications between devices, monitoring equipment, and many offer "smart" type technologies. These ADSs are offering more advantages and functions than legacy systems, but they do so at higher costs and complexity.

15.6.2 Subtransmission Feeder Protection Requirements

Although simple overcurrent instantaneous and timed overcurrent relays are mainly used for subtransmission feeder protections, different utilities use different feeder protection schemes. A utility's feeder protection scheme is based on the specifics of the feeder (radial, non-radial, etc.) and upon the company's operating philosophy and objectives.

One such feeder protection scheme is presented below, depicted in Figure 15.24. This scheme is characterized as a fuse-saving overcurrent feeder protection system with automatic reclosing. However, other variations schemes are also used, and the scheme that is deployed is a function of a distribution utility's business needs.

This specific feeder protection scheme discussed here is representative of typical distribution overcurrent protection and automatic reclosing. It is intended to illustrate feeder protection application, and it is based on meeting the following philosophy:

(a) The feeder protection should ensure that transient faults, either on the main trunk or the laterals, would cause no more than a momentary interruption to any part of the feeder … this is referred to as a fuse saving scheme.
(b) Any permanent fault in a three-phase distribution transformer or its associated secondary circuit, or on a single-phase lateral, shall not cause an extended outage.
(c) The feeder timed overcurrent protection must coordinate with protections on the supply side of the feeder breaker (bus, transformer, and HV line, if applicable) as well as the fuses or protective devices associated with the three-phase taps off the main trunk, and the single-phase laterals.
(d) The timed protection setting should not be approximately greater than three seconds.
(e) For 4-Wire applications, the relationship, between the operating times of the lateral fuses and the feeder protective relays, should be selected such that, in the event of a fault on the lateral:
   1. The minimum melting time (plus margin) of the fuse shall be greater than the total clearing time for the breaker's first trip operation;
   2. The fuse total clearing time shall be less than the time required to initiate tripping for the breaker's second trip operation following automatic reclosure.

15.6.3 Subtransmission Protection

For radial feeders, since there is only one direction in which current flows, non-directional overcurrent protections are usually used. For non-radial feeders, which will be discussed separately, utilities use directional overcurrent or distance protection.

15.6.3.1 Protection Functions

The feeder protection scheme example, depicted in Figure 15.24 is implemented using legacy measuring relays, and the logic scheme uses discrete auxiliary relays and timers. This format was intentionally selected to make it is easier for the reader to understand and follow. This logic is easily converted to Boolean logic and can be implemented using digital relays, optimized for feeder protection applications.

As an overview and referring to Figure 15.24a, the feeder protection scheme presented in this example uses two instantaneous overcurrent elements, and one timed overcurrent for phase protection, and a similar set for ground protection. The two instantaneous overcurrents are designated

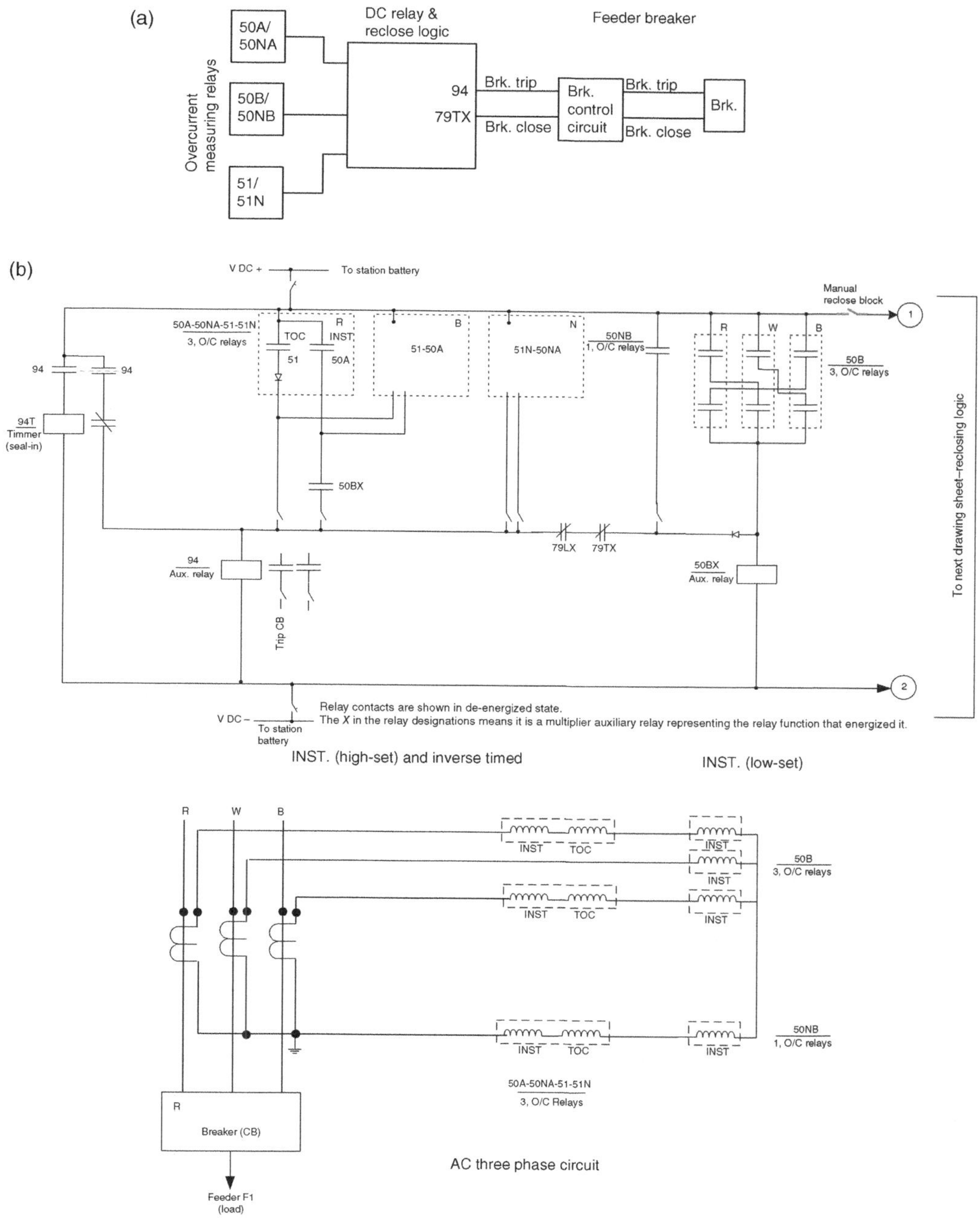

Figure 15.24 (a) Feeder protection and automatic reclose scheme overview. (b) DC and AC diagrams for the feeder protection scheme example. (c) DC diagram – for the feeder protection scheme example continued.

(c)

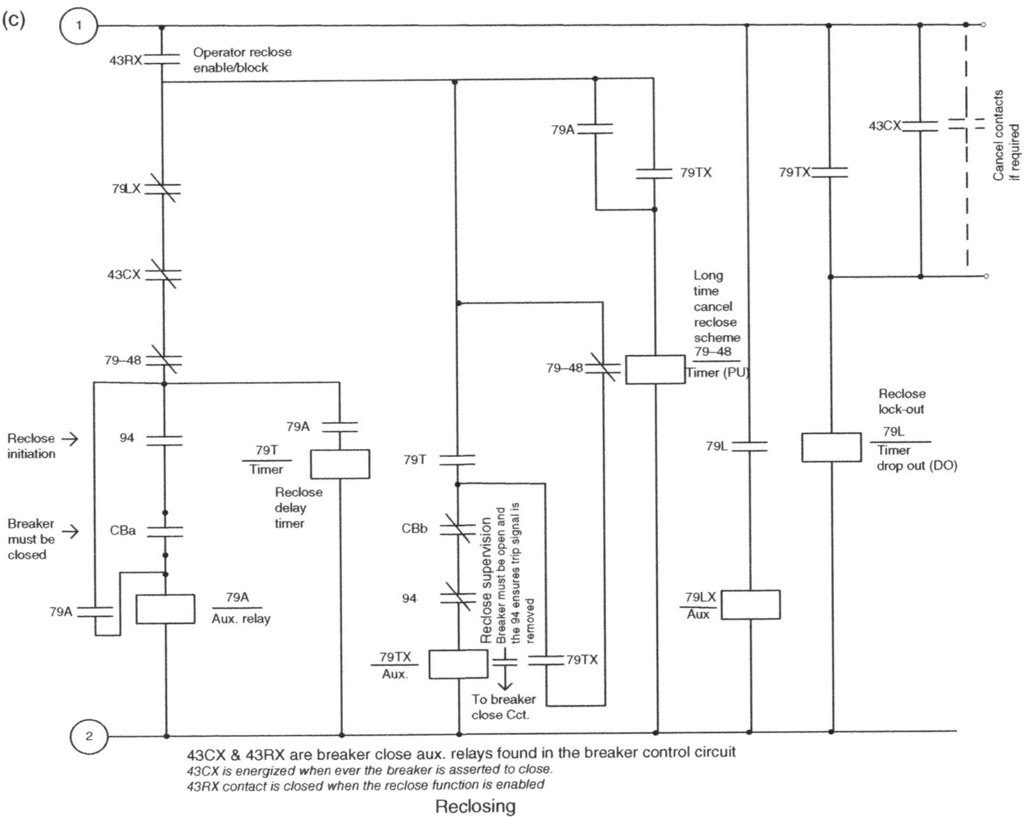

Figure 15.24 *(Continued)*

as 50A and 50B for phase and 50NA and 50NB for ground, respectively. The timed overcurrents are designated as 51 and 51N. The 50A element is further designated as high-set and the 50B as low-set. This designation is intended to distinguish the two instantaneous overcurrent elements by settings and, as further discussed below, by function.

This example also employs a reclosing scheme that automatically recloses the feeder breaker after a protection trip (94) and a set time delay (reclosing delay timer 79T). The reclosing logic initiates a close via auxiliary relay 79TX that interfaces with the feeder breaker control circuit.

15.6.3.1.1 Instantaneous Overcurrent (Low-Set) Protection (50B/50NB)

Refer to Figures 15.24b,c.

The instantaneous low-set protection represents one of two instantaneous protections used for this feeder scheme. The instantaneous low-set is intended to operate for all feeder transient faults; therefore, the use of qualifier low-set, and is used with an automatic reclosing scheme.

It consists of three instantaneous overcurrent elements, one per phase (50B) and one ground element (50NB). This protection is set low to cover the entire feeder and should operate first when a fault first occurs. This first operation of the feeder protection initiates the reclosing scheme (via auxiliary relay 94). When reclosing is initiated, this protection is blocked via reclosure relay (via 79TX and lockout relay 79LX set typically for 10 seconds). The rationale being that a fault that persists after reclosing is probably permanent and requires coordinated isolation.

Since this protection has a low setting, it also needs to be blocked for manual or operator-initiated feeder energization. This is accomplished via contact 43CX, lockout timer 79L, and relay 79LX for manual feeder energization. The reason is that during any type of line energization or re-energization inrush current may cause it to operate. During live line maintenance, reclosing is dangerous to personnel and is often blocked (via contact 43RX or blocking switch). In this case, the low set instantaneous protection remains functional to provide high-speed clearance as an added safety feature.

15.6.3.1.2 Timed Overcurrent Protection (51/51N)

Extremely inverse timed overcurrent relays are used for the timed protection. They are designed to provide a backup to the instantaneous protections. Following the blocking of the instantaneous low-set protection after reclosure, the extremely inverse characteristics of the timed protection provide the time coordination with the downstream protective devices such as fuses, reclosers, and other current relays.

This protection in the example is provided by the timed element of the overcurrent relay. Two relays are used, one connected in each of the red and blue phases. The instantaneous element of these relays is used in the instantaneous high-set (50A) protection.

15.6.3.1.3 Instantaneous Overcurrent High-Set Protection (50A/50NA)

Since the timed overcurrent protection should coordinate with load/fault protections (overcurrent relays, fuses, etc. that are further "downstream" on the feeder), it may be impossible to clear all faults in the protected section in acceptable time. Therefore, instantaneous high-set protections are used to protect the portion of the feeder closest to the substation so that fast clearance can be achieved for close-in faults. High-speed clearance after reclosure is essential to protect the substation itself from the possible cumulative damage caused by permanent, large, close-in faults.

15.6.3.2 Circuit Operation and Tripping

Refer to Figure 15.24b.

The DC logic portion of the overcurrent feeder protections consists of several measuring relay contacts in parallel. When any of these relays operate, the trip auxiliary relay (94) is energized to trip the circuit breaker and initiate the automatic reclosing scheme. It seals itself in and energizes the seal-in timer (94T). When the timer times out, the seal-in of the trip relay (94) is terminated if and when the measuring relays have reset.

The purpose of the seal-in is to protect the sensitive measuring relay contacts that are not rated for the necessary breaking current resulting from having energized the heavy-duty auxiliary relay (94). The measuring relay contacts will remain closed as long as the fault is not yet cleared via the opening of the feeder breaker. Without the seal-in, these contacts will be damaged when the breaker trips. The seal-in timer has to be set long enough to allow the measuring relay contacts to fully open following the opening of the feeder breaker.

15.6.3.2.1 Phase Instantaneous Arrangement – 4 Wire System

The phase instantaneous low-set overcurrent protection (50B) is cross-connected to require the operation of at least two of the phase relays to trip the feeder. The phase instantaneous high-set protection (50A) is supervised by the contacts of the 50BX relay to achieve the same operation.

This arrangement allows phase and ground setting flexibility because it prevents phase instantaneous tripping for ground faults (for example on tapped single-phase distribution feeders). Correct annunciation/targeting for phase and ground faults is also achieved.

15.6.4 Automatic Reclosing

15.6.4.1 General

Approximately 80% of faults on feeders or distribution systems are transient. Of the remaining 20%, 10% are semi-permanent and the additional 10% are permanent. Eighty to ninety percent of all these faults are ground faults. Feeders are predominately overhead line construction, especially in rural areas, and, therefore, are prone to environmentally induced fault conditions. Clashing of conductors caused by wind, lightning strikes, tree branches falling, animals, etc. causes these faults.

It has become common practice to provide at least a single-shot reclosing scheme for feeders. As the name implies, reclosing is the reclosure of the protective relay's associated feeder breaker at a preset interval after the breaker has been tripped. Normally, little or no damage results if the transient faults are promptly cleared by the protective relays and breaker. Therefore, reclosing has the following advantages:

(1) Minimizes interruption in supply to the customer by restoring the feeder to service as quickly as possible.
(2) Minimizes damage at the fault.
(3) Reduces operating costs. (Fewer man-hours in repairing damage).
(4) Prevents blowing of fuses at tapped DS; therefore, considerable cost savings in both man-hours and replacement fuses.

Semi-permanent faults cause initial tripping but can be cleared by subsequent time delayed tripping. Permanent faults require location and repair before supply can be restored.

If any reclosure of the breaker is successful, the scheme resets to its normal state. If the fault is non-transient, the reclose scheme recloses the breaker a preset number (multi-shot) and then locks out. The time intervals between reclosures are independently adjustable. Most feeders have at least a single-shot reclosing scheme and are the most extensively used; it gives considerable benefit in clearing transient faults and imposes a minimum duty on circuit breakers. Multi-shot schemes have the further ability to clear semi-permanent faults and to deal with successive lightning strikes; however, additional tripping duty is imposed on circuit breakers. By immediate single-shot reclosing, service is restored for the majority of faults such that customers do not realize noticeable interruption.

15.6.4.2 Automatic Reclosing Circuit Operation

Refer to Figure 15.24c.

15.6.4.2.1 Brief Summary of Intention

The feeder circuit breaker instantaneous overcurrent relay is set to trip the breaker for all feeder faults and does so before any of the tapped DS HV fuses operate. When the breaker opens, the instantaneous relay is blocked from operating for any subsequent faults. If the fault is transient, service is restored to the entire feeder and the circuit is reset. If the fault is permanent, the breaker recloses onto a fault and the time delay overcurrent relay must now operate since the instantaneous relay has been blocked after the first shot. The time delay characteristics should be that it is slower than the tapped DS fuses and coordinates with other downstream devices. This results in allowing the downstream fuses to blow and isolate the faulty section, leaving the remainder of the feeder in service.

15.6.4.2.2 Reclose Interfaces with Breaker Control

Feeder protections interact with the feeder breaker control circuits to achieve and meet their particular feeder protection philosophy. In summary, two control relays are normally used to block reclosure dependent upon the operating condition and, as such, interface with the feeder protections. The following is a brief overview of these breaker control relays and the rationale for them. In this case, they are designated as CB-43RX and CB-43CX; however, they may be designated differently by other utilities.

15.6.4.2.2.1 Permanent Block Automatic Breaker Reclose (43RX) Operations personnel require that during live-line maintenance of the feeder, the station operator should be able to block automatic reclosure. This facility is normally achieved by the provision of a two-position push-button on the control panel, and in the case of the remotely controlled station, via supervisory; to switch a throw-over relay (designated in this case as 43RX). The two positions on the push-button are named "Block" and "Reclose." This feature also provides flexibility for other purposes such as tree trimming, work on adjacent circuits, etc.

The CB-43RX is a reclose selection-latching relay. Its contact (in the feeder reclosing circuit) closes when reclosure is selected. If block reclosure is selected, the contact opens to disable the reclosure circuit.

15.6.4.2.2.2 Block Automatic Breaker Reclose When Closed (43CX) A 43CX relay is energized whenever the breaker is closed. This relay prevents an automatic reclosure during a breaker closing operation (and for some time afterwards, via lockout timer 79L and relay 79LX) or when the breaker is under maintenance. This relay does not seal-in or latches.

15.6.4.2.3 Initiation

The reclosing scheme is initiated by a contact from the feeder protection (trip relay, 94), refer to Figure 15.24c. A feeder breaker interlock (CB "a" auxiliary switch) is provided to prevent the reclosing scheme from being initiated if the feeder breaker was previously open (for example tripped by another protection – bus protection or manually tripped).

The initiation signal can only be expected to exist for a short time; therefore, a seal-in (reclose seal-in relay, 79A) is provided. The reclose seal-in relay (79A) energizes and starts the timing of the reclose delay timer (79T) and the reclose cancel timer (79-48).

15.6.4.2.4 Supervision and Reclosure

If the breaker is in the closed position, no reclosure signal can be sent to the breaker. A breaker form "B" auxiliary switch is used to accomplish this (in series with the 79TX reclosure auxiliary relay).

For a radial feeder, a reclosing scheme with timed supervision is used (reclosing delaying timer, 79T). Timed supervision is usually adequate for radial feeders. The time delay should allow the load to decay or to be disconnected by automatic controls. The decay time may be substantial if the feeder supplies a significant amount of industrial motor or capacitive load. For non-radial feeders, supervision other than timed alone may have to be used.

A contact of the trip relay (94) in series with the reclosure relay (79TX) prevents any reclosure attempt before the trip signal is removed. This could occur if the 79T relay was set at minimum (no intentional time delay), and the resultant time was less than the trip relay seal-in (94T) time.

The reclosure relay (79TX) seals itself in, initiates breaker closure, and energizes the lockout timer (79L). The low set instantaneous protection is blocked initially by the reclose relay (79TX) contact, and for some time after, it is reset by the lockout timer multiplier (79LX). The reclosure relay (79TX) also ensures that the reclose long time reset timer (79-48) remains energized since the reclose seal-in relay (79A) is reset by the lockout timer multiplier (79LX) and control relay 43CX.

The reclosure relay (79TX) is normally reset by the breaker form "b" auxiliary switch, or in the event, the breaker fails to close, by the cancel reclose relay (79-48). For normal operation, 79TX upon dropping out resets 79-48 timer.

15.6.4.2.5 Lock-Out

The lockout timer (79L) initiates a time delay when de-energized. Its lock-out multiplying relay (79LX) blocks reclosure whenever 79L is energized and for a set time-period after 79L is de-energized.

Thus, reclosing can be canceled by any normally open contact closure (for example from another protection) or reclosing can be delayed from the temporary closure of any external contact, such as 43CX.

There is a lockout period (via 79TX, 79L, and 79LX) following a reclosure so that the single-shot reclosing scheme cannot be initiated if a retrip occurs. This indicates that the fault is probably permanent. For the multi-shot reclosing scheme, the lock-out (block initiation) occurs when the last reclose signal in the multi-shot reclosing sequence is sent to the breaker.

A lockout period is also activated (via 43CX) when the breaker is manually closed, to block the reclosing scheme. If the breaker trips on feeder energization, the fault is likely permanent.

The lockout-multiplying relay (79LX) resets the initiation portion of the reclosing scheme.

15.6.4.2.6 Cancel

There is a maximum time following initiation for which the reclosing scheme can remain energized. After that, the reclosing scheme (relays 79A, 79T, and 79TX) will reset via the reclose cancel relay (79-48X). This feature is often referred to as "long time cancel." For example, if a reclose signal is sent to the breaker but the breaker fails to reclose, then the reclosing scheme will reset after a set period. If the breaker recloses successfully, the reclose cancel (or long time cancel) feature will cease to be in effect, the reclose relay (79TX) resets via the breaker form "b" auxiliary switch, and the reclose seal-in relay (79A) resets by the lockout timer multiplier (79LX).

15.6.4.2.7 Other Functions

The reclosing scheme provides a normally closed contact (reclosure relay 79TX) which blocks the low set instantaneous overcurrent protection during feeder automatic reclosure. The contact of the lockout auxiliary relay (79LX) blocks the low set instantaneous approximately 100 ms after the reclosure relay (79TX) operates and for a set time after the reclosure relay has reset (the breaker has closed). This allows fault clearance by the high set and timed protections following an automatic reclosure.

15.6.5 Relay Settings

This section provides a summary of general feeder protection settings and coordination practices. It must be emphasized that the following discussion does not attempt to provide settings for specific feeders. To do so requires detailed feeder information. However, it describes and summarizes the general practices used for radial feeder protections. Since protection engineering is both a science and a craft, the final settings will be based on judicious use of these principles, customer requirements, and sound engineering and economic judgment.

The purpose of feeder protection systems is to detect faults on the feeder and initiate the isolation of the faulty element(s) from the power system. These systems must be reliable, to this end, the process of setting the relays – selectivity and sensitivity- is extremely important.

The integrity of the bulk power system must be maintained for feeder faults. The following discussion of settings is presented with the understanding that feeder settings must be coordinated with the protections of the supplying bus and Transformer Station.

To achieve selectivity, protection devices must coordinate. Coordination refers to the sequence of operation of protective devices, usually in series, during a fault where fault current flows through both devices, such that the intended device operates to clear the fault. The objective is to clear all transient faults without a prolonged outage and to minimize the extent of the outage following a permanent fault.

In addition to providing feeder settings/coordinations, protection practitioners are also asked to assess the suitability of feeder relaying. Suitability is reviewed when the normal feeder configuration is altered for planned work or emergencies. The operator requires readily available information on relay settings and short circuit values for the desired configurations.

Relay setting criteria are based on normal feeder operation, with margin, and some operating allowances, where possible. However, to maintain continuity of service, since feeders interface directly with customers, protection device setting criteria are allowed to deviate from normal, for short periods of time. These deviations will be referenced as minimum acceptable – less margin from normal – and emergency only – abnormal short-time operating conditions to maintain continuity until corrective action is taken.

Therefore, as a convention, settings by default are based on normal conditions, and where applicable, minimum acceptable settings will be enclosed in round brackets ( ), whereas emergency settings will be enclosed in square brackets [ ]. It should be noted that the margins presented here are examples, and the actual minimum and short-time margins used are utility dependent, based on their risk tolerance.

15.6.5.1 Three-Wire vs. Four Wire Feeders

The protection philosophy is typically the same for three-wire and four-wire feeders. Four-wire feeders in addition to supplying three-phase loads, also supply single-phase loads. The main difference between them is the requirement for the fourth wire to connect single-phase loads.

Historically, the main trunk of the feeder (three-wire subtransmission) supplied three-phase loads and Distribution Transformer Stations (DS) which in turn supplied, generally, single-phase loads. Due to large cost savings of distribution equipment and maintenance, four-wire subtransmission feeders dominated. Four-wire feeders in addition to supplying three-phase loads, also through pole top/pad mount distribution transformers are used to supply single-phase loads. While the advent of four-wire feeders permitted economies in distribution, it has introduced some problems: increased momentary interruptions due to increased exposure occasioned by the laterals, more severe voltage disturbances due to the higher ground currents, and additional relay setting coordination requirements due to single-phase close-in laterals.

In addition to the fourth neutral wire, four-wire feeders generally have low impedance ground sources compared to three-phase feeders. For the latter, a minimum ground current is required only to detect remote-end ground faults, whereas for four-wire, it must also be able to carry the continuous unbalance currents.

15.6.5.2 General Information/Data Requirements

The following information/data should be gathered before beginning to set and coordinate feeder protection devices:

(1) An operating diagram of the feeder of interest is required. The diagram should provide normal conditions and temporary operating conditions for which calculations and settings are required.

(2) Supply station minimum and maximum fault values will need to be acquired or calculated, to determine source equivalent impedances. In addition, supply station transformer impedances are also required.
(3) The supply station's protections should be reviewed to determine any backup protections with which the feeder must coordinate with. Setting information should be gathered for the backup relays.
(4) Feeder loading information, normally from the Operating Group, will be required – maximum peak, unbalance, and cold load peak if available. In addition, the maximum operating voltage is required.
(5) Feeder topology and data are required on the following:
Conductor sizes/configuration or impedances; location, ratings, and impedances of Distribution Transformers (DS); fuse specifications; recloser and; any other protection device specifications which are used for the particular feeder of interest.
(6) Time current curves (TCCs) should be gathered for all equipment.
(7) Protection settings information and TCCs.
(8) Fault current information for minimum and maximum conditions is required for faults at different locations and different operating conditions.

Review of protection device's response under varying operation conditions should ensure operation within ratings and use of appropriate TCCs that meet the requirements of the specific application for which it was intended. The interrupting current ratings of the device should be also checked that is greater than or equal to the maximum fault currents the device may be called upon to interrupt. In addition, the continuous current/voltage rating of the device must be greater than or equal to the maximum load current/voltage the device may have to carry/withstand. Lastly, the response TCCs should provide adequate protection, loadability, and coordinate with both source-side (upstream) and load-side (downstream) protective devices.

Reference is made to allow for a margin several times in this section. Margin allows for errors in fault calculations, system data, and arc resistance. A "Two Times Rule" is normally used in industry as an accepted value. A "Minimum Two Times Rule" states that: the fault current at the end of the zone of protection of an overcurrent protective device must be at least two times the minimum pickup current of the device. This rule defines minimum requirements for fault sensitivity of the protection. Moreover, a coordination time interval of 0.4 (0.3) seconds is normally used.

15.6.5.3 Phase Protection

Feeder phase protections are intended to operate for phase-to-phase and three-phase faults; however, phase relays may also operate for ground faults involving the same phase(s). It is common practice to cross-connect the low-set instantaneous overcurrent protection DC output contacts – RW, WB, and BW for 4-wire feeders. In addition, through a multiplier auxiliary relay (50BX), the phase high-set protections are also supervised to ensure the operation of phase faults only. This allows different sensitivities for phase and ground faults (4-wire feeders have laterals) and aids postfault analysis. It should be noted, however, that the low-set timed protection is not supervised by 50BX, and it could operate for ground faults.

Phase relay settings must meet the following basic criteria:

- clear minimum three-phase and phase-to-phase faults
- must not pick up on load current
- should not trip for faults on the secondaries of the DS or lateral transformers.

15.6.5.3.1 Phase Instantaneous Low-Set Protection – 50B

The instantaneous low-set relay pickup should be set to cover, with margin, the whole feeder. This may involve extending the setting through fused transformer banks, and past the fusing of distribution feeders. When automatic reclosure is not in service, these protections may be blocked.

Information required	Setting principles	Setting criteria
• Maximum feeder load • Minimum feeder remote end 3-PH fault current • Primary through current for a maximum 3-PH fault on the secondary of the tapped station with the highest 3PH fault, (normally the closest tapped station, and hereafter referred to as the critical bank)	• Is intended to operate for all feeder transient phase faults. This protection is blocked after the first operation – assuming single-shot reclosure – and when the feeder is first energized. • The pickup setting principles are normally the same as the low-set timed overcurrent protection, therefore, resulting in the same pickup • The pickup should be low to provide an adequate margin for instantaneous clearing of minimum feeder end phase-to-phase faults, and; high enough not to pick up under maximum load, and secondary tapped station faults if possible	• The pickup should be equal to or greater than 2.0, (1.5), [1.2] times the maximum feeder load • The minimum feeder end short circuit must be equal to or greater than 2.0, (2), [1.5] times the pickup • If possible, the pickup should be equal to or greater than 1.1 (1.1) [1.0] times the primary through current for a maximum 3-PH fault on the secondary of the critical bank • The pickup is typically, equal to the pickup of the low-set timed protection

15.6.5.3.2 Load Impact on Overcurrent Protection Settings

The premise that has been made until now is that an adequate margin exists between the minimum feeder end short circuit current, to the maximum load current. A fault-to-load ratio (FTLR) of not less than 4:1, (3:1) is generally used; this is required to ensure that the overcurrent protection does not pick up for load current, but yet can be set to cover the entire feeder with margin. In addition, this ratio will ensure correct operation during a cold load pickup, and allow for variations in system operating conditions.

Many feeders meet this requirement naturally or by utility-imposed operating limits. However, some feeders are heavily loaded or are of such length that the FTLR criterion cannot be met. Many utilities today, due to economic reasons, are adding more and more load to existing feeders to eliminate or postpone the cost of constructing a new feeder. As a consequence, there are more feeders which cannot support this 4:1 (3:1) FTLR.

It should be noted that only phase overcurrent relays/settings are affected, as they cannot distinguish between loads and fault current; ground relays will not respond to balanced loading conditions. To overcome this loading problem, a recloser or other protection devices that sectionalizes the feeder can be added or a phase distance relay can be used in place of an overcurrent relay; or if a microprocessor type relay is being used, some form of native load supervision logic can be adopted.

For feeder applications where the FTLR is not met and a directional mho distance relay is used, the distance relay would replace the three-phase overcurrent relays (50B). The distance relay set to cover the entire feeder with margin will allow relatively high load current to flow because, in comparison with the feeder impedance, the load represents a large impedance outside its operating characteristic.

The phase timed relays settings must also allow for the load as well as provide line-end fault sensitivity. To accomplish this, the elements are "torque-controlled" by the distance relay. Torque-controlled phase timed relays will not be enabled for operation until the torque-controlling relay (distance) asserts. This differs from trip supervision because the low set phase relays can operate or be picked up before the supervising element asserts. Torque controlling allows proper time coordination with downstream protection devices and permits low settings without tripping for load currents.

15.6.5.3.3 Instantaneous (Low-Set) Distance (MHO)

For applications where overcurrent relays cannot be used due to loadability issues.

Information required	Setting principles	Setting criteria
• Positive sequence impedance from the supply station to the furthest point to be protected • Impedance from the supply station to the secondary of the critical bank • The maximum feeder load, including current and phase angle	• An impedance setting such that it sees feeder end faults • A setting such that the relay's zone does not reach through the critical bank, or operate under a maximum load condition	• The relay setting should be equal to 1.5, (1.25), [1.25] times the positive sequence feeder impedance to the furthest point to be protected; an MTA closest to the line angle shall be chosen • The relay setting should be equal to or less than 0.85 (0.85) times the impedance from the supply station to the LV side of the critical bank • The maximum load should be outside the relay, operating characteristic, with margin

Three-phase loads can be represented on an *R-X* Diagram, see Figure 15.25, as impedance at the load power factor angle. The maximum load point should fall outside the relay operating

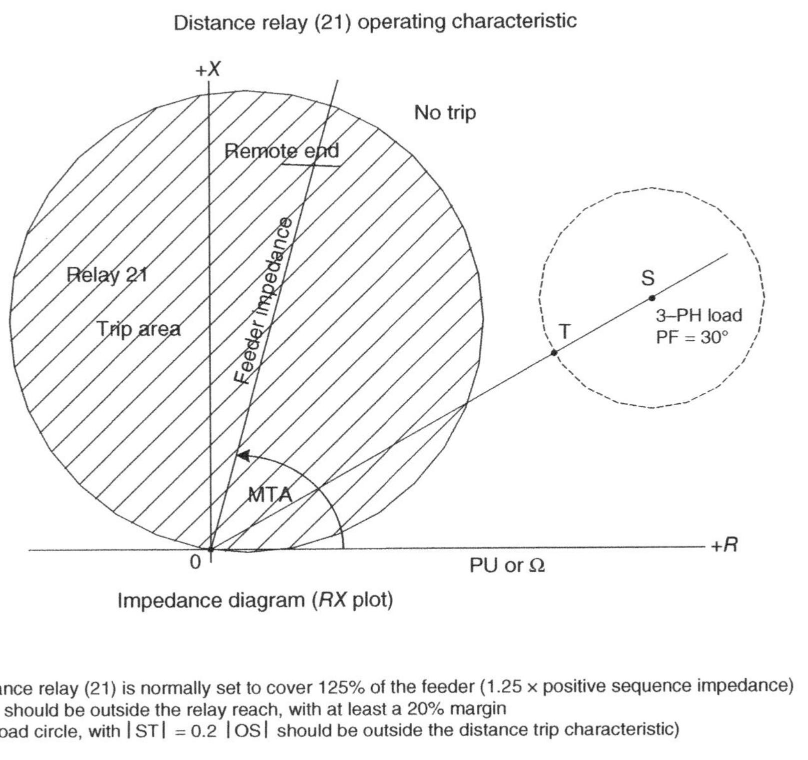

Figure 15.25 Distance relay (21) operating characteristic.

characteristic. As regards to the margin requirement, the circle drawn around the maximum load point with a radius equal to approximately 20% of the distance between the origin of the *R-X* coordinates and the load point should not intersect the relay characteristic. A load angle of 30° (0.87 PF) can be assumed if this information is not available.

15.6.5.3.4 Phase Instantaneous (High-Set) Protection – 50A

Information required	Setting principles	Setting criteria
• Maximum feeder load • Maximum three-phase short circuit current on the primary of tapped stations • Fuse curves	• This protection is intended to clear close-in permanent maximum faults to mitigate damages that could otherwise occur in a timed trip; it is not required to protect 100% of the line • It should be set high enough to ensure the feeder can be energized and not locked out for permanent faults beyond the primary fuses of the closest tapped station. This protection should not trip until the fuse has had a chance to clear the fault, otherwise, the operator would not be able to put the feeder back in service without tripping. • The pickup should be above, with margin, feeder cold load inrush on energization	• The pickup should be greater than the maximum inrush/cold load pickup current, with margin, as a rule-of-thumb, 4-6 times load • The pickup should be equal to or greater than 1.3 (1.3) [1 or less to meet the 0.1 s rule] times the current for a maximum 3-PH fault at the HV side of the first tapped station

In some cases, the first tapped station may be close and the relay cannot be set high enough to eliminate this station or possibly several stations from the relay's zone. Lower settings are acceptable provided the primary fuses operate before the feeder breaker opens. Therefore, the relay operating time plus the breaker opening time must be greater than the total clearing time of any fuse in the relay's zone. To ensure a reasonable margin, the time that can be used is 0.1 seconds (100 ms).

The pickup should be set to avoid response to faults beyond the high voltage fuses at the first tapped transformer. Without the problem of a close-in transformer, the pickup can often be set high enough for coordination yet still cover a substantial portion of the feeder.

15.6.5.3.5 Phase Timed Overcurrent Protection – 51

Using the criteria below, the pickup should be set based on the normal minimum feeder end three-phase fault current. A sufficient margin exists to cover phase-to-phase faults, which is equal to 86.6% ($\sqrt{3}/2$) of the three-phase fault current. The minimum feeder end fault current is achieved with minimum source conditions.

Information required	Setting principles	Setting criteria
• Maximum feeder load • Minimum feeder remote end three-phase short-circuit current • Settings of the phase upstream backup relays	• The 51 protection is intended to trip for permanent faults, back up downstream protection devices such as fuses and reclosers, and should allow coordination with tapped stations to maintain service to the remaining customers. Extremely inverse characteristics are normally used.	• The pickup should be equal to or greater than 2.0, (1.5), [1.2] times the maximum feeder load. • The minimum feeder end short circuit should be equal to or greater than 2.0, (2), [1.5] times the pickup

Information required	Setting principles	Setting criteria
• Maximum 3-PH fault current on the primary of tapped stations • The fuse curves employed on the primary of tapped stations	• It should not trip on load – the pickup should be high enough to ensure that abnormal operating conditions will not trip the feeder. This criterion is relaxed if the phase timed overcurrent relay is torque controlled. • Should trip for all feeder phase faults. It should also be set low enough to ensure that the phase relays pickup for minimum remote-end phase-to-phase faults with margin. • Should allow tapped station fuses to clear an HV tapped station fault. This allows continuity of service for other feeder customers.	• The pickup should be equal to or less than 0.8 (0.8) [0.8] times the pickup of the phase bus backup relay. This should naturally occur unless the feeder is extremely short.
Time Settings	• Fast operating times are required to clear minimum feeder end phase-to-phase faults at appropriate times and to coordinate with phase bus backup relays at the load station. • However, it should be also set to coordinate with tapped station fuses	• The time to clear a minimum feeder end fault is typically equal to or less than 1.5 s, (2.5 s), [3.0 s] • The phase timed relay should operate equal to or greater than 0.4 s, (0.4 s), [0.3 s] faster than the bus backup for all possible fault levels. • The phase timed relay must operate equal to or greater than 0.4 s, (0.4 s), [0.3 s] slower than a tapped station fuse for maximum three-phase fault current on the primary of the tapped station

15.6.5.4 Ground Protection

15.6.5.4.1 Instantaneous (Low-Set) Overcurrent Protection (50NB)

15.6.5.4.1.1 Three-Wire Low Set Instantaneous Overcurrent Protection

Information required	Setting principles	Setting criteria
• Minimum feeder end $I_{L\text{-}g}$ • Pickup of ground bus backup relay	• This protection is required to clear transient faults and is blocked after the initial trip for subsequent breaker reclosing • The setting should be approximately the same as the pickup of the ground timed overcurrent relay	• The pickup should be 0.1, (0.25), [0.3] times the minimum feeder end $I_{L\text{-}g}$ • The pickup should be equal to or less than 0.8 (0.8) times the ground bus backup pickup

15.6.5.4.1.2 Four-Wire Instantaneous (Low-Set) Overcurrent Protection

Information required	Setting principles	Setting criteria
• Minimum feeder end $I_{L\text{-}g}$ • Pickup of ground bus backup relay	• This protection is required to clear transient faults and is blocked after the initial trip for subsequent breaker reclosing • The pickup should be above any prefault unbalance current; unbalance current is normally kept below 20% • It should be set low enough to ensure that it picks up for minimum remote-end line-to-ground faults with margin	• The minimum feeder end line-to-ground fault should be equal or greater than 2, (1.5), [1.5] times the pickup • The pickup must be greater than or equal to 2.0, (1.5), [1.5] times the maximum load unbalance current • The pickup must be less than or equal to 0.8, (0.8), [0.8] times the ground bus backup pickup

15.6.5.4.2 Timed Overcurrent Protection – 51N
15.6.5.4.2.1 Three-Wire Timed Overcurrent Protection

Information required	Setting principles	Setting criteria
• Minimum feeder end $I_{L\text{-}g}$ • Pickup of ground bus backup relay	• This relay will generally not see secondary ground faults at tapped stations (Delta-Y connected) • Most existing 3-wire low-set timed relays are set low, generally between 60 and 100 A, which is sufficient to cover the longer feeders • The ground low set timed can be set low because little unbalance current exits.	• the pickup is usually in the order of 100A primary and should be 0.1, (0.25), [0.3] times the minimum feeder end $I_{L\text{-}g}$ • The pickup should be equal to or less than 0.8 (0.8) times the ground bus backup pickup **Timing:** • Time to clear a minimum feeder end $I_{L\text{-}g}$ fault must be equal to or less than 1.5 s, (2 s), [3 s] • The ground timed relay must operate equal to or greater than 0.5 s, (0.4 s), [0.3 s] faster than the bus backup for all possible fault levels.

15.6.5.4.2.2 Four-Wire Timed Overcurrent Protection

Information required	Setting principles	Setting criteria
• Minimum feeder end $I_{L\text{-}g}$ • Pickup of ground bus backup relay	• Low enough setting to see all faults on the feeder • High setting to coordinate with tapped station fuses • Must not trip for feeder load unbalance • Should clear minimum feeder end faults quickly • Should coordinate with tapped station fuses	• The pickup should be less than 0.8 (0.8) times the ground bus backup • The minimum feeder end fault $I_{L\text{-}g}$ must be greater than or equal to two times the pickup • The pickup must be greater than or equal to 2, (1.5), [1.5] times the maximum prefault feeder load unbalance. `Timing:` • Time to clear a minimum feeder end $I_{L\text{-}g}$ fault must be equal to or less than 1.5 s, (2 s), [3 s] • The ground timed relay must operate equal to or greater than 0.5 s, (0.4 s), [0.3 s] faster than the bus backup for all possible fault levels • The ground low set timed relay should operate equal to or greater than 0.5 s, (0.4 s), [0.3 s] slower than tapped station fuses

15.6.5.4.3 Ground Instantaneous (High-Set) Overcurrent 50NA

Information required	Setting principles	Setting criteria
• Maximum $I_{L\text{-}g}$ on the primary of tapped stations • Tapped station fuse curves	• This protection is intended to clear close-in permanent maximum faults to mitigate damages that could otherwise occur in a timed trip; it is not required to protect 100% of the line • It should be set high enough to ensure the feeder can be energized and not locked out for permanent faults beyond the primary fuses of the closest tapped station • This protection should not trip until the fuse has had a chance to clear the fault, otherwise, the operator would not be able to put the feeder back in service without tripping	• The pickup should be greater than the maximum unbalance current with margin • The pickup should be equal to or greater than 1.3 (1.3) [1 or less meeting the 0.1 s rule] times the current for a maximum l_g t at the HV of the first tapped station

In some cases, the first tapped station may be close and the relay cannot be set high enough to eliminate this station or possibly several stations from the relay's zone. Lower settings are acceptable provided the primary fuses blow before the feeder breaker opens. Therefore, the relay operating time plus the breaker opening time must be greater than the total clearing time of any fuse in the relay's zone. To ensure a reasonable margin, the time that can be used is 0.1s (100 ms).

15.6.5.4.4 Upstream Protection Devices

Feeder protection must coordinate not only with downstream devices but also with upstream protections.

Mostly, feeder protections are required to coordinate only with the bus backup protection. However, legacy systems may require additional coordination with transformer backups and some grounding transformer protections.

The need to coordinate with which upstream protection devices is application-dependent. It will be assumed, in this case, that the feeder protection needs only to coordinate with an upstream bus backup protection.

15.6.5.4.4.1 Bus Backup Protection At a minimum, the feeder protections must coordinate with the bus backup protection. Normally an "A" Group bus differential or a Bus blocking Scheme is used, and a "B" Group phase (51B-Bus) and ground (51NB-Bus) overcurrent backup are used on the load station LV buses such as 27.6 and 44 kV. The overcurrent backups are required since typically on legacy systems, no feeder breaker failure protections are employed. The bus backup protections are set based on the following type of principles:

15.6.5.4.4.1.1 51B-Bus Pickup: greater than 1.2 times the maximum emergency loading of the Station – the emergency loading capability of the in-service banks under minimum operating conditions and preferably equal to 0.5 times the minimum three-phase bus fault.

Timing: Normally, set to trip in 1.2 seconds under a maximum three-phase bus fault condition; providing, at a minimum, a 0.4 seconds coordination interval between the bus and the feeder.

15.6.5.4.4.2 51NB-Bus Pickup: greater than the unbalance current permissible by the grounding device(s) preferably less than 0.5 times the minimum line-to-ground bus fault – one bank/grounding device out-of-service.

Timing: normally set to trip in 1.2 seconds under a maximum line-to-ground bus fault condition; providing, at a minimum, a 0.4 seconds coordination interval between the bus and the feeder.

15.7 Impact of Distributed Generators (DGs) on Distribution Feeder Protection

Many distribution feeders are radial, which has only one source of power to the customer load, load customers, and feeder faults are completely isolated when the load station feeder breaker is opened. However, the recent shift, in support of renewable energy, to connect DG (wind, solar, etc.) at the distribution level, has altered some of these radial configurations. It should be noted that DGs are also referred to as DERs.

This section is intended to briefly summarize some of the impacts that DGs are having on distribution feeders and consequently on the feeder protection described above. It is not intended to address all the protection issues and solutions but to introduce the reader to the basic issues and salient points.

15.7.1 Significant Protection Issues Caused by Distributed Generation

15.7.1.1 The Feeder Becomes Non-Radial

The introduction of a DG or multiple DGs on a given feeder causes it to be transformed from a single source configuration to a multi-source configuration; refer to Figure 15.26 below. In effect, the DGs are operated in parallel to the connected system generators.

15.7.1.2 Islanding

In theory, with sufficient connected DG capacity that matches the load requirements, the power flow from the system could be net-zero. In this case, the aggregate of DGs can sustain the load without contribution from the system and the load then becomes a self-sustaining island.

Referring to Figure 15.27, with Breaker $M1$ open, the DG has the potential to continue to supply the feeder load. The length of time it can and the capacity to do so is a function of the size of the DG or the aggregate size of several connected DGs on the feeder. With small DGs less than 1–10%

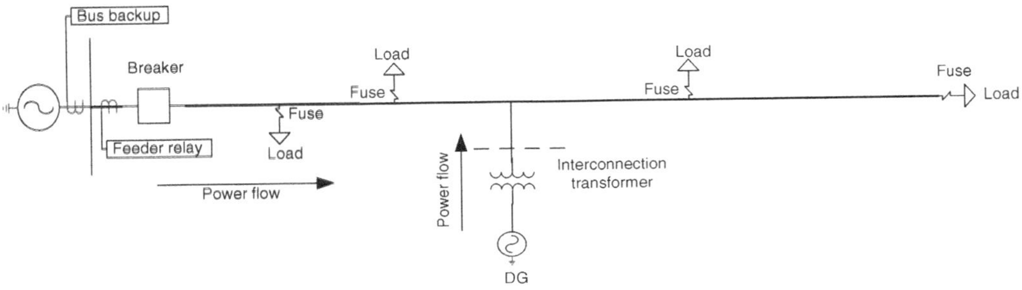

Figure 15.26 A feeder-connected DG.

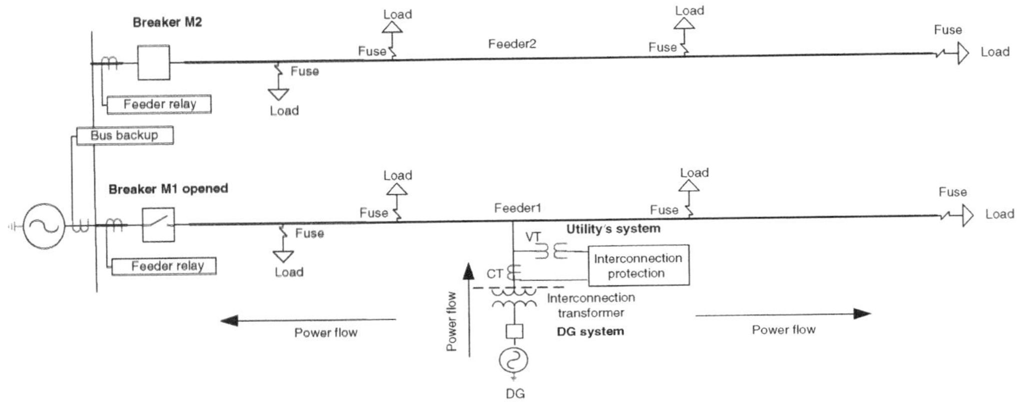

Figure 15.27 DG islanding.

of feeder capacity, this concern diminishes, but for large DGs or aggregates approaching 30–50% of the feeder capacity they can potentially maintain voltage and frequency on the feeder even when separated from the System.

Utilities must ensure delivery of electric power to their customers within the prescribed quality of service limits, such as voltage and frequency, to prevent damage and to maintain proper operation of customer utilization equipment.

Islanded operation of DGs with utility loads external to the DG site is not generally permitted due to the impact on the quality of service that the utility is required to provide.

Normally, the DGs will operate as a part of the interconnected system – in parallel. A disturbance on the distribution system could lead to the generator becoming islanded, as the feeder is designed to isolate all in-zone fault disturbances by opening the feeder breaker, i.e. the DGs supply power to one or more of the utilities' customers. Islanding could lead to damaging the customer's equipment caused by irregularities in power quality.

Once the feeder breaker trips (opens) due to a feeder fault or other abnormal condition, most DGs will island, however, their ability to do so is dependent upon several factors, such as the generator technology type, the size of the DG, the aggregate amount of DGs connected to the feeder, the DG ride-through capabilities, and the ratio of aggregate DG capacity to connected load.

To prevent islanding, the DGs are required to have installed anti-islanding protection. These anti-islanding systems must be reliable and are required to separate the DGs from the utilities' feeder. Anti-islanding protection can be divided into telecommunication and non-telecommunications based. The non-communications-based schemes can sometimes be used for small DGs where the rate of change of frequency or the like at the DG site is proven to remove the DG connection.

One reliable method of anti-islanding protection is the addition of a transfer trip communications facility to remotely trip off the DGs' generation upon opening of the distribution feeder main circuit breaker or circuit recloser. It also has the advantage of speed especially, in high-speed reclosing applications.

15.7.1.3 Internal Faults (Faults on the Feeder Connecting the DGs)

DGs can contribute considerable variation in fault current on the connected feeder affecting feeder overcurrent devices as well as feeder sectionalizing fuses, load fuses, and reclosers causing these feeder overcurrent devices to not operate as intended.

The DG as well will be required to detect all faults on the feeder sequentially following breaker *M*1 opening and trip its own HVI (high voltage interrupter).

Referring to Figure 15.28, for a phase fault at F_1 on Feeder 1, the DG provides an additional energy source that increases fault current magnitudes. This additional fault current can interfere with the operation of the existing protective devices on the feeder.

The 50/51 feeder protections coverage for feeder end faults can become compromised due to the DG connection on the feeder. For a fault at point F_1, the DG on Feeder 1 will increase the total fault current $I_F = I_S + I_{DG}$. However, the increased total current over a common section of feeder to the fault will also increase the nodal voltage at the junction where the currents sum. The magnitude of fault current I_S coming from the station will be reduced, desensitizing the overcurrent or impedance relays at the feeder relay location. This effect (referred to as the apparent effect) increases as a function of the DG size, connection location, and overall feeder length. More sensitive relay settings are required to detect feeder end faults, which will reduce the load-carrying capability. The effect is illustrated in Figure 15.29 below which shows fault infeeds from two sources.

As an example, Figure 15.29b shows the apparent effects for a range of DG sizes, connected at various locations along a 44 kV feeder that is 57 km long. The reduction in current starts for DG connected near the TS. When the DG is connected at the very end of the feeder, there is no apparent effect because at that location there is no common section of feeder over which the separate sources

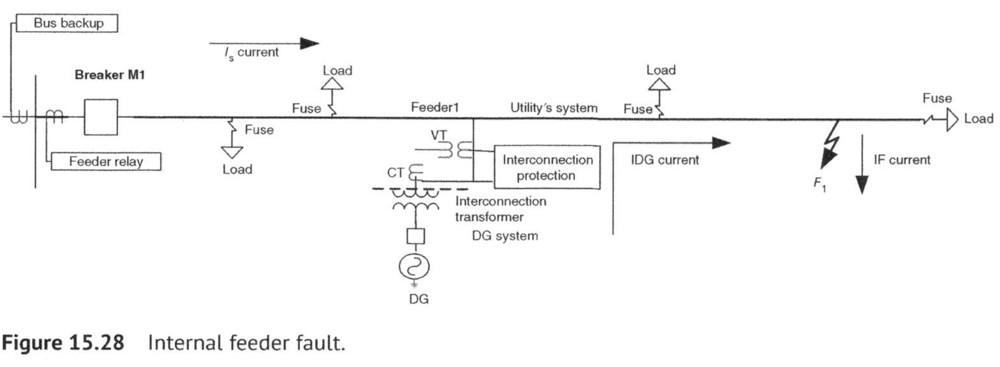

Figure 15.28 Internal feeder fault.

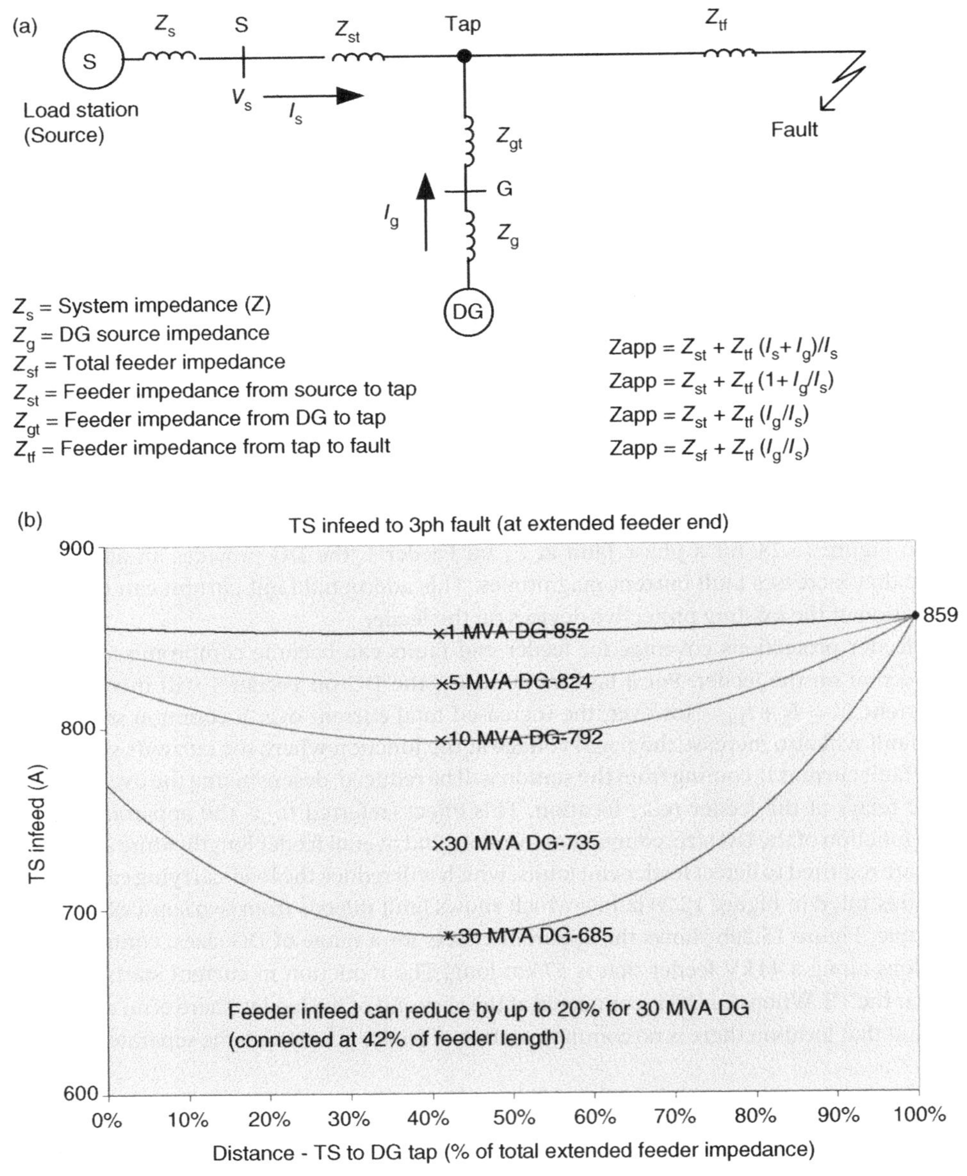

Figure 15.29 (a) Apparent effects due to the connected DG. (b) Apparent effect on a 44 kV, 57 km feeder.

are summing. For this example, the feeder end fault currents are less and the current reduction is largest (20%) at 42% of the feeder length.

The increased fault currents due to DG installations will cause the loss of fuse coordination for distribution loads.

15.7.1.3.1 DG Step-Up Transformer Configuration

The method that the DG configures its step-up transformer significantly affects the system ground protections.

Should the DG step-up transformer be configured Wye-ground – Delta with the Wye-ground on the feeder system side, a second ground source is introduced that has the effect of reducing the ground current sensed by the system feeder overcurrent devices thereby altering their coordination values. This is particularly apparent for resistive ground faults. Furthermore, the higher total ground fault currents as a result of possibly multiple ground sources when more than one DG is connected has the effect of encroaching upon sectionalizing feeder fuse minimum melt characteristics. Each time a ground fault would occur on the feeder the minimum melt characteristic may be altered until one time the fuse melts on load. For utilities where hundreds of thousands of such fuses are employed years later, they may find themselves with a large number of fuses operating for no obvious reason thereby affecting operations and continuity of supply.

The purpose for low-set phase and ground overcurrent protections is to trip system feeder breaker before the minimum melt characteristics of fuses are encroached upon. Infeeds from DG depending on DG penetration will reduce the time till fuse minimum melt is reached. In general, the grounding at the DG facility should not disrupt coordination of the distribution system ground fault protection.

Should the DG step-up transformer be configured Delta on the feeder system side, a temporary $\sqrt{3}\,V_{L\text{-}N}$ transient overvoltage (TOV) is introduced on the other two phases. This happens the moment the system feeder breaker opens thereby eliminating all ground sources while the DG is still connected. In general, the method of grounding at the DG facility should not cause voltage disturbances including overvoltages.

Based on the competing issues, it may be advisable as an overall solution to transfer trip all DG HVI's keyed from the system low set instantaneous protection and have them locked out upon system feeder breaker reclosure thereby allowing normal coordination to be achieved. In this manner, fuse minimum melt characteristics are not affected either.

In this solution, the DG step-up transformers would be configured Wye-ground on the system feeder side. The grounding should be via a suitably sized current-limiting device such as a grounding reactor or high impedance grounding transformer. The reactor or grounding transformer impedance should be chosen to not introduce TOV but yet allow suitable ground fault infeed at the system protections to allow them to operate correctly. Unfortunately, this solution can be expensive.

15.7.1.4 External Faults (Faults on Feeders Adjacent to One Connected to DG)

DGs will contribute to fault current to the feeder they are connected to, to transformer station faults, and to adjacent parallel feeders that emanate from the same bus. Refer to Figure 15.30.

For a fault at location F_2 on Feeder 2, the DG on feeder 1 will contribute fault current to F_2 via Breaker $M1$ and in a direction towards F_2. As distribution feeders are mostly radial, it normally only carries power unidirectional from the HV system to the loads on the LV feeder system. The feeder protections have been designed such that fault current and power on a radial circuit can only flow in one direction only – from the feeder breaker towards the feeder end.

If the DG fault current contribution is equal to or above the overcurrent threshold settings of the feeder protection relays associated with Breaker $M1$ (the unfaulted feeder), Feeder 1 will trip for out of zone Feeder 2 faults.

The 50/51 feeder protections coverage can become compromised due to the DG fault current for out of zone faults with the following impact:

(1) Protection settings will have to be reviewed and modified to ensure complete feeder fault coverage with all DGs in service.

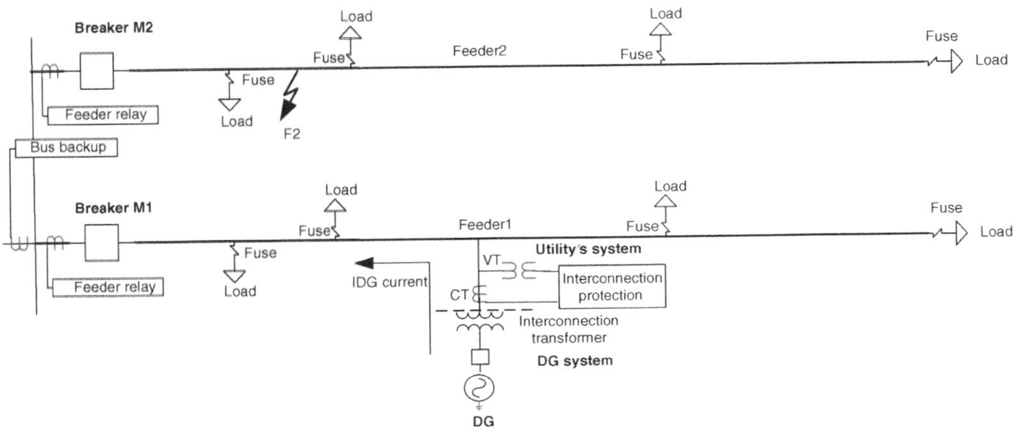

Figure 15.30 External feeder zone faults.

(2) Misoperations by feeder relays for external faults can be minimized by employing directional relaying.

15.7.1.5 Automatic Reclosing and Synchronism

Automatic reclosure introduces problems when DGs are connected to the utilities' system. Normally, the DG must detect faults on the feeder and disconnect as quickly as possible to prevent damage to customer equipment and personnel injuries. For transitory faults, the DG must disconnect from the feeder within the feeder breaker recloser time interval plus some dead time margin, otherwise, the DG, utilities', and other customer equipment could be damaged due to out of phase switching. Reclosure settings require readjustment so that there is a balance between preventing equipment damage and degraded customer supply.

To mitigate the reclosure of the feeder without damaging the DG, a communication signal may be used between the DG and the feeder's reclose protection. This signal, referred to in this textbook as DGEO (distributed generator end open), represents the status of the DG connection. If the DG is disconnected from the feeder, this signal is transmitted to the feeder reclose protection permitting reclosure to occur without synchronism, as in Figure 15.31.

Sensing feeder substation voltage is another method to check whether generation is off-line before attempting an automatic or manual reclose. However, sensing line voltage with a single phase-to-ground connected VT is not adequate because the line may still be energized with a phase-to-ground fault on the sensing phase. A phase-to-phase VT is needed to determine whether

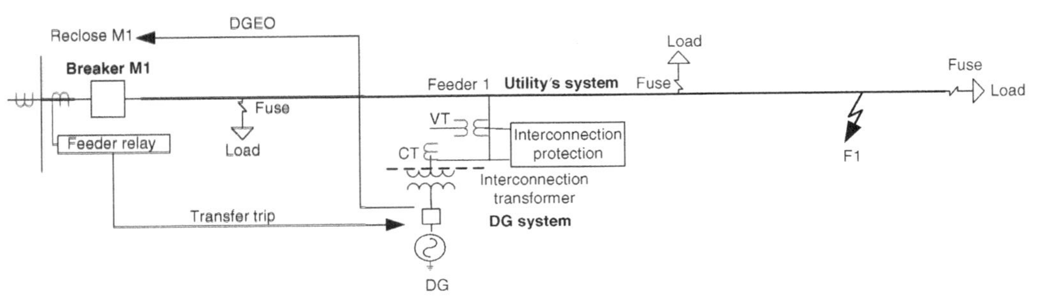

Figure 15.31 DG sends a DGEO signal once it is disconnected.

the line is energized or de-energized. Moreover, a voltage sensing scheme may not work if multiple DGs are connected to the same feeder which can support normal voltage levels.

15.7.1.6 Inrush Current – Interconnection Transformers

When DG interconnection step-up transformers are placed on potential onto a live distribution feeder, the phase low set instantaneous protections of the system protections may operate on the inrush current.

15.7.1.7 Summary of Protection Impacts Due to DGs

- Feeder protection may need to incorporate directionality to prevent sympathetic tripping on external faults. Such directionality can be achieved by the use of a distance feeder protection device, among others.
- Feeder protections may use low-set phase and ground instantaneous distance protection elements instead of traditional overcurrent elements to cater for transient faults and provide a measure of fuse saving for faults beyond the feeder main trunk. These elements may be set to respond to all remote-end faults and would be blocked after the first reclose. Distance elements would permit higher feeder load flows, provide load encroachment and would be less sensitive to system variations for fault currents.
- Phase distance instantaneous and ground instantaneous elements may be used to cater for close-in faults. Timed (51/51N) elements, set to cover far-end faults, would provide coordinated tripping with fuses, reclosers and any customer relays after the first reclosure. The impedance (21) function may be used in place of the overcurrent function (50) to discriminate between limited remote fault current and maximum load current in cases where the ratio of far-end fault current to maximum feeder load is less than 4:1.
- This results in each feeder protection employing three forward zones of protection (both phase and ground) utilizing impedance elements. Load supervision logic may be applied where required. The reach of each element can be set as follows:
  - **Zone 1 (With load supervision/encroachment enabled).** This zone may be set short of any tapped load or lateral fuse, or of the first recloser. This is considered as the high-set protection and would provide instantaneous tripping for all faults in the protected zone.
  - **Zone 2 (With load supervision/encroachment enabled).** This zone may be set to see 125% of the apparent impedance past the end of the feeder, or of the first recloser. The zone 2 instantaneous provides (i) bus block send initiation to the LV bus blocking scheme, if used (ii) transient fault fuse saving protection for feeder faults, (iii) torque control of 51 inverse timed overcurrent element may be set to provide timed coordinated tripping with downstream devices for permanent faults.

    Before an auto-reclose or manual close operation of the feeder breaker, the zone 2 instantaneous tripping could be blocked, however, the zone 2 element should continue to supervise and torque control a 51 inverse timed overcurrent element.
  - **Zone 3 (With load encroachment enabled).** This zone may be set to see the maximum apparent impedance to the end of the feeder. It would provide a definite timed trip of three seconds (absolute maximum time to clear a feeder fault). This zone may be set to cater for the failure of any downstream protection device and should be introduced as a backup protection.
- DGEO may be required for each DG to send a DGEO signal back to the feeder protection indicating whether the DG is off-line. This would be essential to ensure that no reclosing of the feeder breaker or recloser occurs after an initial fault is cleared until all downstream DGs are offline.

- Transfer Tripping to DGs; the feeder protection would initiate a trip of all downstream DGs from the low set instantaneous protections so that when reclosing takes place there is no longer any fault current contribution from the feeder-connected DGs. The reason is that following the low set instantaneous relay operation and subsequent reclosure should the fault persist; the inverse time overcurrent relays would be relied upon to provide coordinated protection with downstream fuses. Without transfer tripping the DGs, it would be virtually impossible to predict what the fault currents would be along the various segments of the feeder. Since the HVI at each DG would remain open when the system feeder breaker is automatically reclosed, all pre-existing feeder relay reclosure and fuse coordination would be maintained.
  The transfer trip of the DG HVI at each DG location would require confirmation at the system feeder protections via a return communication channel that the HVI is, in fact, open prior to the automatic reclosing of the system feeder breaker.
  This communication channel would also double to block the instantaneous phase low set overcurrent or distance elements at the system feeder protections to ensure non-operation when the DG interconnection step-up transformers are placed on potential onto live distribution feeders. This feature would eliminate the possibility of the system feeder breaker being tripped while the DG is placed into service onto a live feeder.
- Feedback from a DG onto an otherwise isolated feeder, besides the protection issues being created, also represents a significant electrical safety hazard to utility personnel as well as to customers and their equipment. By using an HVI located at the DG side of the point of common coupling (PCC) to the system feeder and ensuring that it opens and stays open is essential for reasons of electrical safety.

15.8 Feeder Protection Application Settings Example

Given the following feeder configuration depicted below. Derive the 44 kV (Bus 2) feeder protection settings for phase functions; 50B/50A/51 and ground functions 50NB/50NA/51N:

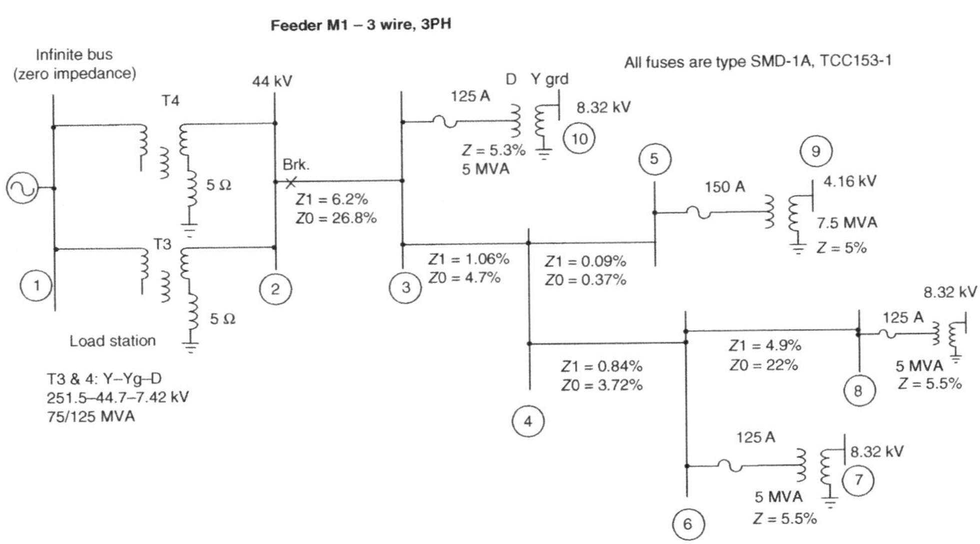

(1) Derive positive and zero-sequence system equivalent impedances for Bus 2, ignoring resistance:

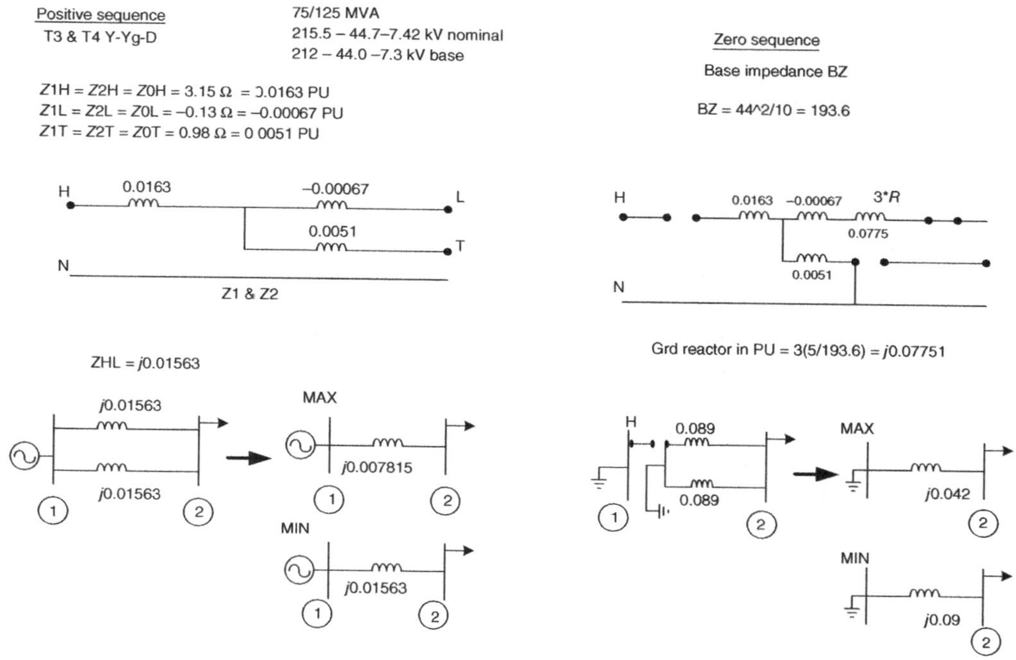

(2) The breaker CT Ratio for this feeder is 800:5 or 160:1. The protection scheme, logic, and reclosing used are depicted in Figures 15.24a,b. The reclose scheme is set to reclose once after one second. The maximum feeder load is 350 A.

(3) Phase Faults – Low Set Instantaneous (50B).
Designed for transient faults, which are the predominant type.
Setting criteria. Set to see all feeder faults, trip breaker, initiates breaker reclose (one second), and is then blocked from an operation. Set it for the minimum most remote 3PH fault with margin and above load.
The minimum 3PH fault at the most remote location which is Bus 8:

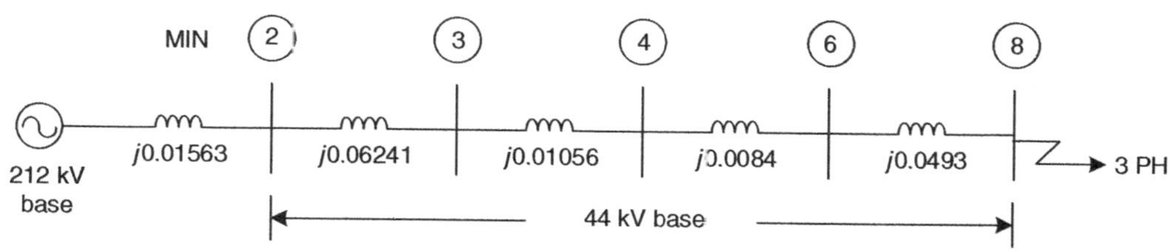

Total Z_1 to Fault Point:

$$0.01563 + 0.06241 + 0.01056 + 0.0084 = 0.0493 = Z_1 \text{ equ.} = 0.1463 \text{ PU}$$

The distribution system MVA base selected is 10 MVA.

I base on the 44 kV side = (10 MVA)/($\sqrt{3}$ × 44 kV) = 131 A

3 PH Fault = VPF/Z_1 = (j1.0)/(j0.1463) = 6.8 PU = 6.8 × 131 (Ibase) = 892 A

Set the low set instantaneous protection to "see" all feeder faults with margin (common practice is approximately 50% of min remote 3PH).

Set at 480 A, which represents 892/480 = 1.85 (1/1.85 = 0.54 times fault).

Check loading: maximum loading for this feeder is 350 A. The pickup setting of 480 A represents 1.4 times the maximum load.

(4) Phase Faults – Timed Protection (51)
It is designed for permanent type faults.
Setting criteria. The overcurrent pick up setting is normally identical to the low set instantaneous protection; it is set to see all feeder faults, trips breaker, and not reclose.
Set it for the minimum most remote 3PH fault with a margin, 480 A.
Time Dial: determine the largest rated fuse that is used on the HV of the distribution transformer stations. Referring to the family of fuse operating curves below, it should be noted that the larger rated fuses take more time to clear for a given fault current. For this example, it is the one used at Bus 5, a 150E fuse.
The largest rated fuse takes more time to clear than the smaller rated fuses. Set 51 for this fuse clearing time pus a coordination margin. This will ensure that the fuse clears HV DS faults before the feeder protection operates and locks out the breaker.
Calculate the 3PH maximum fault at Bus 5:

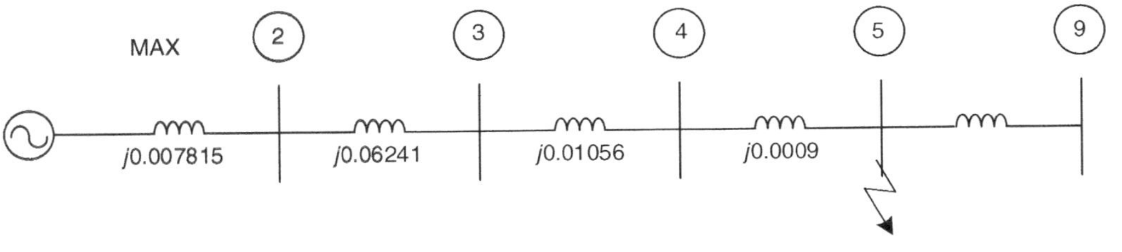

Total Z_1 to Fault Point:

0.007815 + 0.06241 + 0.01056 + 0.0009 = Z_1 equ. = 0.0816 PU

I base on the 44 kV side = (10 MVA)/($\sqrt{3}$ × 44 kV) = 131 A

3 PH Fault = VPF/Z_1 = (j1.0)/(j0.0816) = 12.24 PU = 12.24 × 131 (Ibase)
= 1604 A (where VPF is the prefault voltage)

The maximum 3PH fault current for an HV DS fault at Bus 5 is 1604 A.

Based on the total clearing time–current characteristics for the 150E fuse, see below, this fuse will operate in approximately 0.33 seconds at 1604 A.

The feeder timed overcurrent element set at 480 A "sees" 1604/480 = 3.3 times the pickup current for the same fault at Bus 5. To coordinate with the fuse, the timed element should trip in 0.33 + 0.4 = 0.74 seconds. A time dial setting of 2 will achieve the required coordination.

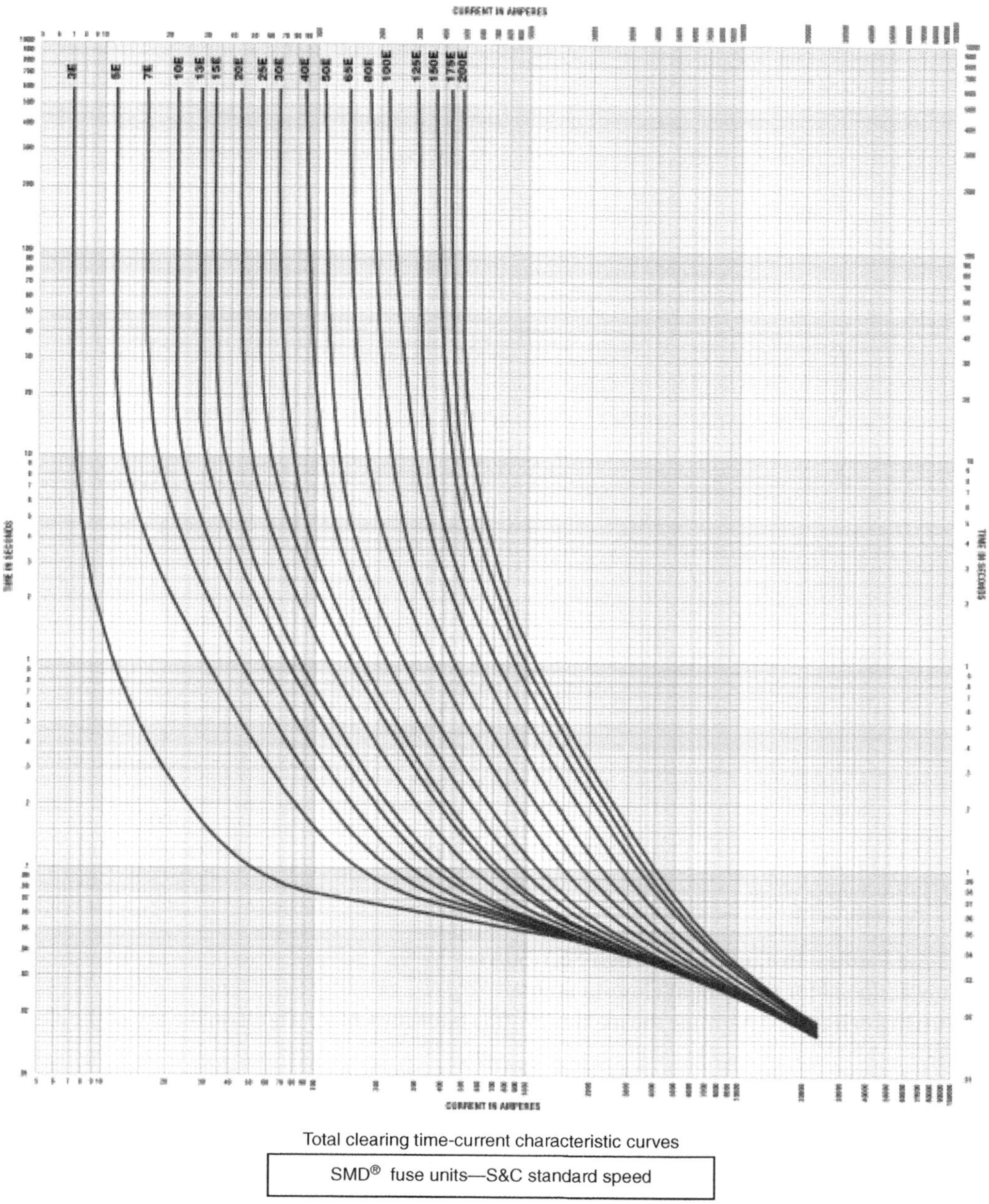

Total clearing TCC fuse curves.
Source: Courtesy of S&C Electric Co.

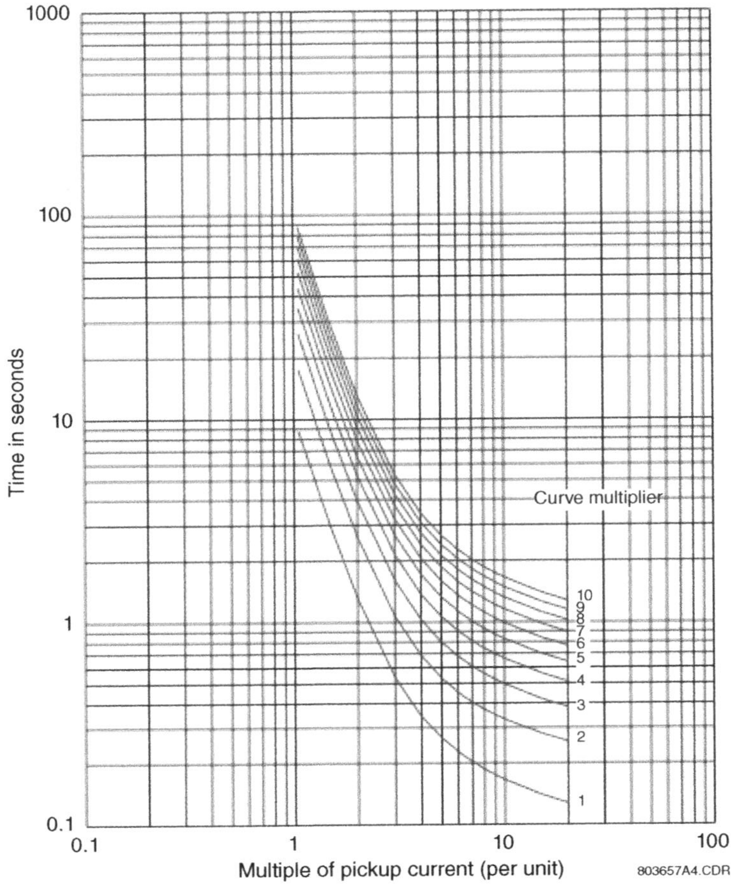

GE relay very inverse curve.
Source: Courtesy of GE.

(5) Phase Faults – Instantaneous (High-Set) (50A)
It is designed for permanent close-in faults.
Setting criteria. Determine the maximum 3PH fault current for a fault on the HV of the closest DS.

Set "short" of that value, trips the breaker, and does not reclose. The first DS is Bus 3; maximum 3PH fault at Bus#3:

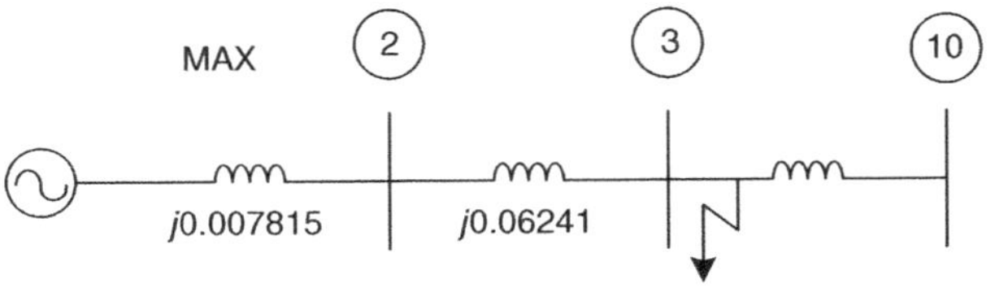

Z_1 equ. = j.07; 3 PH Fault = $j1.0/j0.07 = 14.2$ PU $= 14.2 \times 131 = 1865$ A

Set the high set instantaneous to approximately $1.3 \times 1865 = 2400$ A. Set short of the first tap to allow the fuse to operate for HV DS faults at the first tap.

(6) Ground Faults – Instantaneous (Low-Set) (50NB)
Designed for transient faults, which are the predominant type.
Setting criteria. Can be set low because it is a 3-wire feeder; there should be no unbalance zero-sequence current. Set the instantaneous low set overcurrent relay between 10% and 25% of the minimum remote end fault.

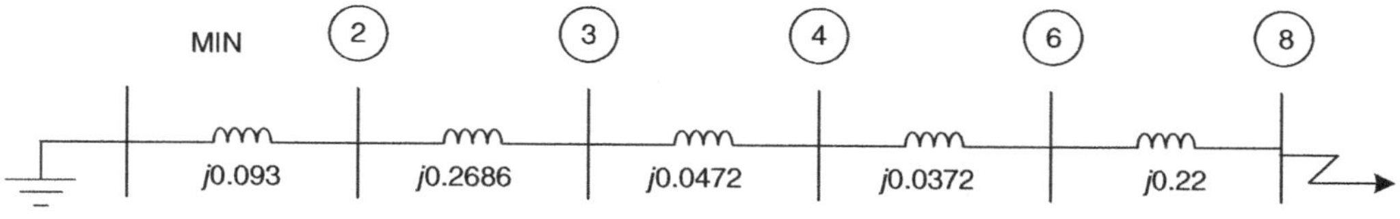

Z_0 total to Fault Point: $0.093 + 0.2686 + 0.0472 + 0.0372 + 0.220 = Z_0$ total $= 0.666$ PU

from 3PH fault above (step3) for the same fault location,

$Z_1 = 0.1463 = Z_2$

$I_0 = j1.0/j(0.1463 + 0.1463 + 0.666); = j1.0/j0.96 = 1$ PU

I_A phase $= 3 \times I_0 = 3 \times 1 = 3$ PU
$= 3 \times$ I Base current
$= 3 \times 131 = 393$ A

The minimum remote end ground fault at bus 8 equals 393 A. Set the instantaneous low set overcurrent relay to $(0.1 - 0.25) \times 393 = (40 - 100)$ A; set it at 70 A.

(7) Ground Faults – Timed (51N)
Setting criteria. Pickup settings should be similar to the low set instantaneous relay. Set at 70 A.

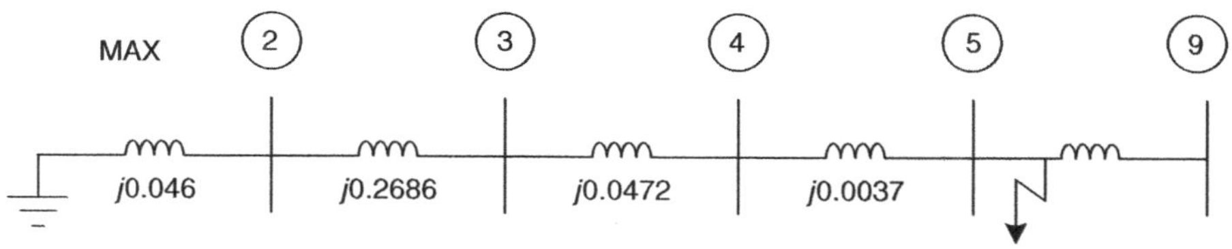

Z_0 total to Fault Point: $0.046 + 0.2686 + 0.0472 + 0.0372 = Z_0$ total $= 0.3655$ PU

from 3PH fault above (step 4)

for the same fault location,

$Z_1 = 0.0816 = Z_2$

$I_0 = j1.0/j(0.0.0816 \text{ x}2 + 0.3655); = j1.0/j0.53 = 1.9$ PU

IA phase $= 3 \times I_0 = 3 \times 1.9 = 5.7$ PU; $= 5.7 \times$ I Base current; $= 5.7 \times 131 = 747$ A

Time Dial: determine the largest rated fuse that is used on the HV of the distribution transformer stations. For this example, that is the one used at Bus 5, a 150E fuse. The maximum L-G fault current for an HV DS fault at Bus 5 is 747 A.

The feeder timed overcurrent element set at 70 A measures 747/70 = 10.7 times the pickup current for the same fault at Bus 5. To coordinate with the fuse, the timed element should trip in 1.5 (fuse) + 0.4 (margin) = 1.9 seconds.

A time dial setting of approximating 12 will achieve the required coordination.

(8) Instantaneous (High-Set) (50NA)

Designed for: permanent close-in faults, trips breaker, no reclose.
The first DS is Bus 3; maximum 3PH fault at Bus#3:

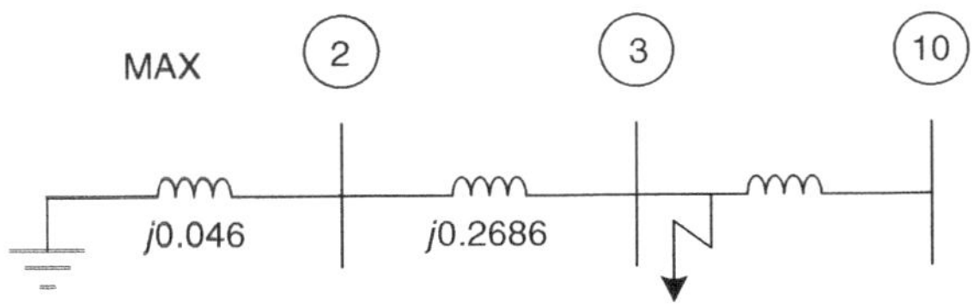

Z_0 total to Fault Point: $0.046 + 0.2686 = Z_0$ total $= 0.3146$ PU

$I_0 = j1.0/j((0.07 \times 2) + 0.3146) = 1/0.4546 = 2.2$ PU; $Z_1 = Z_2 = 0.07$ from above (step 5)

$I_A = 3 \times 2.2 \times 131 = 865$ A

Set the high set instantaneous to approximately $1.3 \times 865 = 1125$ A.

References

1 IEEE Institute of Electrical and Electronic Engineers.
2 CEA, "Application Guide For Distribution Fusing," 1998 Canadian Electricity Association [from Page 6–7].
3 CEA, "Application Guide For Distribution Fusing," 1998 Canadian Electricity Association [from Table 4.1].
4 IEEE C37.60 - 2019 Standard C37.60 Standard for High-Voltage Switchgear and Control Gear.
5 IEEE Standard C37.61 Standard Guide for the Application, Operation and Maintenance of Automatic Circuit Reclosers, 1973.
6 CEATI "Engineering Guide for Distribution Overcurrent Protection", 2009 Table 2–6.
7 CEATI "Engineering Guide for Distribution Overcurrent Protection", 2009 Table 2–7.
8 CEATI "Engineering Guide for Distribution Overcurrent Protection", 2009 Table 4–9.
9 IEEE Standard C57.109 Guide for Liquid-Immerse Transformers, C37.91 - 2021.
10 IEEE Standard C37.91 Guide for Protecting Power Transformers, C57.109 -2018.
11 IEEE Standard C37.91-2021, Guide for Protecting Power Transformers.
12 CEATI Engineering Guide for Distribution Overcurrent Protection 2009, (4–1).
13 CEA Application Guide for Distribution Fusing, August 1998, Table 10.2.

Index

d

i

k

l

m

n

o

Books in the IEEE Press Series on Power and Energy Systems

Series Editor: Ganesh Kumar Venayagamoorthy, Clemson University, Clemson, South Carolina, USA.

The mission of the IEEE Press Series on Power and Energy Systems is to publish leading-edge books that cover a broad spectrum of current and forward-looking technologies in the fast-moving area of power and energy systems including smart grid, renewable energy systems, electric vehicles and related areas. Our target audience includes power and energy systems professionals from academia, industry and government who are interested in enhancing their knowledge and perspectives in their areas of interest.

1. *Electric Power Systems: Design and Analysis, Revised Printing*
   Mohamed E. El-Hawary
2. *Power System Stability*
   Edward W. Kimbark
3. *Analysis of Faulted Power Systems*
   Paul M. Anderson
4. *Inspection of Large Synchronous Machines: Checklists, Failure Identification, and Troubleshooting*
   Isidor Kerszenbaum
5. *Electric Power Applications of Fuzzy Systems*
   Mohamed E. El-Hawary
6. *Power System Protection*
   Paul M. Anderson
7. *Subsynchronous Resonance in Power Systems*
   Paul M. Anderson, B.L. Agrawal, J.E. Van Ness
8. *Understanding Power Quality Problems: Voltage Sags and Interruptions*
   Math H. Bollen
9. *Analysis of Electric Machinery*
   Paul C. Krause, Oleg Wasynczuk, and S.D. Sudhoff
10. *Power System Control and Stability, Revised Printing*
    Paul M. Anderson, A.A. Fouad
11. *Principles of Electric Machines with Power Electronic Applications*, Second Edition
    Mohamed E. El-Hawary
12. *Pulse Width Modulation for Power Converters: Principles and Practice*
    D. Grahame Holmes and Thomas Lipo
13. *Analysis of Electric Machinery and Drive Systems*, Second Edition
    Paul C. Krause, Oleg Wasynczuk, and S.D. Sudhoff
14. *Risk Assessment for Power Systems: Models, Methods, and Applications*
    Wenyuan Li
15. *Optimization Principles: Practical Applications to the Operations of Markets of the Electric Power Industry*
    Narayan S. Rau

16. *Electric Economics: Regulation and Deregulation*
Geoffrey Rothwell and Tomas Gomez
17. *Electric Power Systems: Analysis and Control*
Fabio Saccomanno
18. *Electrical Insulation for Rotating Machines: Design, Evaluation, Aging, Testing, and Repair*
Greg C. Stone, Edward A. Boulter, Ian Culbert, and Hussein Dhirani
19. *Signal Processing of Power Quality Disturbances*
Math H. J. Bollen and Irene Y. H. Gu
20. *Instantaneous Power Theory and Applications to Power Conditioning*
Hirofumi Akagi, Edson H. Watanabe and Mauricio Aredes
21. *Maintaining Mission Critical Systems in a 24/7 Environment*
Peter M. Curtis
22. *Elements of Tidal-Electric Engineering*
Robert H. Clark
23. *Handbook of Large Turbo-Generator Operation and Maintenance*, Second Edition
Geoff Klempner and Isidor Kerszenbaum
24. *Introduction to Electrical Power Systems*
Mohamed E. El-Hawary
25. *Modeling and Control of Fuel Cells: Distributed Generation Applications*
M. Hashem Nehrir and Caisheng Wang
26. *Power Distribution System Reliability: Practical Methods and Applications*
Ali A. Chowdhury and Don O. Koval
27. *Introduction to FACTS Controllers: Theory, Modeling, and Applications*
Kalyan K. Sen, Mey Ling Sen
28. *Economic Market Design and Planning for Electric Power Systems*
James Momoh and Lamine Mili
29. *Operation and Control of Electric Energy Processing Systems*
James Momoh and Lamine Mili
30. *Restructured Electric Power Systems: Analysis of Electricity Markets with Equilibrium Models*
Xiao-Ping Zhang
31. *An Introduction to Wavelet Modulated Inverters*
S.A. Saleh and M.A. Rahman
32. *Control of Electric Machine Drive Systems*
Seung-Ki Sul
33. *Probabilistic Transmission System Planning*
Wenyuan Li
34. *Electricity Power Generation: The Changing Dimensions*
Digambar M. Tagare
35. *Electric Distribution Systems*
Abdelhay A. Sallam and Om P. Malik
36. *Practical Lighting Design with LEDs*
Ron Lenk, Carol Lenk

37. *High Voltage and Electrical Insulation Engineering*
Ravindra Arora and Wolfgang Mosch
38. *Maintaining Mission Critical Systems in a 24/7 Environment, Second Edition*
Peter Curtis
39. *Power Conversion and Control of Wind Energy Systems*
Bin Wu, Yongqiang Lang, Navid Zargari, Samir Kouro
40. *Integration of Distributed Generation in the Power System*
Math H. Bollen, Fainan Hassan
41. *Doubly Fed Induction Machine: Modeling and Control for Wind Energy Generation Applications*
Gonzalo Abad, Jesús López, Miguel Rodrigues, Luis Marroyo, and Grzegorz Iwanski
42. *High Voltage Protection for Telecommunications*
Steven W. Blume
43. *Smart Grid: Fundamentals of Design and Analysis*
James Momoh
44. *Electromechanical Motion Devices, Second Edition*
Paul Krause, Oleg Wasynczuk, Steven Pekarek
45. *Electrical Energy Conversion and Transport: An Interactive Computer-Based Approach, Second Edition*
George G. Karady, Keith E. Holbert
46. *ARC Flash Hazard and Analysis and Mitigation*
J.C. Das
47. *Handbook of Electrical Power System Dynamics: Modeling, Stability, and Control*
Mircea Eremia, Mohammad Shahidehpour
48. *Analysis of Electric Machinery and Drive Systems, Third Edition*
Paul C. Krause, Oleg Wasynczuk, S.D. Sudhoff, Steven D. Pekarek
49. *Extruded Cables for High-Voltage Direct-Current Transmission: Advances in Research and Development*
Giovanni Mazzanti, Massimo Marzinotto
50. *Power Magnetic Devices: A Multi-Objective Design Approach*
S.D. Sudhoff
51. *Risk Assessment of Power Systems: Models, Methods, and Applications, Second Edition*
Wenyuan Li
52. *Practical Power System Operation*
Ebrahim Vaahedi
53. *The Selection Process of Biomass Materials for the Production of Bio-Fuels and Co-Firing*
Najib Altawell
54. *Electrical Insulation for Rotating Machines: Design, Evaluation, Aging, Testing, and Repair, Second Edition*
Greg C. Stone, Ian Culbert, Edward A. Boulter, and Hussein Dhirani
55. *Principles of Electrical Safety*
Peter E. Sutherland

56. *Advanced Power Electronics Converters: PWM Converters Processing AC Voltages*
Euzeli Cipriano dos Santos Jr., Edison Roberto Cabral da Silva
57. *Optimization of Power System Operation, Second Edition*
Jizhong Zhu
58. *Power System Harmonics and Passive Filter Designs*
J.C. Das
59. *Digital Control of High-Frequency Switched-Mode Power Converters*
Luca Corradini, Dragan Maksimoviæ, Paolo Mattavelli, Regan Zane
60. *Industrial Power Distribution, Second Edition*
Ralph E. Fehr, III
61. *HVDC Grids: For Offshore and Supergrid of the Future*
Dirk Van Hertem, Oriol Gomis-Bellmunt, Jun Liang
62. *Advanced Solutions in Power Systems: HVDC, FACTS, and Artificial Intelligence*
Mircea Eremia, Chen-Ching Liu, Abdel-Aty Edris
63. *Operation and Maintenance of Large Turbo-Generators*
Geoff Klempner, Isidor Kerszenbaum
64. *Electrical Energy Conversion and Transport: An Interactive Computer-Based Approach*
George G. Karady, Keith E. Holbert
65. *Modeling and High-Performance Control of Electric Machines*
John Chiasson
66. *Rating of Electric Power Cables in Unfavorable Thermal Environment*
George J. Anders
67. *Electric Power System Basics for the Nonelectrical Professional*
Steven W. Blume
68. *Modern Heuristic Optimization Techniques: Theory and Applications to Power Systems*
Kwang Y. Lee, Mohamed A. El-Sharkawi
69. *Real-Time Stability Assessment in Modern Power System Control Centers*
Savu C. Savulescu
70. *Optimization of Power System Operation*
Jizhong Zhu
71. *Insulators for Icing and Polluted Environments*
Masoud Farzaneh, William A. Chisholm
72. *PID and Predictive Control of Electric Devices and Power Converters Using MATLAB®/Simulink®*
Liuping Wang, Shan Chai, Dae Yoo, Lu Gan, Ki Ng
73. *Power Grid Operation in a Market Environment: Economic Efficiency and Risk Mitigation*
Hong Chen
74. *Electric Power System Basics for Nonelectrical Professional, Second Edition*
Steven W. Blume
75. *Energy Production Systems Engineering*
Thomas Howard Blair
76. *Model Predictive Control of Wind Energy Conversion Systems*
Venkata Yaramasu, Bin Wu

77. *Understanding Symmetrical Components for Power System Modeling*
J.C. Das

78. *High-Power Converters and AC Drives, Second Edition*
Bin Wu, Mehdi Narimani

79. *Current Signature Analysis for Condition Monitoring of Cage Induction Motors: Industrial Application and Case Histories*
William T. Thomson, Ian Culbert

80. *Introduction to Electric Power and Drive Systems*
Paul Krause, Oleg Wasynczuk, Timothy O'Connell, Maher Hasan

81. *Instantaneous Power Theory and Applications to Power Conditioning, Second Edition*
Hirofumi, Edson Hirokazu Watanabe, Mauricio Aredes

82. *Practical Lighting Design with LEDs, Second Edition*
Ron Lenk, Carol Lenk

83. *Introduction to AC Machine Design*
Thomas A. Lipo

84. *Advances in Electric Power and Energy Systems: Load and Price Forecasting*
Mohamed E. El-Hawary

85. *Electricity Markets: Theories and Applications*
Jeremy Lin, Jernando H. Magnago

86. *Multiphysics Simulation by Design for Electrical Machines, Power Electronics and Drives*
Marius Rosu, Ping Zhou, Dingsheng Lin, Dan M. Ionel, Mircea Popescu, Frede Blaabjerg, Vandana Rallabandi, David Staton

87. *Modular Multilevel Converters: Analysis, Control, and Applications*
Sixing Du, Apparao Dekka, Bin Wu, Navid Zargari

88. *Electrical Railway Transportation Systems*
Morris Brenna, Federica Foiadelli, Dario Zaninelli

89. *Energy Processing and Smart Grid*
James A. Momoh

90. *Handbook of Large Turbo-Generator Operation and Maintenance, 3rd Edition*
Geoff Klempner, Isidor Kerszenbaum

91. *Advanced Control of Doubly Fed Induction Generator for Wind Power Systems*
Dehong Xu, Dr. Frede Blaabjerg, Wenjie Chen, Nan Zhu

92. *Electric Distribution Systems, 2nd Edition*
Abdelhay A. Sallam, Om P. Malik

93. *Power Electronics in Renewable Energy Systems and Smart Grid: Technology and Applications*
Bimal K. Bose

94. *Distributed Fiber Optic Sensing and Dynamic Rating of Power Cables*
Sudhakar Cherukupalli, and George J Anders

95. *Power System and Control and Stability,* Third Edition
Vijay Vittal, James D. McCalley, Paul M. Anderson, and A. A. Fouad

96. *Electromechanical Motion Devices: Rotating Magnetic Field-Based Analysis and Online Animations,* Third Edition
Paul Krause, Oleg Wasynczuk, Steven D. Pekarek, and Timothy O'Connell

97. *Applications of Modern Heuristic Optimization Methods in Power and Energy Systems*
Kwang Y. Lee and Zita A. Vale

98. *Handbook of Large Hydro Generators: Operation and Maintenance*
Glenn Mottershead, Stefano Bomben, Isidor Kerszenbaum, and Geoff Klempner

99. *Advances in Electric Power and Energy: Static State Estimation*
Mohamed E. El-hawary

100. *Arc Flash Hazard Analysis and Mitigation,* Second Edition
J.C. Das

101. *Maintaining Mission Critical Systems in a 24/7 Environment,* Third Edition
Peter M. Curtis

102. *Real-Time Electromagnetic Transient Simulation of AC-DC Networks*
Venkata Dinavahi and Ning Lin

103. *Probabilistic Power System Expansion Planning with Renewable Energy Resources and Energy Storage Systems*
Jaeseok Choi and Kwang Y. Lee

104. *Power Magnetic Devices: A Multi-Objective Design Approach*, Second Edition
Scott D. Sudhoff

105. *Optimal Coordination of Power Protective Devices with Illustrative Examples*
Ali R. Al-Roomi

106. *Resilient Control Architectures and Power Systems*
Craig Rieger, Ronald Boring, Brian Johnson, and Timothy McJunkin

107. *Alternative Liquid Dielectrics for High Voltage Transformer Insulation Systems: Performance Analysis and Applications*
Edited by U. Mohan Rao, I. Fofana, and R. Sarathi

108. *Introduction to the Analysis of Electromechanical Systems*
Paul C. Krause, Oleg Wasynczuk, and Timothy O'Connell

109. *Power Flow Control Solutions for a Modern Grid using SMART Power Flow Controllers*
Kalyan K. Sen and Mey Ling Sen

110. *Power System Protection: Fundamentals and Applications*
John Ciufo and Aaron Cooperberg

111. *Soft-Switching Technology for Three-phase Power Electronics Converters*
Dehong Xu, Rui Li, Ning He, Jinyi Deng, and Yuying Wu

Printed and bound by CPI Group (UK) Ltd, Croydon, CR0 4YY

21/09/2026

14965842-0001